CHAPMAN & HALL/CRC COMPUTER and INFORMATION SCIENCE SERIES

Handbook of Bioinspired Algorithms and Applications

CHAPMAN & HALL/CRC
COMPUTER and INFORMATION SCIENCE SERIES

Series Editor: Sartaj Sahni

PUBLISHED TITLES

HANDBOOK OF SCHEDULING: ALGORITHMS, MODELS, AND PERFORMANCE ANALYSIS
Joseph Y.-T. Leung

THE PRACTICAL HANDBOOK OF INTERNET COMPUTING
Munindar P. Singh

HANDBOOK OF DATA STRUCTURES AND APPLICATIONS
Dinesh P. Mehta and Sartaj Sahni

DISTRIBUTED SENSOR NETWORKS
S. Sitharama Iyengar and Richard R. Brooks

SPECULATIVE EXECUTION IN HIGH PERFORMANCE COMPUTER ARCHITECTURES
David Kaeli and Pen-Chung Yew

SCALABLE AND SECURE INTERNET SERVICES AND ARCHITECTURE
Cheng-Zhong Xu

HANDBOOK OF BIOINSPIRED ALGORITHMS AND APPLICATIONS
Stephan Olariu and Albert Y. Zomaya

CHAPMAN & HALL/CRC COMPUTER and INFORMATION SCIENCE SERIES

Handbook of Bioinspired Algorithms and Applications

Edited by

Stephan Olariu

Old Dominion University
Norfolk, Virginia, U.S.A.

Albert Y. Zomaya

University of Sydney
NSW, Australia

Published in 2006 by
Chapman & Hall/CRC
Taylor & Francis Group
6000 Broken Sound Parkway NW, Suite 300
Boca Raton, FL 33487-2742

No claim to original U.S. Government works
Printed in the United States of America on acid-free paper
10 9 8 7 6 5 4 3 2 1

International Standard Book Number-10: 1-58488-475-4 (Hardcover)
International Standard Book Number-13: 978-1-58488-475-0 (Hardcover)

Library of Congress Cataloging-in-Publication Data

Catalog record is available from the Library of Congress

Taylor & Francis Group
is the Academic Division of T&F Informa plc.

Visit the Taylor & Francis Web site at
http://www.taylorandfrancis.com

and the CRC Press Web site at
http://www.crcpress.com

Preface

The *Handbook of Bioinspired Algorithms and Applications* seeks to provide an opportunity for researchers to explore the connection between biologically inspired (or bioinspired) techniques and the development of solutions to problems that arise in a variety of problem domains. The power of bioinspired paradigms lies in their capability in dealing with complex problems with little or no knowledge about the search space, and thus is particularly well suited to deal with a wide range of computationally intractable optimizations and decision-making applications.

Vast literature exists on bioinspired approaches for solving an impressive array of problems and there is a great need to develop repositories of "how to apply" bioinspired paradigms to difficult problems. The material of the handbook is by no means exhaustive and it focuses on paradigms that are "bioinspired," and therefore, chapters on fuzzy logic or simulated annealing were not included in the organization. There was a decision to limit the number of chapters so that the handbook remains manageable within a single volume.

The handbook endeavors to strike a balance between theoretical and practical coverage of a range of bioinspired paradigms and applications. The handbook is organized into two main sections: Models and Paradigms and Application Domains, and the titles of the various chapters are self-explanatory and a good indication to what is covered. The theoretical chapters are intended to provide the fundamentals of each of the paradigms in such a way that allows the readers to utilize these techniques in their own fields. The application chapters show detailed examples and case studies of how to actually develop a solution to a problem based on a bioinspired technique. The handbook should serve as a repository of significant reference material, as the list of references that each chapter provides will become a useful source of further study.

Stephan Olariu
Albert Y. Zomaya

Acknowledgments

First and foremost we would like to thank and acknowledge the contributors of this book for their support and patience, and the reviewers for their useful comments and suggestions that helped in improving the earlier outline of the handbook and presentation of the material. Professor Zomaya would like to acknowledge the support from CISCO Systems and members of the Advanced Networks Research Group at Sydney University. We also extend our deepest thanks to Jessica Vakili and Bob Stern from CRC Press for their collaboration, guidance, and, most importantly, patience in finalizing this handbook. Finally, we thank Mr. Mohan Kumar for leading the production process of this handbook in a very professional manner.

Stephan Olariu
Albert Y. Zomaya

Editors

Stephan Olariu received his M.Sc. and Ph.D. degrees in computer science from McGill University, Montreal, in 1983 and 1986, respectively. In 1986 he joined the Old Dominion University where he is a professor of computer science. Dr. Olariu has published extensively in various journals, book chapters, and conference proceedings. His research interests include image processing and machine vision, parallel architectures, design and analysis of parallel algorithms, computational graph theory, computational geometry, and mobile computing. Dr. Olariu serves on the Editorial Board of *IEEE Transactions on Parallel and Distributed Systems, Journal of Parallel and Distributed Computing, VLSI Design, Parallel Algorithms and Applications, International Journal of Computer Mathematics*, and *International Journal of Foundations of Computer Science.*

Albert Y. Zomaya is currently the CISCO Systems chair professor of internetworking in the School of Information Technologies, The University of Sydney. Prior to that he was a full professor in the Electrical and Electronic Engineering Department at the University of Western Australia, where he also led the Parallel Computing Research Laboratory from 1990 to 2002. He served as associate, deputy, and acting head in the same department, and held visiting positions at Waterloo University and the University of Missouri–Rolla. He is the author/co-author of 6 books and 200 publications in technical journals and conferences, and the editor of 6 books and 7 conference volumes. He is currently an associate editor for 14 journals, the founding editor of the *Wiley Book Series on Parallel and Distributed Computing*, and the editor-in-chief of the *Parallel and Distributed Computing Handbook* (McGraw-Hill 1996). Professor Zomaya was the chair of the *IEEE Technical Committee on Parallel Processing* (1999–2003) and currently serves on its executive committee. He has been actively involved in the organization of national and international conferences. He received the 1997 Edgeworth David Medal from the Royal Society of New South Wales for outstanding contributions to Australian science. In September 2000 he was awarded the *IEEE Computer Society's Meritorious Service Award.* Professor Zomaya is a chartered engineer (CEng), a fellow of the *IEEE, a fellow of the Institution of Electrical Engineers* (U.K.), and member of the ACM. He also serves on the boards of two startup companies. His research interests are in the areas of high performance computing, parallel algorithms, networking, mobile computing, and bioinformatics.

Contributors

Enrique Alba
Department of Languages and Computer Science
University of Málaga
Campus de Teatinos
Málaga, Spain

Abdullah Almojel
Ministry of Higher Education
Riyadh, Saudi Arabia

Sanghamitra Bandyopadhyay
Machine Intelligence Unit
Indian Statistical Institute
Kolkata, India

Nilanjan Banerjee
Center for Research in Wireless Mobility and Networking
Department of Computer Science & Engineering
The University of Texas at Arlington
Arlington, Texas

Mohamed Belal
Helwan University
Cairo, Egypt

Utpal Biswas
Department of Computer Science and Engineering
University of Kalyani
Kalyani, India

Azzedine Boukerche
SITE
University of Ottawa
Ottawa, Canada

Anthony Brabazon
Faculty of Commerce
University College Dublin
Dublin, Ireland

Jürgen Branke
Institute AIFB
University of Karlsruhe
Karlsruhe, Germany

Forbes Burkowski
School of Computer Science
University of Waterloo
Waterloo, Ontario, Canada

S. Cahon
Laboratoire d'Informatique Fondamentale de Lille
Lille, France

J. Francisco Chicano
Department of Languages and Computer Science
University of Málaga
Málaga, Spain

Ernesto Costa
Evolutionary and Complex Systems Group
Centro de Informática e Sistemas da Universidade de Coimbra
Pinhal de Marrocos
Coimbra, Portugal

Carlos Cotta
Department of Languages and Computer Science
University of Málaga
Campus de Teatinos
Málaga, Spain

Kris Crnomarkovic
Advanced Networks Research Group
School of Information Technologies
The University of Sydney
Sydney, Australia

Sajal K. Das
Center for Research in Wireless Mobility and Networking
Department of Computer Science & Engineering
The University of Texas at Arlington
Arlington, Texas

Tiago Ferra de Sousa
Escola Superior de Tecnologia
Instituto Politecnico de Castelo Branco
Castelo Branco, Portugal

Francisco Fernández de Vega
Grupo de Evolución Artificial
Centro Universitario de Mérida
Universidad de Extremadura
Mérida, Spain

C. Dhaenens
Laboratoire d'Informatique Fondamentale de Lille
Lille, France

Bernabe Dorronsoro
Central Computing Services
University of Málaga
Campus de Teatinos
Málaga, Spain

Hoda El-Sayed
Bowie State University
Bowie, Maryland

Mohamed Eltoweissy
Virginia Tech
Falls Church, Virginia

Muddassar Farooq
Informatik III
University of Dortmund
Dortmund, Germany

Marcos Fernández
Instituto de Robótica
Universidad de Valencia
Polígono de la Coma
Paterna (Valencia), Spain

Gianluigi Folino
Institute of High Performance Computing and Networks
Rende (CS), Italy

Agostino Forestiero
Institute of High Performance Computing and Networks
Rende (CS), Italy

Jafaar Gaber
UTMB
France

Mario Giacobini
Information Systems Department
University of Lausanne
Lausanne, Switzerland

Michael Guntsch
Institute AIFB
University of Karlsruhe
Karlsruhe, Germany

Salim Hariri
High Performance Distributed Computing Laboratory
The University of Arizona
Tuscon, Arizona

Piotr Jedrzejowicz
Department of Information Systems
Faculty of Business Administration
Gdynid Maritime University
Gdynia, Poland

Kennie H. Jones
NASA Langley Research Center
Hampton, Virginia

L. Jourdan
Laboratoire d'Informatique Fondamentale de Lille
Lille, France

Kathia Regina Lemos Jucá
Federal University of Santa Catarina
Florianopolis, Brazil

M. Khabzaoui
Laboratoire d'Informatique Fondamentale de Lille
Lille, France

Bithika Khargaria
High Performance Distributed Computing Laboratory
The University of Arizona
Tuscon, Arizona

Peter Korošec
Computer Systems Department
Jožef Stefan Institute
Ljubljana, Slovenia

Barbara Koroušić-Seljak
Computer Systems Department
Jožef Stefan Institute
Ljubljana, Slovenia

Zhen Li
The Applied Software Systems Laboratory
Rutgers, The State University of New Jersey
Camden, New Jersey

Kenneth N. Lodding
NASA Langley Research Center
Hampton, Virginia

Mi Lu
Department of Electrical Engineering
Texas A&M University
College Station, Texas

Francisco Luna
Department of Languages and Computer Science
ETS Ingeniería Informática
University of Málaga
Málaga, Spain

Gabriel Luque
Department of Languages and Computer Science
ETS Ingeniería Informática
University of Málaga
Málaga, Spain

Ujjwal Maulik
Department of Computer Science and Engineering
Jadavpur University
Kolkata, India

N. Melab
Laboratoire d'Informatique Fondamentale de Lille
Lille, France

M. Mezmaz
Laboratoire d'Informatique Fondamentale de Lille
Lille, France

Michelle Moore
Department of Computing and Mathematical Sciences
Texas A&M University-Corpus Christi
Corpus Christi, Texas

Pedro Morillo
Instituto de Robótica
Universidad de Valencia
Polígono de la Coma
Paterna (Valencia), Spain

Anirban Mukhopadhyay
Department of Computer Science and Engineering
University of Kalyani
Kalyani, India

Mrinal Kanti Naskar
Department of Electronics and Telecommunication Engineering
Jadavpur University
Kolkata, India

Antonio J. Nebro
Department of Languages and Computer Science
ETS Ingeniería Informática
University of Málaga
Málaga, Spain

Ana Neves
Escola Superior de Tecnologia
Instituto Politécnico de Castelo Branco
Castelo Branco, Portugal
and
Evolutionary and Complex Systems Group
Centro de Informática e Sistemas da Universidade de Coimbra
Pinhal de Marrocos, Portugal

Alioune Ngom
Computer Science Department
University of Windsor
Windsor, Ontario, Canada

Mirela Sechi Moretti Annoni Notare
Barddal University
Florianopolis, Brazil

Stephan Olariu
Old Dominion University
Norfolk, Virginia

Michael O'Neill
Department of Computer Science & Information Systems
University of Limerick
Limerick, Ireland

Juan Manuel Orduña
Departamento de Informática
Universidad de Valencia
Burjassot (Valencia), Spain

Gregor Papa
Computer Systems Department
Jožef Stefan Institute
Ljubljana, Slovenia

Manish Parashar
The Applied Software Systems Laboratory
Rutgers, The State University of New Jersey
Camden, New Jersey

Zhiquan Frank Qiu
Intel Corporation
Chandler, Arizona

Borut Robič
Faculty of Computer and Information Science
University of Ljubljana
Ljubljana, Slovenia

Abhishek Roy
Center for Research in Wireless Mobility and Networking
Department of Computer Science & Engineering
The University of Texas at Arlington
Arlington, Texas

Hartmut Schmeck
Institute AIFB
University of Karlsruhe
Karlsruhe, Germany

Franciszek Seredynski
Polish-Japanese Institute of Information Technologies
Koszykowa
Warsaw, Poland
and
Institute of Computer Science
Polish Academy of Sciences
Ordona
Warsaw, Poland

Jurij Šilc
Computer Systems Department
Jožef Stefan Institute
Ljubljana, Slovenia

Arlindo Silva
Escola Superior de Tecnologia
Instituto Politécnico de Castelo Branco
Castelo Branco, Portugal
and
Centro de Informatica e Sistemas da Universidade de Coimbra
Pinhal de Marrocos, Portugal

João Bosco Mangueira Sobral
Federal University of Santa Catarina
Florianopolis, Brazil
and
Evolutionary and Complex Systems Group
Centro de Informática e Sistemas da Universidade de Coimbra
Pinhal de Marrocos, Portugal

Tiago Sousa
Escola Superior de Tecnologia
Instituto Politécnico de Castelo Branco
Castelo Branco, Portugal
and
Evolutionary and Complex Systems Group
Centro de Informática e Sistemas da Universidade de Coimbra
Pinhal de Marrocos, Portugal

Giandomenico Spezzano
Institute of High Performance Computing and Networks
Rende (CS), Italy

Michael Stein
Institute AIFB
University of Karlsruhe
Germany

Ivan Stojmenović
Department of Computer Science
School of Information Technology and Engineering
University of Ottawa
Ottawa, Ontario, Canada

Anna Święcicka
Department of Computer Science
Bialystok University of Technology
Bialystok, Poland

Javid Taheri
Advanced Networks Research Group
School of Information Technologies
The University of Sydney
Sydney, Australia

El-Ghazali Talbi
Université des Sciences et Technologies de Lille
Cité Scientifique, France

Domenico Talia
University of Calabria
DEIS
Rende, Italy

Marco Tomassini
Information Systems Department
University of Lausanne
Lausanne, Switzerland

Ashraf Wadaa
Old Dominion University
Norfolk, Virginia

Horst F. Wedde
Informatik III
University of Dortmund
Dortmund, Germany

B. Wei
Laboratoire d'Informatique Fondamentale de Lille
Cité scientifique, France

Benjamin Weinberg
Université des Sciences et Technologies de Lille
Cité Scientifique, France

Larry Wilson
Old Dominion University
Norfolk, Virginia

Xin-She Yang
Department of Engineering
University of Cambridge
Cambridge, United Kingdom

Y. Young
Civil and Computational Engineering Centre
School of Engineering
University of Wales Swansea
Swansea, United Kingdom

Albert Y. Zomaya
Advanced Networks Research Group
School of Information Technologies
The University of Sydney
Sydney, Australia

Joviša Žunić
Computer Science Department
Cardiff University
Cardiff, Wales, United Kingdom

Contents

SECTION I Models and Paradigms

SECTION II Application Domains

I

Models and Paradigms

1

Evolutionary Algorithms

Enrique Alba
Carlos Cotta

1.1 Introduction

One of the most striking features of Nature is the existence of living organisms adapted for surviving in almost any ecosystem, even the most inhospitable: from abyssal depths to mountain heights, from volcanic vents to polar regions. The magnificence of this fact becomes more evident when we consider that the life environment is continuously changing. This motivates certain life forms to become extinct whereas other beings evolve and preponderate due to their adaptation to the new scenario. It is very remarkable that living beings do not exert a conscious effort for evolving (actually, it would be rather awkward to talk about consciousness in amoebas or earthworms); much on the contrary, the driving force for change is controlled by supraorganic mechanisms such as natural evolution.

Can we learn — and use for our own profit — the lessons that Nature is teaching us? The answer is a big YES, as the optimization community has repeatedly shown in the last decades. "*Evolutionary algorithm*" is the key word here. The term evolutionary algorithm (EA henceforth) is used to designate a collection of optimization techniques whose functioning is loosely based on metaphors of biological processes.

This rough definition is rather broad and tries to encompass the numerous approaches currently existing in the field of evolutionary computation [1]. Quite appropriately, this field itself is continuously evolving; a quick inspection of the proceedings of the relevant conferences and symposia suffices to demonstrate the impetus of the field, and the great diversity of the techniques that can be considered "evolutionary."

This variety notwithstanding, it is possible to find a number of common features of all (or at least most of) EAs. The following quote from Reference 2 illustrates such common points:

> The algorithm maintains a collection of potential solutions to a problem. Some of these possible solutions are used to create new potential solutions through the use of *operators.* Operators act on and produce collections of potential solutions. The potential solutions that an operator acts on are selected on the basis of their quality as solutions to the problem at hand. The algorithm uses this process repeatedly to generate new collections of potential solutions until some stopping criterion is met.

This definition can be usually found in the literature expressed in a technical language that uses terms such as *genes, chromosomes, population,* etc. This jargon is a reminiscence of the biological inspiration mentioned before, and has deeply permeated the field. We will return to the connection with biology later on.

The objective of this work is to present a gentle overview of these techniques comprising both the classical "canonical" models of EAs as well as some modern directions for the development of the field, namely, the use of parallel computing, and the introduction of problem-dependent knowledge.

1.2 Learning from Biology

Evolution is a complex fascinating process. Along history, scientists have attempted to explain its functioning using different theories. After the development of disciplines such as comparative anatomy in the middle of the 19th century, the basic principles that condition our current vision of evolution were postulated. Such principles rest upon Darwin's Natural Selection Theory [3], and Mendel's work on genetic inheritance [4]. They can be summarized in the following points (see Reference 5):

- Evolution is a process that does not operate on organisms directly, but on *chromosomes.* These are the organic tools by means of which the structure of a certain living being is encoded, that is, the features of a living being are defined by the decoding of a collection of chromosomes. These chromosomes (more precisely, the information they contain) pass from one generation to another through reproduction.
- The evolutionary process takes place precisely during reproduction. Nature exhibits a plethora of reproductive strategies. The most essential ones are *mutation* (that introduces variability in the gene pool) and *recombination* (that introduces the exchange of genetic information among individuals).
- Natural selection is the mechanism that relates chromosomes with the adequacy of the entities they represent, favoring the proliferation of effective, environment-adapted organisms, and conversely causing the extinction of lesser effective, nonadapted organisms.

These principles are comprised within the most orthodox theory of evolution, the Synthetic Theory [6]. Although alternate scenarios that introduce some variety in this description have been proposed — for example, the Neutral Theory [7], and very remarkably the Theory of Punctuated Equilibria [8] — it is worth considering the former basic model. It is amazing to see that despite the apparent simplicity of the principles upon which it rests, Nature exhibits unparalleled power in developing and expanding new life forms.

Not surprisingly, this power has attracted the interest of many researchers, who have tried to translate the principles of evolution to the realm of algorithmics, pursuing the construction of computer systems with analogous features. An important point must be stressed here: evolution is an undirected process, that is, there exists no scientific evidence that evolution is headed to a certain final goal. On the contrary, it can be regarded as a reactive process that makes organisms change in response to environmental variations. However, it is a fact that human-designed systems do pursue a definite final goal. Furthermore, whatever this goal might be, it is in principle, desirable to reach it quickly and efficiently. This leads to the distinction between two approaches to the construction of natureinspired systems:

1. Trying to reproduce Nature principles with the highest possible accuracy, that is, simulate Nature.
2. Using these principles as inspiration, adapting them in whatever required way so as to obtain efficient systems for performing the desired task.

Both approaches concentrate nowadays on the efforts of researchers. The first one has given rise to the field of Artificial Life (e.g., see Reference 9), and it is interesting because it allows re-creating and studying numerous natural phenomena such as parasitism, predator/prey relationships, etc. The second approach can be considered more practical, and constitutes the source of EAs. Notice anyway that these two approaches are not hermetic containers, and have frequently interacted with certainly successful results.

1.3 Nature's Way for Optimizing

As mentioned above, the standpoint of EAs is essentially practical — using ideas from natural evolution in order to solve a certain problem. Let us focus on optimization and see how this goal can be achieved.

1.3.1 Algorithm Meets Evolution

An EA is a stochastic iterative procedure for generating tentative solutions for a certain problem $\mathcal{P}$. The algorithm manipulates a collection P of individuals (the *population*), each of which comprises one or more chromosomes. These chromosomes allow each individual represent a potential solution for the problem under consideration. An encoding/decoding process is responsible for performing this mapping between chromosomes and solutions. Chromosomes are divided into smaller units termed *genes*. The different values a certain gene can take are called the *alleles* for that gene.

Initially, the population is generated at random or by means of some heuristic seeding procedure. Each individual in P receives a *fitness* value: a measure of how good the solution it represents is for the problem being considered. Subsequently, this value is used within the algorithm for guiding the search. The whole process is sketched in Figure 1.1.

As it can be seen, the existence of a set $\mathcal{F}$ (also known as *phenotype space*) comprising the solutions for the problem at hand is assumed. Associated with $\mathcal{F}$, there also exists a set $\mathcal{G}$ (known as *genotype space*). These sets $\mathcal{G}$ and $\mathcal{F}$ respectively constitute the domain and codomain of a function g known as the *growth* (or *expression*) function. It could be the case that $\mathcal{F}$ and $\mathcal{G}$ were actually equivalent, being g a trivial identity function. However, this is not the general situation. As a matter of fact, the only requirement posed on g is subjectivity. Furthermore, g could be undefined for some elements in $\mathcal{G}$.

After having defined these two sets $\mathcal{G}$ and $\mathcal{F}$, notice the existence of a function ι selecting some elements from $\mathcal{G}$. This function is called the *initialization* function, and these selected solutions (also known as *individuals*) constitute the so-called *initial population*. This initial population is in fact a pool

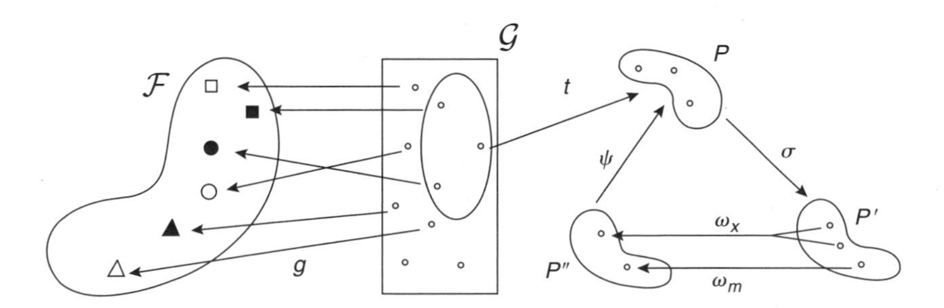

FIGURE 1.1 Illustration of the evolutionary approach to optimization.

```
Evolutionary-Algorithm:

1. P ← apply ι on G to obtain μ individuals (the initial population);
2. while Termination Criterion is not met do
   (a) P' ← apply σ on P; /* selection */
   (b) P'' ← apply ω_1, ..., ω_k on P'; /* reproduction */
   (c) P ← apply ψ on P and P''; /* replacement */
   endwhile
```

FIGURE 1.2 Pseudocode of an evolutionary algorithm.

of solutions onto which the EA will subsequently work, iteratively applying some evolutionary operators to modify its contents. More precisely, the process comprises three major stages: *selection* (promising solutions are picked from the population by using a selection function σ), *reproduction* (new solutions are created by modifying selected solutions using some reproductive operators ω_i), and *replacement* (the population is updated by replacing some existing solutions by the newly created ones, using a replacement function ψ). This process is repeated until a certain termination criterion (usually reaching a maximum number of iterations) is satisfied. Each iteration of this process is commonly termed a *generation.*

According to this description, it is possible to express the pseudocode of an EA as shown in Figure 1.2. Every possible instantiation of this algorithmic template[1] will give rise to a different EA. More precisely, it is possible to distinguish different EA families, by considering some guidelines on how to perform this instantiation.

1.3.2 The Flavors of Evolutionary Algorithms

EAs, as we know them now, began their existence during the late 1960s and early 1970s (some earlier references to the topic exist, though; see Reference 10). In these years — and almost simultaneouly — scientists from different places in the world began the task of putting Nature at work in algorithmics, and more precisely in search of problem-solving duties. The existence of these different primordial sources originated the rise of three different EA models. These classical families are:

- *Evolutionary Programming* (EP): This EA family originated in the work of Fogel et al. [11]. EP focuses on the adaption of individuals rather than on the evolution of their genetic information. This implies a much more abstract view of the evolutionary process, in which the behavior of individuals is directly modified (as opposed to manipulating its genes). This behavior is typically modeled by using complex data structures such as finite automata or as graphs (see Figure 1.3[a]). Traditionally, EP uses asexual reproduction — also known as mutation, that is, introducing slight changes in an existing solution — and selection techniques based on direct competition among individuals.
- *Evolution Strategies* (ESs): These techniques were initially developed in Germany by Rechenberg [12] and Schwefel [13]. Their original goal was serving as a tool for solving engineering problems. With this goal in mind, these techniques are characterized by manipulating arrays of floating-point numbers (there exist versions of ES for discrete problems, but they are much more popular for continuous optimization). As EP, mutation is sometimes the unique reproductive operator used in ES; it is not rare to also consider recombination (i.e., the construction of new solutions by combining portions of some individuals) though. A very important feature of ES is the utilization of self-adaptive mechanisms for controlling the application of mutation. These mechanisms are aimed at optimizing the progress of the search by evolving not only the solutions for the problem being considered, but also some parameters for mutating these solutions (in a typical situation,

[1] The mere fact that this high-level heuristic template can host a low-level heuristic, justifies using the term *metaheuristic*, as it will be seen later.

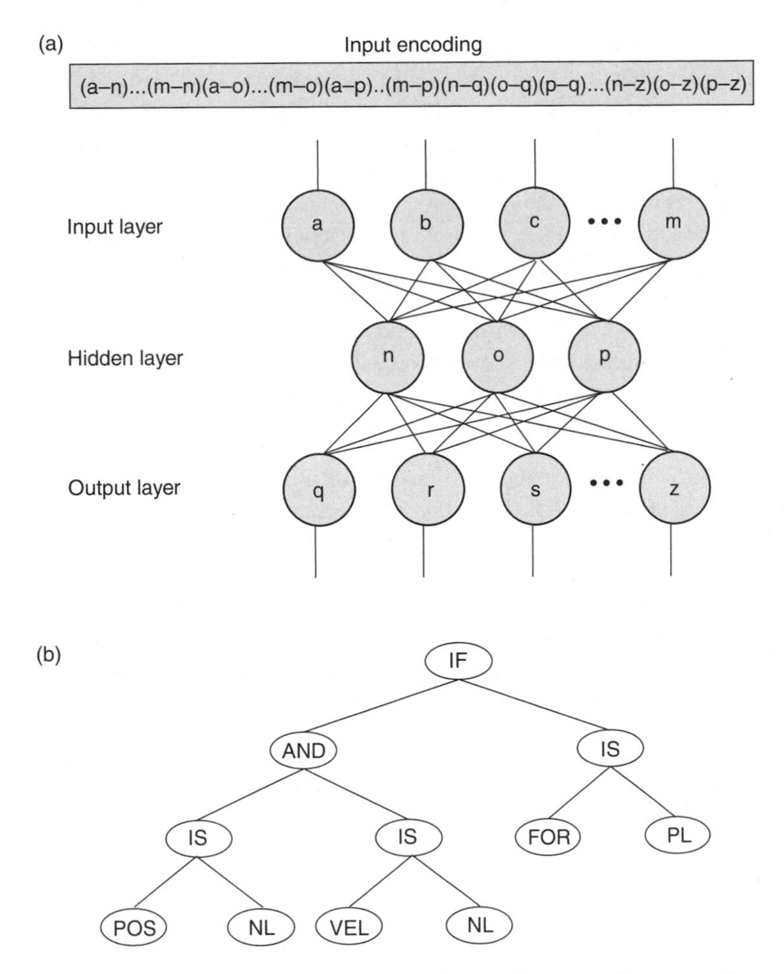

FIGURE 1.3 Two examples of complex representations. (a) A graph representing a neural network. (b) A tree representing a fuzzy rule.

an ES individual is a pair $(\vec{x}, \vec{\sigma})$, where $\vec{\sigma}$ is a vector of standard deviations used to control the Gaussian mutation exerted on the actual solution $\vec{x}$).

- *Genetic Algorithms* (GAs): GAs are possibly the most widespread variant of EAs. They were conceived by Holland [14]. His work has had a great influence in the development of the field, to the point that some portions — arguably extrapolated — of it were taken almost like dogmas (i.e., the ubiquitous use of binary strings as chromosomes). The main feature of GAs is the use of a recombination (or *crossover*) operator as the primary search tool. The rationale is the assumption that different parts of the optimal solution can be independently discovered, and be later combined to create better solutions. Additionally, mutation is also used, but it was usually considered a secondary background operator whose purpose is merely "keeping the pot boiling" by introducing new information in the population (this classical interpretation is no longer considered valid though).

These families have not grown in complete isolation from each other. On the contrary, numerous researchers built bridges among them. As a result of this interaction, the borders of these classical families tend to be fuzzy (the reader may check [15] for a unified presentation of EA families), and new variants have emerged. We can cite the following:

- *Evolution Programs* (EPs): This term is due to Michalewicz [5], and comprises those techniques that, while using the principles of functioning of GAs, evolve complex data structures, as in EP.

Nowadays, it is customary to use the acronym GA — or more generally EA — to refer to such an algorithm, leaving the term "traditional GA" to denote classical bit-string based GAs.

- *Genetic Programming* (GP): The roots of GP can be traced back to the work of Cramer [16], but it is undisputable that it has been Koza [17] the researcher who promoted GP to its current status. Essentially, GP could be viewed as an evolution program in which the structures evolved represent computer programs. Such programs are typically encoded by trees (see Figure 1.3[b]). The final goal of GP is the automatic design of a program for solving a certain task, formulated as a collection of (input, output) examples.
- *Memetic Algorithms* (MAs): These techniques owe their name to Moscato [18]. Some widespread misconception equates MAs to EAs augmented with local search; although such an augmented EA could be indeed considered a MA, other possibilities exist for defining MAs. In general, a MA is problem-aware EA [19]. This problem awareness is typically acquired by combining the EA with existing algorithms such as hill climbing, branch and bound, etc.

In addition to the different EA variants mentioned above, there exist several other techniques that could also fall within the scope of EAs, such as *Ant Colony Optimization* [20], *Distribution Estimation Algorithms* [21], or *Scatter Search* [22] among others. All of them rely on achieving some kind of balance between the exploration of new regions of the search space, and the exploitation of regions known to be promising [23], so as to minimize the computational effort for finding the desired solution. Nevertheless, these techniques exhibit very distinctive features that make them depart from the general pseudocode depicted in Figure 1.2. The broader term *metaheuristic* (e.g., see Reference 24) is used to encompass this larger set of modern optimization techniques, including EAs.

1.4 Dissecting an Evolutionary Algorithm

Once the general structure of an EA has been presented, we will get into more detail on the different components of the algorithm.

1.4.1 The Fitness Function

This is an essential component of the EA, to the point that some early (and nowadays discredited) views of EAs considered it as the unique point of interaction with the problem that is intended to be solved. This way, the fitness function measured how good a certain tentative solution is for the problem of interest. This interpretation has given rise to several misconceptions, the most important being the equation "fitness = quality of a solution." There are many examples in which this is simple not true [19], for example, tackling the satisfiability problem with EAs (i.e., finding the truth assignment that makes a logic formula in conjunctive normal form be satisfied). If quality is used as fitness function, then the search space is divided into solutions with fitness 1 (those satisfying the target formula), and solutions with fitness 0 (those that do not satisfy it). Hence, the EA would be essentially looking for a needle in a haystack (actually, there may be more than one needle in that haystack, but that does not change the situation). A much more reasonable choice is making fitness equal to the number of satisfied clauses in the formula by a certain solution. This introduces a gradation that allows the EA "climbing" in search of near-optimal solutions.

The existence of this gradation is thus a central feature of the fitness function, and its actual implementation is not that important as long this goal is achieved. Of course, implementation issues are important from a computational point of view, since the cost of the EA is typically assumed to be that of evaluating solutions. In this sense, it must be taken into account that fitness can be measured by means of a simple mathematical expression, or may involve performing a complex simulation of a physical system. Furthermore, this fitness function may incorporate some level of noise, or even vary dynamically. The remaining components of the EA must be defined accordingly so as to deal with these features of the fitness function, for example, using a nonhaploid representation [25] (i.e., having more than one chromosome) so as to have a genetic reservoir of worthwhile information in the past, and thus be capable of tackling dynamic changes in the fitness function.

Notice that there may even exist more than one criterion for guiding the search (e.g., we would like to evolve the shape of a set of pillars, so that their strength is maximal, but so that their cost is also minimal). These criteria will be typically partially conflicting. In this case, a *multiobjective* problem is being faced. This can be tackled in different ways, such as performing an aggregation of these multiple criteria into a single value, or using the notion of Pareto dominance (i.e., solution x dominates solution y if, and only if, $f_i(x)$ yields a better or equal value than $f_i(y)$ for all i, where the f_i's represent the multiple criteria being optimized). See References 26 and 27 for details.

1.4.2 Initialization

In order to have the EA started, it is necessary to create the initial population of solutions. This is typically addressed by randomly generating the desired number of solutions. When the alphabet used for representing solutions has low cardinality, this random initialization provides a more or less uniform sample of the solution space. The EA can subsequently start exploring the wide area covered by the initial population, in search of the most promising regions.

In some cases, there exists the risk of not having the initial population adequately scattered all over the search space (e.g., when using small populations and/or large alphabets for representing solutions). It is then necessary to resort to systematic initialization procedures [28], so as to ensure that all symbols are uniformly present in the initial population.

This random initialization can be complemented with the inclusion of heuristic solutions in the initial population. The EA can thus benefit from the existence of other algorithms, using the solutions they provide. This is termed *seeding*, and it is known to be very beneficial in terms of convergence speed, and quality of the solutions achieved [29,30]. The potential drawback of this technique is having the injected solutions taking over the whole population in a few iterations, provoking the stagnation of the algorithm. This problem can be remedied by tuning the selection intensity by some means (e.g., by making an adequate choice of the selection operator, as it will be shown below).

1.4.3 Selection

In combination with replacement, selection is responsible for the competition aspects of individuals in the population. In fact, replacement can be intuitively regarded as the complementary application of the selection operation.

Using the information provided by the fitness function, a sample of individuals from the population is selected for breeding. This sample is obviously biased towards better individuals, that is good — according to the fitness function — solutions should be more likely in the sample than bad solutions.[2]

The most popular techniques are fitness-proportionate methods. In these methods, the probability of selecting an individual for breeding is proportional to its fitness, that is,

$$p_i = \frac{f_i}{\sum_{j \in P} f_j}, \tag{1.1}$$

where f_i is the fitness[3] of individual i, and p_i is the probability of i getting into the reproduction stage. This proportional selection can be implemented in a number of ways. For example, *roulette-wheel selection* rolls

[2]At least, this is customary in genetic algorithms. In other EC families, selection is less important for biasing evolution, and it is done at random (a typical option in evolution strategies), or exhaustively, that is, all individuals undergo reproduction (as it is typical in evolutionary programming).

[3]Maximization is assumed here. In case we were dealing with a minimization problem, fitness should be transformed so as to obtain an appropriate value for this purpose, for example, subtracting it from the highest possible value of the guiding function, or taking the inverse of it.

a dice with $|P|$ sides, such that the ith side has probability p_i. This is repeated as many times as individuals are required in the sample. A drawback of this procedure is that the actual number of instances of individual i in the sample can largely deviate from the expected $|P| \cdot p_i$. *Stochastic Universal Sampling* [31] (SUS) does not have this problem, and produces a sample with minimal deviation from expected values.

Fitness-proportionate selection faces problems when the fitness values of individuals are very similar among them. In this case, p_i would be approximately $|P|^{-1}$ for all $i \in P$, and hence selection would be essentially random. This can be remedied by using fitness scaling. Typical options are (see Reference 5):

- *Linear scaling:* $f_i' = a \cdot f_i + b$, for some real numbers a, b.
- *Exponential scaling:* $f_i' = (f_i)^k$, for some real number k.
- *Sigma truncation:* $f_i' = \max(0, f_i - (\bar{f} - c \cdot \sigma))$, where $\bar{f}$ is the mean fitness of individuals, σ is the fitness standard deviation, and c is a real number.

Another problem is the appearance of an individual whose fitness is much better than the remaining individuals. Such *super-individuals* can quickly take over the population. To avoid this, the best option is using a nonfitness-proportionate mechanism. A first possibility is *ranking selection* [32]: individuals are ranked according to fitness (best first, worst last), and later selected — for example, by means of SUS — using the following probabilities:

$$p_i = \frac{1}{|P|}\left[\eta^- + (\eta^+ - \eta^-)\frac{i-1}{|P|-1}\right], \tag{1.2}$$

where p_i is the probability of selecting the ith best individual, and $\eta^- + \eta^+ = 2$.

Another possibility is using *tournament selection* [33]. In this case, a direct competition is performed whenever an individual needs to be selected. To be precise, α individuals are sampled at random, and the best of them is selected for reproduction. This is repeated as many times as needed. The parameter α is termed the *tournament size*; the higher this value, the stronger the selective pressure. These unproportionate selection methods have the advantage of being insensitive to fitness scaling problems and to the sense of optimization (maximization or minimization). The reader is referred to, for example, References 34 and 35 for a theoretical analysis of the properties of different selection operators.

Regardless of the selection operator used, it was implicity assumed in the previous discussion that any two individuals in the population can mate, that is, all individuals belong to an unstructured centralized population. However, this is not necessarily the case. There exists a long tradition in using structured populations in EC, especially associated to parallel implementations. Among the most widely known types of structured EAs, *distributed* (dEA) and *cellular* (cEA) algorithms are very popular optimization procedures [36].

Decentralizing a single population can be achieved by partitioning it into several subpopulations, where component EAs are run performing sparse exchanges of individuals (dEAs), or in the form of neighborhoods (cEAs). The main difference is that a dEA has a large subpopulation, usually much larger than the single individual that a cEA has typically in every component algorithm. In a dEA, the subpopulations are loosely coupled, while for a cEA they are tightly coupled. Additionally, in a dEA, there exist only a few subpopulations, while in a cEA there is a large number of them.

The use of decentralized populations has a great influence in the selection intensity, since not all individuals have to compete among them. As a consequence, diversity is often better preserved.

1.4.4 Recombination

Recombination is a process that models information exchange among several individuals (typically two of them, but a higher number is possible [37]). This is done by constructing new solutions using the information contained in a number of selected *parents*. If it is the case that the resulting individuals (the *offspring*) are entirely composed of information taken from the parents, then the recombination is said to

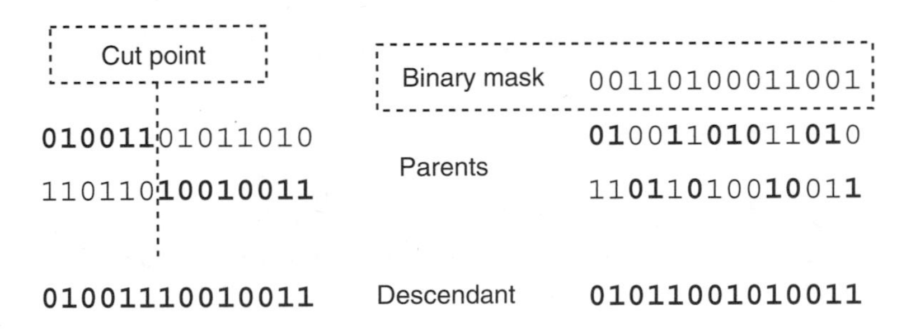

FIGURE 1.4 Two examples of recombination on bitstrings: single-point crossover (left) and uniform crossover (right).

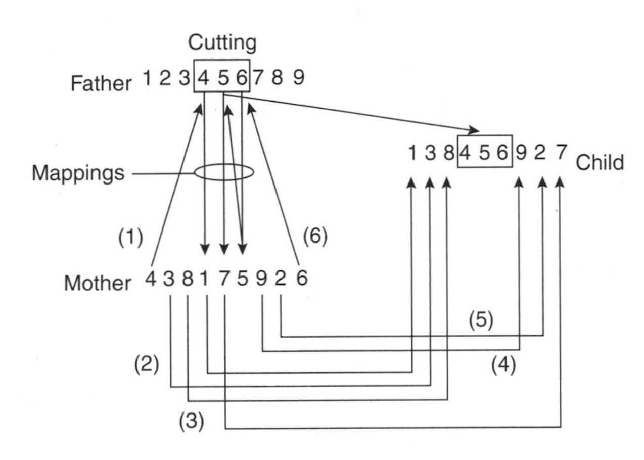

FIGURE 1.5 PMX at work. The numbers in brackets indicate the order in which elements are copied to the descendant.

be *transmitting* [38,39]. This is the case of classical recombination operators for bitstrings such as *single-point crossover*, or *uniform crossover* [40], among others. Figure 1.4 shows an example of the application of these operators.

This property captures the a priori role of recombination: combining good parts of solutions that have been independently discovered. It can be difficult to achieve for certain problem domains though (the *Traveling Salesman Problem* (TSP) is a typical example). In those situations, it is possible to consider other properties of interest such as *respect* or *assortment*. The former refers to the fact that the recombination operator generates descendants carrying all features common to all parents; thus, this property can be seen as a part of the *exploitative* side of the search. On the other hand, *assortment* represents the exploratory side of recombination. A recombination operator is said to be *properly assorting* if, and only if, it can generate descendants carrying any combination of compatible features taken from the parents. The assortment is said to be *weak* if it is necessary to perform several recombinations within the offspring to achieve this effect.

The recombination operator must match the particulars of the representation of solutions chosen. In the GA context, the representation was typically binary, and hence operators such as those depicted in Figure 1.4 were used. The situation is different in other EA families (and indeed in modern GAs too). Without leaving GAs, another very typical representation is that of permutations. Many ad hoc operators have been defined for this purpose, for example, order crossover (OX) [41], partially mapped crossover (PMX; see Figure 1.5) [42], and uniform cycle crossover (UCX) [43] among others. The reader may check [43] for a survey of these different operators.

When used in continuous parameter optimization, recombination can exploit the richness of the representation, and utilize a variety of alternate strategies to create the offspring. Let $(x_1, \ldots, x_n)$ and

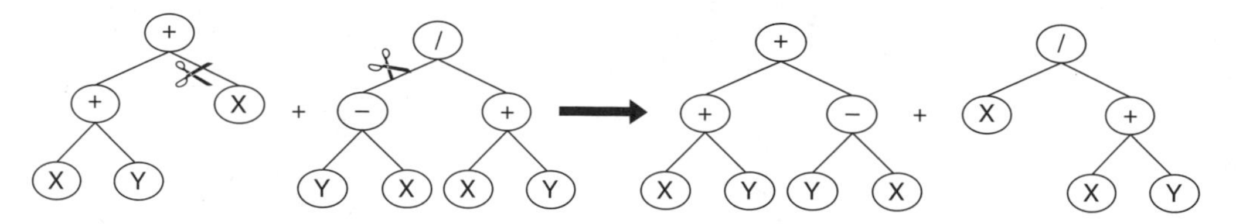

FIGURE 1.6 An example of branch-swapping recombination, as it is typically used in GP.

$(y_1, \ldots, y_n)$ be two arrays of real valued elements to be recombined, and let $(z_1, \ldots, z_n)$ be the resulting array. Some possibilities for performing recombination are the following:

- *Arithmetic recombination*: $z_i = (x_i + y_i)/2, 1 \leq i \leq n$.
- *Geometric recombination*: $z_i = \sqrt{x_i y_i}, 1 \leq i \leq n$.
- *Flat recombination*: $z_i = \alpha x_i + (1 - \alpha)y_i, 1 \leq i \leq n$, where α is a random value in $[0, 1]$.
- *BLX-α recombination* [44]: $z_i = r_i + \beta(s_i - r_i), 1 \leq i \leq n$, where $r_i = \min(x_i, y_i) - \alpha|x_i - y_i|$, $s_i = \max(x_i, y_i) + \alpha|x_i - y_i|$, and β is a random value in $[0, 1]$.
- *Fuzzy recombination*: $z_i = Q(x_i, y_i), 1 \leq i \leq n$, where Q is a fuzzy connective [45].

In the case of self-adaptive schemes as those typically used in ES, the parameters undergoing self-adaption would be recombined as well, using some of these operators. More details on self-adaption will follow in next subsection.

Solutions can be also represented by means of some complex data structure, and the recombination operator must be adequately defined to deal with these (e.g., References 46 to 48). In particular, the field of GP normally uses trees to represent LISP programs [17], rule-bases [49], mathematical expressions, etc. Recombination is usually performed here by swapping branches of the trees involved, as exemplified in Figure 1.6.

1.4.5 Mutation

From a classical point of view (atleast in the GA arena [50]), this was a secondary operator whose mission is to *keep the pot boiling*, continuously injecting new material in the population, but at a low rate (otherwise, the search would degrade to a random walk in the solution space). EP practitioners [11] would disagree with this characterization, claiming a central role for mutation. Actually, it is considered the crucial part of the search engine in this context. This later vision has nowadays propagated to most EC researchers (atleast in the sense of considering mutation as important as recombination).

As it was the case for recombination, the choice of a mutation operator depends on the representation used. In bitstrings (and in general, in linear strings spanning Σ^n, where Σ is arbitrary alphabet) mutation is done by randomly substituting the symbol contained at a certain position by a different symbol. If a permutation representation is used, such a procedure cannot be used for it would not produce a valid permutation. Typical strategies in this case are swapping two randomly chosen positions, or inverting a segment of the permutation. The interested reader may check [51] or [5] for an overview of different options.

If solutions are represented by complex data structures, mutation has to be implemented accordingly. In particular, this is the case of EP, in which, for example, finite automata [52], layered graphs [53], directed acyclic graphs [54], etc., are often evolved. In this domain, it is customary to use more than one mutation operator, making for each individual a choice of which operators will be deployed on it.

In the case of ES applied to continuous optimization, mutation is typically done using Gaussian perturbations, that is,

$$z_i = x_i + N_i(0, \sigma_i), \tag{1.3}$$

where σ_i is a parameter controlling the amplitude of the mutation, and $N(a, b)$ is a random number drawn from a normal distribution with mean a and standard deviation b. The parameters σ_i usually undergo self-adaption. In this case, they are mutated prior to mutating the x_i's as follows:

$$\sigma_i' = \sigma_i \cdot e^{N(0,\tau')+N_i(0,\tau)}, \tag{1.4}$$

where τ and τ' are two parameters termed the *local* and *global learning rate*, respectively. Advanced schemes have been also defined in which a covariance matrix is used rather than independent σ_i's. However, these schemes tend to be unpractical if solutions are highly dimensional. For a better understanding of ES mutation see Reference 55.

1.4.6 Replacement

The role of replacement is keeping the population size constant.[4] To do so, some individuals from the population have to be substituted by some of the individuals created during reproduction. This can be done in several ways:

- *Replacement-of-the-worst*: the population is sorted according to fitness, and the new individuals replace the worst ones from the population.
- *Random replacement*: the individuals to be replaced are selected at random.
- *Tournament replacement*: a subset of α individuals is selected at random, and the worst one is selected for replacement. Notice that if $\alpha = 1$ we have random replacement.
- *Direct replacement*: the offspring replace their parents.

Some variants of these strategies are possible. For example, it is possible to consider the *elitist* versions of these, and only perform replacement if the new individual is better than the individual it has to replace.

Two replacement strategies (*comma* and *plus*) are also typically considered in the context of ES and EP. Comma replacement is analogous to replacement of the worst, with the addition that the number of new individuals $|P''|$ (also denoted by λ) can be larger than the population size $|P|$ (also denoted by μ). In this case, the population is constructed using the best μ out of the λ new individuals. As to the plus strategy, it would be the elitist counterpart of the former, that is, pick the best μ individuals out of the μ old individuals plus the λ new ones. The notation (μ, λ) — EA and $(\mu + \lambda)$ — EA is used to denote these two strategies.

It must be noted that the term "elitism" is often used as well to denote replacement-of-the-worst strategies in which $|P''| < |P|$. This strategy is very commonly used, and ensures that the best individual found so far is never lost. An extreme situation takes place when $|P''| = 1$, that is, just a single individual is generated in each iteration of the algorithm. This is known as *steady-state* reproduction, and it is usually associated with faster convergence of the algorithm. The term *generational* is used to designate the classical situation in which $|P''| = |P|$.

1.5 Fields of Application of EAs

Evolutionary algorithms have been thoroughly used in many domains. One of the most conspicuous fields in which these techniques have been utilized is combinatorial optimization (CO). This way, EAs have been used to solve classical *NP* — hard problems such as the Travelling Salesman Problem [57–59], the Multiple Knapsack Problem [60,61], Number Partitioning [62,63], Max Independent Set [64,65], and Graph Coloring [66,67], among others.

Other nonclassical — yet important — CO problems to which EAs have been applied are scheduling (in many variants [43, 68–71]), timetabling [72,73], lot-sizing [74], vehicle routing [75,76], quadratic assignment [77,78], placement problems [79,80], and transportation problems [81].

[4] Although it is not mandatory to do so [56], it is common practice to use populations of fixed size.

Telecommunications is another field that has witnessed the successful application of EAs. For example, EAs have been applied to the placement of antennas and converters [82,83], frequency assignment [84–86], digital data network design [87], predicting bandwidth demands in ATM networks [88], error code design [89,90], etc. See also Reference 91.

Evolutionary algorithms have been actively used in electronics and engineering as well. For example, work has been done in structure optimization [92], aeronautic design [93], power planning [94], circuit design [95] computer-aided design [96], analogue-network synthesis [97], and service restoration [98] among other areas.

Besides the precise application areas mentioned before, EAs have been also utilized in many other fields such as, for example, medicine [99,100], economics [101,102], mathematics [103,104], biology [105–107], etc. The reader may try querying any bibliographical database or web search engine for "evolutionary algorithm application" to get an idea of the vast number of problems that have been tackled with EAs.

1.6 Conclusions

EC is a fascinating field. Its optimization philosophy is appealing, and its practical power is striking. Whenever the user is faced with a hard search/optimization task that she cannot solve by classical means, trying EAs is a must. The extremely brief overview of EA applications presented before can convince the reader that a "killer approach" is in her hands.

EC is also a very active research field. One of the main weaknesses of the field is the absence of a conclusive general theoretical basis, although great advances are being made in this direction, and in-depth knowledge is available about certain idealized EA models.

Regarding the more practical aspects of the paradigm, two main streamlines can be identified: parallelizing and hybridizing. The use of decentralized EAs in the context of multiprocessors or networked systems can result in enormous performance improvement [108], and constitutes an ideal option for exploiting the availability of distributed computing resources. As to hybridization, it has become evident in the last years that it constitutes a crucial factor for the successful use of EAs in real-world endeavors. This can be achieved by hard-wiring problem-knowledge within the EA, or by combining it with other techniques. In this sense, the reader is encouraged to read other essays in this volume to get valuable ideas on suitable candidates for this hybridization.

Acknowledgments

This work has been partially funded by the Ministry of Science and Technology (MCYT) and Regional Development European Found (FEDER) under contract TIC2002-04498-C05-02 (the TRACER project) `http://tracer.lcc.uma.es`.

References

[1] T. Bäck, D.B. Fogel, and Z. Michalewicz. *Handbook of Evolutionary Computation.* Oxford University Press, New York, 1997.

[2] T.C. Jones. Evolutionary Algorithms, Fitness Landscapes and Search. Ph.D. thesis, University of New Mexico, 1995.

[3] C. Darwin. *On the Origin of Species by Means of Natural Selection.* John Murray, London, 1859.

[4] G. Mendel. Versuche über pflanzen-hybriden. *Verhandlungen des Naturforschendes Vereines in Brünn*, 4: 3–47, 1865.

[5] Z. Michalewicz. *Genetic Algorithms + Data Structures = Evolution Programs.* Springer-Verlag, Berlin, 1992.

[6] J. Huxley. *Evolution, the Modern Synthesis.* Harper, New York, 1942.

[7] M. Kimura. Evolutionary rate at the molecular level. *Nature*, 217: 624–626, 1968.

[8] S.J. Gould and N. Elredge. Punctuated equilibria: The tempo and mode of evolution reconsidered. *Paleobiology*, 32: 115–151, 1977.

[9] C.G. Langton. Artificial life. In C.G. Langton, Ed., *Artificial Life 1*. Addison-Wesley, Santa Fe, NM, 1989, pp. 1–47.

[10] D.B. Fogel. *Evolutionary Computation: The Fossil Record*. Wiley-IEEE Press, Piscataway, NJ, 1998.

[11] L.J. Fogel, A.J. Owens, and M.J. Walsh. *Artificial Intelligence Through Simulated Evolution*. John Wiley & Sons, New York, 1966.

[12] I. Rechenberg. *Evolutionsstrategie: Optimierung technischer Systeme nach Prinzipien der biologischen Evolution*. Frommann-Holzboog Verlag, Stuttgart, 1973.

[13] H.P. Schwefel. *Numerische Optimierung von Computer–Modellen mittels der Evolutionsstrategie*, Vol. 26 of *Interdisciplinary Systems Research*. Birkhäuser, Basel, 1977.

[14] J.H. Holland. *Adaptation in Natural and Artificial Systems*. University of Michigan Press, Ann Harbor, MI, 1975.

[15] T. Bäck. *Evolutionary Algorithms in Theory and Practice*. Oxford University Press, New York, 1996.

[16] M.L. Cramer. A representation for the adaptive generation of simple sequential programs. In J.J. Grefenstette, Ed., *Proceedings of the First International Conference on Genetic Algorithms*. Lawrence Erlbaum Associates, Hillsdale, NJ, 1985.

[17] J.R. Koza. *Genetic Programming*. MIT Press, Cambridge, MA, 1992.

[18] P. Moscato. On Evolution, Search, Optimization, Genetic Algorithms and Martial Arts: Towards Memetic Algorithms. Technical report Caltech Concurrent Computation Program, Report 826, California Institute of Technology, Pasadena, CA, USA, 1989.

[19] P. Moscato and C. Cotta. A gentle introduction to memetic algorithms. In F. Glover and G. Kochenberger, Eds., *Handbook of Metaheuristics*. Kluwer Academic Publishers, Boston, MA, 2003, pp. 105–144.

[20] M. Dorigo and G. Di Caro. The ant colony optimization meta-heuristic. In D. Corne, M. Dorigo, and F. Glover, Eds., *New Ideas in Optimization*. Maiden head, UK, 1999, pp. 11–32.

[21] P. Larrañaga and J.A. Lozano. *Estimation of Distribution Algorithms. A New Tool for Evolutionary Computation*. Kluwer Academic Publishers, Boston, MA, 2001.

[22] M. Laguna and R. Martí. *Scatter Search. Methodology and Implementations in C*. Kluwer Academic Publishers, Boston, MA, 2003.

[23] C. Blum and A. Roli. Metaheuristics in combinatorial optimization: Overview and conceptual comparison. *ACM Computing Surveys*, 35: 268–308, 2003.

[24] F. Glover and G. Kochenberger. *Handbook of Metaheuristics*. Kluwer Academic Publishers, Boston, MA, 2003.

[25] R.E. Smith. Diploid genetic algorithms for search in time varying environments. In *Annual Southeast Regional Conference of the ACM*. ACM Press, New York, 1987, pp. 175–179.

[26] C.A. Coello. A comprehensive survey of evolutionary-based multiobjective optimization techniques. *Knowledge and Information Systems*, 1: 269–308, 1999.

[27] C.A. Coello and A.D. Christiansen. An approach to multiobjective optimization using genetic algorithms. In C.H. Dagli, M. Akay, C.L.P. Chen, B.R. Fernández, and J. Ghosh, Eds., *Intelligent Engineering Systems Through Artificial Neural Networks*, Vol. 5. ASME Press, St. Louis, MO, 1995, pp. 411–416.

[28] C.R. Reeves. Using genetic algorithms with small populations. In S. Forrest, Ed., *Proceedings of the Fifth International Conference on Genetic Algorithms*. Morgan Kaufmann, San Mateo, CA, 1993, pp. 92–99.

[29] C. Cotta. On the evolutionary inference of temporal Boolean networks. In J. Mira and J.R. Álvarez, Eds., *Computational Methods in Neural Modeling*, Vol. 2686 of *Lecture Notes in Computer Science*. Springer-Verlag, Berlin, Heidelberg, 2003, pp. 494–501.

[30] C. Ramsey and J.J. Grefensttete. Case-based initialization of genetic algorithms. In S. Forrest, Ed., *Proceedings of the Fifth International Conference on Genetic Algorithms*. Morgan Kaufmann, San Mateo, CA, 1993, pp. 84–91.

[31] J.E. Baker. Reducing bias and inefficiency in the selection algorithm. In J.J. Grefenstette, Ed., *Proceedings of the Second International Conference on Genetic Algorithms*. Lawrence Erlbaum Associates, Hillsdale, NJ, 1987, pp. 14–21.
[32] D.L. Whitley. Using reproductive evaluation to improve genetic search and heuristic discovery. In J.J. Grefenstette, Ed., *Proceedings of the Second International Conference on Genetic Algorithms*. Lawrence Erlbaum Associates, Hillsdale, NJ, 1987, pp. 116–121.
[33] T. Bickle and L. Thiele. A mathematical analysis of tournament selection. In L.J. Eshelman, Ed., *Proceedings of the Sixth International Conference on Genetic Algorithms*. Morgan Kaufmann, San Francisco, CA, 1995, pp. 9-16.
[34] E. Cantú-Paz. Order statistics and selection methods of evolutionary algorithms. *Information Processing Letters*, 82: 15–22, 2002.
[35] K. Deb and D. Goldberg. A comparative analysis of selection schemes used in genetic algorithms. In G.J. Rawlins, Ed., *Foundations of Genetic Algorithms*. San Mateo, CA, 1991, pp. 69–93.
[36] E. Alba and J.M. Troya. A survey of parallel distributed genetic algorithms. *Complexity*, 4: 31–52, 1999.
[37] A.E. Eiben, P.-E. Raue, and Zs. Ruttkay. Genetic algorithms with multi-parent recombination. In Y. Davidor, H.-P. Schwefel, and R. Männer, Eds., *Parallel Problem Solving from Nature III*, Vol. 866 of *Lecture Notes in Computer Science*. Springer-Verlag, Berlin, Heidelberg, 1994, pp. 78–87.
[38] C. Cotta and J.M. Troya. Information processing in transmitting recombination. *Applied Mathematics Letters*, 16: 945–948, 2003.
[39] N.J. Radcliffe. The algebra of genetic algorithms. *Annals of Mathematics and Artificial Intelligence*, 10: 339–384, 1994.
[40] G. Syswerda. Uniform crossover in genetic algorithms. In J.D. Schaffer, Ed., *Proceedings of the Third International Conference on Genetic Algorithms*. Morgan Kaufmann, San Mateo, CA, 1989, pp. 2–9.
[41] L. Davis. *Handbook of Genetic Algorithms*. Van Nostrand Reinhold Computer Library, New York, 1991.
[42] D.E. Goldberg and R. Lingle, Jr. Alleles, loci and the traveling salesman problem. In J.J. Grefenstette, Ed., *Proceedings of an International Conference on Genetic Algorithms*. Lawrence Erlbaum Associates, Hillsdale, NJ, 1985.
[43] C. Cotta and J.M. Troya. Genetic forma recombination in permutation flowshop problems. *Evolutionary Computation*, 6: 25–44, 1998.
[44] L.J. Eshelman and J.D. Schaffer. Real-coded genetic algorithms and interval-schemata. In D. Whitley, Ed., *Foundations of Genetic Algorithms 2*. Morgan Kaufmann Publishers, San Mateo, CA, 1993, pp. 187–202.
[45] F. Herrera, M. Lozano, and J.L. Verdegay. Dynamic and heuristic fuzzy connectives-based crossover operators for controlling the diversity and convengence of real coded genetic algorithms. *Journal of Intelligent Systems*, 11: 1013–1041, 1996.
[46] E. Alba, J.F. Aldana, and J.M. Troya. Full automatic ann design: A genetic approach. In J. Cabestany, J. Mira, and A. Prieto, Eds., *New Trends in Neural Computation*, Vol. 686 of *Lecture Notes in Computer Science*. Springer-Verlag, Heidelberg, 1993, pp. 399–404.
[47] E. Alba and J.M. Troya. Genetic algorithms for protocol validation. In H.M. Voigt, W. Ebeling, I. Rechenberg, and H.-P. Schwefel, Eds., *Parallel Problem Solving from Nature IV*. Springer-Verlag, Berlin, Heidelberg, 1996, pp. 870–879.
[48] C. Cotta and J.M. Troya. Analyzing directed acyclic graph recombination. In B. Reusch, Ed., *Computational Intelligence: Theory and Applications*, Vol. 2206 of *Lecture Notes in Computer Science*. Springer-Verlag, Berlin, Heidelberg, 2001, pp. 739–748.
[49] E. Alba, C. Cotta, and J.M. Troya. Evolutionary design of fuzzy logic controllers using strongly-typed GP. *Mathware & Soft Computing*, 6: 109–124, 1999.

[50] D.E. Goldberg. *Genetic Algorithms in Search, Optimization and Machine Learning.* Addison-Wesley, Reading, MA, 1989.
[51] A.E. Eiben and J.E. Smith. *Introduction to Evolutionary Computing.* Springer-Verlag, Berlin, Heidelberg, 2003.
[52] C.H. Clelland and D.A. Newlands. PFSA modelling of behavioural sequences by evolutionary programming. In R.J. Stonier and X.H. Yu, Eds., *Complex Systems: Mechanism for Adaptation.* IOS Press, Rockhampton, Queensland, Australia, 1994, pp. 165–172.
[53] X. Yao and Y. Liu. A new evolutionary system for evolving artificial neural networks. *IEEE Transactions on Neural Networks*, 8: 694–713, 1997.
[54] M.L. Wong, W. Lam, and K.S. Leung. Using evolutionary programming and minimum description length principle for data mining of bayesian networks. *IEEE Transactions on Pattern Analysis and Machine Intelligence*, 21: 174–178, 1999.
[55] H.-G. Beyer. *The Theory of Evolution Strategies.* Springer-Verlag, Berlin, Heidelberg, 2001.
[56] F. Fernandez, L. Vanneschi, and M. Tomassini. The effect of plagues in genetic programming: A study of variable-size populations. In C. Ryan et al., Eds., *Genetic Programming, Proceedings of EuroGP'2003*, Vol. 2610 of *Lecture Notes in Computer Science.* Springer-Verlag, Berlin, Heidelberg, 2003, pp. 320–329.
[57] S. Chatterjee, C. Carrera, and L. Lynch. Genetic algorithms and traveling salesman problems. *European Journal of Operational Research*, 93: 490–510, 1996.
[58] D.B. Fogel. An evolutionary approach to the traveling salesman problem. *Biological Cybernetics*, 60: 139–144, 1988.
[59] P. Merz and B. Freisleben. Genetic local search for the TSP: New Results. In *Proceedings of the 1997 IEEE International Conference on Evolutionary Computation.* IEEE Press, Indianapolis, USA, 1997, pp. 159–164.
[60] C. Cotta and J.M. Troya. A hybrid genetic algorithm for the 0–1 multiple knapsack problem. In G.D. Smith, N.C. Steele, and R.F. Albrecht, Eds., *Artificial Neural Nets and Genetic Algorithms 3.* Springer-Verlag, Wien New York, 1998, pp. 251–255.
[61] S. Khuri, T. Bäck, and J. Heitkötter. The zero/one multiple knapsack problem and genetic algorithms. In E. Deaton, D. Oppenheim, J. Urban, and H. Berghel, Eds., *Proceedings of the 1994 ACM Symposium of Applied Computation proceedings.* ACM Press, New York, 1994, pp. 188–193.
[62] R. Berretta, C. Cotta, and P. Moscato. Enhancing the performance of memetic algorithms by using a matching-based recombination algorithm: Results on the number partitioning problem. In M. Resende and J. Pinho de Sousa, Eds., *Metaheuristics: Computer-Decision Making.* Kluwer Academic Publishers, Boston, MA, 2003, pp. 65–90.
[63] D.R. Jones and M.A. Beltramo. Solving partitioning problems with genetic algorithms. In R.K. Belew and L.B. Booker, Eds., In *Proceedings of the Fourth International Conference on Genetic Algorithms.* Morgan Kaufmann, San Mateo, CA, 1991, pp. 442–449.
[64] C.C. Aggarwal, J.B. Orlin, and R.P. Tai. Optimized crossover for the independent set problem. *Operations Research*, 45: 226–234, 1997.
[65] M. Hifi. A genetic algorithm-based heuristic for solving the weighted maximum independent set and some equivalent problems. *Journal of the Operational Research Society*, 48: 612–622, 1997.
[66] D. Costa, N. Dubuis, and A. Hertz. Embedding of a sequential procedure within an evolutionary algorithm for coloring problems in graphs. *Journal of Heuristics*, 1: 105–128, 1995.
[67] C. Fleurent and J.A. Ferland. Genetic and hybrid algorithms for graph coloring. *Annals of Operations Research*, 63: 437–461, 1997.
[68] S. Cavalieri and P. Gaiardelli. Hybrid genetic algorithms for a multiple-objective scheduling problem. *Journal of Intelligent Manufacturing*, 9: 361–367, 1998.
[69] D. Costa. An evolutionary tabu search algorithm and the NHL scheduling problem. *INFOR*, 33: 161–178, 1995.
[70] C.F. Liaw. A hybrid genetic algorithm for the open shop scheduling problem. *European Journal of Operational Research*, 124: 28–42, 2000.

[71] L. Ozdamar. A genetic algorithm approach to a general category project scheduling problem. *IEEE Transactions on Systems, Man and Cybernetics, Part C (Applications and Reviews)*, 29: 44–59, 1999.

[72] E.K. Burke, J.P. Newall, and R.F. Weare. Initialisation strategies and diversity in evolutionary timetabling. *Evolutionary Computation*, 6: 81–103, 1998.

[73] B. Paechter, R.C. Rankin, and A. Cumming. Improving a lecture timetabling system for university wide use. In E.K. Burke and M. Carter, Eds., *The Practice and Theory of Automated Timetabling II*, Vol. 1408 of *Lecture Notes in Computer Science*. Springer-Verlag, Berlin, 1998, pp. 156–165.

[74] K. Haase and U. Kohlmorgen. Parallel genetic algorithm for the capacitated lot-sizing problem. In Kleinschmidt et al., Eds., *Operations Research Proceedings*. Springer-Verlag, Berlin, 1996, pp. 370–375.

[75] J. Berger and M. Barkaoui. A hybrid genetic algorithm for the capacitated vehicle routing problem. In E. Cantú-Paz, Ed., *Proceedings of the Genetic and Evolutionary Computation Conference 2003*, Vol. 2723 of *Lecture Notes in Computer Science*. Springer-Verlag, Berlin, Heidelberg, 2003, pp. 646–656.

[76] J. Berger, M. Salois, and R. Begin. A hybrid genetic algorithm for the vehicle routing problem with time windows. In R.E. Mercer and E. Neufeld, Eds., *Advances in Artificial Intelligence. 12th Biennial Conference of the Canadian Society for Computational Studies of Intelligence*. Springer-Verlag, Berlin, 1998, pp. 114-127.

[77] P. Merz and B. Freisleben. Genetic algorithms for binary quadratic programming. In W. Banzhaf et al., Eds., *Proceedings of the 1999 Genetic and Evolutionary Computation Conference*, Morgan Kaufmann, San Francisco, CA, 1999, pp. 417–424.

[78] P. Merz and B. Freisleben. Fitness landscape analysis and memetic algorithms for the quadratic assignment problem. *IEEE Transactions on Evolutionary Computation*, 4: 337–352, 2000.

[79] E. Hopper and B. Turton. A genetic algorithm for a 2d industrial packing problem. *Computers & Industrial Engineering*, 37: 375–378, 1999.

[80] R.M. Krzanowski and J. Raper. Hybrid genetic algorithm for transmitter location in wireless networks. *Computers, Environment and Urban Systems*, 23: 359–382, 1999.

[81] M. Gen, K. Ida, and L. Yinzhen. Bicriteria transportation problem by hybrid genetic algorithm. *Computers & Industrial Engineering*, 35: 363–366, 1998.

[82] P. Calegar, F. Guidec, P. Kuonen, and D. Wagner. Parallel island-based genetic algorithm for radio network design. *Journal of Parallel and Distributed Computing*, 47: 86–90, 1997.

[83] C. Vijayanand, M.S. Kumar, K.R. Venugopal, and P.S. Kumar. Converter placement in all-optical networks using genetic algorithms. *Computer Communications*, 23: 1223–1234, 2000.

[84] C. Cotta and J.M. Troya. A comparison of several evolutionary heuristics for the frequency assignment problem. In J. Mira and A. Prieto, Eds., *Connectionist Models of Neurons, Learning Processes, and Artificial Intelligence*, Vol. 2084 of *Lecture Notes in Computer Science*. Springer-Verlag, Berlin, Heidelberg, 2001, pp. 709–716.

[85] R. Dorne and J.K. Hao. An evolutionary approach for frequency assignment in cellular radio networks. In *1995 IEEE International Conference on Evolutionary Computation*. IEEE Press, Perth, Australia, 1995, pp. 539–544.

[86] A. Kapsalis, V.J. Rayward-Smith, and G.D. Smith. Using genetic algorithms to solve the radio link frequency assignment problem. In D.W. Pearson, N.C. Steele, and R.F. Albretch, Eds., *Artificial Neural Nets and Genetic Algorithms*. Springer-Verlag, Wien New York, 1995, pp. 37–40.

[87] C.H. Chu, G. Premkumar, and H. Chou. Digital data networks design using genetic algorithms. *European Journal of Operational Research*, 127: 140–158, 2000.

[88] N. Swaminathan, J. Srinivasan, and S.V. Raghavan. Bandwidth-demand prediction in virtual path in atm networks using genetic algorithms. *Computer Communications*, 22: 1127–1135, 1999.

[89] H. Chen, N.S. Flann, and D.W. Watson. Parallel genetic simulated annealing: A massively parallel SIMD algorithm. *IEEE Transactions on Parallel and Distributed Systems*, 9: 126–136, 1998.

[90] K. Dontas and K. De Jong. Discovery of maximal distance codes using genetic algorithms. In *Proceedings of the Second International IEEE Conference on Tools for Artificial Intelligence.* IEEE Press, Herndon, VA, 1990, pp. 805–811.

[91] D.W. Corne, M.J. Oates, and G.D. Smith. *Telecommunications Optimization: Heuristic and Adaptive Techniques.* John Wiley, New York, 2000.

[92] I.C. Yeh. Hybrid genetic algorithms for optimization of truss structures. *Computer Aided Civil and Infrastructure Engineering,* 14: 199–206, 1999.

[93] D. Quagliarella and A. Vicini. Hybrid genetic algorithms as tools for complex optimisation problems. In P. Blonda, M. Castellano, and A. Petrosino, Eds., *New Trends in Fuzzy Logic II. Proceedings of the Second Italian Workshop on Fuzzy Logic.* World Scientific, Singapore, 1998, pp. 300–307.

[94] A.J. Urdaneta, J.F. Gómez, E. Sorrentino, L. Flores, and R. Díaz. A hybrid genetic algorithm for optimal reactive power planning based upon successive linear programming. *IEEE Transactions on Power Systems,* 14: 1292–1298, 1999.

[95] M. Guotian and L. Changhong. Optimal design of the broadband stepped impedance transformer based on the hybrid genetic algorithm. *Journal of Xidian University,* 26: 8–12, 1999.

[96] B. Becker and R. Drechsler. Ofdd based minimization of fixed polarity Reed-Muller expressions using hybrid genetic algorithms. In *Proceedings of the IEEE International Conference on Computer Design: VLSI in Computers and Processor.* IEEE, Los Alamitos, CA, 1994, pp. 106–110.

[97] J.B. Grimbleby. Hybrid genetic algorithms for analogue network synthesis. In *Proceedings of the 1999 Congress on Evolutionary Computation.* IEEE, Washington D.C., 1999, pp. 1781–1787.

[98] A. Augugliaro, L. Dusonchet, and E. Riva-Sanseverino. Service restoration in compensated distribution networks using a hybrid genetic algorithm. *Electric Power Systems Research,* 46: 59–66, 1998.

[99] M. Sipper and C.A. Peña Reyes. Evolutionary computation in medicine: An overview. *Artificial Intelligence in Medicine,* 19: 1–23, 2000.

[100] R. Wehrens, C. Lucasius, L. Buydens, and G. Kateman. HIPS, A hybrid self-adapting expert system for nuclear magnetic resonance spectrum interpretation using genetic algorithms. *Analytica Chimica ACTA,* 277: 313–324, 1993.

[101] J. Alander. Indexed Bibliography of Genetic Algorithms in Economics. Technical report 94-1-ECO, University of Vaasa, Department of Information Technology and Production Economics, 1995.

[102] F. Li, R. Morgan, and D. Williams. Economic environmental dispatch made easy with hybrid genetic algorithms. In *Proceedings of the International Conference on Electrical Engineering,* Vol. 2. International Academic Publishers, Beijing, China, 1996, pp. 965–969.

[103] C. Reich. Simulation if imprecise ordinary differential equations using evolutionary algorithms. In J. Carroll, E. Damiani, H. Haddad, and D. Oppenheim, Eds., *ACM Symposium on Applied Computing 2000.* ACM Press, New York, 2000, pp. 428–432.

[104] X. Wei and F. Kangling. A hybrid genetic algorithm for global solution of nondifferentiable nonlinear function. *Control Theory & Applications,* 17: 180–183, 2000.

[105] C. Cotta and P. Moscato. Inferring phylogenetic trees using evolutionary algorithms. In J.J. Merelo, P. Adamidis, H.-G. Beyer, J.-L. Fernández-Villacañas, and H.-P. Schwefel, Eds., *Parallel Problem Solving from Nature VII,* Vol. 2439 of *Lecture Notes in Computer Science.* Springer-Verlag, Berlin, 2002, pp. 720–729.

[106] G.B. Fogel and D.W. Corne. *Evolutionary Computation in Bioinformatics.* Morgan Kaufmann, San Francisco, CA, 2003.

[107] R. Thomsen, G.B. Fogel, and T. Krink. A clustal alignment improver using evolutionary algorithms. In David B. Fogel, Xin Yao, Garry Greenwood, Hitoshi Iba, Paul Marrow, and Mark Shackleton, Eds., *Proceedings of the Fourth Congress on Evolutionary Computation (CEC-2002)* Vol. 1. 2002, pp. 121–126.

[108] E. Alba. Parallel evolutionary algorithms can achieve super-linear performance. *Information Processing Letters,* 82: 7–13, 2002.

2

An Overview of Neural Networks Models

Javid Taheri
Albert Y. Zomaya

2.1 Introduction

Artificial Neural Networks have been one of the most active areas of research in computer science over the last 50 years with periods of intense activity interrupted by episodes of hiatus [1]. The premise for the evolution of the theory of artificial Neural Networks stems from the basic neurological structure of living organisms. A cell is the most important constituent of these life forms. These cells are connected by "synapses," that are the links that carry messages between cells. In fact, by using synapses to carry the pulses, cells can activate each other with different threshold values to form a decision or memorize an event. Inspired by this simplistic vision of how messages are transferred between cells, scientists invented a new computational approach, which became popularly known as Artificial Neural Networks (or Neural Networks for short) and used it extensively to target a wide range of problems in many application areas.

Although the shape or configurations of different Neural Networks may look different at the first glance, they are almost similar in structure. Every neural network consists of "cells" and "links." Cells are the computational part of the network that perform reasoning and generate activation signals for other cells, while links connect the different cells and enable messages to flow between cells. Each link is usually a one directional connection with a weight which affects the carried message in a certain way. This means, that a link receives a value (message) from an input cell, multiplies it by a given weight, and then passes it to the output cell. In its simplest form, a cell can have three states (of activation): +1 (TRUE), 0, and −1 (FALSE) [1].

2.2 General Structure of a Neural Network

Cells (or neurons) can have more sophisticated structure that can handle complex problems. These neurons can basically be linear or nonlinear functions with or without biases. Figure 2.1 shows two simple neurons, unbiased and biased.

2.2.1 Single- and Multi-Layer Perceptrons

The single-layer perceptron is one of the simplest classes of Neural Networks [1]. The general overview of this network is shown in Figure 2.2 while the network has n inputs and generates only one output. The input of the function $f(\cdot)$ is actually a linear combination of the network's inputs. In this case, W is a vector of neuron weights, X is the input vector, and y is the only output of the network defined as follows:

$$y = f(W \cdot X + b),$$

$$W = (w_1 \quad w_2 \quad \ldots \quad w_n),$$

$$X = (x_1 \quad x_2 \quad \ldots \quad x_n)^{\mathrm{T}}.$$

The above-mentioned basic structure can be extended to produce networks with more than one output. In this case, each output has its own weights and is completely uncorrelated to the other outputs. Figure 2.3

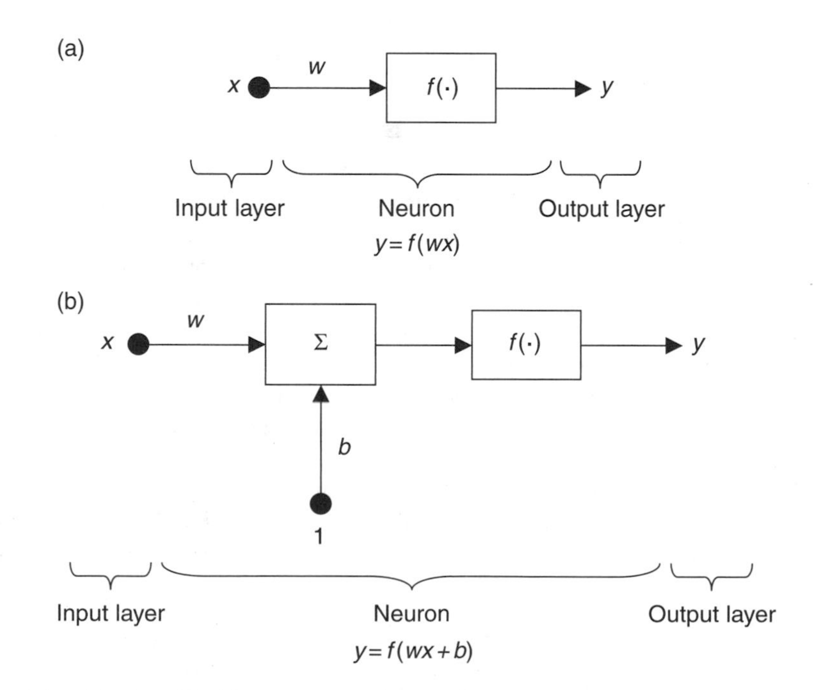

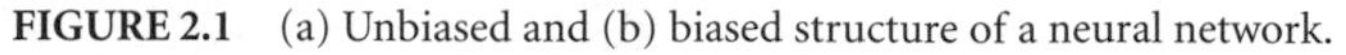

FIGURE 2.1 (a) Unbiased and (b) biased structure of a neural network.

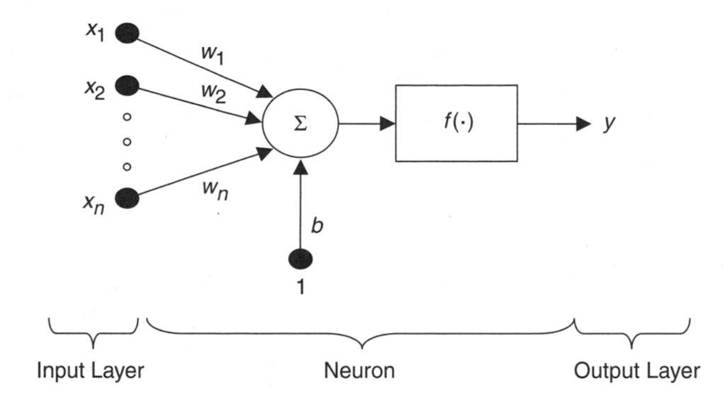

FIGURE 2.2 A single output single-layer perceptron.

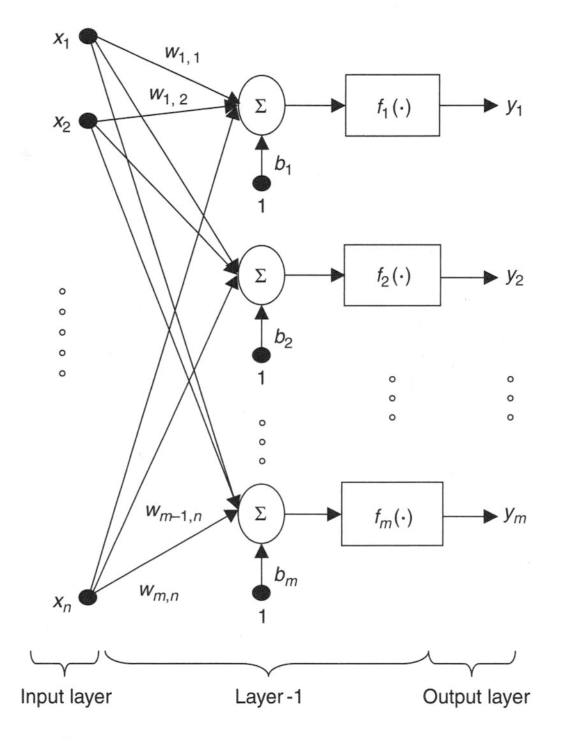

FIGURE 2.3 A multi output single-layer perceptron.

shows an instant of such a network with the following formulas:

$$
\begin{aligned}
Y &= F(W \cdot X + B), \\
W &= \begin{bmatrix} w_{1,1} & w_{1,2} & \dots & w_{1,n} \\ w_{2,1} & & & \\ \vdots & & & \\ w_{m,1} & & \dots & w_{m,n} \end{bmatrix}, \\
X &= (x_1 \quad x_2 \quad \dots \quad x_n)^{\mathrm{T}}, \\
Y &= (y_1 \quad y_2 \quad \dots \quad y_m)^{\mathrm{T}}, \\
B &= (b_1 \quad b_2 \quad \dots \quad b_m)^{\mathrm{T}}, \\
F(\cdot) &= (f_1(\cdot) \quad f_2(\cdot) \quad \dots \quad f_m(\cdot))^{\mathrm{T}},
\end{aligned}
$$

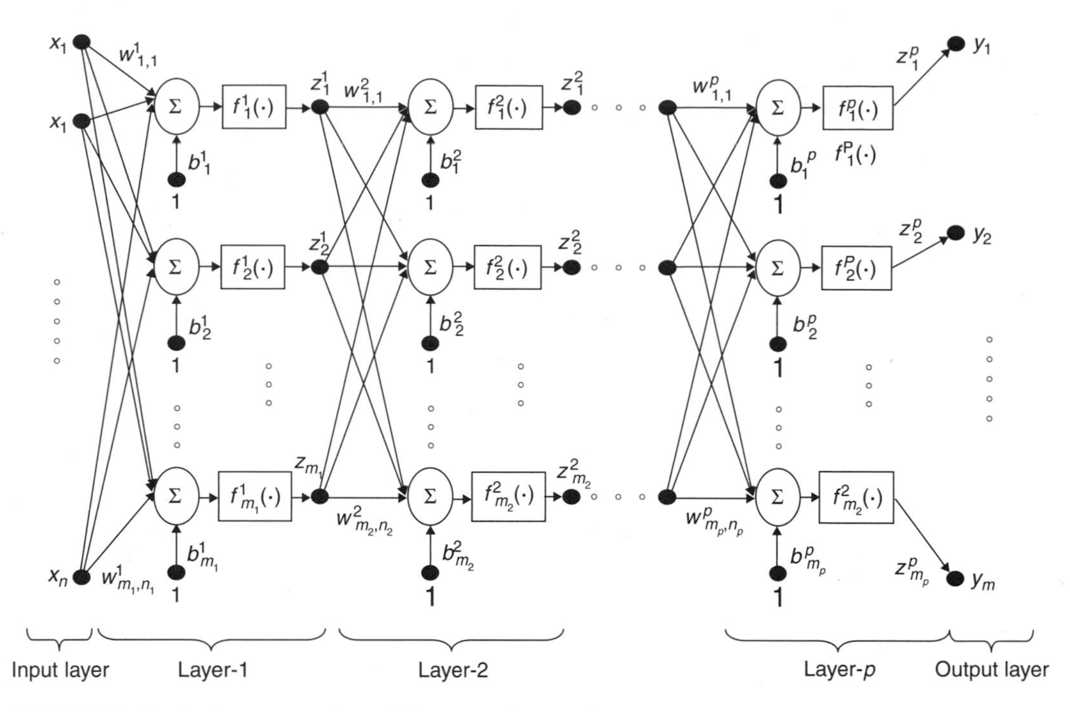

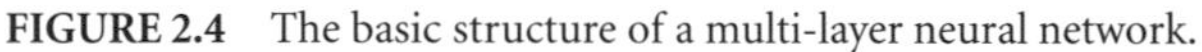

FIGURE 2.4 The basic structure of a multi-layer neural network.

where n is the number of inputs, m the number of outputs, W the weighing matrix, X the input vector, Y the output vector, and $F(\cdot)$ the array of output functions.

A multi-layer perceptron can simply be constructed by concatenating several single-layer perceptron networks. Figure 2.4 shows the basic structure of such network with the following parameters [1]: X is the input vector, Y the output vector, n the number of inputs, m the number of outputs, p the total number of layers in the network, m_i the number of outputs for the ith layer and, n_i the number of inputs for the ith layer.

Note that in this network, every internal layer of the network can have its own number of inputs and outputs only by considering the concatenation rule, that is, $n_i = m_{i-1}$. The output of the first layer is calculated as follows:

$$Z^1 = F^1(W^1 \cdot X + B^1),$$

$$W^1 = \begin{bmatrix} w^1_{1,1} & w^1_{1,2} & \cdots & w^1_{1,n} \\ w^1_{2,1} & & & \\ \vdots & & & \\ w^1_{m_1,1} & & \cdots & w^1_{m_1,n} \end{bmatrix},$$

$$X = (x_1 \quad x_2 \quad \ldots \quad x_n)^{\mathrm{T}},$$

$$B^1 = (b^1_1 \quad b^1_2 \quad \ldots \quad b^1_{m_1})^{\mathrm{T}},$$

$$Z^1 = (z^1_1 \quad z^1_2 \quad \ldots \quad Z^1_{m_1})^{\mathrm{T}},$$

$$F^1(\cdot) = (f^1_1(\cdot) \quad f^1_2(\cdot) \quad \ldots \quad f^1_{m_1}(\cdot))^{\mathrm{T}}.$$

Consequently the output of the second layer would be:

$$Z^2 = F^2(W^2 \cdot Z^1 + B^2),$$

$$W^2 = \begin{bmatrix} w_{1,1}^2 & w_{1,2}^2 & \cdots & w_{1,n}^2 \\ w_{2,1}^2 & & & \\ \vdots & & & \\ w_{m_2,1}^2 & & \cdots & w_{m_2,m_1}^2 \end{bmatrix},$$

$$B^2 = (b_1^2 \quad b_2^2 \quad \ldots \quad b_{m_2}^2)^{\mathrm{T}},$$

$$Z^2 = (z_1^2 \quad z_2^2 \quad \ldots \quad z_{m_2}^2)^{\mathrm{T}},$$

$$F^2(\cdot) = (f_1^2(\cdot) \quad f_2^2(\cdot) \quad \ldots \quad f_{m_2}^2(\cdot))^{\mathrm{T}},$$

and finally the last layer formulation can be presented as follows:

$$Y = Z^p = F^p(W^p \cdot Z^{p-1} + B^p),$$

$$W^p = \begin{bmatrix} w_{1,1}^p & w_{1,2}^p & \cdots & w_{1,n}^p \\ w_{2,1}^p & & & \\ \vdots & & & \\ w_{m_1,1}^p & & \cdots & w_{m_p,m_{p-1}}^p \end{bmatrix},$$

$$B^p = (b_1^p \quad b_2^p \quad \ldots \quad b_{m_p}^p)^{\mathrm{T}},$$

$$Z^p = (z_1^p \quad z_2^p \quad \ldots \quad z_{m_p}^p)^{\mathrm{T}},$$

$$F^p(\cdot) = (f_1^p(\cdot) \quad f_2^p(\cdot) \quad \ldots \quad f_{m_p}^p(\cdot))^{\mathrm{T}}.$$

Notice that in such networks, the complexity of the network raises in a fast race based on the number of layers. Practically experienced, each multi-layer perceptron can be evaluated by a single-layer perceptron with comparatively huge number of nodes.

2.2.2 Function Representation

Two of the most popular uses of Neural Networks are to represent (or approximate) functions and model systems. Basically, a neural network would be used to imitate the behavior of a function by generating relatively similar outputs in comparison to the real system (or function) over the same range of inputs.

2.2.2.1 Boolean Functions

Neural networks were first used to model simple Boolean functions. For example, Figure 2.5 shows how a neural network can be used to model an AND operator, while Figure 2.6 gives the truth table. Note that, "1" stands for "TRUE" while "−1" represents a "FALSE" value. The network in Figure 2.5 actually simulates a linear (function) separator, which simply divides the decision space into two parts.

2.2.2.2 Real Valued Functions

In this case, the network weights must be set so that it can generate continuous outputs of a real system. The generated network is also intended to act as an extrapolator that can generate output data for inputs that are different from the training set.

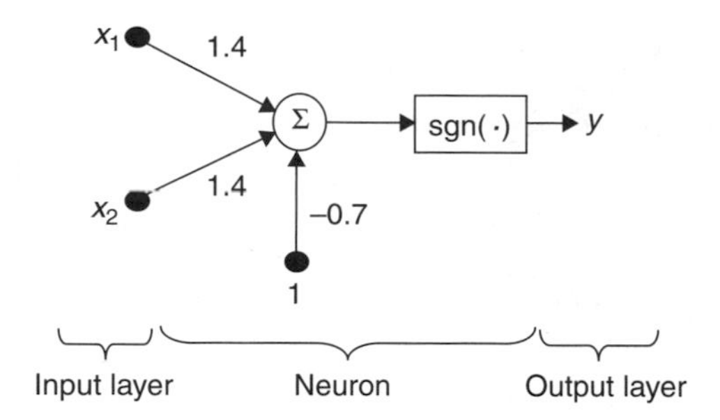

FIGURE 2.5 A neural network to implement the logical AND.

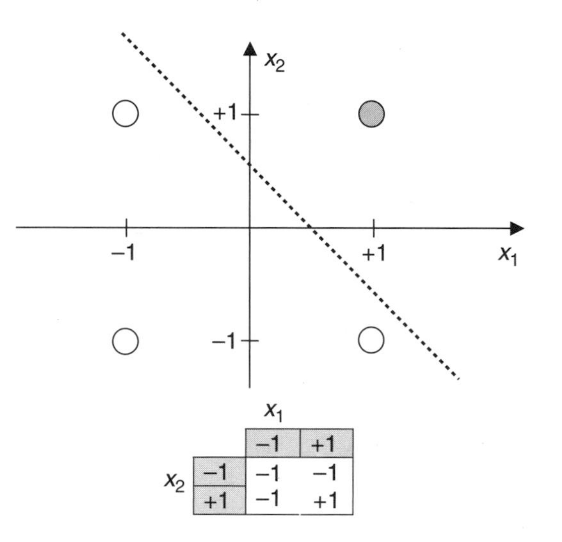

		x_1	
		−1	+1
x_2	−1	−1	−1
	+1	−1	+1

FIGURE 2.6 Implementation of the logical AND of Figure 2.5.

2.3 Learning in Single-Layer Models

The main, and most important, application of all Neural Networks is their ability to model a process or learn the behavior of a system. Toward this end, several algorithms were proposed to train the adjustable parameters of a network (i.e., W). Basically, training a neural network to adjust the W's is categorized into two different classes: supervised and unsupervised [2–6].

2.3.1 Supervised Learning

The main purpose of this kind of training is to "teach" a network to copy the behavior of a system or a function. In this case, there is always a need to have a "training" data set. The network topology and the algorithm that the network is trained with are highly inter-related. In general, a topology of the network is chosen first and then an appropriate training algorithm is used to tune the weights (W) [7,8].

2.3.1.1 Perceptron Learning

As mentioned earlier, the perceptron is the most basic form of Neural Networks. Essentially, this network tries to classify input data by mapping it onto a plane. In this approach, to simplify the algorithm, suppose that the network's input is restricted to $\{+1, 0, -1\}$, while the output can be $\{+1, -1\}$. The aim of the algorithm is to find an appropriate set of weights, W, by sampling a training set, T, that will capture the mapping that associates each input to an output, that is,

$$W = (w_0 \quad w_1 \quad \dots \quad w_n),$$
$$T = \{(R^1, S^1), (R^2, S^2), \dots, (R^L, S^L)\},$$

where n is the number of inputs, R^i is the ith input data, S^i represents the appropriate output for the ith pattern, and, L is the size of the training set. Note that, for the above vector W, w_n is used to adjust the bias in the values of the weights. The Perceptron Learning can be summarized as follows:

Step 1: Set all elements of the weighting vector to zero, that is, $W = (0 \quad 0 \quad \cdots \quad 0)$.
Step 2: Select training pattern at random, namely kth datum.
Step 3: IF the current W has not been classified correctly, that is, $W \cdot R^k \neq S^k$, then, modify the weighing vector as follows: $W \leftarrow W + R^k S^k$.
Step 4: Repeat steps 1 to 3 until all data are classified correctly.

2.3.1.2 Linear Auto-Associators Learning

An auto-associate network is another type of network which has some type of memory. In this network, the input and output nodes are basically the same. Hence, when a datum enters the network, it passes through the nodes and converges to the closest memorized data, which was previously stored in the network during the training process [1].

Figure 2.7 shows an instance of such network with five nodes. It is worthwhile mentioning that the weighing matrix of such network is not symmetrical. That is, $w_{i,j}$ which relate the node "i" to node "j" may have different value than $w_{j,i}$. The main key of designing such network is the training data. In this case, the assumption is to have orthogonal training data or at least approximately orthogonal, that is,

$$\langle T_i, T_j \rangle \approx \begin{cases} 0 & i \neq j, \\ 1 & i = j, \end{cases}$$

where T_i is the ith training data and $\langle \cdot \rangle$ is the inner product of two vectors. Based on the above assumption the weight matrix for this network is calculated as follows where $\otimes$ stands for outer product of two vectors:

$$W = \sum_{i=1}^{N} T_i \otimes T_i.$$

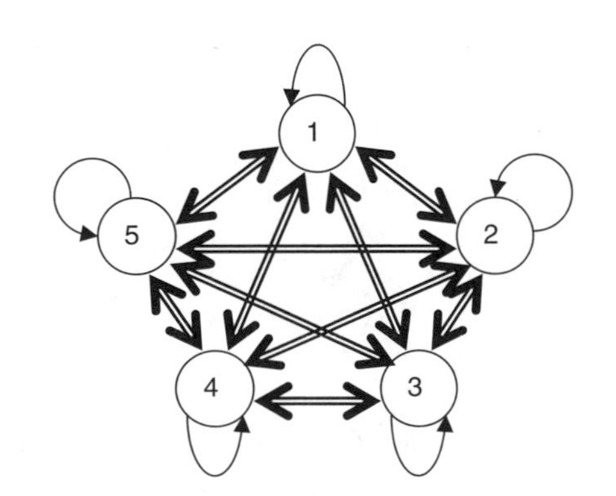

FIGURE 2.7 A sample linear auto-associate network with five nodes.

As it can be seen, the main advantage of this network is in its one-shot learning process, by considering orthogonal data. Note that, even if the input data are not orthogonal in the first place, they can be transferred to a new space by a simple transfer function.

2.3.1.3 Iterative Learning

Iterative learning is another approach that can be used to train a network. In this case, the network's weights are modified smoothly, in contrast to the one-shot learning algorithms. In general, network weights are set to some arbitrary values first, then, trained data are fed to the network. In this case, in each training cycle, network weights are modified smoothly. Then, the training process proceeds until achieving an acceptable level of acceptance for the network. However, the training data could be selected either sequentially or randomly in each training cycle [9–11].

2.3.1.4 Hopfield's Model

A Hopfield neural network is another example of an auto-associative network [1,12–14]. There are two main differences between this network and the previously described auto-associate network. In this network, self-connection is not allowed, that is, $w_{i,i} = 0$ for all nodes. Also, inputs and outputs are either 0 or 1. This means that the node activation is recomputed after each cycle of convergence as follows:

$$S_i = \sum_{j=1}^{N} w_{i,j} \cdot u_j(t), \tag{2.1}$$

$$u'_j = \begin{cases} 1 & \text{if } S_i \geq 0, \\ 0 & \text{if } S_i < 0. \end{cases} \tag{2.2}$$

After feeding a datum into the network, in each convergence cycle, the nodes are selected by a uniform random function, the input are used to calculate Equation (2.1) and then followed by Equation (2.2) to generate the output. This procedure is continued until the network converges.

The proof of convergence for this network uses the notion of "energy." This means that an energy value is assigned to each state of the network and through the different iterations of the algorithm, the overall energy is decreased until it reaches a steady state.

2.3.1.5 Mean Square Error Algorithms

These techniques emerged as an answer to the deficiencies experienced by using Preceptrons and other simple networks [1,15]. One of the most important reasons is the inseparability of training data. If the data used to train the network are naturally inseparable, the training algorithm never terminates (Figure 2.8).

The other reason for using this technique is to converge to a better solution. In Perceptron learning, the training process terminates right after finding the first answer regardless of its quality (i.e., sensitivity of the answer). Figure 2.9 shows an example of such a case. Note that, although the answer found by the Perceptron algorithm is correct (Figure 2.9[a]), the answer in Figure 2.9(b) is more robust. Finally, another reason for using Mean Square Error (MSE) algorithms, which is crucial for most neural network algorithms, is that of speed of convergence.

The MSE algorithm attempts to modify the network weights based on the overall error of all data. In this case, assume that network input and output data are represented by T_i, R_i for $i = 1, \ldots, N$, respectively. Now the MSE error is defined as follows:

$$E = \frac{1}{N} \sum_{i=1}^{N} (W \cdot T_i - R_i)^2.$$

Note that, the stated error is the summation of all individual errors for the all the training data. Inspite of all advantages gained by this training technique, there are several disadvantages, for example, the network might not be able to correctly classify the data if it is widely spread apart (Figure 2.10). The other

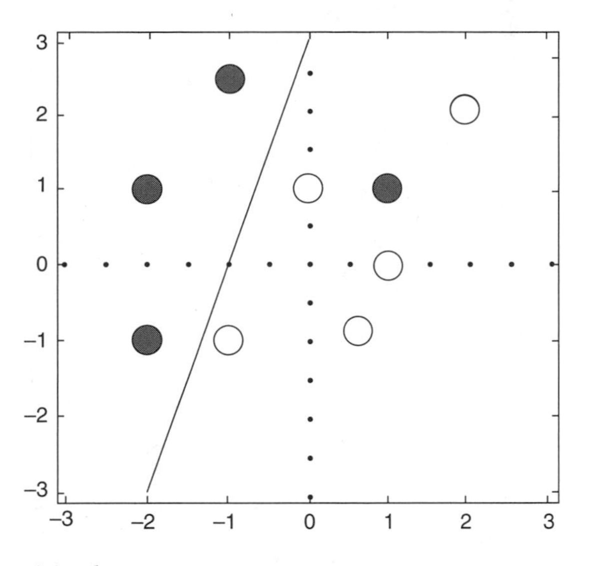

FIGURE 2.8 Inseparable training data set.

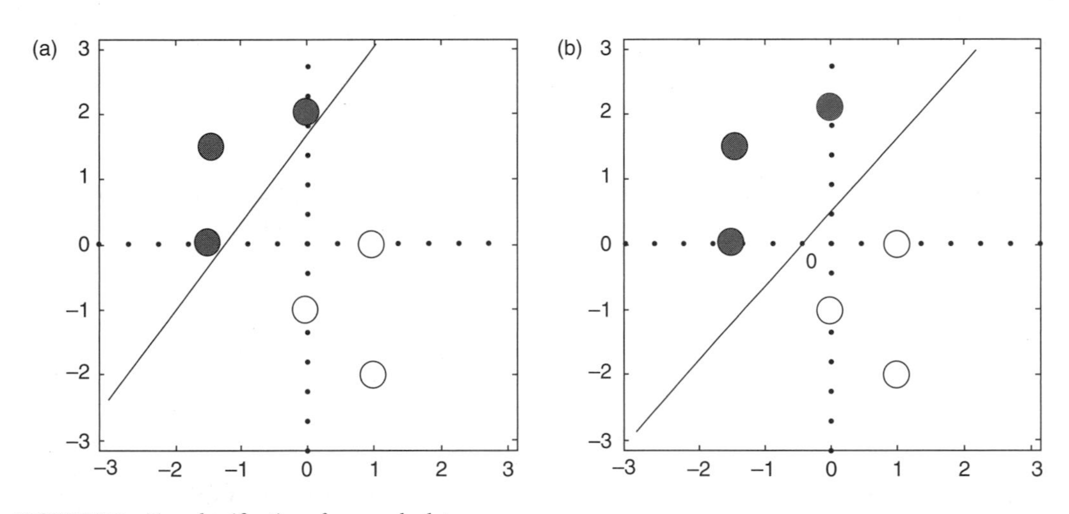

FIGURE 2.9 Two classifications for sample data.

disadvantage is that of the speed of convergence which may completely vary from one set of data to another.

2.3.1.6 The Widow–Hoff Rule or LMS Algorithm

In this technique, the network weight is modified after each iteration [1,16]. A training datum is selected randomly, then, the network weights are modified based on the corresponding error. This procedure continues until converging to the answer. For a randomly selected kth entry in the training data, the error is calculated as follows:

$$\varepsilon = (W \cdot T_k - R_k)^2.$$

The gradient vector of this error would be:

$$\nabla\varepsilon = \left\langle \frac{\partial\varepsilon}{\partial W_0} \quad \frac{\partial\varepsilon}{\partial W_1} \quad \cdots \quad \frac{\partial\varepsilon}{\partial W_N} \right\rangle.$$

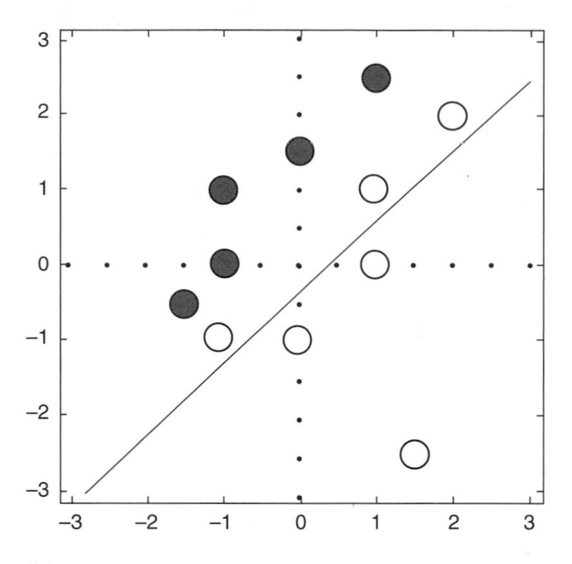

FIGURE 2.10 A data set with far apart solutions.

Hence,

$$\frac{\partial \varepsilon}{\partial W_j} = 2(W \cdot T_k - R_k) \cdot T_k.$$

Based on the Widow–Hoff algorithm, the weights should be modified opposite the direction of the gradient. As a result, the final update formula for the weighting matrix W would be:

$$W' = W - \rho \cdot (W \cdot T_k - R_k) \cdot T_k.$$

Note that, ρ is known as the learning rate and it absorbs the multiplier of value "2."

2.4 Unsupervised Learning

This class of networks attempts to cluster input data without the need for the traditional "learn by example" technique that is commonly used for Neural Networks. Note that, clustering applications tend to be the most popular type of applications that these networks are normally used for. The most popular networks in this class are: K-means, Kohonen, ART1, and ART2 [17–21].

2.4.1 K-Means Clustering

This is the simplest technique used for classifying data. In this technique, a network with a predefined number of clusters is considered, then, each datum is assigned to one of these clusters. This process continues until all data are checked and classified properly. The following algorithm shows how this algorithm is implemented:

Step 1: Consider a network with K clusters.
Step 2: Assign all data to one of the above clusters, with respect to the distance between the center of the cluster and the datum.
Step 3: Modify the center of the assigned cluster.
Step 4: Check all data in the network to ensure proper classification.
Step 5: If a datum has to be moved from one cluster to another, then, update the center of both clusters.
Step 6: Repeat steps 4 and 5 until no datum is wrongly classified.

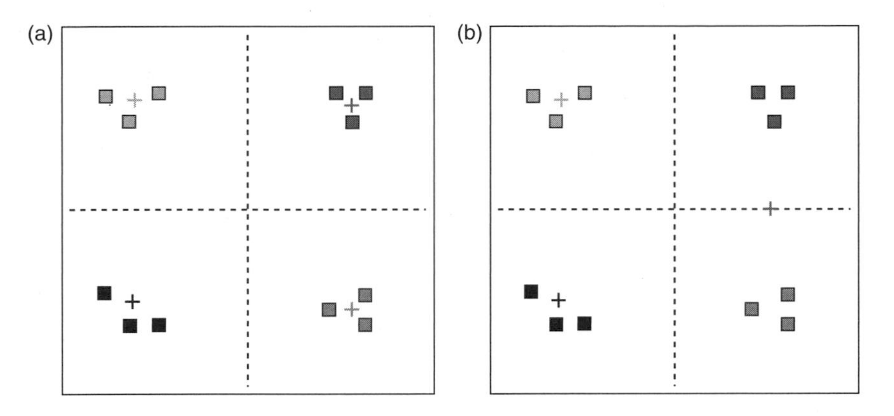

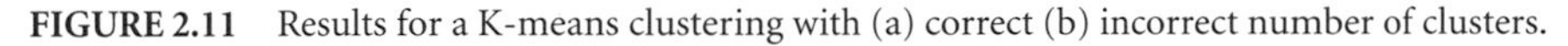

FIGURE 2.11 Results for a K-means clustering with (a) correct (b) incorrect number of clusters.

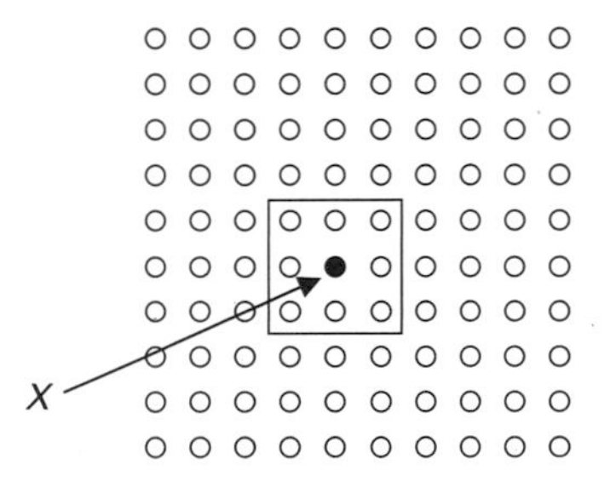

FIGURE 2.12 Output topology of a Kohonen network.

Figure 2.11 shows an instance of applying such network for data classification with the correct and incorrect number of clusters.

2.4.2 Kohonen Clustering

This classification method clusters input data based on how the topological representation of the data. The outputs of the network are arranged so that each output has some neighbors. Thus, during the learning process, not only one output, but a group of close outputs are modified to classify the data. To clarify the situation, assume that a network is supposed to learn how a set of data is to be distributed in a two-dimensional representation (Figure 2.12).

In this case, each point is a potential output with a predefined neighborhood margin. For example, the cell marked as "X" and eight of its neighbors are given. Therefore, whenever this cell gets selected for update, all its neighbors are included in the process too. The main idea behind this approach for classifying the input data is analogus to some biological facts. In a mammalian brain, all vision, auditory, tactile sensors are mapped into a number of "cell sheets." Therefore, if one of the cells is activated all cells close to it will be affected, but at different levels of intensity.

2.4.3 ART1

This neural classifier, known as "Adaptive Resonance Theory" or ART, deals with digital inputs ($T_i \in \{0, 1\}$). In this network, each "1" in the input vector represents information while a "0" entry is considered noise or unwanted information. In ART, there is no predefined number of classes before the start of classification; in fact, the classes are generated during the classification process.

Moreover, each class prototype may include the characteristics of more than a training datum. The basic idea of such network relies on the similarity factor for data classification. In summary, every time a datum is assigned to a cluster, firstly, the nearest class with this datum is found, then, if the similarity of this datum and the class prototype is more than a predefined value, known as a vigilance factor, then, the datum is assigned to this class and the class prototype is modified to have more similarity with the a new data entry [1,22,23].

The following procedure shows how this algorithm is implemented. However, the following needs to be noted before outlining the algorithm:

1. $\|X\|$ is the number of 1's in the vector X.
2. $X \cdot Y$ is the number of common 1's between these vectors X and Y.
3. $X \cap Y$ is the bitwise AND operator applied on vectors X and Y.

Step 1: Let β be a small number, n be the dimension of the input data; and ρ be the vigilance factor $(0 \leq \rho < 1)$.
Step 2: Start with no class prototype.
Step 3: Select a training datum by random, T_k.
Step 4: Find the nearest unchecked class prototype, C_i, to this datum by minimizing $(C_i \cdot T_k)/(\beta + \|C_i\|)$.
Step 5: Test if C_i is sufficiently close to T_k by verifying if $(C_i \cdot T_k)/(\beta + \|C_i\|) > (\|T_k\|/(\beta + \rho))$.
Step 6: If it is not similar enough, then, make a new class prototype and go to step 3.
Step 7: If it is sufficiently similar check the vigilance factor: $(C_i \cdot T_k/\|T_k\|) \geq \rho$.
Step 8: If vigilance factor is exceeded, then, modify the class prototype by $C_i = C_i \cap T_k$ and go to step 3.
Step 9: If vigilance factor is not exceeded, then, try to find another unchecked class prototype in step 4.
Step 10: Repeat steps 3 to 9 until none of the training data causes any change in class prototypes.

2.4.4 ART2

This is a variation to ART1 with the following differences:

1. Data are considered continuous and not binary.
2. The input data is processed before passing it to the network. Actually, the input data is normalized, then, all elements of the result vector that are below a predefined value are set to zero and the vector normalized again. The process is used for noise cancellation.
3. When a class prototype is found for a datum, the class prototype vector is moved fractionally toward the selected datum. As a result, contrary to the operation of ART1, the weights are moved smoothly toward a new datum. The main reason for such a modification is to 'memorize' previously learnt rules.

2.5 Learning in Multiple Layer Models

As mentioned earlier, multi-layer Neural Networks consist of several concatenated single-layer networks [1,24–26]. The inner layers, known as hidden layers, may have different number of inputs and outputs. Because of the added complexity the training process becomes more involved. This section presents two of the most popular multi-layer neural network are presented.

2.5.1 The Back Propagation Algorithm

Back propagation algorithm is one of the most powerful and reliable techniques that can be used to adjust the network weights. The main idea of this approach is to use gradient information of a cost function to modify the network's weights.

However, using this approach to train multi-layer networks is a little different from single-layer networks. In general, multi-layer networks are much harder to train than single-layer ones. In fact, convergence of such networks is much slower and very error sensitive.

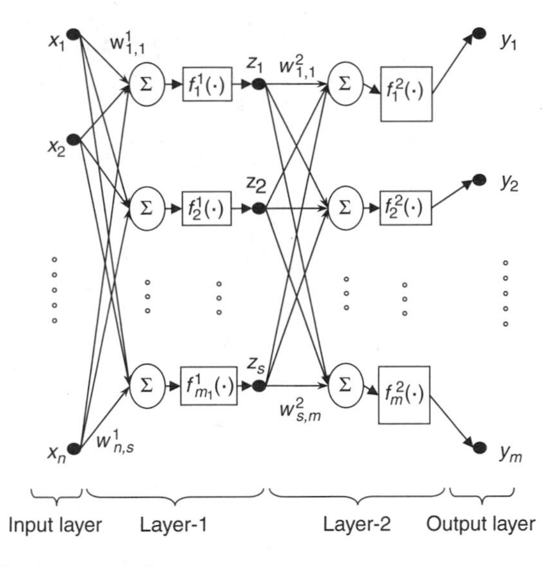

FIGURE 2.13 A single hidden layer network.

In this approach, an input is presented to the network and allowed to "forward" propagate through the network and the output is calculated. Then, the output will be compared to a "desired" output (from the training set) and an error calculated. This error is then propagated "backward" into the network and the different weights updated accordingly. To simplify describing this algorithm, consider a network with a single hidden layer (and two layers of weights) given in Figure 2.13.

In relation to the above network, the following definitions apply. Of course, the same definitions can be easily extended to larger networks.

T_i, R_i for $i = 1, \ldots, L$: The training set of input and outputs, respectively.
N, S, M: The size of the input, hidden, and output layers, respectively.
W^1: Network weights from the input layer to the hidden layer.
W^2: Network weights from the hidden layer to the output layer.
X, Z, Y: Input and output of the hidden layer, and the network output, respectively.
$F^1(\cdot)$: Array of network functions for the hidden layer.
$F^2(\cdot)$: Array of network functions for the output layer.

It is important to note that, in such network, different combinations of weights might produce the same input/output relationship. However, this is not crucial as long as the network is able to "learn" this association. As a result, the network weights may converge to different sets of values based on the order of the training data and the algorithm used for training although their stability may differ.

2.5.2 Radial Basis Functions

Radial Basis Function (RBF) Neural Network is another popular multi-layer neural network [27–31]. The RBF network consists of two layers, one hidden layer and one output layer. In this network, the hidden layer is implemented by radial activation functions while the output layer is simply a weighted sum of the hidden layer outputs.

The RBF neural network is able to model complex mappings, which Perceptron Neural Networks can only accomplish by means of multiple hidden layers. The outstanding characteristics of such network makes it applicable for variety of applications, such as, function interpolation [32,33], chaotic time serious modeling [34,35], system identification [36–38], control systems [39,40], channel equalization [41–43], speech recognition [44,45], image restoration [46,47], motion estimation [48], pattern classification [49], and data fusion [50].

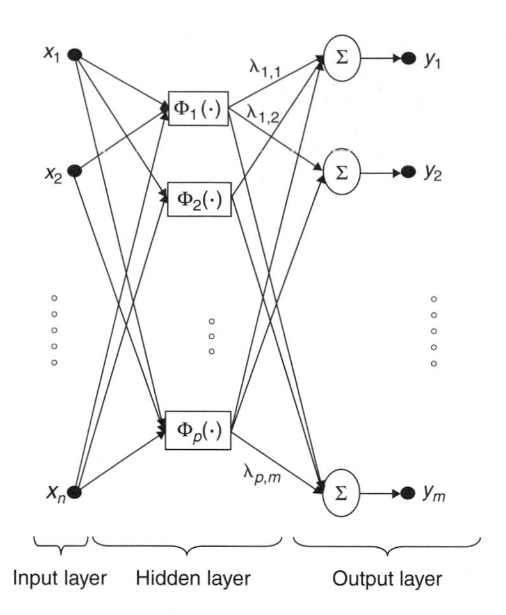

FIGURE 2.14 The basic structure of a RBF network.

The main topology of this network as is shown in Figure 2.14. Many functions were introduced for possible use in the hidden layer; however, radial functions (Gaussian) remain the most effective to use for data or pattern classification. The Gaussian functions are defined as follows:

$$\Phi_i(X) = \exp\lfloor -(X - \mu_j)^{\mathrm{T}}\Gamma_j^{-1}(X - \mu_j)\rfloor,$$

where $j = 1, 2, \ldots, L$ and L represents the number of nodes in the hidden layer, X is the input vector, μ_j and Γ_j are the mean vector and covariance matrix of the jth Gaussian function, respectively. In some approaches, a polynomial term is appended to the above expression while in others the functions are normalized to the sum of all Gaussian components as in the Gaussian mixture estimation. Geometrically, a radial basis function in this network represents a bump in the N-dimensional space where N is the number of entries (input vector size). In this case, the μ_j represents the location of this bump in the space and Γ_j models its shape.

Because of the nonlinear behavior of this network, training procedure of the RBF network (as in multi-layer networks) is approached in a completely different manner to that of single-layer networks. In this network, the aim is to find the center and variance factor of all hidden layer Gaussian functions as well as the optimal weights for the linear output layer. In this case, the following cost function is usually considered as the main network objective:

$$\mathrm{Min}\left(\sum_{i=0}^{N}([Y(T_i) - R_i]^{\mathrm{T}} \cdot [Y(T_i) - R_i])\right),$$

where N is the number of inputs in the training data set, $Y(X)$ is the output of the network for input X and, $\langle T_k, R_k\rangle$ is the kth training data pair. So, the actual output of the network is a combination of a nonlinear computation followed by a linear operation. Therefore, finding an optimal set of weights for hidden layers and output layer parameters is hardly achievable.

In this case, several approaches were used to find the optimal set of weights, however, none of these can provide any guarantees that optimality can be achieved. For example, many approaches suggest that the hidden layer parameters are set randomly and the training procedure is just carried on for the output linear components. In contrast, in some other cases, the radial basis functions are homogenously distributed over the sample space before finding the output linear weights. However, the back propagation algorithm seems to be the most suitable approach for training such a network.

2.6 A Sample of Neural Network Applications

This section briefly reviews a number of application areas in which Neural Networks were used effectively, and this is by no means an exhaustive list of applications.

2.6.1 Expert Systems

One popular application is the use of Neural Networks as expert systems. Several definitions were presented to clearly distinguish this kind of systems from other approaches [55–57]. Generally, an expert system is defined as a system than can imitate the action of a human being for a given process. This definition does not restrict the design of such systems by traditional Artificial Intelligence approaches. Therefore, a variety of such systems can be built by using Fuzzy Logic, Neural Networks, and Neuro-Fuzzy techniques. In most of these systems there is always a knowledge-based component that holds information about the behavior of the system as simple rules followed by operators (usually in Fuzzy Systems) or a large database collected from the system performance that a neural network can be trained to emulate.

2.6.2 Neural Controllers

Neural controllers are a specific class of the expert systems that deal with the process of regulating a linear or nonlinear system. There are two methods to train such system, supervised and unsupervised. In the supervised approach, another controller usually exists and the neural controller is trained to imitate its behavior. In such case, the neural controller is connected in parallel to the other controller and during the process, by sampling inputs and outputs, the network is trained to generate similar outputs for similar inputs of the real controller. This process is known as on-line training. In contrast, in the case of off-line training a database of the real controller inputs and outputs can be employed to train the network [58–60].

2.6.3 Decision Makers

In this specific class, which can also be viewed as an expert system, a neural network is used to make critical decisions in unexpected situations. One such popular application is in financial markets such as stock market applications. One of main characteristics of such systems that distinguish them from simple expert systems is their stability. In fact, these systems must be able to produce acceptable output for untrained situations. Therefore, a sufficiently "rich" data set must be used for the training process [61–63].

2.6.4 Robot Path Planning

Another complex scenario in which Neural Networks were used with some promise is that of robot path planning. In this case, the robot tries to navigate its way to reach a target location. The situation can be made more complicated by adding obstacles in the environment or even other mobile robots. Normally, this situation is modeled as an optimization problem in which some cost function is minimized (e.g., minimize the distance that the robot needs to travel) while satisfying certain constraints (e.g., no collisions) [64–66].

2.6.5 Adaptive Noise Cancellation

Neural networks have been used very effectively to filter noise. In this case, the target signal (in the training set) is the non-noisy signal that the input should be generating. The network must learn how to make imitate the noise and in the process manage to neutralize it. Many approaches were introduced in the literature over the years and some of these were deployed in real environments [67–69].

2.7 Conclusion

In this chapter, a general overview of the artificial Neural Networks was presented. These networks vary in their sophistication from the very simple to the more complex. As a result, their training techniques vary as well as their capabilities and suitability for certain applications. Neural networks have attracted a lot of interest over the last few decades and it is expected they will be an active area of research for years to come. Undoubtedly, more robust neural techniques will be introduced in the future that could benefit a wide range of complex application.

References

[1] Gallant, S.I. *Neural Network Learning and Expert Systems*, MIT Press, Cambridge, MA, 1993.

[2] Karayiannis, N.B. and Venetsanopoulos, A.N. Efficient learning algorithms for Neural Networks (ELEANNE). *IEEE Transactions on Systems, Man, and Cybernetics*, 23, 1993, 1372–1383.

[3] Hassoun, M.H. and Clark, D.W. An adaptive attentive learning algorithm for single-layer Neural Networks. In *Proceedings of the IEEE International Conference on Neural Networks*, Vol. 1, July 24–27, 1988, pp. 431–440.

[4] Ulug, M.E. A single layer fast learning fuzzy controller/filter, Neural Networks. In *Proceedings of the IEEE World Congress on Computational Intelligence*, Vol. 3, June 27–July 2, 1994, pp. 1662–1667.

[5] Karayiannis, N.B. and Venetsanopoulos, A.N. Fast learning algorithms for Neural Networks. *IEEE Transactions on Circuits and Systems II: Analog and Digital Signal Processing*, 39, 1992, 453–474.

[6] Hrycej, T. Back to single-layer learning principles. In *Proceedings of the International Joint Conference on Neural Networks*, Seattle, Vol. 2, July 8–14, 1991, p. 945.

[7] Healy, M.J. A logical architecture for supervised learning, Neural Networks. In *Proceedings of the IEEE International Joint Conference on Neural Networks*, Vol. 1, November 18–21, 1991, pp. 190–195.

[8] Brandt, R.D. and Feng, L. Supervised learning in Neural Networks without feedback network. In *Proceedings of the IEEE International Symposium on Intelligent Control*, September 15–18, 1996, pp. 86–90.

[9] Gong, Y. and Yan, P. Neural network based iterative learning controller for robot manipulators. In *Proceedings of the IEEE International Conference on Robotics and Automation*, Vol. 1, May 21–27, 1995, pp. 569–574.

[10] Park, S. and Han, T. Iterative inversion of fuzzified Neural Networks. *IEEE Transactions on Fuzzy Systems*, 8, 2000, 266–280.

[11] Zhan, X., Zhao, K., Wu, S., Wang, M., and Hu, H. Iterative learning control for nonlinear systems based on Neural Networks. In *Proceedings of the IEEE International Conference on Intelligent Processing Systems*, Vol. 1, October 28–31, 1997, pp. 517–520.

[12] Chen, C.J., Haque, A.L., and Cheung, J.Y. An efficient simulation model of the Hopfield Neural Networks. In *Proceedings of the International Joint Conference on Neural Networks*, Vol. 1, June 7–11, 1992, pp. 471–475.

[13] Galan-Marin, G. and Munoz-Perez, J. Design and analysis of maximum Hopfield networks. *IEEE Transactions on Neural Networks*, 12, 2001, 329–339.

[14] Nasrabadi, N.M. and Li, W. Object recognition by a Hopfield neural network. *IEEE Transactions on Systems, Man and Cybernetics*, 21, 1991, 1523–1535.

[15] Xu, J., Zhang, X., and Li, Y. Kernel MSE algorithm: a unified framework for KFD, LS-SVM and KRR. In *Proceedings of the International Joint Conference on Neural Networks*, Vol. 2, July 15–19, 2001, pp. 1486–1491.

[16] Hayasaka, T., Toda, N., Usui, S., and Hagiwara, K. On the least square error and prediction square error of function representation with discrete variable basis. In *Proceedings of the Workshop on Neural Networks for Signal Processing*, Vol. VI. IEEE Signal Processing Society, September 4–6, 1996, pp. 72–81.

[17] Park, D.-C. Centroid neural network for unsupervised competitive learning. *IEEE Transactions on Neural Networks*, 11, 2000, 520–528.

[18] Pedrycz, W. and Waletzky, J. Neural-network front ends in unsupervised learning. *IEEE Transactions on Neural Networks*, 8, 1997, 390–401.

[19] Park, D.-C. Development of a neural network algorithm for unsupervised competitive learning. In *Proceedings of the International Conference on Neural Networks*, Vol. 3, June 9–12, 1997, pp. 1989–1993.

[20] Hsieh, K.-R. and Chen, W.-T. A neural network model which combines unsupervised and supervised learning. *IEEE Transactions on Neural Networks*, 4, 1993, 357–360.

[21] Dajani, A.L., Kamel, M., and Elmastry, M.I. Single layer potential function neural network for unsupervised learning. In *Proceedings of the International Joint Conference on Neural Networks*, Vol. 2, June 17–21, 1990, pp. 273–278.

[22] Georgiopoulos, M., Heileman, G.L., and Huang, J. Properties of learning in ART1. In *Proceedings of the IEEE International Joint Conference on Neural Networks*, Vol. 3, November 18–21, 1991, pp. 2671–2676.

[23] Heileman, G.L., Georgiopoulos, M., and Hwang, J. A survey of learning results for ART1 networks. In *Proceedings of the IEEE International Conference on Neural Networks, IEEE World Congress on Computational Intelligence*, Vol. 2, June 27–July 2, 1994, pp. 1222–1225.

[24] Song, J. and Hassoun, M.H. Learning with hidden targets. In *Proceedings of the International Joint Conference on Neural Networks*, Vol. 3, June 17–21, 1990, pp. 93–98.

[25] Kwan, H.K. Multilayer feedbackward Neural Networks. In *Proceedings of the International Conference on Acoustics, Speech, and Signal Processing*, Vol. 2, April 14–17, 1991, pp. 1145–1148.

[26] Shepanski, J.F. Fast learning in artificial neural systems: multilayer perceptron training using optimal estimation. In *Proceedings of the IEEE International Conference on Neural Networks*, Vol. 1, July 24–27, 1988, pp. 465–472.

[27] Karayiannis, N.B. and Randolph-Gips, M.M. On the construction and training of reformulated radial basis function Neural Networks. *IEEE Transactions on Neural Networks*, 14, 2003, 835–846.

[28] Leonard, J.A. and Kramer, M.A. Radial basis function networks for classifying process faults. *IEEE Control Systems Magazine*, 11, 1991, 31–38.

[29] Li, R., Lebby, G., and Baghavan, S. Performance evaluation of Gaussian radial basis function network classifiers. In *Proceedings of the IEEE*, SoutheastCon, April 5–7, 2002, pp. 355–358.

[30] Heimes, F. and van Heuveln, B. The normalized radial basis function neural network. In *Proceedings of the IEEE International Conference on Systems, Man, and Cybernetics*, Vol. 2, October 11–14, 1998, pp. 1609–1614.

[31] Craddock, R.J. and Warwick, K. Multi-layer radial basis function networks. An extension to the radial basis function. In *Proceedings of the IEEE International Conference on Neural Networks*, Vol. 2, June 3–6, 1996, pp. 700–705.

[32] Carr, J.C., Fright, W.R., and Beatson, R.K. Surface interpolation with radial basis functions for medical imaging. *IEEE Transactions on Medical Imaging*, 16, 1997, 96–107.

[33] Romyaldy, M.A., Jr. Observations and guidelines on interpolation with radial basis function network for one dimensional approximation problem. In *Proceedings of the 26th Annual Conference of the IEEE Industrial Electronics Society*, Vol. 3, October 22–28, 2000, pp. 2129–2134.

[34] Leung, H., Lo, T., and Wang, S. Prediction of noisy chaotic time series using an optimal radial basis function neural network. *IEEE Transactions on Neural Networks*, 12, 2001, 1163–1172.

[35] Katayama, R., Kajitani, Y., Kuwata, K., and Nishida, Y. Self generating radial basis function as neuro-fuzzy model and its application to nonlinear prediction of chaotic time series. In *Proceedings of the Second IEEE International Conference on Fuzzy Systems*, Vol. 1, March 28–April 1, 1993, pp. 407–414.

[36] Warwick, K. and Craddock, R. An introduction to radial basis functions for system identification. A comparison with other neural network methods. In *Proceedings of the 35th IEEE Decision and Control Conference*, Vol. 1, December 11–13, 1996, pp. 464–469.

[37] Lu, Y., Sundararajan, N., and Saratchandran, P. Adaptive nonlinear system identification using minimal radial basis function Neural Networks. In *Proceedings of the IEEE International Conference on Acoustics, Speech, and Signal Processing*, Vol. 6, May 7–10, 1996, pp. 3521–3524.

[38] Tan, S., Hao, J., and Vandewalle, J. A new learning algorithm for RBF Neural Networks with applications to nonlinear system identification. In *Proceedings of the IEEE International Symposium on Circuits and Systems*, Vol. 3, April 28–May 3, 1995, pp. 1708–1711.

[39] Ibayashi, T., Hoya, T., and Ishida, Y. A model-following adaptive controller using radial basis function networks. In *Proceedings of the International Conference on Control Applications*, Vol. 2, September 18–20, 2002, pp. 820–824.

[40] Dash, P.K., Mishra, S., and Panda, G. A radial basis function neural network controller for UPFC. *IEEE Transactions on Power Systems*, 15, 2000, pp. 1293–1299.

[41] Deng, J., Narasimhan, S., and Saratchandran, P. Communication channel equalization using complex-valued minimal radial basis function Neural Networks. *IEEE Transactions on Neural Networks*, 13, 2002, 687–696.

[42] Lee, J., Beach, C.D., and Tepedelenlioglu, N. Channel equalization using radial basis function network. In *Proceedings of the IEEE International Conference on Neural Networks*, Vol. 4, June 3–6, 1996, pp. 1924–1928.

[43] Lee, J., Beach, C.D., and Tepedelenlioglu, N. Channel equalization using radial basis function network. In *Proceedings of the IEEE International Conference on Acoustics, Speech, and Signal Processing*, Vol. 3, May 7–10, 1996, pp. 1719–1722.

[44] Sankar, R. and Sethi, N.S. Robust speech recognition techniques using a radial basis function neural network for mobile applications. In *Proceedings of IEEE Southeastcon*, April 12–14, 1997, pp. 87–91.

[45] Ney, H. Speech recognition in a neural network framework: discriminative training of Gaussian models and mixture densities as radial basis functions. In *Proceedings of the IEEE International Conference on Acoustics, Speech, and Signal Processing*, Vol. 1, April 14–17, 1991, pp. 573–576.

[46] Cha, I. and Kassam, S.A. Nonlinear image restoration by radial basis function networks. In *Proceedings of the IEEE International Conference on Image Processing*, Vol. 2, November 13–16, 1994, pp. 580–584.

[47] Cha, I. and Kassam, S.A. Nonlinear color image restoration using extended radial basis function networks. In *Proceedings of the IEEE International Conference on Acoustics, Speech, and Signal Processing*, Vol. 6, May 7–10, 1996, pp. 3402–3405.

[48] Bors, A.G. and Pitas, I. Optical flow estimation and moving object segmentation based on median radial basis function network. *IEEE Transactions on Image Processing*, 7, 1998, 693–702.

[49] Gao, D. and Yang, G. Adaptive RBF Neural Networks for pattern classifications. In *Proceedings of the International Joint Conference on Neural Networks*, Vol. 1, May 12–17, 2002, pp. 846–851.

[50] Fan, C., Jin, Z., Zhang, J., and Tian, W. Application of multisensor data fusion based on RBF Neural Networks for fault diagnosis of SAMS. In *Proceedings of the 7th International Conference on Control, Automation, Robotics and Vision*, Vol. 3, December 2–5, 2002, pp. 1557–1562.

[51] Tou, J.T. and Gonzalez, R.C. *Pattern Recognition*, Addison-Wesley, Reading, MA, 1974.

[52] Lo, Z.-P., Yu, Y., and Bavarian, B. Derivation of learning vector quantization algorithms. In *Proceedings of the International Joint Conference on Neural Networks*, Vol. 3, June 7–11, 1992, pp. 561–566.

[53] Burrascano, P. Learning vector quantization for the probabilistic neural network. *IEEE Transactions on Neural Networks*, 2, 1991, 458–461.

[54] Karayiannis, N.B. and Randolph-Gips, M.M. Soft learning vector quantization and clustering algorithms based on non-Euclidean norms: multinorm algorithms. *IEEE Transactions on Neural Networks*, 14, 2003, 89–102.

[55] Medsker, L. Design and development of hybrid neural network and expert systems. In *Proceedings of the IEEE International Conference on Neural Networks, IEEE World Congress on Computational Intelligence*, Vol. 3, June 27–July 2, 1994, pp. 1470–1474.

[56] Kurzyn, M.S. Expert systems and Neural Networks: a comparison, artificial Neural Networks and expert systems. In *Proceedings of the First International Two-Stream Conference on Neural Networks*, New Zealand, November 24–26, 1993, pp. 222–223.
[57] Hudli, A.V., Palakal, M.J., and Zoran, M.J. A neural network based expert system model. In *Proceedings of the Third International Conference on Tools for Artificial Intelligence*, November 10–13, 1991, pp. 145–149.
[58] Wang, W.-Y., Cheng, C.-Y., and Leu, Y.-G. An online GA-based output-feedback direct adaptive fuzzy-neural controller for uncertain nonlinear systems. *IEEE Transactions on Systems, Man and Cybernetics, Part B*, 34, 2004, 334–345.
[59] Zhang, Y., Peng, P.-Y., and Jiang, Z.-P. Stable neural controller design for unknown nonlinear systems using backstepping. *IEEE Transactions on Neural Networks*, 11, 2000, 1347–1360.
[60] Nelson, A.L., Grant, E., and Lee, G. Developing evolutionary neural controllers for teams of mobile robots playing a complex game. In *Proceedings of the IEEE International Conference on Information Reuse and Integration*, October 27–29, 2003, pp. 212–218.
[61] Rothrock, L. Modeling human perceptual decision-making using an artificial neural network. In *Proceedings of the International Joint Conference on Neural Networks*, Vol. 2, June 7–11, 1992, pp. 448–452.
[62] Mukhopadhyay, S. and Wang, H. Distributed decomposition architectures for neural decision-makers. In *Proceedings of the 38th IEEE Conference on Decision and Control*, Vol. 3, December 7–10, 1999, pp. 2635–2640.
[63] Rogova, G., Scott, P., and Lolett, C. Distributed reinforcement learning for sequential decision making. In *Proceedings of the Fifth International Conference on Information Fusion*, Vol. 2, July 8–11, 2002, pp. 1263–1268.
[64] Taheri, J. and Sadati, N. Fully modular online controller for robot navigation in static and dynamic environments. In *Proceedings of the 2003 IEEE International Symposium on Computational Intelligence in Robotics and Automation*, Vol. 1, July 16–20, 2003, pp. 163–168.
[65] Sadati, N. and Taheri, J. Genetic algorithm in robot path planning problem in crisp and fuzzified environments. In *Proceedings of the IEEE International Conference on Industrial Technology*, Vol. 1, December 11–14, 2002, pp. 175–180.
[66] Sadati, N. and Taheri, J. Solving robot motion planning problem using Hopfield neural network in a fuzzified environment. In *Proceedings of IEEE International Conference on Fuzzy Systems*, Vol. 2, May 12–17, 2002, pp. 1144–1149.
[67] Bambang, R. Active noise cancellation using recurrent radial basis function Neural Networks. In *Proceedings of the Asia-Pacific Conference on Circuits and Systems*, Vol. 2, October 28–31, 2002, pp. 231–236.
[68] Chen, C.K. and Chiueh, T.-D. Multilayer perceptron Neural Networks for active noise cancellation. In *Proceedings of the IEEE International Symposium on Circuits and Systems*, Vol. 3, May 12–15, 1996, pp. 523–526.
[69] Tao, L. and Kwan, H.K. A neural network method for adaptive noise cancellation, Circuits and Systems. In *Proceedings of the IEEE International Symposium on Circuits and Systems*, Vol. 5, May 30–June 2, 1999, pp. 567–570.

3

Ant Colony Optimization

Michael Guntsch
Jürgen Branke

3.1 Biological Background

Although only 2% of all insect species are social, they comprise more than 50% of the total insect biomass globally [1], and more than 75% in some areas like the Amazon rain forest [2]. By social, we mean that these insects, including all ants and termites and some subspecies of bees and wasps, live in colonies composed of many interacting individuals. Insect colonies are capable of solving a number of optimization problems that none of the individual insects would be able to solve by itself. Some examples are finding short paths when foraging for food, task allocation when assigning labor to workers, and clustering when organizing brood chambers, all of which are problems that have counterparts in real world optimization problems.

In order for a swarm of insects to cooperate in problem solving, some form of communication is necessary. This communication between the individuals of a colony can be more or less direct, depending on the exact species. When a social bee has found a food source, it communicates the direction and distance of the location where it found the food to the other bees by performing a characteristic dance [3]. This is a very direct communication, as the other bees must perceive the dance the one bee is performing in order to be able to locate the food source. Other forms of direct communication include stimulation by physical contact or the exchange of food or liquid.

Indirect communication between the individuals of a colony is more subtle and requires one individual to modify the environment in such a way that it will alter the behavior of individuals passing through this modified environment at a later time. One scenario where this type of environmentally induced action exists in nature is when termites construct a nest that has a very complicated structure and exhibits properties like temperature control [4]. Whenever a construction phase has ended, the surroundings of the worker have changed, and the next phase of working is encouraged, which in turn results in new surroundings, and so forth. Another example for indirect communication is the laying of pheromone trails performed by certain species of ants. An ant foraging for food will mark its path by distributing an amount of pheromone on the trail it is taking, encouraging (but not forcing) ants who are also foraging for food to follow its path. The principle of modifying the environment in order to induce a change in behavior as a means of communication is called stigmergy and was first proposed in Reference 5.

Stigmergy is the basis for the organization in many ant colonies. Although the ants have a queen, this is a specialized ant which is only responsible for laying eggs and does not have any governing function. Instead, the ants of a colony are self-organized. The term self-organization (SO) is used to describe the complex behavior which emerges from the interaction of comparatively simple agents. Its origins lie in the fields of physics and chemistry, where SO is used to describe microscopic operations resulting in macroscopic patterns, see Reference 6. Through SO, the ants are able to solve the complex problems which they encounter on a daily basis. The benefits of SO as a basis for problem solving are especially apparent in its distributed and robust character. Effectively, an ant colony can maintain a meaningful behavior even if a large number of ants are incapable of contributing for some amount of time.

To better understand the mechanism behind an ant colony's ability to converge to good solutions when looking for a short path from the nest to a food source, some experiments were undertaken in References 7 and 8. In Reference 8 a nest of the Argentine ant *Linepithema humile* was given two paths of identical length that it could take to reach a food source, and after some time had passed, it was observed that the ants had converged to one of the paths, following it practically to the exclusion of the alternative. To test whether this type of ant would converge to the shorter of two alternate paths, an experimental setup similar to the one depicted in Figure 3.1 was evaluated in Reference 7.

The Argentine ant is practically blind, so it has no means of directly identifying a shorter path. However, despite this deficiency, a swarm of these ants is capable of finding the shorter path connecting the nest to the foraging area containing the food, as the experiment shows. Initially, all ants are located at the nest site. A number of ants start out from the nest in search of food, each ant laying pheromone on its path, and reach the first fork at point A. Since the ants have no information which way to go, that is, no ant has walked before them and left a pheromone trail, each ant will choose to go either right or left with equal probability. As a consequence, about one half of the foraging ants will take the shorter route, the others the longer route to intersection B. The ants which were on the shorter track will reach this intersection first, and have to decide which way to turn. Again, there is no information for the ants to use as orientation, so half of the ants reaching intersection B will turn back toward the nest, while the rest continues toward the foraging area containing the food. The ants on the longer branch between intersections A and B, unaffected by the other ants they met head-on, arrive at intersection B and will also split up; however,

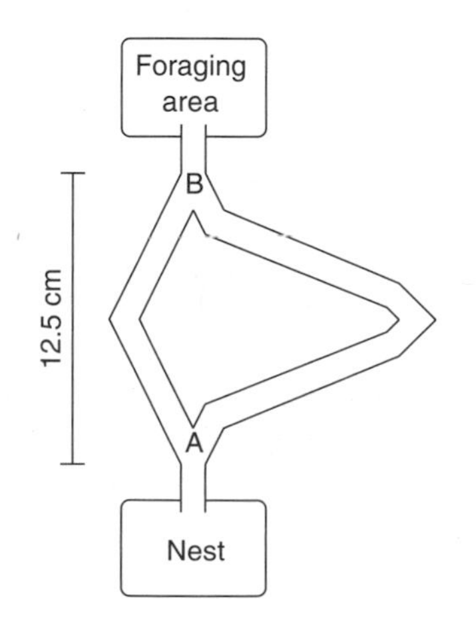

FIGURE 3.1 Single bridge experimental setup.

since the intensity of the pheromone trail heading back toward the nest is roughly twice as high as that of the pheromone trail heading for the foraging area, the majority will turn back toward the nest, arriving there at the same time as the other ants which took the long way back. Interestingly, since more ants have now walked on the short branch between intersections A and B in comparison to the long one, future ants leaving the nest will now already be more inclined to take the short branch, which is a first success in the search for the shortest path.

The ants which continued toward the foraging area pick up some food to carry back to the nest. Arriving at intersection B, the ants will prefer the short branch by the same argument as used above for ants starting out fresh from the nest. Since the amount of pheromone at intersection A on the path back to the nest is (roughly) equal to the sum of the pheromone amounts on the two branches leading away from the nest, the shortest complete path from the foraging area back to the nest is also the most likely to be chosen by the returning ants. Since the ants are continually distributing pheromone as they walk, the short path is continually reinforced by more and more ants, until the amount of pheromone placed thereon in relation to the alternative routes is so high that practically all ants use the shortest path, that is, the system converges to the shortest path through self-reinforcement.

One point that we have neglected to mention so far is that the pheromone used by ants to mark their trails slowly evaporates over time. This does not render the arguments used to explain the double bridge experiment any less valid, it simply makes some of the math used for explanation less rigorous than implied. Indeed, due to the evaporation of pheromone, a path that has not been chosen for some time, invariably the long one, will contain almost no traces of pheromone after a sufficient amount of time, further increasing the likelihood of ants taking the short path identified by the continually updated pheromone.

In the rest of this chapter, we explain how the concept of marking paths with pheromones can be exploited to construct algorithms capable of solving highly complex combinatorial optimization problems, which was first proposed in Reference 9. The following section outlines the structure of Ant Colony Optimization (ACO) algorithms using the Traveling Salesman Problem as an example. Afterwards, the design decisions which must be made when implementing an ACO algorithm for a particular problem are discussed. Finally, a number of applications where ACO algorithms lead to good results are surveyed.

3.2 An ACO Algorithm for the TSP

Consider the problem of finding the shortest tour connecting a number of points $v \in V$ in a complete, weighted graph $G = (V, V \times V, d)$ with $d : V \times V \rightarrow \mathbb{R}^+$ indicating the distance between nodes. This problem, also known as the Traveling Salesman Problem (TSP), is a generalization of the problem solved by the real ants looking for the shortest path between the food source and the nest.

An ACO algorithm for the TSP proceeds iteratively. In every iteration, a number of ants independently construct tours by moving progressively from the current city to one that has not been visited yet, completing the tour after all cities have been visited. The decision of which city to move to from the current location is made using pheromone information τ_{ij}, which denotes whether going from city i to city j led to good solutions in the past, and heuristic information η_{ij}, which signifies an immediate benefit of moving from city i to city j. For both types of information, larger values are better. In case of the TSP, we set $\eta_{ij} = 1/d_{ij}$ to indicate that short distances are preferable to long ones. Unlike biological ants, artificial ants perform an *a posteriori* update on the global pheromone information depending on how good the respective constructed tour is in comparison to the competing ants of the same iteration. Algorithm 3.1 shows a standard ACO algorithm for the TSP.

As can be seen in Algorithm 3.1, the artificial ants used to solve this problem are augmented in a number of ways in comparison to their biological inspiration. The algorithm starts by initializing all pheromone values to a level $\tau_0 > 0$, which is necessary since the ants make decisions according to relative pheromone values and not absolute ones. We will discuss the merits of the different values for τ_0 in Section 3.3. For now, we use $\tau_0 = 1$.

Algorithm 3.1 Basic ACO Algorithm for the TSP

1: initialize pheromone values $\forall i, j \in [1, n] : \tau_{ij} \mapsto \tau_0$
2: **repeat**
3: **for** each ant $l \in \{1, \ldots, m\}$ **do**
4: initialize selection set $S \mapsto \{1, \ldots, n\}$
5: randomly choose starting city $i_0 \in S$ for ant l
6: move to starting city $i \mapsto i_0$
7: **while** $S \neq \emptyset$ **do**
8: remove current city from selection set $S \mapsto S \setminus \{i\}$
9: choose next city j in tour with probability $p_{ij} = \dfrac{\tau_{ij}^{\alpha} \cdot \eta_{ij}^{\beta}}{\sum\limits_{h \in S} \tau_{ih}^{\alpha} \cdot \eta_{ih}^{\beta}}$
10: update solution vector $\pi_l(i) \mapsto j$
11: move to new city $i \mapsto j$
12: **end while**
13: finalize solution vector $\pi_l(i) \mapsto i_0$
14: **end for**
15: **for** each solution $\pi_l, l \in \{1, \ldots, m\}$ **do**
16: calculate tourlength $f(\pi_l) \mapsto \sum\limits_{i=1}^{n} d_{i\pi_l(i)}$
17: **end for**
18: **for all** (i, j) **do**
19: evaporate pheromone $\tau_{ij} \mapsto (1 - \rho) \cdot \tau_{ij}$
20: **end for**
21: determine best solution of iteration $\pi^+ = \arg\min\limits_{l \in [1,m]} f(\pi_l)$
22: **if** π^+ better than current best $\pi^\star$, i.e. $f(\pi^+) < f(\pi^\star)$ **then**
23: set $\pi^\star \mapsto \pi^+$
24: **end if**
25: **for all** $(i, j) \in \pi^+$ **do**
26: reinforce $\tau_{ij} \mapsto \tau_{ij} + \Delta/2$
27: **end for**
28: **for all** $(i, j) \in \pi^\star$ **do**
29: reinforce $\tau_{ij} \mapsto \tau_{ij} + \Delta/2$
30: **end for**
31: **until** condition for termination met

After the initialization of the pheromone matrix, m ants per iteration each independently construct a solution for the TSP instance. The solution of ant l is stored in a permutation vector π_l, which contains the edges that were traversed during tour construction, that is, ant l moved from city i to city $\pi_l(i)$ in the tour. In order to ensure that each solution constructed is feasible, each ant maintains a selection set S which, while the ant constructs a solution, contains all cities that still need to be visited to complete the tour. The starting city for the ants is chosen randomly, since a tour is a circle visiting all cities and therefore the starting position is arbitrary. Afterwards, as long as there are still cities that need to be visited, the ant chooses the next city in the tour in a probabilistic fashion according to Equation (3.1), which is called the *random proportional transition rule*:

$$p_{ij} = \frac{\tau_{ij}^{\alpha} \cdot \eta_{ij}^{\beta}}{\sum_{h \in S} \tau_{ih}^{\alpha} \cdot \eta_{ih}^{\beta}}. \tag{3.1}$$

The artificial ant tends to move to the city which has the best combination of pheromone information, signifying a promising overall solution quality, and heuristic information, which equates to a short distance to the immediate successor in the tour. The pheromone and heuristic information can be weighted with the parameters $\alpha, \beta \in \mathbb{R}^+$ in order to gauge the influence of the respective information on the decision process. For the TSP, using $\alpha = 1$ and $\beta = 5$ yields good results, see Section 3.3 for more details.

Once the artificial ant has visited all the cities initially in S, it returns to the starting city, thereby completing the tour. After m ants have completed their individual tours in this fashion, the solutions are evaluated and the pheromone information is updated. The update consists of two parts: evaporation and intensification. The purpose of evaporation is to diminish all pheromone values by a relative amount, and is accomplished by multiplying all pheromone values with a factor $(1 - \rho)$, where ρ is called the evaporation rate. For the intensification, the best ant of the iteration (with solution π^+) and the best ant found by the algorithm over all iterations (with solution $\pi^\star$) are used, each updating the pheromone values corresponding to the edges traversed on the tour with $\Delta/2$. In conjunction with $\tau_0 = 1$, setting $\Delta = \rho$ usually works well.

After a number of iterations, the exact number correlating to how low the pheromone evaporation is set, the ants will cease finding new solutions because the pheromone matrix has converged to one solution. It is customary to stop the ACO algorithm after a number of iterations chosen in accordance with the evaporation rate ρ and problem size n in order to facilitate the aggregation and comparison of results.

3.3 Design Decisions

In Section 3.2, we introduced a basic ACO algorithm for the TSP. All ACO algorithms follow a basic layout which consists of ants repeatedly constructing solutions, which is modeled as a walk through a decision graph, and updating the pheromone values in order to influence the behavior of future ants [10]. Thus, the TSP is a very intuitive application for ACO algorithms. In this section, we discuss the design decisions made for Algorithm 3.1 in more detail and show alternatives used for other problem classes.

3.3.1 Matrix Interpretation

In Section 3.2, we stated that the pheromone value τ_{ij} denotes how opportune it was in the past to move from city i to city j in the tour. When pheromone values denote a relationship between two items of a problem instance, for example, two cities in the TSP, we say that the pheromone matrix is encoded in an *item* $\times$ *item* fashion. The reason for using an *item* $\times$ *item* encoding for the TSP is that the tourlength is a function of the edges traversed in a tour, and hence the pheromone values guide the ants toward solutions with low overall tourlength. This type of encoding is also used for many other problems, for example, scheduling problems with setup costs, where the solution quality is derived from a predecessor–successor relationship. However, for problems like the Quadratic Assignment Problem (QAP), where facilities are allocated to sites, the goal being to minimize the sum of distance-weighted flows between the facilities (see Reference 11), it makes no difference whether a facility i was positioned before or after another facility j. The only information which is relevant for the evaluation function is the location of the site that facility i is assigned to. Hence, for problems where items need to be positioned optimally, whether in a geometric sense as for the QAP or in a schedule, for example, when minimizing tardiness, an *item* $\times$ *place* encoding is employed, with τ_{ij} denoting how good it was in the past to put item i on place j.

Not all problems allow for the pheromone values to be arranged sensibly in a matrix fashion. In Reference 12, ACO algorithms are used for constructing a shortest super-sequence to a given number of strings, that is, the shortest string which contains all given strings as a sequence. The characters of each string are marked with pheromone information to increase the likelihood of being chosen as the next character when the ants construct a supersequence. Marking individual nodes (or items) with pheromone values is proposed in Reference 13 when using ACO algorithms to solve subset problems like the Multiple Knapsack Problem (MKP), where the optimal subset (in the form of a binary vector stating which items are in the subset) to a given number of items is sought which satisfies a number of linear constraints and maximizes a linear function. The higher the pheromone value assigned to an item is, the more likely its inclusion in the knapsack becomes.

3.3.2 Solution Construction

As we mentioned in Section 3.2, ants construct solutions by making a number of consecutive local decisions which result in a global solution. For the TSP and most other *item* × *item* problems, the exact sequence of these decisions is the result of the individual decisions made, as the ant continually moves to the successor to make the next decision. However, for *item* × *place* problems, the decision sequence is arbitrary, unless there are precedence constraints. In the absence of such constraints, it is usually best to randomly permute the decision sequence every time an ant constructs a solution in order to avoid a bias in the construction process, see Reference 14. Although it is generally preferable to proceed in a random sequence, for some problems, for example, when scheduling jobs to minimize total tardiness, it is necessary to start at place 1 and proceed in order so that heuristic information is available (see below); however, even in these cases, a combination with a random decision sequence yields better results [15]. For the QAP, where so far no effective heuristic guidance for the ants has been discovered, a random decision sequence should always be employed.

The basic rule by which ants make decisions is the random proportional transition rule defined by Equation (3.1), in which the existence of heuristic information is already implied. If no heuristic information is used, the formula is reduced to

$$p_{ij} = \frac{\tau_{ij}^{\alpha}}{\sum_{h \in S} \tau_{ih}^{\alpha}}. \tag{3.2}$$

In References 16 and 17, a modification to the random proportional transition rule is proposed which allows for greater control over the balance between exploration and exploitation. A parameter $q_0 \in [0, 1)$ denotes the probability for choosing the best combination of pheromone and heuristic information perceived by the ant, that is, for choosing j with

$$j = \arg\max_{h \in S} \tau_{ih}^{\alpha} \cdot \eta_{ih}^{\beta}, \tag{3.3}$$

instead of proceeding probabilistically according to Equation (3.1), which is done with probability $(1-q_0)$. This *pseudo-random proportional rule* allows for a calibration of the amount of exploitation (q_0-case) versus biased exploration ($(1-q_0)$-case) performed by the ants, similar to the temperature setting in Simulated Annealing or to reinforcement learning, specifically Q-learning [18]. In most cases, however, it is sufficient to use only the random proportional transition rule defined by Equation (3.1), that is, to set $q_0 = 0$.

For an instance of size n, constructing a solution takes $O(n^2)$ steps for an ant, since it must make n decisions and has $O(n)$ candidates for each decision. A common method for reducing this computational burden for very large problem instances is using so called candidate lists. For example, when using an ACO algorithm for the TSP, it could make sense to define a candidate list L_i for each city i which contains the $\mathcal{L}$ nearest neighbors of city i. When the ant makes a decision, it uses $S_i = S \cap L_i$ instead of S as the selection set, unless $S_i = \emptyset$, in which case the ant reverts to using S. On average, this will greatly reduce the computational expense of constructing a solution while still maintaining good solution quality, see also Reference 19.

Equations (3.1), (3.2), or (3.3) are usually applied in permutation problems. For some problem classes, however, extensions or alternatives exist. For scheduling problems which minimize total tardiness, using the sum over all pheromone values up to and including i, that is,

$$\tau_{ij}' = \sum_{l=1}^{i} \tau_{lj}, \tag{3.4}$$

instead of using only the pheromone value τ_{ij}, is studied in Reference 20. This type of pheromone evaluation more accurately reflects that a job which was not chosen for a particular place in the schedule despite a high pheromone value should be scheduled soon afterwards.

In Reference 21, an ACO algorithm for the Maximum Clique Problem (MCP) is proposed, where new nodes for the clique C being constructed are chosen by evaluating how strong the pheromone-based connection between nodes already in C and the candidate nodes in S are, that is, the probability to add node i is

$$p_i = \frac{\tau_{iC}^{\alpha}}{\sum_{h \in S} \tau_{hC}^{\alpha}} \tag{3.5}$$

with $\tau_{iC} = \sum_{j \in C} \tau_{ij}$ and S being updated in such a way that $C' = C \cup \{i\}$ is still a clique for all $i \in S$. Correspondingly, a pheromone update is undertaken on all edges of the largest final clique of the iteration.

In addition to the pheromone information, ants have the possibility of exploiting heuristic information, if available. Although not a strict necessity, heuristic guidance usually plays an important role for attaining a good solution quality. Two known exceptions to this rule are the QAP, where no beneficial heuristic guidance seems to exist, and the MCP mentioned above, for which results in Reference 21 showed that employing a straightforward heuristic guidance scheme ultimately led to a worse solution quality.

For the problem classes for which beneficial heuristic guidance does exist, we distinguish between constant and dynamic heuristic values. For the TSP, the heuristic values $\eta_{ij} = 1/d_{ij}$ are computed once and remain constant. For other problems like the Probabilistic TSP (see References 22 and 23) or the single machine total weighted tardiness problem (SMTWTP), the η_{ij} values are functions of past decisions which the ant has made, that is, they need to be recalculated during every solution construction. For example, when considering which job j to schedule at place i when minimizing the total tardiness of all jobs, the heuristic values

$$\eta_{ij} = \frac{1}{\max(T + p_j, d_j) - T} \tag{3.6}$$

are used in Reference 20, where d_j is the due date and p_j the processing time of job j, and $T = \sum_{h=1}^{i-1} p_{\pi(h)}$ is the sum of processing times of all jobs that have already been scheduled. Even when heuristic values are dynamic, the effort for their computation should be restricted to constant time if possible, since otherwise the amount of time necessary for constructing a solution becomes prohibitively large. Another disadvantage of the heuristic information for the SMTWTP is that it cannot be combined with a random decision sequence.

Finally, it is possible to influence the ant's decisions by tuning the pheromone and heuristic weights α and β. For most applications, setting $\alpha = \beta = 1$ is sufficient. However, tuning these parameters can lead to better performance for some problem classes. For the TSP, choosing $\beta > 1$ has been shown to yield good results, for example, using $\beta = 2$ in References 17 and 24 or $\beta = 5$ in References 25 and 26. Using a steadily decreasing value of β has also been applied successfully for the Resource Constrained Project Scheduling Problem (RCPSP) in Reference 27. Using values of $\alpha > 1$ has been shown in Reference 21 to achieve quicker convergence at the cost of a lower solution quality in the long run.

3.3.3 Pheromone Update

The purpose of the pheromone update is to focus the search process of the ants on a promising portion of the solution space, which is then explored more extensively in the hope of finding the optimal solution. The field of update strategies is perhaps the most studied part of ACO algorithms, and a variety of methods to update the pheromone values have been proposed. In this section, we briefly explain the conceptually different approaches which have been suggested to update pheromone values, and also explain in detail some of the more successful strategies.

At the most abstract level, the pheromone update should accomplish a positive reinforcement of those search space regions which seem promising and a negative reinforcement of all others. The principal mechanisms used for this are pheromone evaporation (for the negative reinforcement), which diminishes all pheromone values by a relative amount each time it is applied, and pheromone intensification (for the positive reinforcement), achieved by adding an update term to selected pheromone values. Formally, an update takes the form

$$\forall i, j \in [1, n] : \tau_{ij} \mapsto (1 - \rho) \cdot \tau_{ij} + \Delta_{ij}, \tag{3.7}$$

where $\rho \in (0, 1]$ is a parameter of the algorithm denoting how much of the pheromone information is lost with every application of evaporation. A high evaporation rate will cause a more rapid convergence which is coupled with less exploration than a low evaporation rate. Thus, the evaporation rate should be tuned in accordance with the number of iterations that the ACO algorithm is allowed to run. Δ_{ij} is an update value, which is 0 if the edge (i, j) was not traversed by the ant and some value greater than 0 if it was. The exact value of Δ_{ij} and especially the strategy when an update is performed is the key difference between most types of ACO algorithm. There are two aspects to consider when characterizing pheromone intensification: which solutions update, and how much is updated by these solutions.

Generally, updates of the pheromone values take place after an iteration of m ants has constructed solutions. In the Ant System (AS), which was introduced in Reference 9 for solving the TSP, every ant of the iteration contributes to the pheromone update. For each ant $l \in [1, m]$, the update value $\Delta_{ij}(l)$, is calculated, and the update is performed with the sum of update values $\Delta_{ij} = \sum_{l=1}^{m} \Delta_{ij}(l)$. Three different methods for determining the individual Δ_{ij} were tested: assigning a constant, using the inverse to the distance d_{ij} between customers i and j, and, performing best and used subsequently, inverse to the length of the entire tour, that is, the solution quality. In addition to the m ants of an iteration being allowed to perform an update, it was also proposed to let a number of so called elitist ants, which represent the best solution found by all ants so far, update the pheromone trail. Using a small number of these elitist ants, inspired by the elitist strategy in Reference 28, intensifies the search near the currently best solution, leading to better results overall.

Further research resulted in the introduction of the Ant Colony System (ACS) [16,17]. Here, an online update of the pheromone values was proposed in order to enforce exploration: each time an ant traversed an edge (i, j), it would reduce the corresponding pheromone value according to

$$\tau_{ij} \mapsto (1 - \rho) \cdot \tau_{ij} + \rho \cdot \tau_0, \tag{3.8}$$

thus encouraging subsequent ants to choose different edges (note that this holds only for $\tau_{ij} \geq \tau_0$). Also, the global update by all ants at the end of an iteration was replaced by one update performed along the best tour found so far, that is, by one elitist ant.

Originating from AS and ACS, many other update schemes have been proposed. In References 24 and 29, the *MAX–MIN* Ant System (MMAS) is introduced, which uses only the best ant of the iteration and an elitist ant for the positive pheromone update and avoids stagnation in the search process by limiting pheromone values to the predetermined interval $[\tau_{\min}, \tau_{\max}]$. Limiting the pheromone values also bounds the minimum and maximum probability with which an edge is selected according to Equation (3.1), if the heuristic values are bounded as well. In Reference 25, a modification of the AS called AS-rank is proposed, where the m ants of an iteration are ranked by their solution quality and, together with a number of elitist ants of maximum rank, update the pheromone trail in proportion to their rank.

Some methods also exist that operate without using evaporation. In Reference 30, pheromone update is accomplished by comparing the solution quality of the ant to the average quality of the m previous ants. If it is better, a positive update is performed along the path; if it is worse, the update is negative. Thus, the update is accomplished in $O(m \cdot n)$ time for m ants, compared to $O(n^2)$ for Equation (3.7). In Reference 31, a population of solutions is maintained from which the pheromone matrix is derived. Updates are performed on the population, with the insertion of a solution being equivalent to a positive update of the pheromone matrix, and a deletion to a negative update (nullifying the previous positive update which was undertaken upon insertion). This update mechanism will be studied in more detail in Section 3.4.

The number of ants m also plays a role in the exact effect of the pheromone updates. The more solutions are constructed before an update is undertaken, the higher the expected quality of the updating solutions. However, the number of ants is a linear factor in the runtime of the algorithm, which is why a trade-off between runtime and quality of the updating solutions must be found. In Reference 17, a method for finding the optimal number of ants is discussed which relies on knowledge of the average size of the pheromone values before and after a change, both being a function of the problem size n. Modeling the local pheromone update as a first-order linear recurrence relation allows m to be expressed as a function of the average pheromone levels before and after an update, and of the initial pheromone values. Although the authors cannot provide these pheromone levels, they argue that experimental results show $m = 10$ to work well, and this is also the case in our own experiments.

3.3.4 Pheromone Initialization

Having discussed a number of different possibilities for updating pheromone information, we now illustrate some possibilities for initializing the pheromone values. Pheromone initialization is usually uniform, that is, all pheromone values τ_{ij} are assigned an initial value τ_0. A notable exception to this rule is studied in Reference 32, where the initialization is performed after a number of preprocessing steps used to identify local optima.

For uniform initialization, the value of τ_0 must be seen in relation to the value for Δ in order to understand its impact on the behavior of the algorithm. Setting $\tau_0 \gg \Delta$ will cause the updates performed during the initial phase of the ACO algorithm to have practically no effect. Instead, the ants are guided almost exclusively by the heuristic information until all pheromone values have dropped to lower values. Conversely, for $\tau_0 \ll \Delta$, there is a high risk of the ACO algorithm converging on the first good solution found, since the first solution to receive some updates will be highly favored in future decisions due to Equation (3.1). Setting

$$\tau_0 = \frac{\Delta}{\rho}, \tag{3.9}$$

which is the maximum value attainable in the long run for any τ_{ij} via Equation (3.7), usually represents a good trade-off between runtime and exploration.

3.3.5 Integration of Local Search

Local search is a possible augmentation for the ACO algorithm which can be used to further improve the solutions found by the ants of an iteration (either all or only those used for pheromone intensification) before the pheromone update is performed. Some common local search methods, which have been especially popular with the TSP and other routing problems, are 2-opt, 3-opt [33], and Lin–Kernighan (LK) [34]. When applicable, using local search represents a trade-off between quickly improving solution quality by exploring the neighborhood of a solution and premature convergence to a local optima. Ideally, the local search performed should be possible in $O(n^2)$ steps for instances of size n, since otherwise it becomes the dominant part of the search process instead of the ants constructing new solutions.

3.4 Population Based Ant Colony Optimization

Population based Ant Colony Optimization (PACO), introduced in Reference 31, represents a conceptually new and different update scheme to the ACO algorithm. Instead of explicitly maintaining a pheromone matrix and using evaporation to gradually reduce old or unnecessary pheromone information, the PACO algorithm maintains a population of solutions $P = \{\pi_i, \ldots, \pi_k\}$, which is updated instead of the pheromone matrix, and from which the pheromone values τ_{ij} can be derived via

$$\tau_{ij} = \tau_0 + \Delta \cdot |\{\pi \in P | (i,j) \in \pi\}|, \tag{3.10}$$

where Δ is the update value used.

After m ants have constructed a solution, the best ant is taken as a candidate to update the population. For the first k iterations, the candidate solution is added to P in all cases. Once $|P| = k$, it becomes necessary to remove a solution from the population in order to maintain a population size of k. Managing the Population as a FIFO queue, that is, removing the oldest solution and adding the best of the iteration, has proven to be a very good update method [31] for population sizes of $k \leq 5$, outperforming MMAS and the standard ACO algorithm on the TSP and the QAP. Other update methods are introduced in Reference 35. It is also possible to define one of the solutions in the population as the elitist solution, which can only be removed if a better solution is added in its place.

Since old information is removed via the deletion of solutions from the population instead of pheromone evaporation, the PACO algorithm can move to a completely new portion of the search space in k iterations. Coupled with the explicit storage of solutions which are available for repair, this makes the algorithm well suited for dynamic problems, see Reference 35. When a number of alternative solutions are sought, for example, for multi-criteria optimization problems, PACO uses the population to store non-dominated solutions, achieving a good approximation of the Pareto front [36,37]. Finally, the population of solutions can be used as an interface with other population driven algorithms like Genetic Algorithms [38]. See Reference 37 for a more detailed study of the PACO algorithm.

3.5 Applications

In this section, we present a survey of some of the noteworthy applications of ACO. Of course, this survey cannot hope to present a complete overview, and the interested reader is referred to References 26, 39, and 40 for additional surveys.

One of the earliest and most intuitive applications of ACO was the TSP [9]. Since all ACO algorithms depend in some fashion on the metaphor of an ant moving through a graph [10], using the TSP to illustrate the basic principles of Ant Algorithms is a logical choice, and it is also used as the introductory example in Reference 39. ACO has delivered good results on many TSP instances, especially when combined with local search [24]. However, due to the existence of very good heuristics like Lin–Kernighan [34] and polynomial time approximation schemes [41] for the Euclidean TSP, ACO algorithms are not the best choice for this problem class. The situation is better for the Sequential Ordering Problem (SOP), an extension of the TSP, where the goal is to find a minimum weight Hamiltonian Path with precedence constraints among the nodes. Here, a form of Ant Algorithm called Hybrid Ant System (HAS-ACO) [42] is currently one of the best algorithms available. Other variations of the standard TSP, like the Probabilistic TSP (PTSP) and Dynamic TSP (DTSP), are also handled well by ACO, using proper heuristic guidance [23] for the PTSP and pheromone resetting strategies [43,44] or the PACO algorithm [35] for the DTSP.

Another problem related to the TSP is the Vehicle Routing Problem (VRP), in which a number of customers must be serviced exactly once, and all vehicles begin and end their respective tours at a depot. The goal is to minimize the number of vehicles while meeting constraints such as capacity per vehicle, maximum tourlength per vehicle, and time windows. Solving this problem with Ant Systems was first proposed in Reference 45, and further research has lead to a unified approach for VRPs [46], where the Ant System is combined with an insertion heuristic from Reference 47.

The QAP, defined in Reference 11 and shown to be NP-hard in Reference 48, is a conceptually different optimization problem compared to the TSP and its derivates in the sense that the pheromone matrix is not interpreted in an *item* $\times$ *item* fashion, but rather as *item* $\times$ *place*. Applying Ant System to the QAP was first undertaken in Reference 49, including a heuristic guidance scheme for the ants when constructing a solution. Adding local search to the AS algorithm was shown to be beneficial in Reference 50. In Reference 51, the Hybrid Ant System (HAS) was introduced and applied to the QAP with good results. The HAS algorithm uses ants to modify solutions instead of building them, and the pheromone values are used to remember beneficial changes.

Another class of problems in which ACO algorithms have seen wide and successful application is in scheduling problems. For scheduling with due dates, for example, the Single Machine Total Weighted Tardiness Problem (SMTWTP), the pheromone matrix is also interpreted in a *item* × *place* fashion. However, in contrast to the QAP, "place" in this case refers to the place in the schedule and not a physical location. An ACO algorithm for the SMTWTP was applied in Reference 52, where ACO found the optimal solution to 125 benchmark problems more often than the other heuristics evaluated. Ant Algorithms have also been applied to somewhat more complex scheduling problems, for example, job shop scheduling [53], flow shop scheduling [54], and, most notably, the Resource Constrained Project Scheduling Problem (RCPSP) [27], where ACO was state of the art at the time of publishing.

Lately, the ACO algorithm has been extended to be able to deal with multi-criteria optimization problems, in particular the Single Machine Total Tardiness with Setup Costs Problem. Here, two criteria exist which must be optimized simultaneously, yet cannot be aggregated into a single optimization function. Rather, the algorithm needs to find a number of solutions which represent different trade-offs between the two (or more) criteria. The PACO algorithm was modified for optimizing multi-criteria problems in Reference 36 with further improvements in Reference 28 yielding an algorithm which performs very well and can deal with an arbitrary number of criteria.

So far, all the problems discussed have been permutation problems, which can be handled quite well by ACO. However, some efforts have been undertaken to apply ACO to areas where solutions are not permutations. As mentioned above, in Reference 12, ACO is successfully applied to the shortest supersequence problem. Also, some partitioning problems, for example, graph coloring [55] and data clustering [56], have been solved with ACO, with varying degrees of success. In Reference 57, ACO is used as a generic algorithm for solving Constraint Satisfaction Problems (CSPs) with promising results.

As a final note, although not being an application, in the recent past it has been shown that under certain conditions, some versions of ACO can provably find the optimal solution to the instance of a problem with a probability arbitrarily close to 1 [58,59]. Although these results have no immediate impact on the applicability of ACO algorithms, they put ACO on the same level as Simulated Annealing or Genetic Algorithms in terms of solution finding capability. Note that with a lower bound greater than 0 on the probability to find the solution or move closer to the solution in a given iteration, any method will find the optimum with a probability arbitrarily close to 1, given enough time.

References

[1] R.H. Arnett. *American Insects: A Handbook of the Insects of America North of Mexico.* Van Nostrand Rehinhold, New York, 1985.

[2] E.J. Fittkau and H. Klinge. On biomass and trophic structure of the central amazonian rain forest ecosystem. *Biotropica*, 5: 2–14, 1973.

[3] K. von Frisch. *The Dance Language and Orientation of Bees.* Harvard University Press, 1967.

[4] M. Lüscher. Air-conditioned termite nests. *Scientific American*, 205: 138–145, 1961.

[5] P. Grassé. La reconstruction du nid et les coordinations interindividuelles chez bellicositermes natalensis et cubitermes sp. la theorie de la stigmergie: essai d'interpretation du comportement des termites constructeurs. *Insectes Sociaux*, 6: 41–81, 1959.

[6] G. Nicolis and I. Prigogine. *Self-Organization in Non-Equilibrium Systems.* John Wiley & Sons, New York, 1977.

[7] S. Goss, S. Aron, J.-L. Deneubourg, and J. Pasteels. Self-organized shortcuts in the argentine ant. *Naturwissenschaften*, 76: 579–581, 1989.

[8] J.-L. Deneubourg, S. Aron, S. Goss, and J. Pasteels. The self-organizing exploratory pattern of the argentine ant. *Journal of Insect Behavior*, 3: 159–168, 1990.

[9] M. Dorigo. *Ottimizzazione, apprendimento automatico, ed algoritmi basati su metafora naturale (Optimization, learning and natural algorithms)* (Italian). Ph.D. thesis. Dipartimento di Elettronica, Politecnico di Milano, Italy, 1992.

[10] M. Dorigo and G.D. Caro. Ant colony optimization: A new meta-heuristic. In P.J. Angeline, Z. Michalewicz, M. Schoenauer, X. Yao, and A. Zalzala, Eds., *Congress on Evolutionary Computation (CEG)*, Vol. 2. IEEE Press, Washington, 1999, pp. 1470–1477.

[11] T.C. Koopmans and M.J. Beckman. Assignment problems and the location of economic activities. *Econometrica*, 25(1): 53–76, 1957.

[12] R. Michel and M. Middendorf. An ACO algorithm for the shortest common supersequence problem. In D. Corne, M. Dorigo, and F. Glover, Eds., *New Ideas in Optimization*, McGraw-Hill, New York, 1999, pp. 51–62.

[13] G. Leguizamón and Z. Michalewicz. A new version of ant system for subset problems. In P.J. Angeline, Z. Michalewicz, M. Schoenauer, X. Yao, and A. Zalzala, Eds., *Congress of Evolutionary Computation*, Vol. 2. IEEE Press, Washington, 1999, pp. 1459–1464.

[14] D. Merkel and M. Middendorf. On the behaviour of ant algorithms: Studies on simple problems. In *Metaheuristics International Conference (MIC)*, July 2001, pp. 573–577.

[15] D. Merkle and M. Middendorf. A new approach to solve permutation scheduling problems with ant colony optimization. In *Applications of Evolutionary Computing — Evo Workshops*, Vol. 2037 of *Lecture Notes on Computer Science*. Springer-Verlag, Heidelberg, 2001, pp. 484–493.

[16] M. Dorigo and L.M. Gambardella. Ant colonies for the traveling salesman problem. *BioSystems*, 43: 73–81, 1997.

[17] M. Dorigo and L.M. Gambardella. Ant colony system: A cooperative learning approach to the traveling salesman problem. *IEEE Transactions on Evolutionary Computation*, 1: 53–66, 1997.

[18] C.J.C.H. Watkins and P. Dayan. Technical note: Q-learning. *Machine Learning*, 8: 279–292, 1992.

[19] T. Stützle and M. Dorigo. ACO algorithms for the traveling salesman problem. In K. Miettinen, M. Makela, P. Neittaanmaki, and J. Periaux, Eds., *Evolutionary Algorithms in Engineering and Computer Science*, Wiley, New Jersey, 1999, pp. 163–183.

[20] D. Merkle and M. Middendorf. An ant algorithm with a new pheromone evaluation rule for total tardiness problems. In *Applications of Evolutionary Computing — Evo Workshops*, Vol. 1803 of *Lecture Notes on Computer Science*. Springer-Verlag, Heidelberg, 2000, pp. 287–296.

[21] S. Fenet and C. Solnon. Searching for maximum cliques with ant colony optimization. In G. Raidl et al., Eds., *Applications of Evolutionary Computing — Evo Workshops*, Vol. 2611 of *Lecture Notes on Computer Science*, Springer, 2003, pp. 236–245.

[22] J. Branke and M. Guntsch. New ideas for applying ant colony optimization to the probabilistic TSP. In *Applications of Evolutionary Computing — Evo Workshops*, Vol. 2611 of *Lecture Notes in Computer Science*. Springer-Verlag, Heidelberg, 2003, pp. 165–175.

[23] J. Branke and M. Guntsch. Solving the probabilistic tsp with ant colony optimization. *Journal of Mathematical Modelling and Algorithms (JMMA)*, 3(4): 403–425, 2004.

[24] T. Stützle and H. Hoos. Max–min ant system. *Future Generation Computer Systems*, 16: 889–914, 2000.

[25] B. Bullnheimer, R. Hartl, and C. Strauss. A new rank based version of the ant system — a computational study. *Central European Journal for Operations Research and Economics*, 7: 25–38, 1999.

[26] D. Corne, M. Dorigo, and F. Glover. *New Ideas in Optimization*. McGraw-Hill, New York, 1999.

[27] D. Merkel, M. Middendorf, and H. Schmeck. Ant colony optimization for resource-constrained project scheduling. *IEEE Transactions on Evolutionary Computation*, 6: 333–346, 2002.

[28] J. Holland. *Adapdation in Natural and Articial Systems*. MIT Press, Ann Arbor, MI, 1975.

[29] T. Stützle and H.H. Hoos. The max–min ant system and local search for the traveling salesman problem. In *IEEE International Conference on Evolutionary Computation*. IEEE Press, Piscataway, NJ, 1997, pp. 309–314.

[30] V. Maniezzo. Exact and approximate nondeterministic tree-search procedures for the quadratic assignment problem. *INFORMS Journal on Computing*, 11(4): 358–369, 1999.
[31] M. Guntsch and M. Middendorf. A population based approach for ACO. In S. Cagnoni et al., Eds., *Applications of Evolutionary Computing — Evo Workshops*, Vol. 2279 of *Lecture Notes on Computer Science*. Springer, 2002, pp. 72–81.
[32] C. Solnon. Boosting ACO with a preprocessing step. In S. Cagnoni, J. Gottlieb, E. Hart, M. Middendorf, and G. Raidl, Eds., *Applications of Evolutionary Computing — Evo Workshops*, Vol. 2279. Springer-Verlag, Kinsale, Ireland, 2002, pp. 161–170.
[33] S. Lin. Computer solutions for the traveling salesman problem. *Bell Systems Technical Journal*, 44(10): 2245–2269, 1965.
[34] S. Lin and B. Kernighan. An effective heuristic algorithm for the traveling salesman problem. *Operations Research*, 21: 498–516, 1973.
[35] M. Guntsch and M. Middendorf. Applying population based ACO to dynamic optimization problems. In *International Workshop on Ant Algorithms ANTS*, Vol. 2463 of *Lecture Notes on Computer Science*. Springer-Verlag, Heidelberg, 2002, pp. 111–122.
[36] M. Guntsch and M. Middendorf. Solving multi-criteria optimization problems with population-based aco. In C. Fonseca, P. Fleming, E. Zitzler, K. Deb, and L. Thiele, Eds., *Evolutionary Multi-Criterion Optimization (EMO)*, Vol. 2632 of *Lecture Notes on Computer Science*. Springer, Berlin, Heidelberg, 2003, pp. 464–478.
[37] M. Guntsch. *Ant Algorithms in Stochastic and Multi-Criteria Environments*, Ph.D. thesis. Institute AIFB, University of Karlsruhe, January 2004. http://www.ubka.uni-karlsruhe.de/cgibin/psview?document=2004/wiwi/3.
[38] J. Branke, C. Barz, and I. Behrens. Ant-based crossover for permutation problems. In E. Cantu-Paz, Ed., *Genetic and Evolutionary Computation Conference*, Vol. 2723 of *Lecture Notes in Computer Science*. Springer, 2003, pp. 754–765.
[39] E. Bonabeau, M. Dorigo, and G. Théraulaz. *Swarm Intelligence*. Oxford University Press, Oxford, 1999.
[40] T. Stützle and M. Dorigo. The ant colony optimization metaheuristic: Algorithms, applications, and advances. In F. Glover and G. Kochenberger, Eds., *Handbook of Metaheuristics*. Kluwer Academic Publishers, Norwell, MA, 2002.
[41] S. Arora. Polynomial time approximation schemes for Euclidean traveling salesman and other geometric problems. *Journal of the ACM*, 45: 753–782, 1998.
[42] L.M. Gambardella and M. Dorigo. An ant colony system hybridized with a new local search for the sequential ordering problem. *IN-FORMS Journal on Computing*, 12: 237–255, 2000.
[43] M. Guntsch, M. Middendorf, and H. Schmeck. An ant colony optimization approach to dynamic TSP. In L. Spector et al., Eds., *Genetic and Evolutionary Computation Conference (GECCO)*. Morgan Kaufmann Publishers, San Francisco, CA, 2001, pp. 860–867.
[44] M. Guntsch and M. Middendorf. Pheromone modification strategies for ant algorithms applied to dynamic TSP. In E. Boers et al., Eds., *Applications of Evolutionary Computing — Evo Workshops*, Vol. 2037 of *Lecture Notes in Computer Science*. Springer-Verlag, Heidelberg, 2000, pp. 213–222.
[45] B. Bullnheimer, R. Hartl, and C. Strauss. An improved ant system algorithm for the vehicle routing problem. Technical report, POM Working Paper No. 10/97, University of Vienna, 1997.
[46] M. Reimann, K. Doerner, and R. Hartl. Analyzing a unified ant system for the vrp and some of its variants. In G. Raidl et al., Eds., *Applications of Evolutionary Computing — Evo Workshops*, Vol. 2611 of *Lecture Notes on Computer Science*. Springer, Heidelberg, 2003, pp. 300–310.
[47] M.M. Solomon. Algorithms for the vehicle routing and scheduling problems with time window constraints. *Operations Research*, 35(2): 254–265, 1987.
[48] S. Sahni and T. Gonzales. P-complete approximation problems. *Journal of ACM*, 23(3): 555–565, 1976.

[49] V. Maniezzo and A. Colorni. The ant system applied to the quadratic assignment problem. *IEEE Transactions on Knowledge and Data Engineering*, 5: 769–778, 1998.

[50] T. Stützle and H. Hoos. Max–min ant system and local search for combinatorial optimization problems. In *Metaheuristics International Conference (MIC)*, Kluwer Academic, Norwell, MA, 1997.

[51] L.M. Gambardella, E.D. Taillard, and M. Dorigo. Ant colonies for the QAP. *Journal of Operations Research Society*, 2: 167–176, 1999.

[52] A. Bauer, B. Bullnheimer, R. Hartl, and C. Strauss. An ant colony optimization approach for the single machine total tardiness problem. In *Congress on Evolutionary Computation (CEC)*, IEEE Press Piscataway, NJ, 1999, pp. 1445–1450.

[53] A. Colorni, M. Dorigo, V. Maniezzo, and M. Trubian. Ant system for job-shop scheduling. *JORBEL — Belgian Journal of Operations Research, Statistics and Computer Science*, 34: 39–53, 1994.

[54] T. Stützle. An ant approach for the flow shop problem. In *6th European Congress on Intelligent Techniques & Soft Computing (EUFIT)*, Vol. 3. Verlag Mainz, Aachen, 1998, pp. 1560–1564.

[55] A. Vesel and J. Zerovnik. How good can ants color graphs? *Journal of Computing and Information Technology — CIT*, 8: 131–136, 2000.

[56] N. Monmarche. On data clustering with artificial ants. In A.A. Freitas, Ed., *Data Mining with Evolutionary Algorithms: Research Directions.* AAAI Press, Orlando, FL, 1999, pp. 23–26.

[57] C. Solnon. Ants can solve constraint satisfaction problems. *IEEE Transactions on Evolutionary Computation*, 6: 347–357, 2002.

[58] W. Gutjahr. ACO algorithms with guaranteed convergence to the optimal solution. *Information Processing Letters*, 82: 145–153, 2002.

[59] T. Stützle and M. Dorigo. A short convergence proof for a class of ACO algorithms. *IEEE Transactions on Evolutionary Computation*, 6: 358–365, 2002.

4

Swarm Intelligence

Mohamed Belal
Jafaar Gaber
Hoda El-Sayed
Abdullah Almojel

4.1 Introduction

Swarm Intelligence (SI) is a computational and behavioral metaphor for solving distributed problems inspired from biological examples provided by social insects such as ants, termites, bees, and wasps and by swarm, herd, flock, and shoal phenomena in vertebrates such as fish shoals and bird flocks.

In other words, SI is based on the principles underlying the behavior of natural systems consisting of many agents, and exploiting local communication forms and highly distributed control. Thus, the SI approach constitutes a very practical and powerful model that greatly simplifies the design of distributed solutions to different kind of problems. In the last few years, SI principles have been successfully applied to a series of applications including optimization algorithms, communications networks, and robotics.

4.2 Swarm Intelligence Overview

The SI approach emphasizes two important paradigms: the highly distributed control paradigm and the emergent strategy-based paradigm.

4.2.1 Emergent Strategy-Based Paradigm

Collective behavior demonstrated by social insects (ants, bees, termites, etc.) often emerges from a small set of simple low-level interactions between individuals, and between individuals and the environment.

The following example illustrates the concept of emergence. To solve a given task, for example, to sort elements scattered on the ground, one can write an algorithm wherein a centralized part distributes the task to achieve between a set of distributed agents. The centralized program, based on the global goal and plans, the current input, and the current state, collects agent results, analyzes them, and decides the actions to be executed next.

One way of achieving the required task without a centralized part is by the addition of the individual efforts of a multitude of agents who do not have any idea of the global objective to be reached; that is, there is the emergence of collective behavior. Deneubourg et al. [1] introduced a model of sorting behavior in ants. They found that simple model ants were able to sort into piles objects initially strewn randomly across a plane. To be precise, near an anthill, one can observe that ants run in all directions to gather corpses, to clean up their nests, or transport their eggs to order them by size etc. One can only imagine that something, such as a special chemical marker, indicates to individual ants where to place their chips, and allows them to distinguish an object already arranged from an object to be arranged. But how are these markers placed and on what criteria? In fact, such interesting collective behavior can be mediated by nothing more than similar, simple individual behavior. For example, F(1) each ant wanders a bit, (2) if an ant meets an object and if it does not carry one, it takes it, and (3) if an ant transports an object and there is a similar object in the same way in front of it, it deposits its load. By following these local strategic rules with only local perceptual capacities, ants display the ability to perform global sorting and clustering of objects.

Swarm Intelligence is a new way to control multiple agent systems. The swarm-type approach to emergent strategy deals with large numbers of homogeneous agents, each of which has fairly limited capabilities on its own. However, when many such simple agents are brought together, globally interesting behavior can emerge as a result of the local interactions of the agents and the interactions between the agents and the environment. A key research issue in such a scenario is determining the proper design of the local control laws that will allow the collection of agents to solve a given problem.

4.2.2 Highly Distributed Control Paradigm

Swarm Intelligence is, intrinsically, a bottom-up approach. Bottom-up approaches are carried out by programming large numbers of independent entities with relatively simple sets of rules. Brought together, constructive behavior emerges, as it does in insects that create complex social behavior and structures from the combined efforts of individuals with extremely limited intelligence. In contrast, the top-down approach is based on the classic centralized method (e.g., the Client/Server approach), wherein central coordination should take place. SI can be applied to fully distributed systems that consist of several autonomous agents working together with local communication and minimal perception capabilities to complete one or more tasks.

4.2.3 Organizing Principles

A study of the SI approach reveals a useful set of organizing principles that can guide the design of efficient distributed applications for different kinds of problems. SI has the following notable features:

Autonomy: The system does not require outside management or maintenance. Individuals are autonomous, controlling their own behavior both at the detector and effector levels in a self-organized way.

Adaptability: Interactions between individuals can arise through direct or indirect communication via the local environment; two individuals interact indirectly when one of them modifies the environment and the other responds to the new environment at a later time. By exploiting such local communication forms, individuals have the ability to detect changes in the environment dynamically. They can then autonomously adapt their own behavior to these new changes. Thus, swarm systems emphasize auto-configuration capabilities.

Scalability: SI abilities can be performed using groups consisting of a few, up to thousands of individuals with the same control architecture.

Flexibility: No single individual of the swarm is essential, that is, any individual can be dynamically added, removed, or replaced.

Robustness: SI provides a good example of a highly distributed architecture that greatly enhances robustness; no central coordination takes place, which means that there is no single point of failure. Moreover, like most biological and social systems, and by combining scalability and flexibility capabilities, the swarm system enables redundancy, which is essential for robustness.

Massively parallel: The swarm system is massively parallel and its functioning is truly distributed. Tasks performed by each individual within its group are the same. If we view each individual as a processing unit, SI architecture can be thought of as single instruction stream–multiple data stream (SIMD) architecture or systolic networks.

Self-organization: Swarm systems emphasize self-organization capabilities. The intelligence exhibited is not present in the individuals, but rather emerges somehow out of the entire swarm. In other words, if we view every individual as a processing unit, solutions to problems obtained are not predefined or preprogrammed but are determined collectively as a result of the running program.

Cost effectiveness: The swarm-type system consists of a finite collection of homogeneous agents, each of which has fairly limited capabilities on its own. Also, each agent has the same capabilities and control algorithm. It is clear that the autonomy and the highly distributed control afforded by the swarm model greatly simplify the task of designing the implementation of parallel algorithms and hardware. For example, for swarm-type multi-robotic systems, robots are relatively simple and their design process effort can be kept minimal in terms of sensors, actuators, and resources for computation and communication.

4.2.4 Swarm Intelligence Communication Forms

SI exploits local communication forms. Interactions between individuals can arise through direct or indirect communication.

4.2.4.1 Indirect Communication

Indirect communication is implicit communication that takes place between individuals via the environment. This is known as Stigmergy communication. The Stigmergy concept describes a class of mechanisms mediating animal–animal interactions through stimuli. When an animal does not explicitly distinguish between its own activity and the activities of others, its actions include modification of its local environment. By sensing its environment, an animal will perform an appropriate action as a response to the new environment at a later time. Thus, interaction takes place in stages through changes in the local environment. Note that the behavior of each insect can then be described as a series of stimulus–response sequences.

There are two forms of Stigmergy. In the Stigmergy Sematectonic communication form, information is communicated through physical modification of the environment. For example, opening a hole in the body of a termitary causes a disruption of the termitary's carefully maintained internal atmosphere (intense gradients in temperature, humidity, carbon dioxide, and oxygen). Sensing some problem in the body of the termitary, termites perform the rebuilding function and attack intruders while repairing the breach in order to restore the termitary's equilibrium.

In the second form of Stigmergy, some signal substance is deposited in the environment that makes no direct contribution to the task being undertaken but is used to influence the subsequent behavior that is task related [2]. For example, for building their nests, termites use highly volatile chemicals called *pheromones*. Termites place tiny balls of mud near other balls of mud that have high pheromone concentrations and, as a consequence, mounds develop. As the mounds grow, pheromones at the bases evaporate and the termites bring the mud to the top, driving the height of some mounds upward of 30 ft and causing adjacent mounds to meet in arches.

Pheromone-based Stigmergy is well developed in ants. Ants are capable of finding the shortest path from a food source to the nest. Also, they are capable of adapting to changes in the environment, and find a new shortest path once the old one is no longer feasible due to an obstacle [3]. Ants deposit a certain amount of pheromone while walking, and each ant probabilistically prefers to follow a direction rich in pheromone rather than a poorer one. Hence, the shorter path will receive a higher amount of pheromone and this will in turn cause a higher number of ants to choose the shorter path. This elementary behavior of real ants explains how they can find the shortest path. The collective behavior that emerges is a form of autocatalytic behavior (or positive feedback), whereby the more the ants follow the trail the more likely they are to do so.

4.2.4.2 Direct Communication

Direct communication is explicit communication that can also take place between individuals. Examples of such interactions are the waggle dance of the honeybee, using antennas, trophallaxis (food or liquid exchange, e.g., mouth-to-mouth food exchange in honeybees), mandibular contact, visual contact, chemical contact (the odor of nearby nest mates), etc.

Direct communication can be implemented by mobile wireless ad hoc networks. Individuals have a very limited memory with the added feature that they are mobile; therefore, they can be considered mobile agents. Indirect interactions through the environment can be thought of as distributed short-term memory. Indeed, agents communicate through pheromone trails. When walking toward the colony or food sources, ants will simply walk toward a high concentration of pheromone. The accumulated pheromone then serves as a distributed shared memory. Note also that we need an analog for indirect interaction through the local environment to implement the autoadaptive mechanism. Such a system can adapt to changes in user behavior and system software through the pheromones. In other words, pheromones will monitor the state of the machines and the network.

4.2.5 The Limitations of Swarm Intelligence

The swarm approach provides a rich source of inspiration and its principles are directly applicable to computer systems. However, although it emphasizes auto-configuration, auto-organization, and adaptability capabilities, the swarm-type approach remains useful for non-time-critical applications involving numerous repetitions of the same activity over a relatively large area, such as finding the shortest path or collecting rock samples on Mars [4]. Indeed, the swarm-type approach deals with the cooperation of large numbers of homogeneous agents. Such approaches usually rely on mathematical convergence results (such as the random walk) that reach the desired outcome over a sufficiently long period of time [4]. Notice that, in addition, the agents involved are homogeneous.

4.3 The Main Applications of Swarm Intelligence

Swarm Intelligence principles have been successfully applied in a variety of problem domains and applications. An example of successful research direction in SI is ant colony optimization (ACO), which focuses on discrete optimization problems. Particle swarm optimization (PSO) is also an efficient and general approach to solve nonlinear optimization problems with constraints. Another example of interesting research direction is swarm robotics, where the focus is on applying SI techniques to the control of large groups of cooperating autonomous robots.

4.3.1 Ant Colony Optimization

Ant colony optimization has been applied successfully to a large number of difficult, discrete optimization problems including the traveling salesman, the quadratic assignment, scheduling, vehicle routing, etc., as well as to routing in telecommunication networks. Ant algorithms are a subset of SI. In other words, ant algorithms can be viewed as multi-agent systems (ant colony), where agents (individual ants) solve required tasks through cooperation in the same way that ants create complex social behavior from the combined efforts of individuals.

4.3.1.1 Basic Ant Algorithm

The basic concept underlying the ant algorithm is inspired by the foraging behavior of real ants. When ants search for food, they start from their nest and move at random toward the food. Ants use highly volatile chemicals called *pheromones* to provide a sophisticated signaling system. While walking, ants deposit quantities of pheromone marking the selected routes that they follow with a trail of the substance. When an ant encounters an intersection, it has to decide which path to follow next. The concentration of pheromone on a certain path is an indication of its usage. An ant chooses a path with a high probability to follow and thereby reinforces it with a further quantity of pheromone. Over time, the concentration of pheromone decreases due to diffusion. This foraging process is an autocatalytic process characterized by a positive feedback loop, where the probability that an ant chooses any given path increases according to the number of ants choosing the path on previous occasions. Ants that take the shortest path will reach the food source first. On their way back to the nest, the ants again have to select a path. After a sufficiently long period of time, the pheromone concentration on the shorter path will be higher than on other longer paths. Thus, all the ants will finally choose the shorter path.

This ant foraging process can be used to find the shortest path in networks. Also, ants are capable of adapting to changes in the environment, and find a new shortest path once the old one is no longer feasible due to some obstacle. Thus, this process is appropriate to mobile ad hoc networks wherein link changes occur frequently [5].

Let $G = (V, E)$ be a connected graph with $N = |V|$ nodes. The simple ant colony optimization meta-heuristic can be used to find the shortest path between a source node v_s and a destination node v_d on the graph G. The path length is defined by the number of nodes on the path. A variable $\varphi_{i,j}$ corresponding to the artificial pheromone concentration is associated with each edge (i, j). An ant located in node v_i uses pheromone $\varphi_{i,j}$ to compute the probability of node v_j being the next hop. This transition probability $p_{i,j}$ is defined as:

$$p_{i,j} = \begin{cases} \dfrac{\varphi_{i,j}}{\sum_{j \in V_i} \varphi_{i,j}} & \text{if } j \in V_i, \\ 0 & \text{if } j \notin V_i, \end{cases} \qquad \text{with} \sum_{j \in V_i} p_{i,j} = 1 \quad \text{for } 1 \leq i \leq N.$$

During the process, ants deposit pheromone on the edges. In the simplest version of the algorithm, the ants deposit a constant amount of pheromone, that is, the amount of pheromone of the edge (i, j) when an ant moves from node v_i to node v_j is updated from the formula:

$$\varphi_{i,j} = \varphi_{i,j} + \Delta\varphi.$$

Moreover, like real ant pheromone, the artificial pheromone concentration should decrease over time. In the simple ant algorithm this is shown by:

$$\varphi_{i,j}(t + \tau) = (1 - q)\varphi_{i,j}(t), \quad \text{where } 0 < q \leq 1.$$

4.3.1.2 The Traveling Salesman Problem

The traveling salesman problem (TSP) is one of the most studied NP-hard problems in combinatorial optimization. In the following, we show how the basic ant algorithm is adapted to solve this problem.

Consider a graph $G = (N, E)$, where N is a set of nodes representing cities and E is a set of arcs fully connecting the nodes. The distance between cities i and j is denoted d_{ij}. The TSP consists of finding a minimal length Hamiltonian circuit on the graph $G = (N, E)$. A Hamiltonian circuit of graph G is a closed tour visiting once and only once all the $n = |N|$ nodes of G, and its length is given by the sum of the lengths of all the arcs of which it is composed.

A swarm of m ants build tours by executing n steps (one step by node). If all the iterations are done in parallel, the m tours will be built in n iterations. The number of ants m at each iteration is kept constant. The addition of new pheromone and pheromone evaporation are executed after all ants have completed their tour, that is, after they have built a complete tour. Each ant has a memory that contains the list of already visited cities. This list is used to define the set of cities that the ant located on city i still has to visit. Recall that a feasible tour visits a city exactly once. Additionally, this allows the ant to cover the same tour (i.e., path) to deposit delayed pheromones on the visited arcs. The probability with which an ant k chooses to go from city i to city j while building its tour at the algorithm iteration t is:

$$p_{i,j}^{(k)}(t) = \begin{cases} \dfrac{a_{i,j}(t)}{\sum_{l\in V_i^{(k)}} a_{i,l}(t)} & \text{if } j \in V_i^{(k)}, \\ 0 & \text{otherwise,} \end{cases}$$

where $V_i^{(k)}$ denotes the set of the neighborhood of node i that ant k has not visited yet.

The ant decision value $a_{i,j}(t)$ is obtained by the composition of the local pheromone trail value with a local heuristic value that supports the nearest neighbors as follows:

$$a_{i,j}(t) = \frac{[\varphi_{i,j}(t)]^{\alpha}[d_{ij}]^{-\beta}}{\sum_{l\in N_i}[\varphi_{i,l}(t)]^{\alpha}[d_{il}]^{-\beta}} \quad \text{for } j \in N_i,$$

where N_i is the set of neighbors of node i, and α and β are two parameters that control the relative weight of the pheromone trail and the heuristic value. A heuristic value should measure or estimate the relevance of adding an arc (i, j). A reasonable heuristic for TSP is $1/d_{ij}$, the inverse of the distance between cities i and j.

After all the ants have completed their tour, pheromone evaporation on arcs is executed. Each ant k deposits a quantity of pheromone

$$\Delta\varphi_{ij}^{(k)}(t) = \begin{cases} \dfrac{1}{L^{(k)}(t)} & \text{if arc } (i,j) \in T^{(k)}(t), \\ 0 & \text{otherwise,} \end{cases}$$

where $T^{(k)}(t)$ is the tour by ant k at iteration t and $L^{(k)}(t)$ is its length. Note that the shorter the tour of ant k, the greater is the amount of pheromone deposited.

The addition of new pheromone and pheromone evaporation are set by the following formula:

$$\varphi_{ij}(t) = (1-q)\varphi_{ij}(t-n) + \sum_{k=1}^{m}\Delta\varphi_{ij}^{(k)}(t),$$

where q is the pheromone decay trail, $0 < q \leq 1$. The initial amount of pheromone $\varphi_{ij}(0)$ is set to the same small positive value on all arcs. The suitable value for q is 0.5, which ensures a tradeoff between sufficient positive feedback and the exploration of new cycles. For α and β, optimal values are $\alpha \approx 1$ and $1 \leq \beta \leq 5$. Note that with $\alpha = 0$, the algorithm corresponds to the classical greedy algorithm, and with $\alpha > 2$, all the agents converge to the same cycle, which is not necessarily optimal.

The comparison of this algorithm with other heuristics such as tabu search and simulated annealing on small TSP problems ($n = 30$ cities), emphasizes its efficiency. The same technique presented here for TSP was applied to solve other optimization problems such as job scheduling, QAP, and routing in networks [2,5–8].

4.3.1.3 Comparison with Other Nature-Inspired Algorithms

A number of modern optimization techniques are inspired by nature. In simulated annealing modeled from the thermodynamic behavior of solids, particles in solution space move under the control of a randomized scheme, with probabilities according to some typically Boltzmann-type distribution. Genetic algorithms (GAs) start with a randomly generated population and use crossover and mutation operators to update it together with fitness function to evaluate the individuals. Neural networks (NNs) are a distributed learning technique in which the knowledge associated with a trained neural network is not stored in any specific location but encoded in a distributed way across its weight matrix.

Ant colony optimization shares many common points with these nature-inspired approaches. ACO, SA, and GA share the same update mechanism with random techniques. Randomness is present in the fuzzy behavior of ants [8]. ACO shares with GA some organizing principles of social population such as interaction and self-organization. ACO shares with NN trained networks the property that knowledge is distributed throughout the network. Moreover, ACO, like NN, exhibits emergence capabilities [8].

4.3.2 Particle Swarm Optimization

Particle swarm optimization algorithms are also a subset of SI. The basic concept of PSO is inspired by the social behavior of bird flocking and fish schooling. More precisely, PSO is a parallel evolutionary computation technique that provides a collaborative population-based search model. Individuals of the population called particles fly around in a multidimensional search space. During flight, each particle adjusts its position according to its own experience and according to the experience of a neighboring particle, moving toward the best position encountered by itself or its neighbors. Thus, the PSO system combines local search methods (through self-experience) with global search methods (through neighboring experience), attempting to balance exploration and exploitation [9,10].

In practice, a PSO algorithm is initialized with a population of random candidate solutions or particles. Two factors characterize a particle status on the search space: its position and its velocity. Additionally, the performance of each particle is measured according to a problem-dependent fitness function (i.e., cost function). Each particle is assigned a randomized velocity and is iteratively moved through the problem space. It is attracted towards the location of the best fitness achieved so far by the particle itself and by the location of the best fitness achieved so far across its neighborhood. Two versions of PSO exist depending on the neighborhood topology used to exchange experience among particles. In the global version of the algorithm, the neighborhood of the particle is the entire population (i.e., the entire swarm). In the local version, the swarm is divided into overlapping neighborhoods of particles.

4.3.2.1 The Standard PSO Algorithm

The basic PSO algorithm can be described by the following equations:

$$v_i(t+1) = av_i(t) + b_1 r_1 (p_i^{(1)} - x_i(t)) + b_2 r_2 (p_i^{(2)} - x_i(t)),$$
$$x_i(t+1) = x_i(t) + v_i(k+1),$$

where $v_i(t)$ denotes the velocity of particle i, which represents the distance to be traveled by this particle from its current position, that is, the difference between two successive particle positions; $x_i(t)$ represents the particle position; $p_i^{(1)}$ represents its own previous best position; and $p_i^{(2)}$ is the best value obtained so far by any particle among the neighbors. In the global version of the algorithm, $p_i^{(2)}$ represents the globally best position among the whole swarm. Particles change their position (or state) in the following manner. At iteration t, the velocity $v_i(t)$ is updated based on its current value affected by a tuning parameter a, and on a term that attracts the particle towards previously found best positions. The strength of attraction is given by the coefficients b_1 and b_2. The particle position $x_i(t)$ is updated using its current value and the newly computed velocity $v_i(t+1)$. The three tuning parameters, a, b_1, and b_2, influence greatly the algorithm performance. The inertia weight a is a user specified parameter, a large inertia weight pressures towards global exploration in a new search area while a small inertia pressures towards fine tuning the current search area. Positive constant acceleration coefficients (or learning factors) b_1 and b_2 control the maximum step size of the particle, usually $b_1 = b_2 = 2$. Suitable selection of the tuning factors a, b_1, and b_2 can provide a balance between the global (i.e., state space exploration), and local search (i.e., state space exploitation). Random numbers r_1 and r_2 are selected in the range $[0, 1]$ and they introduce useful randomness for state space exploitation.

In the following, we show how this basic PSO algorithm can be adapted to solve an optimization problem. The task-mapping problem (TMP) is one of the most studied NP-hard problems in distributed computing.

4.3.2.2 PSO for Task Assignment Problems

An important issue in distributed computing is the efficient assignment of computations into different processing elements. Given the graph of clustered tasks and the graph of the target distributed architecture, one should find a mapping by placing the highly communicative tasks on adjacent nodes of the processor network. It is well known that TMP is NP-hard. Salman et al. [9] use the PSO approach to address this problem.

For each problem, particles should be designed such that potential solutions can be represented. This is the key issue in designing a PSO algorithm. For a TMP, each particle is represented by a vector of length equal to M, which corresponds to the number of vertices in the task-graph. The value of each element in the particle is an integer in the range $[1, N]$, where N is the number of processors in the target architecture. For example, for a task-graph with eight vertices and three processors, the particle representation (1, 2, 1, 2, 2, 3, 1, 3) means that tasks 1, 3, and 7 are assigned to processor 1; tasks 2, 4, and 5 are assigned to processor 2; and tasks 6 and 8 are assigned to processor 3. Thus, a particle position corresponds to an M-coordinate position in an M-dimensional search space. According to the PSO equations presented here, each particle is iteratively moved (i.e., flies) through an M-dimensional search space [9].

4.3.2.3 Comparison with Other Nature-Inspired Algorithms

Particle swarm optimization and genetic algorithm have some common features. Both algorithms start with a group of a randomly generated population and use fitness values to evaluate the individuals. Moreover, both update the population and search for the optimum with random techniques. However, PSO does not use genetic operators like crossover and mutation. Particles update themselves with internal velocities (i.e., the difference between two successive particle positions). Note also that a PSO algorithm uses a tracking memory that accelerates its convergence to the best solution, even in the local version in most cases [11].

The performance of a PSO algorithm presented in Reference 9 is evaluated in comparison with a GA on randomly generated mapping problem instances. The results showed that the quality of the PSO algorithm solution is better than that of the GA's in most of the test cases and, in addition, the PSO algorithm runs faster than that of the GA's.

Further, the PSO approach seems to be a promising method to train ANN. Indeed, PSO can be used to replace the back-propagation learning algorithm in ANN. For example, Reference 12, showed that PSO is faster and gets better results in most cases than the evolutionary approach.

4.4 Conclusion

Swarm Intelligence is a rich source of inspiration for our computer systems. Specifically, SI has many features that are desirable for distributed computing. These include auto-configuration, auto-organization, autonomy, scalability, flexibility, robustness, emergent behavior, and adaptability. These capabilities suggest a wide variety of applications that can be solved by SI principles. We believe that the emergence paradigm and the highly distributed control paradigm will be fruitful to new technologies, such as nanotechnology, massively parallel supercomputers, embedded systems, and scalable systems for deep space applications.

References

[1] J.L. Deneubourg, S. Goss, N. Franks, A. Sendova-Franks, C. Detrain, and L. Chrétien. The dynamics of collective sorting, robot-like ants and ant-like robots. In *Simulation of Animal*

Behaviour: From Animal to Animals (J.A. Meyter and S. Wilson, Eds.), MIT Press, Cambridge, MA, 1991, pp. 356–365.

[2] T. White. Swarm intelligence and problem solving in telecommunications. *Canadian Artificial Intelligence Magazine*, spring, 1997.

[3] R. Beckers, J.L. Deneubourg, and S. Goss. Trails and U-turns in the selection of the shortest path by the ant Lasius niger. *Journal of Theoretical Biology*, 159, 397–415, 1992.

[4] Lynne E. Parker. ALLIANCE: An architecture for fault tolerant multi-robot cooperation. *IEEE Transactions on Robotics and Automation*, 14, 220–240, 1998.

[5] Mesut Günes and Otto Spanio. Ant-routing-algorithm for mobile multi-hop ad-hoc networks. In *Network Control and Engineering for Qos, Security and Mobility II*, Kluwer Academic Publishers, 2003, pp. 120–138.

[6] Eric Bonabeau and Guy Theraulaz. *Intelligence Collective*, Edition HERMES, Paris, 1994.

[7] Marc Dorigo, Eric Bonabeau, and Guy Theraulaz. Ant algorithms and stigmergy. *Future Generation Computer Systems*, 16, 851–871, 2000.

[8] B. Denby and S. Le Hégarat-Mascle. Swarm intelligence in optimization problems. *Nuclear Instruments and Methods in Physic Research Section A*, 502, 364–368, 2003.

[9] Ayed Salman, Imtiaz Ahmad, and Sabah Al-Madani. Particle swarm optimization for task assignment problem. *Microprocessors and Microsystems*, 26, 363–371, 2002.

[10] Ioan Cristian Trelea. The particle swarm optimization algorithm: Convergence analysis and parameter selection. *Information Processing Letters*, 85, 317–325, 2003.

[11] X. Hu, R. Eberhart, and Y. Shi. Particle swarm with extended memory for multiobjective optimization. In *Proceedings of the IEEE Swarm Intelligence Symposium*, 2003, Indianapolis, IN, USA.

[12] L. Messerschmidt and A.P. Engelbrecht. Learning to play games using a PSO-based competitive learning approach. In *Proceedings of the 4th Asia-Pacific Conference on Simulated Evolution and Learning*, 2002.

5

Parallel Genetic Programming: Methodology, History, and Application to Real-Life Problems

Francisco Fernández de Vega

5.1 Introduction to Genetic Programming

Software industry costs are mainly influenced by human resources required for developing software, because software development is still supported by human expertise. Although prices for hardware have dropped during the last 50 years, costs for hiring computer engineers have steadily increased in the same period. Hence the advent of new techniques for automatic software development would be welcomed by the software industry.

Scientists have been interested for long in the search for techniques capable of automatic software development. The first attempts to endow computers with the ability to learn can be traced back to the 1950s. The term machine learning, which was coined at that time, embodied the idea of computer algorithms that can learn by themselves through experience [1].

The finding of reliable methods for automatic programming would be a revolution for computer science, and would completely change our perception of the software industry.

Although the proposal described in the 1950s was of interest, until very recently results have not unveiled the potential that techniques of machine learning can attain. This potential has been clearly shown by a new technique that can be considered part of "evolutionary algorithms" (EAs), genetic programming (GP) [2].

Genetic programming is aimed at evolving computer programs. It begins with a basic description of the problem to be solved, after which the initialization process takes place; GP proceeds by automatically generating a set of candidate solutions for the problem to be solved. Each of the candidate solutions takes the shape of a computer program. Finally, GP enters a loop that evaluates each of the solutions, selects the best one — according to a measurement criteria — and produces a new set of candidate solutions employing the information contained in those selected solutions that acts as parents for the next generation.

Not only GP but also the set of techniques comprised within the EA field resembles the natural evolution of species in nature, as described by the Natural Evolution Theory [3].

Among the techniques that arose under the umbrella of natural evolution, genetic algorithms [4], evolutionary programming [5], and evolution strategies [6,7] have matured and demonstrated their usefulness.

During the last few years, GP has demonstrated not only its capability of automatically developing software modules [8], but has also been employed for designing industrial products with outstanding quality — such as electronic circuits that have been recently patented [9].

Although GP has proved its usefulness, the computational resources required for solving complex problems may become huge. In such instances, improvement of the technique is sought by using concepts borrowed from the parallel processing area. Not only GP but also EA have incorporated some degree of parallelization when difficult problems are to be solved.

This chapter focuses on parallel genetic programming. We first provide some basic ideas about how GP works, and then how it can be parallelized. Finally, after reviewing the history of the field, we show some problems — benchmark and real-life problems — that are solved by Parallel GP.

5.1.1 How GP Works

5.1.1.1 The GP Algorithm

The GP algorithm is quite similar to other EAs. The idea behind this group of techniques is to improve a set of candidate solutions for the problem at hand by means of a series of iterations. Thus, by applying several genetic operators, partial solutions from different individuals in the population will combine to produce an optimal or at least a useful solution for the problem to be solved. One of the main features of GP is that every individual from the population is a computer program — a candidate program that tries to solve the problem we are facing. At the end of the evolutionary process, the best individual from the population will be the program that solves the problem in the best way among all the programs evaluated so far.

The main steps in the algorithm are the following:

1. Initialize the population of candidate solutions (individuals) for the problem to be solved.
2. Evaluate all of the individuals in the population and assign them a fitness value.
3. Select individuals in the population who will become parents. A selection algorithm is employed in this step.
4. Apply genetic operations to the selected individuals in order to create new ones — descendants.
5. Introduce these new individuals into the new population.
6. If the population is not saturated go to step 3.
7. If the termination criterion is reached, then present the best individual as the output. Otherwise, replace the existing population with the new population and go to step 3.

In the following sections, we will describe the different steps of the algorithm; but before that, an introduction to program representation is offered.

5.1.1.2 Terminal and Function Sets

As stated earlier, GP populations are made up of computer programs. Therefore, each individual is a program, and usually, these programs are encoded be means of tree-like structures. The reason for employing tree structures is simply a tradition that has been inherited from the times when Koza described the technique [2]. Recently, other data structures have been successfully employed [10].

Each individual is made up of *functions* — internal nodes — and *terminals* — the leaves of the tree (see Figure 5.1). Both sets have to be decided according to the problem to be solved. The intuition about the general shape that a solution for the problem might adopt usually helps GP designer for the selection of these sets. For instance, if we try to discover the shape of a function by means of GP when solving a symbolic regression problem, probably, we should include arithmetic functions in the *function set.* Nevertheless, no guarantee about the suitability of the set is frequently provided.

The terminal set is usually made up of variables and constant values that are significant for the problem at hand.

Thus, the first concern for GP designers is to appropriately define the *function* and *terminal* sets: even when the solution to the faced problem is not known, one should take care that a solution can be found using the functions and terminals selected.

For making things easier, we will now describe a very simple problem, and we will see how terminal and function sets can be decided, and how the problem is then solved.

For instance, suppose we want to solve this problem: find an algorithm that accepts an integer number as the input and returns double the value of the input. Although the problem is extremely easy, it will be illustrative for demonstrating the power of GP. The technique is capable of developing the algorithm without any help from human programmers.

The first step is to build both terminal and function sets useful for the problem at hand. In this problem, we know that the function that the algorithm has to implement is simply $F(n) = (n + n)$, so we know that the function set should include an operator for computing addition, so that $F = \{+\}$, and the terminal set must include a parameter for the input $T = \{n\}$. Once the appropriate sets have been established, GP can develop by means of the evolution of an individual like that depicted in the left part of Figure 5.2.

But of course, a different selection for the function and terminal sets might also lead to a solution for the problem. Consider the following sets: $F = \{*\}$, and $T = \{2, n\}$. This time we have not only included a parameter in the terminal set, but also a constant value. Again, GP could obtain a solution like that depicted in the right part of Figure 5.2.

Although one could think that any function and terminal set will be useful for solving the problem, this is not always the case. If we choose the sets $F = \{-\}$ and $T = \{n\}$, no matter which function we might build $(n, n - n, n - n - n, \ldots)$ using these sets, the solution would never be found. Notice that we use a binary subtraction operator; if we add the equivalent unary operator, the solution could be built as (n–n).

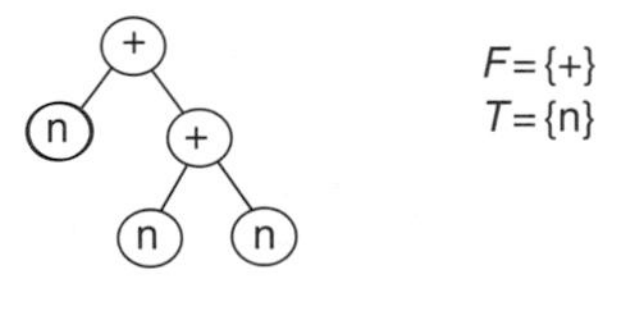

FIGURE 5.1 Individuals in GP are usually encoded by means of tree structures.

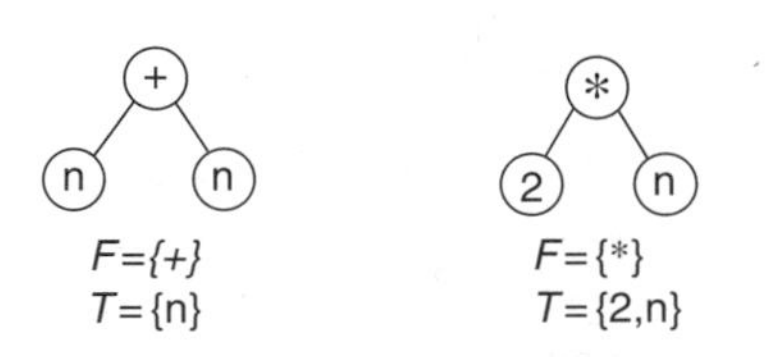

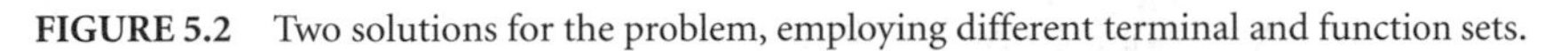
FIGURE 5.2 Two solutions for the problem, employing different terminal and function sets.

Even when the designer does not know the optimal solution, he should be able to extract some information from the high-level specification of the problem, which helps him to define the appropriate function and terminal sets.

5.1.1.3 Fitness Function

It is not only the task of defining the *terminal* and *function sets* that is crucial for obtaining good results when using GP, a good *fitness function* is also required. The idea behind any EA is that good parents will probably have similar or sometimes better children, and after a number of generations, descendants will be good enough for solving the problem. But behind this idea, there is a need for a measure that correctly classifies individuals from the population as good or bad. If the function in charge of measuring and classifying individuals — who would later become parents — does not take into account good traits for the problem, and instead assigns high fitness values to individuals with a poor performance, the algorithm will probably never find a solution for the problem.

So, the fitness function is in charge of evaluating individuals from the population and assigning them a value according to the performance obtained when solving the problem. High fitness values will thus favor individuals for being selected when generating descendants. On the other hand, if a minimization problem is faced, low fitness values are preferred instead.

For the problem described above, a good fitness function would be in charge of evaluating an individual — computer program — and comparing the values returned by that program with those values that should really be obtained, and are computed by using the objective function $F(n) = \{n + n\}$. Therefore, the fitness function will compute an *error value*. This error value is considered the *fitness value* for the individual; low fitness values are thus preferred for this problem.

The selection of a good fitness function is easy for the example shown here, because we know the objective function. However, this is not a typical case, and the election of a good fitness function is a difficult task, given that no clue about the shape of the solution for the problem is known.

After the fitness function has been defined, the GP algorithm possesses the capability of choosing good individuals for breeding purposes. Good individuals have higher probabilities of producing good and sometimes better descendants than bad individuals. Nevertheless, the GP algorithm choose's good individuals with a given probability, so that worse ones also have opportunities for transmitting their genetic material to descendants.

The process of transmitting genetic material to descendants is not merely performed by copying selected individuals. A set of genetic operations are applied instead, so that the information transmitted is somehow altered and descendants will thus differ from parents while at the same time they will inherit their features. Therefore, in the process of creating new candidate solutions, several genetic operators act as the source for the variation that is required.

5.1.1.4 Genetic Operations

Genetic operators are the variation mechanisms that generate new candidate solutions, similar to their parents but including some differences. If parents are good, they are allowed to breed new individuals that share some features with them but that are not completely identical to them. Possibly, some of these offspring can have better fitness than their parents. Some of them will also have worse fitness values, but evolution will discard them as generations are computed. On the other hand, sometimes bad individuals are selected for breeding, and may produce individuals that will help to solve the problem by introducing new genetic material in the population.

A couple of genetic operators are usually in charge of this task: *crossover* and *mutation*. Crossover takes two parents and mixes them up with a given probability so that new individuals are generated. Mutation takes an individual and randomly changes a part of it with a certain probability.

In GP, when tree structures are employed, crossover exchanges two randomly selected subtrees (see Figure 5.3) of the parents, while mutation randomly modifies a subtree from an individual (see Figure 5.4).

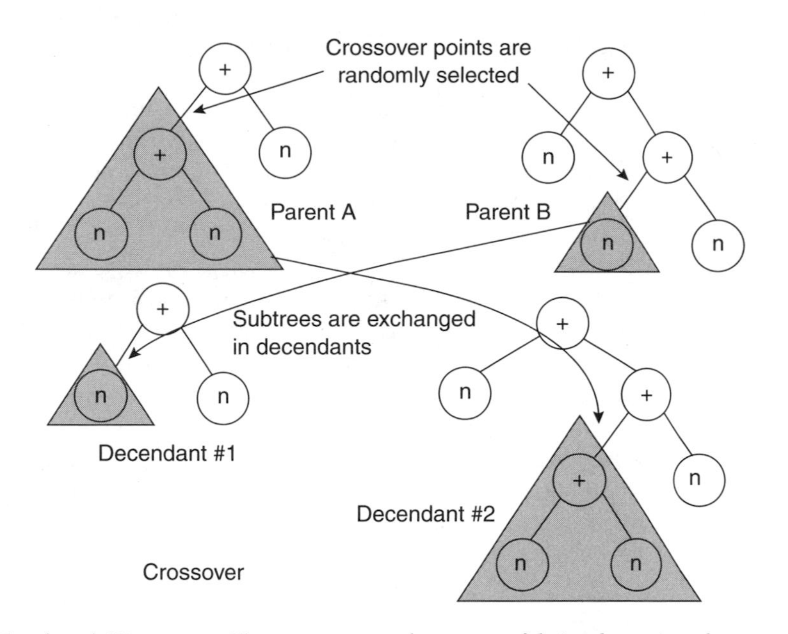

FIGURE 5.3 Tree-based GP crossover. The two parents exchange one of their subtrees in order to generate two new offspring.

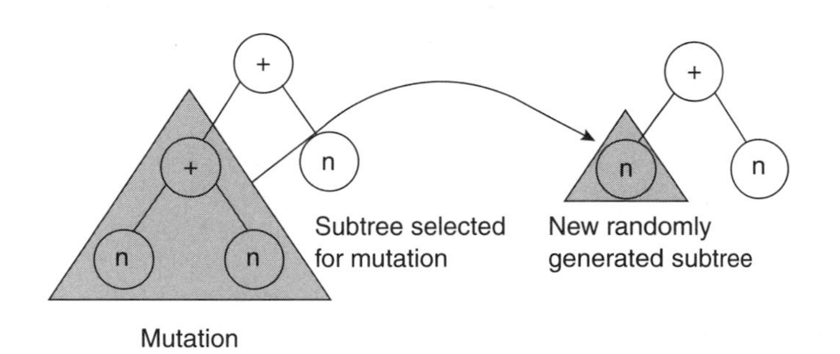

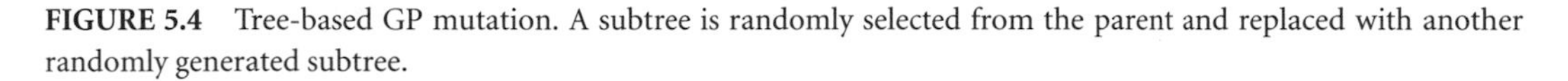

FIGURE 5.4 Tree-based GP mutation. A subtree is randomly selected from the parent and replaced with another randomly generated subtree.

5.1.1.5 Termination Criterion

The GP algorithm may be stopped when a satisfactory solution has been found, or after a given number of generations have been computed.

Although the first criteria is the preferred one, given that it means that the solution has been found, the second one is also employed because computational resources available are always limited. Frequently, the process has to be stopped when no more time for computing is available. This latter case is a problem for GP researchers, and also for any research dealing with EAs. The fact is that EAs are useful for difficult problems, and sometimes the difficulty of the problem is so large that the computational resources required for solving it are not available.

Researchers have been fighting this problem during the last decade and have tried to improve the algorithms for tackling more difficult problems. One of the more successful way of improving algorithms is by adding some degree of parallelization. Parallel EAs thus allow us to employ parallel architectures and distributed computing resources. In the next section we describe the development of a parallel version of GP.

5.2 Models of Parallel and Distributed GP

Typically, the reason for parallelizing any computer algorithm is to achieve time saving. In EAs there is a second reason: the implementation of the new parallel EA may improve the convergence process, so that a smaller number of iterations will be required for finding a solution of similar quality as when using the classic algorithm. Although this second improvement is not always present, some smart decisions when developing the parallel algorithm will provide these two advantages.

The necessity for a parallel implementation is frequently crucial, given that a large set of individuals has to be evaluated for many generations, and large amounts of computational resources are consequently required. In GP, parallelization is even more important: researchers have demonstrated that the size of individuals tends to grow when generations are progressively computed and evolved [11]. Therefore, computational resources required for the GP algorithm will not be constant along generations, but will increase as they are computed.

When parallelism is added to any EA, different models can be employed, depending upon the operations that are parallelized. Usually, parallelization can be applied at the level of the population and at the level of the fitness evaluation. In the following sections, we briefly describe these models, concentrating on GP; for a more detailed discussion, see Reference 12.

5.2.1 Parallelizing at the Fitness Level

The simplest way of applying parallelism to the GP algorithm, is to distribute the evaluation of individuals among several processors (see Figure 5.5). The main algorithm — *the master* — is computed in one processor, and the individuals are sent to other processors that in turn will evaluate and return the fitness values to the main process. This model is useful when the evaluation process is the most time-consuming step of the algorithm, which is usually the case in many real-life problems.

A well-known feature of GP individuals is the differences in size and complexity when compared to other individuals belonging to the same population. This requires the application of a load balancing policy (see for instance Reference 13). Load balancing can automatically be obtained if steady-state reproduction is used instead of generational reproduction.

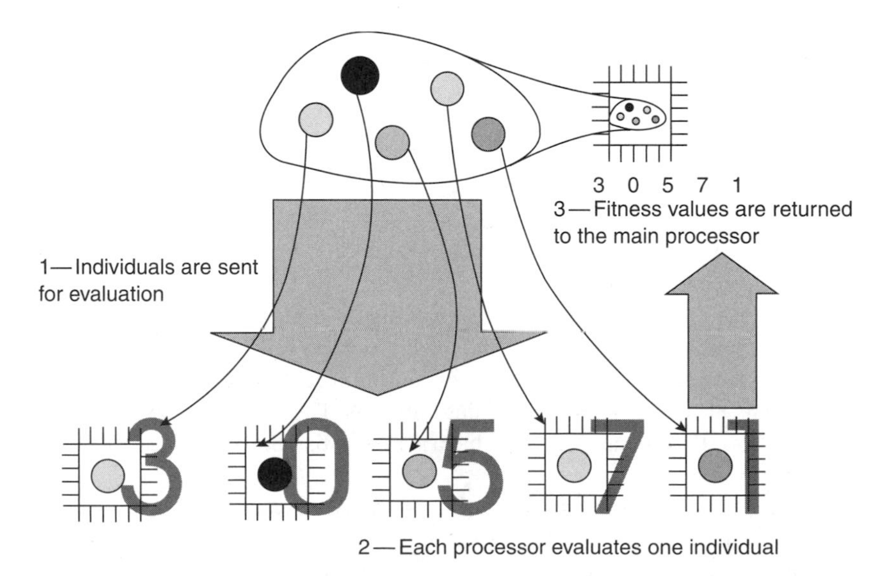

FIGURE 5.5 Parallelizing at the fitness level. Different processors or workstations are in charge of evaluating individuals, while the main processor runs the GP algorithm.

We have to bear in mind that the previously described parallel algorithm is basically the same as the sequential version, the only difference is the parallel evaluation of individuals.

In the next section we describe other ways of developing parallel GP, but with some changes to the basic algorithm, which will also help to solve the problem at hand more quickly.

5.2.2 Parallelizing at the Population Level

Populations of any species in nature feature a spatial distribution that usually depends on the orography of the landscape. They are said to be grouped in *demes*, semi-independent groups of individuals that are relatively isolated and scarcely interact with other neighboring demes by migrating some of their individuals.

This idea can be applied to GP, and different models have been proposed during the last few years. Although a detailed description of the algorithms can be found in Reference 14, we present here the main features of the new models.

5.2.2.1 The Island Model

The idea of having several semi-independent groups of individuals can also be applied to GP (see Figure 5.6). We could thus build a parallel algorithm by distributing the whole population of individuals among the processors available. Each of the processors will thus be in charge of applying evolution to its deme, and sometimes will exchange good individuals with other neighboring processors. This model is usually called the *Island Model* [15,16].

The idea is that each of the subpopulation focuses on a different area of the search space, because the convergence process within each of the deme takes a different path; and this is an important difference from the classic algorithm. Researchers have found that this difference helps to solve problems tackled by GP in a smaller number of steps, which is a new advantage for the parallel model. Researchers have also found that this model helps to add diversity to populations, which is a good feature for avoiding premature convergence.

Different connection topologies have been employed during the last few years. The most common ones are rings, two- and three-dimensional meshes, stars and hypercubes, although recently a *random* communication topology has also been defined in which a given subpopulation sends its emigrants to another randomly chosen subpopulation [17]. The most common replacement policy replaces the worst k individuals in the receiving population with k immigrants, who are the best k individuals of their original island.

We may notice that the new parallel GP algorithm requires several new parameters:

1. Number of subpopulations
2. Frequency of exchange
3. Number of exchanged individuals
4. The communication topology
5. Replacement policy

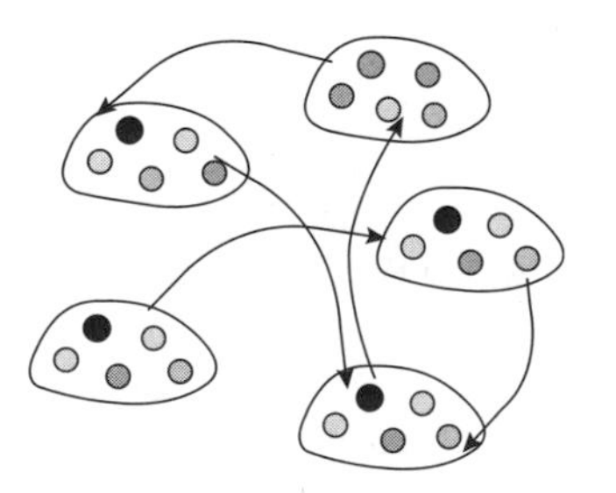

FIGURE 5.6 General island topology.

Some of these important parameters have been studied recently [14]. Researchers have found that a good choice for the previous parameters is to send 10% of individuals from each subpopulation to an adjacent one every 10 generations. The communication topology does not affect significantly the convergence process. On the other hand, a compromise between the number of subpopulations and the total number of individuals employed for solving the problem has to be adopted: if each of the subpopulations is made up of a small number of individuals, the exploration phase performed by each of the subpopulation will not be satisfactory.

A difference may be established between the parallel model we use and the parallel architecture we employ to run that algorithm. When dealing with distributed EAs, such as island GP, one can run the algorithm on both distributed memory multiprocessor machines and also on the network of workstations (NOWs).

In these architectures, the address spaces of each processor are separated and communication between processors must be implemented through some form of message passing. NOWs are widely used because of their low cost and ubiquity, although their performances are limited by communication latencies and heterogeneous workload distribution. In the case of parallel EAs, given that the communication step is not the most time-consuming part of the algorithm, and the migration step is rarely performed, these kinds of low-cost architecture are useful enough.

The migrations between the different demes can be implemented, for example, using the Message Passing Interface Standard (MPI) with synchronous communication operations, that is, each island runs a standard generational GP and individuals are exchanged at fixed synchronization points between generations. Implementation details can be found in Reference 18.

Researchers have also found that the Island Model may obtain better results than the panmictic model — classic model — even when it is run on a standard sequential machine [19]. The improvement does not come from the parallelization of the operations, because only one processor is employed, but from the change in the model. The Island Model introduces the migration step that helps to conserve diversity during the run, which is good for the finding of better solutions.

Other spatial distributions are also available for parallel EAs. For instance, one could distribute each of the individuals from the population on a two-dimensional grid (other dimensions are also possible). The idea is that each of the individuals interact only with their direct neighbors. Therefore, reproduction and mating take place locally for each of the individuals.

The model allows the slow diffusion of information from good individuals across the grid, and semi-isolated niches of individuals arise in different areas of the space.

5.3 The Length of Individuals and Measurement of Results

As described earlier, GP individuals may feature different sizes and complexities because variable-size individuals are employed. This means that the evaluation of a population of individuals does not necessarily always require the same amount of computing resources, even when the number of individuals is always the same.

There is a second factor that directly influences the way we should measure results: Researchers have found that the length of individuals progressively increases in a given experiment as generations are computed [20]. Therefore, even for a particular experiment, different generations require different computing time to be evaluated. So, if one wants to compare a technique with another one, the comparison has to be carefully carried out. This is the case for parallel GP: we are not only interested in solving a particular problem, but also want to know the power of the new technique when compared with the more common implementation of GP.

Traditionally, researchers present results comparing fitness values — quality of solutions — and the number of generations required to find that particular solution. In the case of GP, a better way of comparing results is by computing the total number of nodes evaluated until a given fitness value is found [17]. This is particularly useful for evaluating parallel GP when compared with GP, because the measure is not biased by the different speeds of each of the processors or computers in charge of evaluating each of the subpopulation. This measure will by itself show us the advantage of the model, and this advantage will be observed even when it is employed in a sequential machine. Of course, in a given architecture and a real-life problem, the time required for obtaining a particular solution is the right measure: researchers try to obtain a solution for a problem as soon as possible.

For computing fitness values, a useful figure is the Mean Best Fitness (MBF, the average over a number of runs) of the best fitness value at the end of the run. Nevertheless, when difficult problems are evaluated, none knows in advance whether the global optimum has been obtained or not. Therefore, the idea is to take the measure when a specified amount of computational effort has been spent.

In the comparisons we show below, we employ the measure described here: MBF versus computing effort (total number of nodes evaluated).

5.4 Parallel GP: The History

A number of researchers have applied ideas borrowed from the parallel processing field to GP. Probably, the first attempt to develop a parallel version of GP was carried out by Tufts in 1993 [21]. The idea described was simply to evaluate simultaneously a number of individuals from the GP population, by using several processors. The problem addressed was a classification over a set of customers, in order to predict credit card attrition.

A couple of years later, Juillè and Pollack described a parallel implementation of GP on a fine-grained SIMD architecture [22]. They employed a MasPar MP-2 Computer. Although they first presented a study on the parallel evaluation of S-expressions on the computer, which basically corresponds to a parallelization based on the fitness level, they also described a proposal based on the Island Model, employing a ring topology for communicating subpopulations. They demonstrated the usefulness of the implementation by using the Tic-Tac-Toe problem.

Koza and Andre [23] described latter an implementation of parallel GP based on the Island Model employing a network of Transputers — single VLSI devices containing 32-bit on-chip processor, on-chip memory, and several independent serial bi-direction physical on-chip communication links. The physical topology was based on a central supervisory process (the boss process) and 64 transputers, each running the basic GP algorithm, and the migration step for sending and receiving individuals. They employed a cellular topology, so that every node was physically connected to four neighbors in the network. The problem selected for the tests was the even-5-parity function. Several migration rates were employed. The main conclusion was that parallelization delivered more than linear speedup in solving the problem. Nevertheless, this result was questioned a couple of years later by Punch [24]. The fact is that the new implementation of the algorithm not only uses a parallel architecture but also modifies the main algorithm, so that the new one helps to improve the performances obtained when solving a problem.

The same year other implementation based on the bulk synchronous programming (BSP) model was presented [25], although this time no super-linear speedup was reported. The results showed that the Island Model achieved better speedups than panmictic GP — the classic version.

Another version of parallel GP was described by Stoffel and Spector [23]. Instead of using tree-based GP, they employed linear programs that were run on a stack-based virtual machine. By means of a symbolic regression problem, authors showed that parallel GP can save computing effort.

In 1997, Oussaidène et al. [13] applied a combination of fitness level parallelization and the island-based approach of parallel GP to trading model induction. They simultaneously tackled the load balancing problem that arises because of the different shapes and sizes of individuals in GP, which affects the performance of the parallel system when different individuals are evaluated on different processors. The proposal was to employ a dynamic scheduling algorithm based on a steady-state version of GP.

In a paper published by Punch [24] in 1998, the super-linear speedup reported by Koza [23] is questioned by a series of tests. The benchmark problems selected there were the royal tree and the ant problem. According to Punch, the features of the problem influences the results obtained. Particularly, the number of solutions in the search space greatly influences the performance of the Island Model. Therefore, the conclusion was that multiple-solution problems would be more amenable to multiple populations than single-solution problems. On the other hand, nondeceptive problems would also be more amenable to multiple populations than deceptive problems.

In the aforementioned papers, different parallel GP models are employed to study the improvement achieved when different benchmark and real-life problems are tackled. Nevertheless, no indepth study on specific parameters of the new models is presented until 2000. For instance, Tongchim and Chongstitvatana [26] studied synchronous and asynchronous versions of parallel GP. Results demonstrated that the parallel asynchronous algorithms obtains better results.

On the other hand, a whole study on the migration policies, topology, and other important parameters for the Island Model is presented for GP in 2003 [14]. Similarly, a study is performed, employing a parallel version of GP based on a cellular approach [27].

Plenty of results dealing with the application of parallel GP to real-life problems, and also papers focusing on any of the important parameters for the new model are available today. Recently, researchers have shown that the technique is even capable of solving some problems and obtaining solutions better than any other techniques previously invented by human being. The results are so impressive that they have even been patented. For instance, Koza [9] describes several applications of parallel GP that had led to solutions to problems; these applications are novel and useful enough to be patented. Particularly, Koza [28] describes some results that can be considered as inventions. He presented several analogue circuits that have been *discovered* by means of parallel GP. Some of those circuits — later patented — can probably be considered as the first inventions developed by computers.

In the following section, we show by means of two examples, how parallel GP can be applied to solving a real-life problem.

5.5 Applications

In this section, we briefly describe several benchmark problems that have been traditionally employed for testing GP performances, and also present two different real-life problems that have been addressed by means of parallel GP.

5.5.1 Typical GP Problems

We briefly describe several benchmark problems that have been traditionally used for experimenting with GP: the even parity problem, the artificial ant on the Santa Fe trail problem [2,20], the symbolic regression problem [24] and the royal tree problem [24]. They are useful for analyzing properties of genetic operators or different implementations of GP, and their description are provided as a reference for readers interested in GP. Although we don't show here results obtained with these problems, interested readers may refer to References 19 and 17. We focus instead on a couple of real-life problems that have been tackled by means of parallel GP.

Even Parity k Problem. The boolean even parity k function of k boolean arguments returns *true* if an even number of its boolean arguments evaluates to true, otherwise it returns *false.* If $k = 4$, then 16 fitness cases must be checked to evaluate the fitness of an individual. The fitness is computed as 16 minus the number of hits over the 16 cases. Thus a perfect individual has fitness 0, while the worst individual has fitness 16. The set of functions to be employed for GP individuals might be the following one: $F = \{AND, OR, NOT\}$. The terminal set in this problem is composed of k different boolean variables $T = \{v_1, \dots, v_N\}$.
Artificial Ant Problem on the Santa Fe Trail. In this problem, an artificial ant is placed on a 32×32 toroidal grid. Some of the cells from the grid contain food pellets. The goal is to find a navigation strategy for the ant that maximizes its food intake. Typically, the function set employed is the following one: $F = \{\text{if} - \text{food} - \text{ahead}\}$ while the terminal set is $T = \{\text{left}, \text{right}, \text{forward}\}$ as described in Reference 2. As fitness function, we use the total number of food pellets lying on the trail (89) minus the

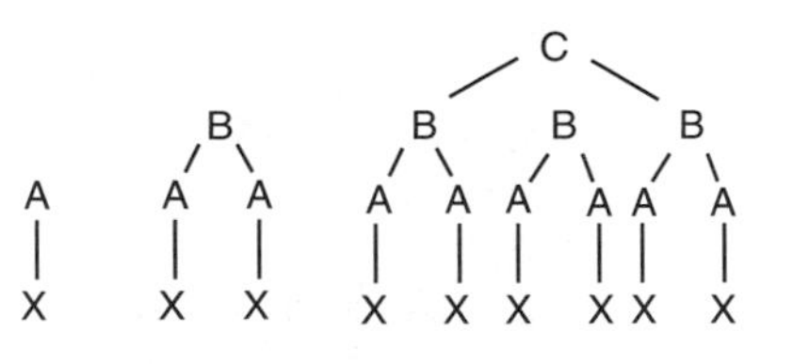

FIGURE 5.7 The Royal Tree Problem. Examples of perfect trees.

amount of food eaten by the ant from the path. This turns the problem into a minimization one, like the previous one.

Symbolic Regression Problem. The problem aims to find a program that matches a given equation. We employ the classic polynomial equation $f(x) = x^4 + x^3 + x^2 + x$, and the input set is composed of 1000 fitness cases. For this problem, the set of functions used for GP individuals is the following: $F = \{*, //, +, -\}$, where // is like / but returns 0 instead of *error* when the divisor is equal to 0, thus allowing syntactic closure. The fitness computes the sum of the square errors at each test point. Again, lower fitness means a better solution.

The Royal Tree Problem. This Problem [24] is commonly used as a standard function for testing the effectiveness of GP. It consists of a single base function that is specialized into as many cases as necessary, depending on the desired complexity of the resulting problem.

A series of functions, *a, b, c*, etc., with increasing arity are defined. (An *a* function has arity 1, a *b* function has arity 2, and so on.) A number of terminals *x*, *y*, and *z* are also defined.

A *level-a* tree is an *a* root node with a single *x* child. A *level-b* tree is a *b* root node with two *level a* trees children. A *level-c* tree is a *c* root node with three *level-b* trees as children. A *level-e* tree has depth 5 and 326 nodes, while a *level-f* tree has depth 6 and 1927 nodes. Perfect trees are defined as shown in Figure 5.7.

The raw fitness of a subtree is the score of its root. Each function calculates its score by adding up the weighted scores of its direct children. If the child is a perfect tree of the appropriate level (for instance, a complete *level-c* tree beneath a *d* node), then the score of that subtree, times a *FullBonus* weight, is added to the score of the root. If the child's root is incorrect, then the weight is *Penalty*. After scoring the root, if the function is itself the root of a perfect tree, the final sum is multiplied by *CompleteBonus*. Typical values used are: FullBonus = 2, PartialBonus = 1, Penalty = 1/3, and CompleteBonus = 2. The score base case is a level-*a* tree, which has a score of 4 (the *a*–*x* connection is worth 1 times the FullBonus, times the CompleteBonus).

5.6 Real-Life Applications

As described in Section 5.4, there is plenty of research dealing with parallel GP. We describe here a couple of real-life problems that have been tackled by using it. Results show that the technique can successfully solve the problems addressed.

5.6.1 Placement and Routing in FPGA

Field Programmable Gate Arrays (FPGAs) are integrated devices used for the implementation of digital circuits by means of a configuration or programming process. Several manufacturers and different kinds of FPGAs are available.

One of the best known is the island-based FPGA (this island has nothing to do with island-based EAs). This model includes three main components: configurable logic blocks (CLBs), input–output blocks (IOBs), and connection blocks (see Figure 5.8). Configurable logic blocks are used to implement all the logic circuitry — they have different configuration possibilities, and are positioned like matrix on the FPGA.

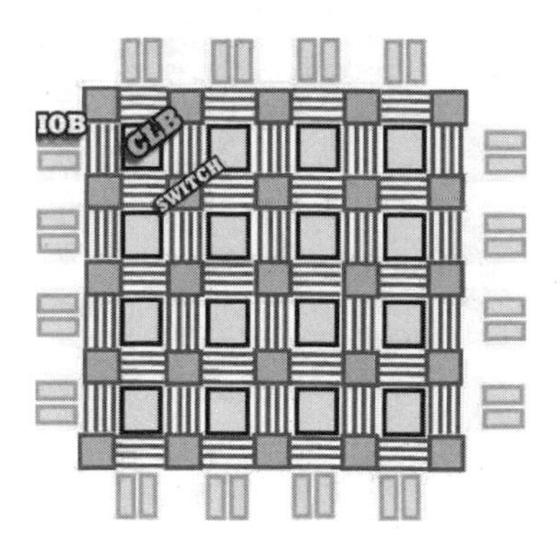

FIGURE 5.8 Island-based FPGA.

The IOBs allow the connection of the circuit implemented by the CLBs with any external system. Finally, the connection blocks (switch-boxes and interconnection lines) are employed for the internal routing of the circuit.

One of the main steps in the FPGA design process is the placement and routing. We present a methodology that is based on parallel GP. The methodology has also been employed for tackling Multi-FPGA Systems Synthesis [29].

The problem we try to solve begins with a circuit description, and the goal is to place components and wires in an FPGA. Genetic programming is thus in charge of encoding circuits, so that a graph — circuit — is described by means of a tree — GP individual. In the following, we describe how graphs are encoded by means of trees.

Although several authors have implemented GP in hardware [30,31], the idea here is completely different: we use GP for implementing circuits on hardware.

5.6.1.1 Circuits Encoding Using Trees

When implementing a circuit on an FPGA, each of the circuit components has to be implemented into a different CLB, and then the CLBs have to be connected according to the circuit topology. Given that circuits are encoded by means of trees, and that evolution will generate new circuits, a fitness function is required for analyzing these circuits, deciding if they are correct or not, and their degree of resemblance with the circuit that is being to implemented.

Any circuit is made up of components and connections. Given that components compute very easy logic functions, any of them can be implemented into any CLB from the FPGA. Therefore, we can describe a given circuit in a way similar to the example depicted in Figure 5.9. This means that we only have to connect CLBs from the FPGA according to the interconnection model that a given circuit implements, and then we can configure each of the CLBs with the function that each component performs in the circuit.

Circuits have to be encoded by means of trees, because we employ tree-based GP. We can first label each component from the circuit with a number, and then assign components' labels to the ends of wires connected to them (as shown in Figure 5.9). Wires can now be disconnected without losing any information. All the wires can be included within a tree by connecting each of the wires as a branch of the tree and keeping them all together in the same tree. The circuit can be easly rebuilt later by using the labels as a guide.

By labeling both extremes of wires, we will have all the information required for reconstructing the circuits. This way of representing circuits allows us to go back and construct the real graph. Moreover, any given tree, randomly generated, will always correspond to a particular graph, regardless of the usefulness of the associated circuit. In this proposal, each node from the tree is representing a connection, and each branch is representing a wire.

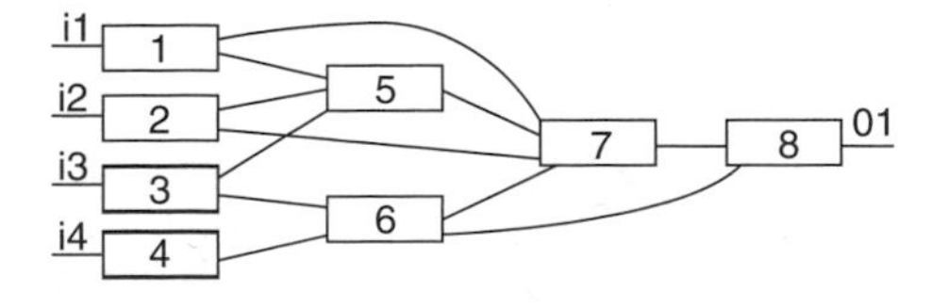

FIGURE 5.9 Representing a circuit with black boxes and labeling connections.

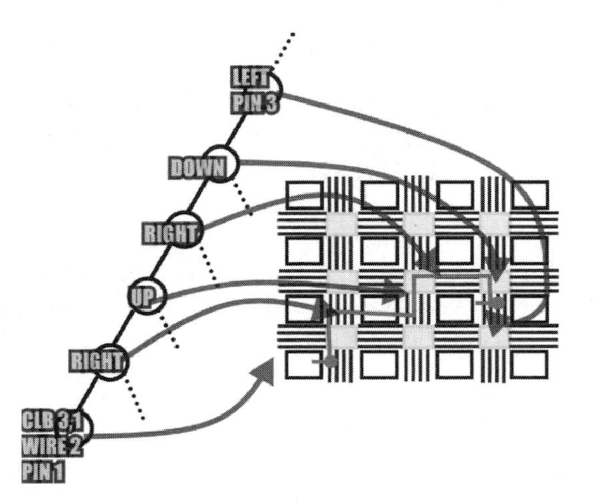

FIGURE 5.10 Mapping an individual into a circuit.

5.6.1.2 GP Sets

The function set for our problem contains only one element: $F = \{\mathrm{SW}\}$, Similarly, the terminal set contains only one element $T = \{\mathrm{CLB}\}$. But SW and CLB may be interpreted differently depending on the position of the node within a tree. Sometimes a terminal node corresponds to an IOB connection, while sometimes it corresponds to a CLB connection in the FPGA. Similarly, an internal node — SW node — sometimes corresponds to a CLB connection (the first node in the branch), while others affect switch connections in the FPGA (internal node in a branch, see Figure 5.10). Each of the nodes in the tree will thus contain different information:

1. If we are dealing with a terminal node, it will have information about the position of CLBs, the number of pins selected, the number of wires to which it is connected, and the direction we are taking when placing the wire.
2. If we are, instead, in a function node, it will have information about the direction we are taking. This information enables us to establish the switch connection, or in the case of the first node of the branch, the number of the pin where the connection ends.

5.6.1.3 Evaluating Individuals

To evaluate an individual we must convert the genotype (tree structure) to the phenotype (circuit in the FPGA), and then compare it to the circuit provided by the partitioning algorithm. We developed an FPGA simulator for this task. This software allows us to simulate any circuit and to check its resemblance to other circuits. Therefore, this software tool is in charge of taking an individual from the population and evaluating every branch from the tree in a sequential way, establishing the connections that each branch specifies. Circuits are thus mapped by visiting each of the useful nodes of the trees and making connections on the virtual FPGA, thus obtaining phenotype.

5.6.1.4 Results

Figure 5.9 graphically depicts one of the test circuits that has been used for validating the methodology.

The main parameters employed were the following: maximum number of generations equal to 500, maximum tree depth equal to 30, steady state, tournament selection of size 10, crossover probability equal to 98%, mutation probability equal to 2%, ramped half-and-half initialization, elitism (i.e., the best individual has been added to the new population at each generation). We employed 5 subpopulations of 500 individuals each, with a period of migration of 10 generations. The GP tool we used is described in Reference 18.

Figure 5.11 shows one of the proposed solutions among those obtained with parallel GP for the circuit. A very important fact is that each of the solutions that GP found possesses different features, such as area of the FPGA used and position of the input/output terminals. This means that the methodology could easily be adapted for managing typical constraints in FPGA placement and routing.

Figure 5.12 presents a comparison of parallel GP and classic GP when applied to the problem of placement and routing on FPGAs. We can see that parallel GP, employing 5 populations and 500 individuals per population, achieved better convergence results than GP with 2500 individuals — the same total amount of individuals: PGP converges more quickly and obtains slightly better results.

The methodology has been successfully applied to Multi-FPGAs System Synthesis [29]. Figure 5.13 shows a picture of the device that was built for testing the methodology.

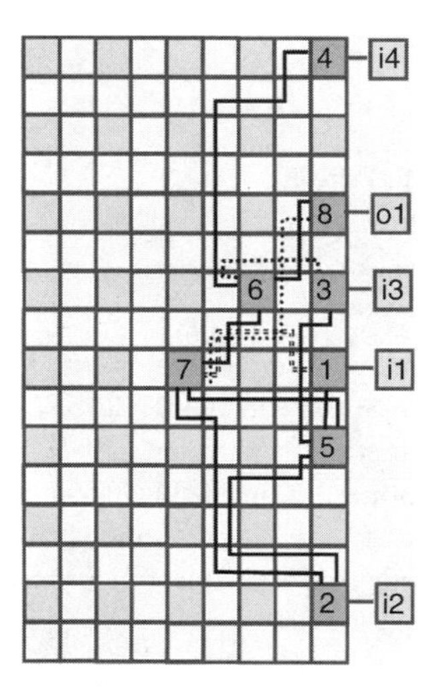

FIGURE 5.11 A solution generated by PGP.

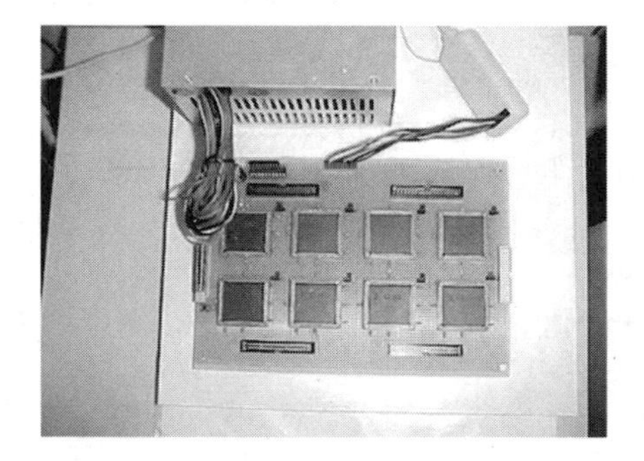

FIGURE 5.12 Comparison between parallel GP — 5 pop., 500 individuals each — and classic GP — 2500 individuals. 50 runs have been performed and averaged for each curve.

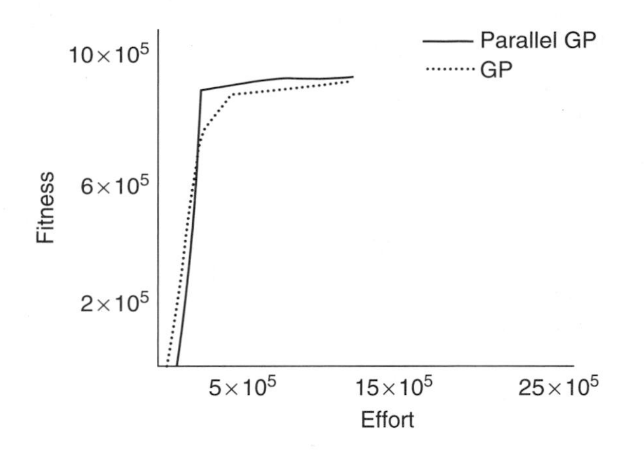

FIGURE 5.13 A Multi-FPGA board designed for testing the methodology.

5.6.2 Medical Knowledge Representation by Means of GP

In this section we present another real-life problem that has been solved by means of parallel GP. We apply GP to the acquisition of medical knowledge. We are focussing on burn diagnosing.

An adaptive system capable of classifying is developed by means of parallel GP. It uses a set of parameters, obtained by specialist doctors, to predict the evolution of a burn according to its initial stages. The system is first trained with a set of parameters and the results of evolutions that have been recorded over a set of clinical cases. Once the system is trained, it is an aid to deciding how new cases will probably evolve. Thanks to the use of parallel GP, an explicit expression of the input parameter is provided, and this explicit expression takes the form of a decision tree, which will be incorporated into software tools that help physicians in their everyday work.

Of course the aim here is not to suppress the task of specialists, but to help them make more accurate diagnoses. Furthermore, developing software tools for medical diagnosis may help nonspecialist doctors when immediate treatment must be applied.

5.6.2.1 The Problem of Burn Diagnosing

When somebody suffers a burn it is necessary to find out its degree, in order to apply the best treatment from the beginning. This classification problem is recognized as being difficult because it is not always clear how much tissue damage there is and how it will evolve.

If we are interested in developing an automatic classifier system capable of deciding the diagnosis of a burn, this system must first be able to extract information from a burn by analyzing a picture of it, and second embody that information in a knowledge system. This knowledge system could take the form of a decision tree. Bearing in mind the feasibility of GP for representing trees, the link between GP and medical knowledge systems is straightforward.

Two different problems must thus be solved: image processing and knowledge representation. We describe here a methodology for representing the knowledge used by a specialist when diagnosing by means of parallel GP. So, the input for the methodology is the features that have been previously extracted by an image processing system. The system must be capable of predicting how a burn will evolve, allowing users to choose the most suitable treatment.

We believe that EAs have an important advantage over other kinds of machine learning techniques such as neural networks: they have the ability to provide information not only about classification, but also about the route taken to reach a decision.

5.6.2.2 Classifying by Means of Decision Trees

Medical diagnosis is a classification problem, in which the search space is made up of a set of points with n-coordinates. Each coordinate is allocated a value for a given symptom. The aim is to find the category to which a point belongs and this data will give us the diagnosis.

Sometimes we are not only interested in obtaining a classifier but also in retrieving important information from the problem's input parameters. Several classifier systems for medical diagnosis have been designed based on decision trees and decision rules [32]. They take the parameters and classify the data.

The decision tree approach [33] falls into the category of inductive learning methods. Taking a set of examples, the aim is to construct a tree that is able to classify new samples within the search space.

Decision trees usually classify the members of a set as either positive or negative examples. In this application of parallel GP, due to the number of possibilities involved in burn diagnosis, we extend the classic approach to multi-class decision-making, as in other medical diagnosis research [34]. In fact a decision tree is a chain of if-then-else constructions that can be seen as a computer program. Each condition is applied to some specific input parameters. The parameters will thus lead us along a branch of the tree, finally reaching a decision, which is the category to which the input data belongs.

We want to generate the decision tree automatically according to a group of clinic cases that make up the training set. This will be done by means of parallel GP.

Bearing in mind that in GP each individual adopts the shape of a tree, a decision tree can be considered to be an individual. If we apply GP to finding a specific decision tree, at the end of the process the best individual takes the form of that decision tree, that is, the genotype of the individual presents the chain of decisions that are necessary for a diagnosis to be made.

When using GP for decision tree extraction, the function set will only be composed of the if-then-else instruction, and some logical operators (see Section 5.3). As a consequence, any tree can be constructed. The terminal set will contain the parameters we have decided to study in each burn.

5.6.2.3 A Case Study: Burns Unit, Virgen del Rocío Hospital. Seville, Spain

In order to apply the methodology, the collaboration of medical specialists from the area of burns treatment was required.

The Burns Unit, at Virgen del Rocío Hospital, was in charge of retrieving information about real cases of burns. A form was provided for several specialists to collect information on cases, which would be useful for testing the problem. Several photographs were taken in each of the cases, and several features of burns were noted in the form.

Thirty one different clinical cases were studied. Each completed form was accompanied by two photographs. No image processing has been done, but photographs are necessary for studying the different parameters that will be automatically retrieved in the future. In this research, following the specialists' indications, just three parameters have been studied.

No photographs are shown here, in order to preserve the privacy of the people who took part in the study, but all of them are stored in the Burns Unit at the Virgen del Rocío Hospital, Seville, Spain.

We are aware that an accurate diagnosis requires the study of additional features, but we believe that these three parameters are enough to see how the methodology works, and to obtain a first sketch for a knowledge system based on GP.

Following specialist doctors' advice, we decided to develop the first version of the knowledge system taking into account just three different parameters from each picture:

1. *Color* (C): Several possible values: White, Black, Yellow, Red, Pink, and combinations
2. *Dryness* (D): Two possibilities: True or False
3. *Ampoule* (A): Two possibilities: Absent or Not

Although studying a wider set of parameters would be more useful, this simple model confirms the validity of the methodology employed. The aim was not to create a tool that can classify 100% of the cases but to demonstrate how easily a doctor's knowledge can automatically be captured by means of GP and decision trees.

Four kinds of diagnosis are possible for a burn: first degree, surface second degree, deep second degree, and third degree.

In order to train the system, a table with parameters, taken by studying the photographs and diagnoses, was built.

5.6.2.4 Results

A set of 31 clinical cases were provided by doctors, and these cases were used to train the system. Each of them was allocated its corresponding parameters together with the result of its evolution over a couple of weeks. This table of values was given to the algorithm in order for it to train.

Due to how GP works, the terminal and function sets were established as $F = \{\text{ifthenelse}, >, =, <, \text{AND}, \text{NOT}\}$ and $T = \{C, D, A\}$.

We ran the algorithm employing 5 populations of 2000 individuals each, with a period of migration of 10 generations. We waited for 60 generations before taking the results. At the end, the best decision tree we obtained was able to classify correctly 27 out of 31 cases. This was due to the presence of several cases with the same parameters but with different evolution values. It was, consequently, impossible to categorize these cases accurately. Figure 5.14 shows the decision tree obtained.

The meanings for nodes are the following: A, Ampoule; D, Dryness; C, Color; D2, Deep second degree; S2, Surface second degree; 3, third degree; P, Pink color; Y, Yellow color.

Bearing in mind that each color has an integer number associated with it, comparisons among them are meaningful. Of course, many different versions of the same tree can be obtained in different executions of the algorithm. For the sake of simplicity, we show just one.

Using this methodology, we are able to represent medical knowledge by means of decision trees. This is just an example about the use of GP for the task proposed, although for obtaining a completely reliable decision tree a much larger set of examples must be used, and training and test sets have to be built and employed in the learning process.

Finally, Figure 5.15 presents a comparison between parallel GP and GP when applied to the problem of medical knowledge representation. We can see that parallel GP employing 5 populations and 2000 individuals per population achieved better convergence results than GP with 10000 individuals (the same total amount of individuals).

We have to consider that the curve shown for PGP is not taking into account the time savings obtained when using a parallel architecture, such as a multiprocessor system or a cluster of computers. If we take both improvements into account — time savings and improvement in the convergence process — parallel GP is far superior to plain GP.

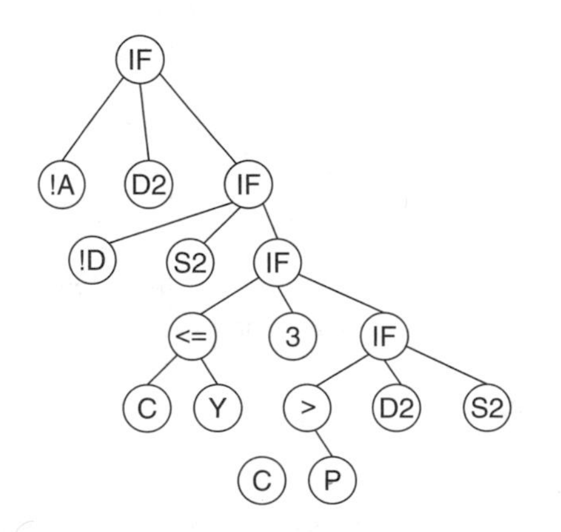

FIGURE 5.14 Burn diagnosing decision tree.

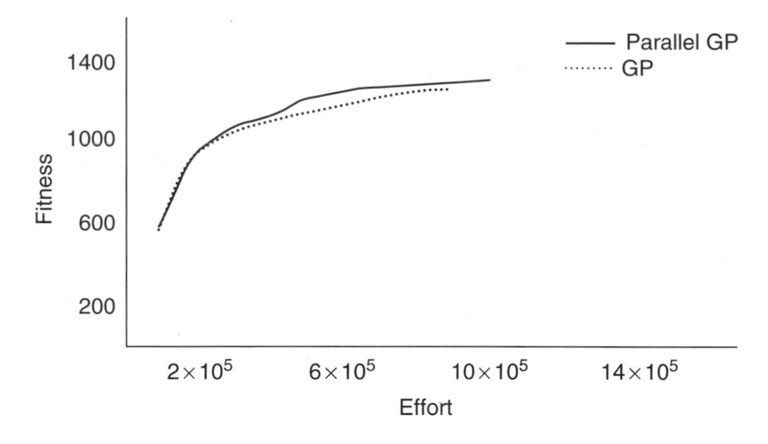

FIGURE 5.15

5.7 Concluding Discussion

In this chapter, we have presented parallel GP and its application to some real-life problems. We have described the methodology and its history. We have also provided a set of benchmark problems commonly used in GP. We have finally shown how parallel GP can be applied to the problem of placing and routing circuits on FPGAs and also to the problem of extracting medical knowledge for burn diagnosing.

Acknowledgment

Part of this research has been possible thanks to Ministerio de Ciencia y Tecnología research projects number TIC2002-04498-C05-01.

References

[1] T. Mitchell. *Machine Learning.* McGraw Hill, New York, 1996.

[2] J.R. Koza. *Genetic Programming.* The MIT Press, Cambridge, MA, 1992.

[3] C. Darwin. *On the Origin of Species by Means of Natural Selection.* John Murray, London, 1859.

[4] John H. Holland. *Adpatation in Natural and Artificial Systems.* University of Michigan Press, Ann Arbor, MI, 1975.

[5] L.J. Fogel, A.J. Owens, and M.J. Walsh. Artificial intelligence through a simulation of evolution. In M. Maxfield, A. Callahan, and L.J. Fogel, Eds., *Biophysics and Cybernetic Systems: Proceedings of the second Cybernetic Sciences Symposium,* Spartan Books, Washington, D.C., 1965, pp. 131–155.

[6] I. Rechenberg. *Evolutionsstrategie: Optimierung technischer Systeme nach Prinzipien der biologischen Evolution.* frommann-holzbog, Stuttgart, 1973. German.

[7] H.P. Schwefel. Evolutionsstrategie und numerische Optimierung. Ph.D thesis, Technische Universitat Berlin, Berlin, 1975.

[8] W.B. Langdon. *Data Structures and Genetic Programming: Genetic Programming + Data Structures = Automatic Programming!* Kluwer Academic Publishers, New York, 1998.

[9] J.R. Koza, F.H. Bennett III, and O. Stiffelman. Genetic programming as a Darwinian invention machine. *Genetic Programming: Proceedings of EuroGP '99, LNCS,* Vol. 1598. Springer-Verlag, May 1999.

[10] K. Stoffel and L. Spector. High-performance, parallel, stack-based genetic programming. In Koza, J.R., et al., Eds., *Genetic Programming 1996: Proceedings of the First Annual Conference.* Stanford University, MIT Press, CA, 1996, pp. 224–229.

[11] W.B. Langdon and R. Poli. Fitness causes bloat. In R. Roy P.K. Chawdhry and R.K. Pant, Eds., *Soft Computing in Engineering Design and Manufacturing*. Springer-Verlag, London, 1997, pp. 13–22.
[12] M. Tomassini. Parallel and distributed evolutionary algorithms: a review. In K. Miettinen, M. Mäkelä, P. Neittanmäki, and J. Périaux, Eds., *Evolutionary Algorithms in Engineering and Computer Science*. John Wiley & Sons, New York, 1999, 113–133.
[13] M. Oussaidéne, B. Chopard, O. Pictet, and M. Tomassini. Parallel genetic programming and its application to trading model induction. *Parallel Computing*, 23: 1183–1198, 1997.
[14] F. Fernández, M. Tomassini, and L. Vanneschi. An empirical study of multipopulation genetic programming. *Genetic Programming and Evolvable Machines*, 4: 21–52, 2003.
[15] J.P. Cohoon, S.U. Hegde, W.N. Martin, and D. Richards. Punctuated equilibria: A parallel genetic algorithm. In J.J. Grefenstette, Ed., *Proceedings of the Second International Conference on Genetic Algorithms*. Lawrence Erlbaum Associates, Mahwah, NJ, 1987, p. 148.
[16] R. Tanese. Parallel genetic algorithms for a hypercube. In J.J. Grefenstette, Ed., *Proceedings of the Second International Conference on Genetic Algorithms*. Lawrence Erlbaum Associates, 1987, pp. 177–183.
[17] F. Fernández de Vega. Distributed Genetic Programming Models with Application to Logic Synthesis on FPGAs. Ph.D. thesis, Computer Science Department, University of Extremadura, Cáceres, Spain, 2001.
[18] F. Fernández, M. Tomassini, L. Vanneschi, and L. Bucher. A distributed computing environment for genetic programming using MPI. In J. Dongarra, P. Kaksuk, and N. Podhorszki, Eds., *Recent Advances in Parallel Virtual Machine and Message Passing Interface*, Vol. 1908 of *Lecture Notes in Computer Science*. Springer-Verlag, Heidelberg, 2000, pp. 322–329.
[19] F. Fernandez, M. Tomassini, and J.M. Sanchez. Experimental study of isolated multipopulation genetic programming. In *IEEE International Conference on Industrial Electronics, Control and Instrumentation*, Nagoya, Japan, 2000. IEEE Press, Washington, 2000, pp. 2672–2677.
[20] W.B. Langdon and R. Poli. *Foundations of Genetic Programming*. Springer-Verlag, Berlin, 2002.
[21] P. Tufts. Parallel case evaluation for genetic programming. In *1993 Lectures in Complex Systems*, Vol. VI of *Santa Fe Institute Studies in the Science of Complexity*, 1993, pp. 591–596.
[22] H. Juille and J.B. Pollack. Parallel genetic programming and fine-grained SIMD architecture. In E.V. Siegel and J.R. Koza, Eds., *Working Notes for the AAAI Symposium on Genetic Programming*. MIT, Cambridge, MA, November 10–12, 1995. AAAI, pp. 31–37.
[23] P.J. Angeline and K.E. Kinnear Jr. (Eds.). *Advances in Genetic Programming 2*. The MIT Press, Cambridge, MA, 1996.
[24] W. Punch. How effective are multiple populations in genetic programming. In J.R. Koza, W. Banzhaf, K. Chellapilla, K. Deb, M. Dorigo, D.B. Fogel, M. Garzon, D. Goldberg, H. Iba, and R.L. Riolo, Eds., *Genetic Programming 1998: Proceedings of the Third Annual Conference*. Morgan Kaufmann, San Francisco, CA, 1998, pp. 308–313.
[25] D. C. Dracopoulos and S. Kent. Bulk synchronous parallelisation of genetic programming. In Jerzy Waśniewski, Ed., *Applied Parallel Computing: Industrial Strength Computation and Optimization; Proceedings of the third International Workshop, PARA '96*, Springer-Verlag, Berlin, Germany, 1996, pp. 216–226.
[26] S. Tongchim and P. Chongstitvatana. Nearest neighbor migration in parallel genetic programming for automatic robot programming. In *Proceedings of the Sixth International Conference on Control, Automation, Robotics and Vision*, Singapore, December 2000.
[27] G. Folino, C. Pizzuti, and G. Spezzano. A scalable cellular implementation of parallel genetic programming. *IEEE Transactions on Evolutionary Computation*, 7: 37–53, 2003.
[28] J.R. Koza, F.H. Bennett III, D. Andre, and M.A. Keane. *Genetic Programming III: Darwinian Invention and Problem Solving*. Morgan Kaufmann, San Francisco, CA, 1999.
[29] F. Fernandez, I. Hidalgo, J. Lanchares, and J.M. Sanchez. A methodology for reconfigurable hardware designed based upon evolutionary computation. *Microprocessors and Microsystems*. Elsevier, 28: 363–371, 2004.

[30] P. Martin. A hardware implementation of a genetic programming system using FPGAs and Handel-C. *Genetic Programming and Evolvable Machines*, 2: 317–343, 2001.

[31] M.I. Heywood and A.N. Zincir-Heywood. Register based genetic programming on fpga computing platforms. In R. Poli, W. Banzhaf, W.B. Langdon, J.F. Miller, P. Nordin, and T.C. Fogarty, Eds., *Proceedings of the European Conference on Genetic Programming*, Vol. 1802 of *Lecture Notes in Computer Science*. Springer-Verlag, London, 2000, pp. 44–59,

[32] J. H. Holmes. Discovering risk of disease with a learning classifier system. In T. Bäck, Ed., *Proceedings of the Seventh International Conference on Genetic Algorithms (ICGA97)*. Morgan Kaufmann, San Francisco, CA, 1997.

[33] J.R. Quinlan. Decision trees and instance-based classifiers. In *The Computer Science and Engineering Handbook*, 1997, pp. 521–535.

[34] M. Kurzynski. The application of unified and combined recognition decision rules to the multistage diagnosis problem. In *Proceedings of the 20th Annual International Conference of the IEEE Engineering in Medicine and Biology Society*, Vol. 3. Hong-Kong, 1998, pp. 1194–1197.

6

Parallel Cellular Algorithms and Programs

Domenico Talia

6.1 Introduction

An *emergent* phenomenon is the large-scale group behavior of a system that does not seem to have any explanation in terms of the single constituent parts only. In other words, *emergence* can be defined by saying that "the whole is greater than the sum of the parts." In emergent systems, we can consider two different levels of description: the *microscopic* level, where all the single components are taken into account; and the *macroscopic* level, where emergent behavior occurs as the synthesis of the complex interaction of the microscopic components. To bring emergent systems out of a speculative horizon it is necessary to experiment and test them. In particular, emergent system simulation on parallel computers is an essential practice for an indepth analysis and evaluation of the accuracy of the proposed models of emergent behavior.

The programming of emergent phenomena and systems using traditional programming models and tools is very difficult and involves long and complex coding. This is mainly because these approaches are based on the design of a system as a whole; hence, design and programming do not start from basic elements. It is better to design emergent systems by means of paradigms that allows for expressing the behavior of single elements and their interactions. The global behavior of these systems then emerges from the evolution and interaction of a massive number of elements; hence, it does not need to be explicitly coded.

The cellular automata (CA) model is a massively parallel computational model that can be effectively used for the investigation and simulation of emergent phenomena and systems. Cellular automata are inherently parallel; therefore, they can be used to model and simulate very large-scale emergent systems on parallel computers [1,2]. Cellular parallel tools allow for the exploitation of the inherent parallelism of CA in the implementation of *natural solvers* that simulate dynamical emergent systems by a massive number of simple agents (cells) that interact locally. Parallel cellular languages and environments provide useful design and programming tools for the development of scalable simulations and models of emergent behavior. This approach is a valid alternative to complex and expensive laboratory experiments and simulations [3].

We discuss here how the basic CA concepts are related to emergent systems, describe parallel CA environments and tools for programming emergence in complex systems, and present some significant programming examples of emergent systems. The remainder of the chapter is organized as follows: Section 6.2 introduces CA and Section 6.3 outlines the main issues in parallel CA programming, describes the main features of the CAMELot environment and its programming language CARPET and discusses some related systems. Section 6.4 shows how different classes of CA can be implemented by using the CARPET language. Section 6.5 presents two examples of emergent phenomena programmed according to the parallel CA model in the CARPET cellular language and gives performance figures for them. Finally, Section 6.6 draws some conclusions.

6.2 Cellular Automata

Cellular automata are discrete dynamical systems in that space, time, and properties can have only a finite number of states [4]. They are an effective model for exploring systems with no central control. Given that the rules are executed in parallel on every cell, we can easily explore systems, using a decentralized control, in which simple local interaction takes place for a large population of cells over some period of time. The basic idea is to describe a complex system and simulate it by the interaction of a massive number of cells following simple rules. Thus, a complex system is not described with complex global equations but the complexity *emerges* from the interaction of simple local rules.

A CA can be defined as a d-dimensional Euclidean space (where $d = 1, 2, 3$ is used in practice), partitioned into cells of uniform size, each one embedding an identical elementary automaton (*ea*). Input for each *ea* is given by the states of the elementary automata in the neighboring cells, where neighborhood conditions are determined by a pattern invariant in time and constant over the cells. At the time $t = 0$, *ea* are in arbitrary states and the CA evolves changing the state of all *ea* at discrete times, according to a local rule. Each cell in the regular spatial lattice can have any one of a finite number of states. As mentioned before, the states of the cells in the lattice are updated according to a local rule called the state transition function. That is, the state of a cell at a given time depends only on its own state in the previous time step and the states of its nearby neighbors at the previous time step [5].

Here, we summarize the basic CA concepts useful for programming emergence.

- *Transition function*: Set of rules that define how the state of each cell changes on the basis of its current state and the states of its neighbor cells.
- *State*: The state of a cellular automaton (*global state*) is completely specified by the values of the variables at each cell (*local state*). The state of a cell is a simple or structured variable (see substate) that takes values in a finite set. The cell state can be either a numeric value or a property. For instance, if each cell represents part of a landscape, then the state might contain the altitude or the type of land.
- *Substate*: If the cell state is represented as a structured variable, substates are the fields of the structure that represent the attributes of the cell state. For example, if each cell represents a particle, its state can be composed of two substates that represent particle mass and speed.

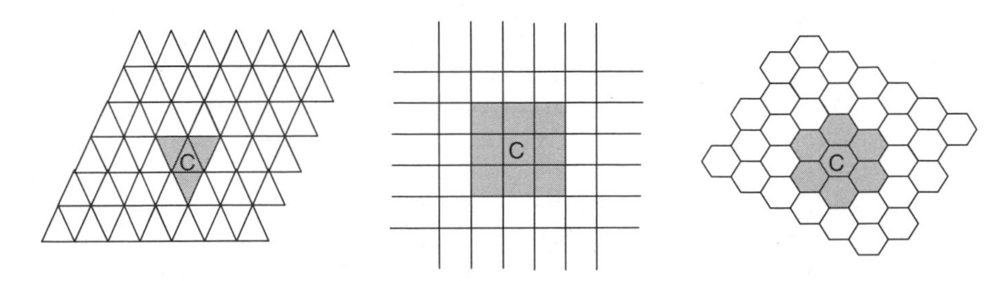

FIGURE 6.1 CA neighborhoods.

- *Neighborhood*: The set of cells that a cell interacts with. The neighborhood of a cell is typically taken to be all immediately adjacent cells. Simple neighborhoods of a cell (C) in a two-dimensional lattice are shown in Figure 6.1.

Let us define a CA as the 4-tuple (E^d, S, N, σ), where

- E^d is a regular lattice (the elements of E^d are called cells).
- S is a finite set of states.
- N a finite set (with $|N| = n$) of neighborhood indices such that for all x in N, all c in E^d : $c + x$ in E^d.
- $\sigma : S^n \rightarrow S$ a transition function.

A configuration $C_t : E^d \rightarrow S$ is a function that associates a state with each cell of the lattice. The effect of the transition function σ is to change the configuration C_t into the new configuration C_{t+1} according to $C_{t+1}(c) = \sigma(\{C_t(i) : i \text{ in } N(c)\})$, where $N(c)$ denotes the set of neighbors of cell c, $N(c) = \{i \text{ in } E^d : c - i \text{ in } N\}$. In standard CA, all cells of the automaton are updated synchronously in parallel; whereas extended CA models define asynchronous updating [6]. In section 6.4, we discuss nonstandard CA models. The state of the entire automaton advances in discrete time steps. Therefore, in CA the transition function plays a role analogous to that of the evolution equation in classical dynamical models. The global behavior of the system is not directly specified but it is determined, in other words, it emerges by the evolution of the states of all cells as a result of multiple interactions. Cellular automata capture the peculiar features of systems that may be seen to evolve exclusively according to the local interactions of their constituent parts, and guarantee computational universality. Furthermore, applied aspects of modeling have been widely investigated from a theoretical viewpoint [4,5].

6.3 Parallel CA Systems

For the implementation of CA on parallel computers two main approaches can be used. One is to write programs that encode the CA rules in a general-purpose parallel programming language such as HPF, HPC++, Linda, or CILK or to use a high-level sequential language such as C, Fortran, or Java with one of the low-level toolkits/libraries currently used to implement parallel applications such as MPI, RMI, PVM, or OpenMP. This approach does not require a parallel programmer to learn new language syntax and programming techniques for cellular programming. However, it is not simple enough to be used by programmers who are not experts in parallel programming, and the code consists of a large number of instructions even if simple cellular models must be implemented.

The other possibility is to use a high-level language specifically designed for CA, in which it is possible to directly express the features and the rules of CA, and then use a compiler to translate the CA code into a program executable on parallel computers. This second approach has the advantage that it offers a programming paradigm that is very close to the CA abstract model and that the same CA description could possibly also be compiled into different codes for various parallel machines. Furthermore, in this approach, parallelism is transparent from the user, so the programmers can concentrate on the specification of the model without worrying about architecture-related issues. In summary, it leads to the writing of software that does express the cellular paradigm in a natural way; these programs are simpler to read, change, and maintain. On the other hand, the regularity of computation and locality of communication allow CA programs to get good performance and scalability on parallel architectures.

Recently, several CA environments have been implemented on current desktop computers. Examples of these systems are CAT, CelLab, CaSim, CDM, Cellsim, DDLab, and Mathematica. A longer list can be found in Reference 7, in which the main features of these systems are outlined. Sequential CA-based systems can be used for educational purposes and very simple simulations, but real-world phenomena simulations generally take a very long time, or in some cases cannot be executed on this class of systems because of memory or computing power limits. Therefore, massively parallel computers are the appropriate computing platform for the execution of CA models when real-life problems must be solved. In fact, for two- and three-dimensional CA of large size the computational load can be enormous. Thus, if CA are to be used for investigating large complex phenomena, their implementation on high performance computers composed of several processors is a must.

In particular, general-purpose distributed-memory parallel computers offer a very useful architecture for a scalable CA machine in terms of speed-up, programmability, and portability. These systems are based on a large number of interconnected processing elements (PE), which perform a task in parallel. According to this approach, in recent years several parallel cellular software environments have been developed.

Significant examples of these parallel cellular systems are CAMELot, StarLogo, NEMO, P-CAM [8], Cellular, ParCel-1, PECANS [9], and DEVS. Some are discussed in Reference 1. Together with these software systems, parallel CA hardware has been developed for a more efficient execution of CA algorithms. Two examples of CA hardware are the CAM-8 machine [10] and the CEPRA FPGA machine. These two systems are special purpose machines that exploit CA parallelism in a very efficient way, although they are specialized machines that do not support general computation models.

Cellular automata parallel systems allow a user to exploit the inherent parallelism of CA to support the efficient simulation of complex systems that can be modeled by a very large number of simple elements (*cells*) with local interaction only. Cellular automata-based languages share several features such as a common computational paradigm and some differences such as, for example, different constructs to specify details of a CA or of mapping and output visualization. Many real-world applications in science and engineering, such as lava-flow simulations, molecular gas simulation, landslide modeling, freeway traffic flow, three-dimensional rendering, soil bioremediation, biochemical solution modeling, and forest fire simulation, have been implemented by using these CA languages. Moreover, parallel CA languages can be used to implement a more general class of fine-grained applications such as finite elements methods, partial differential equations, and systolic algorithms.

The main issues that influence the way in which CA languages support the design of applications on high performance architectures are:

1. The programming approach: The unit of programming is the single cell of the automaton.
2. The cellular lattice declaration: It is based on the definition of the lattice dimension and the lattice size.
3. The cell state definition and operations: Cell state is defined as a single variable or a record of typed variables; cell state access and update operations are needed.
4. The neighborhood declaration and use: Neighborhood concept is used to define interaction among cells in the lattice.
5. The Parallelism exploitation: The unit of parallelism is the cell, and parallelism, like communication, is implicit.
6. The cellular automata mapping: Data partitioning and process-to-processor mapping is implicit at the language level.
7. The output visualization: Automaton global state, as the collection of the cell states, is shown as it evolves.

Many of these issues are taken into account in parallel CA systems and similar or different solutions are provided by parallel CA languages. By discussing these concepts, we intend to illustrate how this class of languages can be effectively used to implement high-performance applications in science and engineering using the massively parallel cellular approach.

Programming Approach. When a programmer starts to design a parallel cellular program, she/he must define the structure of the lattice that represents the abstract model of a computation in terms of cell-to-cell interaction patterns. Then it must concentrate on the unit of computation that is a single cell of the automaton. The computation to be performed must be specified as the evolution rule (transition function) of the cells that compose the lattice. Thus, as against other approaches, a user does not specify a global algorithm that contains the program structure in an explicit form. The global algorithm consists of all the transition functions of all cells that are executed in parallel for a certain number of iterations (steps).

It is worth to notice that in some CA languages it is possible to define transition functions that change in time and space to implement inhomogeneous CA computations. Thus, after defining the dimension (e.g., one-, two-, or three-dimensional) and the size of the CA lattice, she/he needs to specify, by the conventional and the CA statements, the transition function of the CA that will be executed by all the cells. Then the global execution of the cellular program is performed as a massively parallel computation in which implicit communication occurs only among neighbor cells that access each other's state.

Cellular Lattice Declaration. As mentioned in the previous section, the lattice declaration defines the lattice dimension and the lattice size. Most languages support two-dimensional rectangular lattices only (e.g., CANL and CDL). However, some of them, such as CARPET and Cellang, allow the definition of one-, two-, and three-dimensional lattices. Some languages allow also the explicit definition of boundary conditions, such as CANL [9], which allows adiabatic boundary conditions where absent neighbor cells are assumed to have the same state as the center cell. Others implement reflecting conditions that are based on mirroring the lattice at its borders. Most languages use standard boundary conditions such as fixed and toroidal conditions.

Cell State. The cell state contains the values of data on which the cellular program works. Thus the global state of an automaton is defined by the collection of the state values of all the cells. While low-level implementations of CA allow to define the cell state as a small number of bits (typically 8 or 16 bits), cellular languages such as CARPET, CANL, DEVS-C++, and CDL allow a user to define cell states as a record of typed variables as follows:

```
cell=(direction : int ; speed : float);
```

where two substates are declared for the cell state. According to this approach, the cell state can be composed of a set of substates that are of integer, real, char, or Boolean type and in some cases (e.g., CARPET), arrays of these basic types can also be used. Together with the constructs for cell state definition, CA languages define statements for state addressing and updating that address the substates by using their identifiers; for example, `cell.direction` indicates the direction substate of the current cell.

Neighborhood. An important feature of CA languages that differentiate them from array-based languages and standard data-parallel languages is that that they do not use explicit array indexing. Thus, cells are addressed with a name or the name of the cells belonging to the neighborhood. In fact the neighborhood concept is used in the CA setting to define interaction among cells in the lattice. In CA languages, the neighborhood defines the set of cells whose state can be used in the evolution rule of the central cell. For example, if we use a simple neighborhood composed of four cells we can declare it as follows

```
neigh cross=(up, down, left, right);
```

and address the neighbor cell states by the ids used in the above declaration (e.g., `down.speed`, `left.direction`). The neighborhood abstraction is used to define the communication pattern among cells. It means that at each time step, a cell sends to and receives from the neighbor cells the state values. In this way implicit communication and synchronization are realized in cellular computing.

The neighbor mechanism is a concept similar to the region construct that is used in the ZPL language [11], in which regions replace explicit array indexing making the programming of vector- or matrix-based computations simpler and more concise. Furthermore, this way of addressing the lattice elements (cells) does not require compile-time sophisticated analysis and complex run-time checks to detect communication patterns among elements.

Parallelism Exploitation. Cellular automata languages do not provide statements to express parallelism at the language level. It turns out that a user does not need to specify what portion of the code must be executed in parallel. In fact, in parallel CA languages the unit of parallelism is a single cell, and parallelism, like communication and synchronization, is implicit. This means that in principle the transaction function of every cell is executed in parallel with the transaction functions of the other cells. In practice, when coarse-grained parallel machines are used, the number of cells N is greater than the number of available processors P, so each processor executes a block of N/P cells that can be assigned to it using a domain decomposition approach.

CA Mapping. Like parallelism and communication, data partitioning and process-to-processor mapping is implicit in CA languages. The mapping of cells (or blocks of them) onto the physical processors that compose a parallel machine is generally done by the run-time system of each particular language and the user usually intervenes in selecting the number of processors or some other simple parameter.

Some systems that run on MIMD computers use load balancing techniques that assign at run-time the execution of cell transition functions to processors that are unloaded, or use greedy mapping techniques that avoid some processor to become unloaded or free during the CA execution for a long period. Examples of these techniques can be found in References 1, 12, and 13.

Output Visualization and Monitoring. A computational science application is not just an algorithm. Therefore it is not sufficient to have a programming paradigm for implementing a complete application. It is also significant to dispose of environments and tools that help a user in all the phases of the application development and execution. Most of the CA languages we are discussing here provide a development environment that allows a user not only to edit and compile the CA programs, but also to monitor the program behavior during its execution on a parallel machine, by visualizing the output as composed of the states of all cells. This is done by displaying the numerical values or by associating colors to these values. Examples of these parallel environments are CAMEL for CARPET, PECANS for CANL, and DEVS for DEVS-C++. Some of these environments provide dynamic visualization of simulations together with monitoring and tuning facilities. Users can interact with the CA environment to change values of cell states, simulation parameters, and output visualization features. These facilities are very helpful in the development of complex scientific applications and make it possible to use these CA environments as real problem-solving environments (PSEs) [14].

In the rest of this section we outline some of the listed issues by discussing the main features of CAMELot, a general-purpose system that can be easily used for programming emergent systems using the CARPET cellular programming language according to a massively parallel paradigm and some related parallel CA environments and languages.

6.3.1 CAMELot—CARPET

CAMELot (*CAMEL open technology*) is a parallel software environment designed to support the parallel execution of cellular algorithms, the visualization of the results, and the monitoring of cellular program execution [15]. CAMELot is an MPI-based portable version of the CAMEL system based on the CARPET language. CARPET (*CellulAR Programming EnvironmenT*) offers a high-level cellular paradigm that offers to a user the main CA features to assist her/him in the design of parallel cellular algorithms without apparent parallelism [6].

A CARPET user can develop cellular programs describing the actions of many simple active elements (implemented by cells) interacting locally. Then, the CAMELot system executes in parallel cell evolution and allows a user to observe the global complex evolution that arises from all the local interactions.

CARPET uses a C-based grammar with additional constructs to describe the rules of the transition function of a single cell. In a CARPET program, a user can define the basic rules of the system to be simulated (by the cell evolution rule), but she/he does not need to specify details about the parallel execution. The language includes

1. A declaration part (**cadef**) that allows to specify:
   - the dimension of the automaton (**dimension**);
   - the radius of the neighborhood (**radius**);
   - the type of the neighborhood (**neighbor**);
   - the state of a cell as a record of substates (**state**);
   - a set of global parameters to describe the global characteristics of the system (**parameter**);
   - a set of constructs for addressing and updating the cell states (e.g., **update, GetX, GetY, GetZ**).

Figure 6.2 shows a simple CA programmed in CARPET that implements the Fredkin's rule. This is a simple rule: a cell becomes *alive* if the number of living cells in its neighborhood is odd; if the number of living cells in its neighborhood is even a cell becomes *dead*. Fredkin's rule is very simple, however it has the fascinating property that any initial pattern of living cells is replicated several times on a larger scale.

CARPET and CAMELot have been used for implementing high-performance real-world simulations based on the emergence paradigm such as lava flow, traffic flow, and forest fire simulations [6]. In Section 6.5 the CARPET language is used to program two significant examples of emergent systems. Its main linguistic features are outlined by describing how it supports the implementation of real emergent applications.

```
#define dead 0
#define alive 1

cadef
{
   dimension 2;
   radius 1;
   state (short value);
   neighbor Neumann[4] ([0,-1]North, [-1,0]West,
                        [0,1]South, [1,0] East);
}
   int sum=0;
 {  for (i=0; i<4; i++)
       sum = sum + Neumann[i]_value;
    if (sum%2 == 0)
     update(cell_value, dead);
    else
     update (cell_value, alive);
 }
```

FIGURE 6.2 The Fredkin's rule written in CARPET.

6.3.2 Other Cellular-Based Parallel Systems

As mentioned before, some cellular-based systems have recently been developed. Among these we discuss some representative examples such as Cellular, DEVS-C++, and ZPL. In particular, ZPL is a language for parallel programming that has not been specifically designed for CA, but it shares with parallel cellular languages the same rationale and some basic ideas.

The Cellular system [16] consists of a CA-based programming language, named Cellang, an abstract virtual CA machine for execution of cellular programs, named *avcam*, and a viewer named *cellview*, which is used to examine the cell values in a graphic format. Cellang combines the classic CA programming paradigm, like CARPET or CDL, with that of agents. The results of an execution can either be viewed graphically, as an output stream of cell locations and values, or passed through a custom filter before being reported. The current implementation of the Cellang compiler generates codes both for sequential computers and shared-memory multiprocessor systems. However, the system has not been implemented on distributed-memory parallel machines or PC clusters.

The NEMO (Neighbourhood Modeling) system is an environment based on CA designed by the Carleton's PARAllel DIGital Modeling (PARADIGM) group [12]. The main component of the NEMO systems is the Cell driver. The Cell driver is the parallel kernel of NEMO that provides the parallel execution of a NEMO application and interacts with the other components. In addition to the Cell driver, NEMO contains the Neighbourhood Analysis (NAN) driver, which is designed to support the parallel computation of spatial statistics on very large lattices, and the Propagation driver (PD), which supports the modeling in parallel of processes in a lattice that spreads along an active boundary. Moreover, NEMO provides a display facility for the visualization of graphical data from the simulation. As against CAMELot and Cellular, NEMO does not offer a high-level language to implement cellular programs although some default functions can be used.

DEVS-C++ is a high-performance environment based on the DEVS (Discrete Event System Specification) formalism that supports the analysis, design, and simulation of discrete event dynamical systems [17]. The DEVS formalism, based on CA theory and discrete event simulation, provides a means of specifying a mathematical object called *system* defined as a lattice of cells. DEVS-C++ is a DEVS parallel implementation that uses the C++ language for programming simulation and models designed by the DEVS formalism. Applications are programmed using a set of specifically designed classes (included in the Container and Devs libraries) provided by the DEVS-C++ environment in an object-oriented programming framework. The Devs class is the basic class to provide methods for the DEVS formalism. Such methods are implemented as virtual methods to be defined by users. A Java version of the system has been recently developed.

ZPL is a data-parallel array programming language designed for fast execution on both sequential and parallel computers [11]. Like CARPET, ZPL is *implicitly parallel*, that is, the programmer does not express the parallelism; the ZPL compiler and run-time system will generate parallel codes and map processes on processors. As cellular languages, ZPL is suitable only for regular parallel computations.

A fundamental concept in ZPL is the notion of a *region*. Fortran and other array languages refer to subarrays using the so-called slice or triple notation, whereas ZPL uses the concept of a region. A region is simply a set of indices, that is,

```
region X = [1..n, 1..n];
```

specifies the standard indices of an *nxn* array. A ZLP program next declares a set of *directions*. Directions are used to transform regions, as in the expression *north of X*. As in cellular programming, array indexing is avoided in ZPL by referring to adjacent array elements using the @ operator, which is similar to the neighborhood mechanism of CARPET. An expression `A@d`, executed in the context of a region X, results in an array of the same size and shape as X composed of elements of A offset in the direction d. ZPL does not have parallel directives as in data-parallel languages such as HPF or other forms of explicit parallelism. This implicit computation can be parceled out to different processors to get parallelism. Thus, parallelism, as in cellular programming, comes simply from the semantics of the array operations.

6.4 Programming Standard and Nonstandard Parallel Cellular Automata

Nonstandard models of CA can be divided between *modifications* and *generalizations* according to terminology introduced by Worsch [7], where several nonstandard automata are described. By modifications of CA we denote computational models that can simulate CA and can be simulated by CA with a linear overhead in time and space (number of cells). On the other hand, generalizations of CA are models that cannot be shown to be simulated by CA in linear time.

In Section 6.4.1, we concentrate on discussing how CARPET can be used for implementing four types of nonstandard CA: inhomogeneous, synchronous, probabilistic, and partitioned CA.

6.4.1 Inhomogeneous CA

Inhomogeneous CA are a generalization of standard CA. In CA we can have spatial inhomogeneity, temporal inhomogeneity, or both. In spatially inhomogeneous CA, there is not a single transition function σ, but there is a set of different transition functions $\sigma_1, \sigma_2, \ldots, \sigma_N$, associated to different cells or regions of a CA in which also different neighborhoods can be defined. This class of automata can be implemented in CARPET using the operations **GetX**, **GetY**, **GetZ** that return the value of the coordinates X, Y, and Z of a cell in the CA lattice. For example, if we want to use a different transition function in a rectangular region of a two-dimensional lattice we can write a code as follows

```
{
    . . . .
    if ((GetX >= 100 && GetX <= 200) && (GetY >= 80 && GetY <= 400)
      { trans-funct1()}
    else
      { trans-funct2()}
    . . . .
}
```

The same approach can be used for different transition functions for cells that belong to the border of lattice to define boundary conditions. For instance, if two border sides are identified by the coordinates $X = 0$ and $X = 400$, to use a transition function for the cells that are placed on that borders we can write

```
    . . . .
if ((GetX == 0 || GetX == 400)
  { border-trans-funct()}
else
  { normal-trans-funct()}
```

In temporal inhomogeneous CA, the transaction function changes during time. This generalization of standard CA is very useful in modeling and simulation of phenomena that consist of more computational phases. Thus, for a certain number of iterations a function σ_{ta} is used, then another function σ_{tb} is used for another time interval, and so on depending on the kind of computation that must be performed.

This class of CA can be easily programmed in CARPET by using the predefined variable **step**. For example, Figure 6.3 shows the structure of a CARPET algorithm for a two-dimensional CA composed of a sequence of three different transition functions. Temporal inhomogeneity can also be implemented using conditions that include the **step** variable and more complex logic expressions.

6.4.2 Asynchronous and Probabilistic CA

Asynchronous CA are only a modification of the standard CA model. In an asynchronous CA, each cell at each time step t can choose nondeterministically between keeping its current state $s(t-1)$ or changing its

```
cadef
{
 dimension 2;
 radius 1;
   ....
}

{
  if (step == 1)
    trans_funct1();
 else
   if (step > 1 && step <= 10 )
      trans_funct2();
   else
      trans_funct2();
}
```

FIGURE 6.3 CARPET code for a temporal inhomogeneous CA.

state according to the transition function σ. This class of CA is useful in the simulation of asynchronous systems, where it is not necessary to update the state of its components at the same time.

Asynchronous CA can be programmed in CARPET by using the **random** function that allows to express nondeterminism in the local transition function of each cell. The **random**(n) function returns a pseudo random integer number between 0 and n. We can use it when the state of a cell must be updated. For example, the following instructions show the nondeterministic update of a substate of a cell (notice that the **randomize** function call creates a different seed for the random number generator, avoiding that the same random value is generated for every cell):

```
cadef
{
  ....

  state ( int substate1, float substate2);
  ...
}
  .....
randomize();
if(random(n)> n/2)
    update(cell_substate2, X)
else
    . . ./* the cell state is not updated */
```

The **random** function can be also used for the implementation of probabilistic CA that in some aspects are similar to asynchronous CA. In a probabilistic CA, given a local configuration c, each cell can enter a new state s with a probability $\sigma(c, s)$ and $\Sigma_{s \in S}\sigma(c, s) = 1$. This type of CA is useful in the simulation of probabilistic phenomena that occur in physics and other sciences. As mentioned before, by the use of the **random** function it is possible to implement this modification of the standard CA model in CARPET. In particular, this can be done by a using a **switch** statement with probability values (computed on the basis of a random number) assigned to the different **case** branches that must contain an **update** operation whose execution will depend on the probability assigned to its own branch.

```
cadef
{
dimension 2;
radius 1;
state (int sub1, sub2, sub3, sub4);
   neighbor VonNeu[4]  ([0,-1]North,[-1,0]West,[0,1]South,[1,0] East);
}
    int sum=0;
{
   sum = VonNeu[1]_sub1+VonNeu[2]_sub2+VonNeu[3]_sub3+VonNeu[4]_sub4;
if (sum <= 2)
    {
      update(cell_sub1, 0);
      update(cell_sub2, 1);
    }
else
    {
      update(cell_sub3, 0);
      update(cell_sub4, 1);
    }
}
```

FIGURE 6.4 A simple two-dimensional partitioned CA written in CARPET.

6.4.3 Partitioned CA

Another significant modification of the standard CA model is represented by the *partitioned* CA. Whereas in a standard CA, a cell can use the whole state of each neighboring cell to compute its next state, in a partitioned CA only the component S_{ni} of a neighbor n_i is used to determine the new state of a cell. Thus, each cell reads only the component i of a cell n_i that belongs to its neighborhood. The transition function of a partitioned CA can be expressed as

$$\sigma : S_{n1} \times S_{n2} \times \cdots \times S_{nk} \quad \rightarrow \quad S_{n1} \times S_{n2} \times \cdots \times S_{nk}$$

where $N = \{n_1, n_2, \ldots, n_k\}$. Partitioned CA is useful in modeling systems where each cell in a neighborhood contributes toward the change of the state of a central cell. Moreover, they can be used to make possible the implementation of transition functions that otherwise would be infeasible for software or hardware limits. In fact, in them the domain of σ is of size $|S|$ instead of $|S|^{|N|}$.

The implementation of partitioned CA in CARPET is quite direct because the language allows the definition of the cell state as a set of substates. Thus we can define the cell state of a partitioned CA as composed of a number of substates equal to the number of neighbors. Figure 6.4 shows a simple CARPET program that implements a two-dimensional partitioned CA.

The state of a cell is updated according to the sum of the ith component of the neighbor cell i. Since a radius 1 von Neumann neighborhood is defined, the cell state is composed of four substates. If we use a Moore neighborhood, a state composed of eight substates must be used.

6.5 Programming Emergent Systems as Massively Parallel CA: Examples

Emergent behavior not only occurs in nature and in life sciences such as biology and medicine, but significant examples of emergent systems can be found in physics, computer science, and engineering. For instance, network synchronization, distributed mobile sensors, cooperative environments, peer-to-peer computing, and distributed control systems are emergent systems that today occur in many real-life problems. We cannot discuss these systems in detail, but we give here two simple but significant examples of how to program emergent systems on parallel machines using a massively parallel cellular paradigm.

In particular, this section presents two examples of emergent systems programmed by using the CARPET language that we implemented on a Linux cluster. The first example is the Q2R Ising model and the second one is an epidemics diffusion model.

6.5.1 The Q2R Ising Model

The Ising model is one of the pillars of statistical mechanics. Each cell represents a spin that can have two values (1/0, up/down), and neighboring cells have an energetic preference to be the same value. As a system of 1/0 spins, it is a model for magnetism: like iron, there is a temperature (the Neel point) above which the magnetization "melts" away. Run at high temperatures to see the melted state, and run at low temperatures to see the magnetized state. The Q2R model is an Ising model with particular properties.

Q2R cellular automata are a good example for addressing an old question: How is thermodynamic irreversibility, as seen in entropy increase, compatible with microscopic reversibility, as in Newton's equations of motion? The Boltzmann–Zermelo argument says that the time for returning to the low-entropy initial state of a large system are far longer than the age of the universe. How can we test this assertion? Molecular dynamics in its usual form is inadequate to address this question, since arithmetic rounding errors as well as discretization of space and time preclude return to the exact initial configuration. Monte Carlo simulations of Ising models avoid such errors but require random numbers and are therefore not reversible in the usual sense. The Q2R update rules of CA for microcanonical Ising models are reversible and the system returns exactly to the initial configuration after an exponentially long time.

In Q2R, all of the spins are flipped simultaneously, while in the canonical Ising model only one randomly selected spin is flipped at a time. The spin-flipping decision is deterministic in Q2R while it is probabilistic in the Metropolis model. Both versions of the Ising model give many comparable results. Computer simulation of Q2R CA helps us to understand some very old fundamental problems.

Here we show the simple implementation of the Q2R model using the CARPET language and some preliminary performance results on a parallel machine. Figure 6.5 shows the CARPET code composed of a very short number of instructions that have been executed on a cluster of PentiumIII connected by Myrinet. The program uses a two-dimensional automaton with a Moore neighborhood composed of eight neighbor cells; to flip the cell spin, the neighbor cells' spins are summed and then, depending on the current value of the cell, the central cell spin is updated.

Table 6.1 shows the performance results (elapsed time in seconds) up to 12 processors for the execution of 100 iteration steps considering four different lattice sizes. As shown in Figure 6.6, the algorithm is scalable despite its simplicity. Relative speedup is very close to ideal. Although the code of each cell is simple, the application makes efficient use of the 12 parallel processors. In fact efficiency (E = speedup/Procs) ranges from 0.9 to 1, this means that 90% to 100% of the computing power of each CPU is exploited.

6.5.2 Epidemic Diffusion Simulation

The dominant features of an epidemic, its distribution over areas of land, and its evolution through time, are the result of the dynamic interactions of the host and pathogen systems, both of which are influenced by numerous complex biological processes. Hence, models must somehow describe the relationships among the individuals, the time, and the space to apply possible infection conditions on them depending on the neighboring infected individuals. Based on these properties, epidemic diffusion is an emergent phenomenon that can be effectively modeled using CA. In a cellular model of a plant epidemic, the host population is divided into small-scale units, each of which is located at a cell on a two-dimensional lattice. Models can be used to develop understanding of quantitative or qualitative behavior, or to make predictions of the dynamics of biological systems.

Figure 6.7 shows the CARPET code that implements a simulation of an epidemic with three types of cells: cells representing healthy cells/creatures, other cells representing sick ones, and blank cells. Healthy

```
#define up 1
#define down 0
cadef
{
  dimension 2;
  radius 1;
  state (short spin);
  neighbor Moore[8] ([0,-1]N, [-1,-1]NW,
                     [-1,0]W, [-1,1]SW,  [0,1]S,
                     [1,1]SE, [1,0]E, [1,-1]NE);
}
int i;
short sum = 0;
{
 for (i = 0; i < 8; i++)
   sum = Moore[i]_spin + sum;
 if (sum == 4)
   { if (cell_spin == down)
          update(cell_spin, up);
     else
         update(cell_spin,down);
   }
}
```

FIGURE 6.5 The Q2R Ising model written in the CARPET language.

TABLE 6.1 Performance of the Q2R Ising Program

Automata size	1 Proc	2 Procs	4 Procs	8 Procs	12 Procs
1152 × 1152	17.2	8.1	4.4	2.2	1.4
2304 × 2304	67.4	37.8	17.4	9.3	5.8
4608 × 4608	275.7	139.8	69.2	33.9	23.8
9216 × 9216	1035.4	561.8	268.8	144.1	94.9

Note: Elapsed time in sec × 100 steps

cells become sick with a chance of 55% when next to a sick cell. Sick cells recover at any given time step with a probability of 80%, but have a 10% chance of dying. The simple behavior of a single cell can be programmed in CARPET as shown in Figure 6.7 and the global behavior of the epidemic implicitly emerges by the parallel execution of a large number of cells that dynamically interact according to a simple neighbor pattern.

Such a CA program can be used to investigate spatial clustering effects in the spread of a simple epidemic or it can be the basic skeleton for a more complex model with many degrees of freedom. Its use makes it very easy to try and test different scenarios and "what if" situations such as effects of vaccination, preventive measures, or the effects of new drugs.

Table 6.2 shows the performance results (elapsed time in seconds) up to 12 processors for the execution of 100 iteration steps considering four different lattice sizes.

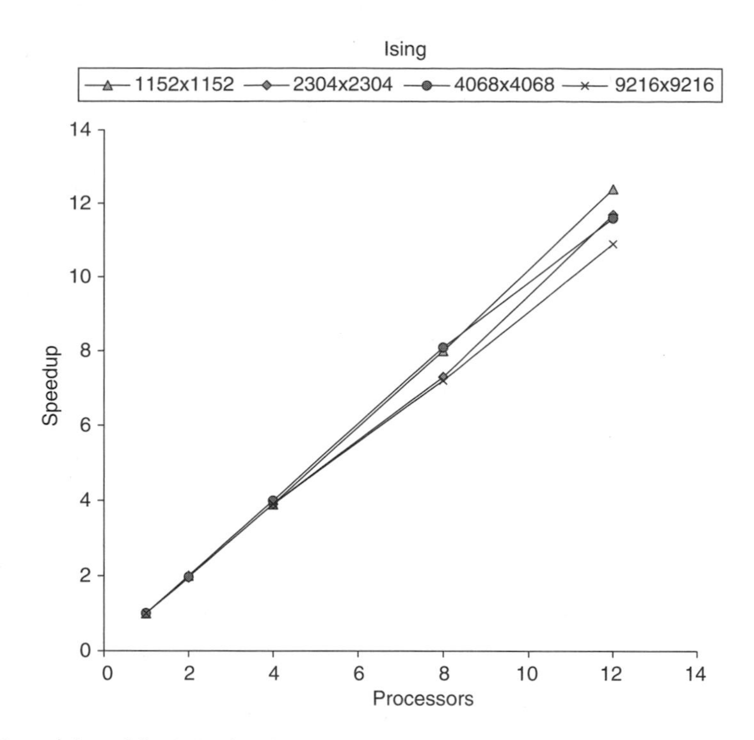

FIGURE 6.6 The speedup of the Q2R Ising CA program.

Figure 6.8 shows the corresponding speedup. As the number of processors increases there is a corresponding increase in the relative speedup, which implies that the computational power of the cluster processors is exploited in an efficient way. In fact, efficiency of the parallel epidemics program ranges from 0.97 to 1, this means that all the CPUs are fully used during the CA program execution.

6.6 Conclusion

A large number of applications can be naturally expressed by combining the emergent system model and the massively parallel paradigm. In many cases, designers do not use this approach because the available tools do not support it. However, parallel implementation of CA-based emergent computation systems and phenomena represent a valid and effective approach in the study of several classes of problems. These kinds of simulation are very helpful in vital scientific areas such as biology, physics, chemistry, medicine, social science, and economy. Cellular automata are a viable abstract model for representing emergent decentralized systems and phenomena in those and other areas.

In this chapter, we discussed how to use parallel CA for complex system development and described the main features of parallel CA environments for developing scalable emergent systems and phenomena. Parallel cellular languages and tools provide a high-level framework for emergent computation and, at the same time, they offer a scalable setting for getting high performance using parallel architectures. While efforts in traditional sequential computer languages and systems design focused on how to express and implement imperative operations and data, the main goal of the cellular paradigm is to offer a massively parallel computational model based on a large number of simple cellular objects and operations that are suitable for defining emergent complex systems.

We showed, through CARPET example programs, how the combination of the CA model with parallel computing techniques and systems could be exploited to efficiently implement emergent computation structures. Finally, modeling and simulation work through parallel cellular methods helps researchers by

```
#define blank   0
#define healthy 1
#define sick    2
cadef
{
  dimension 2;
  radius 1;
  state (short status);
  neighbor Moore[8] ([0,-1]N,[-1,-1]NW,[-1,0]W,[-1,1]SW, [0,1]S,
                    [1,1]SE,[1,0]E,[1,-1]NE);
  parameter (probS 0.55, probR 0.8, probD 0.1);
}
 int i; float probX; short cond = 0;
{
  for (i = 0; i < 8 && cond == 0; i++)
    if (cell_status == healthy &&  Moore[i]_status == sick)
    { probX = random(1);
      if (probX <= probS)
        update(cell_status, sick);
      cond = 1;
    }
    if (cell_status == sick)
    { probX = random(1);
      if (probX <= probR)
        update(cell_status, healthy);
      else
       { probX = random(1);
         if(probX <= probD)
           update(cell_status,blank);
       }
    } }
```

FIGURE 6.7 The simple epidemics model written in the CARPET language.

TABLE 6.2 Performance of the Epidemics Program

Automata size	1 Proc	2 Procs	4 Procs	8 Procs	12 Procs
1152 × 1152	19.3	9.5	4.9	2.4	1.6
2304 × 2304	77.5	43.4	19.1	10.2	6.7
4608 × 4608	316.4	154.2	80.5	39.5	26.9
9216 × 9216	1220.9	627.2	313.2	166.5	104.3

Note: Elapsed Time in sec × 100 steps

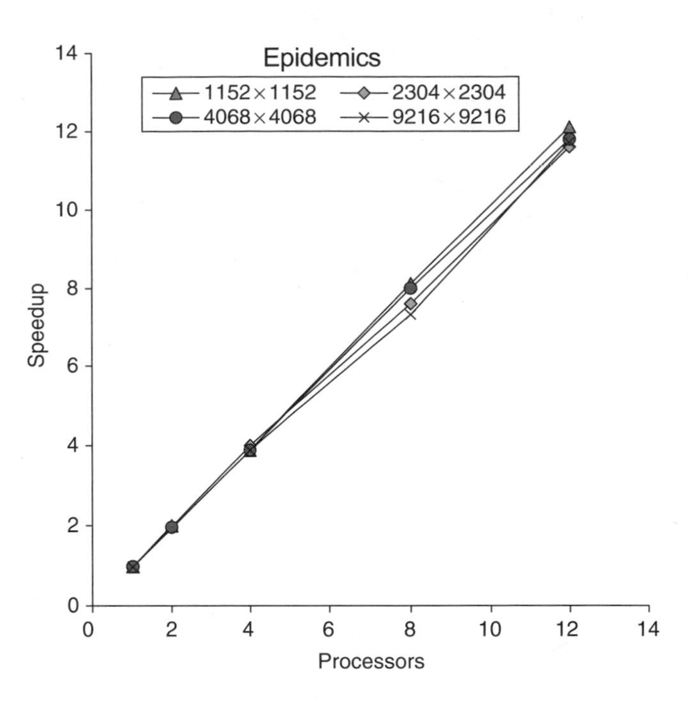

FIGURE 6.8 Speedup of the epidemics CA program.

setting up a cross-disciplinary framework and assisting the designers during software development from the design phase to the execution, tuning, and validation phases. Therefore, these environments can be used as virtual laboratories where scientists may work cooperatively by programming and experimenting as in a real laboratory, getting data and knowledge on the modeled systems.

A Linux version of the CAMELot system is available online at the webpage `www.icar.cnr.it/spezzano/camelot/carpet.html`. Programmers, scientists, students, and professionals may download and use it to simulate complex systems according to the CA model.

References

[1] D. Talia, "Cellular processing tools for high-performance simulation," *IEEE Computer*, 33, 44–52, 2000.

[2] J.R. Weimar, *Simulation with Cellular Automata*, Logos-Verlag, Berlin, 1997.

[3] M. Sipper, "The Emergence of Cellular Computing," *IEEE Computer*, 32, 18–26, 1999.

[4] J. von Neumann, *Theory of Self Reproducing Automata*, University of Illinois Press, IL, 1966.

[5] S. Wolfram, "Computation theory of cellular automata," *Communicates in Math. Physics*, 96, 15–57, 1984.

[6] D. Talia, "Implementing Standard and Nonstandard Parallel Cellular Automata in CARPET," In *Proceedings of the 9th Euromicro Workshop on Parallel and Distributed Processing (PDP 2001)*, Mantova, Italy, IEEE Computer Society Press, Washington, 2001, pp. 243–249.

[7] T. Worsch, "Simulation of cellular automata," *Future Generation Computer Systems*, 16, 157–170, 1999.

[8] A. Schoneveld and J.F. de Ronde, "P-CAM: a framework for parallel complex systems simulations," *Future Generation Computer Systems*, 16, 217–234, 1999.

[9] L. Carotenuto, F. Mele, M. Mango Furnari, and R. Napolitano, "PECANS: A parallel environment for cellular automata modeling," *Complex Systems*, 10, 23–41, 1996.

[10] T. Toffoli and N. Margolus, *Cellular Automata Machines: A New Environment for Modeling.* The MIT Press, Cambridge, MA, 1986.

[11] B.L. Chamberlain et al., "The Case for high level parallel programming in ZPL," *IEEE Computational Science & Engineering*, 5, 76–86, 1998.

[12] C. Hecker, D. Roytenberg, J.-R. Sack, and Z. Wang, "System development for parallel cellular automata and its applications," *Future Generation Computer Systems*, 16, 235–247, 1999.

[13] M. Cannataro, S. Di Gregorio, R. Rongo, W. Spataro, G. Spezzano, and D. Talia, "A Parallel Cellular automata environment on multicomputers for computational science," *Parallel Computing*, 21 803–824, 1995.

[14] E. Gallopoulos, E.N. Houstis, and J.R. Rice, "Workshop on problem-solving environments: findings and recommendations," *ACM Computing Surveys*, 27, 277–279, 1995 .

[15] G. Spezzano and D. Talia, "Programming cellular automata algorithms on parallel computers," *Future Generation Computer Systems*, 16, 203–216, 1999.

[16] J.D. Eckart, "Cellang 2.0: Reference Manual," *ACM Sigplan Notices*, 27, 107–112, 1992.

[17] P.Z. Zeigler et al., "The DEVS environment for high-performance modeling and simulation," *IEEE Computational Science & Engineering*, 4, 61–71, 1997.

7

Decentralized Cellular Evolutionary Algorithms

Enrique Alba
Bernabe Dorronsoro
Mario Giacobini
Marco Tomassini

7.1 Introduction

This chapter focuses on the class of algorithms called cellular evolutionary algorithms (cEAs). Here, we present the canonical algorithm and suggest the interesting variants targeted to solve complex problems accurately with a minimum effort for customization. These techniques, also called diffusion or fine-grained models, have been popularized, among others, by the early work of Gorges-Schleuter [1] and Manderick and Spiessens [2]. The basic idea is to add some structure to the population of tentative solutions. The pursued effect is to improve on the diversity and exploration capabilities of the algorithm while, still admitting an easy combination with local search and other search techniques to improve on exploitation.

These structured models are based on a spatially distributed population in which genetic operations may only take place in a small neighborhood of each individual. Usually, individuals are arranged on a regular grid of dimensions, $d = 1, 2$, or 3. cEAs are a kind of decentralized EA model [3]. They are not a parallel implementation of an EA; in fact, although parallelism could be used to speed up the search, we do not address parallel implementations in this work. In addition, it is worth remarking that, although SIMD (single instruction stream–multiple data stream) machine implementations were popular a decade ago, this is no longer true, and today the best distributed implementation of a cEA should make use of domain decomposition on clusters of networked machines.

Although, fundamental theory is still an open research line for cEAs, they have been empirically reported as being useful in maintaining diversity, and promoting slow diffusion of solutions through the grid. Part of their behavior is due to a lower selection pressure compared with that of *panmictic* EAs (here panmictic means that, any individual may mate with any other individual in the population). The influence of the selection method [4,5], neighborhood [6], and grid topology on the efficiency of cEAs in comparison with other EAs [7] have been investigated in detail, and tested on different applications, such as combinatorial and numerical optimization.

Cellular evolutionary algorithms can be seen as stochastic cellular automata (CA) [8,9] where the cardinality of the set of states is equal to the number of points in the search space. CAs, as well as cEAs, usually assume a *synchronous* or "parallel" update policy, in which all the cells are updated simultaneously. However, this is not the only option available. Indeed, several works on *asynchronous* CAs have shown that sequential update policies have a marked effect on their dynamics [10,11]. Furthermore, the shape of the structure in which individuals evolve has a deep impact on the performance of the cEA. The algorithm admits a special, easy modulation of its shape that can sharpen the exploration or the exploitation capabilities of the canonical technique, as shown in Reference 7. Thus, it is interesting to investigate asynchronous cEAs and nonsquare shaped cEAs, in order to analyze their problem solving capabilities, which is the subject of the second part of this chapter.

This work is organized as follows. Section 7.2 contains some background on synchronous and asynchronous cEAs. In Section 7.3, we discuss the ability of cEAs for changing their behavior depending on the population shape. In Section 7.4, we illustrate, quantitatively, the modifications on the selection intensity due to asynchronicity and population shape. We deal with a set of discrete and continuous benchmark problems in Sections 7.5 and 7.6, respectively, with the goal of analyzing the actual computational power of the algorithms. Finally, Section 7.7 discusses our conclusions, as well as some comments on future work.

7.2 Synchronous and Asynchronous cEAs

In this section we summarize the canonical behavior of cEAs. A cEA starts with the cells (individuals) in a random state and proceeds by successively updating them using evolutionary operators, until a termination condition is met. Updating a cell in a cEA means selecting two parents in the individual's neighborhood (including the individual itself), applying genetic operators to them, and finally replacing the individual if an offspring has a better fitness or using another replacement policy.

Cells can be updated *synchronously* or *asynchronously*. In synchronous (parallel) update all the cells change their states simultaneously, while in asynchronous, or sequential, update cells are updated one at a time in some order. There exist many ways for sequentially updating the cells of a cEA (an excellent discussion of asynchronous update in cellular automata, which are essentially the same system as a cEA, is available in Reference 10). We consider four asynchronous update methods: *fixed line sweep* (LS), *fixed random sweep* (FRS), *new random sweep* (NRS), and *uniform choice* (UC).

- In *fixed line sweep*, the simplest method, the n grid cells are updated sequentially $(1, 2, \ldots, n)$ line after line.

- In *fixed random sweep*, the next cell to be updated is chosen with uniform probability without replacement; this will produce a certain update sequence $(c_1^j, c_2^k, \ldots, c_n^m)$, where c_q^p means that cell number p is updated at time q and $(j, k, \ldots, m)$ is a permutation of the n cells. The same permutation is then used for all update cycles.
- The *new random sweep* method works like FRS, except that a new random cell permutation is used for each sweep through the array.
- In *uniform choice*, the next cell to be updated is chosen at random with uniform probability and with replacement. This corresponds to a binomial distribution for the update probability.

The concept of *generation* that is customarily used in EAs and in synchronous cEAs has to be replaced by *time step* in the asynchronous cases. A time step is defined as updating n times sequentially, which corresponds to updating *all* the n cells in the grid for LS, FRS, and NRS, and possibly less than n different cells in the UC method, since some cells might be updated more than once. It should be noted that, with the exception of LS, the other asynchronous updating policies are stochastic, representing an additional source of nondeterminism besides that of the genetic operators.

7.3 New cEA Variants Based on a Modified Ratio

After explaining the basic algorithm and the asynchronous variants in Section 7.2, we now proceed to characterize the population grid itself. For this goal, we use the "radius" definition given in Reference 7, which is refined from the seminal one appeared in Reference 6 to account for non-square grids. The grid is considered to have a radius equal to the dispersion of n^* points in a circle centered in $(\bar{x}, \bar{y})$ (Equation [7.1]). This definition always assigns different numerical values to different grids.

$$\text{rad} = \sqrt{\frac{\sum (x_i - \bar{x})^2 + \sum (y_i - \bar{y})^2}{n^*}}, \qquad \bar{x} = \frac{\sum_{i=1}^{n^*} x_i}{n^*}, \quad \bar{y} = \frac{\sum_{i=1}^{n^*} y_i}{n^*}. \tag{7.1}$$

Although it is called a "radius," rad measures the dispersion of n^* patterns. Other possible measures for symmetrical topologies would allocate the same numerical value to different topologies (which is undesirable). Two examples are, the radius of a circle surrounding a rectangle containing the topology, or an *asymmetry coefficient.* The definition not only characterizes the grid shape but also provides a radius value for the neighborhood. As proposed in Reference 6, the grid-to-neighborhood relationship can be quantified by the ratio between their radii (Equation [7.2]).

$$\text{ratio}_{\text{cEA}} = \frac{\text{rad}_{\text{Neighborhood}}}{\text{rad}_{\text{Topology}}}. \tag{7.2}$$

When solving a given problem with a constant number of individuals ($n = n^*$, for making fair comparisons) the topology radius will increase as the grid gets thinner (Figure 7.1[b]). Since the neighborhood is kept constant in size and shape throughout this chapter (we always use linear5 (L5), Figure 7.1[a]), the ratio will be smaller as the grid gets thinner.

This ratio value directly influences the behavior of the algorithm. During the search for reducing the ratio means reducing the global selection intensity on the population (see Section 7.4), thus promoting *exploration,* that is, the importance of such a ratio measure. This is expected to allow a higher diversity in the genotype that could improve the results in difficult problems (such as in multimodal or epistatic tasks). On the other hand, the search performed inside each neighborhood is guiding the *exploitation* of the algorithm. In this chapter we study how the ratio affects the search efficiency over a variety of domains. Changing the ratio during the search is a unique feature of cEAs that can be used to shift from

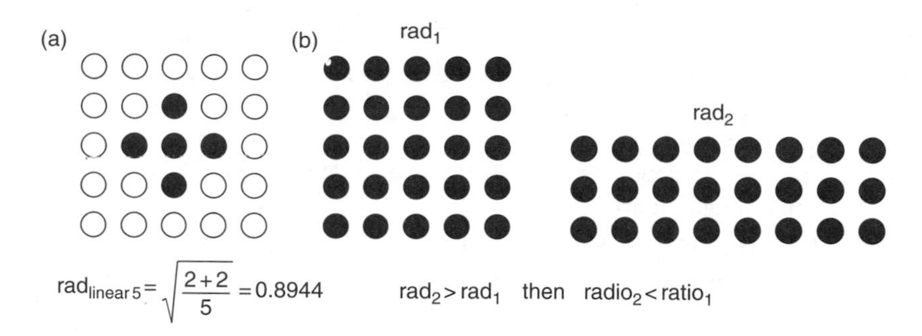

FIGURE 7.1 (a) Radius of neighborhood L5. (b) $5 \times 5 = 25$ and $3 \times 8 \approx 25$ grids; equal number of individuals with two different ratios.

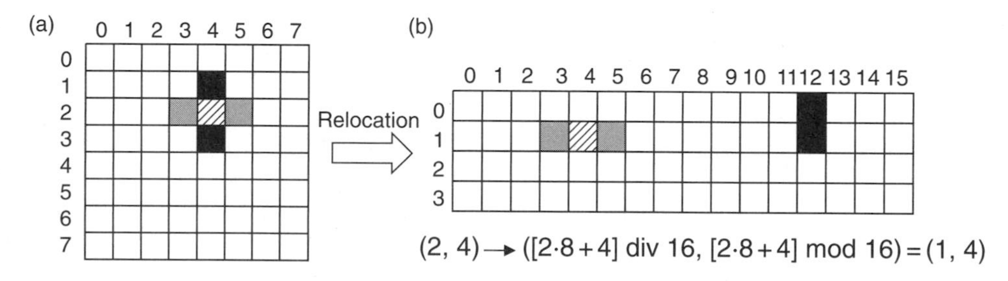

FIGURE 7.2 Relocation of an individual and its neighbors when the grid changes from (a) 8×8 to other (b) with shape 4×16.

exploration to exploitation at a minimum complexity without introducing another new algorithm family in the literature.

Many techniques for managing the exploration/exploitation trade-off are possible. Among them, it is worth to mention the heterogeneous EAs [12,13], in which algorithms with different features run in separate subpopulations and collaborate in order to avoid premature convergence. A different alternative is using *Memetic Algorithms* [14], in which *local search* is combined with the genetic operators in order to promote local exploitation.

Since, a shift between exploration and exploitation can be made by changing the shape of the population (and thus its radius), one can think on changing it during the search. Hence, we theoretically consider the population as a list of length $m \cdot n$, such that, the first row of the $m \times n$ grid is composed by the first n individuals of the list, the second row is made up with the next n individuals, and so on. Therefore, when performing a change from a $m \times n$ grid to a new $m' \times n'$ grid (being $m \cdot n = m' \cdot n'$) the individual placed at position (i, j) will be relocated as follows:

$$(i, j) \rightarrow ([i * n + j] \text{ div } n', \; [i * n + j] \text{ mod } n'). \tag{7.3}$$

We call this redistribution method *contiguous*, because the new grid is filled up by following the order of appearance of the individuals in the list. Figure 7.2 contains an example of a grid change from 8×8 to 4×16. It can be shown, how an individual in position (2, 4) is relocated at (1, 4), changing its neighbors placed at its north and south positions, and keeping close to those placed at its east and west positions. Hence, the change in the grid shape can be seen as an actual migration of individuals among neighborhoods, which will introduce additional diversity into the population for the forthcoming generations. Note that in this chapter we only use static ratios, that is, grid and neighborhood shapes that do not change during the run.

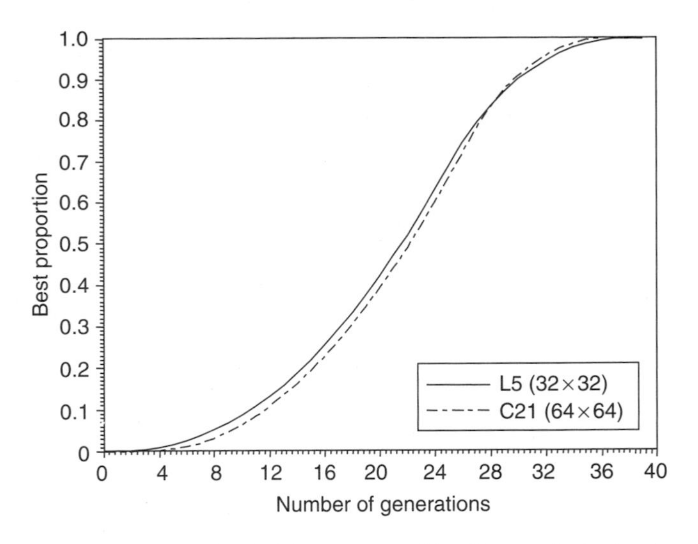

FIGURE 7.3 Growth curves of the best individual for two cEAs with different neighborhood and population shapes, but similar ratio values. The vertical axis represents the proportion of population consisting of best individual as a function of time.

7.4 Selection Pressure, Grid Shape, and Time

Selection pressure is related to the concept of *takeover time*, which is the time taken for a single best individual to colonize the whole population with copies of itself under the effects of selection only [15]. Shorter takeover time mean a more intense selection.

Algorithms with similar ratio show a similar selection pressure, as stated in Reference 5. In Figure 7.3, we plot such a similar behavior for two algorithms with different neighborhood and population radii, but having two similar ratio values. The algorithms plotted are those using a L5 neighborhood with a 32 × 32 population, and a compact21 (C21), neighborhood with a population of 64 × 64 individuals. In the C21 neighborhood a central cell is surrounded by two cells in all directions, including the diagonals, and the four corner cells are cut out.

Hence, it may be very interesting to see how the shape of the grid influences the search of the algorithm. Thus, we study the selection pressure for synchronous cGAs with different grid shapes. In Figure 7.4 we plot the selection pressure for different cGAs using L5 neighborhood and six possible grid shapes for a population of 1024 individuals. Note that the selection pressure induced in synchronous rectangular grids falls under the curve for a synchronous square grid (32 × 32 population), which means that thinner grids favor a more explorative style of search.

If we now keep the shape of the grid constant (say a square), but allow the cell update mode to change, we observe a similar effect on the selection pressure: the global selection pressure induced by the various asynchronous policies fall between the low synchronous limit and the high panmictic bound (see Figure 7.5, [16]). Thus, by varying the update policies it is possible to influence the explorative or exploitative character of the search.

7.5 Experiments in Discrete Optimization

In this section we present the set of discrete problems that we have chosen, to study the behavior of our algorithms. The selected benchmark is representative because it contains many different interesting features found in optimization, such as epistasis, multimodality, deceptiveness, use of constraints, parameter identification, and problem generators. These are important ingredients in any work trying to evaluate algorithmic approaches with the objective of getting reliable results, as stated by Whitley et al. [17].

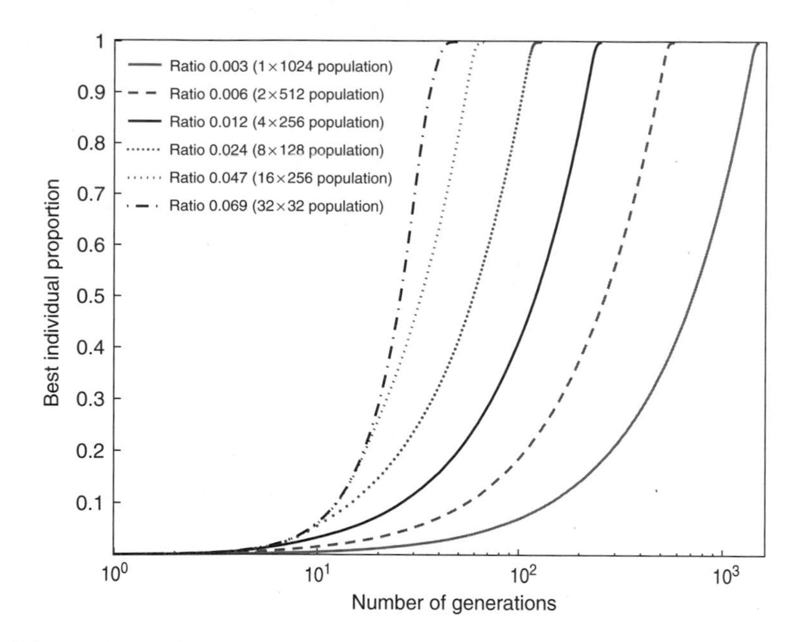

FIGURE 7.4 Takeover times with tournament selection using a L5 neighborhood in a population of 1024 individuals with different grid shapes; mean values over 100 runs. The vertical axis represents the proportion of population consisting of best individual as a function of time. Horizontal axis is in logarithmic scale.

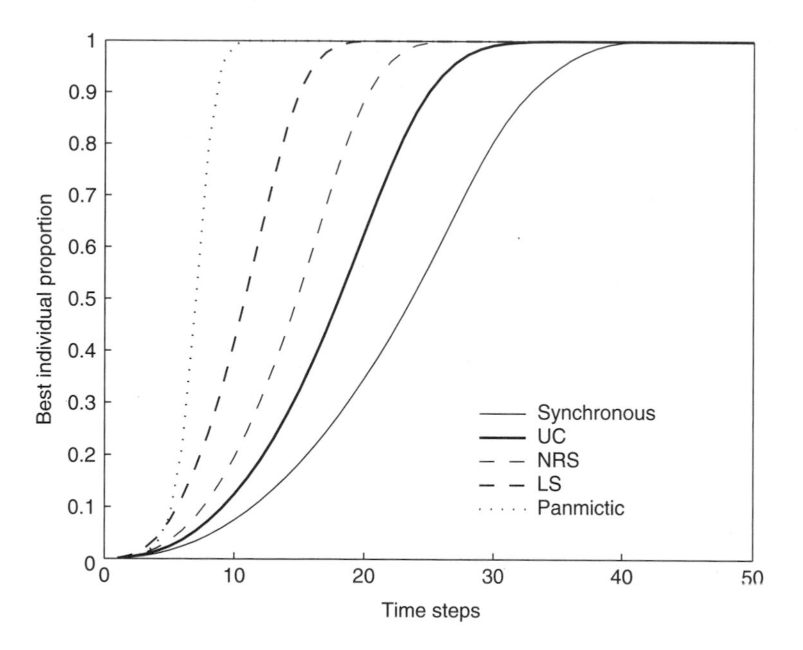

FIGURE 7.5 Takeover times with tournament selection using a L5 neighborhood in a 32×32 grid; mean values over 100 runs. The vertical axis represents the proportion of population consisting of best individual as a function of time.

We experiment with the set of problems studied in Reference 7 which includes the massively multimodal deceptive problem (MMDP), the frequency modulation sounds (FMSs), and the multimodal problem generator P-PEAKS; we then extend this basic three-problem benchmark with error correcting code design (ECC), maximum cut of a graph (MAXCUT), the minimum tardy task problem (MTTP), and the satisfiability problem (SAT).

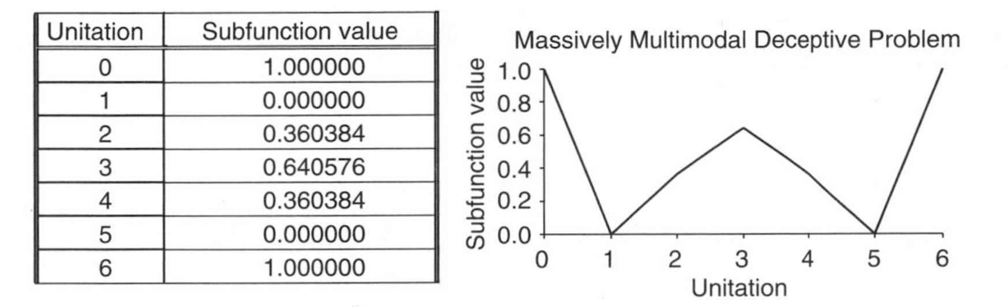

Unitation	Subfunction value
0	1.000000
1	0.000000
2	0.360384
3	0.640576
4	0.360384
5	0.000000
6	1.000000

FIGURE 7.6 Basic deceptive bipolar function (s_i) for MMDP.

The choice of this set of problems is justified by both their difficulty and their application domains (parameter identification, telecommunications, combinatorial optimization, scheduling, etc.). This gives us a high level of confidence in the results obtained, although the evaluation of conclusions is consequently more laborious than with a small test suite.

The problems selected for this benchmark are explained in Sections 7.5.1 to 7.5.7. We include the explanations in the chapter to make it self-contained and to avoid the typical small information lacks that could preclude other researchers from reproducing the results. Finally, in Section 7.5.8 we present and analyze the results.

7.5.1 Massively Multimodal Deceptive Problem (MMDP)

The MMDP is a problem that has been specifically designed to be difficult for an EA [18]. It is made up of k deceptive subproblems (s_i) of six bits each, whose value depends on the number of ones (*unitation*) that a binary string has (see Figure 7.6). It is easy to see (graphic of Figure 7.6) that these subfunctions have two global maxima and a deceptive attractor in the mid-point.

In MMDP, each subproblem s_i contributes to the fitness value according to its *unitation* (Figure 7.6). The global optimum has a value of k and it is attained when every subproblem is composed of zero or six ones. The number of local optima is quite large (22^k), while there are only 2^k global solutions. Therefore, the degree of multimodality is regulated by the k parameter. Here, we use a considerably large instance of $k = 20$ subproblems. The instance we try to maximize for solving the problem is shown in Equation (7.4), and its maximum value is equal to k.

$$f_{\text{MMDP}}(\vec{s}) = \sum_{i=1}^{k} \text{fitness}_{s_i}. \tag{7.4}$$

7.5.2 Frequency Modulation Sounds (FMS)

The FMS problem [19] is defined as, determining the six real parameters, $\vec{x} = (a_1, w_1, a_2, w_2, a_3, w_3)$, of the frequency modulated sound model given in Equation (7.5) for approximating it to the sound wave given in Equation (7.6) (where $\theta = 2 \cdot \pi/100$). The parameters are defined in the range $[-6.4, +6.35]$, and we encode each parameter into a 32 bit substring in the individual.

$$y(t) = a_1 \cdot \sin(\omega_1 \cdot t \cdot \theta + a_2 \cdot \sin(\omega_2 \cdot t \cdot \theta + a_3 \cdot \sin(\omega_3 \cdot t \cdot \theta))), \tag{7.5}$$

$$y_0(t) = 1.0 \cdot \sin(5.0 \cdot t \cdot \theta - 1.5 \cdot \sin(4.8 \cdot t \cdot \theta + 2.0 \cdot \sin(4.9 \cdot t \cdot \theta))). \tag{7.6}$$

The goal is to minimize the sum of square errors given by Equation (7.7). This problem is a highly complex multimodal function having strong epistasis, with optimum value 0. Due to the extreme difficulty of solving this problem with high accuracy, without applying local search or specific operators for continuous optimization (such as gradual GAs [13]), we stop the algorithm when the error falls below 10^{-2}. Hence, our objective for this problem will be to minimize Equation (7.7)

$$f_{\mathrm{FMS}}(\vec{x}) = \sum_{t=0}^{100} (y(t) - y_0(t))^2. \tag{7.7}$$

7.5.3 Multimodal Problem Generator (P-PEAKS)

The P-PEAKS problem [20] is a multimodal problem generator. A problem generator is an easily parameterizable task, which has a tunable degree of epistasis, thus admitting to derive instances with growing difficulty. In addition, using a problem generator removes the opportunity to hand-tune algorithms to a particular problem, therefore allowing a larger fairness when comparing algorithms. With a problem generator, we evaluate our algorithms on a high number of random problem instances, since a different instance is solved each time the algorithm runs, the predictive power of the results for the problem class as a whole is increased.

The idea of P-PEAKS is to generate P random N-bit strings that represent the location of P peaks in the search space. The fitness value of a string is the number of bits the string has in common with the nearest peak in that space, divided by N (as shown in Equation [7.8]). By using a small/large number of peaks we can get weakly/strongly epistatic problems. In this chapter we have used an instance of $P = 100$ peaks, of length $N = 100$ bits each, which represents a medium/high epistasis level [7]. The maximum fitness value for this problem is 1.0.

$$f_{\mathrm{P\text{-}PEAKS}}(\vec{x}) = \frac{1}{N} \max_{1 \le i \le p} \{N - \text{Hamming } D(\vec{x}, \text{Peak}_i)\}. \tag{7.8}$$

7.5.4 Error Correcting Code Design Problem (ECC)

The ECC problem was presented in Reference 21. We will consider a three-tuple (n, M, d), where n is the length of each codeword (number of bits), M is the number of codewords, and d is the minimum Hamming distance between any pair of codewords. Our objective will be to find a code that has a value for d as large as possible (reflecting greater tolerance to noise and errors), given previously fixed values for n and M. The problem we have studied is a simplified version of that in Reference 21. In our case, we search half of the codewords ($M/2$) that will compose the code, and the other half is made up by the complement of the codewords computed by the algorithm.

The fitness function to be maximized is:

$$f_{\mathrm{ECC}} = \frac{1}{\sum_{i=1}^{M} \sum_{j=1, i \ne j}^{M} d_{ij}^{-2}}, \tag{7.9}$$

where d_{ij} represents the Hamming distance between codewords i and j in the code C (made up of M codewords, each of length n). Here, we consider an instance where $M = 24$ and $n = 12$. The search space is of size $\binom{4096}{24}$, which is approximately 10^{87}. The optimum solution for $M = 24$ and $n = 12$ has a fitness value of 0.0674 [22].

7.5.5 Maximum Cut of a Graph (MAXCUT)

The MAXCUT problem looks for a partition of the set of vertices (V) of a weighted graph $G = (V, E)$ into two disjoint subsets V_0 and V_1 so that the sum of the weights of the edges with one endpoint in V_0 and the other in V_1 is maximized. For encoding the problem, we use a binary string, $(x_1, x_2, \ldots, x_n)$, of length n where each digit corresponds to a vertex. If a digit is 1 then the corresponding vertex is in set V_1;

if it is 0 then the corresponding vertex is in set V_0. The function to be maximized [23] is:

$$f_{\mathrm{MAXCUT}}(\vec{x}) = \sum_{i=1}^{n-1} \sum_{j=i+1}^{n} w_{ij} \cdot [x_i \cdot (1 - x_j) + x_j \cdot (1 - x_i)]. \tag{7.10}$$

Note that w_{ij} contributes to the sum only if nodes i and j are in different partitions. While one can generate different random graph instances to test the algorithm, here we have used the case "cut20.09," with 20 vertices and a probability of 0.9 of having an edge between any two randomly chosen vertices. The maximum fitness value for this instance is 56.740064.

7.5.6 Minimum Tardy Task Problem (MTTP)

The MTTP [24] is a task-scheduling problem wherein each task i from the set of tasks $T = \{1, 2, \ldots, n\}$ has a length l_i (the time it takes for its execution), a deadline d_i (before which a task must be scheduled, and its execution completed), and a weight w_i. The weight is a penalty that has to be added to the objective function in the event that the task remains unscheduled. The lengths, weights, and deadlines of the tasks are all positive integers. Scheduling the tasks of a subset S of T is to find the starting time of each task in S, such that at most one task at a time is performed and each task finishes before its deadline.

We characterize a one-to-one scheduling function g defined on a subset of tasks $S \subseteq T : S \mapsto Z^+ \cup \{0\}$, so that for all tasks $i, j \in S$ has the following properties:

1. A task cannot be scheduled before the previous one is completed: $g(i) < g(j) \Rightarrow g(i) + l_i \leq g(j)$.
2. Every task finishes before its deadline: $g(i) + l_i \leq d_i$.

The objective function for this problem is to minimize the sum of the weights of the unscheduled tasks. Therefore, the optimum scheduling minimizes Equation (7.11):

$$f_{\mathrm{MTTP}}(\vec{x}) = \sum_{i \in T-S} w_i. \tag{7.11}$$

The schedule of tasks S can be represented by a vector $\vec{x} = (x_1, x_2, \ldots, x_n)$ containing all the tasks ordered by its deadline. Each $x_i \in \{0, 1\}$, where if $x_i = 1$ then task i is scheduled in S, while if $x_i = 0$ means that task i is not included in S. The fitness function is the inverse of Equation (7.11), as described in Reference 23. We have used in this study an instance called "mttp20," with size 20, and maximum fitness value of 0.02439.

7.5.7 Satisfiability Problem (SAT)

The SAT problem has received much attention by the scientific community since it plays a central role in NP-completeness [25]. The SAT problem was the first that was demonstrated to belong to the NP class of problems.

The SAT problem consists of assigning values to a set of n boolean variables $x = (x_1, x_2, \ldots, x_n)$ such that they satisfy a given set of clauses $c_1(\vec{x}), \ldots, c_m(\vec{x})$, where $c_i(\vec{x})$ is a disjunction of literals, and a literal is a variable or its negation. Hence, we can define SAT as a function $f : B^n \to B, B = \{0, 1\}$ like:

$$f_{\mathrm{SAT}}(\vec{x}) = c_1(\vec{x}) \wedge c_2(\vec{x}) \wedge \cdots \wedge c_m(\vec{x}). \tag{7.12}$$

An instance of SAT, $\vec{x}$, is called satisfiable if $f_{\mathrm{SAT}}(\vec{x}) = 1$, and unsatisfiable otherwise. A k-SAT instance consists of clauses with length k. When $k \geq 3$ the problem is NP-hard [25]. In this chapter we will consider an instance of 3-SAT made up of 430 clauses and 100 variables. This instance belongs to the well-known phase transition of hard SAT instances. The fitness function is a linear function of the number of satisfied clauses. In this, we use the so-called *stepwise adaptation of weights* (SAWs) [26]:

$$f_{\text{SAW}}(\vec{x}) = w_1 \cdot c_1(\vec{x}) + \cdots + w_m \cdot c_m(\vec{x}). \tag{7.13}$$

This function weighs the values of the clauses with $w_i \in B$ in order to give more importance to those clauses which are not yet satisfied by the current best solution. These weights are adjusted dynamically according to $w_i = w_i + 1 - c_i(\vec{x}^*)$, $\vec{x}^*$ being the current fittest individual.

7.5.8 Experimental Analysis

Although a full-length study of the problems presented in Section 7.5.7 is beyond the scope of this work, we present results comparing synchronous and asynchronous cEAs, and also cEAs having different values of the ratio, always with a constant neighborhood shape (L5). Note that it is not our aim to compare cEAs performance with state-of-the-art algorithms and heuristics for combinatorial and numerical optimization. To end this, we should at least tune the parameters and include local search capabilities in the algorithm, which is not the case. Thus, the results only pertain to the relative performance of the different cEA update methods and ratios among themselves.

In this section we present the results of solving some problems using JCell v1.5, our custom simulation program written in Java, with three different static ratios. The configuration of the algorithm for the binary encoded problems is shown in Table 7.1, and the static ratios used are shown in Table 7.2.

The following tables show the results for the problems mentioned preciously: MMDP (Table 7.3), FMS (Table 7.4), P-PEAKS (Table 7.5), ECC (Table 7.6), MAXCUT (Table 7.7), MTTP (Table 7.8), and SAT (Table 7.9). In these tables we report the average of the final best fitness of the algorithm, the average number of evaluations to obtain the optimum value (if obtained), and the hit rate (percentage of successful runs). Therefore, we are analyzing the final distance to the optimum (especially interesting when the optimum is not found), the effort of the algorithm, and the expected efficacy of the algorithm, respectively. In order to get reliable results, we have performed 100 independent runs for any algorithm and for every problem in the test suite.

From the inspection of these tables some conclusions can be clearly drawn. First, the studied asynchronous algorithms tend to need a smaller number of generations than the synchronous ones to locate an optimum, in general. Moreover, the differences among asynchronous and synchronous algorithms

TABLE 7.1 Parameterization Used in the Algorithm for the Binary Encoded Problems

Population size	400 individuals
Selection of parents	Binary tournament + binary tournament
Recombination	DPX, $p_c = 1.0$
Bit mutation	Bit-flip, $p_m = 1/L$ ($10/L$ for FMS)
Individual length	L
Replacement	Rep_if_Better

DPX indicates standard double point crossover.

TABLE 7.2 Studied Ratios

Name (shape of population)	Value of ratio
Square (20 × 20 individuals)	0.11
Rectangular (10 × 40 individuals)	0.075
Narrow (4 × 100 individuals)	0.031

TABLE 7.3 MMDP with a Maximum of 1000 Generations

Algorithm	Average solution (best = 20)	Average generations	Hit rate (%)
Square	19.813	214.2	57
Rectangular	19.824	236.1	58
Narrow	19.842	299.7	61
LS	19.518	343.5	23
FRS	19.601	209.9	31
NRS	19.536	152.9	28
UC	19.615	295.7	36

TABLE 7.4 FMS Problem with a Maximum of 3000 Generations

Algorithm	Average solution (best $\geq$ 100)	Average generations	Hit rate (%)
Square	90.46	437.4	57
Rectangular	85.78	404.3	61
Narrow	80.76	610.9	63
LS	81.44	353.4	58
FRS	73.11	386.2	55
NRS	76.21	401.5	56
UC	83.56	405.2	57

TABLE 7.5 P-PEAKS Problem with a Maximum of 100 Generations

Algorithm	Average solution (best = 1)	Average generations	Hit rate (%)
Square	1.0	51.8	100
Rectangular	1.0	50.4	100
Narrow	1.0	53.9	100
LS	1.0	34.8	100
FRS	1.0	38.4	100
NRS	1.0	38.8	100
UC	1.0	40.1	100

TABLE 7.6 ECC Problem with a Maximum of 500 Generations

Algorithm	Average solution (best = 0.0674)	Average generations	Hit rate (%)
Square	0.0670	93.9	85
Rectangular	0.0671	93.4	88
Narrow	0.0673	104.2	94
LS	0.0672	79.7	89
FRS	0.0672	82.4	90
NRS	0.0672	79.5	89
UC	0.0671	87.3	86

TABLE 7.7 MAXCUT Problem with a Maximum of 100 Generations

Algorithm	Average solution (best = 56.74)	Average generations	Hit rate (%)
Square	56.74	11.3	100
Rectangular	56.74	11.0	100
Narrow	56.74	11.9	100
LS	56.74	9.5	100
FRS	56.74	9.7	100
NRS	56.74	9.6	100
UC	56.74	9.6	100

TABLE 7.8 MTTP with a Maximum of 50 Generations

Algorithm	Average solution (best = 0.02439)	Average generations	Hit rate (%)
Square	0.02439	8.4	100
Rectangular	0.02439	8.3	100
Narrow	0.02439	8.9	100
LS	0.02439	5.9	100
FRS	0.02439	6.2	100
NRS	0.02439	6.3	100
UC	0.02439	6.3	100

TABLE 7.9 SAT Problem with a Maximum of 3000 Generations

Algorithm	Average solution (best = 430.0)	Average generations	Hit rate (%)
Square	429.54	703.1	79
Rectangular	429.67	706.3	84
Narrow	429.61	763.7	81
LS	429.52	463.2	78
FRS	429.67	497.7	85
NRS	429.49	610.5	75
UC	429.50	725.5	76

are statistically significant (with two exceptions), thus indicating that the asynchronous versions perform more efficiently with respect to cEAs with a changing ratio.

On the contrary, synchronous algorithms perform like asynchronous or even better in terms of the percentage of solutions found (hit rate), while the quality of solutions found by the algorithms does not always have significant differences (the exceptions are probably due to the difference on the hit rate).

Another interesting result is that, we can define two classes of problems: those solved by any method to optimality (100% hit rate) and those in which no 100% rate is achieved at all. Problems seem to be amenable for cEAs directly, or to need some (yet unstudied) help, for example, by including local search.

In order to summarize the large set of results and get some useful conclusion, we present a ranking with the best algorithms by following three different metrics: average best final solution, average number of generations on success, and hit rate. Table 7.10 shows the three mentioned rankings. These rankings have

TABLE 7.10 Ranking of the Algorithms with Discrete Problems

Average solution		Average generations		Hit rate (%)	
1 Narrow	10	1 LS	14	1 Narrow	6
1 Rectangular	10	2 NRS	16	2 Rectangular	10
3 Square	14	3 FRS	18	3 FRS	14
4 FRS	15	4 UC	30	4 LS	15
5 LS	18	5 Rectangular	33	5 Square	17
5 UC	18	6 Square	37	6 UC	19
7 NRS	21	7 Narrow	48	7 NRS	21

been computed by adding the position (from better to worse: 1, 2, 3, . . .) that algorithms are allocated for the previous results presented from Table 7.3 to Table 7.9, according to the three criteria.

As we would expect after the previous comments, according to the average final best fitness and hit rate criteria, synchronous algorithms with any of the three studied ratios are, in general, more accurate than all the asynchronous ones for our test problems, with a special leading position for narrow population grids. On the other hand, asynchronous versions clearly outperform any of the synchronous algorithms, in terms of the average number of generations (efficiency), with a trend toward LS as being the best ranked flavor of cEA for our discrete test suite.

7.6 Experiments in Continuous Optimization

We will extend the work of the previous sections by testing all the algorithms with some continuous functions in order to get a more extensive study. This study may be interesting for analyzing the behavior of the algorithms in continuous optimization, in contrast to the study performed on discrete optimization. The functions selected for the study are the three typical multimodal numerical benchmarks: Rastrigin's (RASTR), Ackley's (ACKL), and fractal (FRAC) function. These three problems are real-coded in the algorithms, while previously the standard binary-coded individuals were used (Section 7.5). That is the cause for our special interest on experimenting with a more traditional global optimization. The codification employed for these three problems has been made by following Michalewicz's implementation [27].

7.6.1 Rastrigin's Function

The generalized RASTR function is a sinusoidally modulated function with a global minimum of zero at the origin. It is a typical example of nonlinear multimodal function. It was first proposed by Rastrigin as a two-dimensional function and has been generalized by Mühlenbein and Schlierkamp-Voosen [28]. This function is a fairly difficult problem due to its large search space and large number of local minima, although the function is separable and the local minima are symmetrical (Equation [7.14]).

$$f_{\text{RASTR}}(\mathbf{x}) = nA + \sum_{i=1}^{n} x_i^2 - A\cos(\omega x_i). \tag{7.14}$$

The constants are given by $A = 10$ and $\omega = 2\pi$. The domain of variables $x_i, i = 1, \ldots, n$, is $-5.12 \leq x_i \leq 5.12$ and $n = 10$. The function has a global minimum at the point $f(\mathbf{0}) = 0$.

7.6.2 Ackley's Function

Ackley's function is a multimodal test function obtained by cosine modulation of an exponential function. Originally proposed by Ackley [29] as a two-dimensional function, it has been generalized by

Bäck et al. [30] (see Equation [7.15]). Contrary to Rastrigin's, Ackley's function is not separable, even though it shows a regular arrangement of the local optima.

$$f_{\mathrm{ACKL}}(\mathbf{x}) = -a \exp\left[-b\left(\frac{1}{n}\sum_{i=1}^{n} x_i^2\right)^{1/2}\right] - \exp\left(\frac{1}{n}\sum_{i=1}^{n}\cos(cx_i)\right) + a + e. \quad (7.15)$$

The constants are $a = 20$, $b = 0.2$, and $c = 2\pi$. The variables x_i, $i = 1, \ldots, n$ are in the domain $-32.768 \leq x_i \leq 32.768$. This function has a global minimum at the point $f(\mathbf{0}) = 0$.

7.6.3 Fractal Function

This function has been taken from Reference 31, were its construction, as well as motivations for introducing it as a test problem, are given. Indeed, the function allows the degree of ruggedness to be controlled, and it is likely to capture features of real-world noisy objective functions.

$$f_{\mathrm{FRAC}}(\mathbf{x}) = \sum_{i=1}^{n}(C'(x_i) + x_i^2 - 1), \quad (7.16)$$

where

$$C'(z) = \begin{cases} \dfrac{C(z)}{C(1)|z|^{2-D}} & \text{if } z \neq 0, \\ 1 & \text{if } z = 0, \end{cases}$$

and

$$C(z) = \sum_{j=-\infty}^{\infty} \frac{1 - \cos(b^j z)}{b^{(2-D)j}}.$$

For the runs, we have chosen the constants $D = 1.85$ and $b = 1.5$. The 20 variables x_i ($x_i = 1, \ldots, 20$) vary in the range $[-5, 5]$. The infinite sum in the function $C(z)$ is practically calculated starting with $j = 0$ and alternating the signs of the j values. The sum is stopped when the relative difference between its previous and present value is lower than 10^{-8}, or when $j = 100$ is reached.

7.6.4 Experimental Analysis

In this section we will study the results of our experiments with the proposed continuous problems, as we did in Section 7.5.8 for the discrete case. We maintain, in this case, the ratios used for the synchronous algorithms with respect to those studied in Section 7.5, as well as the asynchronous update criteria. On the other hand, we needed a specific configuration for the genetic operators and theirs probabilities in the case of real-encoded problems. This codification is given in detail in Table 7.11.

The following tables present the results of our experiments with RASTR (Table 7.12), ACKL (Table 7.13), and FRAC (Table 7.14) problems. Like in the case of discrete problems, these tables contain values for the average of the final best fitness, the average generations needed for finding it, and the hit rate. These three values are calculated over 100 independent runs. For the three real-coded problems a run is stopped successfully as soon as an individual is found with fitness within 0.1 from the optimum.

The results obtained with continuous problems are not as clear as in the discrete case. Regarding the average number of generations needed to find an optimal solution, asynchronous algorithms do not always perform a lesser number of generations with respect to synchronous ones; for example, ACKL and FRAC problems. Differences among synchronous and asynchronous algorithms are usually significant.

TABLE 7.11 Parameterization Used in the Algorithm for the Real-Encoded Problems

Population size	400 individuals
Selection of parents	Binary tournament + binary tournament
Recombination	AX, $p_c = 1.0$
Bit mutation	Uniform, $p_m = 1/2L$
Individual length	L
Replacement	Rep_if_Better

AX stands for standard arithmetic crossover.

TABLE 7.12 RASTR Problem with a Maximum of 700 Generations

Algorithm	Average solution (best $\leq$ 0.1)	Average generations	Hit rate (%)
Square	0.0900	323.8	100
Rectangular	0.0883	309.8	100
Narrow	0.0855	354.2	100
LS	0.0899	280.9	100
FRS	0.0900	289.6	100
NRS	0.0906	292.2	100
UC	0.0892	292.4	100

TABLE 7.13 ACKL Problem with a Maximum of 500 Generations

Algorithm	Average solution (best $\leq$ 0.1)	Average generations	Hit rate (%)
Square	0.0999	321.7	78
Rectangular	0.0994	293.1	73
Narrow	0.1037	271.9	65
LS	0.0932	302.0	84
FRS	0.0935	350.6	92
NRS	0.0956	335.5	87
UC	0.0968	335.0	85

TABLE 7.14 FRAC Problem with a Maximum of 100 Generations

Algorithm	Average solution (best $\leq$ 0.1)	Average generations	Hit rate (%)
Square	0.0224	75.2	94
Rectangular	0.0359	62.8	78
Narrow	0.1648	14.6	16
LS	0.0168	69.7	98
FRS	0.0151	71.5	100
NRS	0.0163	73.6	98
UC	0.0138	72.8	96

TABLE 7.15 Ranking of the Algorithms with Continuous Problems

Average solution		Average generations		Hit rate (%)	
1 UC	8	1 LS	7	1 FRS	3
2 LS	9	2 Narrow	9	5 NRS	5
2 FRS	9	2 Rectangular	9	4 LS	7
4 NRS	13	4 FRS	13	6 UC	8
4 Rectangular	13	5 UC	14	7 Square	11
6 Narrow	15	6 NRS	15	3 Rectangular	13
7 Square	16	7 Square	17	2 Narrow	15

This result tells us about a larger efficiency of changing ratio cGAs, which contrast with the findings for discrete problems. On the other hand, contrary to that observed in the case of discrete problems, the success rates of asynchronous algorithms are greater than those of synchronous, in general. Contrary to the results of Section 7.5.8, where either all the algorithms get a 100% hit rate or none finds the solution in every run for any problem, the FRS cEA is the unique algorithm which is able to find the solution in all the executions for FRAC problem.

In order to summarize these results, and following the structure of Section 7.5.8, we present in Table 7.15 a ranking with the best algorithms in all the problems by means of the average solution found, the number of generations needed to find an optimal solution, and the success rate. It can be seen in this table that there exists a trend of asynchronous algorithms that perform better than synchronous ones in terms of the average solution found and the success rate, while synchronous changing-ratio algorithms seem to be more efficient than most asynchronous ones, in general (square and LS are the exceptions).

7.7 Conclusions

In the first part of this chapter we have described several asynchronous update policies for the population of a cEA, followed by some ratio policies, all of them inducing a different kind of search in the cEA. One can tune the selection intensity of a cEA by choosing the update policy and grid ratio without having to deal with additional numerical parameter settings. This is a clear advantage of the algorithms that allows users to utilize existing knowledge instead of inventing a new class of heuristic.

Our conclusion is that cEAs can be easily induced to promote exploration or exploitation by simply changing the update policy or the ratio of the population. This opens new research lines to decide efficient manners of shifting from one given policy/ratio to another in order for the optimum to be reached with a smaller effort when compared with the basic cEA or other types of EAs.

In a later part of the chapter we have applied our extended cEAs to a set of both discrete and continuous test problems. Although our goal has not been to compete with state-of-the-art specialized heuristics, the results clearly show that cEAs are very efficient optimization techniques, that could be further improved by being hybridized with local search techniques [32]. The results on the test problems largely confirm, with some small exceptions, that the solving abilities using the various update/ratio modes are directly linked to their induced selection pressures, showing that exploitation plays an important role. It is clear that the role of exploration might be more important on even harder problem instances, but this aspect can be addressed in our algorithms by using more explorative settings, as well as by using different cEA strategies at different times during the search, dynamically [33].

We conclude that, with respect to discrete problems, asynchronous algorithms are more efficient than synchronous; with statistically significant values for most problems. On the other hand, if we pay attention to the success (hit) rate, it can be concluded that the synchronous policies for the different evaluated ratios outperform the asynchronous algorithms: slightly in terms of the average final fitness, and clearly in terms of probability of finding a solution (i.e., frequency of optimum location).

On the contrary, if we pay attention to the experiments on continuous problems we can get different (somewhat complementary) conclusions. Asynchronous algorithms outperform synchronous ones in the cases of average solutions found and hit rate (significant differences in many cases), while in the average number of generations synchronous algorithms are, in general, more efficient than asynchronous ones.

As future research, we plan to address a single problem of large difficulty and to characterize selection intensity analytically for all the models.

Acknowledgments

This work has been partially funded by the Ministry of Science and Technology (MCYT) and Regional Development European Fund (FEDER) under contract TIC2002-04498-C05-02 (the TRACER project) `http://tracer.lcc.uma.es`. We thank J. Kaempf for performing a part of the computer simulations for the real-valued problems. M. Giacobini gratefully acknowledges financial support by the Fonds National Suisse pour la Recherche Scientifique under contract 200021-103732/1.

References

[1] M. Gorges-Schleuter. ASPARAGOS an asynchronous parallel genetic optimisation strategy. In J.D. Schaffer, Ed., *Proceedings of the Third International Conference on Genetic Algorithms*, Morgan Kaufmann, San Francisco, CA, 1989, pp. 422–427.

[2] B. Manderick and P. Spiessens. Fine-grained parallel genetic algorithms. In J.D. Schaffer, Ed., *Proceedings of the Third International Conference on Genetic Algorithms*, Morgan Kaufmann, San Francisco, CA, 1989, pp. 428–433.

[3] E. Alba and M. Tomassini. Parallelism and evolutionary algorithms. *IEEE Transactions on Evolutionary Computation*, 6: 443–462, 2002.

[4] M. Gorges-Schleuter. An analysis of local selection in evolution strategies. In *Genetic and Evolutionary Conference, GECCO99*, Vol. 1, Morgan Kaufmann, San Francisco, CA, 1999, pp. 847–854.

[5] J. Sarma and K.A. De Jong. An analysis of local selection algorithms in a spatially structured evolutionary algorithm. In T. Bäck, Ed., *Proceedings of the Seventh International Conference on Genetic Algorithms*, Morgan Kaufmann, San Francisco, CA, 1997, pp. 181–186.

[6] J. Sarma and K.A. De Jong. An analysis of the effect of the neighborhood size and shape on local selection algorithms. In H.M. Voigt, W. Ebeling, I. Rechenberg, and H.P. Schwefel, Eds., *Parallel Problem Solving from Nature (PPSN IV)*, Vol. 1141 of *Lecture Notes in Computer Science*, Springer-Verlag, Heidelberg, 1996, pp. 236–244.

[7] E. Alba and J.M. Troya. Cellular evolutionary algorithms: Evaluating the influence of ratio. In M. Schoenauer et al., Eds., *Parallel Problem Solving from Nature, PPSN VI*, Vol. 1917 of *Lecture Notes in Computer Science*, Springer-Verlag, Heidelberg, 2000, pp. 29–38.

[8] M. Tomassini. The parallel genetic cellular automata: Application to global function optimization. In R.F. Albrecht, C.R. Reeves, and N.C. Steele, Eds., *Proceedings of the International Conference on Artificial Neural Networks and Genetic Algorithms*, Springer-Verlag, Heidelberg, 1993, pp. 385–391.

[9] D. Whitley. Cellular genetic algorithms. In S. Forrest, Ed., *Proceedings of the Fifth International Conference on Genetic Algorithms*, Morgan Kaufmann Publishers, San Mateo, CA, 1993, p. 658.

[10] B. Schönfisch and A. de Roos. Synchronous and asynchronous updating in cellular automata. *BioSystems*, 51: 123–143, 1999.

[11] M. Sipper, M. Tomassini, and M.S. Capcarrere. Evolving asynchronous and scalable non-uniform cellular automata. In *Proceedings of International Conference on Artificial Neural Networks and Genetic Algorithms (ICANNGA97)*, Springer-Verlag KG, Vienna, 1998, pp. 67–71.

[12] P. Adamidis and V. Petridis. Co-operating populations with different evolution behaviours. In *Proceedings of the Third IEEE Conference on Evolutionary Computation*, IEEE Press, Washington, 1996, pp. 188–191.

[13] F. Herrera and M. Lozano. Gradual distributed real-coded genetic algorithms. *IEEE-EC*, 4: 43–62, 2000.

[14] P. Moscato. Memetic algorithms. In P.M. Pardalos and M.G.C. Resende, Eds., *Handbook of Applied Optimization*, Oxford University Press, Oxford, 2000, pp. 157–167.

[15] D.E. Goldberg and K. Deb. A comparative analysis of selection schemes used in genetic algorithms. In G.J.E. Rawlins, Ed., *Foundations of Genetic Algorithms*, Morgan Kaufmann, San Francisco, CA, 1991, pp. 69–93.

[16] M. Giacobini, E. Alba, and M. Tomassini. Selection intensity in asynchronous cellular evolutionary algorithms. In E. Cantú-Paz et al., Eds., *Proceedings of the Genetic and Evolutionary Computation Conference GECCO'03*, Springer-Verlag, Berlin, 2003, pp. 955–966.

[17] D. Whitley, S. Rana, J. Dzubera, and K.E. Mathias. Evaluating evolutionary algorithms. *Artificial Intelligence*, 85:245–276, 1997.

[18] D.E. Goldberg, K. Deb, and J. Horn. Massively multimodality, deception and genetic algorithms. In R. Männer and B. Manderick, Eds., *Parallel Problem Solving from Nature, PPSN II*, North-Holland, 1992, pp. 37–46.

[19] S. Tsutsui and Y. Fujimoto. Forking genetic algorithm with blocking and shrinking modes. In S. Forrest, Ed., *Proceedings of the fifth International Conference of Genetic Algorithms*, Morgan Kaufmann, San Mateo, CA, 1993, pp. 206–213.

[20] K.A. De Jong, M.A. Potter, and W.M. Spears. Using problem generators to explore the effects of epistasis. In T. Bäck, Ed., *Proceedings of the Seventh ICGA*, Morgan Kaufmann, San Francisco, CA, 1997, pp. 338–345.

[21] F.J. MacWilliams and N.J.A. Sloane. *The Theory of Error-Correcting Codes.* North-Holland, Amsterdam, 1977.

[22] H. Chen, N.S. Flann, and D.W. Watson. Parallel genetic simulated annealing: A massively parallel SIMD algorithm. *IEEE Transactions on Parallel and Distributed Systems*, 9: 126–136, 1998.

[23] S. Khuri, T. Bäck, and J. Heitkötter. An evolutionary approach to combinatorial optimization problems. In *Proceedings of the 22nd ACM Computer Science Conference*, ACM Press, Phoenix, AZ, 1994, pp. 66–73.

[24] D.R. Stinson. *An Introduction to the Design and Analysis of Algorithms*, 2nd ed. (1987). The Charles Babbage Research Center, Winnipeg, Manitoba, Canada, 1985.

[25] M. Garey and D. Johnson. *Computers and Intractability: A Guide to the Theory of NP-Completeness.* Freeman, San Franciso, CA, 1979.

[26] A. Eiben and J. van der Hauw. Solving 3-SAT with adaptive genetic algorithms. In *Proceedings of the Fourth IEEE Conference on Evolutionary Computation*, IEEE, Piscataway, NJ, 1997, pp. 81–86.

[27] Z. Michalewicz. *Genetic Algorithms + Data Structures = Evolution Programs*, 3rd ed. Springer-Verlag, Heidelberg, 1996.

[28] H. Mühlenbein and D. Schlierkamp-Voosen. The science of breeding and its application to the breeder genetic algorithm (bga). *Evolutionary Computation*, 1: 335–360, 1993.

[29] D.H. Ackley. *A Connectionist Machine for Genetic Hillclimbing.* Kluwer, Boston, MA, 1987.

[30] T. Bäck, G. Rudolf, and H.-P. Schwefel. Evolutionary programming and evolution strategies: Similarities and differences. In D.B. Fogel and W. Atmar, Eds., *Proceedings of the Second Conference on Evolutionary Programming*, Evolutionary Programming Society, La Jolla, CA, 1993, pp. 11–22.

[31] T. Bäck. *Evolutionary Algorithms in Theory and Practice: Evolution Strategies, Evolutionary Programming, Genetic Algorithms.* Oxford University Press, New York, 1996.

[32] G. Folino, C. Pizzuti, and G. Spezzano. Parallel hybrid method for SAT that couples genetic algorithms and local search. *IEEE Transactions on Evolutionary Computation*, 5: 323–334, 2001.

[33] E. Alba and B. Dorronsoro. The exploration/exploitation tradeoff in dynamic cellular evolutionary algorithms. *IEEE Transactions on Evolutionary Computation*, 9(2): 126–142, 2005.

8

Optimization via Gene Expression Algorithms

Forbes Burkowski

8.1 Introduction

The extreme diversity of life forms on our planet is a testament to the many ways in which millions of different species have adapted to the huge number of biological niches that are available ranging, as they do, over extremes in temperature and pressure in air, land, and water. Viewed in an abstracted formulation, we may consider evolutionary adaptation of species within a biological niche to be a robust search and optimization mechanism (Fogel, 1994), the phenotype changing over time so that the species survives in its competitive environment. Under the Darwinian view, evolutionary processes are involved with reproduction, mutation, competition, and selection. The bioinspired strategies of genetic algorithms and evolutionary optimization, in general, strive to capture these biologically robust search mechanisms by formulating mathematical models that search function domains using algorithmic procedures that mimic some of the evolutionary processes of the cell, for example, mutation, and crossover. Before discussing these procedures, let us go over a superficial but quick review of the main biological events in the cell.

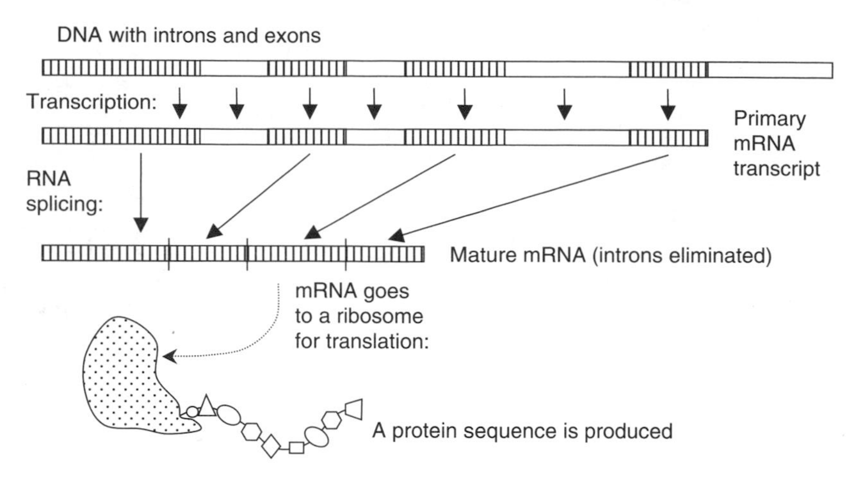

FIGURE 8.1 The central dogma.

8.2 The Central Dogma of Biology

During the 1950s Watson and Crick proposed the central dogma of biology. The basic idea was that information within a cell flowed from DNA to RNA to protein. Protein was not able to influence DNA. We now know that this is too simple. For example, a virus such as HIV can change the DNA of a host cell to achieve the replication of the virus.

Nonetheless, as a generalization, the central dogma is a reasonable assumption. Figure 8.1 illustrates the main events. The interested reader may consult almost any undergraduate text on genetics or biomolecular chemistry to get the details, for example, Lewin (1997). A single DNA molecule is comprised of a sequence of nucleotides: Adenine, Guanine, Cytosine, and Thymine typically denoted as A, G, C, and T. The sequence can be subdivided into intron and exon subsequences. Exons contain nucleotides that specify the content of protein, while introns (sometimes called "junk" DNA) contain nucleotide subsequences with a significance that is largely unknown. During the operation of transcription, some particular sequence of introns and exons making up a gene will be transcribed to form a primary mRNA transcript, which essentially contains the same code sequence as the DNA except that thymine is replaced with a different nucleotide called Uracil, that is, U replaces T in the reproduced string. Later a splicing event removes all intron nucleotides and the mature mRNA is ready for translation by a ribosome, a very large assembly of molecules that takes the available amino acids in the cell and links them into long chains that make up a protein. The ordering of the amino acids in the protein sequence is dictated by three letter codons originally within the exons of the DNA. The now somewhat outmoded "one gene, one protein" hypothesis states that, a single gene in the DNA is responsible for the specification of one type of protein in the cell.

As seen in this setting, mutation is an event that typically introduces a single nucleotide substitution in the DNA of a gamete cell that will participate in reproduction (sperm or egg). In crossover, there is an interchange of material between two different molecules of DNA, each containing the same sequence of genes but typically with slightly different nucleotide sequences within these genes.

Of course, nature achieves its complexity using strategies that go well beyond simple mutation and crossover of genes. For example, while codons in the DNA will prescribe the sequence of amino acids in protein, other mechanisms at the molecular level work on some proteins to modify them. This posttranslational modification and alternate splicing strategies that leave out or reorder exons allow the cell to have many more different proteins than the original "one gene, one protein" hypothesis indicated. Another example of increased complexity is the huge evolutionary leap that took place when some early primordial cell became endowed with mitochondria, the energy producing organelles is now present in all cells. This event is considered to have started as a symbiotic relationship between two cellular structures, one enveloping the other. In support of this view, we note that mitochondria have their own DNA, which does

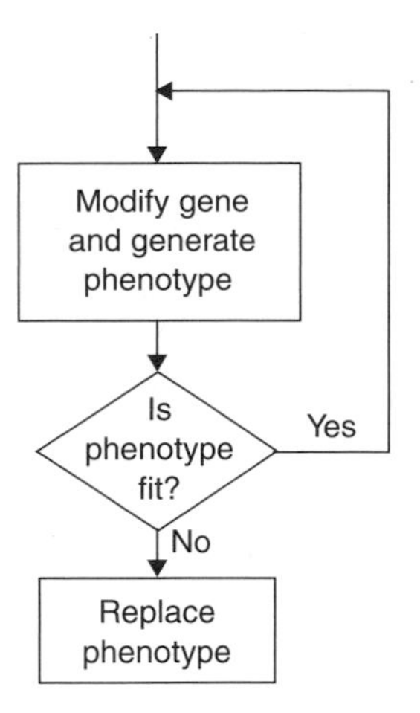

FIGURE 8.2 The reproductive cycle.

not exactly follow the same encoding as regular DNA. This symbiotic combination illustrates a mechanism that achieves a remarkable advance in complexity without any reliance on mutation and crossover per se.

8.3 Borrowing from Nature

Ever since the seminal work by John Holland in 1975 (Holland, reprinted 1992), the evolutionary computation research community has experimented with a wide range of mathematical models for function optimization. Since the full mechanism of biological evolution is daunting in its formidable complexity, these mathematical models must necessarily pick some limited feature of the entire process and this source of "bioinspiration" becomes the central theme of the model. For example, the binary representation of a gene is formatted to represent a feasible solution of some optimization problem and a fitness function is applied to the gene to assess how well the feasible solution meets the requirements of the problem. If the gene has produced a high fitness evaluation then that gene stays in a population of such genes, otherwise it is replaced. Figure 8.2 provides a summary of the basic idea. Creation of new genes and selection strategies to chose the surviving genes are basic to the procedures involved with genetic algorithms. As optimization strategies, genetic algorithms take inspiration from the evolutionary strategies of the cell and incorporate concepts and mathematical objects that show some relationship with biological genes, mutation, crossover, etc. We will assume that the reader is familiar with these ideas.

Researchers have borrowed many other features from evolutionary biology. For example, research developing genetic algorithms inspired by gene duplication (Sawai and Adachi, 1999) and gene signaling (Goldberg et al., 1989). Although not discussed in Section 8.3, most genes occur in a cell with duplication. This is known as diploidy and the corresponding mathematical exploration of this within a genetic algorithm framework has been studied (Bagley, 1967; Smith, 1988). Recalling that introns do not directly specify amino acid sequences, the presence of noncoding areas in a binary gene for genetic algorithms has also been considered (Wu and Lindsay, 1995).

The main content of this chapter presents yet another mathematical model that is based on a biological feature, namely that of gene expression.

8.4 The Gene Expression Algorithm

The many thousands of proteins produced by living cells collectively make up the phenotype of the individual. The external environment will subject this individual to competitive pressures that effectively serve to determine whether that individual will reproduce or be eliminated. In general, genes are *expressed*

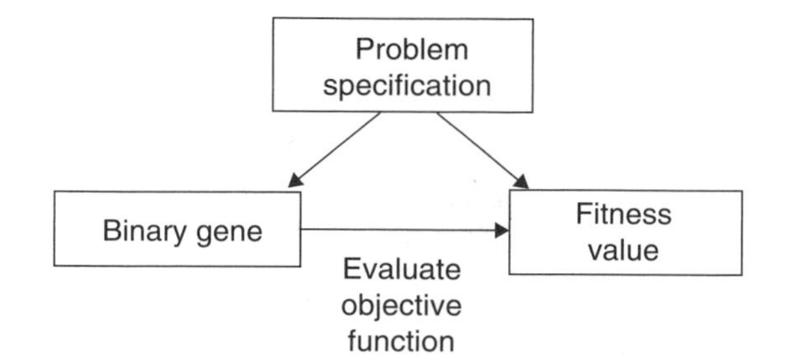

FIGURE 8.3 Fitness evaluation from a binary gene.

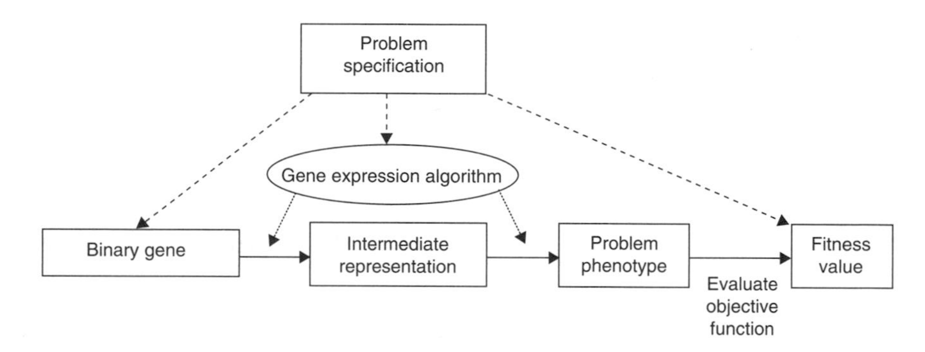

FIGURE 8.4 Fitness evaluation from a phenotype.

by being chosen for transcription leading ultimately to some particular protein with possible modification. The important issue is that, fitness is essentially due to the capabilities of protein making up the phenotype. More significantly, fitness is not an *immediate* property of the genes that have been expressed.

In the traditional genetic algorithm (Goldberg, 1989: 60), the phenotype of an individual is the decoded parameter or parameters of the binary gene. We use Figure 8.3 to describe a somewhat simplified version of the algorithm, showing how the problem specification will dictate the format of a binary gene. A subsequent calculation using an objective function evaluates the fitness of that gene. In this scenario, the fitness function is applied directly to the binary gene. As noted by Mitchell (1996) p. 6:, there is often no notion of a phenotype in the context of genetic algorithms, although more recently, various works have presented various models to incorporate them.

When a genetic algorithm involves a *gene expression* strategy, the fitness function is applied to a computed phenotype that is more than a simple alternate representation of the binary gene. Typically, the computation of this phenotype is fairly complicated and is achieved by an algorithm that is called a gene expression algorithm. As shown in Figure 8.4, the gene expression strategy utilizes an intermediate stage that produces a final phenotype, which is then subjected to fitness evaluation. To make the biological inspiration more obvious, we simplify the transformation of cellular information as follows:

$$\text{DNA gene} \rightarrow \text{mRNA} \rightarrow \text{protein for the phenotype.}$$

In our computational environment, we will have various binary representations that run in parallel with this natural model:

$$\text{gene} \rightarrow \text{intermediate representation} \rightarrow \text{phenotype.}$$

The intermediate representation may be absent or may exist as a sequence of intermediate transformations depending on the requirements of the algorithm. The most significant component of our strategy is that, we use a *gene expression algorithm* to produce a binary phenotype representation that is ready for fitness evaluation. The phenotype meets various feasibility requirements and, in its construction, the gene expression algorithm utilizes various parameters that eventually take on values that are determined through the use of a genetic algorithm. Another description of the gene expression strategy would be the use of optimization heuristics that are dependent on parameters derived via evolutionary computation.

It is important to understand how an algorithm designer might approach a problem with the intention of utilizing a gene expression algorithm. In the traditional deployment of a genetic algorithm, the algorithm designer would have a binary string acting as a gene and then would exercise the genetic algorithm with an appropriate fitness function applied to such a gene. With a gene expression algorithm, some appropriate data structure (usually a binary string) acts as a phenotype structure which provides an input for the fitness function. The "cleverness" of the algorithm designer is then challenged by two goals:

1. Specify the format of a binary gene (suitable for mutation and crossover operations) that can then be transformed into the phenotype structure.
2. Design the gene expression algorithm that does this transformation.

8.5 Advantages in Using Gene Expression

As will be demonstrated in the sections to follow, using a gene expression algorithm has significant advantages:

- *Separation of Design Concerns*: The binary representation of a feasible solution for an optimization problem may be unsuitable for genetic algorithm operators, such as crossover. Using the gene expression strategy, the parameters of the optimization heuristics can be encoded in a binary representation that is completely different from the binary representation of the phenotype to be used for a fitness evaluation. This facilitates the design of representations that are appropriate for their utilization: the gene for crossover, the phenotype for fitness evaluation.
- *Generation of Feasible Phenotypes*: Frequently, the designer of a genetic algorithm must take special pains to ensure that a crossover operation generates a child gene that is a valid member of the search space. Since a gene expression algorithm is responsible for the generation of a phenotype, we can demand that the constraint of feasibility be a part of the gene expression algorithm, the onus of this responsibility being taken away from the reproductive subsystem that must build a valid gene.
- *Using Optimization Heuristics to Extend Genetic Algorithms*: Researchers studying optimization and search strategies for hard NP-complete problems have developed sophisticated "problem specific" heuristics that, in typical applications, generate very reasonable approximations. Gene expression algorithms can often take advantage of these techniques thus incorporating solution strategies that are not typically a part of the traditional approach provided by a simple genetic algorithm.
- *Using Genetic Algorithms to Extend Optimization Heuristics*: We may gain a mutual benefit by combining problem specific heuristics with a genetic algorithm approach. Often, the specification of the heuristic is highly deterministic but will perform better when a stochastic element is provided, in this case, through the use of an evolutionary computation environment.

8.6 Prior Research in Gene Expression Algorithms

In addition to the work of Julstrom that is mentioned later, we discuss two other research efforts dealing with gene expression algorithms. Ferreira (2001) introduces a methodology called gene expression programming. The main idea is that a binary gene is transformed into an expression tree. The expression tree (ET) is an infix expression for a program that may be evaluated when actual values are assigned to the variables located in the leaves of the binary ET. During a training phase, a fitness function compares the evaluation done by the tree with prescribed target values. The expression algorithm determines how the ET is constructed from any valid gene made up of variable names and arithmetic operators. The strategy is applied to problems such as symbolic regression, planning, Boolean concept learning, and cellular automata rules.

Another interesting set of gene expression papers has been written by Hillol Kargupta. He uses the idea of *genetic code-like transformations.* Kargupta (1999) emphasizes the importance of learning functions from data. This would have application in areas such as inductive learning, statistics, data mining, and optimization. In this approach, a function is learnt or induced by generating the coefficients in its Fourier expansion. In general, a function (e.g., a fitness function) defined on an n-bit binary string, requires $O(2^n)$ Fourier coefficients. Hence, this inductive procedure is computationally inefficient. However, it is sometimes possible to find very reasonable approximations of a function if the set of Fourier coefficients has a smaller subset of "large" coefficients that are polynomial in number. In such a case, it must be demonstrated that the *power spectrum*, defined by the set of coefficients, has a high concentration of energy in a few coefficients ($O(n^k)$ in number) with an exponential drop-off in the magnitude of all other nonzero coefficients. This is equivalent to the demand that, the "small" coefficients ($O(2^n)$ in number) be exponentially small so that, cumulatively, they do not amount to any significant sum. The contribution of this chapter to gene expression resides in the observation that for some functions with $O(2^n)$ large coefficients, it is possible to use a "genetic code-like" transformation of the data that will transform the function into a new function that has only $O(n^k)$ large coefficients. The strategy is considered to involve a gene expression algorithm because we assume that the given function is defined on a phenotype space and the gene expression algorithm is used to derive a possibly degenerate transformation that establishes a mapping between a phenotype string and its gene image, which typically has a length that is a small integer multiple of the phenotype string length. This is similar to a natural system in which the coding part of the gene has a string length that is three times the length of the amino acid sequence that it encodes. After such a transformation of data, the fitness function can be seen to have the gene space as its domain and if the genetic code-like transformation is defined in the appropriate manner, then the fitness function will have $O(n^k)$ large Fourier coefficients. These are two other papers (Kargupta and Park, 2000; Kargupta and Ghosh, 2002) that deal with there related issues.

8.7 Gene Expression for the TSP

In the rest of this chapter we review the work presented by Burkowski (2003) which demonstrates the application of the gene expression algorithm to the traveling salesperson problem (TSP). We will assume that the reader is familiar with the basic notion of the TSP. We will also limit our attention to the Euclidean case of finding a minimum length tour through n cities assuming both, the triangle inequality and the symmetry of the intercity distances.

It should be stressed that we are not advocating the use of this algorithm as a serious performance contender in the solution of a TSP. There are currently many highly successful optimization strategies to derive near-optimal solutions of the TSP for problem instances involving several hundred or even thousands of cities. When compared with *any* algorithm, that is a variant of a genetic algorithm, they are typically much faster and produce better results. Our choice of the TSP as a problem to be solved by a gene expression algorithm is mainly being done to illustrate the techniques employed by this methodology with the goal of showing that it can meet the demands of a difficult problem.

Numerous papers have been written in this field of research and a fine bibliography can be found in Reinelt (1994). Reinelt provides the reader with a useful categorization of TSP heuristics: *Construction heuristics* are used to rapidly derive a somewhat coarse approximation to an optimal tour while *improvement heuristics* are used to modify a given tour in an effort to shorten its length, hopefully deriving an optimal tour. The improvement heuristics include the well-known 2-Opt, 3-Opt, and Lin–Kernighan heuristics commonly used in practice.

8.7.1 Construction Heuristics

Among the construction heuristics, we can find various *insertion* strategies that typically start with a short subtour going through a small subset of the cities and subsequently employ some technique to enlarge the tour by iteratively adding more cities until a full tour is generated. As described by Ausiello et al. (1999)

we can derive a tour T from a set of cities C by doing tour extensions of a tour that initially starts as some single city in C:

```
begin
  Select any city c in C;
  T := <c>;
  C := C - {c};
  while C != ϕ  do
  begin
    Let c = f(C, T) in C be the city
    meeting the insertion criterion;
   Insert c into T;
   C := C- {c};
  end
end
```

There is a variety of choice for the city selection function `f`. For example, we may select a city using one of the following predefined criteria:

- *Nearest Neighbor*: Select the city that is nearest to the last city inserted in `T`.
- *Nearest Insertion*: Select the city that has the minimal distance from any city already included in the subtour `T`.
- *Farthest Insertion*: Select the city whose distance from the nearest city in `T` is maximal.
- *Cheapest Insertion*: Select the city whose insertion in `T` involves the minimum increase in path length.

8.7.2 Genetic Algorithms and the TSP

The application of genetic algorithms to the TSP has been met with mixed success by many research studies. A key issue is the quest for an effective gene representation. A good introduction to these issues can be found in the work done by Michalewicz (1992). Further research in Tao and Michalewicz (1998) report on the inver-over operator for fast, high quality solutions to the TSP. In Nagata and Kobayashi (1997) a technique called edge assembly crossover is used.

An interesting strategy in applying GAs to the TSP has been done by Julstrom (1999). He uses an insertion heuristic that is guided by selection priorities, which are specified by means of a binary gene. This is an interesting strategy because the gene is not attempting to directly describe a feasible solution. Instead, it essentially defines a plan of action that leads to a feasible solution. Of course, in a natural setting the DNA of an animal is similarly utilized. DNA is used to specify a collection of proteins that function in the environment of a cell, leading to the characteristics of a phenotype that then exhibits an observable fitness. Our optimization strategies mimic this flow of information: the facilitation of gene expression followed by phenotype generation leading to fitness evaluation.

8.7.3 The TSP Gene Expression Algorithm

Our construction heuristic is similar in spirit to the insertion strategies mentioned earlier; however, admits more flexibility in which it involves more choice at each step of the construction. This extra choice at each step allows us to formulate a strategy that involves a stochastic component. This is done by using a *priority* vector that will specify the sequence of choices made in the complete construction of a feasible tour. The stochastic part of this strategy arises because such a priority vector is derived from a gene in a population that is subjected to an evolutionary process. We now describe this in more detail using the terminology introduced earlier:

- The gene representation will be an inversion vector (to be defined later).
- The intermediate representation is a priority vector.
- The phenotype representation is a sequence of cities defining a tour.

We start by describing the conversion of a priority vector to a tour and then describe the inversion vector.

8.7.3.1 A Priority Vector is an Intermediate Representation

We assume that, our TSP instance involves n cities labeled $1, 2, 3, \ldots, n$. We define an intermediate representation p to be a vector $p = [a_1, a_2, \ldots, a_n]$. Such a vector is intended to represent some permutation of the set $\{1, 2, 3, \ldots, n\}$. In other words, p will be the numbers $1, 2, 3, \ldots, n$ in some "scrambled" order. As described later, the entries in this vector will represent the priority order in which a set of subtours is coalesced to form a final full tour.

8.7.3.2 Merging of Subtours to Derive the Phenotype

The phenotype corresponding to this intermediate representation will be a Hamiltonian tour represented by a sequence of cities. It will have the same appearance as a priority vector but the significance of its components will be very different. To build the phenotype corresponding to some particular priority vector, we go through the following $n + 1$ steps:

Step(0): Build a set of subtours initialized to be n elementary subtours each containing a single city. We then process the entries of the priority vector in a consecutive order.

Step(i): Find the subtour containing the city labeled a_i and merge this subtour with its closest neighboring subtour.

On completion of the final step, we will have a full n-city tour. To derive the phenotype, we simply read out the cities in the final tour that is just constructed. To provide more details about step(i) we describe the merge of subtours as follows: Two subtours are merged by making two cuts (removing an edge from each subtour), thus opening up both subtours, and then reconnecting them to make one large subtour (Figure 8.5).

Given an arbitrary subtour S_A, the closest neighboring subtour S_C is defined to be the subtour providing the lowest merge cost. The merge cost of any two subtours is the minimal merge cost calculated by

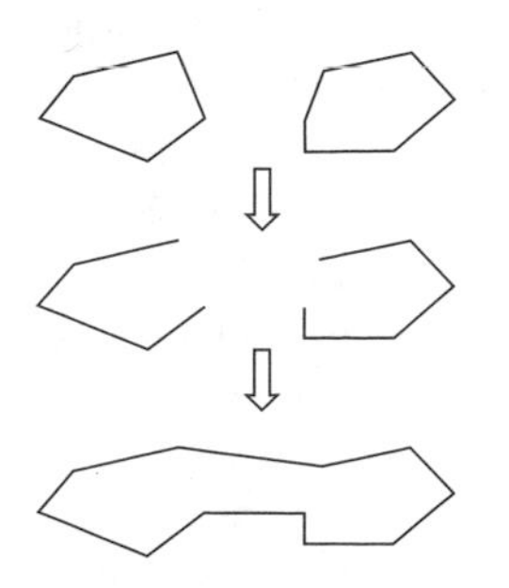

FIGURE 8.5 Merge of two subtours.

evaluating all possible cut pairs, one cut from each subtour. We define the cost of a merge as follows:

$$\text{cost of merge} = \text{total length of edges added} - \text{total length of edges deleted.}$$

Merging two *single* city subtours will produce a simple two city loop and merging a single city subtour with a multi-city subtour is essentially the insertion strategy described earlier.

It should be noted that, this strategy does more than simply generating feasible "single loop" tours. Each tour constructed is typically a reasonable approximation. Experimentation with the TSPLIB test suite (Reinelt, 1991) shows that an arbitrary priority vector (without any improvements facilitated by the genetic algorithm described later) will generate a tour that is roughly within 15% to 20% of optimal.

8.7.3.3 Evaluation of Fitness

Once the phenotype (i.e., a full tour) is constructed, we calculate the sum of all intercity distances covered by the tour and obtain the fitness calculation associated with the given priority vector.

8.8 A New Gene Representation

We now describe the binary gene representation. To the best of our knowledge, this is a novel data representation for a TSP problem and we expect that future refinements of this strategy should produce some interesting avenues of exploration.

8.8.1 The Inversion Vector

Let us consider $a_1, a_2, \ldots, a_n$ to be a permutation of the set $\{1, 2, \ldots, n\}$. The set is deliberately chosen to be a set of integers because we wish to have the ability to perform order comparisons on the given elements. As in Knuth (1998)[1] we define an *inversion vector* $b_1, b_2, \ldots, b_n$ of the permutation $a_1, a_2, \ldots, a_n$ by letting b_j be the number of elements to the left of j that are greater than j. For example, the permutation:

$$a_1, a_2, \ldots, a_n = 1\,7\,6\,9\,5\,8\,3\,4\,2$$

generates the inversion vector:

$$b_1, b_2, \ldots, b_n = 0\,7\,5\,5\,3\,1\,0\,1\,0.$$

To illustrate why this is true, note that $b_5 = 3$ because there are 3 elements in the permutation, namely 7, 6, and 9 that are greater than 5 and situated to the left of 5.

Because of the manner in which the b_j are defined, we will always have the following equations:

$$\begin{aligned} &0 \leq b_1 \leq n-1, \\ &0 \leq b_2 \leq n-2, \\ &\quad\vdots \\ &0 \leq b_{n-1} \leq 1, \\ &b_n = 0. \end{aligned} \tag{8.1}$$

[1]We have adopted the terminology *inversion* to pay proper respect to Knuth who has a prior use of this term in the context of a permutation. For Knuth inversion is used in the sense of a two element swap. The reader should not confuse our use of the term with the notion of inversion as used in genetics.

8.8.2 Recovering the Permutation

As noted by Marshall Hall (1956), there is a one-to-one correspondence between the set of all permutations of $\{1, 2, \ldots, n\}$ and the set of inversion vectors that are constrained to follow the dictates of (1). In fact, if we are given the inversion vector we can easily recover the original permutation by successively determining the relative placement of the elements, considering them in the order of largest to smallest. In the given example, we would write down 9 (the largest number in the city set), then place 8 to the right of 9, since the inversion vector indicates that there is a number to the left of 8 that is larger than 8. Similarly, we place 7 to the left of 9 since there is a 0 in the 7 position of the inversion vector, indicating that there are no elements to the left of 7 that are larger than 7. The 1 in position 6 of the inversion vector will cause the 6 to be written between the 7 and the 9. The permutation elaborated up to this point will be 7 6 9 8. By continuing in this fashion, until all the numbers are written out, we can recover the original permutation. The following table gives the order of element insertion and the results:

Next element to be put into the permutation	Result
9	9
8	9 8
7	7 9 8
6	7 6 9 8
5	7 6 9 5 8
4	7 6 9 5 8 4
3	7 6 9 5 8 3 4
2	7 6 9 5 8 3 4 2
1	1 7 6 9 5 8 3 4 2

8.8.3 Using the Inversion Vector

Why go through the trouble of expressing a permutation as an inversion vector? We do this because the representation is wonderfully compatible with a crossover operation. Note that any two properly formed inversion vectors will follow the constraints dictated by Equations (8.1). Furthermore, a crossover operation between these two inversion vectors will preserve the relative positions of entries in the inversion vectors produced by the crossover. Hence, a child is immediately a legitimate representation of some inversion vector since it too will adhere to the constraints imposed by Equation (8.1). By following the recovery instructions described in Section 8.10, the inversion vector of the child can be converted to a priority vector and we can commence the generation of a phenotype.

Most importantly, there is no need for any further "postcrossover" modifications of the representation such as we see with the adjacency representation or the PMX representation (Michalewicz, 1992: 168–172). It should be noted that the ordinal representation discussed by Michalewicz is also "crossover compatible"; however, in practice, the representation shows poor experimental results since there is little evidence of good parents producing good children due to the disruptive effects of crossover with this type of representation. If the reader indulges in some "back-of-the-envelope" experimentation, you will discover that the inversion vector representation does exhibit some degree of priority preservation from parent to child when a crossover is applied. This is fairly easy to appreciate since the inversion vector, by counting elements larger than a given element, will tend to place a city in a particular comparison relation with the other cities. This is a quality that is not supported by the ordinal representation.

Since a suitable TSP tour representation has been something of a Holy Grail in the GA community, we initially went through the trouble of designing an experiment that used the inversion vector as the representation of the permutation of cities in a tour. Although results were somewhat encouraging, there was still too much disruption, and experiments demonstrated a disappointing lack of inheritance of high

TABLE 8.1 Experimental Results (Population Size = 256)

Problem instance	Opt. length	Tour length	Quality of tour	Number of tour evals.	CPU time sec
att48	10,628	10,653	1.0024	640	9
berlin52	7,542	7,718	1.0233	1,024	12
ch130	6,110	6,336	1.0370	640	49
ch150	6,528	6,910	1.0585	768	89
eil101	629	647	1.0286	896	40
eil51	426	432	1.0140	896	9
kroA100	21,282	22,043	1.0358	640	26
lin105	14,379	14,747	1.0256	1,024	43
tsp225	3,919	4,002	1.0212	1,536	420

TABLE 8.2 Experimental Results (Population Size = 1024)

Problem instance	Opt. length	Tour length	Quality of tour	Number of tour evals.	CPU time sec
att48	10,628	10,628	1.0	2,560	25
berlin52	7,542	7,542	1.0	3,072	39
ch130	6,110	6,227	1.0191	6,144	483
ch150	6,528	6,711	1.0280	4,096	503
eil101	629	641	1.0191	3,584	169
eil51	426	427	1.0023	3,072	34
kroA100	21,282	21,831	1.0258	4,608	199
lin105	14,379	14,379	1.0	6,144	266
tsp225	3,919	3,952	1.0084	8,704	2,745

quality paths. As a consequence, we have chosen to work with the permutation as representing not the order of the cities but rather the order of merging subtours as described earlier.

8.9 Experimental Results

All experiments were done using problem instances taken from TSPLIB (Reinelt, 1991). The quality of the final approximations is presented as the ratio of final tour length divided by the published optimal tour length. A population size of 256 members was used. Each new generation of 128 genes replaced the least fit half of the population. A roulette strategy was used for selection of parent genes. If two successive generations produced no improvement in the best fitness evaluation then the production of a new generation was stopped, mutation was not used. Results are presented in Table 8.1. See Table 8.2 for the same experiment run with a population size of 1024. Processing times (600 MHz Pentium III) are in seconds.

The objective of the experiment was both to test the feasibility of the gene expression model and to verify the utility of the inversion vector strategy. We contend that our results are very encouraging and provide empirical evidence that the gene expression approach is a viable strategy.

Figures 8.6 through 8.10 show the progress of the algorithm in the solution of the Berlin52 problem from TSPLIB (Reinelt, 1991). In all figures except for the last one, the final tour is provided as a lighter line. In Figure 8.6, we see that the first ten merge operations produce short subtours that are in line with the final tour. Figures 8.7 to 8.9 show the solution as a growing set of subtours appearing as small loops that undergo various merging operations. Figure 8.10 is the final tour represented in dark black lines.

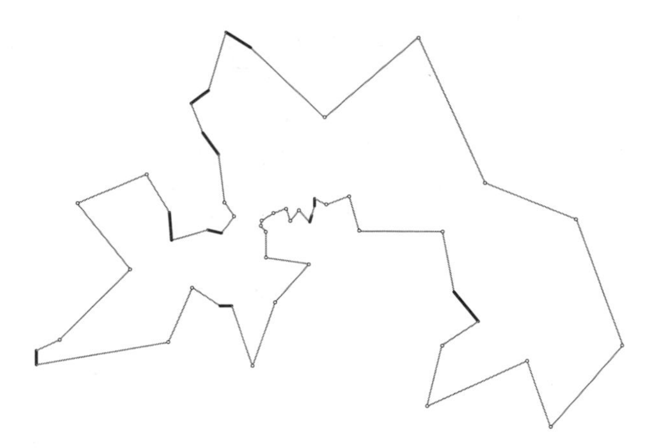

FIGURE 8.6 After 10 merge operations.

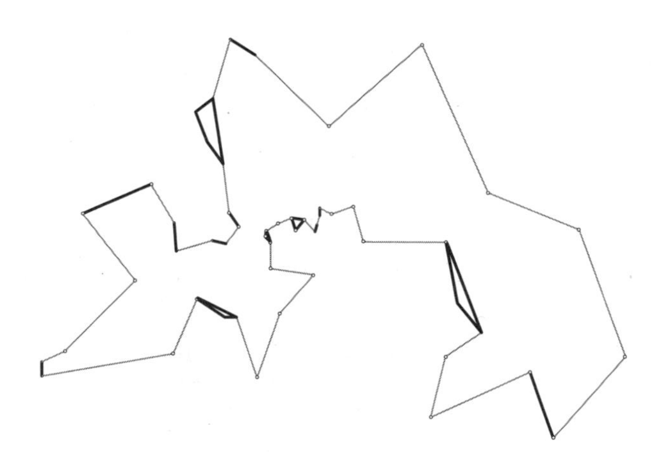

FIGURE 8.7 After 20 merge operations.

FIGURE 8.8 After 30 merge operations.

8.10 Final Discussion

In this chapter, we have described a mathematical model for an optimization strategy that is inspired by gene expression in a living cell. We have argued that, there are various advantages in using separate representations for genes and phenotypes with the specific intention that the gene will provide a set of

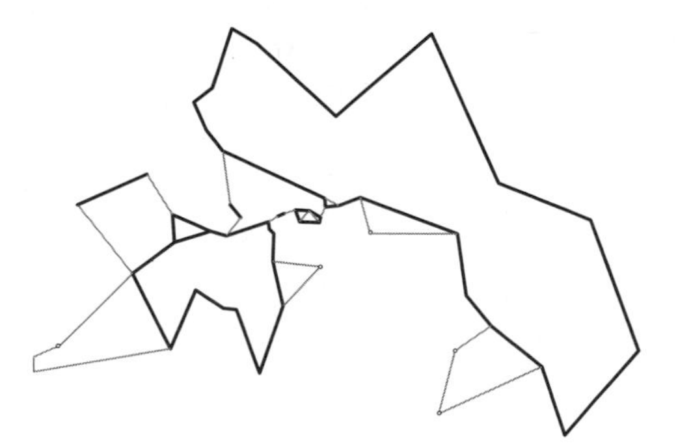

FIGURE 8.9 After 40 merge operations.

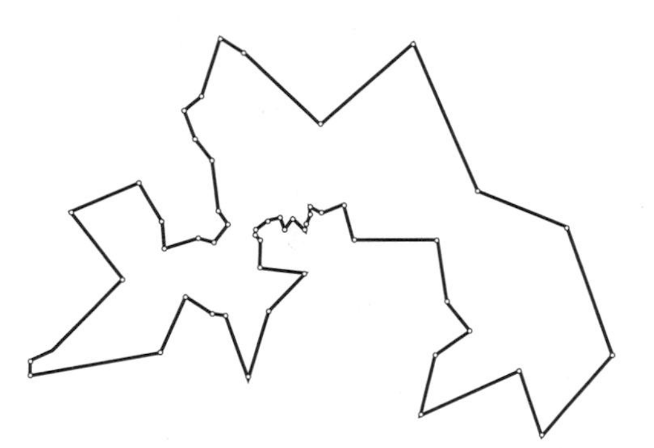

FIGURE 8.10 The optimal tour.

parameters that define the construction of a phenotype that will then be subjected to a fitness evaluation. In particular:

- We have a gene representation that is compatible with crossover and a phenotype representation with the usual appearance of a Hamiltonian path.
- All phenotypes generated by the gene expression algorithm are valid representations suitable for fitness evaluation.
- Our experiments show that, utilization of a sophisticated algorithm such as subtour merging gives the evolutionary computation a "head start" in the construction of good approximations to the optimal tour.
- Our experiments further demonstrate that a TSP construction heuristic will do better when combined with a subsystem for evolutionary computation.

We have referred to the heuristic strategy used to construct a phenotype as a gene expression algorithm since gene expression in the living cell is responsible for utilizing the gene when protein products contributing to the phenotype need to be synthesized. While there is no guarantee that such a gene expression algorithm can ever produce an optimal solution for an NP-complete problem our empirical evidence shows that optimal solutions for a TSP instance can at times be constructed using a subtour merging heuristic adapted to work with a genetic algorithm. This at least holds the promise of providing a viable evolutionary computation environment for future avenues of research.

References

Ausiello G., Crescenzi P., G. Gambosi, V. Kann, A. Marchetti-Spaccamela, and M. Protasi (1999) *Complexity and Approximation.* Springer-Verlag, Berlin.

J.D. Bagley (1967) The behaviour of adaptive systems which employ genetic and correlation algorithms. *Dissertation Abstracts International*, 28, 5106B (University Microfilms No. 68-7556).

F.J. Burkowski (2003) Proximity and priority: Applying a gene expression algorithm to the traveling salesperson problem, paper presented at *NIDISC'03 (The Sixth International Workshop on Nature Inspired Distributed Computing)* Nice, France, April 22–26, 2003.

C. Ferreira (2001) Gene expression programming: A new adaptive algorithm for solving problems. *Complex Systems*, 13, 87–129.

D.B. Fogel (1994) An introduction to simulated evolutionary optimization. *IEEE Transactions on Neural Networks*, 5, 3–14.

D. Goldberg (1989) *Genetic Algorithms in Search, Optimization, and Machine Learning*. Addison-Wesley Publishing, Reading, MA.

D.E. Goldberg, B. Korb, and K. Deb (1989) Messy genetic algorithms: Motivation, analysis, and first results. *Complex Systems*, 3, 493–530.

J.H. Holland (1992) *Adaptation in Natural and Artificial Systems*. MIT Press, Cambridge, MA.

B. Julstrom (1999) Coding TSP tours as permutations via an insertion heuristic. In *SAC '99, Proceedings of the 1999 ACM Symposium on Applied Computing*. ACM Press, New York, pp. 297–301.

H. Kargupta (1999) A striking property of genetic code-like transformations. School of EECS Technical report EECS-99-004, Washington State University, Pullman, WA.

H. Kargupta and S. Ghosh (2002) Toward machine learning through genetic code-like transformations. *Genetic Programming and Evolvable Machines*, 3, 231–258.

H. Kargupta and B.H. Park (2000) Gene expression and fast construction of distributed evolutionary representation. *Evolutionary Computation*, 9, 45–68.

D. Knuth (1998) *The Art of Computer Programming, Sorting and Searching*, Vol. 3, 2nd ed. Addison-Wesley, Reading, MA.

B. Lewin (1997) *Genes VI*. Oxford University Press, New York.

Z. Michalewicz (1992) *Genetic Algorithms + Data Structures = Evolution Programs*. Springer-Verlag, Berlin.

M. Mitchell (1996) *An Introduction to Genetic Algorithms*, MIT Press, Cambridge, MA.

Y. Nagata and S. Kobayashi (1997) Edge assembly crossover: A high-power genetic algorithm for the traveling salesman problem. In *Proceedings of the Seventh International Conference on Genetic Algorithms*, T. Bäck (Ed.), pp. 450–457.

G. Reinelt (1991) TSPLIB — A Traveling Salesman Problem Library, *ORSA Journal on Computing*, 3, 376–384. See also: http://softlib.rice.edu/softlib/tsplib/

G. Reinelt (1994) *The Traveling Salesman*. Springer-Verlag, Berlin.

H. Sawai and A. Adachi (1999) Genetic algorithms inspired by gene duplication. *Congress on Evolutionary Computation*, July. 1999, IEEE Press, Washington, D.C., pp. 480–487.

R.E. Smith (1988) An investigation of diploid genetic algorithms for adaptive search of non-stationary functions, TCGA report No. 88001, University of Alabama, The Clearinghouse for Genetic Algorithms, Tuscaloosa.

G. Tao and Z. Michalewicz (1998) Inver-over operator for the TSP. In *Proceedings of the 5th Parallel Problem Solving from Nature*, T. Baeck, A. Eiben, M. Schoenauer, and H. Schwefel (Eds.), Lecture Notes in Computer Science. Springer-Verlag, Amsterdam, pp. 803–812.

A. Wu and R. Lindsay (1995) Empirical studies of the genetic algorithm with non-coding segments. *Journal a Evolutionary Computation*, 3, 121–147.

A. Wu and R. Lindsay (1996) A survey of intron research in genetics. *Parallel Problem Solving from Nature (PPSN IV)*, H. Voigt, W. Ebeling, I. Rechenberg, and H. Schwefel (Eds.), Springer-Verlag, Berlin, pp. 101–110.

9

Dynamic Updating DNA Computing Algorithms

Zhiquan Frank Qiu
Mi Lu

9.1 Introduction

It has been clearly shown that DNA computing can be used to solve those problems that are currently intractable on even the fastest electronic computers. For methods to design the algorithms for DNA computing, however, is not straightforward. To develop efficient DNA computing algorithms requires a strong background in both DNA molecule and computer engineering. All of these algorithms need to start over from the very beginning when their initial conditions change. It is very frustrating, especially when the initial condition change is very small. The existing models based on which a few DNA computing algorithms were developed are not able to accomplish the dynamic updating.

People have been talking about the huge memory made possible through DNA computing due to the fact that each strand can be treated as both storage media and processor for a long time. Currently, there is no existing application that has used this huge memory because although it is easy to read from this memory, it is extremely hard to store data in it. Memories can only be ready after data has been stored.

A new DNA computing model is introduced based on which new algorithms are developed to solve the 3-coloring problem. These algorithms are presented as vehicles to demonstrate the advantages of the new model. They can be expanded to solve other NP-complete problems. They have the advantage of dynamic updating, so answers can be changed when the initial conditions are modified. The new model takes advantage of this huge memory by generating "lookup tables" during the process of implementing the algorithms. When the initial conditions change, the answers are changed accordingly. The new model can be used to solve computationally intense problems both efficiently and attractively.

9.1.1 Motivation

A strand of DNA is composed of four different base nucleotides: A (adenine), C (cytosine), G (guanine), and T (thymine). When attached to deoxyribose, these base nucleotides can be strung together to form a strand. Because DNA strand can be used to encode information and DNA bio-operations are completely based on the interactions between strands, each DNA strand can be counted as a processor as well as storage media. Numerous strands are involved in DNA bio-operations and the interactions between one another occur simultaneously. This, then, can be viewed as a realization of massive parallel processing.

Since Adleman [1] solved a 7-vertex instance of the Hamiltonian Path Problem, a well-known representative of NP-complete problems, the major goal of subsequent research in the area of DNA computing has been to develop new techniques for solving NP-complete problems. NP-complete problems are those problems for which no polynomial-time algorithm has yet been discovered, in contrast to polynomial-time algorithms whose worst-case run time is $O(n^k)$ for some constant k, where n is the size of the problem.

Consider that 1 litre of water can hold 10^{22} DNA strands. The potential computing power is significant, and this recognition raises the hope of solving problems currently intractable on electronic computers. Rather than using electronic computers upon which the time needed to solve NP-complete problems grows exponentially with the size of the problem, DNA computing technology can be used to solve these problems within a time proportional to the problem size. An NP-complete problem that may take thousands of years for current electronic computers to solve would take a few months, if the existing DNA computing techniques were adopted.

As indicated in several articles [2–7], most DNA computing algorithms are based on certain developed DNA computing models. The most popular models are the sticker based model [8,9], the surface-based model [10,11], and the self-assembly based model [12,13]. The problem with the sticker-based model is that the stickers annealed to the long strand may fall off during bio-operations, thus causing a very high rate of error. The limitation of the surface based model is that the scale of computation is severely restricted by the two-dimensional nature of surface based computations. The shortcoming of the self-assembly based model is that it makes use of biological operations that are not yet matured.

While the theory of molecular computation has developed rapidly, most of these algorithms usually take months to solve problems that may take thousands of years to solve with electronic computers. The problem is that when the initial condition changes, the algorithms have to start over again. Here, a new DNA computing model, which can eliminate this problem, is introduced. Based on this model, algorithms can be designed to dynamically update the answer. When the initial condition changes, the new algorithms can continue with the current process, and the solution for the new problem can be generated by a few extra processes. In addition, this new model can also be used to solve several similar problems simultaneously.

9.1.2 Our New Model

Our new model uses only the DNA biological operations that are matured [1,6]. The following are the basic principle operations: synthesis, ligation, separation, combination, and detection that are picked for the new model.

Synthesis $I(P, \pi)$: this operation is used to generate a pool of coded strands, P, following the predefined criteria π. Different applications code strands differently use the four base nucleotides: A, G, T, and C. A group of strands are defined as one set and a pool is defined as the container holding a set of strands. If the criteria are the colors of a node in a graph, then a pool of strands coding all the possible colors for the nodes is expected after synthesis. In the graph coloring problem, the colors of a number of nodes are encoded by the strands. A few consecutive nucleotides on the strand coded for the color of one node form a region. For example, in Figure 9.1, one strand consists of three regions such that $s = \{RBR\}$ where (CCAAG), (AATTC), and (CCAAG) each represents the color for one node as R(Red), B(Blue), and R(Red), respectively. Ligation $L(P_3, P_1, P_2)$: this operation is used to bind strands in P_1 with strands in P_2. Each code s_{1i} in P_1, is ligated to every other code s_{2j} in P_2. Assuming before the ligation, strands in P_1 represent the codes $\{s_{1i} | i = 1, 2, \ldots, c, \text{where } s_{1i} \in P_1\}$ and those in P_2 represent the code

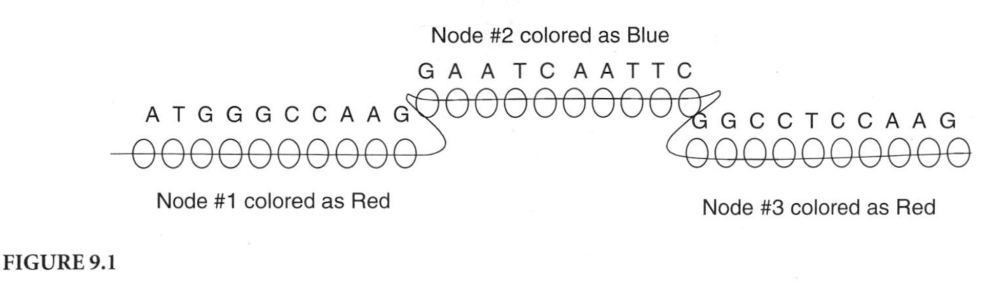

FIGURE 9.1

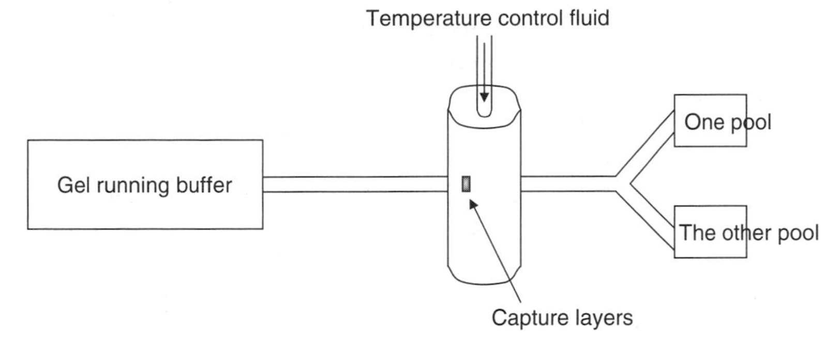

FIGURE 9.2

$\{s_{2j} | j = 1, 2, \ldots, c$, where $s_{2j} \in P_2\}$, after the ligation, the ligated strands are stored in P_3 and they represent the code $\{s_{1i} | i = 1, 2, \ldots, c$, where $s_{1i} \in P_1\}$ where $s_k = s_{1i}s_{2j}$, for $k = i + (j - 1) \times c$.

In separation operation, many identical short strands defined as probes are attached to magnetic beads. These probes are then sent into the pool containing the strands to be separated. Every probe will be paired up with a complementary strand and together they form a double helix. Such pair-ups occur only under the Watson–Crick complement rule: C only pairs with G and T only pairs with A. For example, in Figure 9.1, if the strands containing the region for node 1 colored as 'R' need to be separated, the DNA short strands TACCCGGTTC should be used as a probe because TACCCGGTTC complements ATGGGCCAAG. Also, the double helix can be separated by heating in order to have the paired strands part from each other without breaking the chemical bonds that hold the nucleotides together inside the single strand. The strands in the pool containing a region that complements the probes will be hybridized to, and captured by, the probe while all those without the region will remain in the pool [8].

A gel-based separation technique for DNA computing [14] has been developed that uses gel-layer probes instead of the bead to capture the strands. The capture layer only retains the strand with a region that complements the probe when it is cooled down, and will let all strands pass when the layer is heated up. The advantage of using gel-based probes over bead-based probes is that the gel-based method is more accurate when capturing DNA molecules. In Figure 9.2, which illustrates the gel-based separation, a set of strands runs from the left side buffer to the right. At each capture layer, the temperature is cold in order to capture the desired strands, and all unwanted strands are passed through into one pool. Then the temperature is raised to let all desired strands in the layer pass into another pool. The strands from the left buffer are separated and stored in two different pools:

Combination B(P, P_1, P_2): to pour two pools, P_1 and P_2, together to form a new pool, P.

Detection D(P): to check if there is any strand left in the pool, P. If the answer is "yes," the strands in the pool may need to be decoded.

The rest of this chapter is organized as follows: Section 9.2 gives an introduction to our new algorithm and how, based on the new model we purposed, it solves the 3-coloring problem. The complexity analysis of the new algorithm is provided in Section 9.3, which shows how dynamic updating can be accomplished. The last section makes the conclusion.

9.2 The New Fundamental Algorithm

This new algorithm for the 3-coloring problem has been developed based on the newly developed DNA computing model. The basic algorithm that will generate the answer to the 3-coloring problem of a given graph will be introduced in this section. The algorithm will be advanced to show how the answer can be dynamically updated and this will be shown in Section 9.2.1.

9.2.1 Coloring Problem

The 3-coloring problem is a special case of the k-coloring problem where $k = 3$, is a well-known representative of the class of NP-complete problems. A new algorithm for solving the 3-coloring problem will be introduced, and will simplify the explanation of our new DNA computing model. These when developed can be expanded to solve the k-coloring problem and hereby can be generalized to solve many other NP-complete problems.

k-coloring problem: k-coloring problems require the coloring of an undirected graph $G = (V, E)$ in such a way that no two adjacent vertices share the same color [15]. The two nodes connected by an edge are referred to as adjacent vertices. The solution is a function c: $V \rightarrow 1, 2, \ldots, k$ such that $c(u) \neq c(v)$ for every edge $(u, v) \in E$. In other words, the numbers $1, 2, \ldots, k$ represent the k colors, and the adjacent vertices must have different colors. The k-coloring problem determines whether k colors are enough to color a given graph [16].

A simple example graph with 10 nodes and 10 edges, $G(10, 10)$, is given in Figure 9.3. From the graph, it is clearly shown that the graph can be colored if $k \geq 3$.

In order to solve this 3-coloring problem, it is necessary to generate a pool of encoded DNA strands to represent all the possible color patterns of the n-node graph where each color pattern is an assignment of colors of nodes [17,18]. For example, for nodes $n_1 n_2 n_3 n_4$, "BBRG" is one pattern, which assigns n_1 with color Blue, n_2 with color Blue, n_3 with color Red and n_4 with color Green. "RGBB" is another pattern with given color Red, Green, Blue, and Blue to $n_1 n_2 n_3 n_4$, respectively. After the strands are generated and stored in one pool, the color patterns with no color conflict need to be separated. Two nodes along an edge are defined as having color conflicts when they have the same color. For the color patters with any color conflicts along some edges of the graph, the corresponding strands must be filtered from the pool.

The new algorithm will be introduced in Section 9.2.2. The dynamically updating algorithm and the advantages of the new algorithms will be described in Section 9.2.3.

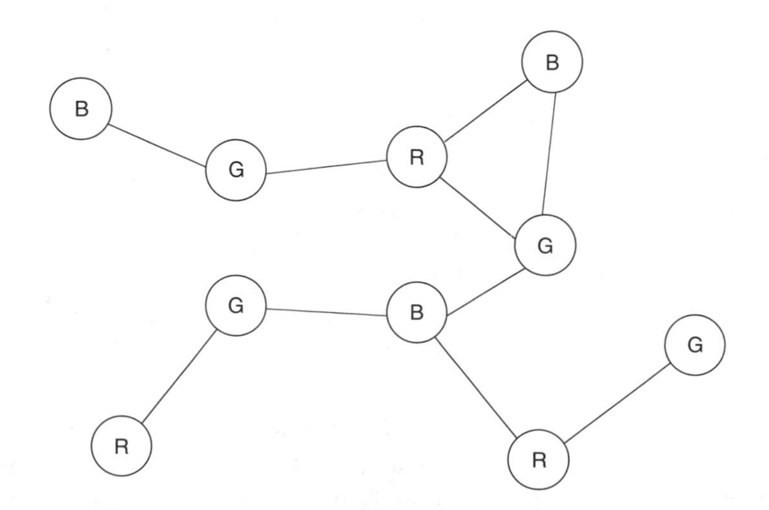

FIGURE 9.3

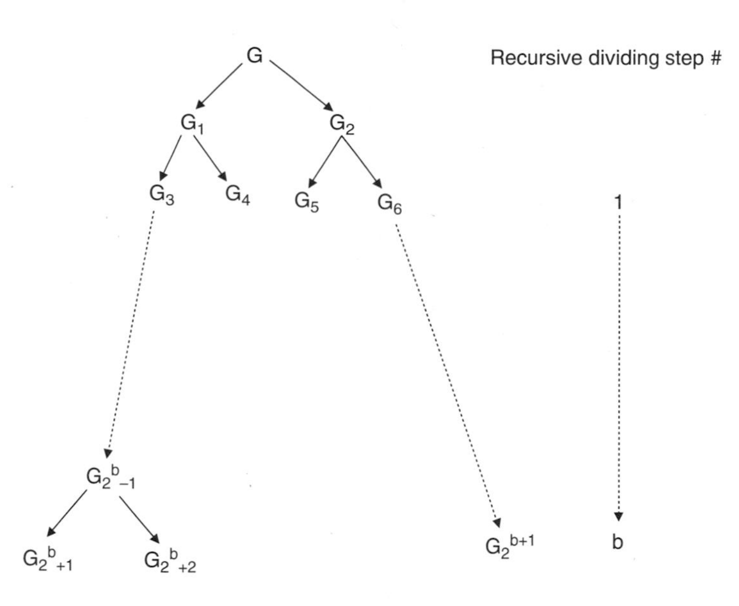

FIGURE 9.4

9.2.2 The New Algorithm

Given a graph $G = (V, E)$, $V = \{v_i \mid i = 1, 2, \ldots, n\}$ is a set of nodes and $E = \{e_j \mid j = 1, 2, \ldots, m\}$ is the corresponding set of edges. The new approach to solving the 3-coloring problem for such graph is divide and merge. Partition graph G into two subgraphs: $G_1 = (V_1, E_1)$ and $G_2 = (V_2, E_2)$ such that $V_1 \cup V_2 = V$, $V_1 \cap V_2 = \phi$ and $|V_1| \approx |V_2|$ by eliminating all edges (u, v) such that $u \in V_1$ and $v \in V_2$. This set of edges, called C, is referred to as the cut set of G [15,19]. The partition process can be executed recursively. That means, subgraphs G_i can be partitioned into G_{2i+1} and G_{2i+2}, until each subgraph contains only one vertex, and n subgraphs exist in total as shown in Figure 9.4.

After partitioning the graph G into n subgraphs, the algorithms start to merge every two subgraphs recursively and in parallel. Before each merge, every subgraph is colored with three colors. During the merge process, the color patterns of the two subgraphs are combined together. The merge operation continues until the original graph G is reestablished. To merge two subgraphs, the edges in the cut sets earlier eliminated for partitioning the subgraphs will be added back. Each addition of such edges will introduce some color conflicts if the nodes linked by the edge are of the same color. The color patterns that worked for the subgraphs may not work for the merged graphs after the subgraphs are combined. Some combined color patterns will be eliminated. This elimination process will continue until the color patterns that are legitimate for the graph are found.

The algorithm to solve the 3-coloring on a sparse graph is shown in Figure 9.5. The first for loop is used to generate n pools of strands representing all possible color patterns for n subgraphs while initially each subgraph contains only one node.

The function of the while loop is first to merge the pairs of the two subgraphs. The bio-operation used to merge two subgraphs is ligation. This step ligates the strands in two pools in order to form longer strands. If the color pattern for the first subgraph is s_i and that for the second graph is s_j, all the s_i should be ligated to s_j. The strands for one color pattern of a subgraph is replicated and each duplicated copy is ligated to those strands representing the color patterns of the other subgraph. All the color patterns of the merged graph will be represented by the ligated longer strands after the merge operation.

Inside the while loop, multiple copies of all strands in every pool need to be prepared for the following round of ligation. The duplication can be accomplished by using the polymerase chain reaction (PCR) process [17,20].

After the merge, some ligated strands encoding the color patterns with color conflicts introduced by those edges in all cut sets are eliminated in the partition step. The next task is to investigate all edges in the

```
Algorithm 1.
For i=1 to n do
In parallel (I(P_i, color of node i))
End
F=n
While f≠1 do
In parallel (Make multiple copies of strands in all pools)
For all odd I do
In parallel ( L (P_i, P_i, P_{i+1} ))
In parallel (relabel all pools 1 to f/2)
End
f=f/2
End
S (P, P_t, P_{f1}, θ),  θ_i is color conflicts along e_i
K = 1;
For i=2 to m do
S (P_t, P_{1t}, P_{1f}, θ), θ_i is color conflicts along e_i
For j=1 to k do
In parallel { S (P_{fj}, P_{jt}, P_{jf}, θ_i) }, θ_i is color conflicts along e_i
End
For j=1 to k do
In parallel do ( (B(P_{fj}, P_{j-1f}, P_{jt}))
End
B (P_t, P_{1t}, θ)
B (P_{fk+1}, P_{kf}, θ)
K=k+1
End
Check if P is empty to return "yes" or "no" accordingly.
```

FIGURE 9.5

cut sets and detect all the color conflicts caused. The task can be accomplished by the separation operation to filter out all strands that contain any color conflict from the pool. Two nodes, *i* and *j*, are connected by an edge. The pool is first separated into three pools while each pool contains the strands coloring node *i* as R, G, and B. The strands having node *j* colored as R, G, and B are filtered out by using the separation operation.

The answer to the 3-coloring problem is "yes" if there is any strand left in the final pool. The answer is "no" and the graph cannot be colored by only three colors if there is no strand left.

9.3 Dynamically Updating the Answers

Once a solution to the 3-coloring problem of the graph is obtained, given minor changes in the initial condition, it is significant to have a method that can quickly update the solution without restarting the algorithm and completely recalculating. What is presented next is an effort to make such dynamically updating solution both realistic and efficient.

For 3-coloring problem, four possible changes may occur with the initial condition: nodes and edges are inserted or removed. Based on the original solution "yes" or "no" to the original graph, different strategies need to be considered to update the answer.

Beginning with the easiest updating strategies, if the original answer is "yes" and an edge or node is removed from the original graph, the answer will remain "yes."

If the original answer is "no," it will remain "no" if a node or edge is added.

If the original answer is "yes," it can be changed to "no" after a node is inserted into the graph. An example is shown in Figure 9.6. The graph shown in Figure 9.6(a) has the answer "yes" for the 3-coloring problem

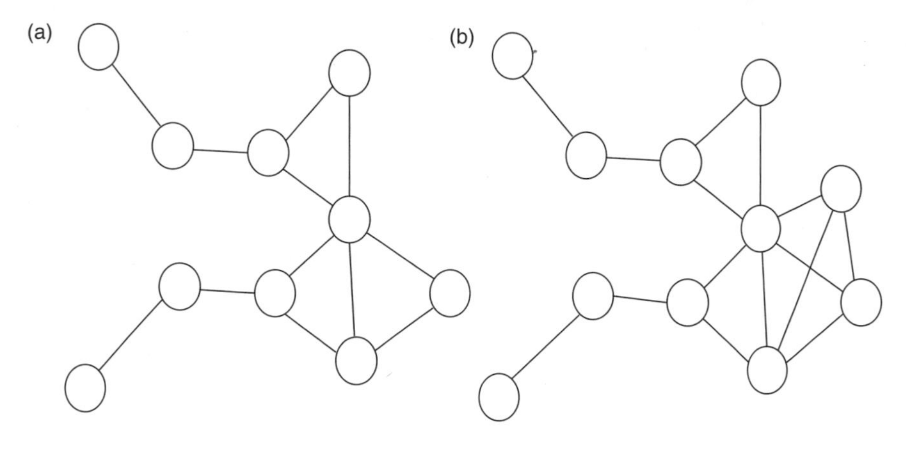

FIGURE 9.6

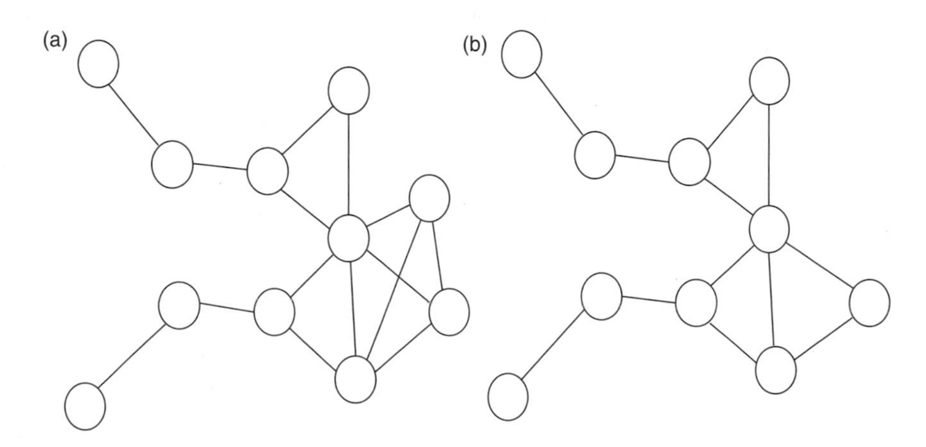

FIGURE 9.7

of the given graph. The answer can be changed to "yes" after one node is inserted to the graph as shown in Figure 9.6(b).

If the original answer is "no," it can be changed to "yes" after a node is removed from the edge. An example is shown in Figure 9.7. Figure 9.7(a) contains the graph with the answer "no" for the 3-coloring problem. The answer is changed to "yes" after one node is removed from the graph as shown in Figure 9.7(b).

If an edge is removed or inserted to the graph, it can be dealt with similarly because at least one edge should be removed if a node is eliminated and at least one edge should be added when a node is inserted.

What is illustrated next shows how to dynamically update a solution when a node or edge is inserted into the graph, following an original answer "yes." The strands in the final set, P_t, are checked for possible new answers. The final set is the only set that can be used because it is the only set that contains the strands representing all possible coloring solution that do not have any color conflicts among all the nodes except the newly added one. Only the color conflicts that occur between the newly added node and nodes connected with it need to be checked based on these sets. Only the newly added edges are checked for color conflicts.

The most difficult case occurs when a node or edge is removed from a graph with an original answer of "no." The answer to the new graph may be either "yes" or "no." To remove a node includes removing both the node itself and all the edges connecting the new node to the graph. What follows is the dynamically updating algorithm for this case. The DNA computing result that reflects an original answer of "no" is

```
Algorithm 2.
For i = 1 to α do
In parallel (S (P_fi, P_newi, P_fi, θ_i), θ_i is color conflicts along exact i # of edges)
End
B (P_new, ϕ, ϕ)
For i = 1 to α do
In parallel ( B (P_new, P_new, P_newi))
End
B (P_new, P_t, P_new)
For j = 1 to β do
S (P_new, P_new, P_newf, w_i), w_j is the colored conflicts based on edge e_j
End
Check if P_new is empty to return "yes" or "no" accordingly.
```

FIGURE 9.8

represented by an empty P_t set with no strand. All other sets represent the coloring patterns of the original graph with the color conflicts. After removing the nodes or edges, some coloring patterns may no longer have conflicts. The task now is to identify those patterns represented by DNA strands. Here, the strand sets that need to be examined is limited. Only those strands representing color patterns with color conflict involving the pair of nodes connected by the edges being removed are checked. Finding the above strand sets takes $O(\alpha)$ steps, where α is the number of edges being removed. This process is much less expensive than recomputing the updated graph from the very beginning when α is not large.

The detailed algorithm needed to find the answer for the new graph with the removed edges, based on the original "no" answer, is illustrated in Figure 9.8.

When only one edge is removed from the original graph, pool P_{f1} is checked. This is because P_{f1} contains all the strands representing all the color combinations for the graph that have no color conflicts along all edges, except one.

Assuming that the two nodes along the edge being removed are n_1 and n_2, the strands that need to be separated from the pool are those that have the two nodes colored as {RR}, {BB}, and $\{GG\}$. That means that only those strands, which have two identically colored nodes are extracted to a new pool, P_{new}. If P_{new} is not empty, the answer to the 3-coloring problem for the new graph is "yes," which is different from the original graph. Otherwise, the "no" answer remains.

When two edges are removed from the graph, both P_{f1} and P_{f2} need to be checked. This is because P_{f2} may contain strands that represent color combinations that have color conflicts along both the edges being removed. P_{f1} may contain strands that represent color combinations of the graph with a color conflict along only one of the two edges being removed. Suppose the two edges being removed are e_1 and e_2. Then, strands that need to be extracted from pool P_{f2} using the separation operation must represent the color combinations of the graph having color conflicts along both edges. Strands that should be extracted from P_{f1} are those representing color combinations with color conflict along either e_1 or e_2. The extracted strands are stored in a new pool, P_{new}. If P_{new} is not empty, the answer to the 3-coloring problem for the new graph is "yes," which is different from the original graph. Otherwise, the answer for the 3-coloring problem to the new graph remains "no."

When α different edges are removed from the original graph, α different pools should be checked. These pools are $P_{f1}, P_{f2}, \ldots, P_{f\alpha}$. For different pools, different operations need to be undertaken. For pool P_{f1}, all strands are left due to the color conflict along one edge. If the edge that caused the conflict is removed, the answer will change to "yes." Because of this, all strands in this pool representing those color combinations with color conflicts along one of the α edges that have been removed should represent the answer to the 3-coloring problem of the new graph. For pool P_{f2}, all strands representing the color combinations have color conflicts along two, and only two, of the edges being removed, representing the answers to the 3-coloring problem of the new graph. For pool P_{ft} where $t \leq \alpha$, all strands representing the color combinations having color conflicts along exactly t different edges being removed will generate the answer to the 3-coloring problem for the new graph. All strands extracted from these sets are stored in a new pool, P_{new}. If P_{new} is not empty, the answer to the 3-coloring problem for the new graph is "yes," and thus different from the original graph. The answer is "no" if P_{new} is empty.

When the graph is changed by both removing and adding edges, multiple processing steps need to be considered. Assuming that the number of edges being removed is α and the number of edges being added is β, the strands with color conflicts along the removed edges should be found first. This will put the strands to be considered for the following operations in one pool, P_{new}, instead of involving several pools. Those α edges should first be considered by using the method introduced above to go through α different pools. Then, P_t is combined with P_{new} and relabeled P_{new}. This is due to the fact that those strands that may generate the "yes" answer are distributed in $\alpha + 1$ different pools. Collecting the strands in one pool will save time and further operations as compared to working on these pools one at a time. If no strands are left in pool P_{new}, then the answer to the new graph is "no." If there are some strands in pool P_{new} after α edges are removed, color conflicts along β edges are checked. This operation can be accomplished in a manner similar to what has been described above for adding edges.

Compared to the existing algorithms, our new method can dynamically update the solution when the initial condition changes for the 3-coloring problem of a graph. It can also solve the 3-coloring problem for many graphs that are similar to each other. The complexity of the existing algorithm is $O(m + n)$, where n is the number of vertices and m is the number of edges [18]. If the updating process is not used, any change to the initial condition will result in a restarting of the process. With our new algorithm, the number of extra processes that need to be undertaken depends upon the significance of the changes. The complexity of the updating process is $O(\alpha + \beta)$, where α is the number of edges being removed. β is the number of edges being added.

When this method is used to solve the 3-coloring problem for multiple graphs that are similar to each other, the time complexity is $O(\theta)$ after the solution for one graph is generated, where θ is the difference between the number of edges of the two graphs.

It is necessary to check the extra space and effort that may be necessary for making dynamic updating available. First, m additional containers are needed to keep all m extra sets of strands. Second, the extra DNA material for generating these sets needs to be contained. Because strands are generated to represent all color combinations for the graph before the separation process takes place, no extra material is necessary as compared with the existing algorithms until the answer is generated for the original graph. The extra material is only necessary if new solution needs to be formed for the modified graph when the edges and/or nodes are added.

When the procedure for approaching a 3-coloring problem of a given graph is finished and a new graph is provided, how can one determine whether to start again from the beginning or to use the dynamic updating method to generate the new answer?

Assuming that the implementation of the algorithms introduced above for the 3-coloring problem of the graph with n nodes and m edges has been finished, the 3-coloring problem of a new graph needs to be solved. This new graph has N nodes and M edges. This graph can be converted from the existing graph by first removing δ nodes and α edges, and then adding γ nodes and β edges. The new graph can be generated by changing the original graph, or it can be treated as a totally new graph. In order to solve the problem for the new graph, N ligation and M separation operations are necessary if the algorithm is being restarted from the beginning. The total time necessary is:

$$T_1 = N \times l + M \times s,$$

where l is the time for each ligation operation and s is the time necessary for each separation operation. Here, combinations are ignored due to their simplicity because the time needed for the combination operations is very short, as compared to the other operations used in DNA computing. When the answer is generated based on the pools already generated using this new, dynamically updating strategy, the time necessary for reaching the answer is

$$T_2 = (\alpha + \beta) \times s + \gamma \times l.$$

In order to take advantage of the new method, the time that is needed must be shorter than restarting the algorithm from the beginning.

$$T_2 \leq T_1,$$
$$(\alpha + \beta) \times s + \gamma \times l \leq N \times l + M \times s,$$
$$(\alpha + \beta) \times s + \gamma \times l \leq (n + \gamma - \delta) \times l + (m + \beta - \alpha) \times s,$$

because $N = n + \gamma - \delta$ and $M = m + \beta - \alpha$. It is easy to get

$$(m - \alpha) \times s + (n - \delta) \times l \geq \alpha \times s,$$

as $n - \delta$ is always >0, the above condition can be tightly restrained as follows:

$$(m - \alpha) \times s > \alpha \times s.$$

So, $\alpha < m/2$. The algorithm needs to be restarted from the beginning only when the change is significant, which means, when more than half of the edges need to be removed to generate the new graph from the original.

Given the above condition, it is clear that there is no need to retain all m sets. At least half of the pools can be destroyed in order to save storage space. This will save the expenses once required for storing m sets of strands and the material needed to work on them.

9.4 Conclusion

A new model for DNA computing is introduced. Based on the new model, our new algorithms for the 3-coloring problem have been presented. The new algorithms have the advantage of dynamic updating, as compared to the existing algorithms. These new algorithms represent a huge improvement over the existing algorithms.

Instead of restarting the DNA computing algorithm from the very beginning every time the initial condition would change, this new method can generate the new solution through a few extra DNA operations based on the existing answer. It can also quickly solve problems similar to those already solved.

No extra material is needed to prepare for the dynamically updating process. The only expense is some extra storage containers for storing the additional pools of DNA strands. As compared to the existing DNA computing algorithms, this new method can achieve a solution much more quickly after the answer for the first problem is generated and it is very financially efficient. This will make DNA computing more attractive to potential users who want to solve the problem that is currently unsolvable.

References

[1] L. Adleman. Molecular computation of solutions to combinatorial problems. *Science*, 1021–1024, 1994.

[2] Y. Gao, M. Garzon, R.C. Murphy, J.A. Rose, R. Deaton, D.R. Franceschetti, and S.E. Stevens Jr. DNA implementation of nondeterminism. In *Proceedings of the Third DMIACS Workshop on DNA Based Computers*, University of Pennsylvania, June 1997, pp. 204–211.

[3] G. Gloor, L. Kari, M. Gaasenbeek, and S. Yu. Towards a DNA solution to the shortest common superstring problem. In *Proceedings of the Fourth International Meeting on DNA Based Computers*, University of Pennsylvania, June 1998, pp. 111–116.

[4] V. Gupta, S. Parthasarathy, and M.J. Zaki. Arithmetic and logic operation with DNA. In *Proceedings of the Third DIMACS Workshop on DNA Based Computers*, University of Pennsylvania, June 1997, pp. 212–222.

[5] P. Kaplan, D. Thaler, and A. Libchaber. Parallel overlap assembly of paths through a directed graph. *In Proceedings of the Third DIMACS Workshop on DNA Based Computers*, University of Pennsylvania, June 1997, pp. 127–141.
[6] R. Lipton. Using DNA to Solve SAT, 1995.
[7] Z.F. Qiu and M. Lu. Arithmetic and logic operations for DNA computers. In *Parallel and Distributed Computing and Networks (PDCN'98), IASTED*, December 1998, pp. 481–486.
[8] S. Roweis, E. Winfree, R. Burgoyne, N. Chelyapov, M. Goodman, P.R. Othemund, and L. Adleman. A sticker based architecture for DNA computation. In *Proceedings of the Second Annual Meeting on DNA Based Computers*, Princeton University, June 1996, pp. 1–27.
[9] S. Roweis, E. Winfree, R. Burgoyne, N. Chelyapov, M. Goodman., P.R. Othemund, and L. Adleman. A sticker based model for DNA computation. *Journal of Computational Biology*, 5: 615–629, 1998.
[10] Q. Liu, Z. Guo, A.E. Condon, R.M. Corn, M.G. Legally, and L.M. Smith. A surface-based approach to DNA computation. In *Proceedings of the Second Annual Meeting on DNA Based Computers*, Princeton University, June 1996, pp. 206–216.
[11] L. Wang, Q. Liu, A. Frutos, S. Gillmor, A. Thiel, T. Strother, A. Condon, R. Corn, M. Lagally, and L. Smith. Surface-based DNA computing operations: Destroy and readout. In *Proceedings of the Fourth International Meeting on DNA Based Computers*, University of Pennsylvania, June 1998, pp. 247–248.
[12] E. Winfree. Proposed techniques. In *Proceedings of the Fourth International Meeting on DNA Based Computers*, University of Pennsylvania, June 1998, pp. 175–188.
[13] E. Winfree, X. Yang, and N.C. Seeman. Universal computation via self-assembly of DNA: Some theory and experiments. In *Proceedings of the Second Annual Meeting on DNA Based Computers*, Princeton University, June 1996, pp. 172–190.
[14] R.S. Braich, C. Johnson, P.W.K. Rothemund, D. Hwang, N. Chelyapov, and L.M. Adleman. Solution of a satisfiablility problem on a gel-based DNA computer. In *Proceedings of the Sixth International Meeting on DNA Based Computers*, June 2000, pp 31–42.
[15] J. Clark and D.A. Holton. *A First Look at Graph Theory.* World Scientific, Singapore, 1991.
[16] T.H. Cormen, C.E. Leisenson, and R.L. Rivest. *Introduction to Algorithms.* MIT Press, Cambridge, MA, 1990.
[17] L. Adleman. On constructing a molecular computer, 1995.
[18] E. Bach and A. Condon. DNA models and algorithms for NP-complete problem. *Journal of Computer and System Sciences*, 57: 172–186, 1996.
[19] N. Christofides. *Graph Theory: An Algorithmic Approach.* Academic Press, New York, 1975.
[20] P.D. Kaplan, G. Cecchi, and A. Libchaber. DNA-based molecular computation: template-template interactions in pcr. In *Proceedings of the Second Annual Meeting on DNA Based Computers*, Princeton University, June 1996, pp. 159–171.

10

A Unified View on Metaheuristics and Their Hybridization

Jürgen Branke
Michael Stein
Hartmut Schmeck

10.1 Introduction

Over the past decades, a multitude of new search heuristics, often called "metaheuristics" have been proposed, many of them inspired by principles observed in nature. Common representatives include evolutionary algorithms (GAs) [1], ant colony optimization (ACO) [2], simulated annealing [3], tabu search [4], or estimation of distribution algorithms [5]. Besides the book at hand, overviews of several such metaheuristics can be found, for example, in References 6 and 7 or 8.

Each of these metaheuristics has been proven successful on a variety of applications. Although there have been attempts to compare their performance, the results are contradicting and inconclusive. There does not seem to be a superior candidate that should generally be preferred over the others. Thus, it is not surprising that recently, there has been a growing interest in hybridization of these metaheuristics (cf. Section 10.2).

In this chapter, we propose a simple unified framework that describes the fundamental principle common to all metaheuristics. The framework focuses on the commonalities rather than the differences between search algorithms. Due to its simplicity and generality, it suggests a natural way for hybridization, basically turning the variety of metaheuristics into one large toolbox from which an algorithm designer can choose those parts that seem most appropriate for the application at hand. The power of the model to unify different metaheuristics will be demonstrated at the example of combining EAs and ACO, and we will report on some preliminary empirical results on the performance of the hybrids so generated.

The chapter is structured as follows: in Section 10.2, we will survey some related work. Then, in Section 10.3, we will describe the proposed unified framework. A specific aspect of that framework, the organization of memory, is discussed in Section 10.4. Section 10.5 demonstrates the application of the model to the hybridization of EAs and ACO. The resulting hybrids are compared empirically in Section 10.6. The chapter concludes with a summary and some suggestions for future work.

10.2 Related Work

There have been numerous attempts to combine aspects of different metaheuristics, usually in the hope of keeping the benefits and avoiding the pitfalls of the pure heuristics. Examples include, but are not limited to, combinations of EAs and simulated annealing [9,10], EAs and tabu search [11], EAs and ACO [12], ACO and tabu search [13], or EAs and particle swarm optimization [14]. Also, there is growing interest in the field of memetic algorithms [15], which focuses on the combination of EAs and local search.

Calegari et al. [16] developed a taxonomy to describe iterative optimization heuristics, but as the goal of a taxonomy is to differentiate rather than to unify, it generally hides the commonalities and opportunities for hybridization. An overview on hybrid approaches involving biologically inspired heuristics, together with a valuable classification and a grammar for hybridization schemes can be found in Reference 17.

An early attempt to create a general framework for designing metaheuristics can be found in Reference 18. The suggested framework is more fine-grained and not quite as general as the framework that we suggest. It attempts to incorporate not only metaheuristics, but also more classical search methods like branch and bound. Implicit assumption of the framework seems to be that the algorithm operates on (partial) solutions, that is, that the memory stores (partial) solutions.

After an extensive description and classification of different metaheuristics, Blum and Roli [19] develop a unified view on the intensification and diversification aspects of algorithmic components. Analyzing the signature of available algorithm components in the developed framework, and identifying suitable combinations, seems to be a promising way toward the development of a systematic design approach for hybrid metaheuristic algorithms.

A framework that is focused on algorithms that use an updated parameterized model to generate candidate solutions was presented in Reference 20. It offers a unified view on ACO, estimation of distribution algorithms (EDAs) and related methods.

10.3 A Unified Framework for Iterative Search Heuristics

Most modern search heuristics like EAs, simulated annealing, ACO, or tabu search, are iterative and repeatedly probe the search space at new locations.

What distinguishes them from random search is primarily that they maintain some sort of memory of the information gathered during the search so far, and that they use this information to select the location where the search space should be tested next. The proposed general model, first presented in Reference 21, follows from this observation and is depicted in Figure 10.1: new solutions are constructed based on information stored in the memory, possibly involving several construction operators that may be applied in parallel (e.g., when different solutions are generated by different operators in each iteration) or sequentially (e.g., when a local optimizer is applied to each solution generated). The construction operators can be rather simple (as e.g., a single bit flip) or rather complicated (e.g., a local optimizer). The new solutions are then evaluated and can be used to update the memory, after which the cycle repeats.

In the following sections, we will show in more detail how this general framework can be used to describe some of the aforementioned metaheuristics:

- *Evolutionary algorithms* store information about the previous search in the form of a set of solutions (population). New solutions are constructed by selecting two solutions (parents), combining them

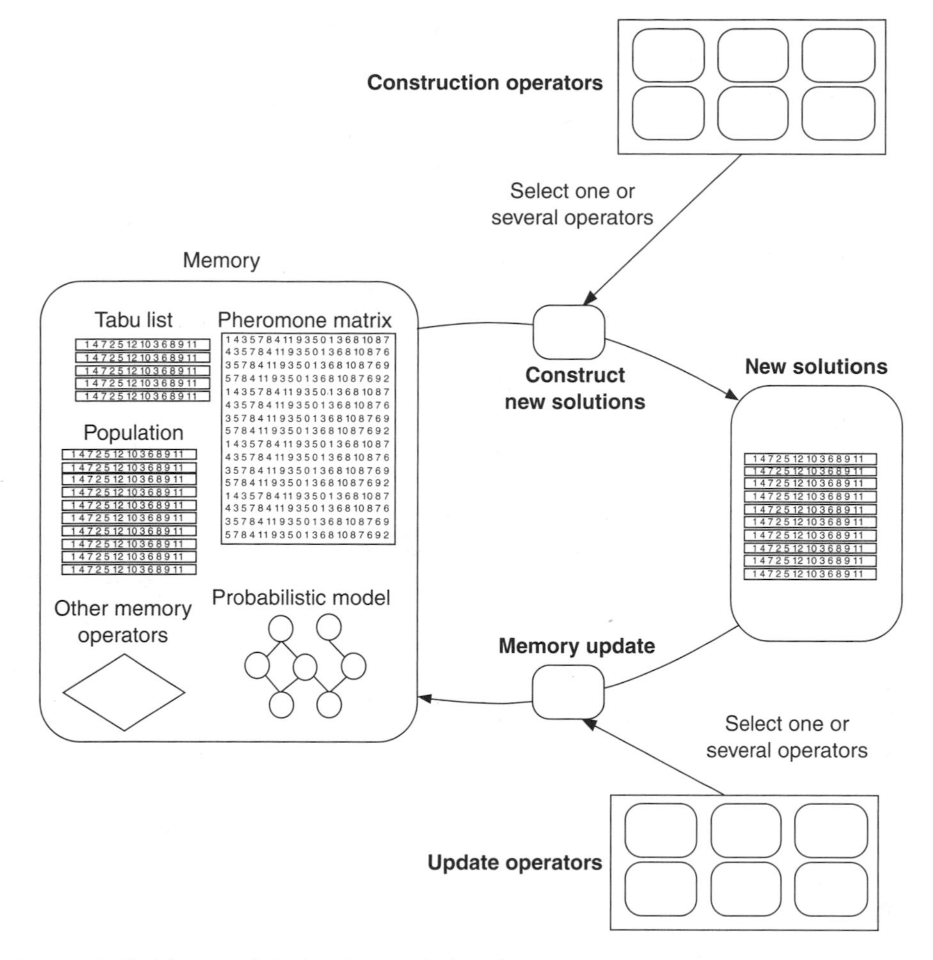

FIGURE 10.1 Unified framework for iterative search algorithms.

in some way (crossover), and performing some local modifications (mutation). Then, the memory is updated by inserting the new solutions into the population. Although there exist a variety of EA variants, they all fit into this general framework. For example, evolution strategies with self-adaptive mutation can be specified by extending the memory to also maintain information about the strategy parameters. Steady state genetic algorithms update the population after every newly generated solution, while genetic algorithms with generational reproduction generate a whole new population of individuals before updating the memory.

- *Simulated annealing* only maintains a single solution in the memory. In addition to that, it keeps track of time by a temperature variable. New solutions are created by local modifications more or less equivalent to the mutation operator in EAs. The new solution replaces the current memory solution depending on the quality difference and the temperature.
- *Tabu search*, just as simulated annealing, creates new solutions based on a single current solution in the memory. Additionally, it maintains so-called tabu-lists to avoid revisiting previous solutions. These tabu lists are generally recently visited solutions or recently performed move operations. New solutions are created by local modifications, while taking into account the tabu list. More advanced tabu search algorithms can comprise of a number of additional memory structures, like for example, a long-term frequency memory that records the number of times a particular component has appeared in a solution.

- *Particle swarm optimization* uses a swarm (set) of particles (current solutions). The search process can be imagined as a parallel search of particles "moving" through the landscape defined by the fitness function. In addition to their locations (solution characteristics), the memory contains for each particle the personal best solution encountered so far and a velocity, which can be seen as a kind of general accumulated search direction. In every iteration, new solutions are generated by moving the particles according to their velocity, and a linear, spring-like attraction to their personal best solution encountered and the overall best solution encountered by any of the swarm's particles. Memory update includes an update of the particle locations, the personal best solutions, and the particles' velocities.
- *Ant colony optimization*, when compared to the approaches outlined earlier, has a completely different way to store information about the search conducted so far. Instead of storing complete solutions, it accumulates information about which partial decisions have been successful when constructing solutions from scratch. For example, for the traveling salesperson problem, it maintains a (so-called pheromone) matrix indicating for each city how desirable it is to visit another city. Using this matrix, new solutions are constructed systematically, starting at a random city, and iteratively selecting the next city probabilistically according to the relative preferences encoded in the matrix. Usually, several new solutions are generated that way, and then the best solution found is used to update the matrix, increasing the probability that future ants will make the same decisions. An elitist ant (best solution found so far) can be modeled by an additional (complete) solution stored in the memory.
- *Estimation of distribution algorithms*, similar to ACO, construct solutions based on a probabilistic model, only that the probabilistic model is not necessarily stored in the form of a matrix. The new solutions are then evaluated, and the information gained is used to update the probabilistic model. The class of EDAs contains a multitude of different approaches that vary primarily in the complexity of the probabilistic model (in particular whether they take variable dependencies into account or not), and in the way the probabilistic model is updated (incrementally or reconstructed every iteration based on the generated samples). Note that many EDA approaches actually do not use a probabilistic model as main memory component, but instead rely on a population of solutions as underlying memory structure, and construct a new probabilistic model in every iteration based on the current population.

Given a description of the different metaheuristics in this general form has many benefits. First, it creates a common language, which allows researchers from different fields to understand each other's approaches easily. Second, it moves the focus from the complete algorithms to the components. And third, it provides the interfaces for the different components to work together.

Based on the presented unified framework, it is almost straightforward to combine different components from different algorithmic paradigms: an algorithm designer can easily select a combination of memorization features, choose a suitable set of construction operators or create new ones that make use of the combined set of selected memorization features, and then decide how the memory is updated with the newly generated information. The framework allows for a lot of freedom: new solutions may be constructed in different ways, using different information from the memory, the solutions thus constructed using one part of the memory may be used to update another part of the memory, and so on.

10.4 Some Thoughts about Memory

Since it is basically the memory that differentiates metaheuristics from random search, the organization of the memory seems to be a crucial aspect. Generally, it would be desirable to store in the memory all the information gathered during the search so far. However, that is usually impossible, not only because of memory space restrictions, but also because extracting relevant information from such a huge memory would be prohibitively slow. Therefore, the information has to be condensed in some way.

Depending on what information is stored in the memory, metaheuristics may be classified into *solution-based* or *model-based* (cf. [20]).

The approaches from the former category primarily keep some of the solutions generated so far. Simulated annealing just stores a single solution; EAs and tabu search store a set of solutions. Although the way these solutions are selected is different for the different algorithms, the implicit assumption always is that the stored solutions sufficiently represent the promising regions of the search space and appropriately reflect the history of the search.

Ant Colony Optimization belongs to the latter category: it assumes that the problem is to make a sequence of decisions, and then accumulates information about the desirability of making a certain decision in a given situation (state). It builds a model of construction methods. The space and complexity limitations are observed by restricting the number of states, and by ignoring interdependencies between decisions. For example, the state usually considered when solving a traveling salesperson problem is the current city, independent of the sequence of cities visited so far. Because usually, only complete *solutions* (corresponding to a combination of decisions) can be evaluated, but the memory stores desirability of *decisions*, ACO has to assign the credit for a good solution to the individual decisions. Currently, this is done in a straightforward way by simply distributing the credit evenly.

The class of EDAs is rather broad, and by using a population of solutions as well as a probabilistic model, different instantiations can be closer to either the solution-based or the model-based memory category. They are more or less decision-based (i.e., construct solutions step by step) but may keep track of variable dependencies through graphical models (e.g., Bayes networks or Gaussian networks).

Obviously, each of the above memorization schemes has its benefits and its drawbacks. Storing complete solutions preserves all the interdependencies between decision variables and thus implicitly takes epistasis[1] into account. However, it discards a lot of the solutions generated. On the other hand, the decision-based memorization scheme used for example, in ACO integrates information about many generated solutions (information is only slowly evaporated), at the expense of losing a lot of information about interdependencies.

From the above considerations, it seems natural that a combination of these two fundamental memorization schemes may be beneficial, which is one of the reasons why we decided to further explore the combination of EAs and ACO using the framework proposed in Section 10.3.

Furthermore, it should be mentioned that besides information about the search space, it seems promising to also store information about algorithmic parameters or meta-information about the search like some characteristics of the fitness values observed over time. Examples for such meta-knowledge are the already mentioned self-adaptive mutation in evolution strategies, or the dependence of the temperature parameter on the fraction of accepted moves in simulated annealing. Tabu search may have several additional memorized features like the frequency of certain moves in the past, etc. Another example, in particular for hybrid approaches, would be to track the performance of different operators in order to decide which operators should be used more often. Again, all those aspects easily fit into the proposed framework.

10.5 Combining Evolutionary Algorithms and Ant Colony Optimization

In this section, we propose a number of EA/ACO hybrids, which attempt to combine the two memorization schemes. Before that, however, let us briefly present the pure EA and ACO we built on. The application considered is the traveling salesperson problem (TSP).

[1] Epistasis, in the field of evolutionary computation, generally refers to the fact that different components of the solution interact.

10.5.1 Basic Evolutionary Algorithm

The standard EA uses a population of n individuals as memory. Individuals are represented as permutations, and the population is initialized randomly. A new individual is constructed by first selecting two parents according to linear ranking selection, applying crossover, and mutating the resulting offspring. As crossover operator, one of the following operators was used:

- Order crossover (OX) selects a random connected part of one parent and fills the remaining places with the missing cities in the order they occur in the other parent [22].
- Edge recombination crossover (ERX) attempts to preserve as many edges as possible. It iteratively constructs a tour, starting from a random city. At each step, it first considers the (up to four) cities that are neighbors (i.e., connected) to the current location in either of the two parents. If at least one of those has not yet been visited, it selects the city that has the fewest yet unvisited other cities as neighbors in the parents. Otherwise, a random successor is selected [23].

Mutation exchanges the position of two randomly selected cities. The mutation operator is called repeatedly with probability p_{m}, that is, an individual is mutated i times with probability $(p_{\mathrm{m}})^i(1-p_{\mathrm{m}})$.

In each cycle, k individuals are constructed (usually $k = 1$) and, to update the memory, the worst individuals in the population are replaced.

10.5.2 Basic Ant Colony Optimization

The memory structure of ACO is an $m \times m$ pheromone matrix with m being the number of cities in the problem. Initially, all values τ_{ij} of the matrix are set to 0.5. New individuals are constructed by starting at a random city i, and choosing the following city j probabilistically according to the relative pheromone values in row i of the pheromone matrix, that is, with probability

$$P_{ij} = \frac{\tau_{ij}^{\alpha}}{\sum_{l \in U} \tau_{il}^{\alpha}},$$

where U being the set of yet unvisited cities and α being a user-defined weighting parameter.

After the k new solutions of one cycle have been constructed, pheromone is evaporated by multiplying every element τ_{ij} of the pheromone matrix with $(1-\rho)$ where ρ is the pheromone evaporation rate. Then, the best individual found in the iteration is used to update the pheromone matrix by adding $1/L$ to all matrix elements τ_{ij} where the edge from city i to city j is part of the corresponding tour with length L.

Furthermore, the algorithm maintains the overall best solution as an elite that is also used to update the matrix in the above way in every cycle.

Note that for many optimization problems, the construction procedure as used by ACO allows to incorporate heuristic knowledge in a very elegant way. In the case of TSP, for example, this can be done by simply preferring close cities in the selection process.

The selection probabilities then become

$$P_{ij} = \frac{\tau_{ij}^{\alpha} \eta_{ij}^{\beta}}{\sum_{l \in U} \tau_{il}^{\alpha} \eta_{il}^{\beta}},$$

where $\eta_{ij} = 1/d_{ij}$ being the reciprocal of the distance d_{ij} between cities i and j and β being the weight for the heuristic information.

Naturally, a direct comparison of algorithms that use heuristic knowledge with algorithms that do not is not fair. Since the focus of this chapter is not to construct the best algorithm to solve TSP, but rather to study the effect of different memorization schemes and algorithmic hybrids, most of the approaches suggested later do not incorporate problem-specific knowledge. Nevertheless, we will also briefly report on the performance of ACO and a hybrid operator when they incorporate heuristic knowledge.

10.5.3 Hybrid Algorithms

50/50: Maybe the most straightforward combination of EAs and ACO is to simply use both basic algorithms to generate a portion of the new solutions each. More specifically, in every cycle, we generate 50% of the k new solutions on the basis of the pheromone matrix, while the remaining 50% are generated using the edge recombination operator. The complete set of k new solutions is then used in the standard way to update the pheromone matrix as well as the population.

Pheromone Completion (PC) crossover: This operator at the same time uses pheromone matrix and population to create a new individual. First, an individual is selected from the population by rank-based selection. A random connected part of that individual is then chosen, and this partial permutation is completed using the standard ACO construction operator, that is, probabilistically according to the pheromone matrix. After k individuals have been created that way, the new individuals are used to update the pheromone matrix as well as the population. Note that this operator is somewhat similar to the approach suggested by Miagkikh and Punch [24]. However, while we are proposing to use separate memory structures for the population and the pheromone matrix, they propose to use a population of "agents," each agent consisting of a solution and an individual pheromone matrix. It is difficult to reason whether using a global pheromone matrix or individual pheromone matrices is more promising. A global pheromone matrix collects more information and will, therefore, perhaps be a better guide in particular in short runs. Individual pheromone matrices, on the other hand, allow for different solutions to be encoded simultaneously, which may be more beneficial for long optimization runs, when diversity is a major issue.

Pheromone-supported Edge Recombination (PsER) crossover: It may happen that all the four edges used by the edge recombination operator to select the next city lead to cities that have already been visited. In these cases, edge recombination selects a city randomly. The pheromone-supported edge recombination operator suggested here instead uses the probabilistic selection based on the pheromone matrix. Again, the resulting individual is subject to mutation, and after k individuals have been created, all of them are used to update both the pheromone matrix as well as the population.

Mutating Ants (MutA): ACO maintains diversity by choosing cities probabilistically in every step, and thus does not seem to require additional mutations. However, a simple change, like swapping two cities (i.e., what mutation does) is very unlikely to be produced by the probabilistic construction procedure. Therefore, we here suggest to mutate the solutions created by the ACO, adding a different kind of change.

Ant-based crossover (ABX): The idea of this hybrid is to combine ACO's sequential construction and elegant integration of heuristic knowledge with the population-based memory of EAs. ABX selects parents from the population just as an ordinary EA. These parents are used to construct a temporary pheromone matrix, by initializing all pheromones to the same small value, and allowing the parents to place additional pheromone on the solutions (paths) they represent. New solutions are then constructed in an ACO way, based on the temporary pheromone matrix and possible heuristic knowledge. Only the best of the generated solutions is returned as child and used to update the population (memory). It is straightforward to extend this idea by running an ACO on the temporary pheromone matrix for a few iterations (allowing the generated solutions to update the temporary pheromone matrix before generating some more solutions). ABX has first been proposed in Reference 25, and for the tests given later, we simply used the same parameter settings that had shown to be successful in that paper: two parents are selected, and the ACO is run on the temporary matrix for 5 iterations with 12 ants each. Only the best solution found is returned as a child.

10.6 Empirical Evaluation

Since ACO is primarily designed for permutation problems, we chose a simple symmetric Euclidean TSP with 100 cities to compare the different algorithms. Three problem instances of varying difficulty have been created: in problem P0, all cities are located equally spaced on a unit circle. To generate problem instances P3 and P6, the location of each of P0's city has been moved in a random direction by a distance of 0.3 or 0.6, respectively. The three resulting problems are visualized in Figure 10.2. Independent of the problem instance, each algorithm was allowed to create and evaluate 200,000 solutions.

Comparing different algorithms is a tricky business, because the results are highly dependent on the parameters used. Therefore, we tested all possible combinations of the parameter settings depicted in Table 10.1.

A comparison of the performance of the different algorithms on all three problems is presented in Table 10.2. The table reports the performance of the best parameter setting for each algorithm. All results are averaged over 30 runs with different random seeds. ACO with incorporated heuristic knowledge ($\beta = 5$), and ABX is also included for comparison.

Looking at the basic algorithms first, it is obvious that the edge recombination crossover yields significantly better results than the order crossover, which again performs better than ACO. The EA is able to find the optimal solution for the simple problem (which is approximately the circumference of a unit

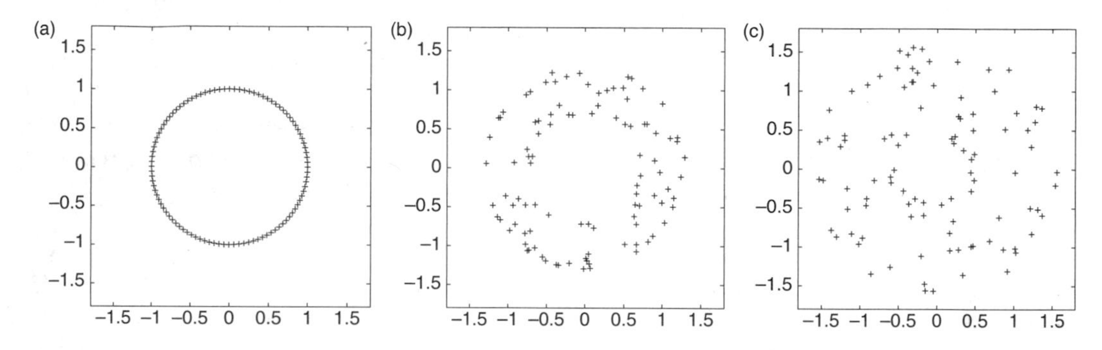

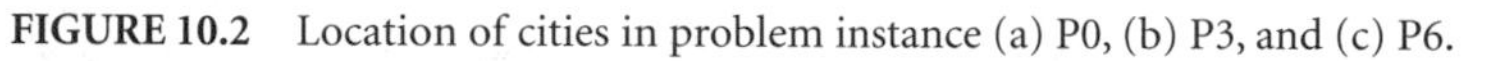
FIGURE 10.2 Location of cities in problem instance (a) P0, (b) P3, and (c) P6.

TABLE 10.1 Parameter Settings Tested

	α	k	ρ	n	p_m	s
ACO	1, 5, 10	10, 20, 50, 100	0.005, 0.01, 0.05	—	—	—
EA with ERX	—	—	—	100, 200, 300	0.3, 0.5	1.25, 1.5, 2.0
EA with OX	—	—	—	100, 200, 300	0.3, 0.5	1.25, 1.5, 2.0
50/50	1, 5	10, 20, 50	0.005, 0.01, 0.05	200, 300	0.3, 0.5	1.5
PsER crossover	1, 5	10, 20, 50	0.005, 0.01, 0.05	200, 300	0.3, 0.5	1.25, 1.5, 2.0
PC crossover	1, 5	10, 20, 50	0.005, 0.01, 0.05	200, 300	0.3, 0.5	1.25, 1.5, 2.0
MutA	1, 5, 10	10, 20, 50, 100	0.005, 0.01, 0.05	—	0.3, 0.5	—

TABLE 10.2 Tour Length of the Best Solution Found by the Different Algorithms

		Problem instance		
	Algorithm	0.0	0.3	0.6
Pure	ACO	14.22	24.54	29.76
	EA with ERX	6.28	18.29	23.22
	EA with OX	12.25	22.30	28.72
Hybrids	50/50	6.58	18.17	23.92
	PC crossover	12.61	22.21	28.56
	PsER crossover	6.28	17.59	22.88
	MutA	8.51	18.05	24.21
With heuristic knowledge	ACO + heuristic	6.28	16.11	21.80
	ABX	6.28	16.11	21.80

circle, i.e., 2π), but deviates from the optimal solution for the other two problems (the optimum is not known, but ACO with heuristic knowledge achieves better results). ACO, on the other hand, seems to be fairly unaffected by the structure of the problem.

As to the hybrids, only the pheromone-supported ER crossover was a clear winner. Simple mutation of the solutions constructed by ACO also performed very well, and consistently outperformed the basic ACO approach, indicating that the algorithm indeed benefited from the mix of operators. The pheromone completion crossover was disappointing (although still better than the basic ACO). However, these results are preliminary, and may be due to the specific problem instances chosen. Certainly, there are problems where ACO prevails, while for other problems, EAs perform better. Yet other problems may require a hybrid approach to be solved.

As expected, heuristic domain knowledge is able to drastically improve performance. The ACO with heuristic knowledge as well as ABX generate equally good (presumably optimal) results on all problem instances, outperforming all methods without heuristic knowledge. Note that besides the idea of ABX, incorporation of domain knowledge into the EA is not as straightforward, and we have not been able to produce similar results, for example, by seeding the population with a heuristic (results not reported). As the results show, the problem instances examined are too simple if heuristic knowledge is incorporated. On the larger problem instances used in Reference 25, ACO and ABX clearly outperformed a standard EA with ERX, and ABX outperformed ACO by ~1.3%.

10.7 Conclusion

In this chapter we have presented a unified framework for iterative search heuristics. According to the framework, each search heuristic maintains some sort of memory of the search history, which is used to construct new solutions, which are then evaluated and used to update the memory. Furthermore, we have argued that different memory schemes have different advantages, and that a search heuristic should benefit from combining different memorization paradigms.

The presented unified framework suggests a natural way for hybridization, and we have demonstrated its usefulness by deriving several interesting combinations of EAs and ACO and conducting a preliminary empirical evaluation of the resulting hybrids. A closer look at the compatibility of different memory schemes, and how they are best combined, is subject to future research. For the algorithm designer, of course, it would be invaluable to know which operators and memory schemes are most promising depending on the application at hand. However, that assumes a useful categorization of problems, and is thus several steps in the future. Overall, we hope that this chapter helps to gain a general understanding of different metaheuristics and of the way they interact.

References

[1] K.A. DeJong. *Evolutionary Computation.* MIT Press, Cambridege, MA, 2002.

[2] E. Bonabeau, M. Dorigo, and G. Theraulaz. *Swarm Intelligence: From Natural to Artificial Systems.* Oxford University Press, Oxford, 1999.

[3] E. Aarts and J. Korst. *Simulated Annealing and Boltzmann Machines.* John Wiley & Sons, New York 1989.

[4] F. Glover. Tabu search — p. I. *ORSA Journal of Computing,* 1: 190–206, 1989.

[5] P. Larrañaga and J.A. Lozano, Eds. *Estimation of Distribution Algorithms.* Kluwer Academic, New York, 2002.

[6] C. Reeves, Ed. *Modern Heuristic Techniques for Combinatorial Optimization.* McGraw-Hill, New York, 1995.

[7] E.L. Aarts and J.K. Lenstra, Eds. *Local Search in Combinatorial Optimization.* Wiley, Chichester, 1997.

[8] Z. Michalewicz and D.B. Fogel. *How to Solve It: Modern Heuristics.* Springer, New York, 1999.

[9] D.E. Brown, C.L. Huntley, and A.R. Spillane. A parallel genetic heuristic for the quadratic assignment problem. In *International Conference on Genetic Algorithms*, Morgan Kaufmann, San Francisco, CA, 1989, pp. 406–415.

[10] S.W. Mahfoud and D.E. Goldberg. Parallel recombinative simulated annealing: A genetic algorithm. *Parallel Computing*, 21: 1–28, 1995.

[11] C. Fleurent and J. Ferland. Genetic and hybrid algorithms for graph coloring. Technical report, Departement d'Informatique, Montreal, Canada, 1994.

[12] V.V. Miagkikh and W.F. Punch. An approach to solving combinatorial optimization problems using a population of reinforcement learning agents. In *Genetic and Evolutionary Computations Conference*, Morgan Kaufmann, San Francisco, CA, 1999, pp. 1358–1365.

[13] E.-G. Talbi, O. Roux, C. Fonlupt, and D. Robilliard. Parallel ant colonies for the quadratic assignment problem. *Future Generation Computer Systems*, 17: 441–449, 2001.

[14] T. Krink and M. Lovbjerg. The lifecycle model: Combining particle swarm optimisation, genetic algorithms and hillclimbers. In J.J. Merelo, P. Adamidis, H.-G. Beyer, J.-L. Fernandez-Villacanas, and H.-P. Schwefel, Eds., *Parallel Problem Solving from Nature*, Vol. 2439 of *Lecture Notes in Computer Science*, Springer, New York, 2002, pp. 621–630.

[15] P. Moscato. Memetic algorithms: A short introduction. In D. Corne, M. Dorigo, and F. Glover, Eds., *New Ideas in Optimization*, McGraw Hill, New York, 1999, chap. 14, pp. 219–234.

[16] P. Calegari, G. Coray, A. Hertz, D. Kobler, and P. Kuonen. A taxonomy of evolutionary algorithms in combinatorial optimization. *Journal of Heuristics*, 5: 145–158, 1999.

[17] E.-G. Talbi. A taxonomy of hybrid metaheuristics. *Journal of Heuristics*, 8: 541–564, 2002.

[18] R. Poli and B. Logan. The evolutionary computation cookbook: Recipes for designing new algorithms. In *Online Workshop on Evolutionary Computation*, 1996, pp. 33–36.

[19] C. Blum and A. Roli. Metaheurisitics in combinatorial optimization: Overview and conceptual comparison. *ACM Computer Survey*, 35: 268–308, 2003.

[20] M. Zlochin, M. Birattari, N. Meuleau, and M. Dorigo. Model-based search for combinatorial optimization. Technical report TR/IRIDIA/2000-15, INRIDIA, Universite Libre de Bruxelles, 2001.

[21] J. Branke, M. Stein, and H. Schmeck. A Unified Framework for Metaheuristics. Technical report 417, University of Karlsruhe, Institute AIFB, Karlsruhe, Germany, 2002.

[22] L. Davis. Applying adaptive algorithms to epistatic domains. In *International Joint Conference on Artificial Intelligence*, Morgan Kaufmann, San Francisco, CA, 1985, pp. 162–164.

[23] D. Whitley, T. Starkweather, and D'A. Fuquay. Scheduling problems and traveling salesman: The genetic edge recombination operator. In J. Schaffer, Ed., *International Conference on Genetic Algorithms*, Morgan Kaufmann, San Francisco, CA, 1989, pp. 133–140.

[24] V.V. Miagkikh and W.F. Punch. Global search in combinatorial optimization using reinforcement learning algorithms. In *Congress on Evolutionary Computation*, IEEE, Piscataway, 1999, pp. 189–196.

[25] J. Branke, C. Barz, and I. Behrens. Ant-based crossover for permutation problems. In E. Cantu-Paz, Ed., *Genetic and Evolutionary Computation Conference*, Vol. 2273 of *Lecture Notes in Computer Science*, Springer, New York, 2003, pp. 754–765.

11

The Foundations of Autonomic Computing

Salim Hariri
Bithika Khargaria
Manish Parashar
Zhen Li

11.1 Introduction

The advances in computing and communication technologies and software tools have resulted in an explosive growth in networked applications and information services that cover all aspects of our life. These services and applications are inherently complex, dynamic, and heterogeneous. In a similar way, the underlying information infrastructure, for example, the Internet, is large, complex, heterogeneous, and dynamic, globally aggregating large numbers of independent computing and communication resources, data stores, and sensor networks. The combination of the two results in application development, configuration, and management complexities that break current computing paradigms, which are based on static behaviors, interactions, and compositions of components and services. As a result, applications, programming environments, and information infrastructures are rapidly becoming brittle, unmanageable, and insecure. This has led researchers to consider alternative programming paradigms and management techniques that are based on strategies used by biological systems to deal with complexity, dynamism, heterogeneity, and uncertainty.

Autonomic computing is inspired by the human autonomic nervous system, which has developed strategies and algorithms to handle complexity and uncertainties, and aims at realizing computing systems and applications capable of managing themselves with minimum human intervention. In this chapter, we first give an overview of the architecture of the nervous system and use it to motivate the autonomic computing paradigm. We then illustrate how this paradigm can be used to build and manage complex applications. Finally, we present an overview of existing autonomic computing systems and applications and highlight two such systems.

11.2 Autonomic Nervous System

The human nervous system is, to the best of our knowledge, the most sophisticated example of autonomic behavior existing in nature today [1]. It is the body's master controller that monitors changes inside and outside the body, integrates sensory input, and effects appropriate response. In conjunction with the endocrine system, which is the body's second important regulating system, the nervous system is able to constantly regulate and maintain homeostasis. Homeostasis is one of the most remarkable properties of highly complex systems. A homeostatic system (e.g., a large organization, an industrial firm, a cell) is an open system that maintains its structure and functions by means of a multiplicity of dynamic equilibriums that are rigorously controlled by interdependent regulation mechanisms. Such a system reacts to every change in the environment, or to every random disturbance, through a series of modifications that are equal in size and opposite in direction to those that created the disturbance. The goal of these modifications is to maintain internal balances.

The manifestation of the phenomenon of homeostasis is widespread in the human system. As an example, consider the mechanisms that maintain the concentration of glucose in the blood within limits — if the concentration should fall below about 0.06%, the tissues will be starved of their chief source of energy; if the concentration should rise above about 0.18%, other undesirable effects will occur. If the blood-glucose concentration falls below about 0.07%, the adrenal glands secrete adrenaline, which causes the liver to turn its stores of glycogen into glucose. This passes into the blood and the blood-glucose concentration drop is opposed. Further, a falling blood-glucose also stimulates appetite causing food intake, which after digestion provides glucose. On the other hand, if the blood-glucose concentration rises excessively, the secretion of insulin by the pancreas is increased, causing the liver to remove the excess glucose from the blood. Excess glucose is also removed by muscles and skin, and if the blood-glucose concentration exceeds 0.18%, the kidneys excrete excess glucose into the urine. Thus, there are five activities that counter harmful fluctuations in blood-glucose concentration [2].

The above example focuses on the maintenance of the blood-glucose concentration within safe or operational limits that have been "predetermined" for the species. Similar control systems exist for other parameters such as systolic blood pressure, structural integrity of the medulla oblongata, severe pressure of heat on the skin, and so on. All these parameters have a bearing on the survivability of the organism, which in this case is the human body. However, all parameters are not uniform in their urgency or their relations to lethality [2]. Parameters that are closely linked to survival and are closely linked to each other so that marked changes in one leads sooner or later to marked changes in the others, have been termed as essential variables by Ashby in his study of the design for a brain [2], which is discussed later.

11.2.1 Ashby's Ultrastable System

Every real machine embodies no less than an infinite number of variables [2], and for our discussion we can safely think of the human system as represented by a similar sets of variables, of which we will consider a few. In order for an organism to survive, its essential variables must be kept within viable limits (see Figure 11.1). Otherwise the organism faces the possibility of disintegration and loss of identity (i.e., dissolution or death) [2].

The body's internal mechanisms continuously work together to maintain its essential variables within their limits. Ashby's definition of adaptive behavior as demonstrated by the human body follows from

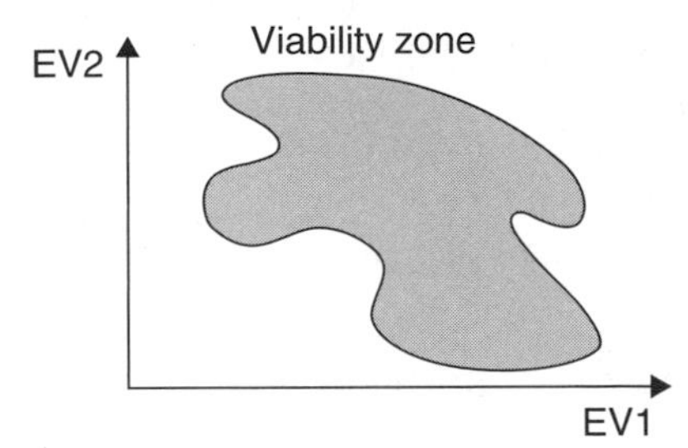

FIGURE 11.1 Essential variables.

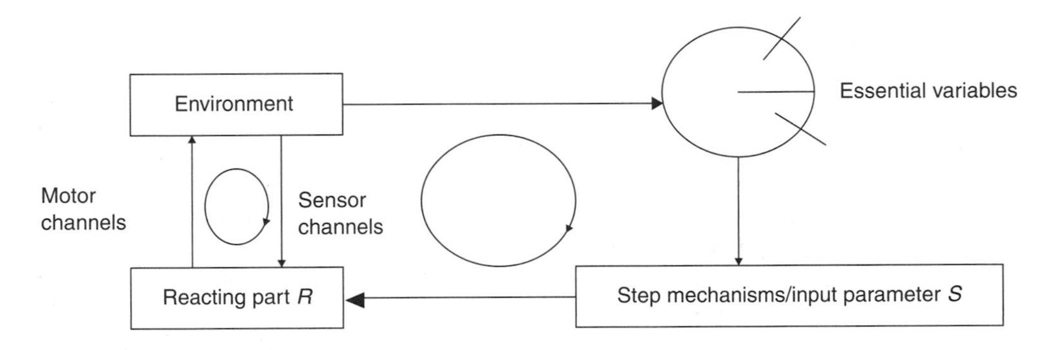

FIGURE 11.2 The ultrastable system architecture [2].

this observation. He states that a form of behavior is adaptive if it maintains the essential variables within physiological limits [2] that define the viability zone. Two important observations can be made:

- The goal of the adaptive behavior is directly linked with the survivability of the system.
- If the external or internal environment pushes the system outside its physiological equilibrium state, the system will always work toward coming back to the original equilibrium state.

Ashby observed that many organisms undergo two forms of disturbance: (1) frequent small impulses to main variables and (2) occasional step changes to its parameters. Based on this observation, he devised the architecture of the Ultrastable system that consists of two closed loops (see Figure 11.2): one that controls small disturbances and a second that is responsible for longer disturbances.

As shown in Figure 11.2, the ultrastable system consists of two subsystems, the environment and the reacting part (*R*). *R* represents a subsystem of the organism that is responsible for overt behavior or perception. It uses the sensor channels as part of its perception capability and motor channels to respond to the changes impacted by the environment. These set of sensors and motor channels constitute the primary feedback between *R* and the environment. We can think of *R* as a set of behaviors of the organism that gets triggered based on the changes affected by the environment. *S* represents the set of parameters that triggers changes in relevant features of this behavior set. Note that in Figure 11.2, *S* triggers changes only when the environment affects the essential variables in a way that causes them to go outside their physiological limits. As mentioned earlier, these variables need to be maintained within physiological limits for any adaptive system/organism to survive. Thus we can view this secondary feedback between the environment and *R* as responsible for triggering the adaptive behavior of the organism. When the changes impacted by the environment on the organism are large enough to throw the essential variables out of their physiological limits, the secondary feedback becomes active and changes the existing behavior sets of the organism to adapt to these new changes. Notice that any changes in the environment tend to push an otherwise stable system to an unstable state. The objective of the whole system is to maintain the subsystems (the environment and *R*) in a state of stable equilibrium. The primary feedback handles finer changes in the environment with the existing behavior sets to bring the whole system to stable equilibrium. The secondary feedback handles coarser and long-term changes in the environment by changing its existing behavior sets and eventually brings back the whole system to stable equilibrium state. Hence, in a nutshell, the environment and the organism always exist in a state of stable equilibrium and any activity of the organism is triggered to maintain this equilibrium.

11.2.2 The Nervous System as a Subsystem of Ashby's Ultrastable System

The human nervous system is adaptive in nature. In this section we apply the concepts underlying the Ashby's ultrastable system to the human nervous system. The goal is to enhance the understanding of an adaptive system and help extract essential concepts that can be applied to the autonomic computing paradigm presented in the following sections.

As shown in Figure 11.3, the nervous system is divided into the Peripheral Nervous System (PNS) and the Central Nervous System (CNS). The PNS consists of *sensory neurons* running from stimulus receptors that inform the CNS of the stimuli and *motor neurons* running from the CNS to the muscles and glands, called effectors, which take action. CNS is further divided into two parts: sensory–somatic nervous system and the autonomic nervous system. Figure 11.4 shows the architecture of the autonomic nervous system as an Ashby utrastable system.

As shown in Figure 11.4, the sensory and motor neurons constitute the sensor and motor channels of the ultrastable system. The triggering of essential variables, selection of the input parameter S and translation of these parameters to the reacting part R constitute the workings of the nervous system. Revisiting the management of blood-glucose concentration within physiological limits discussed earlier, the five mechanisms that get triggered when the essential variable (i.e., concentration of glucose in blood)

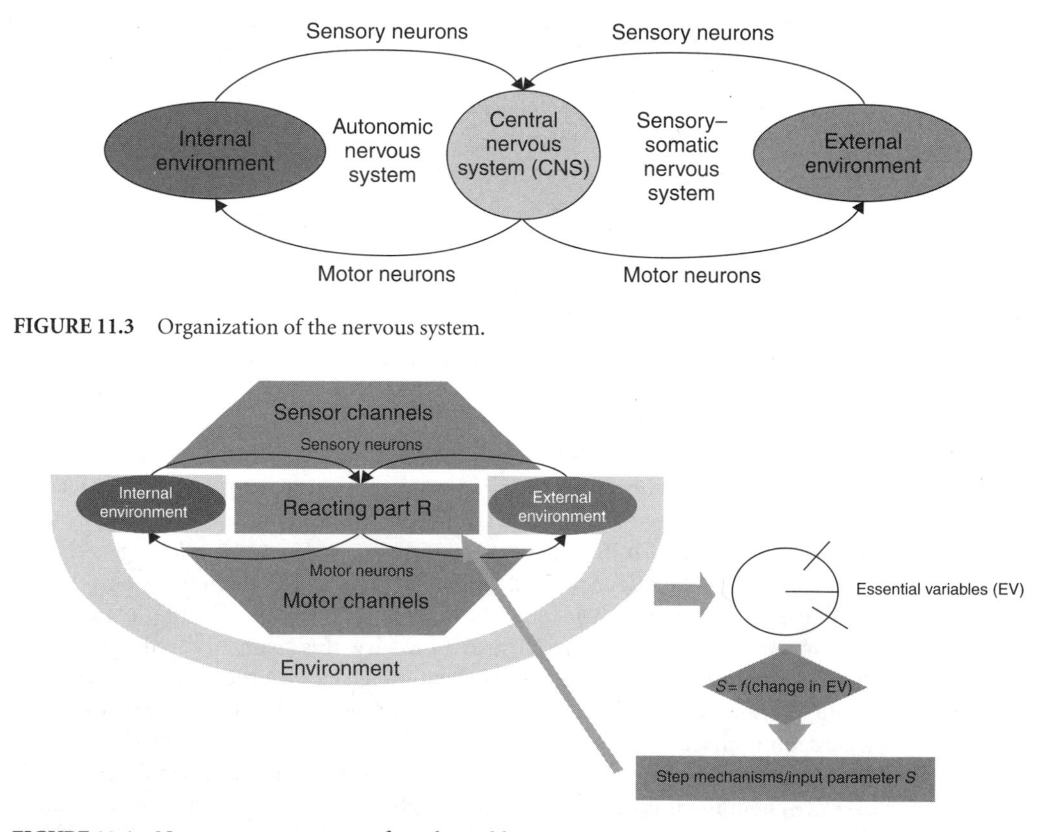

FIGURE 11.3 Organization of the nervous system.

FIGURE 11.4 Nervous system as part of an ultrastable system.

goes out of the physiological limits change the normal behavior of the system such that the reacting part R works to bring the essential variable back within limits. It uses its motor channels to effect changes so that the internal environment and the system (organism) come into the state of stable equilibrium. It should be noted that the environment here is divided into the internal environment and external environment. The internal environment represents changes impacted internally within the human system and the external environment represents changes impacted by the external world. However, the goal of the organism is to maintain the equilibrium of the entire system where all the subsystems (the organism or system itself, and the internal and external environments) are in stable equilibrium.

11.3 Autonomic Computing Paradigm

An autonomic computing paradigm, modeled after the autonomic nervous system, must have a mechanism whereby changes in its essential variables can trigger changes in the behavior of the computing system such that the system is brought back into equilibrium with respect to the environment. This state of stable equilibrium is a necessary condition for the survivability of the organism. In the case of an autonomic computing system, we can think of survivability as the system's ability to protect itself, recover from faults, reconfigure as required by changes in the environment, and always maintain its operations at a near optimal performance. Its equilibrium is impacted by both the internal environment (e.g., excessive CPU utilization) and the external environment (e.g., protection from an external attack). The autonomic computing system requires: (1) sensor channels to sense the changes in the internal and external environment, and (2) motor channels to react to and counter the effects of the changes in the environment by changing the system and maintaining equilibrium. The changes sensed by the sensor channels have to be *analyzed* to determine if any of the essential variables has gone out of their viability limits. If so, it has to trigger some kind of *planning* to determine what changes to inject into the current behavior of the system such that it returns to the equilibrium state within the new environment. This planning would require *knowledge* to select the right behavior from a large set of possible behaviors to counter the change. Finally, the motor neurons *execute* the selected change. "Sensing," "Analyzing," "Planning," "Knowledge," and "Execute" are in fact the keywords used to identify an autonomic system [3]. We use these concepts to present the architecture of an autonomic component that represents the smallest unit of an autonomic application or a system with self-managing capabilities.

11.3.1 Architecture of an Autonomic Component

An autonomic component (see Figure 11.5) is the smallest unit of an autonomic application or system. Multiple autonomic components (units) can be composed to form an autonomic application or system, which is a self-contained software module or system with specified input/output interfaces and explicit context dependencies. It also has embedded mechanisms for self-management responsible for providing functionalities, exporting constraints, managing its own behavior in accordance with context and policies, and interacting with other components. For example, an autonomic component consists of the following parts:

Managed Element: This is the smallest unit of the application and it contains executable code (e.g., numerical model of a physical process) and a data structure that defines the executable code's attributes (e.g., its purpose, operation, input and output requirements, criteria for when and how to control it). At runtime, the managed element can be affected in different ways, for example, it can encounter a failure during execution, it can be externally attacked, or it may slow down and affect the performance of the entire application.

Environment: The environment represents all the factors that can impact the managed element. The environment and the managed element can be two subsystems forming a stable system. Any change in the environment causes the whole system to go from a stable state to an unstable state. This change is then offset by reactive changes in the managed element causing the system to move

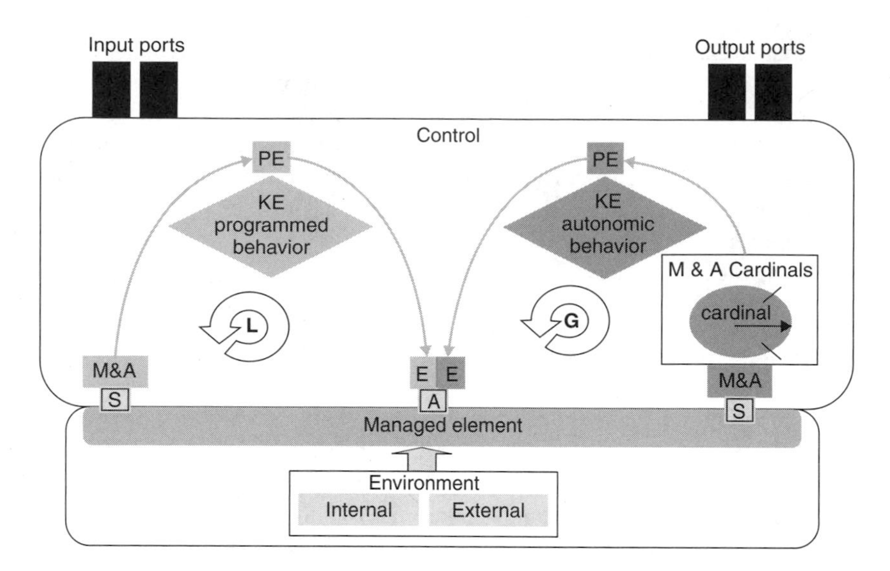

FIGURE 11.5 An automatic component.

back from the unstable state to a different stable state. Notice that the environment consists of two parts — internal and external. The internal environment consists of changes internal to the managed element, which can be looked at as reflecting the state of the application. The external environment can be thought of as reflecting the state of the execution environment.

Control: Each autonomous component has its own manager that: (1) accepts user-specified requirements (fault tolerance, performance, security, etc.) (2) interrogates the data structure that characterizes the executable code (3) senses the state of the overall computation (4) determines the nature and instantaneous state of the overall computational environment and (5) uses this information to control the operation of its associated executable code within the overall system in order to effectively achieve the user-specified requirements. This process is accomplished on-the-fly and continuously throughout the execution of the overall computation. As is evident from Figure 11.5, the control part consists of two control loops — the local loop and the global loop.

The *local loop* can only handle known environment states. Its knowledge engine contains the mapping of environment states to behaviors. For example, when the load on the local system goes above the threshold value, the local control loop will work toward balancing the load by either controlling the local resources available to the managed element or by reducing the size of the problem handled by this element. This will work only if the local resources can handle the computational requirements. However, the local loop is *blind* to the overall behavior of the entire application or system and thus cannot achieve the desired global objectives. In a scenario where the entire system is affected, the local loop will continue repeating local optimization that may lead to degradation in performance and result in unadapted or chaotic behavior. At some point, one of the essential variables of the system (in this case, a performance cardinal) overshoots its limits. This is when the global loop comes into action.

The *global loop* can handle unknown environment states and may involve machine learning. It uses four cardinals for the monitoring and analysis of the managed elements. These are performance, configuration, protection, and security. These cardinals are like the essential variables described in Ashby's ultrastable system. This control loop acts toward changing the existing behavior of the managed element such that it can adapt itself to changes in the environment. For example, in load-balancing, the desired behavior of the managed element (as directed by the local loop) requires its local load to be within prescribed limits. However, the local loop might not be able to maintain the local load within these acceptable limits, which in turn might degrade the performance of the overall system. Consequently, this change in the overall performance cardinal triggers the global loop, which then selects an alternate behavior pattern from the pool of behavior patterns for the managed element. This analysis and planning uses its knowledge engine. Finally, the new plan is executed to adapt the behavior of the managed element to the new environment conditions.

Input and Output Ports: Many interacting autonomous components may be composed to form a complex application. These autonomic components use the input and output ports for such a composition.

11.4 Autonomic Computing Systems

An autonomic computing system can be a collection of autonomic components, which can manage their internal behaviors and relationships with others in accordance to high-level policies. The principles that govern all such systems have been summarized as eight defining characteristics [4]:

Self-Awareness: an autonomic system knows itself and is aware of its state and its behaviors.

Self-Protecting: an autonomic system is equally prone to attacks and hence it should be capable of detecting and protecting its resources from both internal and external attack and of maintaining overall system security and integrity.

Self-Optimizing: an autonomic system should be able to detect suboptimal behaviors and intelligently perform self-optimization functions.

Self-Healing: an autonomic system must be aware of potential problems and should have the ability to reconfigure itself to continue to function smoothly.

Self-Configuring: an autonomic system must have the ability to dynamically adjust its resources based on its state and the state of its execution environment.

Contextually Aware: an autonomic system must be aware of its execution environment and be able to react to changes in the environment.

Open: an autonomic system must be portable across multiple hardware and software architectures, and consequently it must be built on standard and open protocols and interfaces.

Anticipatory: an autonomic system must be able to anticipate, to the extent possible, its needs and behaviors and those of its context, and be able to manage itself proactively.

Sample self-managing system/application behaviors include installing software when it detects that the software is missing (self-configuration), restarting a failed element (self-healing), adjusting current workload when it observes an increase in capacity (self-optimization), and taking resources offline if it detects an intrusion attempt (self-protecting).

Each of the attributes listed above are active research areas toward realizing autonomic systems and applications. Generally, self-management is addressed in four primary system/application aspects, that is, configuration, optimization, protection, and healing. Further, self-management solutions typically consist of the steps outlined earlier: (1) the application and underlying information infrastructure provide information to enable context and self-awareness (2) system/application events trigger analysis, deduction, and planning using system knowledge and (3) plans are executed using the adaptive capabilities of the system. An autonomic system implements self-managing attributes using the control loops described earlier to collect information, makes decisions, and adapt, as necessary.

Autonomic components need to collaborate to achieve coherent autonomic behaviors at the application level. This requires a common set of underlying capabilities including representations and mechanisms for solution knowledge, system administration, problem determination, monitoring and analysis, and policy definition, enforcement, and transaction measurements [5]. For example, a common solution knowledge capability captures installation, configuration, and maintenance information in a consistent manner, and eliminates the complexity introduced by heterogeneous tools and formats. Common administrative console functions ranging from setup and configuration to solution runtime monitoring and control provide a single platform to host administrative functions across systems and applications, allowing users to manage solutions rather than managing individual systems/applications. Problem determination is one of the most basic capabilities of an autonomic element and enables it to decide on appropriate actions when healing, optimizing, configuring, or protecting itself. Autonomic monitoring is a capability that provides an extensible runtime environment to support the gathering and filtering of data obtained through sensors. Complex analysis methodologies and tools provide the power and flexibility required to perform a range of analyses of sensor data, including deriving information about resource configuration, status, offered workload, and throughput. A uniform approach to defining the policies is necessary to support adaptations and govern decision-making required by the autonomic system. Transaction measurements are needed to understand how the resources of heterogeneous systems combine into a distributed transaction execution environment. Using these measurements, analysis and plans can be derived to change resource allocations to optimize performance across these multiple systems as well as determine potential bottlenecks in the system.

11.5 Illustrative Example: Distributed Cellular DEVS Dynamic Forest Fire Simulation

One important feature of autonomic applications is their ability to change the components and the structure that interconnects these components at runtime as required by the self-control and management algorithms. For example, in fire forest simulation, you might need to change the type of components used to simulate the fire depending on the current atmosphere conditions (rain, dry, speed of wind), and the vegetation type at each computational grid. The runtime system framework will then modify the data structures of the autonomic components in order to add or remove components.

The forest fire simulation model has been developed using the concepts of cellular automata. A cellular automaton is an array of identically programmed automata, or "cells," which interact with one another. Essentially, it is a one-dimensional (1D) string of cells, a 2D Grid or a 3D solid and has three important features — state, neighborhood, and its program. Just as every living cell contains all of the instructions for its replication and operation, each individual cell in a cellular automaton can be programmed with a set of rules that defines how its state changes in response to its current state and that of the neighbors. In the forest fire model, the forest is represented as a 2D cell space composed of cells of dimensions $l \times b$ (l: length, b: breadth). For each cell there are eight major wind directions N, NE, NW, S, SE, SW, E, W as shown in Figure 11.6 and Figure 11.7

A group of such individual cells will together constitute, a Virtual Computational Unit (VCU). The weather and vegetation conditions are assumed to be uniform within a cell, but may vary in the entire cell space. A cell interacts with its neighbors along all the eight directions as listed earlier, using input and

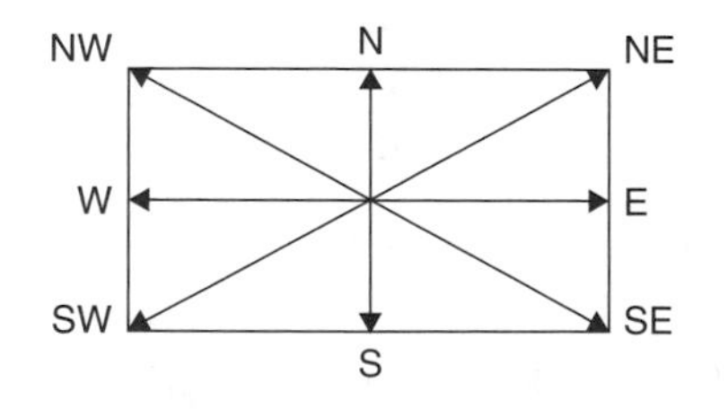

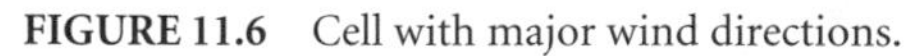
FIGURE 11.6 Cell with major wind directions.

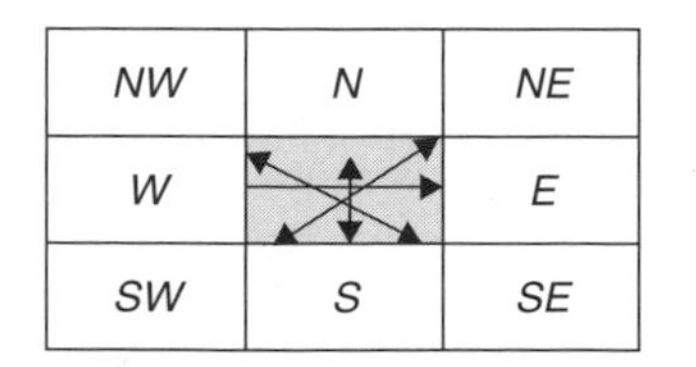

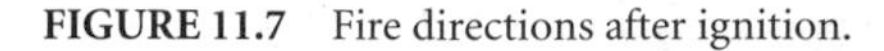
FIGURE 11.7 Fire directions after ignition.

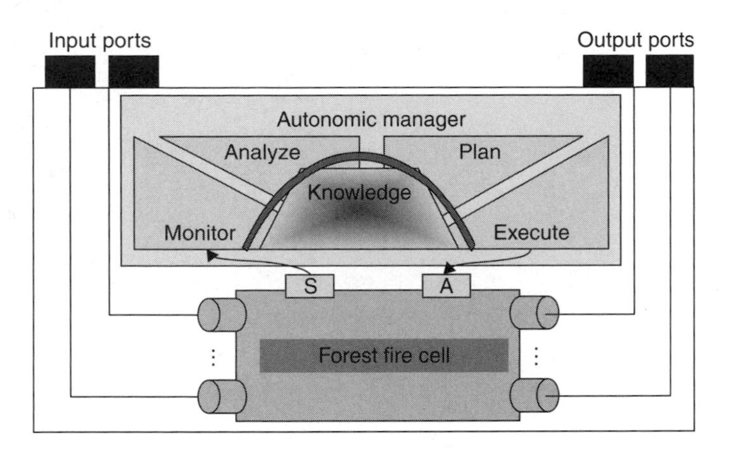

FIGURE 11.8 A forest fire cell expressed as an autonomic component.

output ports (the DEVS-Java model). A cell is programmed to undergo state changes from "unburned" to "burning" if it is hit by an igniter or gets a notification message to compute its fire-line intensity value. The cell changes state from "unburned" to "burning" only if the computed fire-line intensity is above a threshold value for "burning." During the "burning" phase, the cell propagates to eight different fire components along the eight directions (refer to Figure 11.5). The direction and value of maximum fire spread is computed using Rothermel's fire spread model [6]. The remaining seven components are then derived using a different decomposition algorithm. Rothermel's model takes into account the wind–speed and direction, the vegetation type, the calorific value of the fuel, and terrain type in calculating the fire spread.

11.5.1 Forest Fire Cell as an Autonomic Component

A single forest fire cell is modeled by an Autonomic Component (Figure 11.8), which consists of:

1. *Specification:* get notification messages from neighbor cell or igniter and send notification messages to the neighbor cells.
2. *States:* $wind_{ew}$, $wind_{ns}$, veg_h, veg_m, veg_l, unburned, burning, burned, blocked.
3. *Actions:* partition cell space horizontally/vertically, allocate/de-allocate resources, migrate component, add/delete coupling, start, terminate, pause, resume, and so on.
4. *Policies:* for self-optimizing, self-healing, self-configuration, and self-protecting.

The set of all possible *actions* on the component, are distributed into different *policies* that can be activated at runtime based on the component current state and the state of the physical resources. The policies are then used to map the *states* to the *actions,* as shown in Figure 11.9. In the following, we will explain few of these mappings with example scenarios.

Case 1: State change from $wind_{ew}$ to $wind_{ns}$ triggers actions guided by the self-optimizing policy.

The entire forest cell space is to be distributed into smaller units consisting of individual forest cells such that each group of forest cells executes on a single compute node. Since this application involves a lot of interactions between neighboring cells, an optimal cell space partitioning plan must ensure that neighbor cells do not spend too much time communicating with each other. One strategy is to partition cells based on wind direction. As is evident from Figure 11.10, the wind direction determines the direction of the communications among neighboring cells in the cell space. Thus, a partitioning strategy based on wind direction will ensure that most of the communication occurs within the cells assigned to a single compute node.

As seen from Figure 11.11, as the wind direction changes from east–west to north–south, the state of the fire cell changes from $wind_{ew}$ to $wind_{ns}$ after a certain time. At the end of the time advance, an output

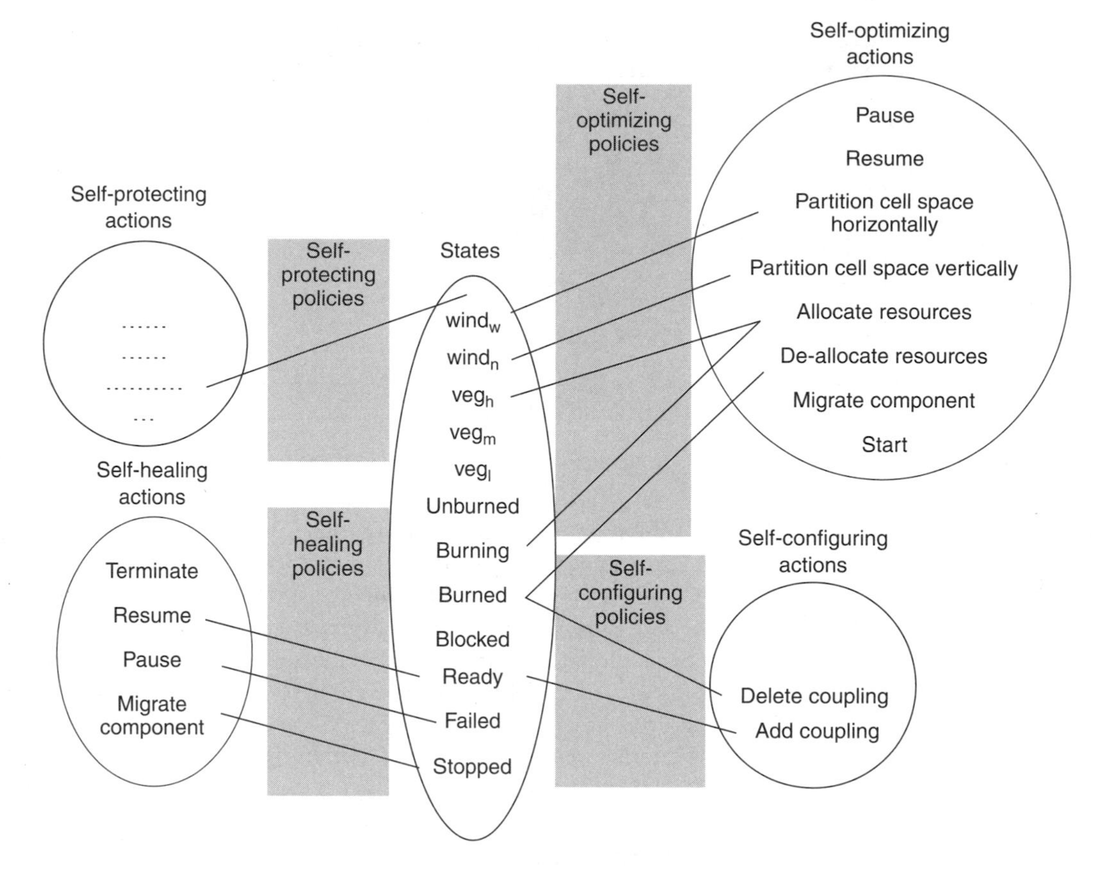

FIGURE 11.9 Mapping from states to actions governed by policies.

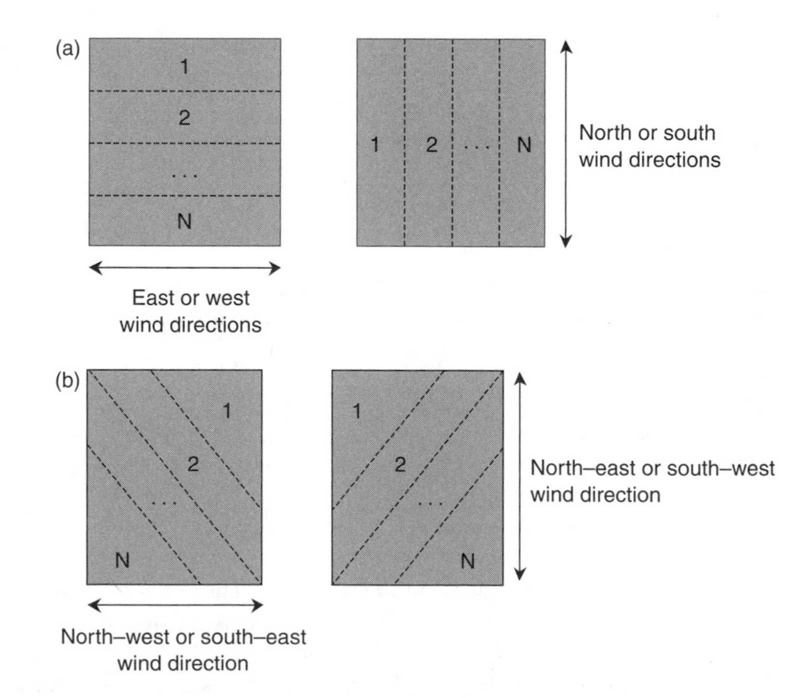

FIGURE 11.10 Partitioning of cell space based on wind direction.

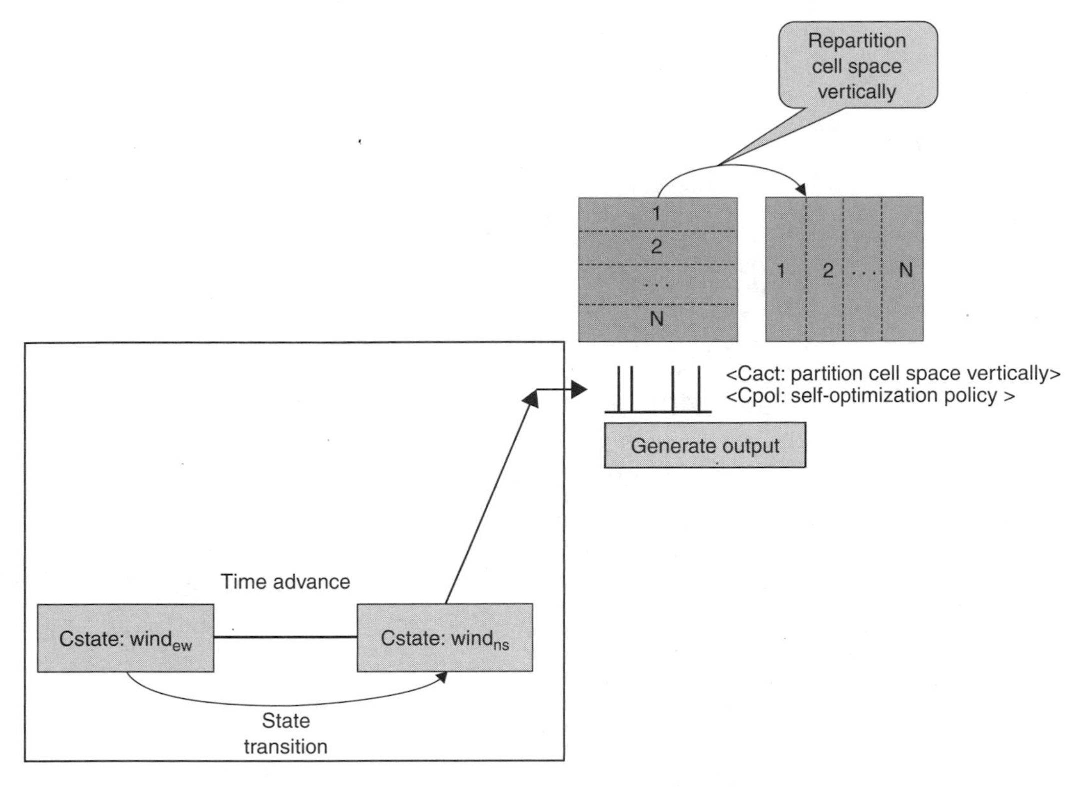

FIGURE 11.11 State transition guided by the self-optimization policy.

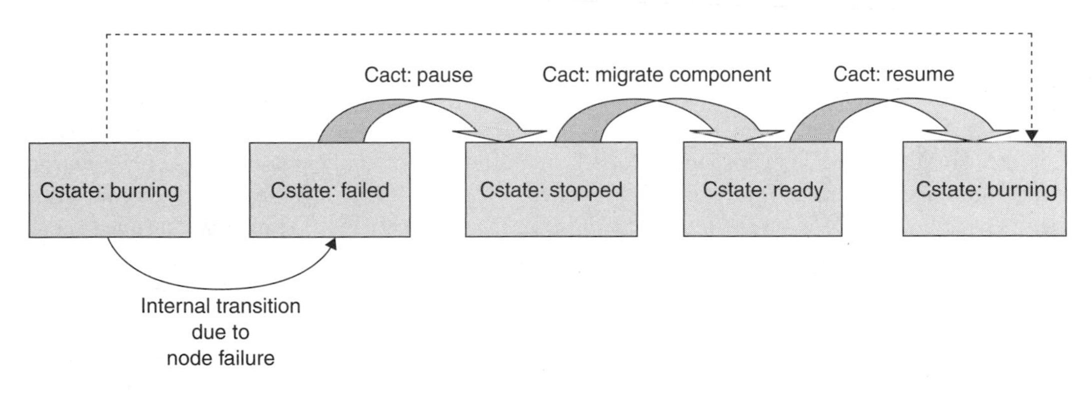

FIGURE 11.12 Sets of state transitions and actions guided by self-healing policies.

is generated, which involves a self-optimization action. In this case, the requested action is to repartition the cell space vertically. Referring to Figure 11.11, we see that this action is mapped with the state wind$_{ns}$ The autonomic manager uses its monitoring and analysis engines to detect this state change. The planning engine then generates the appropriate action as shown in Figure 11.11. The appropriate action (mapping function) is a part of the knowledge stored in the knowledge engine. Finally, the execution engine executes this action.

Case 2: State change from burning to failed triggers actions guided by the self-healing policy.

A scenario where a component stops running due to a failure in the node is shown in Figure 11.12. A series of state changes generates the corresponding actions as shown in Figure 11.12. Finally the component resumes execution from the last check-pointed state and its state changes into *burning*.

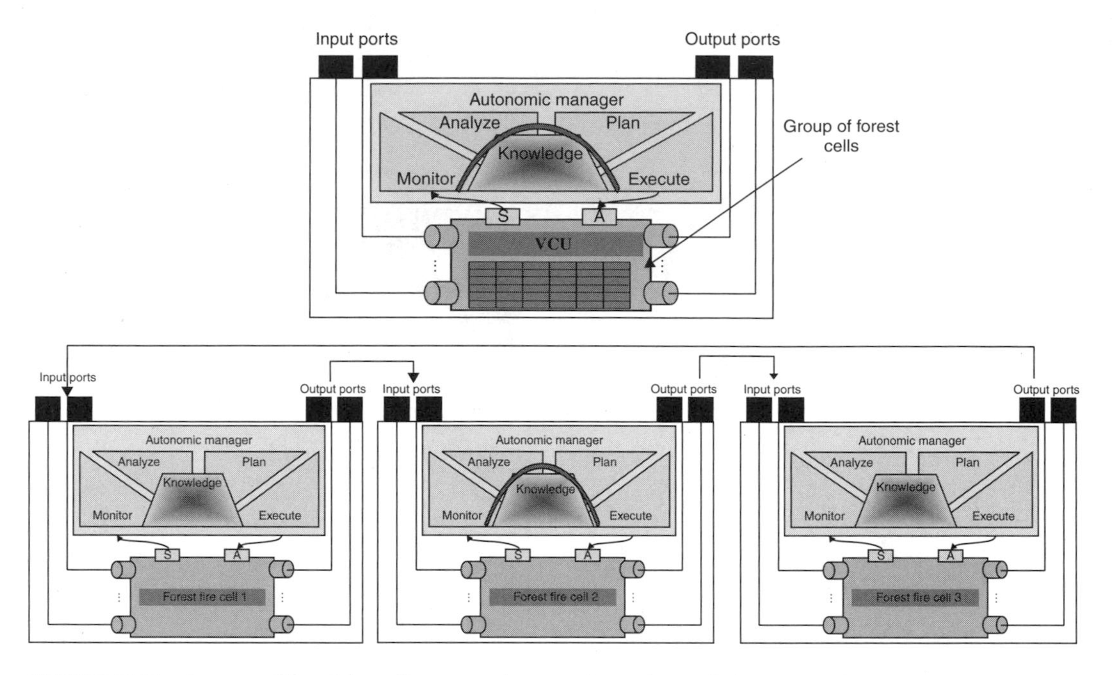

FIGURE 11.13 A group of forest fire cells expressed as an autonomic coupled component.

11.5.2 Forest Fire Cell Space as Autonomic Coupled Component

The forest fire cell space is nothing but a collection of forest cells that are connected in a particular manner to create the entire forest application. In terms of our autonomic component formalisms, we can view the forest cell space as a collection of autonomic forest cells (Figure 11.13).

A forest fire autonomic coupled component consists of: (1) *specification:* a union of a group individual forest cell autonomic components with connected input, output ports. (2) *actions:* repartition cell space, migrate, pause, resume. (3) *global policies:* for self-optimization, self-configuration, and so on to manage the autonomic capabilities of the coupled component which, in our case, is the collection of forest fire cells. The interaction between the *states*, *actions*, and *policies* are very much similar as explained earlier. In what follows, we will explain this interaction in the context of the forest fire application.

Load balanced execution is crucial to the performance of the forest fire application because the entire application proceeds at the speed of the slowest problem piece (VCU) in the entire cell space. In Figure 11.14, we depict a scenario where the complexity of VCU1 (number of forest cells) is reduced by removing some cells from that VCU and adding them to another VCU2 that is lightly loaded.

VCU1 starts at *loadHigh.* Note that this state is a union of the states of the cells in that VCU. Hence *loadHigh is* the union of *burning1*, *burning2*, *unburned3*, ..., *burning*$_n$. This could be a possible scenario because cells perform the major amount of computation during the *burning* state. Hence, if the complexity of VCU1 is high and in addition, most of the cells in it are *burning*, then it can go to the *loadHigh* state. The self-optimization policy for this VCU then selects the defined individual cell action to *migrate cells.* This causes the VCU to transition to state *paused.* The next action is to *select cells for migration.* Again, the self-optimizing policy uses its criteria to guide the selection of cells for migration. For example, it might pick cells with the objective of minimal increase in communication overhead after the migration. At this point the VCU goes to state *reconfigure.* The corresponding action is to *remove couplings* for the selected cells. After this action is applied on the VCU, it goes back to state *ready* when it is ready to run again with the load reduced. Its execution is then *resumed* when it changes state to the *loadMid state* and starts execution.

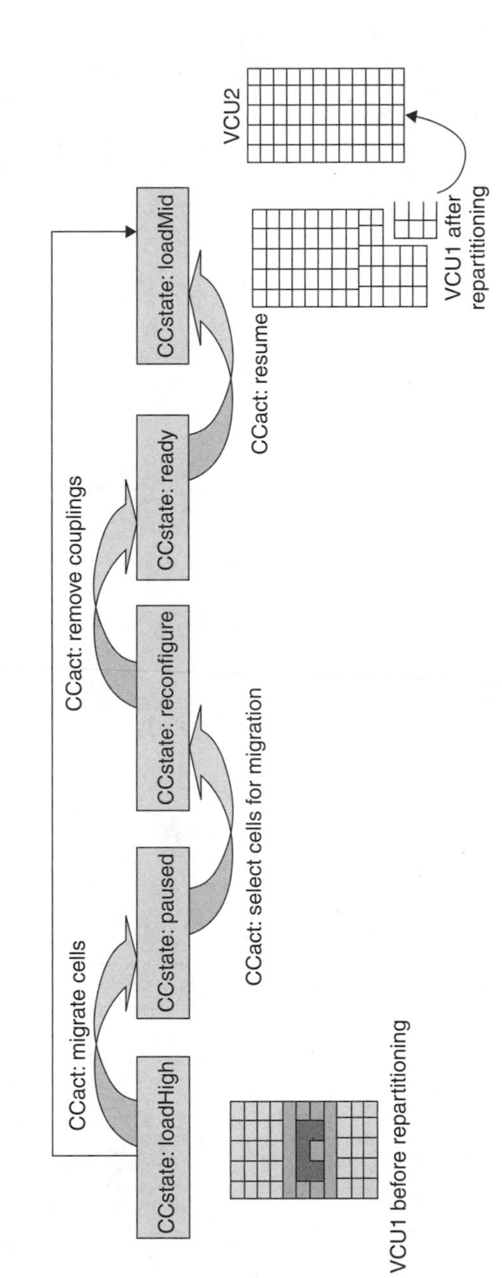

FIGURE 11.14 Sets of state transitions and actions guided by self-optimizing policies bring a component from heavily loaded to lightly loaded state.

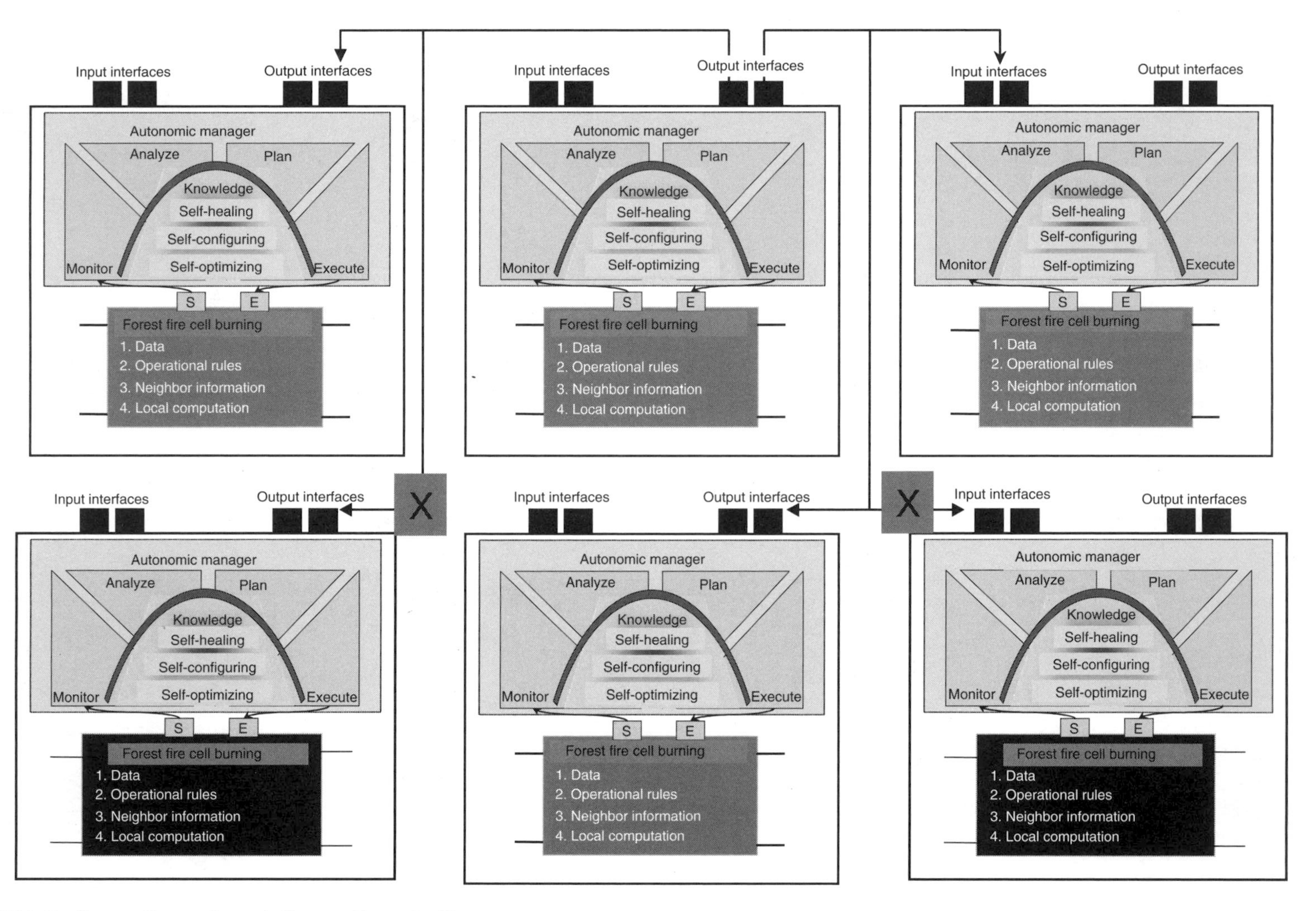

FIGURE 11.15 Forest cell space dynamically reconfigures itself.

11.5.3 Dynamic Composition and Self-Management of Forest Fire Application

As discussed previously, we might encounter scenarios where the cells need to be dynamically deleted from or added to the cell space. In this section, we will discuss how our framework achieves this objective.

As shown in Figure 11.15, two cells have *burned* and the other cells are *burning*. Hence, the couplings of the burned cells are to be dynamically removed from this group of cells at runtime (denoted by the "cross" in Figure 11.15). The monitoring engine senses this state change from *burning* to *burned*. It notifies the analysis engine and planning engine. The planning engine then looks up the appropriate action for this state by using the information stored in the self-configuration policy. In this case, the action is to *remove the couplings* of the *burned* cells. Now, the execution engine deletes the specific ports and then sets the new state of the VCU as *ready* and the VCU *resumes* execution.

11.6 The Autonomic Computing Landscape

There have been a number of research efforts in both academia and industry addressing autonomic computing concepts and investigating the issues outlined earlier. Existing projects and products can be broadly classified as: (1) Systems that incorporate mechanisms to address autonomic properties for problem determination, autonomic monitoring, complex analysis, policies for autonomic managers, and transaction measurements. (2) Systems that investigate models, programming paradigms and development environments to support the development of autonomic applications and systems. Systems in the former category are summarized in Table 11.1, while systems in the latter category are summarized in Table 11.2.

TABLE 11.1 Systems Incorporating Autonomic Properties

System	Application area	Addressed autonomic issues addressed/Key issues
OceanStore [7,8]	Global, consistent, highly-available persistent data storage	Self-healing, self-optimization, self-configuration, self-protection Policy-based caching, routing substrate adaptation, autonomic replication, continuous monitoring, testing, and repairing
Storage Tank [9]	Multi-platform, universally accessible storage management	Self-optimization, self-healing Policy-based storage and data management, server redirection and log-based recovery
Océano [10]	Cost effective scalable management of computing resources for software farms	Self-optimization, self-awareness Autonomic demands distribution, constant component monitoring
SMART DB2 [11]	Reduction of human intervention & cost for DB2	Self-optimization, self-configuration Autonomic index determination, disaster recovery, continuous monitoring of DB2's health and alerting the DBA
AutoAdmin [12]	Reduction of total cost of ownership (TCO)	Self-tuning, self-administration Usage tracking, index tuning and recommending based on workload
Sabio [13]	Autonomic classification of documents	Self-organization, self-awareness Group documents according to the word and phrase usage
Q-Fabric [14]	System support for continuous online management	Self-organization Continuous online quality management through "customizability" of each application's QoS

TABLE 11.2 Systems Support Development of Autonomic Applications and Systems

System	Focus	Autonomic issues addressed
KX (Kinesthetics eXtreme) [15]	Retrofitting automicity	Enabling autonomic properties in legacy systems
Anthill [16]	P2P systems based on Ant colonies	Complex adaptive behavior of P2P systems
Astrolabe [17]	Distributed information management	Self-configuration, monitoring and to control adaptation
Gryphon [18]	Publish/subscribe middleware	Large communication
Smart Grid [19]	Autonomic principles applied to solve Grid problems	Autonomic Grid computing
Autonomia [20]	Model and infrastructure for enabling autonomic applications	Autonomic applications
AutoMate [21]	Execution environment for autonomic applications	Autonomic applications

Two projects, AutoMate [21] and Autonomia [20], belonging to the second category, directly investigate the key issues of autonomic component/service definition and construction, autonomic application construction, execution and management, and autonomic middleware services. These systems are briefly described below.

11.6.1 AutoMate — Enabling Autonomic Applications

Project AutoMate (TASSL, Rutgers University) [21] investigates autonomic strategies to deal with the challenges of complexity, dynamism, heterogeneity, and uncertainty in Grid computing systems, and enables systems and applications that are capable of managing (i.e., configuring, adapting, optimizing, protecting, healing) themselves. The overall goal of Project AutoMate is to investigate conceptual models and implementation architectures that can enable the development and execution of such self-managing Grid applications. Specifically, it investigates programming models, frameworks, and middleware services that support the definition of autonomic elements, the development of autonomic applications as the dynamic and opportunistic composition of these autonomic elements, and the policy, content, and context driven definition, execution, and management of these applications.

A schematic overview of AutoMate is presented in Figure 11.16. AutoMate builds on emerging Grid/P2P middleware services to define and manage virtual organizations. Its key components include the Accord [22] programming system, the Rudder [23] decentralized coordination framework and agent-based deductive engine, and the Meteor [24] content-based middleware providing support for content-based routing, discovery and associative messaging. Project AutoMate additionally includes the Sesame context-based access control infrastructure, the DAIS cooperative-protection services and the Discover collaboratory [25] services for collaborative monitoring, interaction, and control.

The core components of AutoMate have been prototyped and are being currently used to enable self-managing applications in science and engineering (e.g., autonomic oil reservoir optimizations, autonomic runtime management of adaptive simulations, etc.) and to enable sensor-based pervasive applications. Further information about AutoMate and its components can be obtained from http://automate.rutgers.edu/.

11.6.2 Autonomia

Autonomia (University of Arizona) provides application developers with the tools required to specify the appropriate control and management schemes, the services to deploy and configure required software and hardware resources, and to run applications. Autonomia can efficiently support the development of pervasive systems and services, and provides an environment to make the control and management

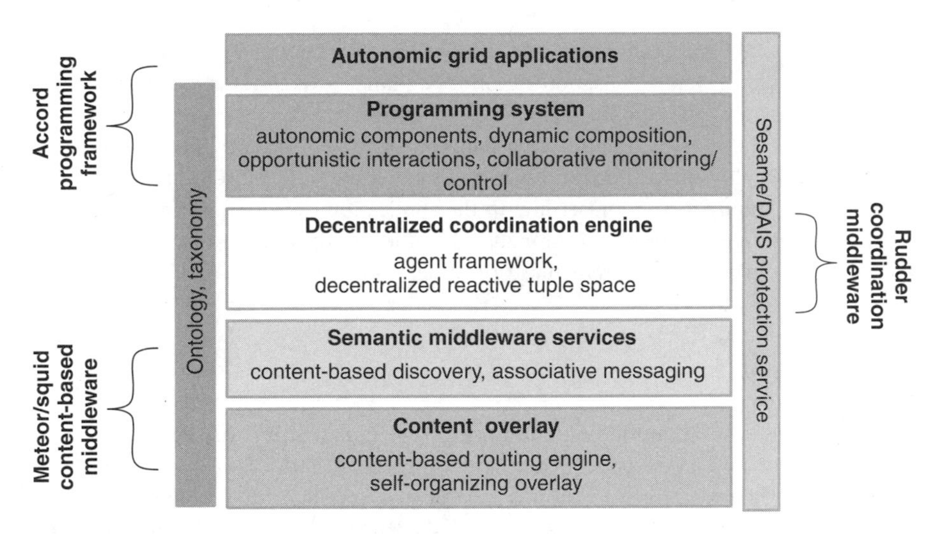

FIGURE 11.16 AutoMate architecture.

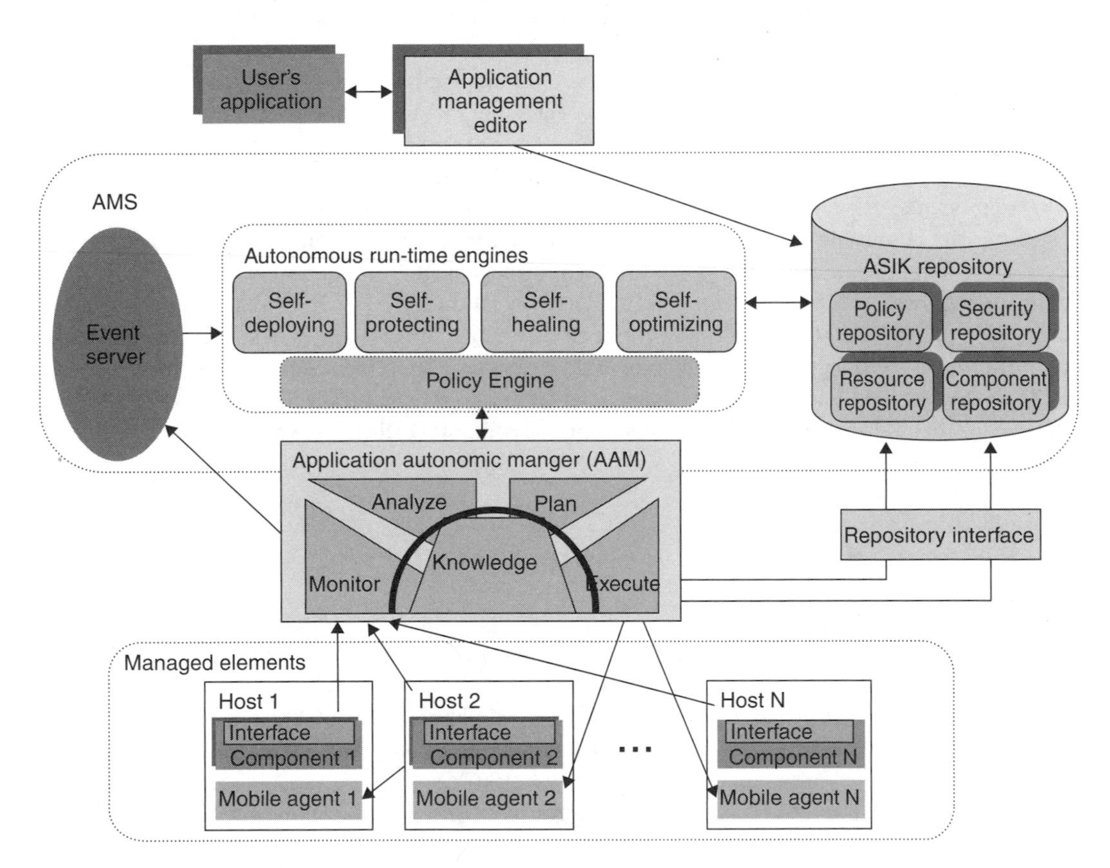

FIGURE 11.17 Autonomia architecture.

of large-scale parallel and distributed applications autonomic. Autonomia provides online monitoring and management to maintain the desired autonomic attributes of applications as well as system services, achieving self-deployment, self-configuration, self-optimization, self-healing, and self-protection by the policy engines.

The main modules of Autonomia include Application Management Editor (AME), Autonomic Middleware Services (AMS) and Application Autonomic Manager (AAM). The AME is a graphical user interface for developing an application using pre-developed and standard components and specifying the management requirements for each application component. The AMS provides common middleware services and tools needed by applications and systems to automate the required control and management functions. The AAM mainly focuses on setting up the application execution environment. It acts as the application administrator that is responsible for allocating the appropriate resources to run the application and maintaining the application requirements at runtime.

11.7 Summary

In this chapter, we presented the autonomic computing paradigm, which is inspired by biological systems such as the autonomic human nervous system and which enables the development of self-managing computing systems. These systems use autonomic strategies and algorithms to handle complexity and uncertainties with minimum human interference, thus shifting the burden of managing systems from people to technologies. An autonomic computing system is a collection of autonomic components, which implement an intelligent control loop to monitor, analyze, plan, and execute using knowledge of the environment.

Several research efforts focused on enabling the autonomic properties address four main areas: (1) A self-healing system could be expected to heal program parts that malfunction. (2) Self-protection of systems to prevent large-scale correlated attacks or cascading failures from permanently damaging valuable information and critical system functions. (3) Self-configuration that involves automatic incorporation of new components and automatic component adjustment to new conditions. (4) Self-optimization on a system level addressing automatic parameter tuning. Projects in both industry and academia have addressed autonomic behaviors at all levels of system management, from the lowest levels of the hardware to the highest levels of software systems and applications. At the hardware level, systems are dynamically upgradable [26], while at the operating system level, active operating system code is replaced dynamically [27]. Efforts have also focused on autonomic middleware, programming systems, and runtime [21, 28]. At the application level, self-optimizing databases and web servers dynamically reconfigure to adapt service performance. The challenges to achieve true autonomic computing still exist, which will be accomplished through a combination of process changes, skills evolution, new technologies and architecture, and open industry standards.

References

[1] Autonomic Nervous System. http://users.rcn.com/jkimball.ma.ultranet/BiologyPages/P/PNS.html#autonomic

[2] W. Ross Ashby. *Design for a Brain*, 2nd ed., Revised, Chapman & Hall Ltd, London, 1960.

[3] J.O. Kephart and D.M. Chess. The vision of autonomic computing. *IEEE Computer*, 36, 41–50, 2003.

[4] IBM Corporation. Autonomic computing concepts. http://www-3.ibm.com/autonomic/library.shtml, 2001.

[5] IBM. An Architectural Blueprint for Autonomic Computing, April 2003.

[6] R. Rothermel. A mathematical model for predicting fire spread in wildland fuels. Research paper INT-115. Ogden, UT: U.S. Department of Agriculture, Forest Service, Intermountain Forest and Range Experiment Station, 1972.

[7] Oceanstore. http://oceanstore.cs.berkeley.edu, July 8, 2002.

[8] J. Kubiatowicz. OceanStore: Global-scale persistent storage. Stanford Seminar Series, Stanford University, 2001.

[9] IBM Almaden Research. IBM storage tank — A distributed storage system WhitePaper. January 24, 2002.
[10] The Océano Project, http://www.research.ibm.com/oceanoproject, IBM Corporation.
[11] Guy M. Lohman and Sam Lightstone. SMART: Making DB2 (More) Autonomic. In *Very Large Data Bases (VLDB) Conference 2002.*
[12] S. Chaudhuri. AutoAdmin: Self-tuning and self-administering databases. http://research.microsoft.com/research/dmx/autoadmin, Microsoft Research Center.
[13] R. Pool. Natural selection, a new computer program classifies documents automatically, 2002.
[14] Q-fabirc. http://www.cc.gatech.edu/systems/projects/ELinux/qfabric.html.
[15] G. Kaiser, P. Gross, G. Kc, J. Parekh, and G. Valetto. An approach to autonomizing legacy systems. In *Workshop on Self-healing, Adaptive and Self-managed Systems*, New York City, NY, June 23, 2002.
[16] Anthill. http://www.cs.unibo.it/projects/anthill/index.html.
[17] Robbert van Renesse, Kenneth Birman, and Werner Vogels. Astrolabe: A robust and scalable technology for distributed system monitoring, management, and data mining. *ACM Transactions on Computer Systems*, 21, 164–206, 2003.
[18] Gryphon. http://www.research.ibm.com/gryphon/gryphon.html
[19] Smart Grid, http://www.ldeo.columbia.edu/res/pi/4d4/testbeds/
[20] S. Hariri, Lizhi Xue, Huoping Chen, Ming Zhang, S. Pavuluri, and S. Rao. Autonomia: an autonomic computing environment. In *Proceedings of the Performance, Computing, and Communications Conference*, IEEE International, April 9–11, 2003.
[21] M. Agarwal, V. Bhat, H. Liu, et al. AutoMate: Enabling autonomic applications on the grid. In *Autonomic Computing Workshop Fifth Annual International Workshop on Active Middleware Services* (AMS'03), June 25–25, 2003.
[22] H. Liu, M. Parashar, and S. Hariri. A component-based programming framework for autonomic applications. In *Proceedings of 1st IEEE International Conference on Autonomic Computing (ICAC-04)*, IEEE Computer Society Press, Washington, 2004, pp. 278–279.
[23] Z. Li and M. Parashar. Rudder: A rule-based multi-agent infrastructure for supporting autonomic grid applications. In *Proceedings of 1st IEEE International Conference on Autonomic Computing (ICAC-04)*," May 2004, pp. 10–17.
[24] N. Jiang, C. Schmidt, V. Matossian, and M. Parashar. *Content-based Middleware for Decoupled Interactions in Pervasive Environments*, Rutgers University, Wireless Information Network Laboratory (WINLAB), Piscataway, NJ, USA, 2004.
[25] V. Bhat and M. Parashar. Discover middleware substrate for integrating services on the grid. In *Proceedings of 10th International Conference on High Performance Computing (HiPC 2003)*, Springer-Verlag, Heidelberg, December 2003, pp. 373–382.
[26] J. Jann, L.M. Browning, and R.S. Burgula. Dynamic reconfiguration: Basic building blocks for autonomic computing on ibm pseries servers. *IBM Systems Journal*, 2003.
[27] J. Appavoo, K. Hui, C.A.N. Soules, R.W. Wisniewski, D.M. Da Silva, O. Krieger, M.A. Auslander, D.J. Edelsohn, B. Gamsa, G.R. Ganger, P. McKenney, M. Ostrowski, B. Rosenburg, M. Stumm, and J. Xenidis. Enabling autonomic behavior in systems software with hot swapping. *IBM Systems Journal*, 2003.
[28] James Kaufman and Toby Lehman. Optimal grid: Grid middleware for high performance computational biology. Research report, IBM Almaden Research Center.
[29] P. Horn. Autonomic Computing: IBM's Perspective on the State of Information Technology, http://www.research.ibm.com/autonomic/, October 2001.
[30] Adaptive Systems. http://www.cogs.susx.ac.uk/users/ezequiel/AS/lectures.

II

Application Domains

12

Setting Parameter Values for Parallel Genetic Algorithms: Scheduling Tasks on a Cluster

Michelle Moore

12.1 Introduction

Optimization problems are widespread and appear frequently in a variety of common, everyday applications. For example, a shipping company handles containers of varying sizes and shapes, and wants to pack the maximum possible number into a fixed space. This packing plan must be generated for each truck. An airport wants to determine the fastest pattern for their fleet of snow plows and dump trucks for clearing snow from the runways. Since heavier snowfall will require more trips by the dump trucks, they also need the plowing pattern generator to include the rate of snowfall in the computations. Finally, consider a school district that needs to reduce the amount of fuel used by its 40 buses, and wants to determine the shortest route that will allow children to be picked up at their homes. The district has a call-in system for children who need not be picked up on a particular day, so the route plan has to be updated every morning.

In each of these examples, the result must be determined quickly. Unfortunately, in order to determine the very best answer in each case, all possible solutions must be considered. For example, in order to determine the best bus route to pick up 50 children for one bus in a fixed area requires that 50! routes be examined. That is 3.04e64 routes for a single bus. For 40 buses of 50 children each, there are 2000! possible schedules! Because of the amount of time taken to find an optimal answer to these types of computationally intractable problems, approximation algorithms are developed to find acceptably good answers in a reasonable amount of time.

The decision to use an approximation algorithm raises additional questions. How close will a particular approximation algorithm's answer be to the optimal answer? Will the approximation algorithm be able to find this answer fast enough? In all the examples given, there is a limit to the amount of time that can be spent to find a solution. The program implementing the algorithm must guarantee a solution before the time limit is reached.

Biologically inspired optimization methods such Genetic Optimization Algorithms (GOAs) offer the advantage of an immediate approximate solution. The available time is then used to refine that solution to bring it as close as possible to the optimal. Of course, in real applications, the actual optimal solution is not known. However, where optimal solutions are known, GOA approximations frequently find them. When the optimal is not found, the approximation is usually very close. GOAs can be written to execute in parallel to speed up the process of finding an optimal or approximate solution. Communication of local results among Parallel GOAs (P-GOAs) can often improve both the quality of the approximations and the speed at which they are found.

The likelihood that a very good solution will be found when using a P-GOA depends largely on the values of the parameters used by the program. These parameters include population size, deme size (the number of individuals on each processor), number of generations, rate of communication, and communication frequency. This chapter discusses a mathematical method to determine appropriate P-GOA values for these parameters for a multiprocessor scheduler. The computations required to determine proper values for the P-GOA parameters are shown in detail. In addition, known optimal schedules are compared with schedules generated using the P-GOA and the results are shown.

The analysis and experimentation discussed here represents ongoing investigation. The results presented here represent the culmination of an initial attempt to find an "all purpose" sizing formula for task scheduling in a computational science programming environment. Using the work by Cantú-Paz [1] and Goldberg et al. [2] as a theoretical and practical foundation, numerous computations were explored, and trials were run using a variety of equations. An "all purpose" formula was not found; however, it is now clear what can be accomplished using this sort of methodology, and what problem qualities suggest a very different approach. The material in some of the sections of this chapter has appeared in previous discussions of the progress in this area [3–6]. It has been repeated here so that this chapter may be read without the need to refer to the earlier work.

The rest of this chapter is organized as follows. Section 12.2 outlines the background of the sample problem and of previous work related to parameter value selection for parallel Genetic Algorithms (GAs). Section 12.3 presents the problem-specific scheduling model and the complexity of the problem. This material has appeared in previous publications. Section 12.4 introduces GAs and the application of the genetic metaphor to the scheduling problem, which has appeared in previous publications. Section 12.5 describes the genetic operators and the parameter variables used in the scheduler. This section is similar to discussions in Reference 6, but provides the critical modifications that produced a successful deme sizing equation for the task scheduling P-GOA. Readers familiar with the previous reports might begin their reading in Section 12.5. Section 12.6 describes the design of the experiments and describes how particular values were chosen for the sizing equations. Section 12.7 details the results of the experiments and summarizes the conclusions.

12.2 Background

A multiprocessor scheduling problem is used to illustrate appropriate calculations for determining several important parameter values for a P-GOA solution approximation. The schedule to be developed by the P-GOA specifies the processor on which each of the tasks required for a particular application is to run. This application schedules tasks on a cluster of homogeneous processors with an Ethernet connection. The task execution times and communication times are known in advance. This models the scheduling requirements that may be encountered when performing distributed database queries or when doing "production runs" in computational science applications with known computation times and varying data. In applications such as these, execution and communication times can be known (or at least estimated very closely) in advance.

Parameter sizing methods were reported by Goldberg [7], Goldberg et al. [2], and others. Precise analysis of these methods for selected problems and processor topologies were presented by Cantú-Paz [1]. In his forward [1], Goldberg states, "I believe it is fair to say that prior to this work, the design of parallel GAs was something of an empirical black art, guided largely by ad hoc trial and error experimentation." The statistical analysis presented in these works had their basis in schema theory. Schema theory attempts to provide an explanation for the behavior of genetic algorithms. Analysis of this explanation of GA behavior resulted in precise formulae for determining, a priori, the values of GA input parameters that would produce a particular solution quality.

A number of objections to schema theory have been published. A recent book by Reeves and Rowe [8] summarizes many of these objections, and provides sufficient bibliographic information to begin a more in-depth investigation into the reasoning of schema theory detractors. They suggest that schema theory-based analysis can be fruitful for specific categories of GAs. However, they insist that the behavior of other types of GAs differs vastly from that described by schema theory. Nevertheless, the benefits of calculating accurate GA parameters in advance, in terms of time saved and solution quality, can be sizable for some applications. In this chapter, effective parameters are found for a task-scheduling problem by applying appropriately adapted analysis.

In Reference 3 these concepts were first applied to this scheduling problem. In References 4 to 6, an in-depth examination of three deme sizing equations developed over time by Cantú-Paz, Goldberg, and others was reported in terms of their applicability to the scheduling problem. Information in Reference 6 outlined how to apply the equations to the cluster-scheduling problem, specifically detailing the appropriate determination of values for the equation variables. This chapter explains refinements developed through analysis of the most promising equations, and provides the rationale for making those refinements. Justification of approximations used in parameter computations and modifications to the original equations are explained.

12.3 The Task-Scheduling Problem

The task-scheduling problem presented here has been the sample problem during the various stages of analysis of the parameter sizing computation development. This definition and description of the scheduling problem has appeared, in slightly varying form in References 3 to 6.

12.3.1 Definitions

A task t_i is represented as a taskpair (e_i, c_i), consisting of an execution time e_i and a communication time c_i. The total number of tasks to schedule is n. The symbol p_i represents one of the m processors available to the scheduler. The makespan of a schedule is the time at which the execution of all taskpairs is completed. The optimal makespan is the shortest possible period in which a given set of taskpairs can execute on the available multiprocessor system. The goal of the scheduler is to produce schedules with makespans as close to optimal as possible within a predictable and practical amount of time. In summary, the set of tasks $\{t_0, \ldots, t_{n-1}\}$ is scheduled to execute on a system of m processors $\{p_0, \ldots, p_{m-1}\}$.

Tasks are created and scheduled by an initial processor designated p_0. The time p_0 uses to create a schedule for the given set of tasks is not considered as a part of the makespan. The time required to send the task assignments to the other processors is assumed constant and is therefore not considered. Any final computations that must occur after all results are communicated back to p_0 are unaffected by the particular schedule, and are not included in the schedule quality evaluations. Finally, messages and processes for control or monitoring are assumed to have no effect on the relative efficiencies of the schedules, so these values are also not considered in the evaluation of schedule quality.

The time required for each task to execute and communicate a result (or estimations of these times) is available in advance. Dependent computations are grouped into a single task so that each task is independent. Consequently, there are no precedence constraints on the tasks. However, the computation portion of each task must complete execution before its corresponding message is sent.

12.3.2 The Computational Model

The target hardware environment is a cluster system using message passing. The cluster consists of m processors $\{p_0, \ldots, p_{m-1}\}$ connected by a shared Ethernet bus. All processors are identical and the communication bandwidth between $\{p_1, \ldots, p_{m-1}\}$ and p_0 is constant. The processor that creates the set of tasks and the schedule is represented as p_0. This process requires no communication time in order to complete its tasks. Processors $\{p_1, \ldots, p_{m-1}\}$ represent the additional processors to which p_0 sends tasks or control messages, and from which p_0 receives the computation results. Tasks scheduled on $\{p_1, \ldots, p_{m-1}\}$ are not considered complete until the result message has left the communication channel. After all n tasks have been completed, p_0 may use the results obtained to execute a final computation. However, since this final execution time is the same regardless of the manner in which the tasks are scheduled, its value is ignored during schedule quality evaluation.

Only one message may be transmitted to or from a processor and only one task may execute on a processor at a time. All schedules and tasks are non-preemptive. If there is a task available to execute, and a processor is available at that time, the task will be scheduled. If the communication channel is available, the message will be transmitted immediately after the associated task is completed. Otherwise, the message will be transmitted as soon as messages associated with any previously executed tasks leave the communication bus.

12.3.3 Complexity of the Problem

In References 8 to 10 and 17, GAs were shown to provide effective approximations for combinatorial optimization problems. The combinatorial optimization problem of finding a schedule on $m > 2$ identical processors that minimizes a finish time for independent tasks consisting of execution times has been shown to be in class NP [11,12,19]. In addition to execution times, e_i, the amount of time, c_i, required to return the result of each computation over the communication channel is also considered. Therefore, in order to schedule n taskpairs $\{t_0, \ldots, t_{n-1}\}$, the values of $\{(e_0, c_0), \ldots, (e_{n-1}, c_{n-1})\}$ must be considered. This problem can be transformed into the above scheduling problem by setting its communication times to zero, and is clearly in NP. When communication times are nonzero, the communication channel is an additional resource that must be scheduled, which increases problem complexity. The quality or "fitness" of the schedule is the total time required for all tasks to complete execution. An exhaustive search of all possible schedules will require time bounded by $\Theta(m^n)$, where n is the number of tasks to schedule, and m is the number of processors to schedule. The addition of a communication bus that must also be "scheduled" increases the complexity of the problem. Since the problem is NP-complete, approximation algorithms for scheduling attempt to create schedules as close to the optimal as possible.

12.3.4 The Optimal Algorithm

In order to accurately evaluate the schedules produced by the P-GOA, optimal schedules were generated for a series of small scheduling problems. It was necessary to keep the problems fairly small due to the complexity of an optimal algorithm. Consider if m is the number of processors and n is the number of tasks, an exhaustive search will require m^n time to evaluate every schedule. If a branch and bound algorithm is used to avoid unfruitful partial schedules, the worst case will occur if data requires all branches to be searched, giving a worst-case complexity of $O(m^n)$. A parallel version of the optimal algorithm executed on r processors has complexity $O(m^n/r)$, again an exponential time complexity measure. Because of the time required to obtain comparison values using an optimal algorithm, the size of the test cases was kept small.

12.3.5 P-GOA Time and Space Complexity

The space requirements of the sequential version of the genetic schedule optimization algorithm are linear,

$$population\_size \times number\_of\_tasks \times 2$$

and the execution time complexity of the sequential version of the genetic schedule optimization algorithm is

$$num\_generations \times population\_size \times number\_of\_tasks \times number\_of\_processors.$$

The scheduler is cost optimal in that the space complexity of the P-GOA is

$$(population\_size \times number\_of\_tasks \times 2)/r$$

and the execution time complexity of the P-GOA is

$$(num\_generations \times population\_size \times number\_of\_tasks \times number\_of\_processors)/r,$$

where r is the number of processors on which the scheduler is executed.

12.4 The Genetic Approach

12.4.1 GAs and Optimization

In 1975, Holland [13] introduced the idea of combining directed randomness with adaptation as a search and optimization technique. Since that time, GAs have been used in a wide array of applications. In addition, the characteristics of GAs have been studied extensively, and diverse theoretical explanations for the ability of these complex systems to generate solutions have appeared [18]. GAs have been used to approximate solutions for a wide variety of NP-complete problems [2,9].

A GA begins with a population of initial coded "guesses" (chromosomes) at a solution. These guesses may be randomly generated or produced using a problem specific heuristic. An application specific fitness or objective function is applied to each chromosome, and the "better" individuals are selected to "survive." Survivor individuals are then merged together to form a new generation (mating). Occasionally, portions of the chromosomes of an individual are randomly altered (mutation). This process of fitness determination, mating, and mutation is repeated for a given number of generations, or until the individuals improve sufficiently to achieve a predefined goal. The simplest algorithms often view an individual as a single chromosome. However, applications with multiple constraints may employ individuals with multiple chromosomes.

12.4.2 The Genetic Metaphor

The P-GOA employs methods and metaphors from nature and genetics to "evolve" an initial population of solutions (schedules in this case) into high quality solutions. The initial population of schedules is randomly generated. The fitness, or quality, of each schedule is evaluated. The "fitter" schedules are saved and used as a "mating pool." Finally, these schedules are paired randomly to create new schedules until there are enough "offspring" to return the population to its original size. The P-GOA divides the population of schedules into local demes and distributes them over the available processors. Periodically, individual schedules are allowed to migrate to other processors in hopes of speeding the rate of schedule improvement.

The P-GOA represents each schedule as a single chromosome. Each individual schedule consists of *lval* genes, where *lval* is the number of tasks to be scheduled. Each position of the chromosome corresponds to a task, and the values at each position correspond to a processor number. A schedule for six tasks on five processors might appear as (1 3 4 2 3 0), indicating that t_0 is executed on p_1, t_1 is executed on p_3, and so on. Chromosomes are also referred to as strings. The cardinality of the string alphabet is the number of digit values that may appear on the chromosome. Many GAs use binary alphabets, but in general, we may say that a chromosome string has a χ-ary alphabet where the cardinality is χ. In the given example, the cardinality of the alphabet (χ) is equal to five.

12.5 Genetic Operators and Parameters

12.5.1 Encoding and Initialization

For scheduling in the P-GOA, each chromosome encodes a schedule solution and each gene represents a scheduled task. The number of genes in a chromosome equals n, which is the number of tasks to be scheduled. In the P-GOA, allele values represent the processor to which a task is assigned. Allele values range from 0 to $m - 1$, where m equals to the number of processors available. For example, given the following taskpair set:

$$\begin{gathered}\{(7,16),(11,22),(12,40),(15,22),(17,23),\\(17,23),(19,23),(20,28),(20,27),(26,27),\\(28,31),(36,37),(31,29),(28,22),(23,19),\\(22,18),(22,17),(29,16),(27,16),(35,15)\},\end{gathered}$$

the optimal schedule on three processors would appear as

$$1\;0\;0\;0\;0\;0\;0\;0\;2\;0\;0\;0\;1\;1\;1\;1\;1\;2\;1\;2.$$

The optimal makespan for this sequence of tasks is 202.

Various population "seeding" techniques were examined in Reference 14. However, for general GA schedulers, random initialization of the population of schedules has proven to be the most effective initialization technique and is used here.

12.5.2 The Sizing Equations and Population Distribution

As mentioned previously, the determination of population and deme sizes for practical applications of parallel GAs required a great deal of what was frequently termed "empiricism," that is, trial and error. Cantú-Paz [1] offered a way to analyze a particular problem so that a very precise population and deme size, an efficient migration strategy, and an effective processor topology could be determined in advance.

The analysis here considers only deme sizing, sets migration to the maximum possible value, and uses a fully connected system.

Consideration of collateral noise (the noise of other partitions when deciding between best and second best building block) is built into the sizing equation. External noise may exist if fitness cannot be directly calculated. This was not the case for the scheduling application, so the sizing computation does not require adjustments for external noise. Each of the processors used in finding the schedule is assigned n_d schedules to evolve locally.

Analysis produced a sizing equation that is identical to the one given in Reference 1. However, many of the variables could not be calculated in the same manner as in Reference 1 and required approximations. In addition, experiments revealed that a scaling factor was needed to compensate for a crucial difference in this application. Unlike the applications in Reference 1, the scheduler does not need to converge. It is only necessary to find one schedule of the required quality. The scaling factor will be discussed in detail in Section 12.6.4. The equation, minus scaling, is given below:

$$\frac{n_d = \mathrm{sqrt}(-\chi^k \ln(1 - P_{\mathrm{bb}}))}{(2p - 1)/p}, \tag{12.1}$$

where

χ = cardinality of the chromosome string alphabet, the number of processors
k = building block size, the number of fixed positions in the schema
P_{bb} = probability that one partition (one building block) in one deme is correct, probability of success per deme $\approx Q/m - \mathrm{sqrt}(\ln(r))/\mathrm{sqrt}(2m)$
Q = required problem solution quality $= (1 - \alpha)m$
α = the probability of making the wrong choice between two competing schema
m = number of partitions in the string, number of building blocks $lval/k$
$lval$ = length of the chromosome (schema), the number of tasks
r = the number of processors used to create the schedule
$p = (1 - q), \approx 1/2 + \psi/\mathrm{sqrt}(2\pi)$
π = the constant value 3.14159
$\psi = d/(\sigma_{\mathrm{bb}}\,\mathrm{sqrt}(2m'))$
d = signal difference for peak to peak (mean to mean) function difference between the best and the second best competing individuals, this affects the likelihood of choosing the better individual
$\sigma_{\mathrm{bb}} = \mathrm{sqrt}(\sigma_{\mathrm{f}}^2)$ the average building block standard deviation
σ_{f}^2 = overall variance of all schemas for each fixed position $= \sum_{i=1}^{m} \sigma_{\mathrm{fi}}^2$

$$\approx \frac{(f_{\max} - f_{\min})^2}{12}, \tag{12.2}$$

$f_{\max}$ = maximum fitness function value possible
$f_{\min}$ = minimum fitness function value possible

12.5.3 Fitness and the Objective Function

Tasks assigned to processors other than p_0, will have an associated message that must be scheduled. The total message time will therefore be longer than the task execution time on $p_{1,\ldots,n-1}$. Tasks assigned to p_0 do not require message time, so the greater of the communication completion time and the execution time on p_0, will be the makespan

$$\text{makespan} = \mathrm{MAX}(p_0, \text{message time}).$$

Because the goal of the scheduler is to minimize the makespan, the fitness of an individual chromosome string is evaluated as

$$\text{fitness} = -1 \times \text{makespan}.$$

12.5.4 Survival, Mating, and Mutation

Each generation, the average of all the individual makespans is calculated. Schedules with makespans less than or equal to the average are placed in the survivor pool. Each individual schedule with a fitness greater than or equal to the average fitness is allowed to reproduce and survive to the next generation. This rank-based selection method combines truncation selection and $(\gamma + \lambda)$ selection. The top $1/\nu$ of the population is selected. Then the $\gamma = 1/\nu$ parents mate to create λ offspring. The union of γ and λ constitute the next generation. This allows very fit individuals to survive and mate repeatedly.

Survivors are randomly chosen, with replacement, to produce offspring until the size of the population returns to its original level. Each mating produces a single offspring. As with sexual reproduction, an offspring is created from the genes of the two parents. Each allele is randomly chosen from one parent or the other. Mating is accomplished using uniform crossover with a 0.5 crossover probability [15,16]. In other words, when creating an offspring from a pair of schedules, the offspring's processor assignment for each task is equally likely to match either parent.

In addition, a 2.5% mutation probability is applied for each allele. This means that after crossover, the GA allows for a .025 probability for each task being reassigned to a randomly chosen processor. This is a relatively high mutation rate, but was chosen to prevent premature convergence from occurring due to small deme sizes. The process of selection, mating, crossover, and mutation occurred for 1000 generations. This is a relatively small number of generations. The goal of the scheduler is to find near optimal schedules in a very short amount of time. A small number of generations, combined with a deme size no larger than necessary, allows this to be accomplished.

12.5.5 Migration Frequency, Policy, and Rate

The initial population of schedules is distributed among the available cluster nodes. After each generation, the more fit individuals of each deme are copied and distributed among the other demes, replacing the least fit individuals. The P-GOA uses the migration rate necessary to fill the positions left open by the unfit individuals. Migration occurs after selection and before mating so that the new migrants can contribute to the next generation. After each migration, the local iterations of selection, mating, crossover, and mutation resume. When the execution of 1000 generations completes, the P-GOA outputs the best schedule found so far, along with its makespan.

Using the results from the migration policy analysis and experiments in Reference 1, migration follows what is called a "best–worst policy," since migrants take the place of the least fit individuals in each deme. The proportion of migrants to the population as a whole is the migration rate. Since migrants are sent to replace individuals below average, the rate of migration decreases as the fitness values begin to converge. Since the individual schedules remain on their original processors and copies are sent to other processors, this process is sometimes referred to as "pollination" rather than migration.

12.6 Experiment Design

The experiments performed by Goldberg et al. [2,7] and Cantú-Paz [1] to verify the given calculations and guidelines, involved problems qualitatively different from the scheduling problem described here. Their experiments involved binary encodings only. The schedule encoding is **m**-ary, where **m** is the number of processors. For the scheduling problem, this is the same as χ-ary where χ is the alphabet size. The fitness of their experimental individuals involved functional calculations that directly evaluated the encodings.

In the scheduling application, fitness is an indirect computation requiring an evaluation of the *meaning* and implications (i.e., effect on communication time) of the encoding. When they compared expected confidence levels with experimental results, the degree of correctness was defined as the percentage of alleles possessing the correct value when the algorithm converged. In the scheduling application, the degree of correctness is defined as the "nearness" to the optimal schedule that can be obtained in a limited amount of time by the best individual. These differences were overcome by problem specific interpretations of the meanings of the equation variables and by employing a scaling factor to adjust the output of the equation.

12.6.1 The Data Sets

The initial data set consisted of taskpair sets, each containing 20 normally distributed, randomly generated (e, c) task pairs. The mean execution time was 25 with a standard deviation of 2%. Three communication means were used, 10, 25, and 40, (std $= 0.02$) to vary the relative effects of communication overhead. This data set was used throughout the experiments with various sizing equations. The final deme sizing equation reported here was also tested using the original data sets as well as data sets of exponentially distributed task execution and mean times, taskpair sets of varying sizes, and data sets with larger or smaller means and standard deviations.

12.6.2 The Test Programs

The parallel scheduling program was written in C using LAM MPI constructs to execute on multiple cluster processors. The migration rate was set to the maximum that could occur for the deme with the most survivors after selection. Migration occurred at every generation.

In order to evaluate the quality of the schedules that were produced, actual optimal schedules were found for the data sets. Because of the exponentially increasing amount of time required to find the comparison optimal schedules, the number of tasks in the tests sets was kept small. A sizing program was written to calculate population sizes that corresponded to varying predicted levels of schedule accuracy. The sizing program generated population and deme sizes using probabilities of success of 0.80, 0.85, 0.90, 0.95, and 0.99. Deme sizes were generated at each confidence level with each of the three communication means. To reduce the effect of stochastic variance, five runs were averaged for each communication mean at each confidence level. The program to find the optimal schedule for each of these cases was also executed five times, and the results were averaged for comparison. An expected quality of 80% meant that when deme size was calculated at the 80% confidence level. They were expected to produce schedules that would allow all tasks to complete execution in an amount of time exceeding the optimal by no more than 20%.

12.6.3 Sizing Equation Variable Value Computation

The graphs shown represent results when the following variable values were set to the given values:

$r = 12$, the number of processors used to create the schedule, the number of cluster nodes running the P-GOA scheduler.

$\chi = 5$, the number of processors to be scheduled, the cardinality of the alphabet, the number of possible allele values.

$k = 1$, the size of a building block.

$lval = 15$, the number of tasks to schedule, the number of genes in the chromosome.

$m =$ the number of partitions, the number of building block sets in the chromosome string, since the building block size is 1, $lval/k = 15$.

$m' = m - 1$.

Signal difference is defined in Reference 1 as the difference in the means of the fitness distribution produced by the best building block and that produced by the second best building block.

The fitness distribution associated with a particular building block in a task schedule cannot be directly determined from the value assigned to the building block. Fitness calculations must consider the meaning of the allele value in terms of previous assignments and the effect on communication time. Consequently, the P-GOA scheduler considers the smallest possible difference in average makespans and the average execution time. The average of these two values is used as an estimate for signal difference.

$$d \approx (1 + \mu_e)/2.$$

This value will vary in relationship with the means of the data set values. Data sets with larger execution times will have larger signal difference values.

The worst possible schedule for a given data set could be generated if tasks were scheduled to maximize the makespan. This worst fitness for the specific scheduling problems is calculated as follows:

$$\begin{aligned} f_{\max} \approx \text{if } \mu_c < \mu_e \quad & lval\, \mu_e + \mu_c \quad \text{which in this example is } 15 \times 25 + 10 = 385, \\ \text{if } \mu_c = \mu_e \quad & lval\, \mu_e + \mu_c \quad \text{which in this example is } 15 \times 25 + 25 = 400, \\ \text{if } \mu_c > \mu_e \quad & lval\, \mu_c + \mu_e \quad \text{which in this example is } 15 \times 40 + 25 = 625. \end{aligned}$$

The minimum makespans were estimated using observations of task placements found by the optimal algorithm

$$\begin{aligned} f_{\min} \approx \text{if } \mu_c \leqslant \mu_e \quad & (1/num\_procs)\; lval\, \mu_e + 2\mu_c \quad \frac{1}{3} \times 15 \times 25 + 10 = 135, \\ \text{if } \mu_c > \mu_e \quad & (1/num\_procs)\; lval\, \mu_c + \mu_e \quad \frac{1}{3} \times 15 \times 40 + 25 = 225. \end{aligned}$$

Next, the probability of retaining the best building block between generations on at least one deme is

$$\begin{aligned} p &\approx \frac{1}{2} + \psi/\text{sqrt}(2\pi) \quad \text{where } \psi = d/(\sigma_{bb}\, \text{sqrt}(2m')), \\ &\approx \frac{1}{2} + (13/(\sigma_{bb}\, \text{sqrt}(28))/\text{sqrt}(2\pi)), \\ &\approx 0.5 + (13/\sigma_{bb} \times 5.292)/0.251, \\ &\approx 0.5 + 2.457/(\sigma_{bb} \times 0.251), \\ &\approx 0.5 + 9.789/\sigma_{bb}, \end{aligned}$$

where

$$\sigma_{bb} = \text{sqrt}(\sigma_f^2)$$

and

$$\sigma_f^2 \approx (f_{\max} - f_{\min})^2/12,$$

then

$$\text{if } \mu_c < \mu_e \quad \text{sqrt}(\sigma_f^2) \approx \text{sqrt}((385 - 135)^2/12) = 72.17$$
$$p \approx 0.5 + 9.789/72.17 = 0.636,$$
$$\text{if } \mu_c = \mu_e \quad \text{sqrt}(\sigma_f^2) \approx \text{sqrt}((400 - 135)^2/12) = 76.50$$
$$p \approx 0.5 + 9.789/76.50 = 0.628,$$
$$\text{if } \mu_c > \mu_e \quad \text{sqrt}(\sigma_f^2) \approx \text{sqrt}((625 - 225)^2/12) = 115.47$$
$$p \approx 0.5 + 9.789/115.47 = 0.585.$$

12.6.4 Scaling Factor

Because the P-GOA does not need to converge on a schedule of the required quality, merely find at least one, the population size can be scaled down. Previous calculations produced population sizes that met or exceeded predicted quality at the 95 to 99% confidence levels. This was the primary goal; however, it was discouraging that the deme sizes produced for the lower confidence levels produced schedules whose quality far exceeded the predicted quality. This "excess" quality was not a problem in practical terms; however, it indicated that the sizing equations were not capturing an important feature in the behavior of the scheduler. It is well known that fitness in GAs tends to improve at a logarithmic rate. Early improvements come quickly, but later ones appear more slowly. Since the deme sizes produced accurate results at the higher range, a scaling factor was needed to make the deme sizes reflect the rate of increase at the lower quality levels. Deme sizes were scaled so that they used the calculated size for the highest confidence level. These sizes were decreased at an exponential rate until they reached the lowest confidence level.

12.7 Results and Conclusions

The sizing equation produced very accurate deme sizes for the scheduling problem where the task sizes were normally distributed. Table 12.1 shows the accuracy levels for each population size for averaged results with all communication means. Figure 12.1 illustrates how closely the actual makespans generated track the makespans predicted by the equation.

Tests run with larger data sets, varying data means, and varying standard deviations gave similar results. The deme sizing equation with the scaling factor applied provides sizes that allow the P-GOA scheduler to produce schedules with makespans very close to a predetermined accuracy. In addition, the very small populations at the lower confidence levels allows for trade-offs between accuracy and speed. Clearly, this deme sizing equation captures the behavior of the P-GOA for this scheduling application.

In fact, Tables 12.2 to 12.4 show the best approximate makespan found for each of the five tests that were run for each communication mean. These approximate makespans are shown with the optimal makespans for each run and with the generation number in which the best approximate makespan was

TABLE 12.1 Accuracy Levels for Each Population Size — Normal Data

Deme size	Actual	Predicted
7	0.805	0.80
10	0.878	0.85
14	0.950	0.90
24	0.975	0.95
50	0.997	0.99

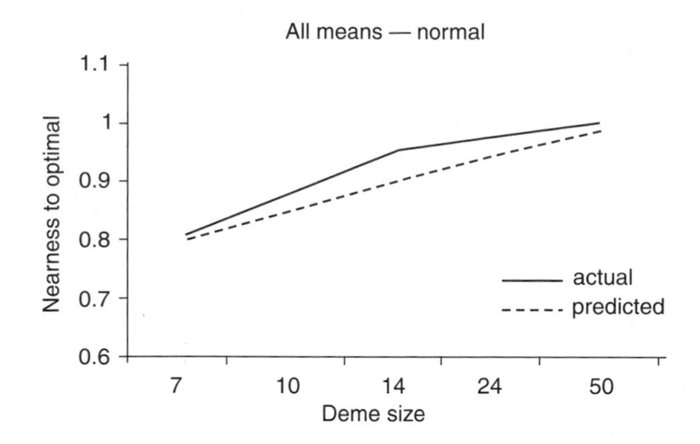

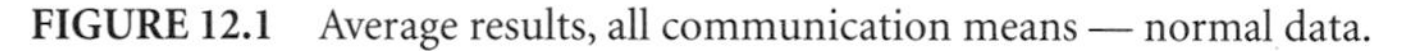

FIGURE 12.1 Average results, all communication means — normal data.

TABLE 12.2

Optimal	Normal 10	Generations
115	115	78
117	117	404
118	119	462
118	118	40
114	114	620

TABLE 12.3

Optimal	Normal 25	Generations
194	195	112
193	195	124
195	197	630
194	194	609
191	191	40

TABLE 12.4

Optimal	Normal 40	Generations
230	231	800
236	237	366
242	242	18
244	245	126
237	237	13

found. The tables illustrate that the P-GOA found the optimal makespan 53.33% of the time. When the optimal was not found, the approximation was very close.

However, experiments with exponentially distributed data yielded very disappointing results. Makespans in all experiments were far from optimal. The distribution of the data is undeniably a strong factor in the applicability of the sizing equations. This is a serious limitation, but it should be kept in perspective. The results of the sizing equation were disappointing, but the scheduler was able to produce schedules that were <3% away from the predicted quality at the high range (99%). In practical terms, this is quite useful and the scheduler is ready for experimental incorporation into a cluster system. Table 12.5 provides the average quality measures compared with the predicted quality at each population size calculated. Figure 12.2 illustrates the disappointing performance.

TABLE 12.5 Accuracy Levels for Each Population Size — Exponential Data

Deme size	Actual	Predicted
7	0.436	0.80
10	0.541	0.85
14	0.668	0.90
24	0.842	0.95
50	0.962	0.99

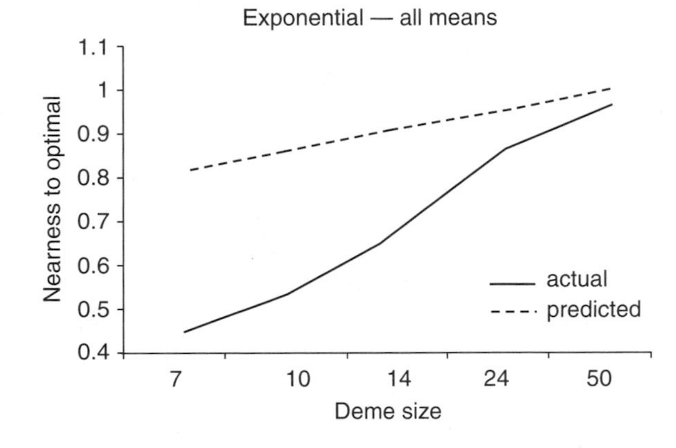

FIGURE 12.2 Average results, all communication means — exponential data.

A great deal of analytical work is needed before an "all purpose" parameter sizing methodology can be found for P-GOAs, but it is encouraging that a "special purpose" parameter sizing methodology has been proven for at least one complex optimization problem. The only data specific variables needed by the P-GOA scheduler are the means of the task execution times and communication times. For the type of ongoing computational science application that this scheduler was designed to work with, this information is either available or easily estimated.

Acknowledgments

Research Assistants John Picarazzi, Shelia Poorman, Brian McCord, Tzintzuni Garcia, William Jackson, Simon San Miguel, Jason Picarazzi, and Lucas Wilson have contributed or are currently contributing to this project. This work is supported by NASA Grant NAG9-1401 and NSF Grant NSF 01-171.

References

[1] Cantú-Paz, E. *Efficient and Accurate Parallel Genetic Algorithms*, Kluwer Academic Publishers, Dordrecht, 2000.

[2] Goldberg, D., Deb, K., and Clark, J. Genetic algorithms, noise, and the sizing of populations. *Complex Systems*, 6, 1992, 332–362.

[3] Moore, M. Parallel genetic algorithms to find near optimal schedules for tasks on multiprocessor architectures. In *Proceedings of Communicating Process Architectures*, Bristol, UK, September 2001, pp. 27–36.

[4] Moore, M. An accurate and efficient parallel genetic algorithm to schedule tasks on a cluster. In *Proceedings of the International Parallel and Distributing Processing Symposium*, Nature Inspired Distributed Computing Workshop, Nice, France, April 2003.

[5] Moore, M. Accurate calculation of deme sizes for a parallel genetic scheduling algorithm. In *Proceedings of Communicating Process Architectures*, Enschede, NL, September 2003, pp. 305–314.

[6] Moore, M. An accurate parallel genetic algorithm to schedule tasks on a cluster. *Parallel Computing*, 30(5–6), 2004, 567–583.

[7] Goldberg, D. Sizing populations for serial and parallel genetic algorithms. In *Proceedings of the Third International Conference on Genetic Algorithms*, Fairfax, VA, USA, June 1989, pp. 70–79.

[8] Reeves, C. and Rowe, K., *Genetic Algorihtms: Principles and Perspectives*, Kluwer Academic Publishers Group, Boston, MA, 2003.

[9] De Jong, K. and Spears, W. Using genetic algorithms to solve NP-complete problems. In *Proceedings of the Third International Conference on Genetic Algorithms*, Fairfax, VA, USA, June 1989, pp. 124–132.

[10] Hou, E., Hong, R., and Ansari, N. Efficient multiprocessor scheduling based on genetic algorithms. In *Proceedings of the 16th Annual Conference of the IEEE Industrial Electronics Society*, Asilomar, CA, USA, November 1990, pp. 1239–1243.

[11] Coffman, E. *Introduction to Deterministic Scheduling Theory, Computer and Job-Shop Scheduling Theory*. John Wiley & Sons, New York, 1976.

[12] Horowitz, E. and Sahni, S. Exact and approximate algorithms for scheduling non-identical processors. *Journal of the ACM*, 23, 1976, 317–327.

[13] Holland, J. *Adaptation in Natural and Artificial Systems*. University of Michigan Press, Ann Arbor, MI, 1975.

[14] Kidwell (Moore), M. Using genetic algorithms to schedule distributed tasks on a bus-based system. In *Proceedings of the Fifth International Conference on Genetic Algorithms*, Urbana-Champaign, IL, USA, July 1993, pp. 368–374.

[15] Eshelman, L., Caruana, R., and Schaffer, J. Biases in the crossover landscape. In *Proceedings of the Third International Conference on Genetic Algorithms*, Fairfax, VA, USA, June 1989, pp. 10–19.

[16] Syswerda, G. Uniform crossover in genetic algorithms. In *Proceedings of the Third International Conference on Genetic Algorithms*, 1989, pp. 2–9.

[17] Goldberg, D. *Genetic Algorithms Search, Optimization, and Machine Learning*. Addison-Wesley, Reading, MA, 1989.

[18] Mitchell, M. *An Introduction to Genetic Algorithms*. MIT Press, Cambridge, MA, 1996.

[19] Salleh, S. and Zomaya, A. *Scheduling in Parallel Computing Systems*. Kluwer Academic Publishers Group, Dordrecht, The Netherlands, 2000.

13

Genetic Algorithms for Scheduling in Grid Computing Environments: A Case Study

Kris Crnomarkovic
Albert Y. Zomaya

13.1 Introduction

The proliferation of the Internet and the availability of powerful computers and high-speed networks as low-cost commodity components are changing the way computing is done today. The interest in coupling geographically distributed computational resources is also growing for solving large-scale problems, leading to what is popularly called Grid Computing. Grids enable the sharing, selection, and aggregation of suitable computational and data resources for solving large-scale data intensive problems in science, engineering, and commerce [1–3].

An important issue for Grid and other Heterogeneous Computing environments is how to assign tasks to resources and order execution of the tasks to maximize some performance criterion of the Grid environment. These procedures are termed matching and scheduling, and taken together, are known as mapping. There are two different types of mapping: static and dynamic. *Static mapping* is performed when the applications are mapped in an offline planning phase, for example, planning the schedule for a set of production jobs. *Dynamic mapping* is performed when the applications are mapped in an online fashion, for example, when tasks arrive at unknown intervals and are mapped as they arrive (the workload is not known a priori) [4,5]. In both cases, this generalized mapping problem has been shown to be NP-hard (e.g., in References 6 to 8).

The goal of this study has been to investigate classes of scheduling algorithms for service-based Grid Environments, that is, where a known request is served to a user or users. Such an environment may involve *request scheduling*, which is a task scheduling architecture that uses a service request as the minimal scheduling unit. A service request is considered to be finer-grained than a job request. A single server would be able to handle multiple requests [9].

This chapter examines some variations of conventional schedulers for dynamic mapping of dependent tasks. Section 13.2 describes further background material. Section 13.3 discusses related work in the literature. Section 13.4 describes enhancements made to a grid simulation toolkit and introduces a genetic mapping heuristic. Section 13.5 gives the results from the study. Section 13.6 describes some further work that might be considered, and Section 13.7 draws some conclusions from the work.

13.2 Background

13.2.1 List Scheduling Heuristics

As described in Section 13.1, this chapter focuses on the dynamic scheduling of tasks on a service-oriented Grid, a network of heterogeneous machines. The service to be scheduled may be represented by a *task graph* or *directed acyclic graph* (DAG). The DAG specifies the tasks that make up the service, as well as the dependencies between tasks [10].

The scheduling of DAGs may be done using a number of broad strategies, including *Clustering Algorithms* [11]. Clustering Algorithms are not considered in this chapter, but involve mapping subsets of tasks that have large inter-task communications to a set of processors with high bandwidth and low latency (e.g., see Reference 12).

An example of a generational algorithm is a *List Scheduler*. A List Scheduler performs mappings based upon a subset of the tasks in the DAG. It only attempts to schedule those tasks that have had all dependency relationships fulfilled. This subset of tasks is called a *meta-task* and consists of tasks that are independent with respect to each other. This greatly simplifies the mapping process and allows it to be done dynamically. An auxiliary algorithm then maps this meta-task to available resources. When a rescheduling event (such as a task finishing) occurs, a new meta-task is constructed and scheduled. The expected execution time of each task on every resource is considered to be known a priori. This assumption is typically made when conducting mapping research [11,13].

The mapping of tasks to resources, whether static or dynamic, is an NP-hard problem. Finding an optimal solution is intractable. Therefore, a heuristic is normally used to choose a mapping.

Some examples [1,4,14,15] of simple mapping heuristics are:

- *OLB (Opportunistic Load Balancing)*: OLB assigns each task, in arbitrary order, to the next available machine, regardless of the task's expected execution time on that machine.
- *UDA (User Defined Assignment)*: UDA assigns each task, in arbitrary order, to the machine with the best expected execution time for that task, regardless of that machine's availability.
- *Fast Greedy*: Each task is assigned to the resource with the minimum completion time. The tasks are assigned in an arbitrary order.
- *Min–Min*: The Min–Min heuristic begins with the set U of all unmapped tasks. For each task in U, the minimum completion time on all machines is calculated. The task with the earliest overall minimum completion time is selected and assigned to the machine that yielded that minimum completion time. This task is removed from U, and the process reiterated until all tasks have been mapped. Min–Min attempts to map as many tasks as possible to their first choice of machine, under the assumption that this will result in a shorter makespan. Min–Min has been reported as superior to many other simple mapping heuristics [14].
- *Max–Min*: Max–Min is similar to Min–Min except that the task with the latest minimum completion time is selected and mapped. Max–Min attempts to minimize the penalties incurred by the scheduling of long-running tasks.

13.2.2 GridSim

GridSim [3,16,17] (http://www.gridbus.org/gridsim/) is a toolkit that supports the modeling and simulation of parallel and distributed computing environments. Entities such as users, applications, resources, and schedulers may be incorporated, primarily to aid in the design and evaluation of scheduling algorithms. The features of GridSim are described in Reference 17.

GridSim embodies a layered and modular architecture [17] to make use of existing infrastructure such as the open-source discrete-event simulator SimJava [18], which itself runs in a java virtual machine. The layered structure is as follows:

- The `gridbroker` package provides high-level support for Schedulers or Grid Resource Brokers. Some of the main classes are:
  - `Broker`, which encapsulates a Scheduler — The `Broker` incorporates a variant of the Fast Greedy heuristic within its default time-optimized job scheduler.
  - `BrokerResource`, which embodies a Resource, as known to the `Broker`.
  - `UserEntity`, which represents a user as known to the `Broker`.
  - `Experiment`, which manages the work done by the `Broker` for the `UserEntity`. It includes a `GridletList` object that represents the tasks to be scheduled.
- The `GridSim` package provides the basic grid infrastructure used by `gridbroker`. It includes the following primary classes:
  - `GridSim`. This class represents a grid entity and gives it communication and simulation support.
  - `Gridlet`. A class that represents a single independent task. It includes attributes such as execution size and size of input and output files. As it models an independent task, there is no support for multiple input and output file sizes.
  - `GridletList`. A class that encapsulates a set of independent tasks. It simply extends the java utility class `LinkedList`, while adding support for sorting component Gridlet objects by execution size.
  - `GridResource`. A class that embodies a grid computing resource. Its attributes are specified in the member class `ResourceCharacteristics`.
  - `ResourceCharacteristics`. A class that embodies the attributes of a grid computing resource. It is used to specify characteristics such as CPU speed, baud rate, availability, and cost.

- The SimJava package is a general-purpose discrete-event simulation package implemented in Java. Simulations in SimJava contain a number of entities each of which runs in parallel in its own thread. An entity's behavior is encoded in Java using its `body()` method [17].
- The Java Virtual Machine (JVM) provides the runtime environment for SimJava, and hence GridSim. Since the JVM essentially interprets code that has been compiled for it, it does not provide the raw performance of a fully compiled language; however, implementations of JVMs exist for multiprocessor systems and clusters [19], and so it does provide scalable support for grid simulations.

13.2.3 Task Graphs for Free

Task Graphs for Free (TGFF) [20] is an open-source package that generates random or semi-random task graphs, or DAGs. Such task graphs are ideal for scheduler simulation research. The task graphs may be generated to have a series-parallel-like structure or may be more purely random.

A generated graph consists of *tasks* and *arcs*. Arcs are the dependencies that link tasks. A definitions file (i.e., a `.tgffopt` file) is created that specifies the parameters for construction of the graph or graphs, as well as the corresponding tables that define task attributes and arc attributes. Upon processing by the TGFF utility, a text file (a `.tgff` file) is produced that fully specifies the task graph [20].

13.2.4 Genetic Algorithms

Genetic Algorithms (GAs) are often used to solve difficult optimization problems heuristically. GAs seek to make use of the fundamental principle of biological evolution, *Natural Selection*. A GA mirrors the same mechanisms by evolving a population of fixed size through a large number of generations. The population consists of randomly generated candidate solutions encoded as a string of bits or numbers, called a *chromosome*. An *objective function* is used to evaluate the chromosomes yielding a *fitness value* that reflects the suitability of the candidate solution. The fitness value is used to select a subset of individuals. This subset is then subjected to the processes of *mutation* and *crossover*, resulting in the next generation of individuals. Mutation and crossover help the GA to escape from local optima in the solution space and find near-optimal solutions to a problem.

Selection of parents for the next generation uses a competitive evaluation of the objective function for members of the population. The degree to which the fit individuals are likely to be selected is called the *selective pressure*. Surprisingly, too high a selective pressure is undesirable, since it may lead to the population converging overly quickly to a local optimum and not finding a global optimum. Various mechanisms have been used to moderate the selective pressure, an example of which is *tournament selection*. In tournament selection, a set of individuals (generally two) is randomly selected from the population. The fittest individual is then selected as a parent [21].

Note that, in tournament selection, the fittest individual in the population may well not proceed to the next generation. However, if the GA is engineered such that such an individual is guaranteed to proceed to the next generation, then the GA is said to have the quality of *elitism*.

Mutation involves a random change to a chromosome and so has the chance of introducing great novelty into the population. However, when the mutation rate is too high, good solutions tend to be overwritten, and the population tends not to *converge* to a solution.

Crossover emulates the genetic mixing effects of biological sex. It allows successful individuals to share their genetic material, and may therefore result in offspring that combine the virtues of both parents. Crossover is a considerably more important optimization mechanism than mutation [21]. Crossover may involve a division in one or many places in the chromosome. A crossover with a single point of division is termed *Single-Point Crossover*.

13.3 Related Work

13.3.1 A Comparison Study of Static Mapping Heuristics

Genetic Algorithms may be used in the static mapping of independent tasks to resources. In [1] the GA was used to find near-optimal mappings using populations of 200 chromosomes. The chromosomes were encoded mappings, represented by a vector of `t` integers, where the `i`th integer represents the `i`th task, and its value represents the machine to which it has been mapped. The GA used the makespan as the objective function. Elitism was used to retain the best solution between generations. The GA used one of three termination criteria, either stopping after 1000 iterations, or after no change in the elite chromosome for 150 generations, or after all chromosomes have converged. The termination criterion that was generally triggered was the second criterion, that is, no change in the elite chromosome for 150 generations.

For each optimization task, the GA was executed eight times: four times with a random initial population, and four times with populations seeded with the Min–Min solution. Note that the best GA solution always came from a population seeded with a Min–Min solution. The GA always improved on the Min–Min solution, providing an improvement of up to 10%.

It was decided to attempt a variant of this optimization technique, for a dynamic generational scheduler operating on a set of dependent tasks. The GA was used to find a near-optimal mapping for each meta-task during each scheduling event. The intention was to improve upon the performance of a Min–Min generational List Scheduler.

13.3.2 A Dynamic Matching and Scheduling Algorithm for HC Systems

In the paper [13], the *critical path length* for a task is defined as the longest path from the task node to an exit node. The paper describes a hybrid remapper, which uses a combination of static and run-time generated information. It uses the critical path length as a motivating factor in its *Minimum Completion Time Static Priority Algorithm*. The concept of a Critical Path was used in the `FAST_GREEDY_CRITICAL` mapping heuristic and the `SUM_CRITICAL_FINISH` objective function.

13.4 Development Work

While GridSim is a versatile and powerful simulation toolkit, it is not perfectly suited to the needs of this project. It was therefore extended in a number of ways.

First, GridSim is designed for the scheduling of independent tasks, that is, there is no support for task dependencies because GridSim is principally designed for job scheduling. Second, GridSim is intended as an application level scheduler, whereas the project requirement was for a scheduler at a service-level. Third, there is no facility in GridSim for flexible configuration of brokers or execution environment. It was therefore extended to provide xml-based support for broker configuration and task and resource specification. (Similarly, TGFF was later integrated, so that the broker might directly parse `.tgff` definition files.)

Various other extensions were also made to the broker, such as the implementation of several alternate mapping heuristics, in particular, one based on a genetic algorithm.

Where possible, all extensions were made by the creation of the new packages:

- `gridscheduler`
- `gridscheduler.ga`

Only one class in the gridsim and gridbroker packages was directly amended. All the required functionality was generally accommodated by writing new subclasses of the existing gridsim or gridbroker classes.

13.4.1 DAG Support

Since GridSim is not designed for the scheduling of dependent tasks, a number of classes required extension to permit scheduling simulation of DAGs. The class GridSim.Gridlet required support to specify its ancestor and dependent tasks. It was decided not to model the data size of these dependencies. As the project was a simulation of a service-based broker, it was accepted that communications would, in fact, be centralized via the broker itself. Therefore, the data sizes of the various dependencies could be represented by the input and output data sizes of the task itself. Gridlet was, therefore, extended by the new gridscheduler class Tasklet. The new class has additional attributes to identify its predecessors and successors, as well as its *critical path length.* The critical path length was an attribute added late in the development phase of the project, and is intended to support a new heuristic. It is a measure of the size of the execution path from the task to the latest subsequent point of termination [13].

Gridlet itself was the only class in the GridSim packages that was directly altered. Some small changes were made to support a new status: PREPARED. This new status indicates that the task has no further dependencies, and may now be executed by the processor. It follows CREATED in the life cycle of a Gridlet, and precedes READY, QUEUED, INEXEC, SUCCESS, and FAILED. READY means that the task has been assigned to a resource. QUEUED means that the task has been committed to a resource and is in its input queue. The meanings of the other statuses are as implied by the labels. The PREPARED status is essential in the operation of a generational scheduler.

Similarly, the class GridSim.GridletList represents a meta-task, a set of independent tasks. While it did not require extension for any fundamental new functionality, it was extended by the class TaskGraph to provide support for a number of methods relating to sets of tasks. The getWaiting() method was provided to return those tasks in the TaskGraph in a state of PREPARED. The setPredecessors() method was added to set the predecessors for all tasks in a TaskGraph, given that all successors had already been set — the rationale being that fewer errors would then be made in the xml definition files of TaskGraphs. This class was made redundant with the introduction of support for TGFF definition files. Two inner classes were created to provide support for sorting the TaskGraph by fanout size or CriticalPathLength. The setCriticalPathLengths() method was added in order to set all critical path lengths for tasks in the TaskGraph. This method delegates to the Tasklet method setAndGetCriticalPathLength(). Setting of the critical path length (cpl) uses the following simple recursive algorithm, which is executed for the initiating task or tasks:

```
cpl = execution size of this task
if this task has dependent tasks
   cpl + = max(cpl of all dependent tasks)
```

The GridSim class ResourceCharacteristics was extended by the class TaskResourceCharacteristics, in order to allow the baud-rate of the resource to be visible by the broker. This visibility was needed so that the broker would be able to include an estimate for the communication delay in its mapping heuristic calculations.

For this reason, the gridbroker class BrokerResource was extended by the class TaskBrokerResource, and the method getExpectedCompletionTime() was overridden to add a term for the communication delay to the estimate. The expected completion time is given by:

$$\begin{aligned}
&\textit{completion time} = \textit{exec} + \max(\textit{avail}, \textit{comms})\\
&\textit{avail} = \text{earliest time at which a resource is available}\\
&\textit{exec} = \text{anticipated execution time for a task}\\
&\textit{comms} = \text{expected communication delay}
\end{aligned}$$

The use of the max( ) function reflects the assumption that communications transmission may proceed in parallel with waiting for the processor to be available.

If the task in question has no predecessors, or is *colocated* with all such predecessors, then the communication delay is considered to be only a nominal amount — reflecting only an initiating communication from the scheduler. A corresponding method `isCoLocated()` was created to test whether the task is colocated. This method is also used by the `TaskBroker` class (see Section 4.2) when tasks are forwarded to resources; if a task is colocated, then the input file size of the `Tasklet` is also set to a nominal value.

A simple and popular model of the communication delay for message passing is:

$$\textit{message time} = \textit{latency} + \textit{message size}/\textit{bandwidth}$$

This model includes a term for the latency of the communication link. However, a bandwidth-only model has been found to lead to better predictions of communication delay, at least in some applications [22]. It was therefore considered not essential to add a network latency component to the `GridSim` communication delay methods.

13.4.2 The TaskBroker

The `gridbroker` class `Broker` was replaced by the new `gridscheduler` class TaskBroker. `Broker` was designed for the scheduling of independent tasks using one of several economic- or deadline-based optimization schemes, based on cost, time, or combined cost/time optimizations. The architecture is described in Reference 17.

For the purposes of this investigation, the economic and deadline support of the Broker class was ignored. The base time optimization scheduling was instead extended to incorporate a Generational Scheduler for DAGs. Time-shared processors have not been supported.

The `scheduleAdviser()` method, which builds a list of those tasks that are to be scheduled, was amended to only schedule `PREPARED` tasks (i.e., those tasks for which all parent tasks have completed successfully). It does this by moving such tasks from the existing `glPreparedList` to the new `glWaitingList` collection (note that `glPreparedList` actually holds those tasks that are `CREATED`, rather than `PREPARED`, and so is something of a misnomer). It then invokes one of two new methods: `scheduleWith_ListInsertion()` or `scheduleWith_ListInsertionFG()`; these methods use various heuristics to map the available tasks to resources.

13.4.3 Mapping Heuristic Extensions

The existing broker used a variant of the Fast Greedy mapping heuristic (Section 13.2.1). In Fast Greedy, each task is assigned to the resource with the minimum completion time. The tasks are assigned in an arbitrary order. The existing broker instead assigned tasks in increasing order of execution size, in an attempt to maximize packing of tasks.

TaskBroker applies various different heuristics. All may be defined in the broker configuration. The configuration parameter <heuristic> defines the desired heuristic.

The `scheduleWith_ListInsertionFG()` method applies various variants of the Fast Greedy heuristic:

- `FAST_GREEDY_BASE`. The base Fast Greedy heuristic.
- `FAST_GREEDY`. Fast Greedy as originally used by GridSim, that is, with tasks mapped in order of execution size.
- `FAST_GREEDY_FANOUT`. Fast Greedy, but with the tasks mapped in reverse order of the number of dependent tasks. This is an attempt to execute those tasks that may have many child tasks first.
- `FAST_GREEDY_CRITICAL`. Fast Greedy, but with the tasks mapped in reverse order of the critical path length. This is an attempt to map those tasks that are in the critical path of a task graph first.

The scheduleWith_ListInsertion( ) method applies the Min_Min heuristic or the GA heuristic. The configuration parameter <minmin_limit> governs when the GA is invoked for any particular scheduling event. When the size of the meta-task (the set of tasks to be scheduled) is less than or equal to the minmin_limit, then Min_Min is used to schedule the meta-task, otherwise the GA is invoked to find a near-optimal mapping.

13.4.4 GA Mapping Heuristic

The GA described in Reference 1 was used as the model for the GA heuristic used in this project to map the meta-task arising at each scheduling event. The GA heuristic was designed to be entirely configurable (see Section 13.4.2).

The functions of the various GA configuration parameters are as follows:

- heuristic. See Section 13.4.3.
- minmin_limit. See Section 13.4.3.
- minmin_seed. Indicates whether the initial population will be seeded with the Min_Min solution.
- population_size. The size of the population selected for breeding.
- max_generations. The GA stops after this number of generations.
- max_winning_run. The GA stops if the same Chromosome is the elite value for this number of generations.
- crossover_rate. The proportion of parent matings that result in crossover, and therefore recombine genetic information.
- mutation_rate. The proportion of Chromosomes that will have a gene randomly changed.
- fertility_rate. The proportion of the population that attempts to breed each generation.
- random_injection. A proportional injection of random Chromosomes injected every generation. Not yet implemented.
- elitism. Whether or not the elite Chromosome is guaranteed to survive to the next generation.
- objective. The objective function used by the GA. In this implementation, the value returned by the objective function is actually the inverse fitness — for example, in the case of MAKESPAN, we wish it minimized. The following objective functions are available:
  - MAX_FINISH (a.k.a. MAKESPAN). The makespan of a mapping, which is the latest time of completion of the mapped tasks (see Section 13.4.1).
  - SUM_TASK_FINISH. The sum of all completion times of the mapped tasks.
  - SUM_PE_FINISH. The sum of all last completion times of all processors.
  - SUM_TASK_CRITICAL. The sum, for all mapped tasks, of the products of their completion times and the log of their critical path lengths.

The classes in the gridscheduler.ga package are:

- Evolution. Encapsulates the scheduler GA itself. Contains the method evolve( ), which performs the complete evolutionary process and returns the solution as a Chromosome.
- Population. The set of Chromosomes that encapsulate the mappings currently under consideration.
- Chromosome. Specifies the processor mappings for all tasks currently ready for execution. A SingleGene object specifies each task mapping. The Chromosome contains a collection of these SingleGene objects.
- SingleGene. A container for a gene within a Chromosome. Each gene specifies the mapping of a task to a processor [resource] and the priority of the mapping to the resource. The priority is encoded as an integer value. If another SingleGene in the Chromosome specifies that its task is to be mapped to the same resource with a higher priority, then the other task will be scheduled before this one.

- `Evaluation`. A container for the result of an evolutionary generation, that is, whether the evolution has completed; why it has completed; how many generations the result took; and the elite `Chromosome` itself. Completion may have occurred for one of three criteria described in Section 13.3.1.
- `Mating`. Encapsulates the genetic pairing of two `Chromosomes`, which may share genetic information to produce two [genetically related] offspring.
- `GaRandom`. A simple class with static methods to return random numbers efficiently.

13.5 Results

13.5.1 Methodology

The GridSim distribution, including its example source files, was used as the starting point for this project. A number of incremental changes were then made in order to provide needed support (see Section 13.4).

With the implementation of the GA mapping heuristic, a large number of tests were carried out using a TGFF generated task graph. These tests indicated that the GA heuristic gave a nearly 8% improvement over a Min–Min heuristic *for the task graph in question.* Interestingly, the GA heuristic displayed a large degree of variability in its performance. In a typical set of 100 simulations for the task graph, the GA heuristic gave an improvement of 7.8%, but with a standard deviation of 5%. However, the vast majority of results were an improvement over the control Min–Min result.

An identical heterogeneous processor arrangement was used for this and all subsequent tests. The configuration file for this is given in Appendix A.

In recognition that this result was for only a single task graph, a script was designed to generate multiple variants of similar task graphs and compare the performance of a trial heuristic against the Min–Min heuristic for all of these task graphs.

Fortunately, TGFF includes excellent support for just this scenario, as it accepts a random number generator seed parameter. This seed affects all the randomized aspects of the generated task graph. Varying the seed while holding all other parameters constant generates *task graph families* containing an arbitrary number of task graphs. Given an identical seed, an identical task graph is produced [20].

The script invokes TGFF with an incremented seed, starting from a value of zero (the default value), generating a family of n task graphs. For each generated task graph, first the Min-Min heuristic is run and then the trial heuristic is run m times. Therefore, each test involved running the trial heuristic $n \times m$ times.

These task graph family tests revealed that the previously recorded 8% improvement had merely been due to a fortunate choice of task graph. The results of other task graphs in the same family led to an overall ambivalent result. Further tests were therefore conducted in an attempt to find circumstances in which the GA might perform with more consistency.

This further testing at first indicated that the GA heuristic does give a significant improvement for a family of task graphs. However, this improvement vanished when further members of the task graph family were evaluated.

13.5.2 Comparisons of the Objective Functions of GA

The various objective functions described in Section 13.4.4 were developed with differing rationales.

- `MAX_FINISH` (a.k.a. `MAKESPAN`). This was the objective function used in Reference 1. It is ideal for a static mapping scenario; but in a generational scheduler it does not allow the GA to discriminate between solutions on the basis of the completion times of all tasks — not just those tasks on the processor that completes last. (The critical path of the mapped meta-task.)
- `SUM_TASK_FINISH`. This function seeks to achieve a better result for all tasks in the meta-task, not just those that execute on the critical path of the mapped meta-task. This function differentiates

between solutions with an identical makespan, but with poorer task completion times on other resources. For example, if one solution mapped those tasks not contributing to the makespan on a single processor, and another solution mapped such tasks to multiple processors — then this function would discriminate between the two solutions. MAKESPAN gives the same score to both solutions.

- SUM_PE_FINISH. This applies a variant of the logic of the previous heuristic, but is likely not to be as discriminating.
- SUM_TASK_CRITICAL. This objective function hopes to introduce some non-locality into the GA. In other objective functions, the position of the task in the task graph plays no part in how well it might be mapped. This function seeks to preferentially map those tasks that are most critical to a total reduction in the makespan of the task graph.

Each GA objective function was tested against 100 task graphs, with each task graph having 100 nodes, and an execution size of 2 to 8 $\times$ 10^9 instructions and a communication size of 0.5 to 1.5 $\times$ 10^5 bytes. Each task graph was a recursive series-parallel graph, with a series length of 1–3 and series width of 1–3. Each task graph also had a local crossover of 20 (see Reference 20 for a full description of parameters).

Appendix B depicts the seed task graph (0) resulting from these parameters.

Other broker configuration parameters were:

```
<broker_properties>
  <heuristic>min_min</heuristic>
  <minmin_limit>1</minmin_limit>
  <minmin_seed>true</minmin_seed>
  <population_size>100</population_size>
  <max_generations>500</max_generations>
  <max_winning_run>50</max_winning_run>
  <crossover_rate>1.0</crossover_rate>
  <mutation_rate>0.01</mutation_rate>
  <fertility_rate>2.0</fertility_rate>
  <random_injection>0.0</random_injection>
  <elitism>true</elitism>
..<max_task_per_PE>1</max_task_per_PE>
</broker_properties>
```

Each test consisted of a family of 10 task graphs, with 20 scheduling runs per task graph, for a total of 800 scheduling runs.

The results indicate that, for this task graph family subset, the GA mapping heuristic gives an average 6% improvement — when the SUM_TASK_CRITICAL objective function is used (Table 13.1). The SUM_TASK_FINISH function performs nearly as well, while the MAKESPAN function gives less than half the same improvement. This result is in line with the expectations of the objective functions. The SUM_TASK_FINISH function, with its contribution of all task finish times, gives a greatly improved result over the MAKESPAN. The SUM_TASK_CRITICAL function, with its non-localized contribution, gives a further improved result.

Note that this improvement, relative to Min–Min, vanished when further members of the task graph family were examined. However, the relative results of the different objective functions still hold.

TABLE 13.1 GA Objective Function Comparisons

Objective function	sum_pe_finish	makespan	sum_task_critical	sum_task_finish
Proportional increase to min-min	0.015	0.024	0.06	0.05

13.5.3 GA Onset Trials

The configuration parameter MINMIN_LIMIT (Section 13.4.3) specifies a meta-task size limit for which the Min–Min heuristic should be used rather than the GA heuristic. This functionality was designed since it is clear that for a single task, the optimal mapping returned by the GA would, at best, simply be that resource that would first complete the task, which is precisely the mapping that a Min–Min heuristic would result in.

Two series of tests were conducted to investigate the effect of increasing the MINMIN_LIMIT. The tests were discontinued at the point where, for a large number of the tasks tested, the GA result was identical to a simple Min–Min heuristic. This occurs when the size of meta-tasks rarely, if ever, exceeds the MINMIN_LIMIT.

In the first series of tests, the task family used in the previous test was tested with the SUM_TASK_FINISH objective function, with MINMIN_LIMIT varied from one to three. Other broker configuration parameters were identical to the previous test.

The second series of tests was similar to the first, with the task graph family having similar characteristics to the previous test, but with a series width of two to eight. The value of the broker parameter MAX_GENERATIONS was 100. Other broker configuration parameters were identical to the previous test. The MINMIN_LIMIT varied between one and five.

Each test consisted of a family of 10 task graphs, with 20 scheduling runs each, for a total of 400 scheduling runs.

The results in Table 13.2 indicate that the favorable results obtained in Section 13.5.2 for the SUM_TASK_FINISH objective function steadily degrade as the MINMIN_LIMIT is increased and more of the mapping is made using Min–Min. This indicates *that for these particular task graphs*, the GA heuristic was more effective than a Min–Min mapping.

The results in Table 13.3 do not demonstrate a consistent pattern. The GA appears not to perform inconsistently given the characteristics of the task graph family.

13.5.4 Large-Scale DAGs

An attempt was made to see whether the GA might be more advantageous for large-scale task graphs.

Each GA objective function was tested against 2 to 5 task graphs, with each task graph having 1000 nodes, with an execution size of 2 to 8×10^9 instructions and communication size of 0.5 to 1.5×10^5 bytes. The task graphs were random, not series parallel. The test has not yet been performed against a larger number of task graphs because of the time taken to perform the test itself.

Each test consisted of a family of 2 to 5 task graphs, with 10 scheduling runs per task graph, for a total of 110 scheduling runs.

The results for SUM_TASK_FINISH and SUM_TASK_CRITICAL appear to approximate that of a Min–Min mapping, while the other objective functions do considerably worse. This again demonstrates the superiority of these two objective functions to SUM_PE_FINISH and MAKESPAN (Table 13.2)

TABLE 13.2 GA Onset

minmin_limit	1	2	3
Proportional increase to minmin	0.05	0.025	0.001

Low-width series parallel.

TABLE 13.3 GA Onset

minmin_limit	1	2	3	4	5
Proportional increase to minmin	−0.017	0.012	0.006	0.019	0.005

Medium degree series parallel.

TABLE 13.4 Large-Scale DAGs

Objective	sum_pe_finish	makespan	sum_task_critical	sum_task_finish
Result versus minmin	−0.094	−0.038	−0.01	−0.006

TABLE 13.5 Comparison of Fast Greedy Heuristic Variants with Min–Min

Task_degree	2–5			8–12			14–17		
Heuristic (fg)	Exec	Fanout	Critical	Exec	Fanout	Critical	Exec	Fanout	Critical
Improvement versus unsorted fg	0.073	0.035	0.051	0.043	0.036	0.029	0.040	0.034	0.033

13.5.5 Fast Greedy Variants

A number of trials were also conducted of the various variants of the Fast Greedy heuristic. Each was tested against a total of 300 task graphs, in 3 families differentiated by task degree, that is, 100 task graphs of low degree, 100 of medium degree, and a further 100 of high degree.

All task graphs had 100 nodes, with an execution size of 2 to 8×10^9 instructions and communication size of 0.5 to 1.5×10^5 bytes. The task graphs were random, not series parallel.

Each test consisted of a family of 100 task graphs, with 1 scheduling run each, for a total of 1500 scheduling runs (including Min–Min runs).

The results indicate that, for all task degrees, performance of the Fast Greedy heuristic is improved by assigning the tasks in order of execution size, order of fanout, or order of critical path size (Table 13.5). The effectiveness of the simple order of execution size case is explained by considering that this approximates to a Min–Min mapping, at least for consistent processors and no communication costs.

13.6 Further Work

13.6.1 Network Latency

As indicated in Section 13.4.1, a generalized model of communication delay for message passing is:

$$\textit{message time} = \textit{latency} + \textit{message size}/\textit{bandwidth}$$

This chapter assumed a latency of zero, which was considered acceptable, given the effectiveness of a bandwidth-only model [22] and GridSim's lack of support for network latency. However, incorporation of non-zero latencies might provide for a more accurate model of a grid environment. This would require some refactoring of several GridSim classes, but particularly IO_Data, Input, and Output. It is expected that this work would have little bearing upon the GA mapping heuristic.

13.6.2 Multiple User Environments

The GA heuristic has only been tested in a single-user environment. It would be of great interest to investigate performance in a true multi-user service-based environment. In such an environment, the load on the grid would be variable, due to tasks being processed for other users. It might then be that in this circumstance, the local nature of the meta-task mapping of the GA heuristic would be of less consequence.

13.6.3 Randomized Heterogeneous Processor Environments

This project has only investigated a single example of a heterogeneous computing environment, in a more complete investigation; a randomized heterogeneous processor environment could be evaluated. The TGFF table generating functionality can be used to generate such randomized grid environments.

13.6.4 Inconsistent Execution

This chapter has only investigated execution of the GA heuristic in a *consistent* processor environment. That is, while the processors have been heterogeneous in performance and available bandwidth, they have behaved consistently for all tasks. However, it would be interesting to evaluate performance in an *inconsistent* environment. This would necessitate adding a `TaskType` member to the `Tasklet` class (which extends `Gridlet`), and supporting differing execution rates for the PE class for different `TaskType` objects.

13.7 Conclusions

This chapter has described several variants of an existing dynamic generational mapping heuristic, the Fast Greedy heuristic. It has been shown that simply sorting the meta-task in order of execution size may make substantial improvements to the Fast Greedy heuristic. Other orderings, such as fanout size and critical-path length, appear not to be as advantageous.

This chapter has also described a new dynamic generational mapping heuristic, the GA mapping heuristic. It has been shown that the total makespan of the heuristic is roughly comparable to that of the Min-Min heuristic, with an added variance. The variance is introduced by the non-deterministic nature of the GA (The execution time of the scheduler itself will, of course, be substantially greater). It is suspected that this result might be enhanced with an improvement in the objective function of the GA. Most evaluated objective functions only make use of the current meta-task, and have no broader knowledge of the task graph. The exception to this was the `MAX-TASK-CRITICAL` function, which attempted to factor the critical path of a task into the fitness value. This GA appeared to be the best performing (along with the `SUM_TASK-FINISH` function). One difficulty in improving the GA is that it may only use the fitness value to choose between solutions. It may not use any other contextual information, and the fitness is only a simple scalar value.

It is also suggested that performance of the GA heuristic would be relatively greater in a multiuser or inconsistent processor environment. In such an environment, the local nature of the GA should be less of a constraint on performance.

References

[1] Braun, T., Siegel, H., et al. A Comparison study of static mapping heuristics for a class of meta-tasks on heterogeneous computing systems. In *8th IEEE Heterogeneous Computing Workshop (HCW '99)*, 1999.

[2] Foster, I. and Kesselman, C., Ed. *The Grid: Blueprint for a New Computing Infrastructure.* Morgan Kaufmann, San Francisco, CA, 1998.

[3] Murshed, M. and Buyya, R. Using the GridSim Toolkit for Enabling Grid Computing Education, http://citeseer.nj.nec.com/574558.html.

[4] Maheswaran, M., Ali, S., et al. Dynamic matching and scheduling of a class of independent tasks onto heterogeneous computing systems. In *8th IEEE Heterogeneous Computing Workshop (HCW '99)*, April 1999, pp. 30–44.

[5] Kim, J., Shivle, S., et al. Dynamic mapping in a heterogeneous environment with tasks having priorities and multiple deadlines. In *International Parallel and Distributed Processing Symposium (IPDPS'03)*, April 2003.

[6] Coffman, E., Jr., Ed. *Computer and Job-shop Scheduling Theory.* John Wiley & Sons, New York, 1976.

[7] Fernandez-Baca, D. Allocating modules to processors in a Distributed System. *IEEE Transaction on Software Engineering*, 15, 1989, 1427–1436.
[8] Ibarra, O. and Kim, C. Heuristic algorithms for scheduling independent tasks on non-identical processors. *Journal of the ACM*, 24, 1977, 280–289.
[9] Novaes, M. and Challenger, J., *Request Level Scheduling for the Grid*. IBM T.J. Watson Research Center, Yorktown Heights, N.Y.
[10] Cirou, B. and Jeannot, E., "Triplet: a clustering scheduling algorithm for heterogeneous systems. *IEEE ICPP International Workshop on Metacomputing Systems and Applications (MSA'2001)*, September 2001, Valencia, Spain.
[11] Freund, R., Gherrity, M., et al. Scheduling resources in multi-user, heterogeneous, computing environments with SmartNet. In *7th IEEE Heterogeneous Computing Workshop (HCW '98)*, March 1998, pp. 184–199.
[12] Taura, K. and Chien, A. A heuristic algorithm for mapping communicating tasks on heterogeneous resources. In *Heterogeneous Computing Workshop*, May 1, 2000.
[13] Maheswaran, M. and Siegel, H. A dynamic matching and scheduling algorithm for heterogeneous computing systems. In *7th IEEE Heterogeneous Computing Workshop (HCW '98)*.
[14] Holenarspiur, P., Yarmolenko, V., et al. Characterisation and Enhancement of Static Mapping Heuristics for Heterogeneous Systems, Technical report OSU-CISRC-2/00-TR07, Department of Computer and Information Science, Ohio State University, 2000.
[15] Siegel, H. and Ali, S. Techniques for mapping tasks to machines in heterogeneous computing systems. *Euromicro Journal of Systems Architecture*, Special Issue on Heterogeneous Distributed and Parallel Architectures: Hardware, Software and Design Tools, 46, 2000, 627–639.
[16] Sherwani, J., Ali, N., et al. Libra: An economy driven job scheduling system for clusters. In *Proceedings of the 6th International Conference on High Performance Computing in Asia-Pacific Region* (HPC Asia 2002), Bangalore, India, 2002.
[17] Buyya, R. and Murshed, M. GridSim: a toolkit for the modeling and simulation of distributed resource management and scheduling for grid computing. *The Journal of Concurrency and Computation: Practice and Experience* (CCPE), 14(13–15), Nov–Dec 2002, 1175–1220.
[18] Howell, F. and McNab, R. Simjava: a discrete event simulation package for java with applications in computer systems modelling. In *First International Conference on Web-based Modelling and Simulation*. San Diego, CA, Society for Computer Simulation, January 1998.
[19] Aridor, Y., Factor, M., and Teperman, A. cJVM: a single system image of a JVM on a cluster. In *Proceedings of the 29th International Conference on Parallel Process (ICPP'99)*, 1999.
[20] Dick, R., Rhodes, D., and Wolf, W. TGFF: Task Graphs for Free. In *Proceedings of the 5th International Workshop Hardware/Software Co-Design/Codes/CASHE'97*, March 1998, pp. 97–100.
[21] Wilkinson, B. and Allen, M., *Parallel Programming*. Prentice Hall, 1999, pp. 372–377.
[22] Dail, H., Berman, F., and Casanova, H. A decoupled scheduling approach for grid application development environments. *Journal of Parallel and Distributed Computing*, 63(5), May 2003, 505–524.

Appendix A. Test Processor Configuration

```
<?xml version="1.0" encoding="utf-8" standalone="yes" ?>
<grid>
 <resource name="resource0">
  <sharing>space</sharing>
  <cost>1</cost>
  <baud>100</baud>
  <machine>
          <pe mips="300"/>
  </machine>
```

```
</resource>
<resource name="resource1">
  <baud>80</baud>
  <machine>
          <pe mips="250"/>
  </machine>
</resource>
<resource name="resource2">
  <baud>150</baud>
  <machine>
          <pe mips="68"/>
  </machine>
</resource>
<resource name="resource3">
  <baud>80</baud>
  <machine>
          <pe mips="500"/>
  </machine>
</resource>
<resource name="resource4">
  <baud>100</baud>
  <machine>
          <pe mips="120"/>
  </machine>
</resource>
</grid>
```

This configuration specifies that the available grid environment consists of five machines, each having a bandwidth ranging from 80 to 150 KBpsec. All are space-sharing architectures, rather than time-sharing, and have a nominal cost. Each of the machines have a single processor element, with speed ratings of 68 to 500 mips.

Appendix B. Sample Task Graph

The `.tgffopt` definition that follows, given a seed value of zero, results in the task graph depicted in Figure 13.B1. Resultant Task sizes are 2000 to 8000 M instructions, and communication costs are 500 to 1500 KB.

```
tg_cnt 1
gen_series_parallel true
series_must_rejoin true
series_len 2 1
series_wid 2 1
series_local_xover 20
task_cnt 100 1
task_degree 2 4
tg_write
eps_write
table_label PROCESS_SIZE
table_cnt 1
```

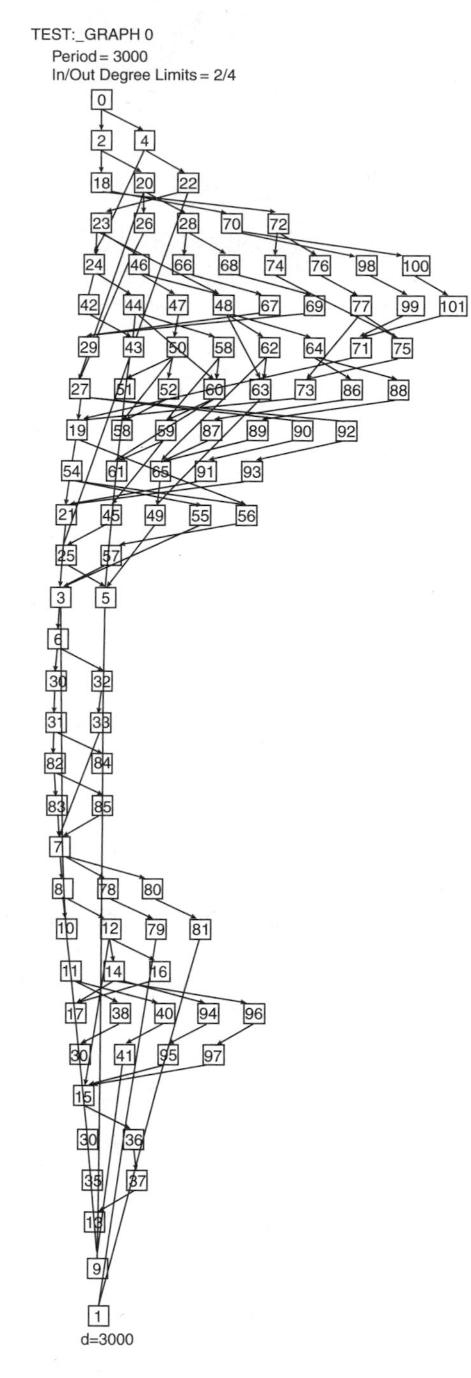

FIGURE 13.B1 Sample task graph.

```
type_attrib exec_time 5000 3000
pe_write
table_label COMMS_SIZE
table_cnt 1
type_attrib comms_size 1000 500
trans_write
```

14

Minimization of SADMs in Unidirectional SONET/WDM Rings Using Genetic Algorithms

Anirban Mukhopadhyay
Utpal Biswas
Mrinal Kanti Naskar
Ujjwal Maulik
Sanghamitra Bandyopadhyay

14.1 Introduction

Most of today's optical networks are hierarchies of SONET rings. This optical fiber communication is employed with WDM technology, where the whole bandwidth of an optical fiber is divided among a number of nonoverlapping wavelengths, each of which is capable of carrying high-speed optical data. In recent years, the bandwidth of a wavelength has increased from 2.5 Gbps (OC-48) to 10 Gbps (OC-192) and is likely to go upto 40 Gbps (OC-768) in the near future [1,2]. Thus, the bandwidth capacity on a wavelength is too large for certain traffic requirements. An approach to provide fractional wavelength capacity is to split a wavelength into multiple time slots and multiplex traffic on the wavelength. Therefore, each wavelength running at the line rate of OC-N can carry several low-speed OC-M ($M < N$) traffic channels in TDM fashion. For example, an OC-48 line can carry 16 OC-3 channels. The resulting networks are called WDM–TDM networks or WDM traffic grooming networks. The ratio of N to the smallest value of M is called the grooming ratio [1–5].

Using WDM technology, multiple rings can be supported on a single fiber ring. In this architecture, each wavelength independently carries a SONET ring. Each SONET ring can further support multiple low-speed streams. At every node, a Wavelength Add/Drop Multiplexer (WADM) adds and drops or bypasses traffic on any wavelength. At each node, there are SONET add/drop multiplexers (SADM) on each wavelength to add/drop low-speed streams. So the number of SADMs per node will increase linearly with the number of wavelengths that a single fiber ring can carry. The cost of SADMs dominates the total cost of the optical network. But in fact, it is not necessary for each node to be equipped with SADMs on each wavelength. An SADM on a wavelength at a node is needed only if there is traffic terminating at this node on this wavelength. So, the problem is to combine different low-speed traffic streams into high-speed traffic streams in such a way that the number of SADMs is minimized [3,4]. This problem is proven to be NP-complete [3]. As far as our knowledge goes, we are the first to propose a genetic algorithm (GA) solution to this problem for unidirectional ring topologies. Here, we have restricted ourselves to consider static traffic pattern only. We have proposed an ILP formulation for the problem and solved it using a GA. We have shown that our algorithm produces better results compared to those found in some recent literature.

The rest of the chapter is organized as follows: Section 14.2 presents an example for SADM minimization problem. Section 14.3 proposes the ILP formulation of the problem. A brief introduction to GA is presented in Section 14.4. Section 14.5 gives an overview of the proposed GA to solve the problem. The results obtained are discussed in Section 14.6. Finally, Section 14.7 concludes the chapter.

14.2 Example of SADM Minimization

In this section, we will present an example to show that careful grooming can reduce the number of SADMs. Consider a ring network with four nodes. Each wavelength can carry two traffic streams. The traffic pattern for this example is unidirectional uniform traffic. Figure 14.1 shows the traffic matrix. Without traffic grooming, every node requires two SADMs, one for each wavelength. Thus, without grooming, a total of $4 \times 2 = 8$ SADMs are required.

The traffic assignment as shown in Table 14.1 will result in Figure 14.2, where we can see that though we have used traffic grooming, still the number of SADMs required is eight. This is due to the lack of careful grooming.

But if we assign traffic as shown in Table 14.2, node 2 will bypass traffic on the wavelength $\lambda 2$ (Figure 14.3); hence saving an SADM, as node 2 will have no SADM for wavelength $\lambda 2$. Thus, in this case, the total number of SADMs required is seven.

$$T = \begin{bmatrix} 0 & 1 & 1 & 1 \\ 0 & 0 & 1 & 1 \\ 0 & 0 & 0 & 1 \\ 0 & 0 & 0 & 0 \end{bmatrix}$$

FIGURE 14.1 The traffic matrix.

TABLE 14.1 Careless Traffic Assignment

Wavelength	Time slots	Traffic
λ1	1 --→	1 → 2, 2→4
	2 ⟶	1 → 3, 3→4
λ2	1 --→	1 → 4
	2 ⟶	2 → 3

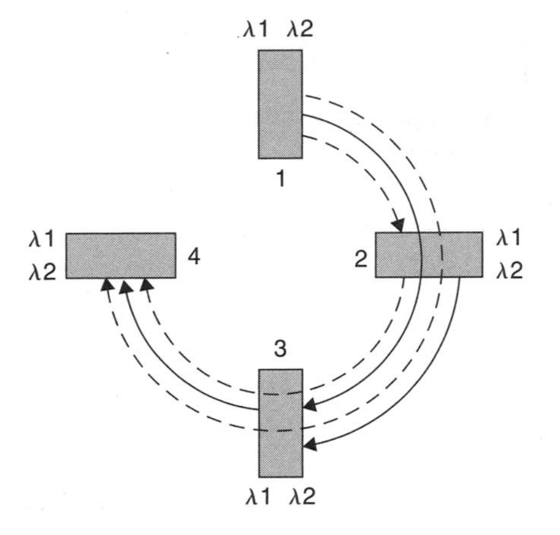

FIGURE 14.2 Traffic configuration requiring eight SADMs.

TABLE 14.2 Careful Traffic Assignment

Wavelength	Time slots	Traffic
λ1	1 --→	1 → 2, 2→4
	2 ⟶	2 → 3, 3 → 4
λ2	1 --→	1 → 4
	2 ⟶	1 → 3

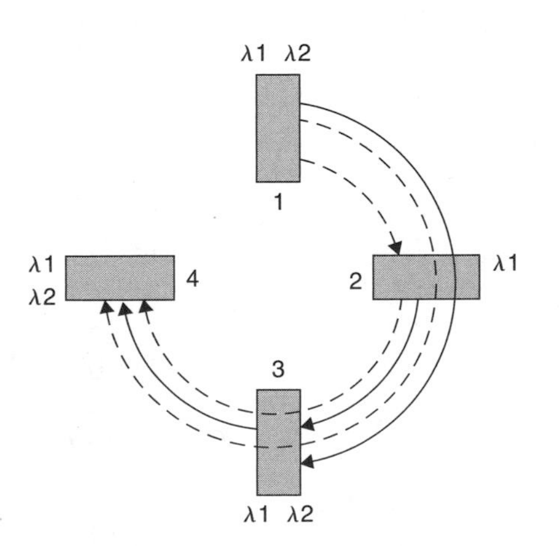

FIGURE 14.3 Traffic configuration requiring seven SADMs.

14.3 Problem Formulation

Here we have considered a unidirectional ring network. Let the number of nodes in the ring be N and L, the total number of wavelengths available in a link; g is the *grooming ratio*, that is, the maximum number of low-speed traffic streams that can be multiplexed onto a single wavelength. An SADM is needed at a node for some wavelength only if some of the traffic contained in that wavelength is originated from, or destined to, that node.

Suppose N nodes in the network are labeled from 0 to $N - 1$ in a clockwise direction. Wavelengths available are 0 to $L - 1$. We are given a set of source–destination node pairs, $R = \{(s_i, d_i) | 0 \leq s_i, d_i \leq N - 1, 0 \leq i \leq |R| - 1\}$, where each pair (s_i, d_i) denotes the requirement of a low-speed, unidirectional traffic stream from node s_i to d_i.

The solution is to find a suitable wavelength assignment $W = \{l_i | 0 \leq l_i \leq L - 1, 0 \leq i \leq |R| - 1\}$, where wavelength l_i is assigned to traffic stream (s_i, d_i), such that total number of SADMs needed is minimized.

For a wavelength assignment $\{l_i\}$, we can find the total number of SADMs needed. Let SADM_{nw} be a binary variable that is true if an SADM is required at node n for wavelength w, where $0 \leq n \leq N - 1, 0 \leq w \leq L - 1$.

$$\text{SADM}_{nw} = \begin{cases} 1 & \text{if node } n \text{ needs SADM for wavelength } w, \\ 0 & \text{if node } n \text{ does not need SADM for wavelength } w. \end{cases}$$

We know that an SADM for wavelength w is needed at node n only if node n is the source or destination of any traffic stream using wavelength w. Hence, we can compute SADM_{nw} for each (n, w) pair in the following way.

$$\text{SADM}_{nw} = \begin{cases} 1 & \text{if } [s_i = n \text{ or } d_i = n] \text{ and } l_i = w, \\ 0 & \text{otherwise,} \end{cases}$$

where $(s_i, d_i) \in R, 0 \leq l_i \leq L - 1, 0 \leq i \leq |R| - 1$. Therefore, the total number of SADMs in the ring is given by $\sum_{n=0}^{N-1} \sum_{w=0}^{L-1} \text{SADM}_{nw}$. The objective is to minimize the number of SADMs, that is, minimize $\sum_{n=0}^{N-1} \sum_{w=0}^{L-1} \text{SADM}_{nw}$.

There are certain constraints that we have to consider. First, $0 \leq \text{SADM}_{nw} \leq 1$, and SADM_{nw} is an integer variable. The other constraint is that no wavelength at any link should be overloaded, that is, the maximum number of traffic streams that pass through it should be less than or equal to the grooming ratio g.

The link from node n to $((n + 1) \mod N)$ is traversed by all low-speed traffic streams $(s_i, d_i) \in R$, where $(s_i \leq n < d_i)$, or $(n < d_i < s_i)$, or $(d_i < s_i \leq n)$. Therefore, the load on the wavelength w on link $n \to ((n + 1) \mod N)$ is given by the cardinality of the set

$$\text{LD}_{nw} = \left\{ \begin{array}{l} (s_i, d_i) \mid [(0 \leq s_i \leq n < d_i \leq N - 1) \text{ or } (0 \leq n < d_i < s_i \leq N - 1) \\ \text{or } (0 \leq d_i < s_i \leq n \leq N - 1)] \text{ and } [l_i = w], \\ (s_i, d_i) \in R, 0 \leq l_i \leq L - 1, 0 \leq i \leq |R| - 1 \end{array} \right\}.$$

Hence, the load constraint will be $\forall n, w(0 \leq n \leq N - 1, 0 \leq w \leq L - 1) \Rightarrow |\text{LD}_{nw}| \leq g$. Thus, the problem of minimization of SADMs in unidirectional SONET/WDM ring via traffic grooming can be given as:

Given integers $N > 0, L > 0, g > 0$ and request set $R = \{(s_i, d_i)\}(0 \leq s_i, d_i \leq N - 1, 0 \leq i \leq |R| - 1)$, find a suitable wavelength assignment $W = \{l_i | 0 \leq l_i \leq L - 1, 0 \leq i \leq |R| - 1\}$, so that,

$$\sum_{n=0}^{N-1} \sum_{w=0}^{L-1} \text{SADM}_{nw} \text{ is minimized.}$$

Subject to:

1. $0 \leq \text{SADM}_{nw} \leq 1$, and SADM_{nw} is an integer variable
2. $\forall n, w\ (0 \leq n \leq N-1, 0 \leq w \leq L-1) \Rightarrow |\text{LD}_{nw}| \leq g$

14.4 Genetic Algorithm

Genetic algorithms, introduced by John Holland, are efficient methods for the solution of many search and optimization problems. Genetic algorithms are robust and stochastic search procedures based on the principle of natural genetics and evolutionary theory of genes [6,7]. They follow the evolutionary process as stated by Darwin. The algorithm starts by initializing a *population* of potential solutions encoded as strings called *chromosomes.* Each solution has some *fitness value,* which indicates the goodness of the encoded solution. Based on the fitness values, the parents that would be used for reproduction are selected (*survival of the fittest*). The new generation is created by applying genetic operators such as *crossover* (exchange of information among parents) and *mutation* (sudden small change in a parent) on selected parents. Thus the quality of population is improved as the number of generations increases. The process continues until some specific criterion is met or the solution converges to some optimized value.

Though GAs sound complicated, one does not have to be a biologist to understand its mechanism. The main advantage of GAs is their robustness, and hence they can be applied to a wide range of problems, for example, VLSI desing [8,9], pattern recognition and image processing [10–12], bioinformatics [13,14]. Genetic algorithms are not guaranteed to find the global optimum solution to a problem in a small number of generations, but they are generally good at finding "reasonably good" solutions to problems "acceptably quickly". In this regard, GAs have been found to outperform many other search processes.

14.5 Genetic Algorithms for Traffic Grooming Problem

In this section, we will present a GA to solve the problem stated in the previous section. We will discuss each step of the algorithm and at last give the final GA for the problem of traffic grooming in unidirectional SONET/WDM ring networks to minimize the number of SADMs.

14.5.1 Solution Representation

Here, we have encoded the solution to integer strings. In the previous section we showed that the input set of source–destination pairs is the set $R = \{(s_i, d_i) | 0 \leq s_i, d_i \leq N-1, 0 \leq i \leq |R|-1\}$, where each pair (s_i, d_i) denotes the requirement of a low-speed, unidirectional traffic stream from node s_i to d_i. We have to find a suitable wavelength assignment such that the total number of SADMs required is minimized. We can represent the solution as an integer string $W = \{l_i | 0 \leq l_i \leq L-1, 0 \leq i \leq |R|-1\}$, where wavelength l_i is assigned to traffic stream (s_i, d_i). The fitness value (objective function in the ILP), that is, the total number of SADMs required, for each solution can be calculated as shown in Section 14.2. We do not explicitly check the validity of a string, that is, whether it satisfies the constraint that a wavelength on a link can support maximum g (*grooming ratio*) numbers of low-speed traffic streams; rather, the fitness function is converted into an unconstrained form, by adding a penalty term greater than $L \times N$ (maximum number of SADMs in the ring) times the number of violations, to the objective function. Thus while minimizing the fitness values, the solutions violating the constraint are automatically ruled out.

Example 14.1

Input set: $R = \{(2,3), (6,2), (1,8), (5,6), (2,4), (3,6)\}$

Solution representation: $W = \{\ 3 \quad 8 \quad 3 \quad 2 \quad 2 \quad 1\ \}$

14.5.2 Initial Population

The initial population is generated by creating a number of individual solutions (strings). Each string is generated by random assignment of the wavelengths to each traffic stream of the input set *R*.

14.5.3 Selection Operator

We use the tournament method of selection, in which a number of strings are picked randomly from the current generation; from among them we choose the best solution to act as a parent. The other parent is always the best solution of the previous generation. Genetic operators are applied on the parents to produce the offspring solutions.

14.5.4 Crossover

The crossover operation is used to exchange genetic materials between two parents. Here we use single-point crossover, which produces two children from two parents. A random crossover point is generated between 0 and the maximum chromosome length of the solution. This point divides each solution into two parts. The corresponding parts of the parent chromosomes are swapped to produce two new offspring.

Example 14.2
Parent1: **1 3 5 2 3 6 7 4 1**
Parent2: 3 4 5 2 2 *5 7 1 5*
Crossover point: 5
Offspring1: **1 3 5 2 3** | *5 7 1 5*
Offspring2: 3 4 5 2 2 | **6 7 4 1**

14.5.5 Mutation

A random mutation point between 0 and maximum chromosome length is selected, and another wavelength is assigned instead of the current wavelength at that point with some probability called the *mutation probability*.

Example 14.3
Chromosome: 1 3 5 **2** 3 6 7 4 1
Mutation point: 4
New assignment to 4th wavelength (2): 8
New chromosome: 1 3 5 **8** 3 6 7 4 1

14.5.6 Acceptance Policy

A new offspring solution obtained by crossover or mutation operation is retained in the population, if its fitness value is better than the worst fitted solution in the population, otherwise the solution is rejected [15].

14.5.7 The Final Genetic Algorithm

Example 14.4

Procedure GA
Begin
Generate initial population as per Section 5.2
Initialize number_of_generations = 0
  While (number_of_generations < max _generation) *do*
    Select the parents for crossover as per Section 5.3

Apply crossover operator as per Section 5.4
Apply mutation operator as per Section 5.5
Accept offspring solution as per section 5.6
Number_of_generations = number_of_generations + 1
End while
Output the solution and its fitness value
End GA

14.6 An Improvement to the Genetic Algorithm for All-To-All Unitary Traffic

In this section, we will present an improvement over the aforementioned GA by introducing an extra constraint. This improvement works best in the case of all-to-all unitary traffic, that is, when exactly one connection is to be established between each pair of nodes. In such a scenario, if the input set R contains a source–destination pair (s, d), then it will also contain the pair (d, s).

The additional constraint is that both the pairs, (s, d) and (d, s), are assigned the same wavelength. The rationale being that this assignment will form a circle and SADM saving is maximum when such a circle is created, because a circle needs two SADMs, one at node s and the other at node d (Figure 14.4). If two different wavelengths are used, then a total of four SADMs are needed, two SADMs at each node (Figure 14.5). Hence, while generating a string, we apply this additional constraint with the expectation of getting a better performance. Let us denote this improved algorithm as GA*.

14.7 Results and Discussions

In this section, we will present the results of our algorithm and make comparisons with two other results found in References 3 and 4. In Reference 3, Modiano presented a heuristic algorithm for the same

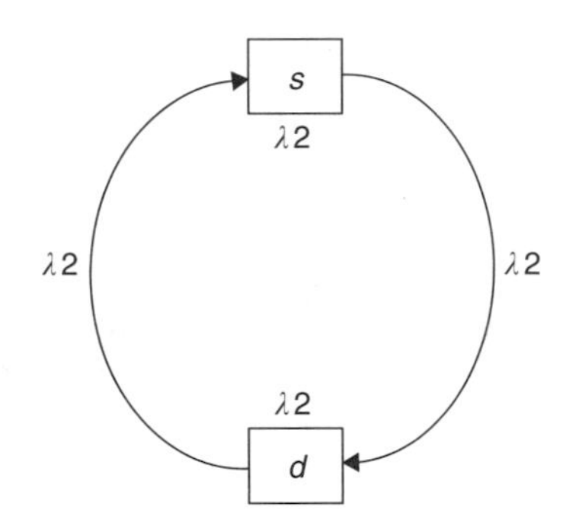

FIGURE 14.4 A total of two SADMs are needed.

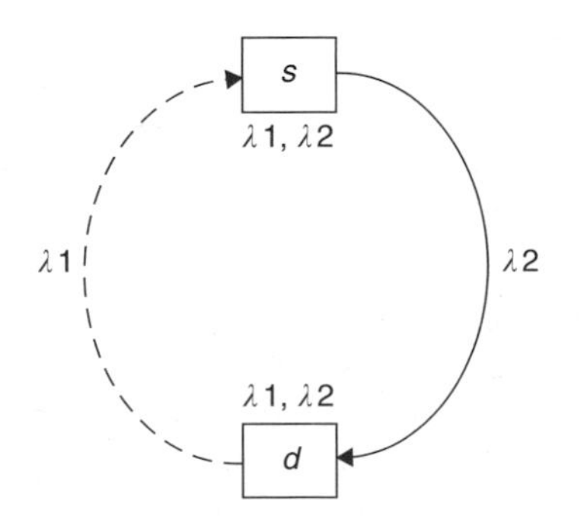

FIGURE 14.5 A total of four SADMs are needed.

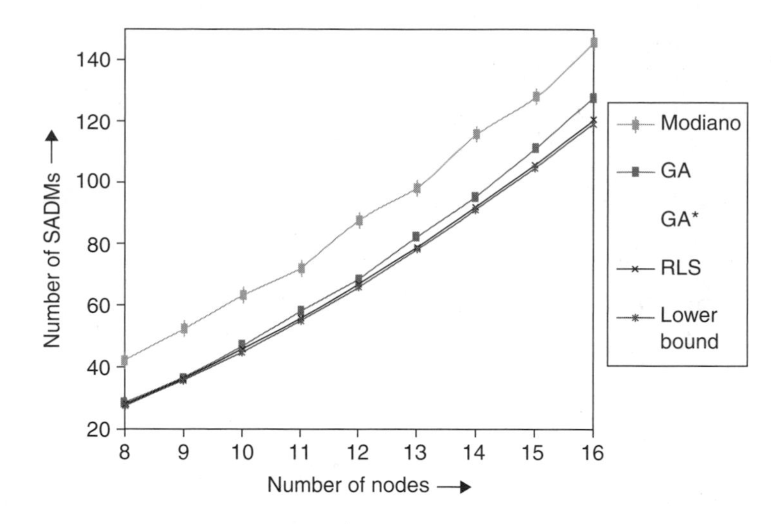

FIGURE 14.6 Performance of algorithms for $g = 4$.

problem, whereas in Reference 4, a Reactive Local Search (RLS) algorithm has been presented. We have tested the performances of the algorithms for all-to-all unitary traffic, that is, traffic requirement is exactly one for each pair of nodes. Hence the input set of traffic requirements R can be calculated as:

Example 14.5
  For $s = 0, 1, 2, 3, \ldots, N - 1$
    For $d = 0, 1, 2, 3, \ldots, N - 1$
      If $s \neq d$ then $R = R \cup \{(s, d)\}$
    End for
  End for

For example, for a SONET ring with three nodes, the input set of connection requirements will be:

$$R = \{(1, 2), (2, 1), (1, 0), (0, 1), (2, 0), (0, 2)\}.$$

We have run our algorithm upto 10,000 generations to get the results, which takes close to 1 min on a 1.7 GHz Pentium IV computer with 128 MB RAM running Windows Me operating system. The execution time does not depend on the number of nodes of the ring.

Figure 14.6 is a comparison of Modiano's algorithm, the RLS algorithm, our GA, and GA* for all-to-all unitary traffic in unidirectional rings. The graphs represent the number of SADMs required for a certain number of nodes. The number of nodes varies from 8 to 16, as most of the SONET ring contains a maximum of 16 nodes. We have shown the graphs for grooming ratio $g = 4$. We have also given the graphs for the optimal number of SADMs required. From the graph it is clear that our algorithm performs better than Modiano's algorithm in all the cases. GA* and RLS give the optimal result for $g = 4$.

Figure 14.7 shows the graphs comparing the four algorithms for grooming ratio $g = 16$. Here we can see that the improved GA, that is, GA* performs better than the other algorithms and its performance is near optimal.

The effectiveness of the proposed improved GA, that is, GA* can be viewed better if we compare the values for the number of SADMs obtained using the algorithm GA* with its lower bound. In Table 14.3, we have shown the optimum solution and the solution obtained by algorithm GA* for different numbers of nodes and for grooming ratio 4 and 16. It is clear that for grooming ratio 4, GA* gives the optimum result. For larger grooming ratios also, GA* provides near-optimal results for a small number of nodes. Undoubtedly, GA* gives the best result among all the algorithms for the case of all-to-all unitary traffic.

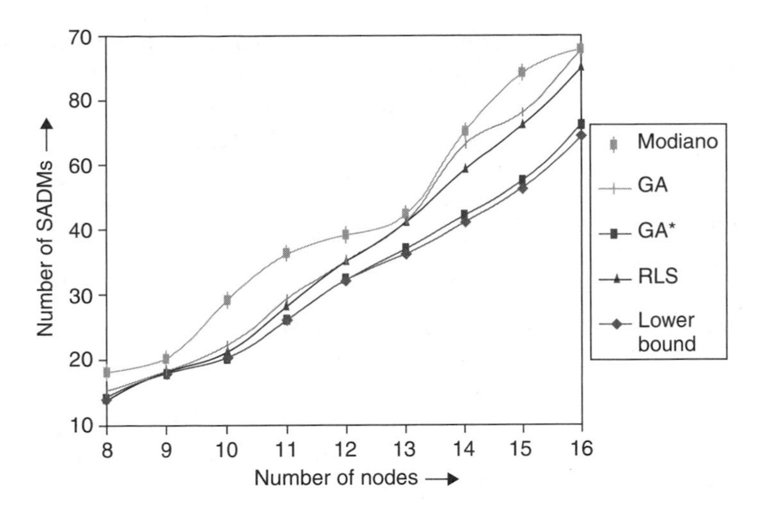

FIGURE 14.7 Performance of algorithms for $g = 16$.

TABLE 14.3 Comparison of results of GA* with optimal value

No. of nodes	Grooming ratio $g = 4$		$g = 16$	
	GA*	Lower bound	GA*	Lower bound
8	28	28	14	14
9	36	36	18	18
10	45	45	20	20
11	55	55	26	26
12	66	66	32	32
13	78	78	37	36
14	91	91	42	41
15	105	105	47	46
16	120	120	56	54

14.8 Conclusions

In this chapter, we have presented a GA for the traffic grooming problem in unidirectional SONET/WDM ring networks to minimize the number of SADMs required. We have shown that our algorithm performs better than other heuristic and reactive search techniques found in recent literature. The GA also takes lesser time to reach the optimal result; this time is independent of the number of nodes as the number of generations has been fixed at 10,000. In most of the cases, the optimum value is found with a much lesser number of generations.

References

[1] A. Mukhopadhyay, J.K. Singh, U. Biswas, and M.K. Naskar, Distributed approaches for dynamic traffic grooming in WDM optical networks. In *Proceedings of International Conference on CODEC-04*, Kolkata, India, January, 2004, p. P-55.

[2] A. Mukhopadhyay, J.K. Singh, U. Biswas, and M.K. Naskar, Improved distributed approaches for dynamic traffic grooming in WDM optical networks. In *Proceedings of Conference on DPN-04*, IIT Kharagpur, India, January, 2004, pp. 92–96.

[3] A.L. Chiu and E.H. Modiano, Traffic grooming algorithms for reducing electronic multiplexing costs in WDM ring networks. *IEEE Journal of Lightwave Technology*, 18, 2–12, 2000.

[4] R. Battiti and M. Brunato, Reactive search for traffic grooming in WDM networks. In S. Palazzo, Ed., *Proceedings of IWDC2001, Lecture Notes in Computer Science*, Springer-Verlag, Heidelberg, 2001.

[5] E.H. Modiano and P.J. Lin, Traffic grooming in WDM networks. *IEEE Communications Magazine*, 39, 124–129, 2001.

[6] D.E. Goldberg. *Genetic Algorithms in Search, Optimization and Machine Learning*. Addison-Wesley, New York, 1989.

[7] M. Mitchell, *An Introduction to Genetic Algorithms*. MIT Press, Cambridge, MA, 1996.

[8] P. Mazumder and E.M. Rudnick, *Genetic Algorithms for VLSI design, Layout & Test Automation*. Prentice Hall, New York, 1999.

[9] S. Roy, S. Bandyopadhyay, U. Maulik, and B.K. Sikdar, A genetic algorithm based state assignment scheme for synthesis of testable FSMs targeting area and delay. *International Journal of Engineering Intelligent Systems*, 10, 45–52, 2002.

[10] U. Maulik and S. Bandyopadhyay, Fuzzy partitioning using real coded variable length genetic algorithm for pixel classification. *IEEE Transactions on Geosciences and Remote Sensing*, 41, 1075–1081, 2003.

[11] U. Maulik and S. Bandyopadhyay, Genetic algorithm based clustering technique. *Pattern Recognition*, 33, 1455–1465, 2000.

[12] S. Bandyopadhyay and U. Maulik, Non-parametric genetic clustering: Comparison of validity indices. *IEEE Transactions on Systems, Man, and Cybernetics Part-C*, 31, 120–125, 2001.

[13] G.B. Fogel and D.W. Corne, *Evolutionary Computation in Bioinformatics*. Morgan Kaufmann, San Francisco, CA, 2003.

[14] S. Bandyopadhyay, An efficient technique for superfamily classification of amino acid sequences: Feature extraction, fuzzy clustering and prototype selection. *Fuzzy Sets and Systems*, 152, 5–16, 2005.

[15] K. Dev and C. Siva Ram Murthy, A genetic algorithm for the knowledge base partitioning problem. *Pattern Recognition Letters*, 16, 873–879, 1995.

15

Solving Optimization Problems in Wireless Networks Using Genetic Algorithms

Sajal K. Das
Nilanjan Banerjee
Abhishek Roy

15.1 Introduction

The surging demand for mobile communications and the emerging field of ubiquitous computing require enormous development in the realm of wireless networking. Although many problems in mobile wireless networking can be successfully solved using techniques borrowed from wireline networks, there exists some problems that are very specific to the wireless domain and often computationally very difficult or sometimes even intractable. This is primarily due to the inherent limitations of wireless communications, such as scarce bandwidth, high bit error rate, or location uncertainty. Most of these problems can be mapped to classical optimization problems, with the goal of optimizing some objective functions, say resource utilization, while satisfying all the constraints imposed by the wireless communication systems. The constraints make most of the problems NP-complete or NP-hard [1]. There are two broad approaches to solve such problems. One is to directly compute the exact solution based on the constraints, using a brute force technique. However, this approach is often infeasible in a large-scale problem domain. The other approach is to use a heuristic-based solution that can be computed in feasible time. Although a heuristic may not yield the optimal solution, a carefully designed heuristic may produce a near-optimal solution with low computational complexity. Various attempts have been made to develop efficient heuristics that range from calculus-based search methods to random search techniques. The paradigm of "evolutionary computing or programming" essentially provides such heuristics for solving computationally difficult problems, by mimicking the evolution process that the nature uses to eliminate weak forms of life.

An important example of evolutionary computing is genetic algorithm (GA), which is a guided, random search technique. In GA, the less fit (or incompetent) solutions are replaced by better solutions produced by applying some genetic operators on the existing ones in the solution space. This evolutionary learning process progressively refines the search space and makes the problem computationally manageable, thus enabling the search procedure to converge quickly and resulting in a near-optimal solution in feasible time. GA has been successfully applied to solve several optimization problems in the wireless domain [2–4]. The common objective of such problems is the optimal usage of scarce and hence costly wireless resources, such as bandwidth. For example, in References 3 and 5, GA has been used to find an optimal location management strategy for cellular networks. The goal here is to determine an optimal policy for location updates of a mobile host, which ensures minimal paging and signaling overhead, thereby conserving the costly wireless bandwidth. Whereas this work considers only a single optimization criteria (the location management cost), a multi-objective GA-based location management framework has been proposed in Referece 6 that considers other optimization factors, such as the load balancing between various mobility routers in an Universal Mobile Telecommunication Systems (UMTS) [7] network. Channel assignment is another challenging problem in cellular networks. The entire spectrum of wireless bandwidth is divided into several channels and assigned to the cells, with the constraint that no two adjacent cells are assigned the same channel, which ensures that there is no co-channel interference. A single cell may have more than one channels. The channel assignment problem can be shown to be equivalent to the graph coloring problem, and hence NP-hard [1]. GA-based channel assignment algorithms [2] provide a scalable mechanism to determine the least number of channels required for a particular cellular network, while satisfying the constraint related to the co-channel interference problem in adjacent cells. Due to the limited resource available in wireless networks, the admission of new calls must be monitored to ensure a minimum quality of service (QoS) for the already admitted calls. This is done by call admission control algorithms, which essentially select, from a large number of possible admission policies, the optimal one with overall service quality improvement and minimum call blocking rate. However, the number of possible policies can be very large such that finding the optimal policy can be computationally intractable. Again, GA provides a heuristic-based approach [4] to yield a near-optimal call admission policy that ensures QoS for the admitted calls while minimizing the call blocking rate. GA has also been used in solving QoS routing problems in networks. It can be shown that unicast routing with two or more QoS constraints can be mapped to a NP-hard problem. Naturally, the problem is also NP-hard for multicast routing. The imprecise network state information in dynamic wireless networks further complicates the problem. In Reference 8, GA has been used to produce near-optimal solutions in computationally feasible time, for multicast routing problem in networks with imprecise state information, while satisfying more than one QoS constraints.

The goal of this chapter is to present a review of GA-based solutions of the above problems. The rest of this chapter is organized as follows. Section 15.2 reviews the basic concept of GAs and their interpretation. The location management problem is discussed in Section 15.3. Section 15.4 presents a GA-based solution for optimal channel allocation in cellular networks. Efficient call admission control in cellular networks is described in Section 15.5, while the design and performance analysis of an efficient multicast routing protocol is described in Section 15.6. Finally, Section 15.7 concludes the chapter.

15.2 Basics of Genetic Algorithms

GAs are stochastic algorithms, whose search methods model a natural phenomenon, genetic evolution. In evolution, the problem each species faces is one of searching for beneficial adaptations to complex changing environments. The "knowledge" that each species has gained is embodied in the makeup of the chromosomes of its members. In this light, GAs can also be classified as a learning technique. They have

```
Generic Genetic Algorithm Framework
g := 0 {generation counter }
Initialize population P(g)
Evaluate population P(g) { i.e., compute fitness values for each member of P(g)}
while (stopping criteria not satisfied) do
  g := g + 1
  Select P(g) from P(g − 1)
  Crossover P(g)
  Mutate P(g)
  Evaluate P(g)
end while
```

FIGURE 15.1 A generic framework for genetic algorithms.

been successfully applied to tackle optimization problems such as scheduling [9], adaptive control [10], game playing [11], cognitive modeling [12], transportation problems [13], traveling salesman problems [14], database query optimization [15], and so on. Although GAs belong to the class of probabilistic algorithms, they differ from random search algorithms because they combine elements of directed and stochastic search. Therefore, GAs are also more robust than purely directed search techniques [16].

As an evolutionary programming model, a GA for a particular problem must have the following five properties:

- A genetic representation of the potential solution space to a symbolic domain, such as a strings of symbols, which form the so-called *chromosomes*.
- A methodology to create an initial population of potential solutions.
- An evolution function that plays the role of the environment, rating solutions in terms of their "fitness."
- Genetic operators that alter the structure of chromosomes.
- Values for various parameters such as population size, probabilities of applying genetic operators, etc.

Figure 15.1 shows a pseudo-code depicting the generic framework of any GA. An initial random population is created and each member of the population is evaluated using a fitness function. Three major genetic operators, selection, crossover, and mutation, are then successively applied to the members of the population in each generation until the stopping criteria is met or there is little change in the quality of the best solution in the entire population. *Selection* is the operation that increases the number of good solutions or chromosomes in a population; usually the best chromosome in each phase of evolution is selected. The *crossover* operation generates new representative chromosomes containing the properties of more than two chromosomes. Finally, *mutation* changes or mutates a portion of a chromosome to introduce a new chromosome in the population. When there is little or no change in the quality (or fitness) of the chromosomes in the population with successive application of GA operators, the best one is selected to represent the final solution.

15.3 Location Management

The location management problem is defined as the tracking of a mobile user in cellular wireless networks. As shown in Figure 15.2, a cellular network is divided into discrete cells, each served by a base station (BS). Mobile users communicate with a BS using wireless links. A number of such BSs are linked to a mobile switching center (MSC), which is connected to the existing wireline networks. The location management problem has two flavors: one on the part of the mobile device and the other on the part of the network system. In a stand-by mode, every mobile device is made to report its location in regular intervals. This process is known as *registration* or *update.* On the arrival of a new call to the mobile device, the network system initiates a search for the device by polling the cells, where it could be possibly

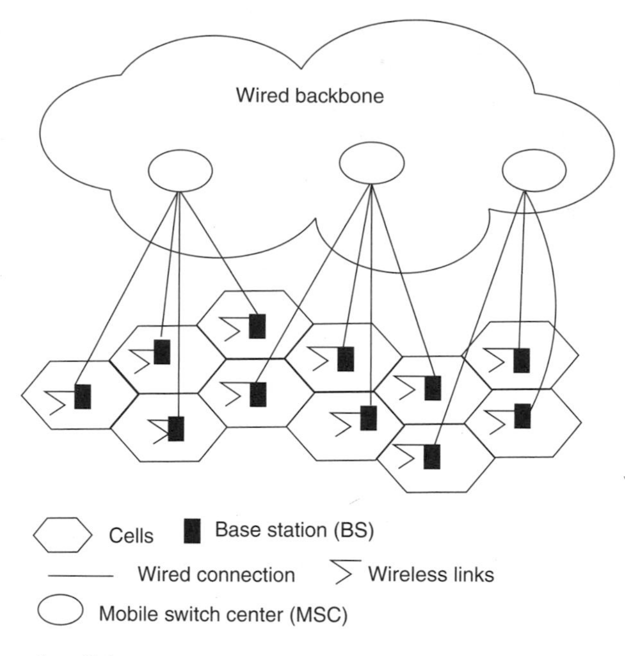

FIGURE 15.2 Overview of a cellular system.

located. This procedure is known as *paging*. By giving a low upper bound on the maximum number of cells that can be polled, the paging process reduces the paging cost, but requires more frequent updates, thus increasing the update cost. On the other hand, a reduction of number of updates essentially decreases the update cost, but increases the location uncertainty and the subsequent paging cost. The essence of an optimal location tracking thus lies in minimizing the combined update and paging costs associated with a particular mobile device. The concept of using a suitable GA-based approach to minimize the average per-user location management cost was first proposed in Reference 3.

A cellular network can be represented by a graph $G(V, E)$, where a node represents the location areas (LA — a logical representation of one or more cells) and an edge represents the access paths between a pair of location areas. At this point, the most interesting question is whether a mobile device will issue an update or not upon entering a new LA. Let δ_i represent an update decision variable for a user in $\text{LA}_i (1 \leq i \leq M)$, such that $\delta_i = 1$ or 0 depending on whether or not an update has been issued. Now, if $\text{Cost}_p(i)$ and Cost_u respectively denote the cost associated with paging LA_i and issuing a single update, then the total location management cost (LMC) is determined by taking a weighted average of the LMCs in the individual LAs. This is mathematically represented by: $\text{LMC} = \sum_{i=1}^{M} \Pi_i \times LM_i^{(\delta_i)}$, where Π_i is the normalized weight associated with the LMC, LM_i, of LA_i. The average LMCs for both cases ($\delta_i = 1$ and $\delta_i = 0$) depend on the individual update cost, paging cost, call arrival rate, and the user's residence time in the LA. Assuming a Poisson arrival with rate λ, geometrically distributed (with parameter p_i) residence time, it has been shown in Reference 3 that $\text{LM}_i^{(1)} = \text{Cost}_u + (\frac{\lambda}{p_i})\text{Cost}_p(i)$ and $\text{LM}_i^{(0)} \approx \text{Cost}_p^0(i) + ((\lambda/p_i) - 1)\text{Cost}_p(i)$. It is now clear that an update strategy $U_s = [\delta_i]$ for the user constitutes a vector of decision variables, having values 0 or 1 for all the LAs. The objective is to obtain optimal strategy U_s^* such that the LMC is minimized. While enumerating all possible update strategies, the state space of the solution increases exponentially with the number of LAs. Therefore, a GA is proposed in Reference 3 to obtain a (near-) optimal solution for the location management problem.

As mentioned in Section 15.2, the first step in a GA-based approach is to map the state space into a symbolic domain. The most obvious way is to represent each bit-string associated with a strategy U_s by a single chromosome (or genome). The length of every chromosome is equal to the number of LAs. A group of strategies is chosen to form the initial population. For faster convergence, the relative proportion of the number of 0s and 1s in the bit-string is chosen based on the call arrival rate and update cost. For a relatively low call arrival rate and high update costs, it is wise to issue less frequent updates, thereby resulting in more 0s than 1s in the chromosomes. Since, at every iteration, the GA inherently attempts to increase the associated fitness values, the fitness function is chosen to be reciprocal of the total LMC, that is, 1/LMC. The roulette wheel spinning selection [16] is used with elitism so that better chromosomes will survive for the next iteration. After this selection, the crossover and mutation functions are executed with probabilities 0.8 and 0.01 respectively. The fitness of the children are now evaluated and the entire process is repeated.

Illustrative Example: At each iteration of the GA, the best chromosome, from the initialization phase till that iteration cycle, is tracked. This gives the optimal (or near-optimal) solution at the termination of the algorithm. The population size for each generation is kept constant at 50 and the number of bits (i.e., the number of LAs) in the chromosome is chosen as 8. The cost function LMC is computed using the steady-state transition probabilities between any two LAs. It was found that the GA converged very fast to the near-optimal solution. The population size is kept constant at 20 and in all the cases the algorithm converges to the optimal solution within 1000 generations. The best and average values of the fitness of the chromosomes as well as the standard deviations are computed for each generation. A sample run of the entire process with different generations having best and average fitness values is shown in Table 15.1, which results in 11011 as the best chromosome having fitness 0.127 corresponding to LMC = 7.87 units. Table 15.2, on the other hand, represents the optimal update strategies for different call arrival rates and update/paging cost ratios.

Recently, the location management problem is also investigated in Reference 5 using a combination of cellular automata (CA) and GAs. The total LMC is estimated as $r\sum_{i \in C} w_{mi} + \sum_{j=0}^{N-1} w_{cj}.v(j)$, where r is the update-to-paging cost ratio, C is total number of reporting cells, that is, the number of cells from which at least one update is issued, w_{mi} represents frequency of movement into a cell i, w_{cj} represents frequency of call arrival within a cell j, and $v(j)$ represents the vicinity (neighborhood) of cell j. Note that

TABLE 15.1 Sample Run of Genetic Algorithms

Generation no.	Best fitness	Average
1	0.102	0.102
3	0.106	0.103
7	0.107	0.105
15	0.120	0.106
20	0.127	0.116
35	0.127	0.116
50	0.127	0.125
100	0.127	0.122
200	0.127	0.126
300	0.127	0.127

TABLE 15.2 Optimal Update Strategies

C_u/C_p	Optimal strategy		
	$\lambda = 0.3$	$\lambda = 0.5$	$\lambda = 0.8$
0.10	00000	11111	11111
0.33	00000	11111	11011
0.50	00000	01101	10011
0.67	00000	01110	10001
0.75	00000	01110	10001
0.90	00000	11010	10001
1.00	00000	11010	10001

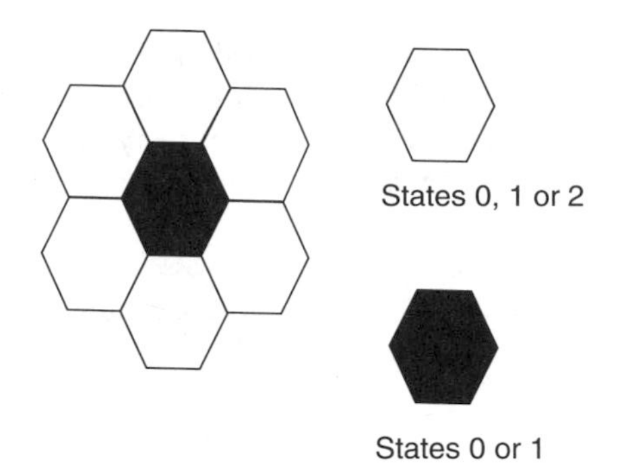

FIGURE 15.3 Location management using cellular automata.

cellular automata is a decentralized, discrete space–time system and the state of every cell is governed by its surrounding cells. Each cellular unit of cellular automata is associated with each cell in the network. Each cell is represented either by "1" or "0" depending on whether or not it is a reporting cell. This leads to two possible states for each cellular unit in the CA. Considering the hexagonal cells of the network, the maximum neighborhood of every cell is taken as 6. For cells with less than 6 neighbors, dummy cells, represented by "2" are added. Thus, as shown in Figure 15.3, each cell itself has 2 possible states (0 or 1) and each of its 6 neighbors has 3 possible states (0, 1, or 2), implying a total number of $3^7 \times 2 = 1458$ possible neighborhood states. Hence, the corresponding rule is of length 1458 bits and there can be a total of 2^{1458} transition functions. Genetic algorithms are used to search for the best one from this exponential number of transition rules. In Reference 5, an initial population of 1000 rules are created with a random value for each rule. At every iteration, a new set of CA test data is generated. The set consists of four randomly generated reporting cells configuration. The fitness function is chosen as the sum of the LMCs of all these four configurations. The selection strategy is used to get the minimum fitness values (minimum location management costs). A two-point crossover [16] with probability 0.8 and a mutation with probability 0.01 is used to achieve better solutions at every iteration. A set of 80 rules (chromosomes) have been used for elitism, while crossover and mutation have been applied on the rest of the 920 rules. Simulation results on different 4×4 and 5×5 cellular networks result in a cost per call arrival between 12.25 and 16 within 200 generations.

A careful look at both of the above mentioned location management strategies reveals that the only objective of both the schemes is to reduce the overall LMC. However, more recently, the problem of location management is combined with load balancing constraints, thereby introducing the notion of multi-objective optimization techniques. Genetic algorithms can be easily extended to solve such multi-objective optimization problems. A multi-objective, hierarchical mobility management optimization for UMTS networks has been proposed in Reference 6, which balances the load among various unique logical areas in the network, such as the LA, and the mobility routers, such as the mobile switch centers (MSCs), Radio Network Controller (RNC), and Serving GPRS Node (SGSN), while minimizing the signaling cost for location update. A schema-based niched Pareto GA [17] has been used that deals with multiple objectives by incorporating the concept of Pareto domination in its selection operator, and applying a niching pressure to spread out its population along the Pareto optimal tradeoff surface. The fitness function or the cost functions used are the RA load balancing and the intra-SGSN signaling cost, which covers the intra-SGSN routing area update and paging cost.

15.4 Channel Assignment

A major challenge associated with cellular wireless networks is the limitation on available bandwidth. In a time division multiple access (TDMA) scheme, the wireless bandwidth is divided into different channels

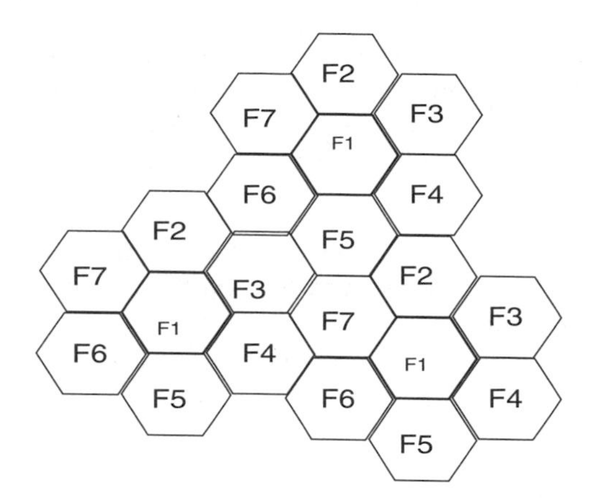

FIGURE 15.4 Channel reuse in cellular systems.

and each cell is allocated to a certain number of channels depending on its traffic density. In the cellular system, the same frequency cannot be used in the adjacent cells, as there will be co-channel interference. The hexagonal cell structure creates a cluster of 7 cells, where the frequencies will differ from each other. Thus, channels used by a cell can be reused by a different cell belonging to a different cluster (sufficiently far apart), so that the interference is bounded by some specific value. This is demonstrated in Figure 15.4, where F_i represents the different frequencies used in the corresponding cell.

The channel assignment technique [2] can be static or dynamic. In a static assignment, a fixed number of channels are allocated to each cell and are estimated by the traffic demand in that cell. On the other hand, in a dynamic allocation strategy, the channels are assigned on demand, and no cell has exclusive control on any channel. Channel assignment strategies need to tackle two types of channel interferences: (i) co-channel interference and (ii) adjacent channel interference. Co-channel interference occurs when a different signal is received in the same channel as the original signal and therefore cannot be eliminated by the receiver itself. Adjacent channel interference, on the other hand, is caused by inadequate/incomplete filtering of unwanted modulation products in frequency modulation systems, improper tuning, or poor frequency control in either the reference channel or the interfering channel, or both. The channel assignment strategy needs to maintain both the co-channel and the adjacent channel interference below a tolerable limit. Most of the research in this area is focussed on finding the minimum bandwidth spectrum to satisfy a given traffic load, while maintaining the interference constraint.

Let the available frequency spectrum consist of $\mathcal{M}$ consecutive channels $f_1, f_2, \ldots, f_{\mathcal{M}}$. Also, let $\mathcal{N}$, d_i, and c_{ij} respectively represent the number of cells, channel demand of cell i, and frequency separation needed between any two channels allotted to a pair of cells i and j. The matrices formed by the elements c_{ij} and d_i, $(1 \leq i, j \leq \mathcal{N})$, are denoted as **C** and **D**, respectively. Intuitively, $c_{ij} = 1$ and $c_{ij} = 2$, respectively mean that the same and adjacent frequencies cannot be assigned to certain channels. The channel assignment strategy now needs to find out a suitable frequency assignment matrix $\mathbf{F} = [f_{mi}]_{\mathcal{M}\times\mathcal{N}}$, where f_{mi} is binary-valued, with 1 indicating that the frequency f_m is assigned to the ith cell and 0 indicating that it is not. The problem now boils down to minimize the value of $\mathcal{M}$ such that the channels can be allocated to all the cells without interference, that is, when $f_{mi} = f_{lj} = 1, |f_{mi} - f_{lj}| \geq c_{ij}$ and $\sum_{m=1}^{\mathcal{M}} f_{mi} = d_i$. It has been shown in Reference 18 that the channel assignment problem is equivalent to the graph coloring problem (NP-hard) [1], even when only the co-channel constraints are considered, that is, when the matrix **C** is binary-valued. The actual channel assignment problem considered here is even more complex and the solution space increases exponentially with the increasing number of cells. The role of evolutionary algorithm now comes into play to obtain a (near-) optimal solution [2] in a polynomial time.

The first step to solve this problem using GA is again to generate the initial pool of encoded strings or chromosomes. Let, $S_1, S_2, \ldots, S_P$ represent the P strings of an initial population. Each S_i is an $\mathcal{M} \times \mathcal{N}$ matrix, where $\mathcal{M}$ and $\mathcal{N}$ are number of frequencies and number of cells, respectively. The elements of the matrix S_i can have values 0, 1, -1, or 9. The interpretation of these four different values are enumerated here:

1. 0: Cell (given by the column number) is not using the frequency (row number), and even if it uses that frequency, there will be no conflict with other existing allocations.
2. 1: Cell is using the particular frequency.
3. -1: Cell is not using the frequency (row number), and cannot use that frequency for possible interference.
4. 9: It is used at the head of all unused channels (rows).

Illustrative Example: Table 15.3 demonstrates an example of a valid solution for the four-node channel allocation problem shown in Figure 15.5. The initial population has been created by using different permutations of the nodes in the chromosomes, e.g., 1, 2, 3, 4, or 1, 3, 2, 4, or 3, 4, 1, 2, or 3, 1, 2, 4. The fitness function is decided based on the total number of channels allocated, that is, the value of $\mathcal{M}$. The objective is to minimize this value of $\mathcal{M}$. For chromosomes with equal $\mathcal{M}$ values, the one with more 0s is selected as the better one. The reason is that the chromosomes having more 0s allow more channels to be added, while satisfying the interference constraint. Now, out of the $P \times \mathcal{N}$ columns, $P \times \mathcal{N} \times \rho_{\Updownarrow}$ columns are selected at random, where $\rho_{\Updownarrow}$ is the probability of mutation. From a selected column, randomly a 0 and 1 is chosen and flipped. An unavailability of a column with 0 or 1 leads to a mutation failure. If the mutation results in a row with all 0 values, a leading 9 is placed, reducing the bandwidth requirement by 1. Simulation results show that the lower bound for the minimum number of frequencies range from $\sim$300 to 500 for different types of demands. The GA approach takes only 1 to 3 generations (fraction of a second) to converge in a near-optimal solution.

TABLE 15.3 Examples of Valid Solutions

	solution		
+1	−1	+1	−1
−1	−1	−1	−1
−1	−1	−1	+1
−1	−1	−1	−1
−1	+1	−1	−1
−1	−1	0	−1
−1	−1	−1	−1
−1	−1	−1	+1
0	−1	−1	−1
0	0	0	−1
0	0	0	−1
0	0	−1	−1
0	−1	−1	+1
9	0	−1	−1
9	0	0	−1

$$C = \begin{bmatrix} 5 & 4 & 0 & 0 \\ 4 & 5 & 0 & 0 \\ 0 & 0 & 5 & 0 \\ 0 & 1 & 2 & 5 \end{bmatrix} \qquad D = \begin{bmatrix} 1 \\ 1 \\ 1 \\ 3 \end{bmatrix}$$

FIGURE 15.5 A four-node channel allocation problem.

15.5 Call Admission Control

An important issue in cellular systems is to maintain a good service quality. Admission control policies are designed to meet this end. A wise admission decision for new calls can improve the overall service quality by guaranteeing QoS to the already admitted calls, while having a minimal average blocking rate for new calls. An admission policy is a binary decision controlling the admission criteria. Thus, any such policy can be represented by a finite number of logic functions forming the policy space. However, the enormously large policy state space often needs some compact coding [4] to restrict the blow up of search space. There is a finite possibility for the coded policies to get stuck at local optimal solutions. Multimodal optimization strategies are required to solve such problems. Genetic algorithms provide such a robust optimization strategy that yields quite satisfactory results even when little is known about the search space.

Following Reference 4, a network system $\mathcal{S}$ with n-dimensional state vector $s = (s_1, \ldots, s_n)$ is considered. The entire set of customers is divided into different classes depending on their relative importance. A state represents the number of customers of each class in the system. It has been shown in Reference 4 that for a system described by a finite dimensional Markovian state vector, the call admission control can be formulated as a Markov Decision Process (MDP). Such a model assumes a set of state-dependent controls. Fixing a particular policy actually specifies the state transition matrix, a set of controls and rewards depending on both the system state and state transitions. At every iteration, the current policy is evaluated and using this evaluation a better policy is obtained. Each step however requires solving a set of n linear equations. Each possible decision for each possible state is also evaluated. For this admission control, the strategy needs $O(n)$ evaluations per iteration. The computational effort becomes excessive when the number of states is very high. Hence, GA has been used to reduce the computational complexity and yield moderately good results. For this, a suitable coding is needed to represent such policy-based decisions suitable for using in GAs.

If η represents an admission-decision and $\eta\Gamma$ denotes the number of bits necessary to represent a policy, then the total number of possible policies is $2^{\eta\Gamma}$. For smaller values, this policy can be directly represented by the string called *direct coding*. However, for any practical system, the enormous size of search space makes such direct coding impossible and intractable. A compact coding strategy of good strings, which is also theoretically capable of coding any arbitrary string, is presented in Reference 4. It introduces two different policy coding procedures. I*n block coding*, a table of entries is built, with each entry corresponding to a K-bit pattern. A number of such K-bit-patterns can be concatenated to form full policy strings of length $\Gamma\eta$. The policy is now specified by the sequence of table indices, rather than sequence of K bits. For any $G < K$ a compression is achieved, and the total number of entries is 2^G. In *program coding* the states of the policies are assumed as inputs to a computer program. The policy coding now specifies the bits of the program. Assuming infinite time and memory, the existence of a universal computer or a Finite State Machine (FSM) can be modeled, which will provide the decision as the output.

These coded strings are represented by the chromosomes. A policy search scheme for two different classes of policies are modeled using Markovian queuing systems [4]. If $\lambda_1, \lambda_2, \varrho_{b_1}^k, \varrho_{b_2}^k$, represents the arrival rates and blocking probability of two different classes of calls, then for any policy k, the average blocking is given by $Pr[\text{block}] = (\lambda_1\varrho_{b_1}^k + \lambda_2\varrho_{b_2}^k)/(\lambda_1 + \lambda_2)$. Minimizing this $Pr[\text{block}]$ is essentially maximizing $(1 - Pr[\text{block}])$. This measure provides the performance of any policy i. For each generation g, the performance perform_i of a policy i in the population is evaluated and the minimum $\text{perform}_{\min}$ is computed. The fitness function of policy i used is given by $\text{fitness}_i = \exp[-\alpha(\text{perform}_i - \text{perform}_{\min})]$, where $\alpha > 0$ is a constant scale factor constant that emphasizes or de-emphasizes deviation from the best population performance, $\text{perform}_{\min}$. Experimental results show that it is useful to increase α with the increase of g. Normal crossover (swapping the portions of parent policies) and mutation (flipping the bits) operations are used to generate the offspring policies. The algorithm is capable of obtaining (near-) optimal solutions within $g = 1000$ generations. Table 15.4 provides some sample results of direct, block, and program codes for 10 and 50 cells respectively. The table demonstrates that the GA approach achieves a near-optimal value of the blocking probability in almost all cases. Obviously, the time, iteration and

TABLE 15.4 Results of Admission Control

Cells	Null policy	Best policy	Direct code (GA)			Block code (GA)			Program code (GA)		
			Pr[B]	bits	G	Pr[B]	bits	G	Pr[B]	bits	G
10	0.121	0.0753	0.076	400	132	0.077	200	100	0.076	24	< 40
50	0.216	0.072	0.097	2652	3600	0.078	510	800	0.072	48	< 40

length of strings varies for different coding strategies. Program codes perform the best in terms of the number of generations and string lengths, followed by the block codes and direct codes.

An adaptive resource allocation and call admission control scheme, based on GA, was proposed in Reference 19 for wireless ATM networks. Multimedia calls have their own distinct QoS requirements (e.g., cell loss rate, delay, jitter, etc.) for each of their substreams. The network usually allocates an appropriate amount of resource that constitutes a certain QoS level, which remains fixed during a call. But such static schemes are inefficient in terms of resource utilization. In the adaptive algorithm proposed in Reference 19, each substream declares a range of acceptable QoS levels (e.g., high, medium, low) instead of just a single one. With the variation of network resources, the algorithm selects the best possible QoS level that each substream can obtain, while achieving maximum utilization of the resources and their fair distribution among the calls. For example, in case of congestion, the algorithm tries to free up some resources by degrading the QoS levels of some of the existing calls. The problem essentially boils down to finding the best QoS levels for all existing calls amidst a large search space. Thus, if three types of streams, namely audio, video, and data are considered with four possible QoS levels — "High," "Medium," "Low," and "No" Component — then a total number of $4^3 = 64$ QoS levels are possible, and if we consider N existing calls, the search space is given by 64^N. For $N = 10$, the dimension of search space will be $64^{10} = 1.153 \times 10^{18}$. A GA has been provided in Reference 19, which is a tool for searching an optimal solution in this huge search space. Simulation results show that the algorithm is capable of searching solutions with a huge gain (250%) in terms of the number of admitted calls, while achieving a resource utilization ranging between 85.87% and 99.985%.

15.6 QoS-based Multicast Routing Protocol

A majority of multimedia applications, such as video on demand and conferencing, depend on efficient multicast protocols. Multicasting is essentially a selective broadcast of information from a source to a given set of destinations. The information is delivered from the source to the destinations using a multicast routing tree computed by multicast routing algorithm. Since the basic unicast routing (route computation from a single source to a single destination) with multiple QoS constraints is an NP-complete problem [20], the computation of a multicast route that satisfies certain QoS constraints is also computationally intractable. In wireless networks, fluctuations in available resources complicates the problem further. Thus, multicast routing in wireless networks, with multiple QoS constraints and imprecise information on the available network resources need some kind of heuristic-based technique to compute an efficient multicast routing tree. In this section, we describe the design and performance analysis of a multicast routing QoS-based multicast routing scheme [21], that relies on the link-state advertisements and uses a GA framework for near-optimal multicast delivery tree computation.

The problem thus is to compute a source-specific tree-based route, spanning the source and a given set of destination nodes, such that the route satisfies multiple QoS parameters, such as minimum bandwidth and end-to-end delay, and also complies with policy requirements such as minimum total bandwidth utilization. Modeling the input network as a graph, with the routers as the nodes and the network links as the edges of the graph, maps the QoS-based multicast routing problem to a constrained Steiner tree computation problem, which is a well-known NP-hard problem [1]. In case of wireless networks, it is rather difficult to obtain accurate values for these QoS parameters at any instant of time due to the time-varying nature of the wireless medium and the dynamism in the available resources. However, the moments of their probabilistic distributions can be obtained by studying the history of their change over a certain period of time, broadcasted with link state advertisements. Proper modeling with probability distributions for each QoS parameter enables us to estimate the probabilistic bound corresponding to a particular value of that parameter. This bound serves as the guarantee provided by the network, for satisfying a particular value of a QoS parameter. Additionally, the tree should admit the maximum number of sessions with such guarantees. This is achieved by efficient resource and traffic management mechanisms.

15.6.1 Underlying Routing Algorithm

The network is represented by a graph $G = (V, E)$, where V is a set of nodes and E is a set of links between node pairs. A logical path between two nodes v_1 and v_n is represented by $v_1, v_2, \ldots, v_n$, where $v_i \in V$ for $1 \leq i \leq n$. There can be multiple such paths between each pair of nodes. The primary objective is to develop an efficient scheme that will find a near-optimal multicast routing tree with source as the root and the destination as the leaves, such that

- The bandwidth provisioning for the specified QoS requirement is guaranteed.
- The end-to-end delay requirement is satisfied with maximal guarantee.
- The amount of bandwidth utilized is minimal.

The GA-based solution is encoded as follows:

- All the possible paths between the source and each destination are first computed and stored in a pool. Note that a path is a sequence of network nodes. Among these paths, we select those paths that conform to the bandwidth requirement.
- From each created pool, one solution is randomly chosen and concatenated with each other to obtain a multicast solution.

In order to achieve optimality in terms of bandwidth availability, bandwidth utilization, and the end-to-end delay guarantees, we devise an optimization function, which the GA can use in its course of action. The transmission delay along each link is assumed to follow an exponential distribution. Hence, the available bandwidth has been assumed to follow Poisson distribution [22] and the end-to-end delay on a path from the source to any particular destination is assumed to follow Erlang-K distribution [22]. Based on these assumptions, the probability distribution function for the end-to-end delay along a path from the source to a single destination is given as $\mathcal{D}_{\text{path}}(t) = (\bar{d}^{\kappa} t^{\kappa-1} e^{-\bar{d}t}/(\kappa - 1)!)$, where $\bar{d}$ is the average delay on each link and κ is the number of links on the path. Similarly, the probability of getting a bandwidth of $\mathcal{B}$ over a link l is given by the probability density function, $\beta_l(\mathcal{B}) = (\bar{b}^{\mathcal{B}} e^{-\bar{b}}/\mathcal{B}!)$, where $\bar{b}$ is the average bandwidth on each link. In addition to these measures, we have considered residual bandwidth as the third measure of optimization. This is given by $\sum_{l \in \mathcal{T}}(\varsigma_l - \sigma_l)$, where ς_l is the capacity of a link $l \in \mathcal{T}$, and σ_l is the bandwidth allocated for all the paths in the multicast tree $\mathcal{T}$, along the link l. The optimization function is chosen as the linear combination of the probabilistic measure of the three QoS parameters. Hence the *fitness* function is given by:

$$F = \Pi_{\text{path} \in \mathcal{T}} \mathcal{D}_{\text{path}}(t) + \Pi_{l \in \mathcal{T}} \beta_l(\mathcal{B}) + \frac{\sum_{l \in \mathcal{T}}(\varsigma_l - \sigma_l)}{\sum_{l \in \mathcal{T}} \varsigma_l}.$$

Genetic algorithm-based operations, namely selection, crossover, and mutation operations are then successively applied to the members of the population until there is little change in the quality of the best solution in the entire population. The best solution replaces the worst solution of the previous population. In crossover operations, the corresponding parts of two randomly selected multicast routing trees are concatenated to obtain new routing trees. In mutation operations, a path in a multicast routing tree is randomly selected and is replaced by another valid path between the same source and destination.

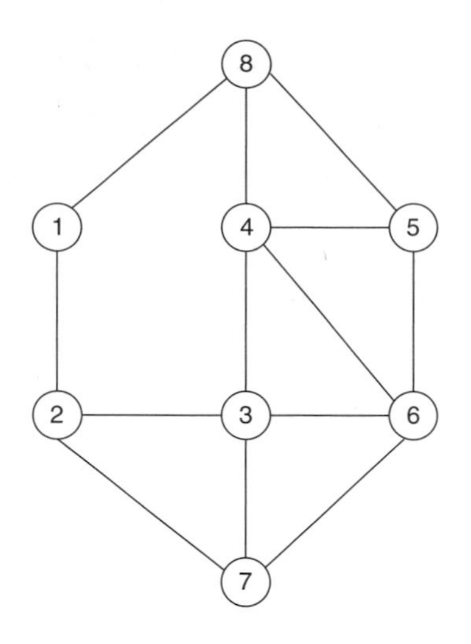

FIGURE 15.6 Example scenario.

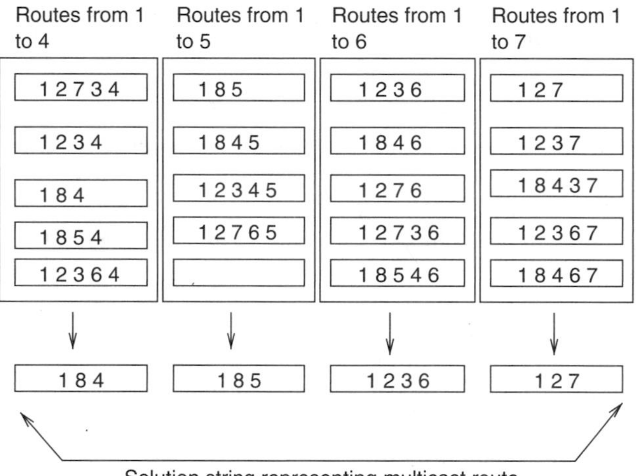

FIGURE 15.7 Multicast solution selection strategy.

Illustrative Example: To illustrate the coding scheme of potential solutions, we study a network of eight nodes, shown in Figure 15.6, where node 1 is the source of multicast delivery and 4, 5, 6, and 7 are the destination nodes. Then the pools of valid paths for each source–destination pair are as shown in Figure 15.7. The initial population, as required by the GA, is created as follows. Each member of the population is formed by randomly selecting a path between each source–destination pair and then concatenating them to represent a multicast tree spanning the source node and the set of destination nodes. Figure 15.8 depicts the multicast delivery tree computed by the underlying routing algorithm for the sample network, shown in Figure 15.8, with node 1 as the source and nodes 4, 5, 6, and 7 as the destination nodes.

15.6.2 Improvement

The GA framework described here, combines the three QoS objectives into a single linear fitness function. This scheme works well when only one solution is needed. But when multiple, mutually conflicting optimization parameters are involved, it cannot yield solutions, which are better than the other with respect to a

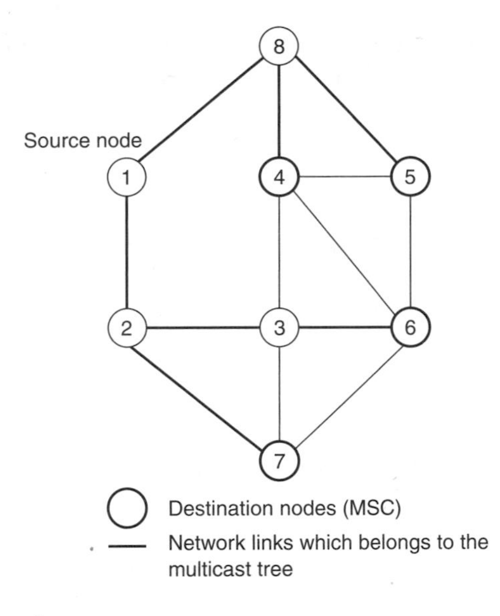

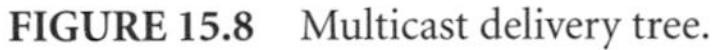

FIGURE 15.8 Multicast delivery tree.

single optimization parameter, but not the superior when all the optimization parameters are considered. In such a case, no solution is dominated by others when all the parameters are considered. These are generally termed as *nondominated* or *pareto-optimal* solutions, generated by using a multi-objective GA. The GA-based multicast routing algorithm has been extended to incorporate a multi-objective QoS-optimization mechanism [8]. The procedure does not combine the three predefined QoS parameters into a single objective function but attempts to optimize each parameter individually, thereby providing a near-optimal and non dominated set of solutions (i.e., multicast trees). The solution set consists of not only those trees that offer best delay, bandwidth requirement, and residual bandwidth guarantee individually, but also a set of trees compromising fairly between the three optimization parameters.

Simulation results demonstrate that the algorithm is capable of obtaining a near-optimal multicast tree in reasonable time. With a session arrival rate of 5–10 multicast sessions, the average session blocking rate is only 2–5%. The multi-objective GA improves the flexibility of this scheme by offering a set of nondominated solutions. The user now has the flexibility to choose his/her favorable solution from this nondominated set. The dynamism and fluctuation of networks always creates resource uncertainty. This might result in the unavailability of a particular QoS-constrained path. The nondominated solutions aids in offering an alternate QoS-guaranteed path, thereby resulting in the graceful degradation scheme of QoS provisioning. With multi-objective GA, it becomes possible to sustain more calls with their minimum level of QoS.

15.7 Conclusion

In this chapter, we have presented a survey of the computationally difficult problems specific to wireless networking that have been solved using GA, a bio-inspired optimization algorithm. The primary objective of each of the problems is to derive a near-optimal solution in a computationally feasible time. GA provided a tool for deriving such solutions. Relevant performance results of the solutions for the problems have been been presented to illustrate the efficiency of GAs in solving the complex problems. Potentially, GA can be used in any of the optimization problems faced in the design and operation of wireless networks. While, in this chapter, we have predominantly discussed problems from the cellular network domain, recent developments in the area of ad hoc and sensor networks are posing new challenges, a majority of which are difficult optimization problems. As a result, there have been some research attempts [23–25] to solve these problems using GA. We hope that this chapter will help in providing a clear idea on the methodology of solving complex, real-life problems in the wireless domain using GAs.

References

[1] M.R. Garey and D.S. Johnson, *Computers and Intractability: A Guide to the Theory of NP-Completeness*, W.H. Freeman and Company, San Francisco, CA, 1983.

[2] G. Chakraborty and B. Chakraborty. A genetic algorithm approach to solve channel assignment problem in cellular radio networks. *IEEE Midnight-Sun Workshop on Soft Computing Methods in Industrial Applications*, June 1999.

[3] S.K. Sen, A. Bhattacharya, and S.K. Das. A selective update strategy for PCS users. *Wireless Networks*, 5, 313–326, 1999.

[4] A. Yener and C. Rose. Genetic algorithms applied to cellular call admission: Local policies. *IEEE Transactions on Vehicular Technology*, 46, 72–79, 1997.

[5] R. Subrata and A.Y. Zomaya. Evolving cellular automata for location management in mobile computing networks. *IEEE Transactions on Parallel and Distributed Computing*, 14, 13–26, 2003.

[6] T. Ozugur, A. Bellary, and F. Sarkar. Multiobjective hierarchical 2G/3G mobility management optimization: niched Pareto genetic algorithm. In *Proceedings of Globecom*, Vol 6, 2001, pp. 3681–3685.

[7] 3GPP - UMTS Standards. http://www.3gpp.org/ftp/tsg_cn/ TSG_CN/TSGN_03/Docs

[8] A. Roy and S.K. Das. QM^2RP: A QoS-based Mobile Multicast Routing Protocol. *ACM/Springer Wireless Networks* (WINET), 10(3), 271–286, 2004.

[9] S.J. Beaty. Genetic algorithms and instruction scheduling. In *Proceedings of the 24th annual international symposium on Microarchitecture*, pp. 206–211

[10] K.Ng and Y.Li. Design of sophisticated fuzzy logic controllers using genetic algorithms. In *Proceedings of the IEEE World Congress on Computational Intelligence*, 1994.

[11] C.T. Sun and M.D. Wu. Multi-stage genetic algorithm learning in game playing. In *Proceedings of the First International Joint Conference of the North American Fuzzy Information Processing Society, the Industrial Fuzzy Control and Intelligent Systems, and the NASA Joint Technology*, 1994, pp. 223–227.

[12] C.L. Karr. Genetic algorithms for modelling, design, and process control. In *Proceedings of the second international conference on Information and knowledge management*, 1993, pp. 233–238.

[13] Z. Michalewich. A genetic algorithm for the linear transportation problem. *IEEE Transactions on Systems, Man, and Cybernetics*, 21, 445–452, 1991.

[14] C.A.R. Jahuria. Hybrid genetic algorithm with exact techniques applied to TSP. In proceedings of the *Second International Workshop on Intelligent Systems Design and Application*, 2002, pp. 119–124.

[15] J.J. yang and R.R. Korfhage. Query optimization in information retrieval using genetic algorithms: Report on the experiments of the TREC project. In *Proceedings of TREC'1. NIST*, 1993, pp. 31–58.

[16] D.E. Goldberg, *Genetic Algorithms in Search, Optimization and Machine Learning*", Addison-Wesley Longman Publishing Co., Reading, MA, 1991.

[17] J. Horn, N. Nafpliotis, and D.E. Goldberg, " A niched pareto genetic algorithm for mutiobjective optimization. In *IEEE Conference on Evolutionary Computation*, New Jersey, vol. 1, 1994, pp. 82–87.

[18] W.K. Hale. Frequency assignment: Theory and applications. *Proceedings of IEEE*, 68, 1497–1514, 1980.

[19] M. Sherif, I. Habib, M. Naghshineh, and P. Kermani. An Adaptive Resource Allocation and Call Admission Control Scheme for Wireless ATM Using Genetic Algorithms. In *Proceedings of Globecom*, 1999, pp. 1500–1504.

[20] S. Chen and K. Nahrstedt. An overview of quality of service routing for next-generation high-speed networks: Problems and solutions. *IEEE Network, Special Issue on Transmission and Distribution of Digital Video*, 12, 64–79, 1998.

[21] N. Banerjee and S.K. Das. Fast determination of QoS-based multicast routes in wireless networks using Genetic Algorithm. *IEEE International Conference on Communications (ICC)*, 8, 2588–2592, 2001.
[22] R. Nelson. Probability, Stochastic Process and Queuing Theory. *Springer-Verlag*, Heidelberg, 1995.
[23] L. Barolli, A. Koyama, and N. Shiratori. A QoS routing method for ad-hoc networks based on genetic algorithm. In *Proceedings of International Workshop on Database and Expert Systems Applications*, 2003, pp. 175–179.
[24] D. Turgut, S.K. Das, R. Elmasri, and B. Turgut. Optimizing clustering algorithm in mobile ad hoc networks using genetic algorithmic approach. In *Proceedings of IEEE GLOBECOM*, 2002, pp. 62–66.
[25] Q. Wu, S.S. Iyengar, N.S.V. Rao, J. Barhen, V.K. Vaishnavi, H. Qi, and K. Chakrabarty. On computing the mobile agent routes for data fusion in a distributed sensor network. *IEEE Transactions on Knowledge and Data Engineering*, 1(6), 740–753, 2004.

16

Medical Imaging and Diagnosis Using Genetic Algorithms

Ujjwal Maulik
Sanghamitra Bandyopadhyay
Sajal K. Das

16.1 Introduction

The last half of the twentieth century has seen a vigorous growth in the field of digital image processing (DIP) and its potential applications. DIP deals with the manipulation and analysis of images that are generated by discretizing the continuous signals. One important area of application that has evolved from the 1970s is that of medical images. Rapid development in different areas of image processing, computer vision, pattern recognition, and imaging technology, and the transfer of technology from these areas to the medical domain has changed the entire way of looking at clinical routine, diagnosis, and therapy. Also, the need for more effective and less (or non) invasive treatment has led to a large amount of research for developing what may be called *computer aided medicine.*

Most modern medical data are expressed as images or other types of digital signals. The explosion in computer technology in recent years introduced new imaging modalities such as x-rays, magnetic resonance imaging (MRI), computer tomography (CT), positron emission tomography (PET), single photon emission computed tomography (SPECT), electrical impedance tomography (EIT), ultrasound, and so on. These images are noninvasive and offer high spatial resolution. Thus the acquisition of a large number of such sophisticated image data has given rise to the development of quantitative and automatic processing and analysis of medical images (as opposed to the manual qualitative assessment done earlier). Moreover, the use of new, enhanced, and efficient computational models and techniques has also become necessary.

A large amount of research is being devoted to the various domains of medical image processing, and some surveys are already published [1,2]. However, in view of the vastness of the field, it has become necessary to specialize any further survey work that is undertaken in this area, so that it can become manageable and can be of more benefit to researchers/users. Some such attempts have already been made, for example, specialization in terms of period of publication [3], image segmentation [4], registration [5], virtual reality, and surgical simulation [6,7].

The area of designing equipments for better imaging and hence improvement in subsequent processing tasks has also received the attention of researchers. The design problem has been viewed as one of optimization, and therefore the use of efficient search strategies has been studied. The application of genetic algorithms, a well-known class of search and optimization strategies, is also one of the important areas that has been investigated in this regard.

Genetic Algorithms (GAs) [8,9] are randomized search and optimization techniques guided by the principles of evolution and natural genetics, and have a large amount of implicit parallelism. They provide near optimal solutions of an objective or fitness function in complex, large, and multimodal landscapes. In GAs, the parameters of the search space are encoded in the form of strings called *chromosomes.* A *fitness function* is associated with each string that represents the degree of *goodness* of the solution encoded in it. Biologically-inspired operators such as *selection*, *crossover*, and *mutation* are used over a number of evolutions (generations) for generating potentially better strings.

The important fallout of (semi-) automated medical image processing tasks is enhanced diagnosis. Several tasks in the area of medical diagnosis have also been modeled as an optimization problem, and researchers have used GAs for solving them. In this chapter, we attempt to provide a state-of-the-art survey in the application of the principles of GAs, an important component of *evolutionary computation*, for improving medical imaging and diagnosis tasks. Section 16.2 describes the basic principles of GAs. Thereafter, the use of GAs in improving equipment design has been studied. Finally, the application of GAs for computer aided diagnosis, including schemes driven by both image and data (consisting of information not derived from images), is provided.

16.2 Preliminaries on Genetic Algorithms

Genetic algorithms [8–10] are efficient, adaptive, and robust search and optimization processes that are usually applied to very large, complex, and multimodal search spaces. They are modeled on the principles of natural genetic systems, in which the genetic information of each individual or potential solution is encoded in structures called chromosomes. They use some domain or problem-dependent knowledge for directing the search in more promising areas of the solution space; this is known as the fitness function. Each individual or chromosome has an associated fitness function, which indicates its degree of goodness with respect to the solution it represents. Various biologically-inspired operators such as selection, crossover, and mutation are applied on the chromosomes to yield potentially better solutions.

16.2.1 Basic Principles and Features

Genetic algorithms emulate biological principles to solve complex optimization problems. It essentially comprises a set of individual solutions or chromosomes (called the population), and some biologically-inspired operators that create a new (and potentially better) population from an old one. According to the theory of evolution, only those individuals in a population who are better suited to the environment are likely to survive and generate offspring, thereby transmitting their superior genetic information to new generations.

The essential components of GAs are the following:

- A representation strategy that determines the way in which potential solutions will be encoded to form string like structures called chromosomes.
- A population of chromosomes.

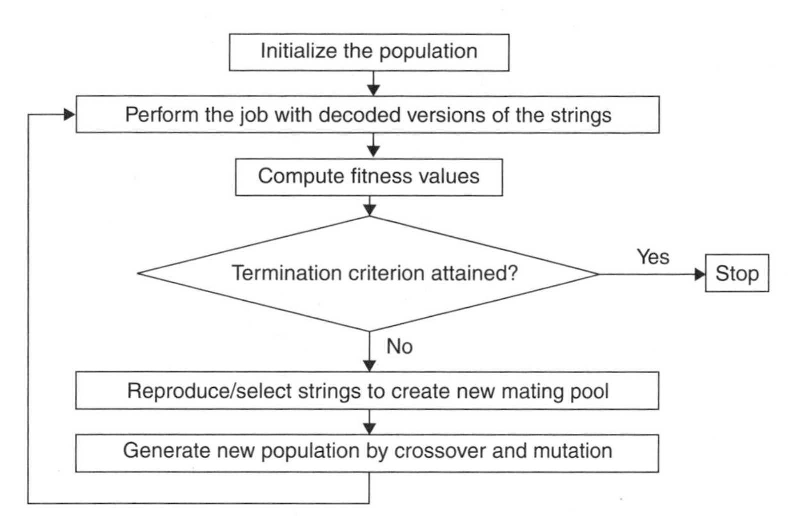

FIGURE 16.1 Basic steps of a genetic algorithm.

- Mechanism for evaluating each string (fitness function).
- Selection/reproduction procedure.
- Genetic operators (crossover and mutation).
- Probabilities to perform genetic operations.

A schematic diagram of the basic structure of a GA is shown in Figure 16.1. The components of GAs are described in the following sections.

16.2.2 Encoding Strategy and Population

To solve an optimization problem, GAs start with the chromosomal representation of a parameter set, which is to be encoded as a finite size string over an alphabet of finite length. Usually, the chromosomes are strings of 0s and 1s. For example, the string

$$1 \quad 0 \quad 0 \quad 1 \quad 1 \quad 0 \quad 1 \quad 0$$

is a binary chromosome of length 8. It is evident that the number of different chromosomes (or strings) is 2^l, where l is the string length. Each chromosome actually refers to a coded possible solution. A set of such chromosomes in a generation is called a population, the size of which may be constant or may vary from one generation to another. A common practice is to choose the initial population randomly.

16.2.3 Evaluation Technique

The fitness/objective function is chosen depending on the problem to be solved, in such a way that the strings (possible solutions) representing good points in the search space have high fitness values. This is the only information (also known as the payoff information) that GAs use while searching for possible solutions.

16.2.4 Genetic Operators

The frequently used genetic operators are selection, crossover, and mutation operators. These are applied to a population of chromosomes to yield potentially new offspring. The operators are described in Sections 16.2.4.1 to 16.2.4.3.

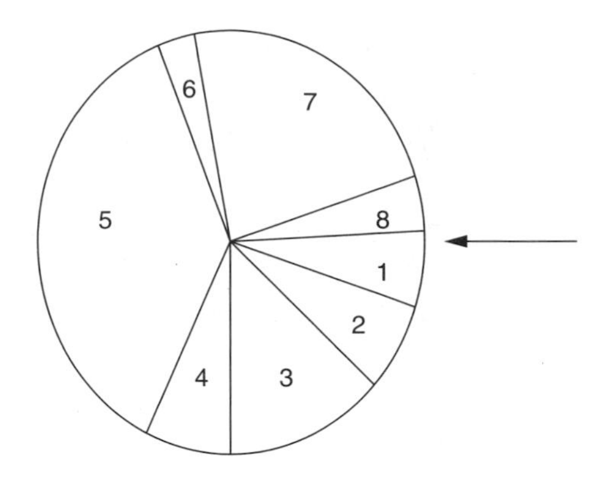

FIGURE 16.2 Roulette wheel selection.

16.2.4.1 Selection

The selection/reproduction process copies individual strings (called parent chromosomes) into a tentative new population (known as mating pool) for genetic operations. The number of copies that an individual receives for the next generation is usually taken to be directly proportional to its fitness value; thereby mimicking the natural selection procedure to some extent. This scheme is commonly called the *proportional selection scheme. Roulette wheel parent selection*, *stochastic universal selection*, and *binary tournament selection* [8,10] are some of the most frequently used selection procedures. Figure 16.2 demonstrates the roulette wheel selection. The wheel has as many slots as the population size P, where the size of a slot is proportional to the relative fitness of the corresponding chromosome in the population. An individual is selected by spinning the roulette, and noting the position of the marker when the roulette stops. Therefore, the number of times that an individual will be selected is proportional to its fitness (or, the size of the slot) in the population. In the commonly used *elitist* model of GAs, thereby providing what is called an *elitist GA* (EGA), the best chromosome seen up to the present generation is retained either in the population, or in a location outside it.

16.2.4.2 Crossover

The main purpose of crossover is to exchange information between randomly selected parent chromosomes by recombining parts of their genetic information. It combines parts of two parent chromosomes to produce offspring for the next generation. *Single-point crossover* is one of the most commonly used schemes. Here, first of all, the members of the selected strings in the mating pool are paired at random. Then each pair of chromosomes is subjected to crossover with a probability μ_c where an integer position k (known as the crossover point) is selected uniformly at random between 1 and $l-1$ ($l > 1$ is the string length). Two new strings are created by swapping all characters from position $(k+1)$ to l. For example, let the two parents and the crossover points be as shown below.

$$1\quad 0\quad 0\quad 1\quad 1|\quad 0\quad 1\quad 0$$

$$0\quad 0\quad 1\quad 0\quad 1|\quad 1\quad 0\quad 0$$

Then, after crossover the offspring will be the following:

$$1\quad 0\quad 0\quad 1\quad 1\quad 1\quad 0\quad 0$$

$$0\quad 0\quad 1\quad 0\quad 1\quad 0\quad 1\quad 0$$

Other common crossover techniques are two-point crossover, multiple point crossover, shuffle-exchange crossover, and uniform crossover [9].

The successful operation of GAs depends, to a great extent, on the coding technique used to represent the problem variables [11,12]. The *building block hypothesis* indicates that GAs work by identifying good building blocks, and by eventually combining them to get larger building blocks [8,13,14]. Unless good building blocks are coded tightly, the crossover operation cannot combine them [15,16]. Thus coding–crossover interaction is important for the successful operation of GAs. The problem of tight or loose coding of problem variables is largely known as the *linkage problem* [17]. Recent work on linkage learning GAs that exploits the concept of gene expression can be found in References 18 to 20.

16.2.4.3 Mutation

Mutation is the process by which a random alteration in the genetic structure of a chromosome takes place. Its main objective is to introduce genetic diversity into the population. It may so happen that the optimal solution resides in a portion of the search space that is not represented in the population's genetic structure. The process will therefore be unable to attain the global optima. In such situations, only mutation can possibly direct the population to the optimal section of the search space by randomly altering the information in a chromosome. Mutating a binary gene involves simple negation of the bit, whereas that for real coded genes are defined in a variety of ways [10,21]. Here, we discuss the binary bit-by-bit mutation, where every bit in a chromosome is subject to mutation with a probability μ_m. The result of applying the bit-by-bit mutation on positions 3 and 7 of a chromosome is shown here.

1 0 **0** 1 1 0 **1** 0

1 0 **1** 1 1 0 **0** 0

16.2.5 Parameters of GA

There are several parameters in GAs that have to be manually tuned and fixed by the programmer. Among these are the population size, probabilities of performing crossover and mutation, and the termination criteria. Several other things must also be determined by the programmer. For example, one must decide whether to use the generational replacement strategy, in which the entire population is replaced by a new population, or the steady state replacement policy where only the less fit individuals are replaced. Most such parameters in GAs are problem dependent, and no guidelines for their choice exist in the literature. Therefore, several researchers have also kept some of the GA parameters variable and/or adaptive [22–24].

As shown in Figure 16.1, the cycle of selection, crossover, and mutation is repeated a number of times till one of the following occurs:

1. The average fitness value of a population becomes more or less constant over a specified number of generations.
2. A desired objective function value is attained by at least one string in the population.
3. The number of generations (or iterations) is greater than some threshold.

16.3 Genetic Algorithms for Equipment Design for Medical Image Acquisition

Magnetic resonance imaging (MRI) is one of the most commonly used medical imaging techniques used for the diagnosis of several ailments including multiple sclerosis, strokes, tumors, and other infections of the brain, bones, spine, or joints, and for visualizing torn ligaments, soft tissues of the body, etc. Magnetic resonance imaging is based on the phenomenon of nuclear magnetic resonance (NMR) discovered by Felix Bloch and Edward Powell, for which they received the Nobel Prize in 1952. In an MRI scan, the patient is placed inside the bore of a magnet, which is usually cubic in shape. The body part to be scanned is placed at the center (or, isocenter) of the magnetic field inside the bore. In this noninvasive procedure, strong magnetic fields along with radio waves are used to visualize the structure of a particular body part.

The design of appropriate equipments for the purpose of good imaging may be considered as the first step in medical image processing. For example, large superconducting solenoids with apertures of typically 1 m, highly uniform (20 ppm) central fields of 1–2T, and low fringe fields (5 Gauss at 5 m) are required in clinical MRI. However, these magnets, which are now available in the clinical market, have deep bores, typically between 1.8 and 2 m length, and have a number of disadvantages such as patient claustrophobia and limited access for intervention. In order to overcome the limitations and evolve good designs for the magnets, researchers have mapped the design problem to one of optimization and have investigated the use of computational methods [25,26], including GAs [27–29], for this purpose.

Analytical techniques have been the preferred approach to design such magnets and gradient sets for MRI. Such technique are computationally efficient but are approximate, particularly away from the axis of symmetry. In Reference 30, an attempt has been made, which uses GA running on massively parallel computers to design an actively shielder whole-body MRI solenoid magnet with a bore of 1 m. The task is to optimize a cost function based on the magnetic field generated by a set of upto 20 circular coils, each having upto 2000 turns. The coils are constrained to be concentric with the magnet, and are arranged in symmetric pairs. A single coil is described by five parameters:

- Two parameters describing its position in the X–Z plane
- Depth of the coil
- Width of the coil
- The direction of the current

A chromosome encodes these five parameters per coil, for upto 20 coils. Since the coils are arranged in pairs that are symmetric about the central X–Y plane of the magnet, only 10 of the coils are independent. Thus a chromosome encodes upto 50 parameters that have to be optimally tuned. The magnetic field is computed using the Biot–Savart law. The fitness function incorporates terms for uniformity of field in the region of interest (ROI), and the smallness of the fringe field. The field is calculated by summing over the contributions from the turns in each coil. Recombination is performed as a two-stage process. In the first stage, a parent subset, which is of half the size of the population, is created by performing binary tournament selection in the top eighth of the population. In the second stage, pairs of chromosomes from the parent subset are picked at random and mated to produce a single offspring. The parents are replaced in the subset, and this process is continued till the next generation is completed. Three types of mutations are considered, which takes care of small perturbations, perturbations to add or remove a coil from the design, and a drastic perturbation for adding extra diversity to the population. The initial result provided in Reference 30 demonstrates the effectiveness of GAs for producing shorter whole-body magnet designs, which are original and innovative.

Two-dimensional ultrasonic arrays provide the possibility of three-dimensional electronic focusing and beam-steering and thus three-dimensional imaging. In simulation studies, it has been demonstrated [31] that reducing the number of elements of a two-dimensional matrix array down to order eight, keeps resolution and leads still to sufficient contrast. These random arrays are usually obtained by generating a random pattern with the desired number of elements. In Reference 32, simulation is presented to show that the imaging quality of a sparse tree can be improved by optimizing the random choice of elements. The optimization is done using GA.

16.4 Genetic Algorithms for Medical Diagnosis

For diagnosis of diseases, several tests are performed on the patient, some of which may involve taking various images such as x-rays, MRI scan, CT scan, etc., and some others that involve pathological and other tests. Data from these tests may become quite large and conflicting, and manual interpretation by collecting all such information may become difficult. This has therefore given rise to the development of

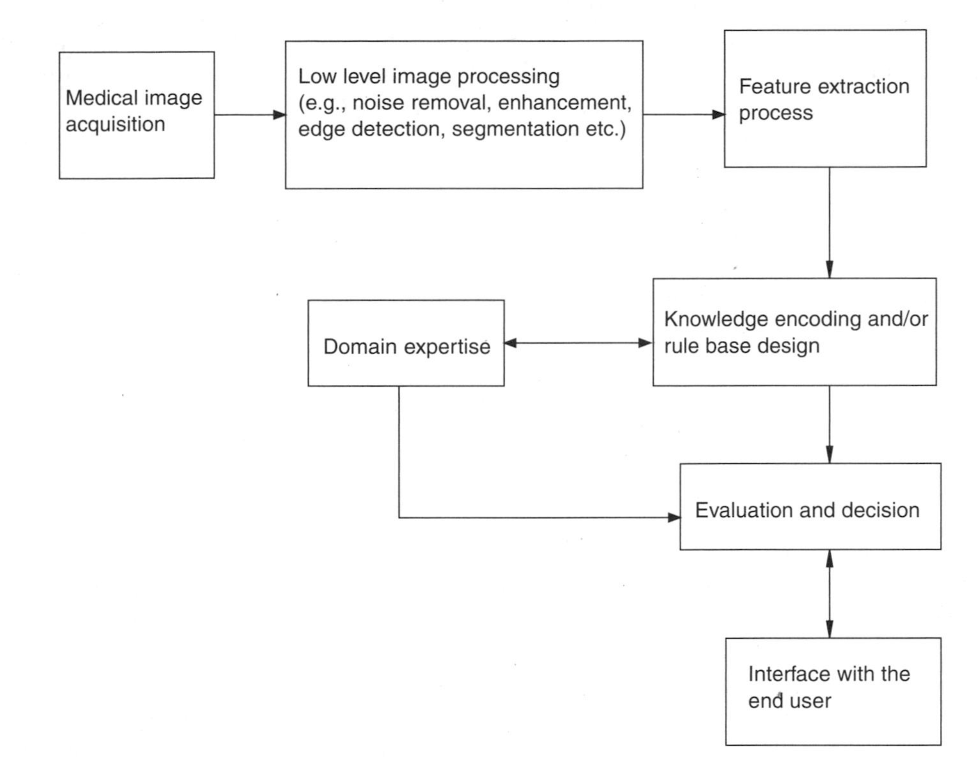

FIGURE 16.3 Block diagram of a CAD system.

computer assisted diagnostic tools, which are intended to help the medical practitioners make sense out of a large amount of data, and make diagnosis and therapy decisions. Figure 16.3 shows a block diagram of a computer aided diagnosis system.

16.4.1 Image Driven Diagnosis

Computer aided detection and classification schemes have the potential of increasing diagnostic accuracy in medical imaging. This may be done, for instance, by alerting radiologists to lesions that they initially overlooked, or assisting in the classification of detected lesions [33]. As with any complicated pattern recognition system, these schemes, generally referred to as computer aided diagnosis (CAD) schemes, typically employ multiple parameters such as threshold values, filter weights, and ROI sizes to arrive at a detection or classification decision. For the system to have a high performance, the values of these parameters need to be set optimally. In general, the optimal set of parameters may change when a component of the imaging chain is modified or changed. When the number of parameters is large, it becomes very difficult to manually determine the optimal choice of parameter values, as some of the values may be correlated in some unknown manner. However, conventional optimizing techniques, including GAs, are designed to optimize a scalar objective function, and the task of optimizing the performance of a diagnostic decision-making system is clearly multiobjective. Therefore, two objectives, increasing the sensitivity and reducing the false-positive rate of the system, are described by a single objective function. This objective function is defined as the weighted sum of the sensitivity and false-positive rate or using area under the receiver operating characteristic (ROC) curve. The use of a niched pareto genetic algorithm (NP-GA) [34] in training two popular diagnostic classifiers for optimizing their performance, has been studied in Reference 35. Unlike conventional classifier techniques that formulate the problem as the solution to a scalar optimization, the NP-GA explicitly addresses the multiobjective nature of the training task. Traditional techniques of classifier training attempt to combine several objective functions into one, so that conventional scalar optimization technique can be utilized. This involves incorporating a priori information into the aggregation method so that the resulting performance of the classifier is satisfactory for the task at hand. It has been shown in Reference 35 that the multiobjective genetic approach removes the ambiguity associated with defining a scalar measure of classifier performance and that it returns a set of optimal solutions that are equivalent in the absence of any information regarding the preference of the objectives, that is, sensitivity and specificity. The a priori knowledge that is used for aggregating the objective functions in conventional classifier training may instead be applied for post-optimization to select from one of the series of solutions returned from multiobjective genetic optimization. This technique is applied in Reference 35 to train a linear classifier and an artificial neural network, using simulated datasets. The performance of the solutions returned from the multiobjective genetic optimization represents a series of optimal pairs, sensitivity and specificity, which can be thought of as operating points on an ROC curve. It is observed that all possible ROC curves for a given dataset and classifier are less than or equal to the ROC curve generated by the NP-GA optimization. In Reference 36, a multiobjective approach for optimizing the performance of two rule-based CAD schemes has been proposed. One of these CAD schemes is designed to detect clustered microcalcifications in digitized mammograms, while the other scheme is developed to detect the breast masses.

Diagnosis and follow up of pigmented skin lesions is an important step toward early diagnosis of skin cancer [37]. For this purpose, digitized epiluminescence microscope (ELM) [38] images of pigmented skin lesions is used. Epiluminescence microscopy is a noninvasive technique that uses an oil immersion to render the skin translucent and make pigmented structure visible. During clinical diagnosis of pigmented skin lesions, one of the main features is the lesion symmetry, which should be evaluated according to its shape, color, and texture. This may be evaluated by drawing two orthogonal axes that maximize the perceived symmetry [39]. The evaluation is binary, that is, the lesion is either symmetrical or asymmetrical. In addition to the small number of possible outcomes, the evaluation is highly subjective and depends on the physicians' experience. As a result, the development of automatic techniques for the quantification of symmetry and the detection of symmetry axes is necessary. Other methods based on principal component technique for computing the axes may be found in References 40 and 41.

Reference 42 has proposed a GA-based technique and an optimization scheme derived from the self-organizing maps theory for the detection of symmetry axes. The notion of symmetry map has been introduced, which allows an object to be mapped to a symmetry space where its symmetry properties can be analyzed. The objective function (ψ) used in Reference 42 is based on a given symmetry measure, which is a function of the mean-square error (MSE) between the original and the reflected images. The MSE is defined as follows:

$$\text{MSE} = E[||\Gamma(x, y) - \Gamma(x', y')||^2] \tag{16.1}$$

where $\Gamma(x, y) : [0, c - 1] \times [0, r - 1] \rightarrow IR^n$ is the vector valued $r \times c$ input image, where the pixel values are in an n dimensional space, (x, y) and (x', y') represent the pixel coordinates and the symmetry coordinates respectively. The input image can be decomposed into symmetric ($\Gamma_s(x, y)$) and asymmetric ($\Gamma_a(x, y)$) components. Therefore, $\Gamma(x, y) = \Gamma_s(x, y) + \Gamma_a(x, y)$. The asymmetric component can be considered as symmetric noise, and the MSE is proportional to its energy. The distortion due to noise can be measured through the peak signal-to-noise ratio (PSNR), and is given by

$$\text{PSNR} = 10 \log_{10}\left(\frac{(N_q - 1)^2}{\text{MSE}}\right), \tag{16.2}$$

where N_q is the number of quantization levels in the image. The fitness function ψ is defined as

$$\psi = 1 - \frac{1}{1 + \text{PSNR}}. \tag{16.3}$$

Real encoding of chromosomes has been used in Reference 42 along with a modified version of simulated binary crossover [43]. The proposed technique is applied for detection and diagnosis of malignant melanoma.

The development of computer supported systems for melanoma diagnosis is of great importance to dermatologists due to its clinical accuracy in identifying malignant melanomas [44,45]. Several techniques for computerized melanoma diagnosis are based on color images making use of image analysis methods to quantify visual features as described by the "ABCD" rule (Asymmetry, irregular Border, varying Color, Diameter) [37,46,47]. Laser profilometry opens up new possibilities to improve tumor diagnostics in dermatology [48].

The recognition task is to classify surface profiles of melanomas and nevi also called moles. Due to the fact that all profiles contain regions with a structure similar to normal skin, each profile is subdivided into 16 non-overlapping quadratic subprofiles and image analysis algorithms are applied to each subprofile separately. Subsequently, feature selection algorithms are applied to optimize the classification performance of the recognition system.

An efficient computer supported technique that uses GAs for diagnosis of skin tumors in dermatology is presented in Reference 49. High resolution skin surface profiles are analyzed to recognize malignant melanomas and nevocytic nevi. In the initial phase, several types of features are extracted by two-dimensional image analysis techniques characterizing the structure of skin surface profiles: texture features based on co-occurrence matrices [50], Fourier features [51], and fractal features [37,52,53]. Subsequently, several feature selection algorithms based on heuristic strategies, greedy technique, and GA are applied to determine suitable feature subsets for the recognition process by describing it as an optimization problem. As a quality measure for feature subsets, the classification rate of the nearest neighbor classifier computed with the leaving-one-out method is used. Among the different techniques used, GAs show the best result. Finally, neural networks with error back-propagation as learning paradigm are trained using the selected feature sets. Different network topologies, learning parameters, and pruning algorithms are investigated to optimize the classification performance of the neural classifier. With the optimized recognition system, a classification performance of 97.7% is achieved.

In the 1980s, microwave imaging was thought to have great potential in developing medical diagnostic tools. However, because of the problem of inverse scattering, good reconstruction of images was found to be difficult. Carosi et al. [54] have used the approach of focused imaging, where only a part of the body is subjected to the investigation, combined with the capabilities of global search techniques like GAs for accurate reconstruction. For experimental purpose, a human abdomen is considered and different electromagnetic sources operating at the working frequency of 433 MHz and 2.54 GHz are used. Numerical investigations are performed to define the optimal dimensions of the reduced investigation domain. To quantitatively evaluate the effects of the reduction of the original investigation domain on the inversion data, suitable relative errors are defined. Once the reduced domain is defined, preliminary reconstructions are performed aiming to evaluate the imaging capability of GAs when a focussed approach is used for tomographic application.

One of the most important facial paralysis diagnosis techniques is quantitative assessment of patient's facial expression motion. Johnson et al. [55] proposed a technique for achieving maximal regional facial motion while the rest of the face is held at rest based on Maximal Static Response Assay of facial nerve function. This requires the placement of removable adhesive dots and a small adhesive ruler on the face at predefined locations. However, many facts, such as misplacement of facial dots, misplacement of the grid, and reading and entry errors, will cause an error in the assay. Helling et al. [56] used region-specific, subtracted, digitized image light reflectance as a two-dimensional marker for the complex three-dimensional surface deformations of the face during expression. The velocity of region-specific facial motion is estimated from the facial motion image sequences using the optimal flow (OF) technique. The computation of the OF field requires estimates of both the spatial gradient and spatial time derivative at each pixel, and this time-consuming process often limits its use, especially in medical application. To overcome this problem, an OF technique based on GA is proposed in Reference 57 to detect facial motions from dynamic image sequences. Experimental results demonstrate that the proposed technique is very useful to diagnose the site of facial paralysis and assess progression or recovery profiles of patients when combined with other diagnosis techniques.

16.4.2 Data Driven Diagnosis

The previous section reported some research on development of diagnostic tools in which one of the inputs considered is an image. In this section, we deal with some other diagnostic tools that use other types of input data, mostly numeric.

Electromyography (EMG) is the recording and study of the electrical activity of voluntary contracting muscles. Clinical EMG findings provide useful information in the electrodiagnostic examination of peripheral nerves and skeletal muscle, and in deciding the level of the lesion in patients suffering from neuromuscular disorders. It is also useful in deciding whether the symptom of muscle weakness in the assessment of neuromuscular disorders is myopathic or neurogenetic in origin. The advantages of automated EMG diagnostic systems can be found in Reference 58. Different approaches have been used to address the problem of automated EMG diagnosis. The utility of artificial neural networks in classifying EMG data trained with back propagation or Kohonen's self-organizing feature map algorithm has recently been demonstrated in References 59 and 60. In Reference 61, a study has been made to investigate how genetics-based machine learning (GBML) can be applied for diagnosing certain neuro muscular disorder based on EMG data. The effect of GBML control parameters on diagnostic performance is also examined. Subsequently, a hybrid diagnostic system is introduced that combines both the neural network and GBML.

In References 62 and 63, a methodology based on GAs for the automatic induction of Bayesian networks from a file containing cases and variables related to the problem of predicting survival in malignant skin melanoma is described. The structure is learned by applying three different techniques: the Cooper and Herskovits metric for a general Bayesian network [64], the Markov blanket approach, and the relaxed Markov blanket method. The methodologies are applied to the problem of predicting survival of people after 1, 3, and 5 years of being diagnosed as having malignant skin melanoma. The induced Bayesian network is used for classifying patients according to their prognosis of survival. These results are compared to those obtained by the Naive–Bayes paradigm. An empirical comparison of Bayesian networks learned using GAs, rule induction, and logistic regression is carried out in Reference 65 where the task is to predict the survival of women suffering from breast cancer. In a more recent attempt, Blanco et al. [66] have studied the problem of learning Bayesian networks using two stochastic, population-based search algorithms: the univariate marginal distribution algorithm and population-based incremental learning. Comparison with the GA-based scheme is also carried out.

The problem of Wisconsin breast cancer diagnosis (WBCD) has been considered in References 67 and 68, by combining fuzzy systems and evolutionary algorithms to design an automatic diagnosis system. The proposed fuzzy-genetic approach produces systems exhibiting two prime characteristics. They attain high classification performance with the possibility of attributing a new confidence measure to the output diagnosis. Moreover, the resulting systems involve a few simple rules, and are, therefore, human interpretable. Another approach for diagnosis of breast cancer by Bayesian networks can be found in Reference 69.

In medicine, prognostic models may be used to assess the likely prognosis of a patient, which, in turn, may determine the treatment of that patient. In Reference 70, a prognostic model based on diffusion GAs is sought to determine whether or not patients suffering from an uncommon form of cancer will survive. The problem considered is a multiobjective one, in which the three objectives are:

- Maximize the correct number of survival predictions
- Maximize the correct number of death predictions
- Minimize the number of factors used

The motivation behind this study is to accurately predict the outcome so that patients who are more likely to die can be identified at diagnosis, and subjected to high dose aggressive chemotherapy (which has several negative side effects). On the other hand, those with a high chance of survival can be spared this treatment and its side effects. Given a set of case histories, a technique is proposed in Reference 70

TABLE 16.1 Different Criteria and Their Ranges Used in Reference 70

Number	Name	Ranges
1	Brain metastases	Yes, No
2	Liver metastases	Yes, No
3	Placental site trophoblastic tumors (PSTT)	Yes, No
4	Prior chemotherapy	Yes, No
5	Pregnancy to term	Yes, No
6	Age	<26, 26–30, 31–41, >41
7	Serum hCG level	<800, 801–29570, 29571–181000, >181000
8	Interval between pregnancy and diagnosis	<4, 5–9, 10–33, >33

that attempts to find the relative weights of the different factors that are used to describe the cases. The eight factors that are used in the prognostic model of Reference 70 and their possible ranges are provided in Table 16.1.

A diffusion GA is used for building the prognostic model that will simultaneously optimize the three objectives given here. A model is represented as a chromosome by encoding the weights associated with each factor. Boolean criteria (1–5 in Table 16.1) have single associated weights, while criteria having a range of values (6–8 in Table 16.1) have one weight per subrange. In addition a combination weight is used in Reference 70, which can incorporate the possibility that a combination of factors might be important. Thus a total of 18 weights are used, 5 for the Boolean criteria, 12 for the subranged criteria, and the combination weight. In the diffusion GA, the individuals in a population are arranged along the vertices of a square lattice. During crossover, each individual chooses one of its four neighbors randomly as the mate. Mutation can either increase or decrease a weight by 10%, or set it to zero. Only if the result of crossover and mutation at a position is better, in the Pareto optimal sense, than the original, the latter will be replaced. Marvin et al. [70] experimented with a population size of 169 placed on a 13×13 grid. The weights are randomly set in the range $[-2000, 2000]$, and the algorithm is executed for 1000 generations. It is found to predict 90% of the survivals and 87% of the deaths, while using 6 of the 8 factors, and 13 of the possible 18 weights when the entire data set is used. For training using 90% data and testing using the remaining 10%, several prognostic models are obtained each coming up with a different compromise among the three objective values. Significantly, the method in Reference 70 enables a simple model to be evolved, one that produces well-balanced predictions and one that is relatively easy for clinicians to use.

Ngan et al. [71] employ evolutionary programming (EP) and genetic programming (GP) in the domain of knowledge discovery in medical systems. Evolutionary Programming is used to learn Bayesian networks, which is known to be an intractable problem. Minimum description length principle is used to measure the goodness of solution in EP followed by the use of GP to learn rules. The entire knowledge discovery system is applied on limb fracture and scoliosis data, where it is able to detect many interesting patterns/rules that were uncovered earlier. Ngan et al. had earlier used GP for discovering comprehensible rules in the medical domain; they used grammar to restrict the seach space, and to ensure the syntactical correctness of the rules [72]. The discovered rules were evaluated within the framework of support confidence proposed for association rule mining. Here, a major limitation was that the grammar was application dependent, and had to be written for each application domain.

Genetic programming is also used to discover comprehensible rules for predicting 12 different diseases using 189 predicting attributes, or measurements [73]. The 12 diseases considered here are stable angina, unstable angina, acute myocardial infarction, aortic dissection, cardiac tamponade, pulmonary embolism, pneumothorax, acute pericarditis, peptic ulcer, esophageal pain, musculoskeletal disorders, and psychogenic chest pain, the characteristic of all of which was chest pain. All the 189 attributes are binary. Genetic programming is used to learn rules expressed in a kind of first-order logic of the form $<Att_i\ Op\ Att_j>$, where Att_i and Att_j are the predicting attributes, and Op is some relational operator. Genetic programming evolves a population of "programs" candidate to the solution of a specific problem. Here, a program is represented in the form of a tree, in which the internal nodes are functions (operators) and the leaf nodes are terminal symbols. In the GP formulation of the problem in Reference 73, the terminal set consists of the 189 attributes, and the function set consists of {AND, OR, NOT}. The GP is executed once for each class, with the appropriate rule for the ith class being evolved in the ith GP run. Thus, each run consists of learning a two-class classification problem, in which the goal is to predict whether a patient has a particular disease (class i) or not (NOT class i). For computing the fitness of a candidate (program or rule), a (labeled) training set is used on which the the rule is tested. The size of the following sets are computed:

- True positives (tp): the rule predicts that the patient has a given disease and the patient does have it.
- False positives (fp): the rule predicts that the patient has a given disease and the patient does not have it.
- True negative (tn): the rule predicts that the patient does not have a given disease and the patient actually does not have it.
- False negative (fn): the rule predicts that the patient does not have a given disease but the patient actually has it.

Thereafter, two measures are computed, the sensitivity (Se) and specificity (Sp):

$$\mathrm{Se} = \frac{\mathrm{tp}}{\mathrm{tp} + \mathrm{fn}} \tag{16.4}$$

$$\mathrm{Sp} = \frac{\mathrm{tn}}{\mathrm{tn} + \mathrm{fp}}. \tag{16.5}$$

The fitness function is taken to be the product of the two, namely,

$$\mathrm{fitness} = \mathrm{Se} \times \mathrm{Sp}. \tag{16.6}$$

For the experiments, the data set consisted of 138 samples (patients), which were partitioned into a training set with 90 samples, and a test set with 48 samples. The GP achieved an accuracy of 77.08% on. the test set. Other related methods that used GP for classification problem can be found in References 74 to 78. On similar lines, in Reference 79, GAs were used to discover comprehensible IF–THEN rules for the diagnosis of dermatological diseases and prediction of the recurrence of breast cancer. Here, a chromosome is of length n, where n is the number of attributes. The i, the gene corresponding to the ith attribute, is divided into three fields: weight (W_i), operator (O_i), and value (V_i). Each gene corresponds to one condition in the IF part of the rule. The GA is executed once for each class, and therefore the THEN part (indicating the class for which the GA was run) is not required to be encoded in the chromosome. The weight value indicated whether the ith attribute, A_i, is at all present in the rule (if $W_i >$ Limit) or not (if $W_i \leq$ Limit). Limit was set to 0.3. The operator O_i could take values from $\{=, \neq\}$, if the corresponding attribute is categorical, and from $\{\geq, <\}$, if the corresponding attribute is continuous. The value V_i could take values from the domain of the attribute A_i. Normal selection and crossover operators are used. Three mutation operators are defined, namely weight mutation, operator mutation, and value mutation. The fitness function of a chromosome was defined as fitness $=$ Se $*$ Sp as in Reference 73. The dermatology data set consists of the differential diagnosis of erythematisquamous. There are six different diagnoses (six classes): psoriasis, seboreic dermatitis, lichen planus, pityriases rosea, chronic dermatitis, and pityriasis rubra pilaris. The data set consists of 366 records with 34 attributes. The breast cancer data consists of 286 records with 9 attributes and 2 classes (recurrence and nonrecurrence of cancer). The accuracy rates achieved in Reference 79 were 95% for the dermatological data and 67% for the cancer data. The resultant rules were also found to be comprehensible, with one rule obtained per class.

16.5 Discussion and Conclusions

This chapter provides a comprehensive survey of the application of GAs to the domain of designing of equipments for medical image acquisition and medical diagnosis. In recent times, with the advent of a variety of sophisticated imaging techniques, use of medical images in clinical diagnosis and therapy has increased manifold. Therefore, attempts at increasing the resolution and quality of such images is an important area of research. This, in turn, leads to research for designing equipments in such a way that the imaging modality becomes faster and more informative. Computer aided diagnosis has also been necessitated due to the large amount of data that is routinely collected for the patients. These problems often turn out to be those of search and optimizations, requiring the use of good optimization tools such as GAs.

The research works reviewed in this chapter have been reported in diverse journals and proceedings like *IEEE Transactions on Medical Imaging, IEEE Transactions on Pattern Analysis and Machine Intelligence, IEEE Transactions on Neural Networks, Artificial Intelligence in Medicine, Medical Image Analysis, Magnetic Resonance Imaging, Evolutionary Computation, International Conference on Evolutionary Computation, International Conference on Genetic Algorithms, ACM SIGMOD Conference on Management of Data*, and so on. Encouraging results have been reported by researchers in this regard. This chapter presented a methodical way in which a large number of such research activities have been compiled and reported within a common platform, namely GA-based techniques.

It may be noted that the main challenges and issues in integrating GAs for solving optimization problems in medical imaging and diagnosis are manifold. First, the encoding strategy must be suitably defined so that it conforms to the building block hypothesis. According to this hypothesis, short low order above average schema, or building blocks, should combine to yield potentially better solutions. Any adhoc encoding strategy may not follow this hypothesis, and hence encoding strategy may not follow this hypothesis, implying GAs may often be found to yield poor results in such situations. Second, the fitness function must be adequately designed. Since fitness computation is often computation intensive, in the medical domain one may need to go for parallel GAs working on massively parallel systems to get real time response. This in turn incorporates an added level of difficulty to the problem. Finally, a still open unsolved issue is the appropriate selection of the operator probabilities and the termination criterion, so as to ensure good performance of the GA-based systems for medical imaging and diagnosis problems.

Acknowledgment

A part of this work was carried out when Dr. U. Maulik visited the University of Texas at Arlington, USA, with the BOYSCAST fellowship provided by Department of Science and Technology, Government of India, during 2001. Dr. S. Bandyopadhyay would like to acknowledge Indian National Science Academy, Government of India sponsored project *Soft computing for medical image segmentation and classification* No. BS/YSP/36/887 for providing partial support to carry out this work.

References

[1] G. Gerig, T. Pun, and O. Ratib. Image analysis and computer vision in medicine. *Computerized Medical Imaging and Graphics*, 18, 85–96, 1994.

[2] N. Ayache. Medical computer vision, virtual reality and robotics. *Image and Computer Vision*, 13, 295–313, 1995.

[3] J.S. Duncan and N. Ayache. Medical image analysis: Progress over two decades and the challenges ahead. *IEEE Transactions on Pattern Analysis and Machine Intelligence*, 22, 85–108, 2000.

[4] T. Mclnerney and D. Terzopolous. Deformable models in medical image analysis: A survey. *Medical Image Analysis*, 1, 91–108, 1996.

[5] J.B.A. Maintz and M.A. Viergever. A survey of medical image restoration. *Medical Image Analysis*, 2, 1–37, 1998.

[6] R. Shahidi, R. Tombropoulos, and R.P.A. Grzeszczuk. Clinical applications of three-dimensional rendering of medical data sets. *Proceedings of IEEE*, 86, 555–568, 1998.

[7] S Lavallee, *Registration for Computer Integrated Surgery: Methodology, and State of Art.* MIT Press, Cambridge, MA, 1996, pp. 77–97.

[8] D.E. Goldberg, *Genetic Algorithms in Search, Optimization and Machine Learning.* Addison-Wesley, New York, 1989.

[9] L. Davis, Ed., *Handbook of Genetic Algorithms.* Van Nostrand Reinhold, New York, 1991.

[10] Z. Michalewicz, *Genetic Algorithms + Data Structures = Evolution Programs.* Springer-Verlag, New York, 1992.

[11] H. Kargupta, K. Deb, and D.E. Goldberg. Ordering genetic algorithms and deception. In *Proceedings of the Parallel Problem Solving from Nature* (R. Manner and B. Manderick, Eds.). North-Holland, Amsterdam, 1992, pp. 47–56.

[12] N.J. Radcliffe. Genetic set recombination. In *Foundations of Genetic Algorithms 2* (L.D. Whitley, Ed.). Morgan Kaufmann, San Mateo, CA, 1993, pp. 203–219.

[13] J.J. Grefenstette, R. Gopal, B. Rosmaita, and D. Van Gucht. Genetic algorithms for the traveling salesman problem. In *Proceedings of the 1st International Conference on Genetic Algorithms* (J.J. Grefenstette, Ed.). Lawrence Erlbaum Associates, Hillsdale, 1985, pp. 160–168.

[14] J.H. Holland, *Adaptation in Natural and Artificial Systems.* The University of Michigan Press, Ann Arbor, MI, 1975.

[15] D.E. Goldberg, K. Deb, and B. Korb. Messy genetic algorithms: Motivation, analysis, and first results. *Complex Systems*, 3, 493–530, 1989.

[16] D.E. Goldberg, K. Deb, and B. Korb. Do not worry, be messy. In *Proceedings of the 4th International Conference on Genetic Algorithms* (R.K. Belew and J.B. Booker, Eds.). San Mateo, CA, Morgan Kaufmann, 1991, pp. 24–30.

[17] D.E. Goldberg, K. Deb, H. Kargupta, and G. Harik. Rapid, accurate optimization of difficult problems using fast messy genetic algorithms. In *Proceedings of the 5th International Conference on Genetic Algorithms* (S. Forrest, Ed.), San Mateo, CA, Morgan Kaufmann, 1993, pp. 56–64.

[18] H. Kargupta and S. Bandyopadhyay. Further experimentations on the scalability of the GEMGA. In *Proceedings of the V Parallel Problem Solving from Nature (PPSN V), Lecture Notes in Computer Science* (T. Baeck, A. Eiben, M. Schoenauer, and H. Schwefel, Eds.), Vol. 1498. Springer-Verlag, Amsterdam, The Netherlands, 1998, pp. 315–324.

[19] S. Bandyopadhyay, H. Kargupta, and G. Wang. Revisiting the GEMGA: Scalable evolutionary optimization through linkage learning. In *Proceedings of International Conference on Evolutionary Computation*, 1998, pp. 603–608.

[20] H. Kargupta and S. Bandyopadhyay. A perspective on the foundation and evolution of the linkage learning genetic algorithms. *The Journal of Computer Methods in Applied Mechanics and Engineering, Special issue on Genetic Algorithms*, 186, 266–294, 2000.

[21] L.J. Eshelman and J.D. Schaffer. Real-coded genetic algorithms and interval schemata. In *Foundations of Genetic Algorithms 2* (L. Whitley, Ed.). Morgan Kaufmann, San Mateo, CA, 1993, pp. 187–202.

[22] J.E. Baker. Adaptive selection methods for genetic algorithms. In *Proceedings of the 1st International Conference on Genetic Algorithms* (J.J. Grefenstette, Ed.). Hillsdale: Lawrence Erlbaum Associates, 1985, pp. 101–111.

[23] M. Srinivas and L.M. Patnaik. Adaptive probabilities of crossover and mutation in genetic algorithm. *IEEE Transaction on System, Man, Cybernetics*, 24, 656–667, 1994.

[24] S. Bandyopadhyay, *Pattern Classification using Genetic Algorithms.* Ph.D. thesis, Machine Intelligence Unit, Indian Statistical Institute, Calcutta, India, 1998.

[25] S. Crosier and D.M. Doddrell. Gradient-coil design by simulated annealing. *Journal of Magnetic Resonance*, 103, 354–357, 1993.

[26] S. Crosier and D.M. Doddrell. Compact MRI magnet design by stochastic optimization. *Journal of Magnetic Resonance*, 127, 233–237, 1997.

[27] B.J. Fisher, N. Dillon, T.A. Carpenter, and L.D. Hall. Design by genetic algorithm of a z gradient set for magnetic resonance imaging of the human brain. *Measurement Science and Technology*, 6, 904–909, 1995.

[28] B.J. Fisher, N. Dillon, T.A. Carpenter, and L.D. Hall. Design of a biplanar gradient coil using a genetic algorithm. *Magnetic Resonance Imaging*, 15, 369–376, 1997.

[29] G.B. Wiliams, B.J. Fisher, C.L.-H. Huang, T.A. Carpenter, and L.D. Hall. Design of biplanar gradient coils for magnetic resonance imaging of the human torso and limbs. *Magnetic Resonance Imaging*, 17, 739–754, 1999.

[30] R.E. Ansorge, T.A. Carpenter, L.D. Hall, N.R. Shaw, and G.B. Williams. Use of parallel supercomputer to design magnetic resonance systems. *IEEE Transactions on Applied Superconductivity*, 10, 1368–1371, 2000.

[31] D.H. Turnbull, A.T. Kerr, and F.S. Foster. Simulation of B-Scan images from two dimensional transducer arrays. In *Proceedings of the 1990 IEEE Ultrasonic Symposium*, 1990, pp. 769–773.

[32] P.K. Weber, R.M. Schmitt, B.D. Tylkowski, and J. Steak. Optimization of random sparse 2-D transducer arrays for 3-D electronic beam steering and focusing. In *Proceedings of the 1994 IEEE, IEEE Ultrasonics Symposium*, 1994, pp. 1503–1506.

[33] M.A. Anastasio, H. Yoshida, R. Nagel, R.M. Nishikawa, and K. Doi. A genetic algorithm-based method for optimizing the performance of a computer-aided diagnosis scheme for detection of clustered microcalcifications in mammograms. *Medical Physics*, 25, 1613–1620, 1998.

[34] J. Horn and N. Nafpliotis. Multiobjective Optimization using the Niched Pareto Genetic Algorithms. *illiGAL report no. 93005*. University of Illinois at Urbana-Champaign, 1993.

[35] M.A. Kupinski and M.A. Anastasio. Multiobjective genetic optimization of diagnostic classifiers with implications for generating receiver operating characteristic curves. *IEEE Transactions on Medical Imaging*, 18, 675–685, 1999.

[36] M.A. Anastasio, M.A. Kupinski, R.M. Nishikawa, and M.L. Giger. A multiobjective approach to optimizing computerized detection schemes. *IEEE Nuclear Science Symposium*, 3, 1879–1883, 1998.

[37] A. Green, N. Martin, J. Pfitzner, M. O'Rourke, and N. Knight. Computer image analysis in the diagnosis of melanoma. *Journal of American Academies of Dermatology*, 31, 958–964, 1994.

[38] Z.B. Argenyi. Dermatoscopy (epiluminescence microscopy) of pigmented skin lesions. *Dermatology Clinical*, 15, 79–95, 1997.

[39] W. Stolz, O. Braun-Falco, M. Landthaler, P. Bilek, and A.B. Cognetta, *Color Atlas of Dermatoscopy*. Blackwell Science, Oxford, 1994.

[40] W.V. Stoecker, W.W. Li, and R.H. Moss. Automatic detection of asymmetry in skin tumors. *Computerized Medical Imagings and Graphics*, 16, 191–197, 1992.

[41] D. Gutkowicz-Krusin, M. Elbaum, P. Szwaykowski, and A.W. Kopf. Can early malignant melanoma be differentiated from a typical melanocytic nerves by in vivo techniques?, Part II, automatic machine vision classification. Skin Research and Technology, Vol. 3, pp. 15–22, 1997.

[42] P. Schmid-Saugeon. Symmetry axis computation for almost-symmetrical and asymmetrical objects: Application to pigmented skin lesions. *Medical Image Analysis*, 4, 269–282, 2000.

[43] K. Deb and A. Kumar. Simulated binary crossover for continuous search space. *Complex Systsems*, 9, 115–148, 1995.

[44] C.M. Balch and G.W. Milton, *Hautmelanome* Springer, Berlin, 1988.

[45] C.M. Grin, A.W. Kopf, B. Welkovich, R.S. Bart, and M.J. Levenstein. Accuracy in the clinical diagnosis of malignant melanoma. *Archives of Dermatology*, 126, 763–766, 1990.

[46] J.E. Golston, W.V. Stoecker, R.H. Moss, and I.P.S. Dhillon. Automatic detection of irregular borders in melanoma and other skin tumors. *Computer Medical Imaging Graphics*, 16, 163–177, 1992.

[47] A.J. Sober and J.M. Burstein. Computerized digital image analysis: An aid for melanoma diagnosis preliminary investigations and brief review. *Journal of Dermatology*, 21, 885–890, 1994.

[48] K.P. Wilhelm, P. Elsner, E. Beradesca, and H. Baibach, *Bioengineering of the Skin: Skin Surface Imaging and Analysis*. CRC Press, Boca Raton, FL, 1997.

[49] H. Handels, T. Rob, J. Kreusch, H.H. Wolff, and S.J. Poppl. Feature selection for optimized skin tumor recognition using genetic algorithms. *Artificial Intelligence in Medicine*, 16, 283–297, 1999.

[50] R.M. Haralick, K. Shanmugam, and I. Dinstein. Texture features for image classification. *IEEE Transactions on System, Man, & Cybernetics*, 3, 610–621, 1973.

[51] D.H. Ballard and M.B. Brown. *Computer Vision*. Prentice-Hall, Englewood Cliffs, NJ, 1982.

[52] K.J. Falconer. *Fractal Geometry, Mathematical Foundations and Applications*. Wiley, Chichester, 1990.

[53] H.O. Peitgen and D. Saupe. *The Science of Fractal Images*. Springer, Berlin, 1988.

[54] S. Carosi, A. Massa, and M. Pastorino. Numerical assessment concerning a focussed microwave diagnostic method for medical application. *IEEE Transactions of Microwave Theory and Techniques*, 48, 1815–1830, 2000.

[55] P.C. Johnson, H. Brown, W.M. Kuzon, J.R. Ballit, J.L. Garrison, and J. Campbell. Simultaneous quantitation of facial movements: The maximal static response assay of facial nerve function. *Annals of Plastic Surgery*, 32, 171–179, 1994.

[56] T.D. Helling and M.J.G. Neely. Validation of objective measures for facial paralysis. *Laryngoscope*, 107, 1345–1349, 1997.

[57] Y. Cui, M. Wan, and J. Li. A new quantitative assessment method of facial paralysis based on motion estimation. In *Proceedings of the 20th Annual International Conference of the IEEE Engineering in Medicine and Biology Society*, Vol. 3. IEEE Press, Hong Kong, 1998, pp. 1412–1413.

[58] L.J. Dorfman and K.C. McGill. AAEE minimonograph #29: Automatic quantitative electromyography. *Muscle Nerve*, 11, 804–818, 1988.

[59] C.N. Schizas, C.S. Pattichis, I.S. Schofield, P.R. Fawcett, and L.T. Middleton. Artificial neural nets in computer-aided macro motor unit potential classification. *IEEE Engineering Medicine and Biology Magazine*, 9, 31–38, 1990.

[60] C.S. Pattichis. *Artificial Neural Networks in Clinical Electromyography*. Ph.D. thesis, Queen Mary and Westfield College, University of London, UK, 1992.

[61] C.S. Pattichis and C.N. Schizas. Genetic-based machine learning for the assessment of certain neuromuscular disorders. *IEEE Transactions on Neural Networks*, 7, 427–439, 1996.

[62] P. Larranaga, B.S.M.Y. Gallego, M.J. Michelena, and J.M. Pikaza. Learning bayesian networks by genetic algorithms. A case study in the prediction of survival in malignant skin melanoma. In *Lecture Notes in Artificial Intelligence*, Vol. 1211. Springer-Verlag, Heidelberg, 1997, pp. 261–272.

[63] B. Sierra and P. Larranaga. Predicting survival in malignant skin melanoma using Bayesian networks automatically induced by genetic algorithms. An empirical comparison between different approaches. *Artificial Intelligence in Medicine*, 14, 215–230, 1998.

[64] F.V. Jensen, *Introduction to Bayesian Networks*. University College of London, UK, 1996.

[65] P. Larranaga, M.Y. Gallego, B. Sierra, L. Urkola, and M.J. Michelena. Bayesian networks, rule induction and logistic regression in the prediction of women survival suffering from breast cancer. In *Lecture Notes in Artificial Intelligence*, Vol. 1323. Springer-Verlag, Heidelberg, 1997, pp. 303–308.

[66] R. Blanco, I. Inza, and P. Larranaga. Learning bayesian networks in the space of structures by estimation of distribution algorithms. *International Journal of Intelligent Systems*, 18, 205–220, 2003.

[67] C.A. Pena-Reyes and M. Sipper. Evolving fuzzy rules for breast cancer diagnosis. In *Proceedings of 1998 International Symposium on Nonlinear Theory and Applications (NOLTA '98)*, Vol. 2. Lausanne, Presses Polytechniques et Universitaires Romandes, pp. 369–372, 1998.

[68] C.A. Pena-Reyes and M. Sipper. A fuzzy-genetic approach to breast cancer diagnosis. *Artificial Intelligence in Medicine*, 17, 131–155, 1999.

[69] C.J. Kahn, L. Roberts, K. Shaffer, and P. Haddawy. Construction of a bayesian network for mammographic diagnosis of breast cancer. *Computers in Biology and Medicine*, 27, 19–29, 1997.

[70] N. Marvin, M. Bower, and J.E. Rowe. An evolutionary approach to constructing prognostic models. *Artificial Intelligence in Medicine*, 15, 155–165, 1999.

[71] S. Ngan, M.L. Wong, W. Lam, K.S. Leung, and J.C.Y. Cheng. Medical data mining using evolutionary computation. *Artificial Intelligence in Medicine*, 16, 73–96, 1999.

[72] S. Ngan, M.L. Wong, and K.S. Leung. Using grammar based genetic programming for data mining of medical knowledge. In *Genetic Programming 1998: Proceedings of 3rd Annual Conference*, Morgan Kaufmann, San Mateo, CA, 1998, pp. 254–259.

[73] C.C. Bojarczuk, H.S. Lopes, and A.A. Freitas. Discovering comprehensible classification rules by using genetic programming: A case study in a medical domain. In *Proceedings of the Genetic and Evolutionary Computation Conference* (W. Banzhaf, J. Daida, A.E. Eiben, M.H. Garzon, V. Honavar, M. Jakiela, and R.E. Smith, Eds.), Vol. 2, (Orlando, FL, USA), pp. 953–958, Morgan Kaufmann, San Mateo, CA, 1999, pp. 13–17.

[74] A. Teller and M. Veloso. Program evolution for data mining. *International Journal of Expert Systems*, 8(3), 213–236, 1995.

[75] B. Marchesi, A.L. Stelle, and H.S. Lopes. Detection of epileptic events using genetic programming. In *Proceedings of the 19th International Conference IEEE/EMBS*, pp. 1198–1201. IEEE Press, Washington, 1997.

[76] J.R. Sherrah, R.E. Bogner, and A. Bouzerdoum. The evolutionary pre-processor: Automatic feature extraction. In *Genetic Programming: Proceedings of 2nd Annual Conference*, Morgan Kaufmann, San Mateo, CA, 1997, pp. 304–312.

[77] N.I. Nikolaev and V. Slavov. Inductive genetic programming with decision trees. In *Proceedings of the 1997 European Conference on Machine Learning (ECML)*, 1997.

[78] L. Martin, F. Moal, and C. Vrain. A relational data mining tool based on genetic programming. In *Principles of Data Mining and Knowledge Discovery: Proceedings of 2nd European Symposium (LNAI)*, Vol. 1510 Springer-Verlag, Heidelberg, 1998, pp. 130–138.

[79] M.V. Fidelis, H.S. Lopes, and A.A. Freitas. Discovering comprehensible classification rules a genetic algorithm. In *Proceedings of the 2000 Congress on Evolutionary Computation CEC00*, La Jolla Marriott Hotel La Jolla, CA, USA, pp. 805–810, IEEE Press, Washington 2000, pp. 6–9.

17

Scheduling and Rescheduling with Use of Cellular Automata

Franciszek Seredynski
Anna Święcicka
Albert Y. Zomaya

17.1 Introduction

Multiprocessor scheduling is one of the most challenging problems in parallel and distributed computing [1]. It is known to be NP-complete in its general form [2]. Researchers have studied restricted forms of the problem by constraining either a parallel program model or a multiprocessor model. However, these special cases do not fully represent real-world systems. To solve the scheduling problem in the general case, a number of heuristics based on different mathematical platforms and metaheuristics based on mechanisms observed in nature, have been introduced. The commonly known heuristics are *list scheduling, critical path*, or *clustering* [3,4]. Recently metaheuristics such as *simulated annealing* (SA), *genetic algorithms* (GAs), *ant colonies, tabu search* (TS), or *neural networks* have been successfully applied [5,6].

The main problem related to heuristics and metaheuristics is the existence of the scheduling overhead represented by the cost of running the scheduler. Here, in the case of metaheuristics on which we focus our attention, the main source of the scheduling overhead is the necessity of calculation of a cost function in subsequent iterations of a scheduling algorithm. One of the main sources of the scheduling overhead is neglecting potential knowledge about the scheduling problem, which could be gained during solving its instances. The prevailing number of scheduling algorithms does not extract, conserve, and reuse any knowledge about the problem while solving instances of the scheduling problem.

The motivation of our work is to develop a framework for designing scheduling algorithms where knowledge about scheduling process can be extracted and reused while solving new instances of the scheduling problem. For this purpose we propose to use a recently emerged and very promising hybrid technique combining cellular automata (CAs) and GA, and a computational paradigm of artificial immune system (AIS) [7].

CAs are discrete dynamical systems made up of a large number of cells, which behave according to local rules. It is an interesting feature of these systems that although cells interact only locally, in a fully distributed manner, a complex global behavior can emerge. For this reason CAs are often used to model real-world phenomena [8]. CAs are also considered as models of highly parallel and distributed computations in multiprocessor and distributed systems [9]. They are used to find solutions of problems such as scheduling and resource management [10].

The main problem related to CAs is a huge space of local CA rules representing possible solutions of a problem. Therefore, most applications of CAs were a result of clever, but time-consuming hand designing rather than an oriented search. Only recent works [11,12] on applying evolutionary computation to search CA rules opened new possibilities. Results described in the literature show that CAs, combined with evolutionary techniques for discovering local rules, can be effectively used to find parallel and distributed solutions of complex global problems such as density classification task and synchronization task [11,12], and location management in mobile computing [13] or cryptography [14]. Recently it has been shown [15] that such a hybrid technique can be applied to discover scheduling algorithms. In this chapter, we extend this methodology and additionally propose to use AIS as a support for reusing discovered CAs scheduling rules to solve new instances of the scheduling problem.

The rest of this chapter is organized as follows. Section 17.2 presents the scheduling problem. Section 17.3 provides a background on CAs. Section 17.4 contains the description of the proposed CA-based scheduling system. Section 17.5 contains experimental results concerning CAs applied to scheduling in two-processor systems. Section 17.6 describes AIS for reusing knowledge conserved in CA rules while solving new instances of the scheduling problem. Section 17.7 presents the extension of the proposed approach on the case of multiprocessor systems consisting of more than two processors. Finally, Section 17.8 contains the conclusions.

17.2 Multiprocessor Scheduling

In our approach, we use some models of a multiprocessor system and a parallel program. They are both represented by corresponding graphs. A multiprocessor system is represented by an undirected unweighted graph $G_s = (V_s, E_s)$, called a *system graph*. V_s is the set of N_s nodes representing processors with their local memories. E_s is the set of edges representing bidirectional channels between processors and defines a topology of the multiprocessor system. Figure 17.1(a) shows an example of a system graph representing a multiprocessor system consisting of two processors P_0 and P_1. It is assumed that all processors that have the same computational power and communication via links do not consume any processor time.

A parallel program is represented by a weighted, directed acyclic graph $G_p = \langle V_p, E_p \rangle$, called a *precedence task graph* or a *program graph*. V_p is the set of N_p nodes of the graph representing elementary tasks. The weight b_k of the node k describes the processing time needed to execute a task k on any processor of the system. E_p is the set of edges of the precedence task graph describing the communication patterns between the tasks. The weight a_{kl} of the edge (k, l) describes the communication time between the pair of tasks k and l when they are located in neighboring processors. If the tasks k and l are located in the same processor than the communication time between them is equal to zero. Figure 17.1(a) shows an example

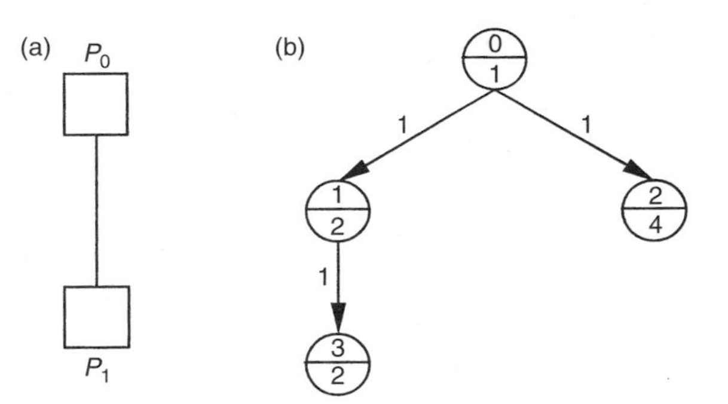

FIGURE 17.1 Examples of a (a) system graph and (b) precedence task graph.

TABLE 17.1 An Example of a Rule Table for a One-Dimensional, Binary State CA

η	000	001	010	011	100	101	110	111
a	0	1	0	1	1	0	1	0

of the system graph corresponding to the two-processor system. Figure 17.1(b) shows an example of the program graph consisting of four tasks with their order numbers from 0 to 3. All communication costs of the program graph are equal to one. Computational costs of tasks are 1, 2, 4, and 2, respectively.

The purpose of the *scheduling* is to distribute the tasks among the processors in such a way that the precedence constraints are preserved and the *response time* T (the total execution time) is minimized. Found optimal schedule is usually represented by a Gantt chart showing allocation of tasks to processors and time when a given task starts and finishes execution.

The response time T for a given precedence task graph depends on an *allocation* of tasks in multiprocessor topology and some *scheduling policy* [15], which defines the order of processing tasks, ready to run in a given processor.

17.3 Cellular Automata (CA)

These [12,16] are dynamical systems in which space and time are discrete. Here, we shall concentrate on one-dimensional CAs.

A one-dimensional CA consists of a spatial lattice of N cells, each of which, at time t, can be in one of k states. In this chapter, we assume that each cell is a Boolean variable: $a_i^t \in \{0, 1\}, i = 0, 1, \ldots, N - 1$. The collection of all local states is called the *configuration*. A CA has a single fixed rule, which is used to update each cell. The rule maps from the states in a neighborhood of a cell to a single state, which is the updated value for that cell. In a one-dimensional CA the neighborhood of a cell includes the cell itself and some *radius* r of neighbors on either side of the cell.

The equations of motion for a CA are often expressed in the form of a *rule table*: a look-up table listing in lexicographical order of each of the neighborhood patterns and the state to which the central cell in that neighborhood is mapped. For example, Table 17.1 presents one possible rule table for a one-dimensional, binary state CA with radius $r = 1$. Each possible neighborhood η is given, along with the output bit a to which the central cell is updated.

The CA starts out with some initial configuration (IC) of cell states. To run the CA, the rule table is applied, usually synchronously, to each neighborhood in the current lattice configuration, respecting the choice of boundary conditions, to produce the configuration at the next time step. The most popular boundary conditions are periodic boundary condition and null boundary condition. Periodic boundary

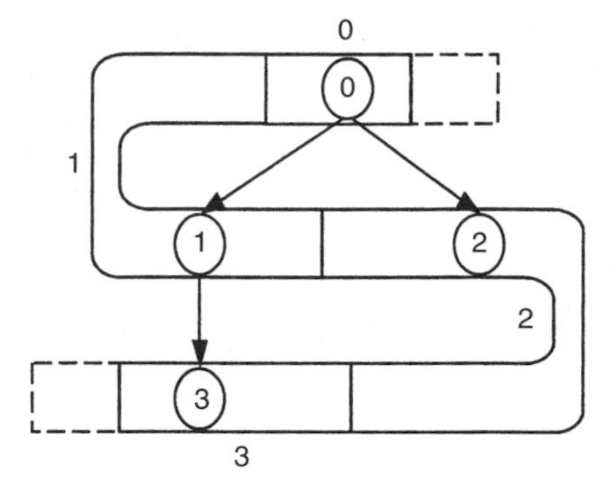

FIGURE 17.2 Program graph and corresponding CA.

condition means that the leftmost cell is considered to be the right neighbor of the rightmost cell and vice versa. Null boundary condition means that "absent" cells are always in state 0.

The behavior of CA is often illustrated using "space-time diagrams" in which the configuration of states in n-dimensional lattice is plotted as a function of time (in most cases space-time diagrams are practical only for $n \leq 2$).

17.4 CA-Based Scheduling Algorithm

To design a scheduling algorithm with the use of CA we assume that a nonlinear structure of a program graph is approximated by a one-dimensional CA of size N_p, in contrast to the approach presented in References 15 and 17, where a two-dimensional irregular structure of CA was used. The way of creating local neighborhood with the use of two-dimensional CA and associating them with program graphs was complex, and resulted in a constant size of the neighborhood that was equal to seven, a long CA rule with the length equal to 250 bits, and a huge space of possible solutions. Therefore, the use of one-dimensional CA seems to be a more simple and promising alternative. The choice of one-dimensional CA means that with each task of a program graph an elementary cell is associated (see Figure 17.2). We assume that the cell 0 is assigned to task 0, the cell 1 is assigned to task 1, etc. Dashed lines in Figure 17.2 mean dummy cells in the state 0, which means null boundary conditions.

Let us consider the problem of scheduling in the two-processor system. In this case, cells of CA corresponding to any program graph will be either in the state 0 or in the state 1. The state 0 or 1 of a cell means that a corresponding task is allocated in the processor P_0 or P_1, respectively. Each configuration of CA corresponds to some allocation of tasks in the system graph.

CA corresponding to the program graph will evolve according to its rule. Initial states of CA correspond to an initial allocation of tasks in the two-processor system. Changing states of CA will result in changing an allocation of tasks in the system, and changing the response time T. We want to know if there exists a rule for CA providing for any initial allocation of tasks converging CA to an allocation which minimizes T.

An architecture of CA-based scheduling system is presented in Figure 17.3. There are three modes of work of the scheduling system: a mode of learning CA scheduling rules, a mode of normal operating and a mode of reusing CA scheduling rules by applying AIS. CAs can update their states (CA mode) sequentially (*seq*), in parallel (*par*) or sequentially with a random order of updating cells' states (*seq–ran*). In sequential mode the CA works asynchronously, that is, at a given moment of time only one cell updates its state. An order of updating states by cells is defined by their order numbers corresponding to tasks in the precedence task graph. A single step of running the CA is completed in N_p steps. In parallel mode all cells update their states synchronously. The third mode works in the same way as sequential mode, but an order of updating cells' states is random.

The purpose of the *learning* mode (see, Figure 17.3, left) is to discover CA rules for scheduling. Randomly generated allocations of program graph tasks into a system graph serve as initial states of CA, which then runs according to a rule from GA population of rules. GA searches a CA rule, which will be able to evolve to a final configuration corresponding to an optimal or suboptimal value of the response time T, for the input allocation of the program tasks. The algorithm for discovering CA rules with use of GA [11,18] is presented.

Learning mode: discovering CA rules using GA

BEGIN

create an initial population of rules of size P;

FOR $l = 1$ **TO** G **DO**

BEGIN

create a set of size I of test problems

FOR $i = 1$ **TO** P **DO**

BEGIN

$T_i^* = 0$;

FOR $j = 1$ **TO** I **DO**

$T_i^* = T_i^* + \text{CA}(rule_i, test_j, CA\ mode, M\ steps)$;

$T_i^* = T_i^*/I$;

END;

sort current population of rules according to T_i^*;

move E of the best individuals to the next population;

FOR $i = 1$ **TO** $\lfloor (P - E)/2 \rfloor$ **DO**

BEGIN

$rule_1^{parent}$ =select();

$rule_2^{parent}$ =select();

$(rule_1^{child}, rule_2^{child})$ =crossover$(rule_1^{parent}, rule_2^{parent})$;

mutation$(rule_1^{child}, rule_2^{child})$;

END;

END;

END;

GA begins with a population of P randomly generated CA rules. The length of the rule is calculated as $L = k^{2r+1}$. We then create a set (different for each generation of GA) of size I of test problems (ICs of CA) by generating a set of random allocations of tasks in multiprocessor system for a given instance of a problem. Each rule i is tested on a whole set of the set problems. The function CA() returns for a rule i the response time T_i^j obtained by CA running M steps under given mode of CA (i.e., *seq*, *seq–ran*, or *par*) on a test problem j. In our experiments, we set $M \approx 4 \times N_p$. M can be calculated by some heuristic based on many observations of behavior of CAs. T_i^j is calculated once, after the last step of running CA, when

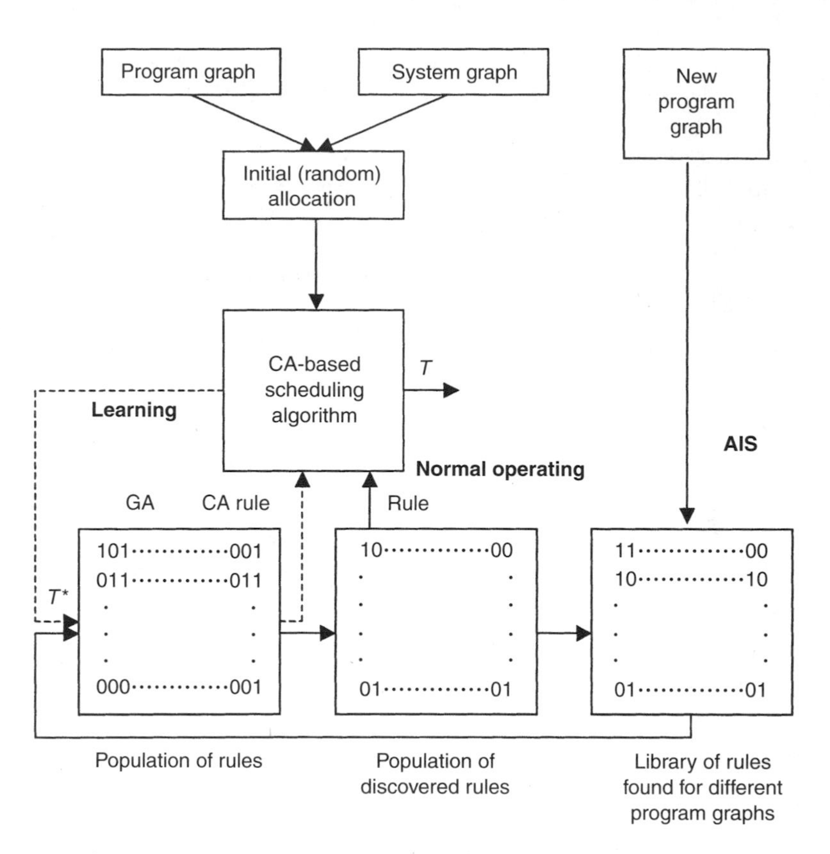

FIGURE 17.3 An architecture of CA-based scheduling system.

the CA final configuration of states corresponds a final allocation of tasks. As a rule fitness of the rule i, a value T_i^* is accepted, which is the sum of values T_i^j averaged over the number I of solved tests, that is, $T_i^* = \sum_{j=1}^{I} T_i^j / I$.

After calculation of the fitness function for all rules, GA operators are used. A number E of the best rules ("*elite*") is copied without modification to the next generation. The remaining $P - E$ rules for the next generation are formed by two-point *crossover* operator applied to randomly chosen pairs of elite rules, which are mutated next with probability p_{m}. This process is continued over a predefined number of generations G and when completed the discovered rules are stored.

In the *normal operating* mode (see Figure 17.3, middle), when a program graph is initially, randomly allocated, CA is initiated and equipped with a rule taken from the set of discovered rules. In this mode we expect that for any initial allocation of tasks of a given program graph, CA will be able to evolve very fast, without time-consuming calculation of T, to a configuration corresponding to the minimal or near minimal value of T.

In the third mode AIS (see Figure 17.3, right), enables a potential reusing of discovered knowledge stored in CA rules. It has an important meaning when trying to reschedule. The concept of AIS, along with experimental results, is presented in Section 17.6.

17.5 Two-Processor Scheduling with CA: Experimental Results

17.5.1 Scheduling Deterministic and Random Program Graphs

A number of experiments with program graphs that available in the literature (see Reference 15) has been conducted. In addition, a number of random graphs were used, with 25 and 50 (in average) tasks. The

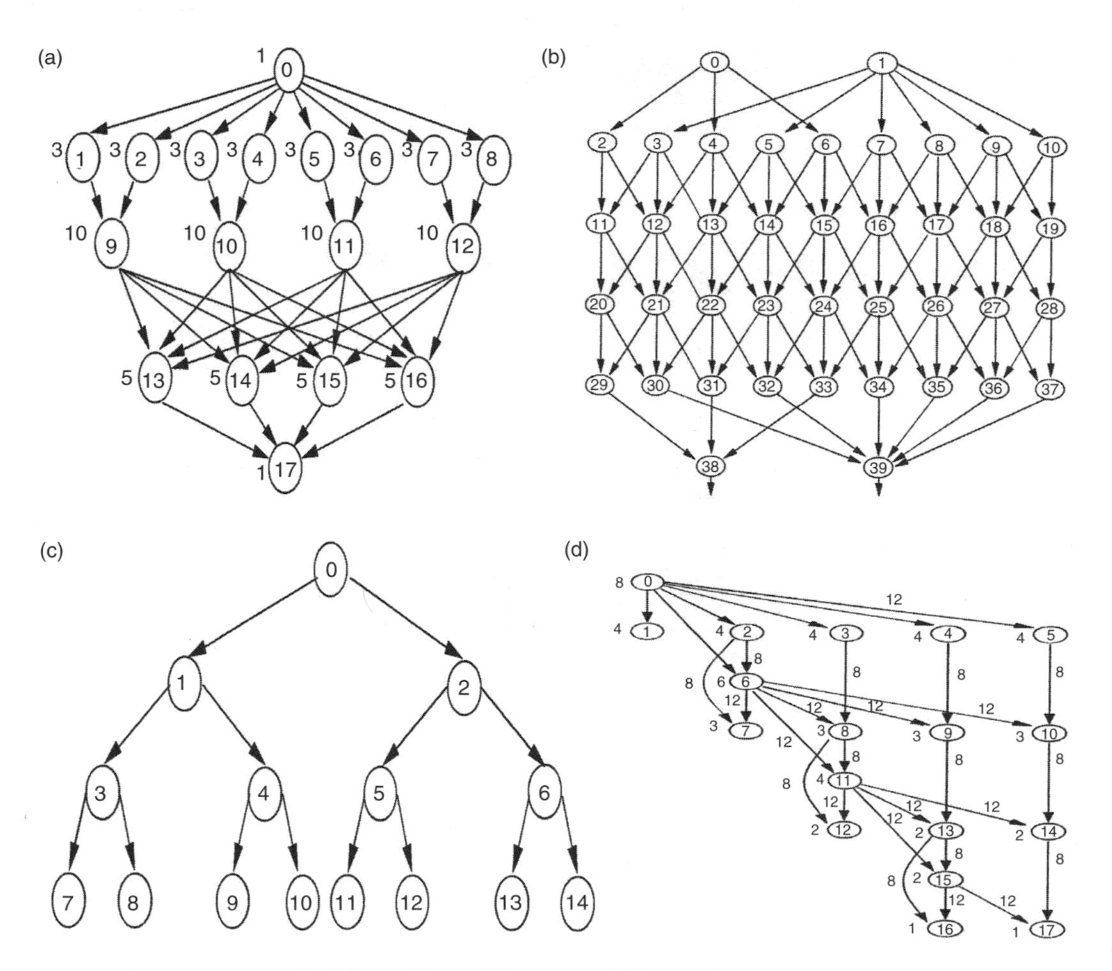

FIGURE 17.4 Program graphs: (a) *g*18, (b) *g*40, (c) *tree*15, and (d) *gauss*18.

experiments were performed on PC, Celeron 745 MHz. We used CAs with a radius of neighborhood $r \in \{1, 2, 3\}$, what corresponds to rule lengths equal to 8, 32, and 128, respectively. We used the following parameters in the experiment: a rules population size $P = 50 \div 200$, elite size $E = 15 \div 100$, a size of test problems $I = 10 \div 50$, a number of CA steps $M = 4 \times N_p$ (N_p — a number of tasks), and mutation probability $p_m = .03$. In this section, we describe in detail the results obtained for the following deterministic program graphs: *g*18, *g*40, *tree*15, and *gauss*18 (see Figure 17.4) and for the randomly generated program graph called *Rnd*25_5.

The first program graph used in experiments is *tree*15. It is a binary tree consisting of 15 tasks. All computational and communication costs are the same and equal to one. The optimal response time T for *tree*15 in the two-processor system is equal to nine. Experiments have shown that for program graphs from family tree it is sufficient to use $r = 1$ to discover scheduling rules. Rules are discovered in a few generations of GA for all three modes of the CA, that is, *seq*, *par*, and *seq–ran*. Figure 17.5 shows typical runs of the CA-based scheduler with the best rule from the final generation, starting from randomly generated initial configuration.

The next experiment is conducted with the program graph *g*40 (see, e.g., Reference 15), which is composed of 40 tasks, with computational and communication costs equal to four and one, respectively. The optimal response time T for *g*40 in the two-processor system is equal to 80. The minimal value of r, which allows in the learning mode for *g*40 converging to optimal value of T is $r = 2$. Figure 17.6(a) shows a typical run of the scheduling system in the learning mode. One can see that for two modes of updating

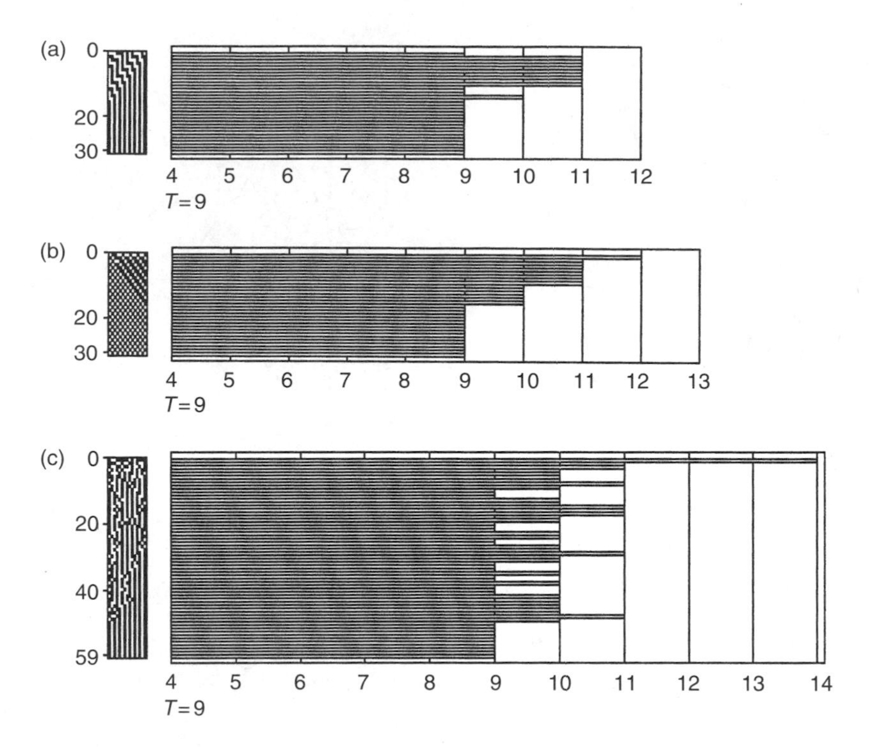

FIGURE 17.5 Space-time diagrams of CA-based scheduler for *tree*15: (a) sequential mode, (b) parallel mode, and (c) sequential mode with a random order of updating cells' states.

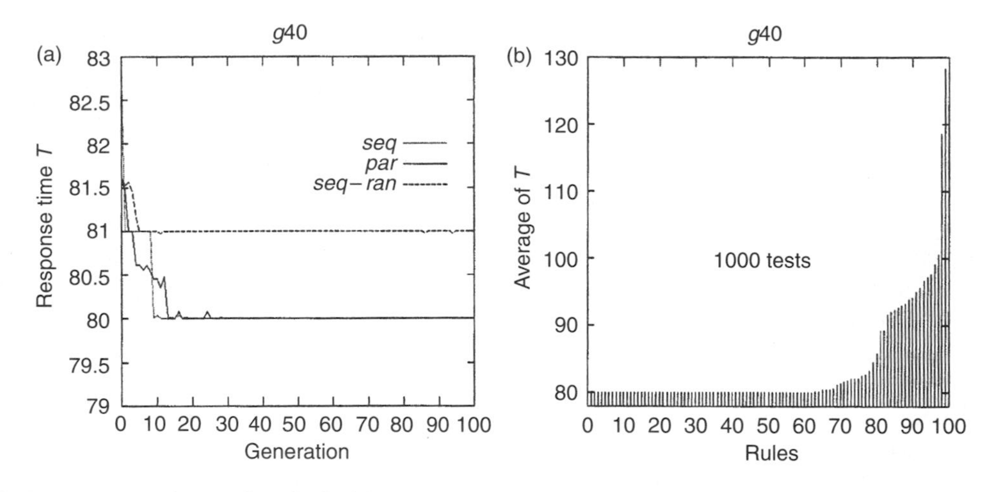

FIGURE 17.6 Running CA-based scheduling system for program graph *g*40: (a) learning mode and (b) normal operating mode.

CA rules, *seq* and *par*, the system discovers rules converging CA to a configuration corresponding to optimal value of $T = 80$, and for *seq–ran* mode the system converges to rules providing suboptimal value of T. It takes <20 generations of GA.

After the run of the system in the learning mode the final population of GA contains rules suitable for CA-based scheduling. We can find out a quality of these rules in the normal operating mode. We generate 1000 random ICs (program tasks' allocations) and use them to schedule each test problem by each of the rules found. Figure 17.6(b) presents the average value of T for each rule over a whole set of test problems. One can see that about 60% of discovered rules are able to find an optimal schedule for each test problem.

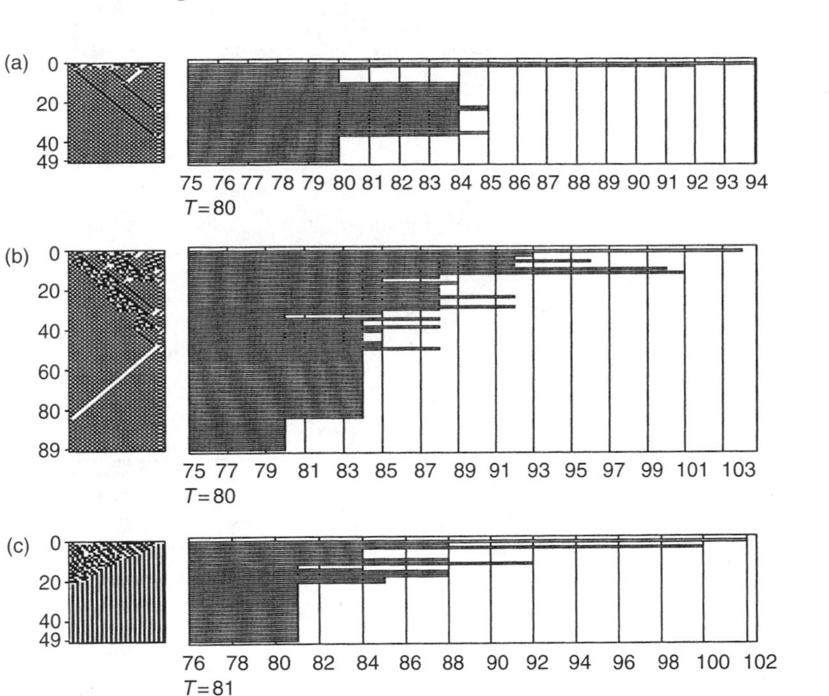

FIGURE 17.7 Space-time diagrams of CA-based scheduler for $g40$: (a) CA in *seq* mode, (b) *par* mode, and (c) *seq–ran* mode.

The scheduling in this mode is performed automatically, very quickly, without calculation of the cost function T, and can be observed by watching a space-time diagram of CA.

Figures 17.7(a)–(c) show space-time diagrams for typical runs of the CA-based scheduler working with the best rule found for $g40$, and performing scheduling in the normal operating mode for a single, randomly generated test problem, with CA working in *seq*, *par*, and *seq–ran* modes, respectively. Left part of each figure presents a space-time diagram of the CA consisting of 40 cells and right part shows graphical values of T corresponding to the allocations found in a given step. Let us consider Figure 17.7(a). One can see that in the step 0 (the first row), cells of the CA are in some initial states corresponding to the initial allocation of tasks (white cell — a corresponding task is allocated to P_0, black cell — a task is allocated to P_1) and the value of T corresponding to this allocation is equal to 94. Then the CA starts to run, changing its states sequentially according to its rule, what results in changing values of T. Each subsequent row of the figure shows a situation after changing states of 40 cells. One can see that the CA needs 38 steps to converge to tasks' allocation corresponding to the minimal value of $T = 80$. During which CA running no calculation of a cost function is performed. The 38 steps of the CA requires $38 \times 40 = 1520$ time steps, in which sequentially performed elementary entries to a simple look-up table are executed to update states of CA cells, and it corresponds to a very low computational cost (see, Table 17.2).

Figure 17.7(b) shows a space-time diagram of CA-based scheduler working in *par* mode. In this case the CA converges to the optimal value of T after 84 steps, which is equivalent to 84 time steps, because 40 updates of the look-up table are performed in parallel in each step. Figure 17.7(c) presents a typical run of the CA-based scheduler working in *seq–par* mode, with the suboptimal value of T equal to 81 found in 20 steps of CA.

The next program graph used in experiments, referred as *gauss*18 (see, e.g., Reference 15), represents the parallel Gaussian elimination algorithm consisting of 18 tasks. The optimal response time T for this program graph in the two-processor system is equal to 44. Despite the lower number of tasks when compared with the $g40$, the *gauss*8 is much more difficult for the learning mode because of its nonregular

TABLE 17.2 Performance of CA-Based Scheduling System

Program graph	Learning mode		Metaheuristics			Normal operating mode
	seq CA	*par* CA	GA	SA	TS	*seq* CA
*tree*15	9, 10	9, 55	9, 0.03, 0.003	9, 3.57	9, 0.086	9, -, 0.002
*g*18	46, 9	46, 38	46, 0.04, 0.006	46, 2.93	46, 0.197	46, -, 0.002
*g*40	80, 50	80, 125	80, 0.08, 0.03	80, 3.29	80, 0.665	80, -, 0.015
*gauss*18	44, 380	46, 1500	44, 1, 0.17	49, 2.32	49, 1.277	44, -, 0.003
*Rnd*25_1 (31)	495,1880	507, 5720	495, 0.4, 0.16	495, 1.74	495, 10.36	495, -, 0.011
*Rnd*25_5 (17)	94, 720	94, 1600	94, 41, 8.2	100, 3.29	95, 14.35	94, -, 0.005
*Rnd*25_10 (21)	62, 25	62, 131	62, 24.37, 0.58	62, 0.07	62, 0.79, -	62, -, 0.006
*Rnd*50_1 (59)	895, 4110	901, 9680	890, 4, 5.32	891, 2.64	890, 55.88	895, -, 0.023
*Rnd*50_5 (62)	225, 4365	238, 12790	203, 220, 290.4	228, 4.29	214, 70.68	225, -, 0.024
*Rnd*50_10 (53)	146, 6480	146, 21350	135, 29.33, 11.44	138, 6.19	145, 97.70	146, -, 0.019

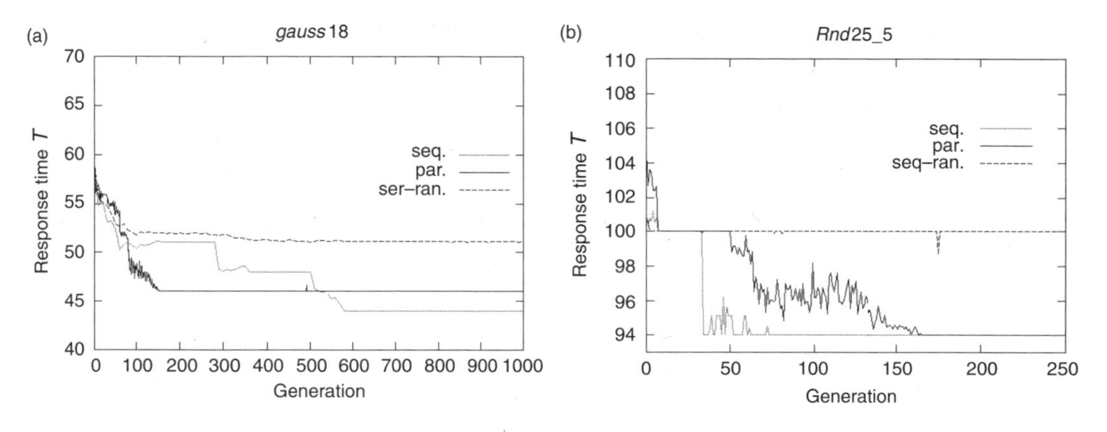

FIGURE 17.8 Learning mode of the scheduling system: (a) for deterministic program graph *gauss*18 and (b) random graph *Rnd*25_5.

structure. The minimal value of r, which allows for *gauss*18 converging in the learning mode to optimal value of $T = 3$, what results in 128-bit long CA rule and a large space of possible rules.

Figure 17.8(a) shows typical runs of the GA in the learning mode for three modes of operating of the CA. One can see that only for *seq* mode of CA the GA is able to find, after about 600 generations, optimal CA rules. However, in the normal operating mode, the best-discovered rule needs only 18 steps to schedule, so the performance in this mode is similar as the performance shown for the *g*40.

Figure 17.9(a) shows a typical run of the CA-based scheduler working in sequential mode with the best rule from the final generation, starting from randomly generated IC. One can see that the CA needs <20 time steps to converge to tasks' allocation corresponding to the minimal value of $T = 44$. Figure 17.9(b) shows space-time diagram of CA-based scheduler working in parallel mode. In this case the CA converges to suboptimal value of $T = 46$. Figure 17.9(c) presents typical run of the CA-based scheduler working in sequential mode with a random order of updating cells' states, with the value of $T = 51$.

Last experiment presented in this section is performed on randomly generated program graph called *Rnd*25_5 (with 17 tasks). The optimal response time T for *Rnd*25_5 in the two-processor system is equal to 94. In the learning mode $r = 3$ is required. Figure 17.8(b) shows typical runs of the GA for three modes of operating of the CA. For *seq* mode, the GA discovers optimal CA rules after about 50 generations. For *par* mode, it takes above 150 generations. For *seq–par* mode, the GA does not discover optimal CA scheduling rules. Figure 17.10 shows typical runs of the CA-based scheduler with the best rule from the final generation, starting from randomly generated IC.

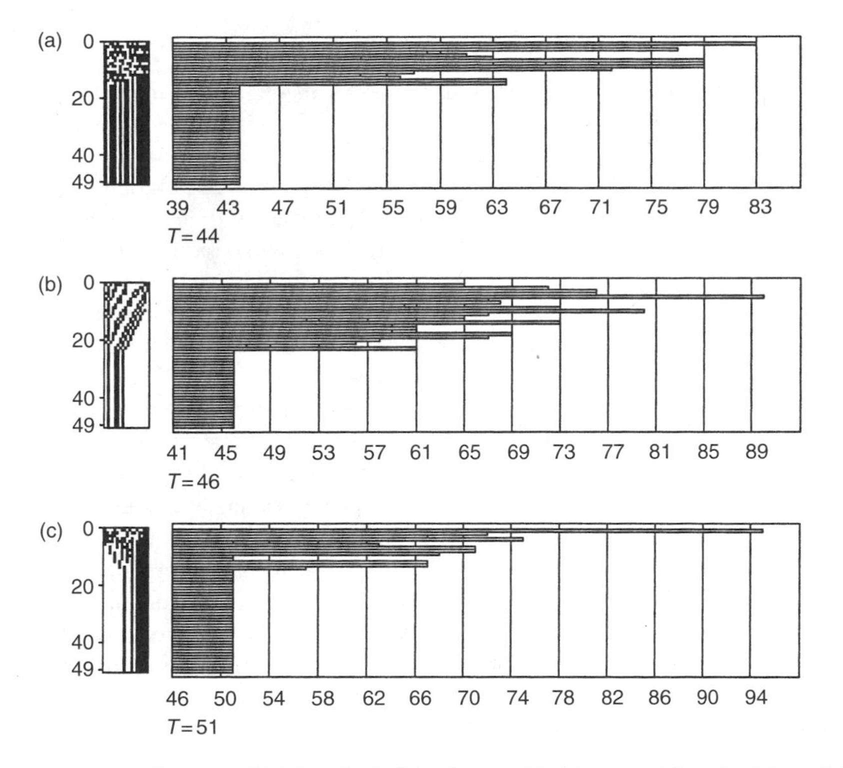

FIGURE 17.9 Space-time diagrams of CA-based scheduler for *gauss*18: (a) sequential mode, (b) parallel mode, and (c) sequential mode with a random order of updating cells' states.

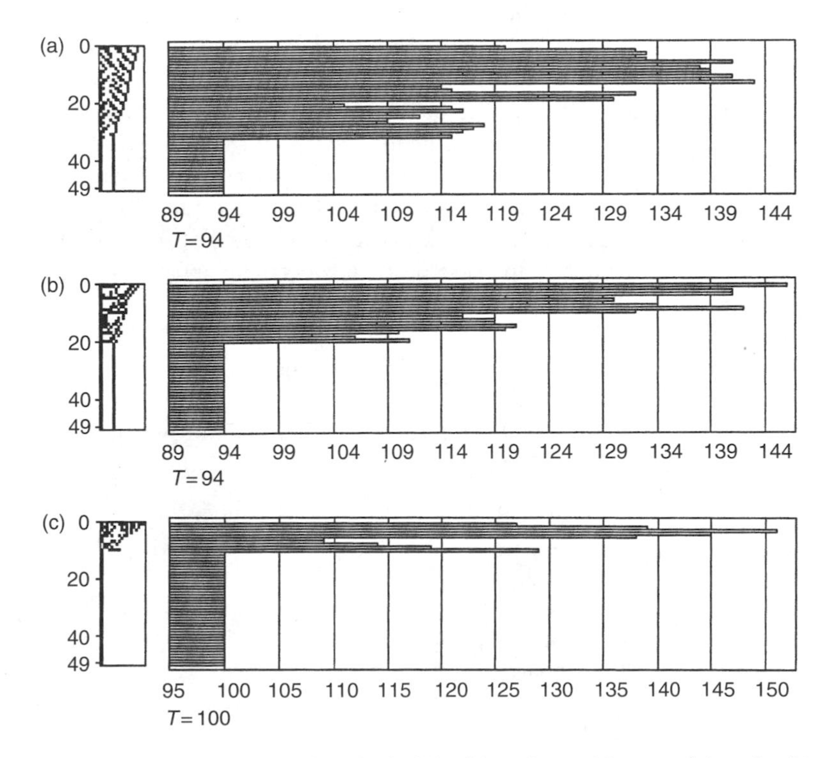

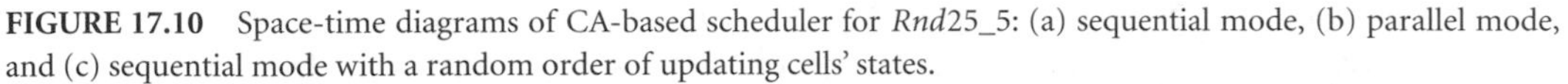

FIGURE 17.10 Space-time diagrams of CA-based scheduler for *Rnd*25_5: (a) sequential mode, (b) parallel mode, and (c) sequential mode with a random order of updating cells' states.

17.5.2 Performance and Complexity Issues

Table 17.2 summarizes results (averaged on ten runs) of experiments conducted with use of deterministic and random program graphs (*Rnd* from 17 to 62 tasks), for the learning mode (*seq* and *par* mode of CA) and normal operating modes (*seq*). The results are compared with the ones obtained with use of other metaheuristics, such as standard GA, SA, and TS. Each result for a given program graph is represented by a pair: the best found response time T, and a required number (in thousands) of calculation of cost function T; for GA and normal operating mode the time (in seconds) of running an algorithm on PC is provided.

We can see that for a prevailing number of program graphs the scheduling system was able to discover CA rules providing optimal or suboptimal value of T. Only for random program graphs we obtained the worst results. These values are validated with use of metaheuristics.

The procedure of discovering CA rules is computationally, relatively expensive, because the learning process requires the presentation of a number of test problems and calculation of the cost function T, which is the main source of the complexity of the learning algorithm. The complexity of the algorithm expressed by a number of the cost function calculations depends (see Table 17.2) not only on a number of tasks, but also on its topology and relation between communication and computational costs in a given program graph. However, this procedure is used once. When learning is completed, discovered rules store the knowledge, which can be reused while solving new instances of the scheduling problem. In general, better results have been obtained for the sequential mode of CA. A reason for this is that a calculation of the cost function T, which in fact is performed by a simulation of execution of a program graph in a system graph, is a sequential algorithm defined by information flow in an oriented program graph. A parallelization of this process is easier and computationally less expensive for graphs with a regular structure. A cost of evolving rules for *par* mode of CA is a few times greater than for *seq* mode.

Metaheuristics were used to find an optimal value of T for different program graphs used in the learning mode. One can see that none of the metaheuristics offer a dominating quality of solutions or performance, which is a consequence of *no free lunch* theorem [19]. Note that the cost of running them defined by a number of calculation of the cost function T is much lower than the cost of rules discovered in the learning mode. However, this cost must be paid regularly each time the metaheuristics are used.

Results of experiments conducted in the normal operating mode (*seq* CA mode) clearly show the advances of the proposed approach. Scheduling is performed without an expensive procedure of calculation of the cost function T and for larger program graphs it may be a few orders of magnitude faster than for example, scheduling performed by GA. The scheduling algorithm is fully distributed — scheduling is the result of local updating states of CA cells. The main component of the algorithm complexity is a simple operation on the look-up table to read a decision concerning updating of a cell state, and this complexity grows linearly with the number of tasks.

Proposed CA-based scheduling algorithm has an interesting feature. Since the coding CA rules does not depend on a program graph, there is a technical possibility of reusing the discovered rules, in the normal operating phase, on other program graphs.

The question that now arises is whether discovered CA rules have abilities of scheduling other program graphs. The goal of experiments described in Section 17.5.3 is to examine this issue.

17.5.3 Reusing Discovered Rules

In the first experiment we used rules discovered for the program graph $g18$ (see Figure 17.4[a] to find solutions of other program graphs constructed on the base of $g18$. The program graph called $g36$ was constructed through linking $g18$ to its final task. The weight of the linking edge was set to zero. The program graph called $g54$ was constructed through linking $g18$ to the final task of $g36$ and setting the weight of the linking edge to zero. In a similar way other program graphs ($g72$, $g90$, and $g108$) were constructed.

Then we used five of the best CA rules found for the program graph $g18$ in the normal operating phase on 1000 random initial allocations of our new program graphs. In this experiment $r = 2$ and two modes

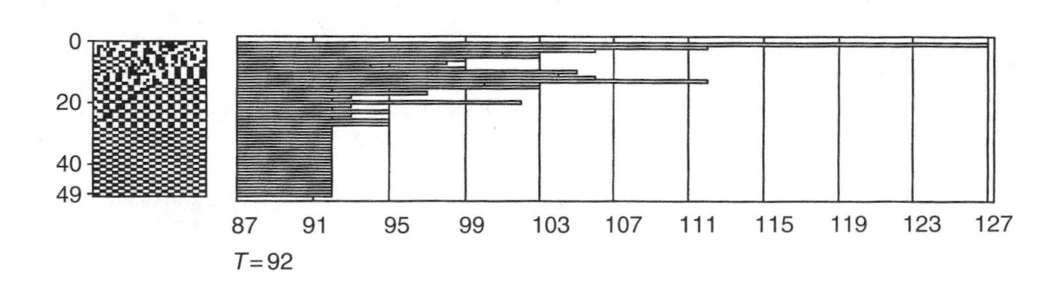

FIGURE 17.11 Space-time diagrams of CA-based scheduler for *g*36: sequential mode.

TABLE 17.3 The Best Response Time *T* Received for Different Program Graphs

Rule\graph	*tree*7	*intree*7	*g*18	*g*40	*gauss*18	*Rnd*25_5
*tree*7	5.0	5.0	48.04	85.9	64.58	119.7
*intree*7	5.0	5.0	47.0	81.0	63.35	124.33
*g*18	5.02	5.08	46.0	84.17	63.93	122.49
*g*40	5.0	5.0	49.01	80.0	61.32	113.65
*gauss*18	5.26	5.11	52.93	92.23	44.0	117.82
*Rnd*25_5	5.74	5.14	51.04	88.07	56.0	94.0
opt.*T*	5	5	46	80	44	94

of operating a CA were used: sequential and parallel mode. The experiments conducted have shown that the best rules obtained for the program graph *g*18 can successfully find an optimal scheduling for each representative of the test. These solutions are obtained without a process of discovering CA rules and without using GA to find optimal schedules. It means that the time required to find optimal solutions for tested program graphs is significantly reduced.

Figure 17.11 shows typical run of the CA-based scheduler, working in sequential mode, for the program graph *g*36, with the best rule discovered for the program graph *g*18. The optimal response time *T* for the program graph *g*36 is equal to 92.

The next question that arises is whether the discovered rules are sensitive to some modifications of our program graph. To find out, discovered rules were used in the normal operating phase (assuming sequential mode of a CA) to find solutions of some other program graphs. These graphs were constructed from *g*18 by introducing some random modifications to it. These modifications included changing the values of the weights of some randomly chosen tasks or (and) edges. This way we obtained 30 new program graphs. In most of the cases the best discovered rules were able to find an optimal (or near optimal) scheduling for each representative of the test. These solutions are obtained without a process of discovering rules and without using GA to find optimal schedules.

The last experiment had a more general character than the two previously described. We wanted to know if the discovered rules had possibilities of scheduling other program graphs. We took the whole population of rules discovered for the following program graphs: *tree*7, *intree*7, *g*18, *g*40, *gauss*18, and *Rnd*25_5. These populations of rules were discovered assuming $r = 3$ and sequential mode of a CA. The course of the experiment was identical as in the experiments described earlier. The populations of rules discovered for these program graphs were tested (in the normal operating mode, assuming $I = 1000$) on other program graphs the following way: "each population of rules on each program graph." Obtained results of the best rules are presented in Table 17.3. These are typical results obtained in a given test. The last row of Table 17.3 contains the optimal response times *T* for tested program graphs.

Results presented in Table 17.3 indicate that in some cases rules discovered for a given program graph have possibilities of scheduling other program graphs. The best rules in the tested populations can be used to obtain optimal (or near optimal) schedules of the program graph *tree*7 and *intree*7. This fact perhaps is not surprising because tree-structured program graphs are the easiest instances. More interesting is that rules discovered for tree-structured program graphs are suitable for scheduling program graphs *g*18 and *g*40. On the other hand, rules discovered for *gauss*18 and *Rnd*25_5 operate on the program graphs (with the exception of trees) weaker than other tested rules.

Presented results indicate that discovered CA rules store knowledge about a way of solving instances of the scheduling problem. Typically, rules specialize in solving the scheduling problem for a specific program graph. Some of them are "more general" than others. They have limited possibilities to schedule other program graphs. These observations lead to a concept of AIS, which would support reusing knowledge stored in discovered CA rules.

17.6 AISs and Rescheduling

17.6.1 A Concept of AIS for Rescheduling

Natural immune systems are adaptive systems in which learning takes place by evolutionary mechanisms similar to biological evolution. AIS is a new computational paradigm based on natural immune system. Different approaches to modeling AISs are presented in the literature [20–22]. We will use a simple GA-based model of AIS studied by Forrest et al. [21]. They use two populations: antigens (foreign material) and antibodies (the cell that perform the recognition). Binary strings represent individuals in these populations and recognition is highly simplified and modeled as string matching. GA is used to evolve population of antibodies that match specific antigens well.

Artificial immune systems (AISs) have been applied in many domain [7], for example, to create anti-virus programs; detect intrusions in computer networks, in pattern recognition, data compression, and optimization. In scheduling domain, an AIS, in conjunction with a GA, was used to produce effective schedules in a factory environment [23].

In our approach rules discovered by the GA are interpreted as antibodies and program graphs — as antigens. A given antibody (a rule) "recognizes" a specific antigen (a program graph) if it can find an optimal or near optimal schedule for it. When a new scheduling instance arrives to the system, the AIS will form a population of antibodies — rules from the stored library, and next run the population with use of GA (the same as in the learning mode) to evolve faster scheduling rules for the new problem.

Figure 17.12 presents an idea of our AIS used for rescheduling. In our approach, rules discovered for different program graphs (the whole populations) are stored in the library. When we want to discover

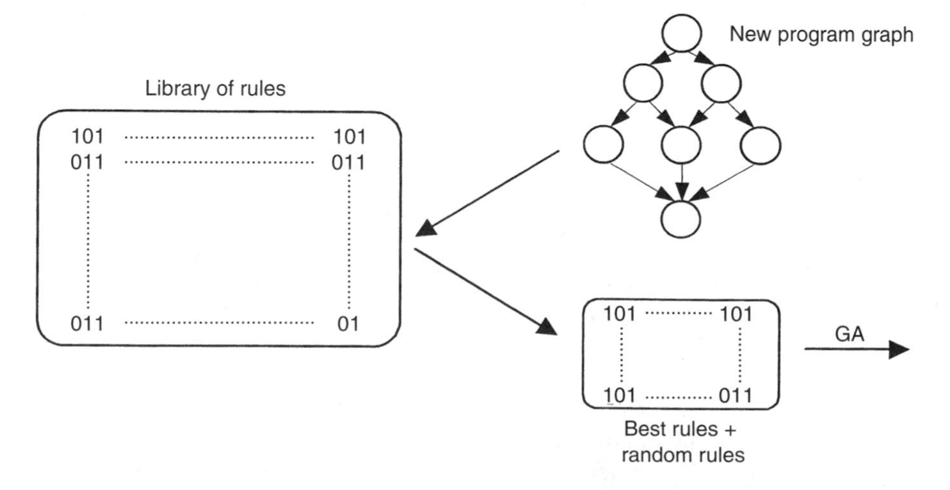

FIGURE 17.12 An idea of the AIS for rescheduling.

rules for a new program graphs, first, we choose from the library some number of the best rules (those which the best "recognize" a new program graph). Then we add to these rules some number of random rules.

An initial population of rules is created as follows: First, we run the CA-based system in the normal operating mode applying stored rules from the library to evaluate schedules T proposed by them. As shown earlier (see Table 17.2) this mode of work is characterized by a very low computational cost. Next, we select the initial population of AIS $d \cdot P$ the best rules, where $d \in (0, 1)$. Remaining $(1 - d) \cdot P$ rules are randomly generated. Finally, the whole population of AIS rules is evolved by GA in the learning mode.

It is important to underline that all rules in the library have the same length (they were discovered assuming $r = 3$). It enables applying genetic operators on rules from different program graphs.

17.6.2 Experimental Results

An experiment is conducted with the program graph called *Rnd*25_1*g*18 shown in Figure 17.13. It was constructed through linking the program graph *g*18 to final tasks of *Rnd*25_1 and setting the weights of linking edges to zero. Both components of *Rnd*25_1*g*18 are known to the system, but their composition is a new instance for scheduling. The optimal response time T for *Rnd*25_1*g*18 is equal to 541. The library of AIS rules is composed of already discovered rules (see Table 17.2).

Preliminarily, we test on *Rnd*25_1*g*18 in the normal operating mode rules found earlier for *Rnd*25_1 and *g*18. We can see (see Figure 17.14) that none of these rules provide an optimal solution. We also run the

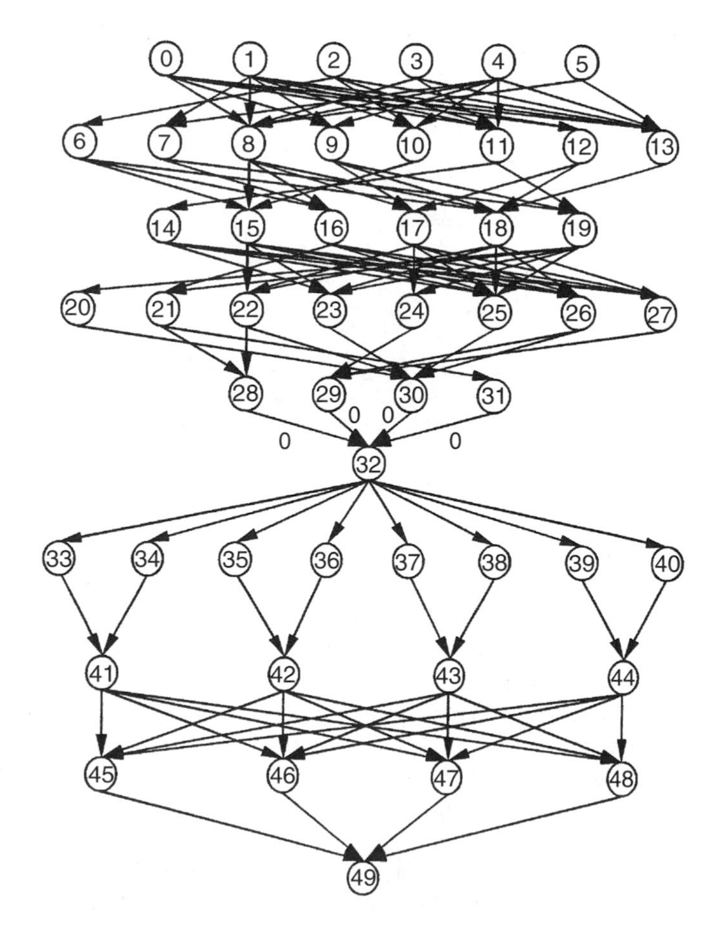

FIGURE 17.13 Program graph *Rnd*25_1*g*18.

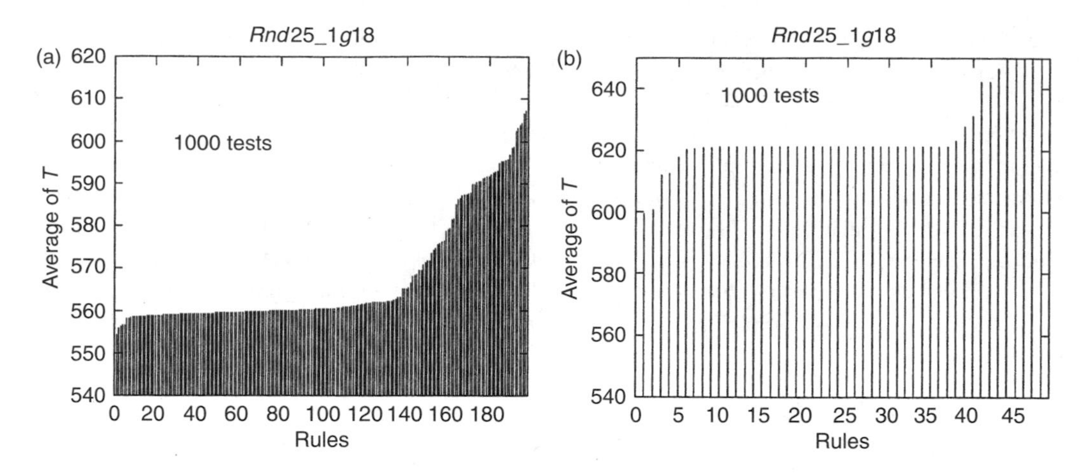

FIGURE 17.14 Normal operating phase for *Rnd*25_1*g*18: (a) population of rules discovered for *Rnd*25_1, (b) population of rules discovered for *g*18.

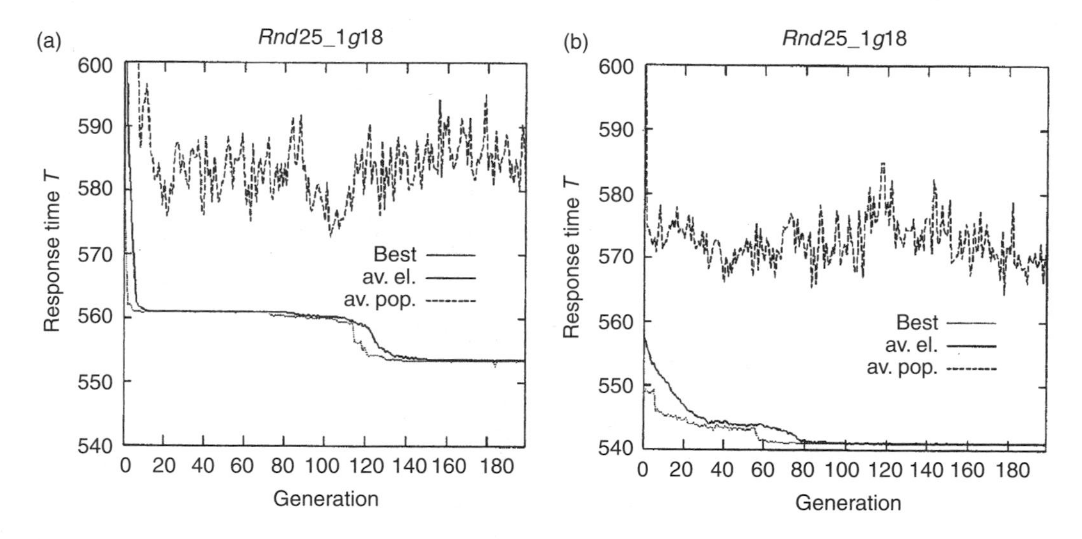

FIGURE 17.15 Learning mode for *Rnd*25_1*g*18: (a) randomly generated initial population (b) AIS with population of rules from the library.

scheduling system in the standard learning mode with randomly created CA rules. We set $P = 200$, $E = 50$, and $p_{\mathrm{m}} = .03$. We can see (Figure 17.15[a]) that the system can discover, during 200 generations, rules providing only suboptimal solutions.

Finally, we run AIS in the learning mode. The initial population of AIS rules was composed of 60% of rules from the library, after evaluating them in the normal operating mode, and the remaining rules were randomly generated. We can see (Figure 17.15[b]) that the behavior of AIS is different when compared with the behavior of the system in the standard learning mode. AIS that uses the knowledge stored in rules library can discover the optimal CA scheduling rule after about 60 generations.

Figure 17.16 shows typical run of the CA-based scheduler with the best-discovered rule, starting from randomly generated IC. One can see that in the step 0, cells of the CA are in some states corresponding to the allocation of tasks and the value of T corresponding to this allocation is >656. Then the CA starts to change its states sequentially, which results in changing values of T. One can see that the CA needs <70 time steps to converge to tasks' allocation corresponding to the minimal value of $T = 541$.

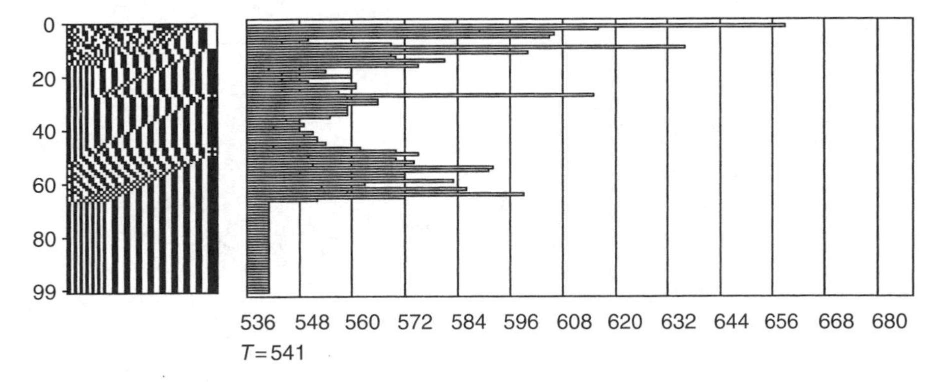

FIGURE 17.16 Space-time diagram of CA-based scheduler for *Rnd*25_1*g*18.

TABLE 17.4 The Best Response Time T for 3, 4, and 8 Processors

Program graph	$k = 3$	$k = 4$	$k = 8$
*tree*15	7.0 (7)	7.0 (7)	7.0 (7)
*g*18	38.0 (36)	27.0 (26)	26.0 (24)
*g*40	57.0 (57)	46.0 (45)	39.0 (33)
*gauss*18	48.0 (44)	52.0 (44)	52.0 (44)

Other conducted experiments have shown the advantage of AIS operating on rules from the library, significantly accelerating the process of discovery rules for new program graphs, which were partially presented earlier to the system.

17.7 Scheduling with More than Two Processors

When considering scheduling problem in the case of more than two processors, we are faced with the major problem — the problem of the length of the rule. The length of the rule L is calculated as $L = k^{2r+1}$. We can see that the length of the rule grows very quickly with k and r. It results in increasing the search space of possible solutions and creates a complex combinatorial optimization problem.

Table 17.4 presents the best response time T obtained for some program graphs (*seq* mode of CA) in the case of 3, 4, and 8 processors (assuming a fully connected system). They are compared with results obtained by a standard GA (in brackets).

One can see that the system can discover CA rules providing the optimal scheduling of *tree*15 for all considered multiprocessor systems. For remaining program graphs, under given setting parameters ($r = 1$, $P = 300, E = 100, I = 50$, and $p_{\mathrm{m}} = .03$), the system discovered rules providing suboptimal solutions.

Figure 17.17 shows a typical run of GA for the program graph *g*40 and multiprocessor system consisting of eight processors. In this case the optimal response time $T = 33$. We can see that GA needs about 200 generations to find CA rules providing suboptimal scheduling with $T = 39$.

Figures 17.18 to 17.20 show typical runs of the CA-based scheduler in the normal operation mode, with the best rule from the final generation, for the program graphs *g*40 and 3-, 4-, and 8-processor systems, respectively starting from randomly generated IC. Note that there is another important feature of the CA-based scheduler in the normal operating mode. The system is not sensitive to the increasing number of processors. Scheduling is performed in a number of steps similar to the case of two-processor system.

The ways of increasing the efficiency of the proposed scheduling system is the subject of our current research. It includes using new definitions of CA rules and applying nonuniform CA [16].

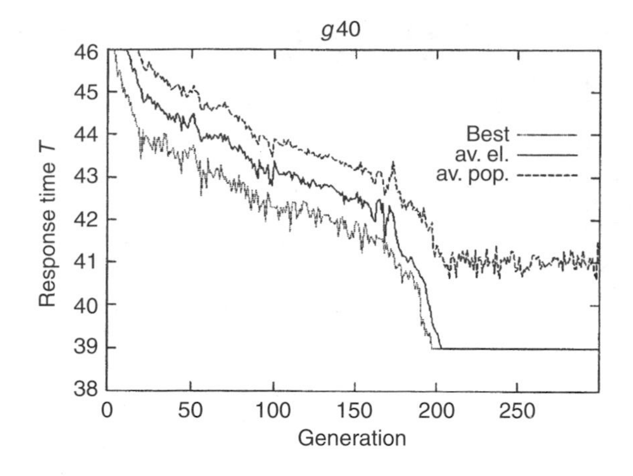

FIGURE 17.17 Learning mode for $g40$ and eight processors.

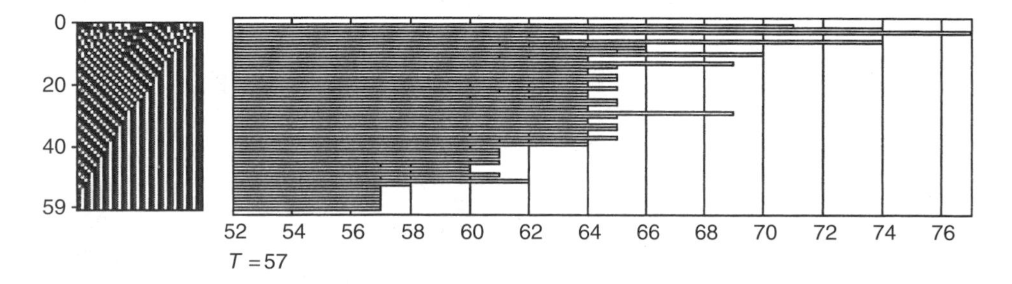

FIGURE 17.18 Space-time diagrams of CA-based scheduler for $g40$ and three processors: sequential mode.

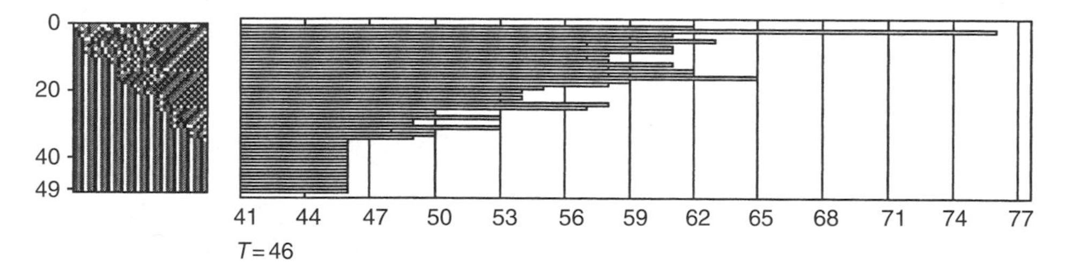

FIGURE 17.19 Space-time diagrams of CA-based scheduler for $g40$ and four processors: sequential mode.

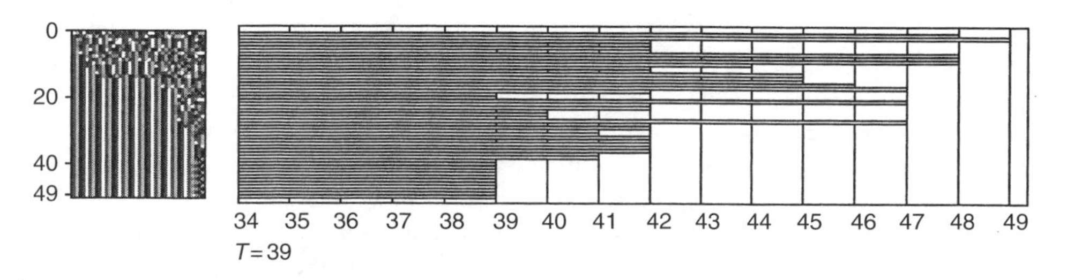

FIGURE 17.20 Space-time diagrams of CA-based scheduler for $g40$ and eight processors: sequential mode.

17.8 Conclusions

In this chapter we have presented the results of research on developing CA-based scheduler, working in sequential and parallel modes. They show that the GA can discover CA rules suitable to solve the scheduling problem for a given instance of the problem. This phase of the algorithm is called the learning phase. In this phase, knowledge about solving a given instance of the scheduling problem is extracted and coded into CA rules. Discovered rules are used in the normal operating phase by CA-based scheduler for automatic scheduling.

However, the main effort was directed to the question of a potential reusing of discovered CA rules. To solve this problem we proposed the immune system approach implemented by GA, which matches the currently available knowledge in discovered and stored CA rules with new instances of the scheduling problem and quickly recognizes familiar building blocks in these instances.

Proposed hybrid technique opens very promising possibilities in developing parallel and distributed scheduling algorithms and reducing their complexity.

Future work will include the expanding of the immune system concept. Perhaps it is possible to keep in the library not the whole chromosomes, as in the current approach, but only selected genes. It might help in further reducing the complexity of the algorithm.

References

[1] J. Błażewicz, K.H. Ecker, G. Schmidt, and J. Węglarz, *Scheduling in Computer and Manufacturing Systems*, Springer-Verlag, Heidelberg, 1994.

[2] M.R. Gary and D.S. Johnson, *Computers and Intractability: A Guide to the Theory of NP-Completeness*. W.H. Freeman and Company, San Francisco, CA, 1979.

[3] H. El-Rewini, T.G. Lewis, and H.H. Ali, *Task Scheduling in Parallel and Distributed Systems*. PTR Prentice Hall, Englewood Cliffs, NJ, 1994.

[4] Y.K. Kwok and I. Ahmad, Benchmarking the task graph scheduling algorithms. In *Proceedings of 1998 IPPS/SPDP Symposium*, Orlando, FL, 1998, pp. 531–537.

[5] S. Saleh and A.Y. Zomaya, Multiprocessor scheduling using mean-field annealing. In *Parallel and Distributed Processing*, Vol. 1388 of *Lecture Notes in Computer Science* Springer-Verlag, Heidelberg, 1998, pp. 288–296.

[6] L. Wang, H.J. Siegel, V.P. Roychowolhury, and A.A. Maciejewski, Task matching and scheduling in heterogeneous computing environments using a genetic-algorithm-based approach. *Journal of Parallel and Distributed Computing*, 47, 8–22, 1997.

[7] L.N. de Castro and J. Timmis, *Artificial Immune Systems: A New Computational Intelligence Approach*, Springer-Verlag, Heidelberg, 2002.

[8] A. Schoneveld, Parallel Complex Systems Simulation, Ph.D. thesis, University of Amsterdam, Holland, 1999 (http://www.science.uva.nl/research /pscs/papers/phd.html).

[9] M. Mitchell, Computation in cellular automata. In T. Gramb, S. Bornholdt, M. Grob, M. Mitchell, and T. Pellizzari (Eds.), *Non-Standard Computation*, Wiley-VCH, Weinhein, Federal Republic of Germany, 1998, pp. 95–140.

[10] B.J. Overeinder, Distributed Event-driven Simulation — Scheduling Strategies and Resource Management, Ph.D. thesis, University of Amsterdam, Holland, 2000 (http://www.science.uva.nl/research /pscs/papers/phd.html).

[11] R. Das, M. Mitchell, and J.P. Crutchfield, A genetic algorithm discovers particle-based computation in cellular automata. In Y. Davidor, H.-P. Schwefel, and R. Männer (Eds.), *Parallel Problem Solving from Nature — PPSN III*, Vol. 866 of *Lecture Notes in Computer Science*. Springer-Verlag, Heidelberg, 1994, pp. 344–353.

[12] M. Sipper, *Evolution of Parallel Cellular Machines*. The Cellular Programming Approach, LNCS 1194, Springer-Verlag, Heidelberg, 1997.

[13] R. Subrata and A.Y. Zomaya, Evolving cellular automata for location management in mobile computing. *IEEE Transactions on Parallel and Distributed Systems*, 14, 13–26, 2003.
[14] M. Tomassini, M. Sipper, and M. Perrenoud, On the generation of high-quality random numbers by two-dimensional cellular automata. *IEEE Transactions on Computers*, 49, 1140–1151, 2000.
[15] F. Seredyński and A.Y. Zomaya, Sequential and parallel cellular automata-based scheduling algorithms. *IEEE Transactions on Parallel and Distributed Systems*, 13, 1009–1023, 2002.
[16] S. Wolfram, Universality and complexity in cellular automata. *Physica D*, 10, 1–35, 1984.
[17] F. Seredyński, Scheduling tasks of a parallel program in two-processor systems with use of cellular automata, *Future Generation Computer Systems*, 14, 351–364, 1998.
[18] Z. Michalewicz, *Genetic Algorithms + Data Structures = Evolution Programs*, Springer-Verlag, Heidelberg, 1992.
[19] D.H. Wolpert and W.G. Macready, No free lunch theorems for optimization. *IEEE Transactions on Evolutionary Computation*, 1, 67–82, 1997.
[20] J.D. Farmer, N.H. Packard, and A.S. Perelson, The immune system, adaptation, and machine learning. *Physica D*, 22, 187–204, 1986.
[21] S. Forrest, B. Javornik, R. Smith, and A.S. Perelson, Using genetic algorithms to explore pattern recognition in the immune system. *Evolutionary Computation*, 1, 191–211, 1993.
[22] A.S. Perelson, Immune network theory. *Immunological Review*, 110, 5–36, 1989.
[23] E. Hart and P. Ross, An immune system approach to scheduling in changing environments. In W. Banzhaf et al. (Eds.), *GECCO-99: Proceedings of the Genetic and Evolutionary Computation Conference*. Morgan Kaufmann, San Mateo, CA, 1999, pp. 1559–1566.
[24] A.A. Khan, C.L. McCreary, and M.S. Jones, A Comparison of multiprocessor scheduling heuristics. In *Proceedings of International Conference on Parallel Processing*, Vol. II, 1994, pp. 243–250.
[25] M. Mitchell and S. Forrest, Genetic algorithms and artificial life. In Ch.G. Langton (Ed.), *Artificial Life. An Overview*. The MIT Press, Cambridge, MA, 1995.
[26] F. Seredyński and A. Święcicka, Immune-like system approach to multiprocessor scheduling, In R. Wyrzykowski, J. Dongarra, M. Paprzycki, and J. Wasniewski (Eds.), *Parallel Processing and Applied Mathematics*, Vol. 2328, of *Lecture Notes in Computer Science*. Springer-Verlag, Heidelberg, 2002, pp. 626–633.

18

Cellular Automata, PDEs, and Pattern Formation

Xin-She Yang
Y. Young

18.1 Introduction

A cellular automaton (CA) is a rule-based computing machine, which was first proposed by von Newmann in early 1950s and systematic studies were pioneered by Wolfram in 1980s. Since a cellular automaton consists of space and time, it is essentially equivalent to a dynamical system that is discrete in both space and time. The evolution of such a discrete system is governed by certain updating rules rather than differential equations. Although the updating rules can take many different forms, most common cellular automata use relatively simple rules (Von Newmann, 1966; Wolfram, 1983). On the other hand, an equation-based system such as the system of differential equations and partial differential equations also describe the temporal evolution in the domain. Usually, differential equations can also take different forms that describe various systems. Now one natural question is what is the relationship between a rule-based

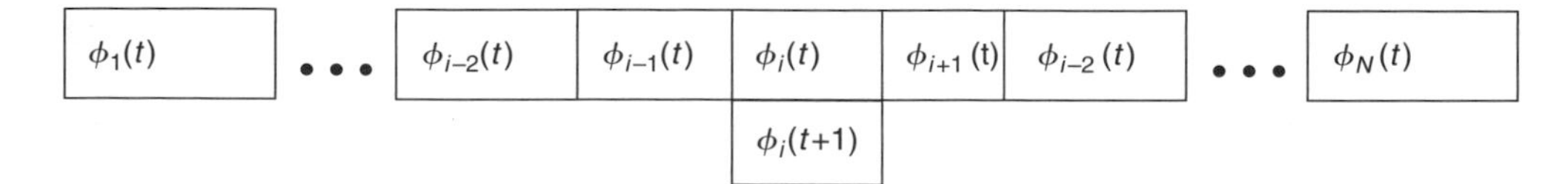

FIGURE 18.1 Diagram of a one-dimensional cellular automaton.

system and an equation-based system? Given differential equations, how can one construct a rule-based cellular automaton, or vice versa? There has been substantial amount of research in these areas in the past two decades. This chapter intends to summarize the results of the relationship among the cellular automata, partial differential equations (PDEs), and pattern formations.

18.2 Cellular Automata

18.2.1 Fundamentals of Cellular Automaton

On a one-dimensional (1D) grid that consists of N consecutive cells, each cell i ($i = 1, 2, \ldots, N$) may be at any of the finite number of states, k. At each time step, t, the next state of a cell is determined by its present state and the states of its local neighbors. Generally speaking, the state ϕ_i at $t + 1$ is a function of its $2r + 1$ neighbors with r cells on the left of the concerned cell and r cells on its right (see Figure 18.1). The parameter r is often referred to as the radius of the neighborhood.

The number of possible permutations for k finite states with a radius of r is $p = k^{2r+1}$, thus the number of all possible rules to generate the state of cells at next time step is k^p, which is usually very large. For example, if $r = 2$, $k = 5$, then $p = 125$ and the number of possible rules is $5^{125} \approx 2.35 \times 10^{87}$, which is much larger than the number of stars in the whole universe.

The rule determining the new state is often referred to as the transition rule or updating rule. In principle, the state of a cell at next time step can be any function of the states of some neighbor cells, and the function can be linear and nonlinear. There is a subclass of the possible rules according to which the new state depends only on the sum of the states in a neighborhood, and this will simplify the rules significantly. For this type of updating rules, the number of possible permutations for k states and $2r + 1$ neighbors is simply $k(2r + 1)$, and thus the number of all possible rules is $k^{k(2r+1)}$. Then for the same parameter $k = 5$, $r = 2$, the number of possible rules is $5^{15} = 3.05 \times 10^{10}$, which is much smaller when compared with 5^{125}. This subclass of sum-rule or totalistic rule is especially important in the popular cellular automata such as Conway's Game of Life, and in the cellular automaton implementation of partial differential equations.

The dynamics and complexity of cellular automata are extremely rich. Stephan Wolfram's pioneering research and mathematical analysis of cellular automata led to his famous classification of 1D cellular automata:

Class 1: The first class of cellular automata always evolves after a finite number of steps from almost all initial states to a homogeneous state where every cell is in the same state. This is something like fixed point equilibrium in the dynamical system.

Class 2: Periodic structures with a fixed number of states occur in the second class of cellular automata.

Class 3: Aperiodic or "chaotic" structures appear from almost all possible initial states in this type of cellular automata.

Class 4: Complex patterns with localized spatial structure propagate in the space as time evolves. Eventually, these patterns will evolve to either homogeneous or periodic. It is suggested that this class of cellular automata may be capable of universal computation.

Cellular automata can be formulated in higher dimensions such as 2D and 3D. One of the most popular and yet very interesting 2D cellular automata using relatively simple updating rules is the

Conway's Game of Life. Each cell has only two states $k = 2$, and the states can be 0 and 1. With a radius of $r = 1$ in the 2D case, each cell has eight neighbors, thus the new state of each cell depends on total nine cells surrounding it. The boundary cells are treated as periodic. The updating rules are: if two or three neighbors of a cell are alive (or 1) and it is currently alive, then it is alive at next time step; if three neighbors of a cell are alive (or 0) and it is currently not alive its next state is alive; the next state is not alive for all the other cases. It is suggested that this simple automaton may have the capability of universal computation. There are many existing computer programs such as Life in Matlab and screen savers on all computer platforms such as Windows and Unix.

18.2.2 Finite-State Cellular Automata

In general, we can define a finite-state cellular automaton with a transition rule $G = [g_{ij,\ldots,l}]$, $(i, j, \ldots, l = 1, 2, \ldots, N)$ from one state $\Phi^t = [\phi^t_{ij,\ldots,l}]$ at time level n to a new state $\Phi^{t+1} = [\phi^{t+1}_{ij,\ldots,l}]$ at a new time step $n + 1$. The value of subscript $(i, j, \ldots, l)$ denotes the dimension, d, of the cellular automaton. Therefore, a CA in the d-dimensional space has N^d cells. For the 2D case, this can be written as

$$G : \Phi^t \mapsto \Phi^{t+1}, \quad g_{ij} : \phi^t_{ij} \mapsto \phi^{t+1}_{ij}, \qquad (i, j = 1, 2, \ldots, N).$$

In the case of sum-rule with a $4r + 1$ neighbors, this becomes

$$\phi^{t+1}_{ij} = G\left(\sum_{\alpha=-r}^{r} \sum_{\beta=-r}^{r} a_{\alpha\beta}\, \phi^t_{i+\alpha, j+\beta}\right), \quad (i, j = 1, 2, \ldots, N),$$

where $a_{\alpha\beta}$ $(\alpha, \beta = \pm 1, \pm 2, \ldots, \pm r)$ are the coefficients. The cellular automata with fixed rules defined this way are deterministic cellular automata. In contrast, there exists another type, namely, the stochastic cellular automata that arise naturally from the stochastic models for natural systems (Guinot, 2002; Yang, 2003).

18.2.3 Stochastic Cellular Automata

When using cellular automata to simulate the phenomena with stochastic components or noise such as percolation and stochastic process, the more effective way is to introduce some probability associated with certain rules. Usually, there is a set of rules and each rule is applied with a probability (Guinot, 2002). Another way is that the state of a cell is updated according to a rule only if certain conditions are met or certain values are reached for some random variables. For example, the rule for 2D a cellular automaton $g(\phi^t_{ij}) = \phi^{t+1}_{ij}$ is applied at a cell only if a random variable $v \leq \Gamma(\phi^t_{ij})$ where the function $\Gamma \in [0, 1]$. At each time step, a random number v is generated for each cell (i, j), and the new state will be updated only if the generated random number is greater than Γ, otherwise, it remains unchanged. Cellular automata constructed this way are called stochastic or probabilistic cellular automata. An example is given later in the next section.

18.2.4 Reversible Cellular Automata

A cellular automaton with an updating rule $\phi^{t+1}_{ij} = g(\phi^t_{ij})$ is generally irreversible in the sense that it is impossible to know the states of a region such as all zeros were the same at a previous time step or not. However, certain class of rules will enable the automata to be reversible. For example, a simple finite difference (FD) scheme for a dynamical system

$$u(t + 1) = g[u(t)] - u(t - 1) \quad \text{or} \quad u(t - 1) = g[u(t)] - u(t + 1),$$

is reversible since for any function $g(u)$, one can compute $u(t+1)$ from $u(t)$ and $u(t-1)$, and invert $u(t-1)$ from $u(t)$ and $u(t+1)$. The automaton rule for 2D reversible automata can be similarly constructed as

$$u_{i,j}^{t+1} = g(u_{i,j}^{t}) - u_{i,j}^{t-1},$$

together with appropriate boundary conditions such as fixed-state boundary conditions (Margolus, 1984).

18.3 Cellular Automata for PDEs

Cellular automata have been used to study many phenomena (Vichniac, 1984; Wolfram, 1984a, 1994; Weimar, 1997; Yang, 2002). In fact, many natural phenomena behave like finite-state cellular automata, and these include self-organized systems such as pattern formation in biological system, insect colonies, and ecosystem; multiple particle system such as the lattice-gas, granular material, and fluids; autocatalytic systems such as enzyme functionality and mineral reactions; and even systems involving society and culture interactions. However, most of these systems have been studied using continuum-based differential equations. We now focus on the formulation of automaton rules from corresponding PDEs.

18.3.1 Rules-Based System and Equation-Based System

Equation-based relationships, often in the form of ordinary differential equations and partial differential equations, form the continuum models of most physical, chemical, and biological processes. Differential equations are suitable and work well for systems with only a small degree of freedom and evolution of system variables in the continuous and smooth manner. There have been vast literatures on analysis and solution technique of the differential models. On the other hand, cellular automata are often considered as an alternative approach and may compliment the existing mathematical basis. The state variables are always discrete, but the numbers of degree of freedom are large (Wolfram, 1984b).

Although continuum models have advantages such as high accuracy and conservation laws, mathematical analysis are usually very difficult and the analytical solutions do not always exist. Only few differential equations have a closed-form solution. In last several decades, the numerical methods have become the essential parts of the solution and of the analysis of almost all problems in engineering, physics, and biology. In fact, computational modeling and numerical computation have become the third component, bridging the gap between theoretical models and experiments. As the computing speed and memory of computers increase, computer simulations have become a daily routine in science and engineering, especially in multidisciplinary research. There are also vast literatures concerning the numerical algorithms, numerical solutions of partial differential equations, and others. These include the following well-established methods: finite difference method, finite element method, finite volume method, cellular automata, lattice-gas, Monte-Carlo method, and genetic algorithms.

Finite difference methods work very well for many problems, but have disadvantages in dealing with irregular geometry. Finite element and finite volume methods can deal with irregular geometry and thus are commonly used and there are quite a few commercial software packages available. Lattice-gas and Monte-Carlo methods are suitable for many problems in physics, especially for systems with multibodies or multiparticles. Cellular automaton is a very interesting and powerful method, and it has gradually become an essential part of the numerical computations owing to its universal computability and the nature of parallel implementation.

18.3.2 Finite Difference Scheme and Cellular Automata

There is a similarity between finite difference scheme and finite-state cellular automata. Finite difference scheme is the discretization of a differential equation on a regular grid of points with the evolution of the states over the discrete time steps. Even the state variable from a differential equation may have continuous values; numerical computation on a computer always lead to the discrete values due to the limited bits of processors or round-off. Similarly, cellular automata are also about the evolution of state variables with finite number of values on a regular grid of cells at different discrete time steps. If the number of states of a cellular automaton is comparable with that of the related finite difference equation, then we can expect the results to be comparable.

To demonstrate this, we choose the 1D heat equation

$$\frac{\partial T}{\partial t} = \kappa \frac{\partial^2 T}{\partial x^2},$$

where T is temperature and κ is the thermal diffusivity. This is a mathematical model that is widely used to simulate many phenomena. The temperature $T(x, t)$ is a real-valued function, and it is continuous for any time $t > 0$ whatever the initial conditions. In reality, it is impossible to measure the temperature at a mathematical point, and the temperature is always the average temperature in a finite representative volume over certain short time. Mathematically, one can obtain a closed-form solution with infinite accuracy in the domain, but physically the temperature would only be meaningful at certain macroscopic levels. No matter how accurate the solution may have at very fine scale, it would be meaningless to try to use the solution at the atomic or subatomic levels where quantum mechanics come into play and, the solution of temperature is invalid (Toffoli, 1984). Thus, numerical computation would be very useful even though it has finite discrete values.

The simplest discretization of the above heat equation is the central difference for spatial derivative and forward scheme for time derivatives, and we have

$$T_i^{n+1} - T_i^n = \frac{\kappa \Delta t}{(\Delta x)^2}(T_{i+1}^n - 2T_i^n + T_{i-1}^n),$$

where i and n are the spatial and time indices. If we choose the time steps and spatial discretization such that $\kappa \Delta t/(\Delta x)^2 = 1$, now we have

$$T_i^{t+1} = (T_{i+1}^t + T_i^t + T_{i-1}^t) - 2T_i^t,$$

which is something like the "mod 2" cellular automata or Wolfram's cellular automata with rule 150 (Wolfram, 1984a; Weimar, 1997).

Cellular automata obtained this way are very similar to the finite difference method. If the state variables are discrete, they are exactly the finite-state cellular automata. However, one can use the continuous-valued state variable for such simulation, in this case, they are continuous-valued cellular automata from differential equations as studied by Rucker's group and used in their the well-known CAPOW program (Rucker, 2003; Ostrov and Rucker, 1996). In some sense, the continuous-valued cellular automata based on the differential equations are the same as the finite difference methods, but there are some subtle differences and advantages of cellular automata over the finite difference simulations due to the CA's properties of parallel nature, artificial life-oriented emphasis on experiment and observation and genetic algorithms for searching large phase space of the rules as proposed by Rucker et al. (1998).

18.3.3 Cellular Automata for Reaction-Diffusion Systems

By extending the discretization procedure from differential equations to derive automaton rules for cellular automata, we now formulate the cellular automata from their corresponding partial differential equations. First, let us start with the reaction-diffusion equation that may form beautiful patterns in the 2D configuration

$$\frac{\partial u}{\partial t} = D\left(\frac{\partial^2 u}{\partial x^2} + \frac{\partial^2 u}{\partial y^2}\right) + f(u),$$

where $u(x, y, t)$ is the state variable that evolves with time in a 2D domain, and the function $f(u)$ can be either linear or nonlinear. D is a constant depending on the properties of diffusion. This equation can also be considered as a vector form for a system of reaction-diffusion equations if let $D = \text{diag}(D_1, D_2)$, $u = [u_1 u_2]^T$. The discretization of this equation can be written as

$$\frac{u_{i,j}^{n+1} - u_{i,j}^n}{\Delta t} = D\left[\frac{u_{i+1,j}^n - 2u_{i,j}^n + u_{i-1,j}^n}{(\Delta x)^2} + \frac{u_{i,j+1}^n - 2u_{i,j}^n + u_{i,j-1}^n}{(\Delta y)^2}\right] + f(u_{i,j}^n),$$

by choosing $\Delta t = \Delta x = \Delta y = 1$, we have

$$u_{i,j}^{n+1} = D[u_{i+1,j}^n + u_{i-1,j}^n + u_{i,j+1}^n + u_{i,j-1}^n] + f(u_{i,j}^n) + (1 - 4D)u_{i,j}^n,$$

which can be written as the generic form

$$u_{i,j}^{t+1} = \sum_{k,l=-r}^{r} a_{k,l} u_{i+k,j+l}^t + f(u_{i,j}^t),$$

where the summation is over the $4r + 1$ neighborhood. This is a finite-state cellular automaton with the coefficients $a_{k,l}$ being determined from the discretization of the governing equations, and for this special case, we have $a_{-1,0} = a_{+1,0} = a_{0,-1} = a_{0,+1} = D$, $a_{0,0} = 1 - 4D$, $r = 1$.

18.3.4 Cellular Automata for the Wave Equation

For the 1D linear wave equation,

$$\frac{\partial^2 u}{\partial t^2} = c^2 \frac{\partial^2 u}{\partial x^2},$$

where c is the wave speed. The simplest central difference scheme leads to

$$\frac{u_i^{n+1} - 2u_i^n + u_i^{n-1}}{(\Delta t)^2} = c^2 \frac{u_{i+1}^n - 2u_i^n + u_{i-1}^n}{(\Delta x)^2}.$$

By choosing $\Delta t = \Delta x = 1$, $t = n$, it becomes

$$u_i^{t+1} = [u_{i+1}^t + u_{i-1}^t + 2(1 - c^2)u_i^t] - u_i^{t-1}.$$

This can be written in the generic form

$$u_i^{t+1} + u_i^{t-1} = g(u^t),$$

which is reversible under certain conditions. This property comes from the reversibility of the wave equation because it is invariant under the transformation: $t \to -t$.

18.3.5 Cellular Automata for Burgers Equation with Noise

One of the important equations arising in many processes such as turbulent phenomenon is the noisy Burgers equation

$$\frac{\partial u}{\partial t} = 2u\frac{\partial u}{\partial x} + \frac{\partial^2 u}{\partial x^2} + \nabla \upsilon,$$

where υ is the noise that is uncorrelated in space and time so that $\langle \upsilon(x,t)\rangle = 0$ and $\langle \upsilon(x,t)\upsilon(x_0,t_0)\rangle = 2D\delta(x-x_0)\delta(t-t_0)$ (Emmerich and Kahng, 1998). This equation with Gaussian white noise can be rewritten as

$$\frac{\partial u}{\partial t} + \xi = 2u\frac{\partial u}{\partial x} + \frac{\partial^2 u}{\partial x^2} + \eta,$$

where both ξ and η are uncorrelated. By introducing the variables $v_i^t = c\,\exp(\Delta x\, u_i^t), \phi_{\mathrm{I}}^t = \beta\,\ln(v_i^t)$, $\alpha = \Delta t/(\Delta x)^2, (1-2\alpha)/c\alpha = \exp(-A/\beta), c^2 = \exp(B/\beta), \xi = \exp(\Phi), \eta = \exp(\Psi)$ and after some straightforward calculations in the limit of β tends zero, we have the automata rule

$$\begin{aligned}\phi_i^{t+1} = \phi_{i-1}^t &+ \max[0, \phi_i^t - A, \phi_i^t + \phi_{i+1}^t - B, \Psi_i^t - \phi_{i-1}^t] \\ &- \max[0, \phi_{i-1}^t - A, \phi_{i-1}^t + \phi_i^t - B, \Phi_i^t - \phi_{i-1}^t],\end{aligned}$$

where we have used the following identity,

$$\lim_{\phi\to+0} \varepsilon \ln(\mathrm{e}^{A/\varepsilon} + \mathrm{e}^{B/\varepsilon} + \cdots) = \max[A, B, \ldots].$$

This forms a generalized probabilistic cellular automaton that is referred to as the noisy Burgers cellular automaton. Burgers equation without noise usually evolves in shock wave, and in the presence of noise, the states of the probabilistic cellular automata may be taken as discrete reminders of those shock waves that were disorganized.

18.4 Differential Equations for Cellular Automata

Computer simulations based on cellular automata and partial differential equations work remarkably well for many different reasons. One possibility is that finite-state cellular automata and finite difference approximations using discrete time and space with a finite precision can represent physical variables very well and thus the models are insensitive to very small space and time scales. It is relatively straightforward to derive the updating rules for cellular automata from the corresponding partial differential equations. However, the reverse is usually very difficult. There is no general method available to formulate the continuum model or differential equations for given rule-based cellular automata despite the obvious importance. Fortunately, there have been some important progress made in this area (Omohundro, 1984; Wolfram, 1984b), and we outline some of the procedures of formulating continuum equations for cellular automata.

18.4.1 Formulation of Differential Equations from Cellular Automata

The mathematical analysis for cellular automata was first pioneered by Wolfram, and has attracted wide attention since 1980s. Wolfram (1983) gave an extensive analysis of statistical mechanics of cellular automata. Later on, Omohundro (1984) provided an instructive procedure of formulating the partial differential equations for cellular automata by using 10 PDE variables in 2D configurations. These ten variables are the state variable $P(x,y,t)$ at present, new state $N(x,y,t)$ and eight variables $U_1, \ldots, U_8$ with eight bump or bell functions. The eight variables have the same format of information as the $P(x,y,t)$. On a 2D grid, these eight functions $S_1, \ldots, S_8$ are shown in Figure 18.2 together with shifted coordinates.

According to Omohundro's formulation, we assume the bumps to be α wide and constant outside β from the transition. If the width of a cell is 1, then $\alpha = 1/5$ and $\beta = 1/100$. The eight functions are

S_2 (−1,+1)	S_3 (0,+1)	S_4 (+1,+1)
S_1	N/P	S_5
S_8 (−1,−1)	S_7 (0,−1)	S_6 (+1,−1)

FIGURE 18.2 Diagram of eight neighbors and the positions of Omohundro's functions.

taken as

$$S_1 = e^{-1/(\beta-x)^2}(-\beta < x < \beta),\ 0\ (|x| \geq \beta), \quad S_2 = \frac{S_1(x\beta/\alpha)}{S_1(0)}, \quad S_3 = \int_{-1}^{x} S_1(x)\mathrm{d}x \bigg/ \int_{-\beta}^{\beta} S_1(x)\mathrm{d}x,$$

$$S_4 = S_3(x-\alpha/2)S_3(\alpha/2-x), \quad S_5 = S_3(x-\alpha)S_3(\alpha-x), \quad S_6 = \sum_{k=-\infty}^{\infty} S_4(x-k),$$

$$S_7 = \sum_{k=-\infty}^{\infty} S_5(x-k), \quad S_8 = \sum_{k=-\infty}^{\infty} S_2(x-k).$$

By using these functions, Omohundro derived the following equation:

$$\frac{\partial N}{\partial t} = -\gamma\left[\frac{\mathrm{d}S_2}{\mathrm{d}x}\frac{\partial N}{\partial x} + \frac{\mathrm{d}S_2}{\mathrm{d}y}\frac{\partial N}{\partial y}\right],$$

where γ is a large constant. Differential equations for other variables can be written in a similar manner although they are more complicated (Omohundro, 1984).

18.4.2 Stochastic Reaction-Diffusion

A stochastic cellular automaton on a 1D ring has N sites with simple rules for the state variable $u_i(t)$ and $S = u_i(t) + u_{i+1}(t)$: (1) $u_i(t+1) = 0$, if $S = 0$; (2) $u_i(t+1) = 1$ with probability p_1, if $S = 1$; (3) $u_i(t+1) = 1$ with probability p_2, if $S = 2$. This Domany–Kinzel cellular automaton is equivalent to the following stochastic reaction-diffusion equation:

$$\frac{\partial v}{\partial t} = D\nabla^2 v + f(v) + \varepsilon\sqrt{v},$$

where v is the concentration of live sites or $u_i(t) = 1$, and ε is a zero-mean Gaussian random variable with unit variance (Ahmed and Elgazzar, 2001). However, there is nonuniqueness associated with the formulation of differential equations from the cellular automata. Bagnoli et al. (2001) demonstrated that the above equation can be obtained from the following rules: (1) $u_i(t+1) = 0$, if $\Sigma = u_{i+1}(t) + u_i(t) + u_{i-1}(t) = 0$; (2) $u_i(t+1) = 1$ with probability p_1, if $\Sigma = 1$; (3) $u_i(t+1) = 2$ with probability p_2, if $\Sigma = 2$; (4) $u_i(t+1) = 1$, if $\Sigma = 3$. This nonuniqueness in the relationship between cellular automata and PDEs require more research.

18.5 Pattern Formation

The behavior and characteristics of cellular automata are very complicated and there is no general or universal mathematical analysis available for the description of such complexity. Even for 1D cellular automata, they are complicated enough as shown by Wolfram classifications. Pattern formation is one of the typical characteristics in cellular automata. The study of pattern complexity and the conditions

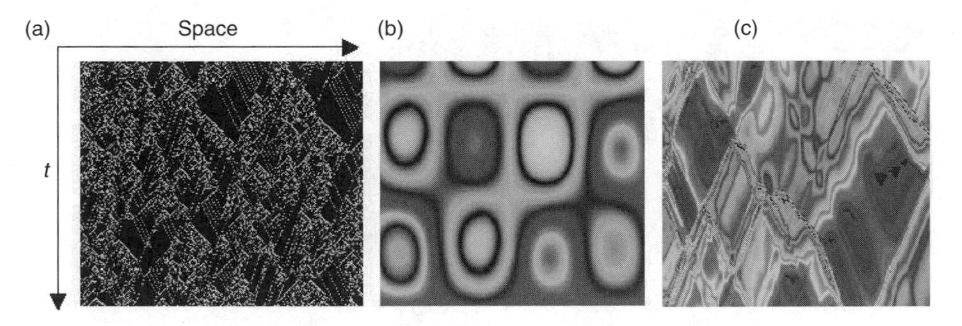

FIGURE 18.3 Pattern formation in cellular automata: (a) 1D CA with disordered initial conditions; (b) CA for 1D wave equation; and (c) nonlinear 1D Sine-Gordon equation.

of its formation is essential in many processes such as biological pattern formation, enzyme dynamics, percolation, and other processes in engineering applications (Turing, 1952; Flake, 1998; Cappuccio et al., 2001; Boffetta et al., 2002). This section focuses on the pattern formation in cellular automata and comparison of CA results with the results using differential equations.

18.5.1 Complexity and Pattern in Cellular Automata

Wolfram (1983) pioneered the studies of complexity and patterns in cellular automata. Complex patterns form in 1D cellular automata even with very simple rules. Figure 18.3(a) shows the typical patterns in 1D cellular automata with random initial conditions for $k = 16$ states and $r = 1$. Figure 18.3(b) is for 1D wave equation with $k = 1024$ states and $r = 1$. Figure 18.3(c) corresponds to the nonlinear wave based on Sine-Gordon equation $u_{tt} - u_{xx} = \alpha \sin(u)$ with $\alpha = -0.1, k = 1024$ states, $r = 2$ and sinusoidal initial conditions. The pattern formations are more complicated in higher dimensions. We compare the results from different methods while we study the pattern formations in the next section.

18.5.2 Comparison of Cellular Automata and PDEs

For a 2D reaction-diffusion system of two partial differential equations:

$$\frac{\partial u}{\partial t} = D_u \nabla^2 u + f(u), \quad \frac{\partial v}{\partial t} = D_v \nabla^2 v + g(v), \quad f(u, v) = \alpha(1 - u) - uv^2,$$

$$g(u, v) = uv^2 - \frac{(\alpha + \beta)v}{1 + (u + v)},$$

where $D_u = 0.05$. The parameters $\gamma = D_v/D_u$, α, β can vary so as to produce the complex patterns. This system can model many systems such as enzyme dynamics and biological pattern formations by slight modifications (Murray, 1989; Meinhardt, 1982, 1995; Keener and Sneyd, 1998; Yang, 2003). Figure 18.4 shows a snapshot of patterns formed by the simulations of the above reaction-diffusion system at $t = 500$ for $\gamma = 0.6, \alpha = 0.01$, and $\beta = 0.02$. The right plot is the comparison of results obtained by three different methods: cellular automata (marked with CA), finite difference method (FD), and finite element method (FE). The plot is for the data on the middle line of the pattern shown on the left.

18.5.3 Pattern Formation in Biology and Engineering

Pattern formation occurs in the many processes in biology and engineering. We will give two examples here to show the complexity and diversity of the beautiful patterns formed. Figure 18.5 shows the 2D and 3D patterns for the reaction diffusion system with $f(u, v) = \alpha(1 - u) - uv^2/(1 + u + v)$, $g(u, v) = uv^2 - (\alpha + \beta)v/(1 + u + v)$.

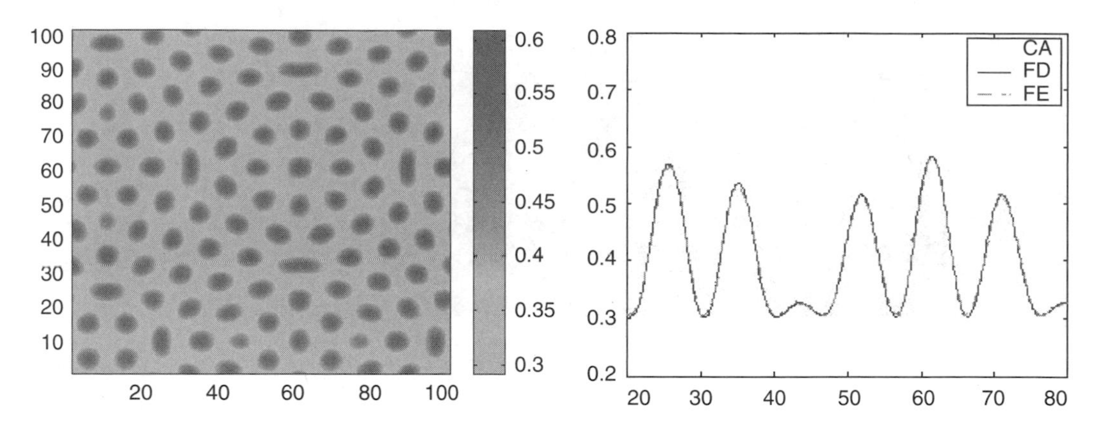

FIGURE 18.4 A snapshot of pattern formation of the reaction-diffusion system (for $\alpha = 0.01, \beta = 0.02$, and $\gamma = 0.6$) and the comparison of results obtained by three different methods (CA-dotted, FD-solid, FE-dashed) through the middle line of the pattern on the left.

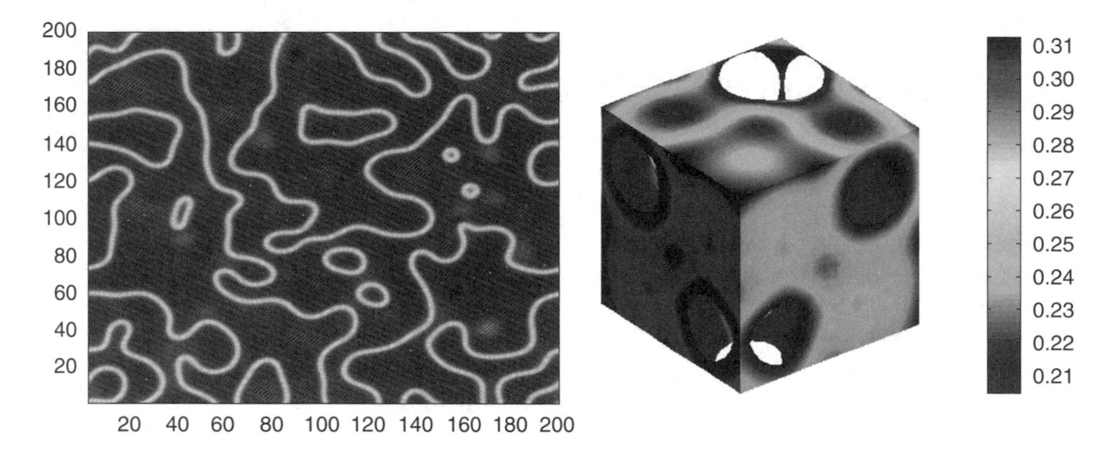

FIGURE 18.5 Pattern formations from 2D random initial conditions with $\alpha = 0.05$, $\beta = 0.01$, and $\gamma = 0.5$ (left) and 3D structures with $\alpha = 0.1$, $\beta = 0.05$, and $\gamma = 0.36$ (right).

Another example is the formation of spiral waves studied by Barkley and his colleagues in detail (Barkley et al., 1990; Margerit and Barkley, 2001)

$$\frac{\partial u}{\partial t} = \nabla^2 u + \frac{u}{\varepsilon^2}(1-u)\left(u - \frac{v+\beta}{\alpha}\right), \qquad \frac{\partial v}{\partial t} = \varepsilon \nabla^2 v + (u - v),$$

where $\varepsilon, \alpha, \beta$ are parameters. Figure 18.6 shows the formation of spiral waves (2D) and scroll waves (3D) of this nonlinear system under appropriate conditions.

The patterns formed in terms of diffusion-reaction equations have been observed in many phenomena. The ring, spots, and stripes exist in animal skin coating, enzymatic reactions, shell structures, and mineral formation. The spiral and scroll waves and spatiotemporal pattern formations are observed in calcium transport, Belousov–Zhabotinsky reaction, cardiac tissue, and other excitable systems.

In this chapter, we have discussed some of important development and research results concerning the connection among the cellular automata and partial differential equations as well as the pattern formation related to both systems. Cellular automata are rule-based methods with the advantages of local interactions, homogeneity, discrete states and parallelism, and thus they are suitable for simulating systems with large degrees of freedom and life-related phenomena such as artificial intelligence and ecosystems.

FIGURE 18.6 Formation of spiral wave for $\alpha = 1, \beta = 0.1$, and $\varepsilon = 0.2$ (left) and 3D scroll wave for $\alpha = 1, \beta = 0.1$, and $\varepsilon = 0.15$ (right).

PDEs are continuum-based models with the advantages of mathematical methods and closed-form analytical solutions developed over many years, however, they usually deal with systems with small numbers of degree of freedom. There is an interesting connection between cellular automata and PDEs although this is not always straightforward and sometimes may be very difficult. The derivation of updating rules for cellular automata from corresponding PDEs are relatively straightforward by using the finite differencing schemes, while the formulation of differential equations from the cellular automaton is usually difficult and nonunique. More studies are highly needed in these areas. In addition, either rule-based systems or equation-based systems can have complex pattern formation under appropriate conditions, and these spatiotemporal patterns can simulate many phenomena in engineering and biological applications.

References

Ahmed, E. and Elgazzar, A.S. On some applications of cellular automata. *Physica A*, **296** (2001) 529–538.

Bagnoli, F., Boccara, N., and Rechtman, R. Nature of phase transitions in a probabilistic cellular automaton with two absorbing states. *Physical Review E*, **63** (2001) 461161–461169.

Barkley, D., Kness, M., and Tuckerman, L.S. Spiral-wave dynamics in a simple model of excitable media: The transition from simple to compound rotation. *Physical Review A*, **42** (1990) 2489–2492.

Boffetta, G., Cencini, M., Falcioni, M., and Vulpiani, A. Predictability: A way to characterize complexity. *Physics Reports*, **356** (2002) 367–474.

Cappuccio, R., Cattaneo, G., Erbacci, G., and Jocher, U. A parallel implementation of a cellular automata based on the model for coffee percolation. *Parallel Computing*, **27** (2001) 685–717.

Emmerich, H. and Kahng, B.N. A random automata related to the noisy Burgers equation. *Physica A*, **259** (1998) 81–89.

Flake, G.W. *The Computational Beauty of Nature*. MIT Press, Cambridge, MA (1998).

Guinot, V. Modelling using stochastic, finite state cellular automata: Rule inference from continuum models. *Applied Mathematical Modelling*, **26** (2002) 701–714.

Keener, J. and Sneyd, J. *Mathematical Physiology*. Springer-Verlag, New York (1998).

Margerit, D. and Barkley, D. Selection of twisted scroll waves in three-dimensional excitable media. *Physical Review Letters*, **86** (2001) 175–178.

Margolus, N. Physics-like models of computation. *Physica D*, **10** (1984) 81–95.

Meinhardt, H. *Models of Biological Pattern Formation*. Academic Press, London (1982).

Meinhardt, H. *The Algorithmic Beauty of Sea Shells*. Springer-Verlag, New York (1995).

Murray, J.D. *Mathematical Biology*. Springer-Verlag, New York (1989).

Omohundro, S. Modelling cellular automata with partial differential equations. *Physica D*, **10** (1984) 128–134.

Ostrov, D. and Rucker, R. Continuous-valued cellular automata for nonlinear wave equations. *Complex Systems*, **10** (1996) 91–117.

Rucker, R. Continuous-valued cellular automata in two dimensions. In Griffeath, D. and Moore, C. (Eds.), *New Constructions in Cellular Automata*. Oxford University Press, (2003). Website for CAPOW98 Software: http://www.cs.sjsu.edu/faculty/rucker/capow/

Toffoli, T. Cellular automata as an alternative to (rather than an approximate of) differential equations in modelling physics. *Physica D*, **10** (1984) 117–127.

Turing, A. The chemical basis of morphogenesis. *Philosophical Transactions of the Royal Society of London B*, **237** (1952) 37–72.

Vichniac, G.Y. Simulating physics with cellular automata. *Physica D*, **10** (1984) 96–116.

von Newmann J. Theory of Self-Reproducing Automata (Edited by A.W. Burks). University of Illinois Press, Urbana, IL (1966).

Weimar, J.R. Cellular automata for reaction-diffusion systems. *Parallel Computing*, **23** (1997) 1699–1715.

Wolfram, S. Statistical mechanics of cellular automata. *Review of Modern Physics*, **55** (1983) 601.

Wolfram, S. Cellular automata as models of complexity. *Nature*, **311** (1984a) 419–424.

Wolfram, S. Universality and complexity in cellular automata. *Physica D*, **10** (1984b) 1–35.

Wolfram, S. *Cellular Automata and Complexity*. Addison-Wesley, Reading, MA (1994).

Yang, X.S. Characterization of multispecies living ecosystems with cellular automata. In Standish, Abbass, and Bedau (Eds.), *Artificial Life VIII*. MIT Press, Cambridge, MA (2002), pp. 138–141.

Yang, X.S. Turing pattern formation of catalytic reaction-diffusion systems in engineering applications. *Modelling and Simulation in Materials Science and Engineering*, **11** (2003) 321–329.

19

Ant Colonies and the Mesh-Partitioning Problem

Borut Robič
Peter Korošec
Jurij Šilc

The complex social behaviors of ants have been much studied by science, and computer scientists are now finding that these behavior patterns can provide models for solving difficult combinatorial optimization problems.

Dorigo and Stützle,
Ant Colony Optimization, MIT Press, 2004.

19.1 Introduction

An ant colony is a "distributed system" that has a highly structured social organization in spite of the simplicity of its individuals. Because of this organization, an ant colony is capable of accomplishing complex natural tasks that far exceed the individual capacity of a single ant. It is not surprising, therefore, that computer scientists have taken inspiration from ants and their behavior in order to design algorithms for solving computationally demanding problems. This chapter is concerned with applying these ideas to solve an important, but *NP*-hard, combinatorial optimization problem: the so-called mesh-partitioning problem.

Many of the problems that arise in mechanical, civil, automobile, and aerospace engineering can be expressed in terms of partial differential equations and solved by using the finite-element method. If a partial differential equation involves a function, f, then the purpose of the finite-element method is to determine an approximation to f. To do this the domain is discretized into a set of geometrical elements consisting of nodes: a process known as meshing. The value of f is then computed for each of these nodes, and the solutions for the other points are interpolated from these values [1].

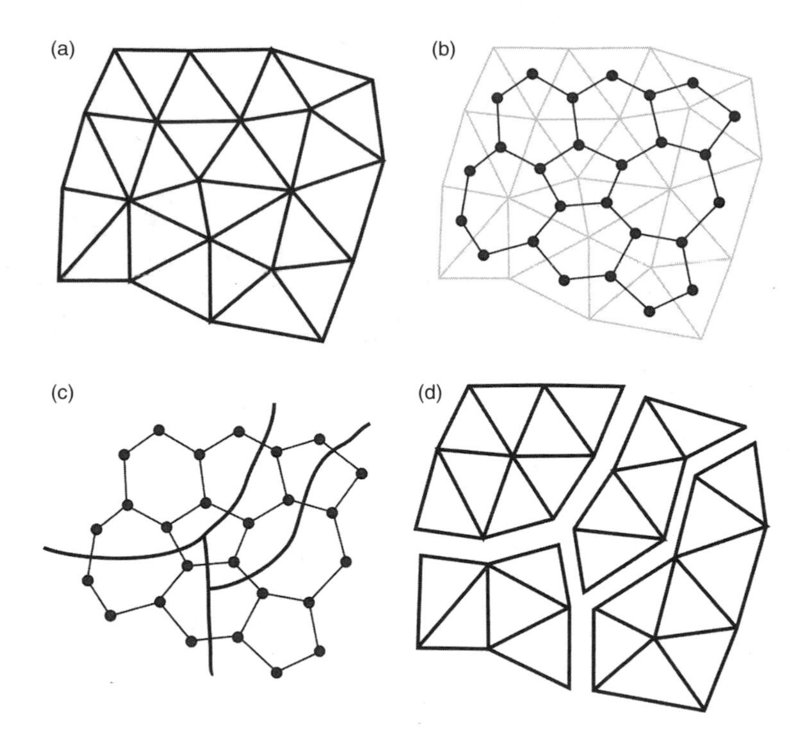

FIGURE 19.1 Mesh partitioning: (a) sample mesh; (b) mesh with induced graph; (c) after graph partitioning; and (d) the resulting partitioned mesh.

However, in real-world engineering problems, meshing is a demanding task because meshes usually have large numbers (e.g., hundreds of thousands) of elements. For this reason, the finite-element method is usually parallelized, which means the mesh is partitioned and distributed among several processors: a process known as *mesh partitioning*. To achieve high computational efficiency, it is important that the mesh partitioning produces workloads that are well balanced and that the interprocessor communication is minimized. This is a combinatorial optimization problem and is a special case of the well-known graph-partitioning problem (see Figure 19.1).

The graph-partitioning problem is defined as follows. Let $G(\boldsymbol{V}, \boldsymbol{E})$ be an undirected graph consisting of a nonempty set $\boldsymbol{V}$ of vertices and a set $\boldsymbol{E} \subseteq \boldsymbol{V} \times \boldsymbol{V}$ of edges. A k-partition $\boldsymbol{D}$ of G comprises k mutually disjointed subsets $\boldsymbol{D}_1, \boldsymbol{D}_2, \ldots, \boldsymbol{D}_k$ (called domains) of $\boldsymbol{V}$, whose union is $\boldsymbol{V}$. The set of edges that connect the different domains of a partition $\boldsymbol{D}$ is called an edge-cut, and is denoted by $\xi(\boldsymbol{D})$. A partition $\boldsymbol{D}$ is considered to be balanced if the sizes of the domains are roughly the same, that is, if

$$\beta(\boldsymbol{D}) = \max_{1 \leq i \leq k} |\boldsymbol{D}_i| - \min_{1 \leq i \leq k} |\boldsymbol{D}_i| \approx 0.$$

The graph-partitioning problem is to find a balanced partition with a minimum edge-cut.

Unfortunately, the graph-partitioning problem is an *NP*-hard optimization problem. This means that it is impossible to find optimum solutions in polynomial time (unless $P = NP$). Consequently, heuristic approaches are normally used for mesh partitioning. Many mesh-partitioning algorithms are described in the literature [2]. Some of these just use geometric information about the mesh and employ a fast, parallel, sorting algorithm in their implementation [3]. Spectral partitioning methods also make use of the mesh-connectivity information and solve the eigenvalue problem to compute the partition [4].

A promising alternative is to use stochastic heuristics, some of which are based on various fundamental principles observed in nature. Recently, a number of studies have shown that such techniques have great potential for solving a wide range of problems [5]. Examples include tabu search [6], simulated

TABLE 19.1 Applications of Ant-Colony Optimization Algorithms

Problem name	Authors	Algorithm name	Year
Traveling salesman	Dorigo, Maniezzo, and Colorni	AS	1991
	Gambardella and Dorigo	Ant-Q	1995
	Dorigo and Gambardella	ACS and ACS-3-opt	1996
	Stützle and Hoos	MMAS	1997
	Bullnheimer, Hartl, and Strauss	AS_{rank}	1997
Quadratic assignment	Maniezzo, Colorni, and Dorigo	AS-QAP	1994
	Gambardela, Taillard, and Dorigo	HAS-QAP	1997
	Stützle and Hoos	MMAS-QAP	1997
	Maniezzo	ANTS-QAP	1998
	Maniezzo and Colorni	AS-QAP	1999
Scheduling problems	Colorni, Dorigo, and Maniezzo	AS-JSP	1994
	Stützle	AS-FSP	1997
	Bauer et al.	ACS-SMTTP	1999
	den Besten, Stützle, and Dorigo	ACS-SMTWTP	1999
Vehicle routing	Bullnheimer, Hartl, and Strauss	AS-VRP	1997
	Gambardella, Taillard, and Agazzi	HAS-VRP	1999
Connection-oriented network routing	Schoonderwoerd et al.	ABC	1996
	White, Pagurek, and Oppacher	ASGA	1998
	Di Carlo and Dorigo	AntNet-FS	1998
	Bonabeau et al.	ABC-smart ants	1998
Connection-less network routing	Di Carlo and Dorigo	AntNet and AntNet-FA	1997
	Subramanian, Druschel, and Chen	Regular ants	1997
	Hausse et al.	CAF	1998
	van der Put and Rothkrantz	ABC-backward	1998
Sequential ordering	Gambardella and Dorigo	HAS-SOP	1997
Graph coloring	Costa and Hertz	ANTCOL	1997
Shortest common supersequence	Michel and Middendorf	AS-SCS	1998
Frequency assignment	Maniezzo and Carbonaro	ANTS-FAP	1998
Generalized assignment	Ramalhinho, Lorenço, and Serra	MMAS-GAP	1998
Multiple knapsack	Leguizamón and Michalewicz	AS-MKP	1999
Optical network routing	Navarro Varela and Sinclair	ACO-VWP	1999
Redundancy allocation	Liang and Smith	ACO-RAP	1999

Source: Adapted from Dorigo, M., Bonabeau, E., and Therauluz, G. *Future Generation Computer Systems*, 16, 858, 2000. With permission from Elsevier.

annealing [7], neural networks [8], and genetic algorithms [9]. These methods are widely applicable and have proven to be very powerful in practice [10].

Ant colonies are yet another such natural phenomenon that has recently given rise to new optimization methods belonging to the group of "Heuristics from Nature," which are implemented in some new bio-inspired algorithms [11,12]. This *ant-colony optimization* has already been used in solving various combinatorial optimization problems, such as the traveling salesman [13], quadratic assignment [14], job-shop scheduling [15], vehicle [16] and network routing [17]. More references to applications of ant-colony optimization can be found in References 18 and 19 (see Table 19.1). For the case of mesh partitioning, however, only a few attempts have been made [20–23].

19.2 Ant-Colony Optimization

Ant-Colony Optimization (ACO) is a metaheuristics approach proposed by Dorigo et al. [12]. The inspiration for ACO is the foraging behavior of real ants. This behavior allows the ants to find the shortest

```
While termination condition not satisfied Do
  ScheduleActivities
    Ants_activity()
    Pheromone_evaporation()
    Deamon_actions()
  EndScheduleActivities
EndWhile
```

FIGURE 19.2 Ant-colony optimization algorithm.

paths between food sources and their nest [24]. Ants deploy a chemical trail (or pheromone trail) as they walk; this trail attracts other ants to take the path that has the most pheromone. This reinforcement process results in the selection of the shortest path: the first ants coming back to the nest are those that took the shortest path twice (from the nest to the source and back to the nest), so that more pheromone is present on the shortest path than on the longer paths immediately after these ants have returned, stimulating nest mates to choose the shortest path.

Ant-colony optimization algorithms (see Figure 19.2) are based on a parameterized probabilistic model (pheromone model) that is used to model the chemical pheromone trails. Artificial ants incrementally construct solutions by adding opportunely defined solution components to a partial solution under consideration. In order to do this, artificial ants perform randomized walks on a completely connected graph $G(\boldsymbol{C}, \boldsymbol{L})$, called a construction graph, whose vertices are the solution components $\boldsymbol{C}$, and the set $\boldsymbol{L}$ composed of the connections. When a constrained combinatorial optimization problem is considered, the problem constraints Ω are built into the ants' constructive procedure in such a way that in every step of the construction process only feasible solution components can be added to the current partial solution.

`Ant_activity()`: In the construction phase an ant incrementally builds a solution by adding solution components to the partial solution constructed so far. The probabilistic choice of the next solution component to be added is done by means of transition probabilities. More specifically, ant n in step t moves from vertex $i \in \boldsymbol{C}$ to vertex $j \in \boldsymbol{C}$ with a probability given by:

$$p_{ij,n}(t) = \begin{cases} \dfrac{\tau_{ij}^{a}(t)\eta_{ij}^{b}}{\sum_{m\in N_{i,n}} \tau_{im}^{a}(t)\eta_{im}^{b}} & j \in N_{i,n}, \\ 0 & j \notin N_{i,n}, \end{cases}$$

where η_{ij} is a priori available heuristic information, a and b are two parameters that determine the relative influence of the pheromone trail $\tau_{ij}(t)$ and heuristic information, respectively, and $N_{i,n}$ is the feasible neighborhood of vertex i. If $a = 0$, then only heuristic information is considered. Similarly, if $b = 0$, then only pheromone information is at work. Once an ant builds a solution, or while a solution is being built, the pheromone is being deposited (on nodes or connections) according to the evaluation of a (partial) solution. This pheromone information will direct the search of the ants in the following iterations. The solution construction ends when an ant comes to the ending vertex (where the food is located).

`Pheromone_evaporation()`: Pheromone-trail evaporation is a procedure that simulates the reduction of pheromone intensity. It is needed in order to avoid a too quick convergence of the algorithm to a suboptimal solution.

`Daemon_actions()`: Daemon actions can be used to implement centralized actions that cannot be performed by single ants. Examples are the use of a local search procedure applied to the solutions built by the ants, or the collection of global information that can be used to decide whether it is useful or not to deposit additional pheromone to bias the search process from a nonlocal perspective.

As we can see from the pseudo code, the **ScheduleActivities** construct does not specify how the three included activities should be scheduled or synchronized. This means it is up to the programmer to specify how these procedures will interact (parallel or independent).

Within the ACO metaheuristic framework the currently best-performing versions in practice are Ant Colony System [13] and MAX–MIN Ant System [25]. Recently, researchers have been dealing with finding similarities between ACO algorithms and Estimation of Distribution Algorithms [26,27]. Furthermore, connections between ACO algorithms and Stochastic Gradient–Descent algorithms are shown in Reference 28.

19.3 Mesh Partitioning with the Ant-Colonies Algorithm

The mesh-partitioning problem, like some other problems, for example, data-mining, text-mining, belong to the wider class of so-called clustering problems, which are concerned with the grouping of objects into homogeneous subgroups. For these problems, ant-based algorithms have also been proposed [29]. However, ant-based clustering differs from ACO in several fundamental respects:

- It draws its inspiration from the clustering behavior observed in real ants (not the foraging behavior as in the ACO).
- It is not metaheuristics, in contrast to ACO; it tackles only the specific task of clustering.
- Unlike ACO it does not make use of artificial pheromones.
- It shows no synergetic effect, that is, its performance is mostly independent of population size.

In contrast to the ant-based clustering approach to the partitioning problem, we discuss three approaches that are based on ACO metaheuristics. Informally, each of these approaches uses multiple ant colonies instead of only one (as is the case in ACO).

19.3.1 Multiple Ant-Colony Algorithms

19.3.1.1 The Basic Algorithm

The main idea of the Basic Multiple Ant-Colony Algorithm (B-MACA) for k-way partitioning was recently proposed by Langham and Grant [22]. There are k colonies of ants that are competing for food, which in this case represents the vertices of the graph (mesh). Eventually, ants gather food to their nests, that is, they partition the mesh into k submeshes. More precisely, the algorithm B-MACA proceeds as follows (see Figure 19.3).

Initially, the graph is mapped onto the grid that represents the ants' habitat (where the ants can move). There are many possibilities for this mapping, one of which is random mapping, as used in our case. Then the ants are evenly distributed to k nest loci on the grid, which are chosen according to a certain strategy, from where they start their foraging and gathering of food. An ant can move in one of three directions (forward, left, and right). The direction is chosen by a probability rule based on pheromone intensity. When an ant attempts to move off the grid it is forced to move left or right with equal probability. When an ant finds food it tries to pick it up. To do this, it checks whether the quantity of the temporarily gathered food in its nest is at the limit (the capacity of storage is limited by the problem's constraints). If the limit has not been reached, then the weight of the food is calculated from the number of cut edges created by assigning the selected vertex to the partition associated with the nest of the current ant; otherwise the ant moves in a randomly selected direction. If the food is too heavy for one ant to pick it up (and not too heavy for a few ants to pick it up) then the ant sends a help signal within a radius of a few cells. So, if other ants are in the neighborhood, they will help this ant to carry the food to the nest locus. On the way back to the nest locus an ant deposits pheromones on the trail that it is making, so the other ants can follow its trail and gather more food from that, or a nearby cell. When an ant reaches the nest locus it drops the

```
Initialize()
While ending condition not satisfied Do
  For all ants of colony Do
    For all colonies Do
      If carrying food Then
        If in nest locus The Drop_Food()
        Else Move_To_Nest()
      Else If food here Then Pick_Up_Food()
      Else If food ahead Then Move_Forrward()
      Else If in nest locus Then Move_To_Away_Pheromone()
      Else If help signal Then Move _To_Help()
      Else Follow_Strongest_Forward_Pheromone()
    EndFor
  EndFor
  For all grid cells Do
    Evaporate_Pheromone()
  EndFor
EndWhile
```

FIGURE 19.3 Algorithm B-MACA.

food in the first possible place around the nest (e.g., in a clockwise direction). After an ant has dropped its food, it starts a new round of foraging. Of course, ants can also gather food from other nest loci. When an ant tries to pick up food from other nest loci it performs the same procedure as when foraging for food, with the exception that when the food is too heavy to be picked up, the ant moves on instead of sending a help signal. In this way the temporary solution is significantly improved.

As was mentioned above, there are some constraints that are imposed on the B-MACA algorithm. The first is the colony's storage-capacity constraint, which is implemented so that no single colony can gather all the food into its nest, and to maintain the appropriate balance between domains. The second constraint ensures that when the pheromone intensity of a certain cell drops below a fixed value, that cell's pheromone intensity is restored to the initial value. In this way we maintain a high exploration level. Other constraints are as follows: there can only be a limited number of vertices put in a single cell; each ant can carry only a limited number of pieces of food; the food that is being brought back to the nest is a kind of a tabu, that is, it is not available to other ants. A short tabu list consisting of the last m pieces of food that were moved helps the algorithm to escape from local minima.

19.3.1.2 The Multilevel Algorithm

One already-established way to speed up and globally improve the partitioning method is the use of multilevel techniques. Here, the basic idea is to group vertices together to form clusters that define a new graph. The next step is to recursively iterate this procedure until the graph size falls below a certain threshold. This is followed by a successive refinement of these coarser graphs. This procedure is known as the multilevel paradigm. The multilevel idea (see Figure 19.4) was first proposed by Barnard and Simon [30] as a method of speeding-up spectral bisection, and improved by Hendrickson and Leland [31] who generalized it to encompass local refinement algorithms.

The implementation consists of two parts: graph contraction and partition expansion. In the first part, a coarser graph $G_{\ell+1}(\boldsymbol{V}_{\ell+1}, \boldsymbol{E}_{\ell+1})$ is created from $G_\ell(\boldsymbol{V}_\ell, \boldsymbol{E}_\ell)$ by finding the largest independent subset of graph edges and then collapsing them. Each selected edge is collapsed and the vertices $u_1, u_2 \in \boldsymbol{V}_\ell$ that are at either end of it are merged into the new vertex $v \in \boldsymbol{V}_{\ell+1}$ with weight

$$|v| = |u_1| + |u_2|.$$

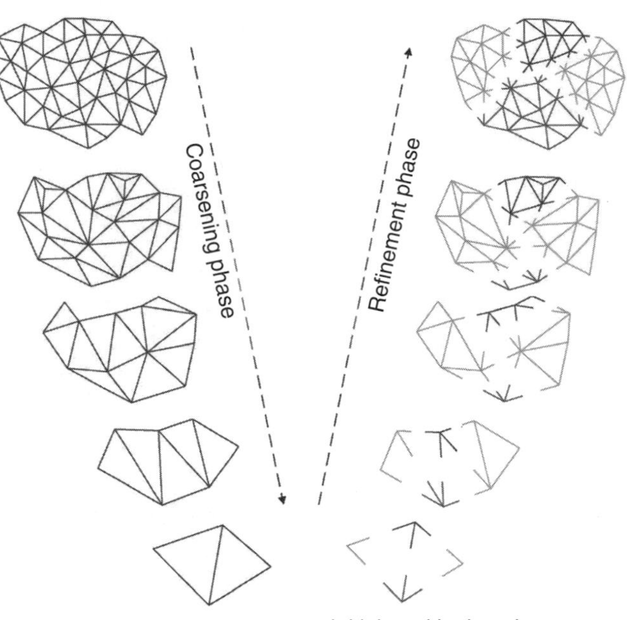

FIGURE 19.4 The three phases of multilevel k-way mesh partitioning.

The edges that have not been collapsed are inherited by the new graph $G_{\ell+1}$ and the edges that have become duplicated are merged and their weight is summed. Because of inheritance, the total weight of the graph remains the same and the total edge weight is reduced by an amount equal to the weight of the collapsed edges, which have no impact on the graph balance or the edge-cut. In the second part, the already-optimized partition (with the algorithm B-MACA) of graph G_ℓ is expanded. The optimized partition must be interpolated onto its parent graph $G_{\ell-1}$. Because of the simplicity of the coarsening in the first part, the interpolation itself is trivial. So, if vertex $v \in \boldsymbol{V}_\ell$ belongs to domain $\boldsymbol{D}_i$, then after refinement the matched pair $u_1, u_2 \in \boldsymbol{V}_{\ell-1}$ that represents v, will also be in $\boldsymbol{D}_i$. In this way the graph is expanded to its original size, and on every level ℓ of our expansion we run our basic ant-colony algorithm. This is referred to as the Multilevel Multiple Ant-Colony Algorithm (M-MACA) approach.

Due to the large graphs and the increased number of levels, the number of vertices in a single cell increases rapidly. To overcome this problem we introduced a method called bucket sort that accelerates and improves the algorithm's convergence by choosing the most "promising" vertex from the cell. The bucket sort, which was first introduced by Fiduccia and Mattheyses [32], has become an essential tool for the efficient and rapid sorting and adjustment of vertices in terms of their gain. The basic idea is that all the vertices with a particular gain g are put together in a "bucket" ranked g. In this way the problem of finding a vertex with maximum gain is converted into finding the nonempty bucket with the highest rank, and then picking a vertex from it. If a chosen vertex migrates from one domain to another, only its gain and the gains of all its neighbors have to be recalculated and put back into appropriate buckets. In our implementation each bucket is represented by a double-linked list of vertices. Because of the multilevel process, it often happens that the potential gain values are dispersed over a wide range. For this reason we have introduced the 2–3 tree. With this we avoided large and sparse arrays of pointers. We store the nonempty buckets in the 2–3 tree, so each leaf in the tree represents a bucket. For even faster searching we have made one 2–3 tree for each colony on every cell that has vertices on it (see Figure 19.5). With this we have increased the speed of the search, as well as the add and delete operations.

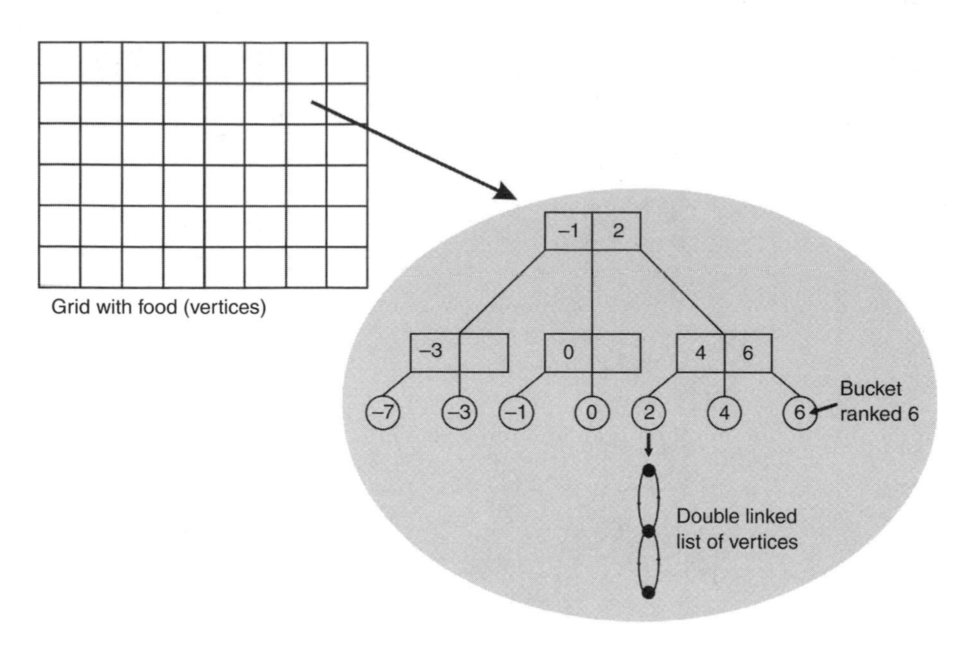

FIGURE 19.5 Bucket sorting.

19.3.1.3 The Hybrid Algorithm

What we have done is to merge vector quantization (VQ) and the basic ant-colony algorithm into a single algorithm called the hybrid ant-colony algorithm (H-MACA) [23]. With the VQ we compute a partition, which is then used as a starting partition for the B-MACA. With the B-MACA we refine this partition so that the best possible result is obtained.

The VQ method [33] is a stochastic approximation method that uses the basic structure of the input vectors to solve a specific problem (e.g., data compression). In other words, the input space is divided into a finite number of distinct regions (domains) and for each region there is a representative vector. When a mapping function (device) receives a new input vector it maps it into a region with which this vector is represented best. This is a simple example of some sort of compression. Of course this is only one possibility of the use of this method. In our case we used it as a mapping device for our mesh-partitioning problem. The mesh vertices are usually locally connected to their neighbors. Now we can treat the position of each mesh vertex as an input vector and each domain in our partition as a region in input space. We try to divide our "mesh" space into domains, so that the size (the number of vertices) of each domain is approximately the same, with as few as possible connections between the domains.

A vector quantizer maps ℓ-dimensional vectors in the vector space $\mathbb{R}^\ell$ into a finite set of vectors

$$\boldsymbol{Y} = \{\boldsymbol{y}_i\colon i = 1, 2, \ldots, k\}.$$

Each vector $\boldsymbol{y}_i$ is called a codeword and the set of all the codewords is called a codebook. Associated with each codeword $\boldsymbol{y}_i$ is a nearest-neighbor region called the Voronoi region, and it is defined by:

$$\boldsymbol{v}_i = \{\boldsymbol{x} \in \mathbb{R}^\ell\colon \|\boldsymbol{x} - \boldsymbol{y}_i\| \leq \|\boldsymbol{x} - \boldsymbol{y}_j\|, \forall i \neq j\}.$$

The set of Voronoi regions partition the entire space $\mathbb{R}^\ell$ such that

$$\left(\bigcup_{i=1}^{k} \boldsymbol{v}_i = \mathbb{R}^\ell\right) \wedge \left(\bigcap_{i=1}^{k} \boldsymbol{v}_i = \emptyset\right).$$

The VQ consists of the following six steps:

Step 1: Determine the number of domains k.

Step 2: Read the input graph and its coordinates. The input vector consists of three elements. The first two are the vertex coordinates and the third is the vertex density. By density we mean how close together its connected neighbors are. The closer they are, the higher is the value for the density.

Step 3: Select k codewords. The initial codewords can be selected randomly from the input vectors or as random points in the input space.

Step 4: Calculate the Euclidian distance between the input vector and the codewords. The input vector is assigned to the domain of the codeword that returns the minimum value according to the function:

$$f(\boldsymbol{x}, \boldsymbol{y}_i) = \varepsilon(\boldsymbol{x}, \boldsymbol{y}_i) - \xi_i - \beta_i,$$

where $\varepsilon(\boldsymbol{x}, \boldsymbol{y}_i)$ represents a function that calculates the Euclidian distance between $\boldsymbol{x}$ and $\boldsymbol{y}_i$, ξ_i represents the change in the edge-cut if $\boldsymbol{x}$ belonged to the i-th domain, and β_i represents the difference between the number of vertices in the largest and the i-th domains.

Step 5: Compute the new set of codewords. We add up all the $\boldsymbol{x}_i$ *vectors* in the i-th domain and divide the summation by the number of input vectors in the domain:

$$\boldsymbol{y}_i = \frac{1}{m} \sum_{j=1}^{m} x_{ij},$$

where m represents the number of input vectors in the i-th domain.

Step 6: Repeat steps 4 and 5 until the values of the codewords converge, usually to a suboptimal solution.

An example of the VQ is shown in Figure 19.6. Here we used a two-dimensional graph ($\ell = 2$), but it can easily be expanded to any other number of dimensions. We can see 45 input vectors that are

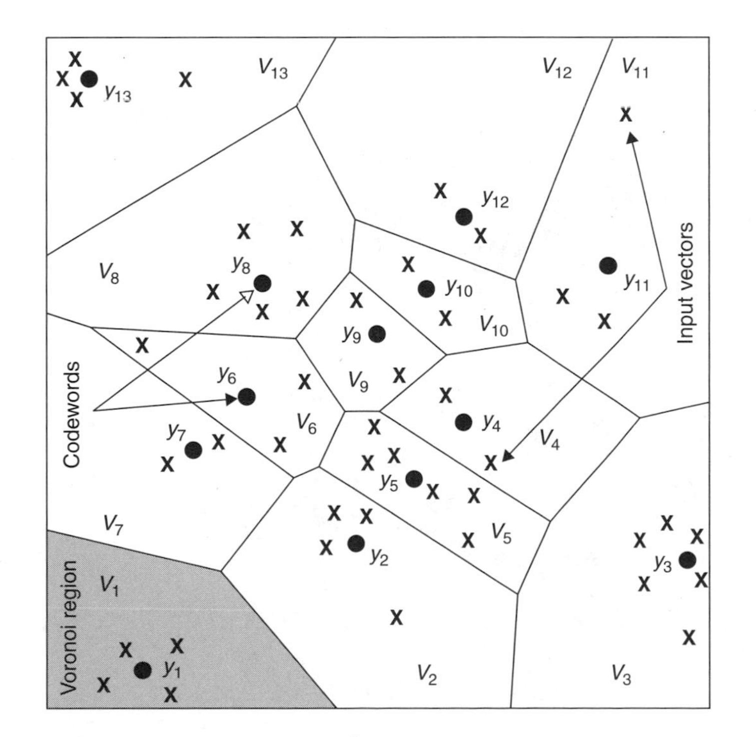

FIGURE 19.6 Vector quantization.

divided into $k = 13$ domains (Voronoi regions $v_1, v_2, \ldots, v_{13}$) and are represented with the codewords $y_1, y_2, \ldots, y_{13}$.

19.3.2 Performance Evaluation

19.3.2.1 Experimental Environment

The B-MACA, M-MACA, and H-MACA were implemented in Borland®Delphi™. The experiments were made on a computer with an AMD Athlon™XP 1800+ processor running the Microsoft®Windows®XP operating system. The implementation also includes a visualization tool to assist the user in selecting the appropriate parameters of the algorithm (see Figure 19.7 and Figure 19.8).

Figure 19.7 shows what the output of the MACA-type algorithm looks like. Here we can see how the algorithm partitioned the graph into four domains and which vertices belong to which partition.

Figure 19.8 consists of two pictures. Figure 19.8(a) represents what the grid looks like after four iterations of the algorithm. As we can see, ants are foraging for food (one food piece represents a graph vertex, which in the case of M-MACA can be composed of many vertices of the initial graph), which is scattered all over the grid. In Figure 19.8(b) we can see that food is laid around the nests and ants are "stealing" food from other nests.

On the right-hand side of the grid you can see "instruments" with which the algorithm is controlled. With it, we are able to set the grid size (Grid size), the number of ants per colony (Number of ants), the desired balance (±diversion), the maximum number of iterations without improvement per level (Number of iterations), the weight of heuristic information (Weight on prob1), the maximum number of gathered food pieces (Fetch number), and the maximum length of the tabu list (Tabu number). All the other data represent the current state of the algorithm.

19.3.2.2 Benchmark Suite

For the purpose of testing, we have randomly generated four graphs. Meanwhile, all the other benchmark graphs were taken from the Graph Partitioning Archive at the University of Greenwich and from the Graph Collection Web page at the University of Paderborn. All the graphs are described in Table 19.2.

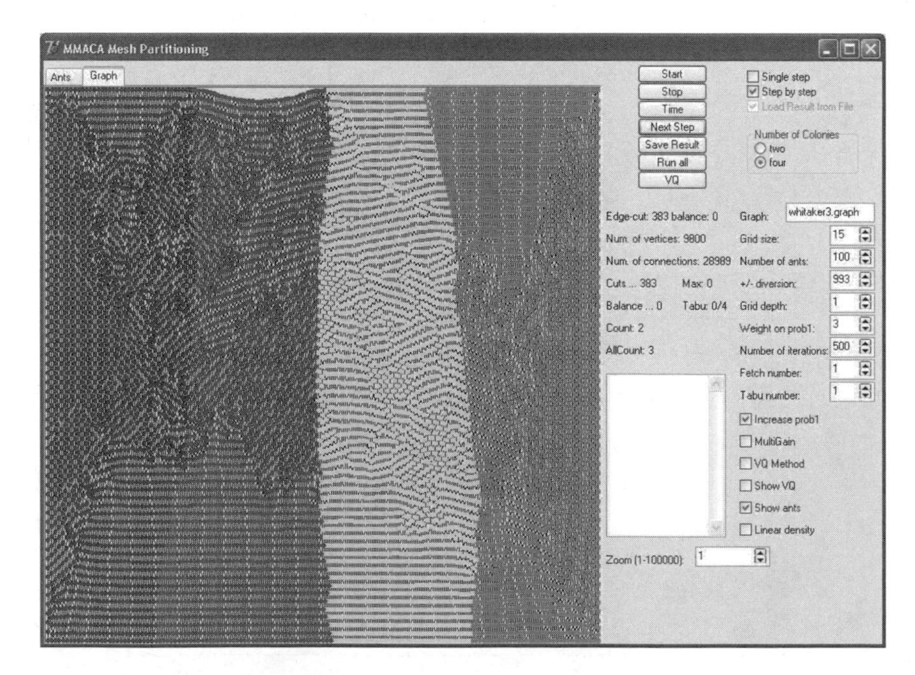

FIGURE 19.7 Mesh-partitioning visualization.

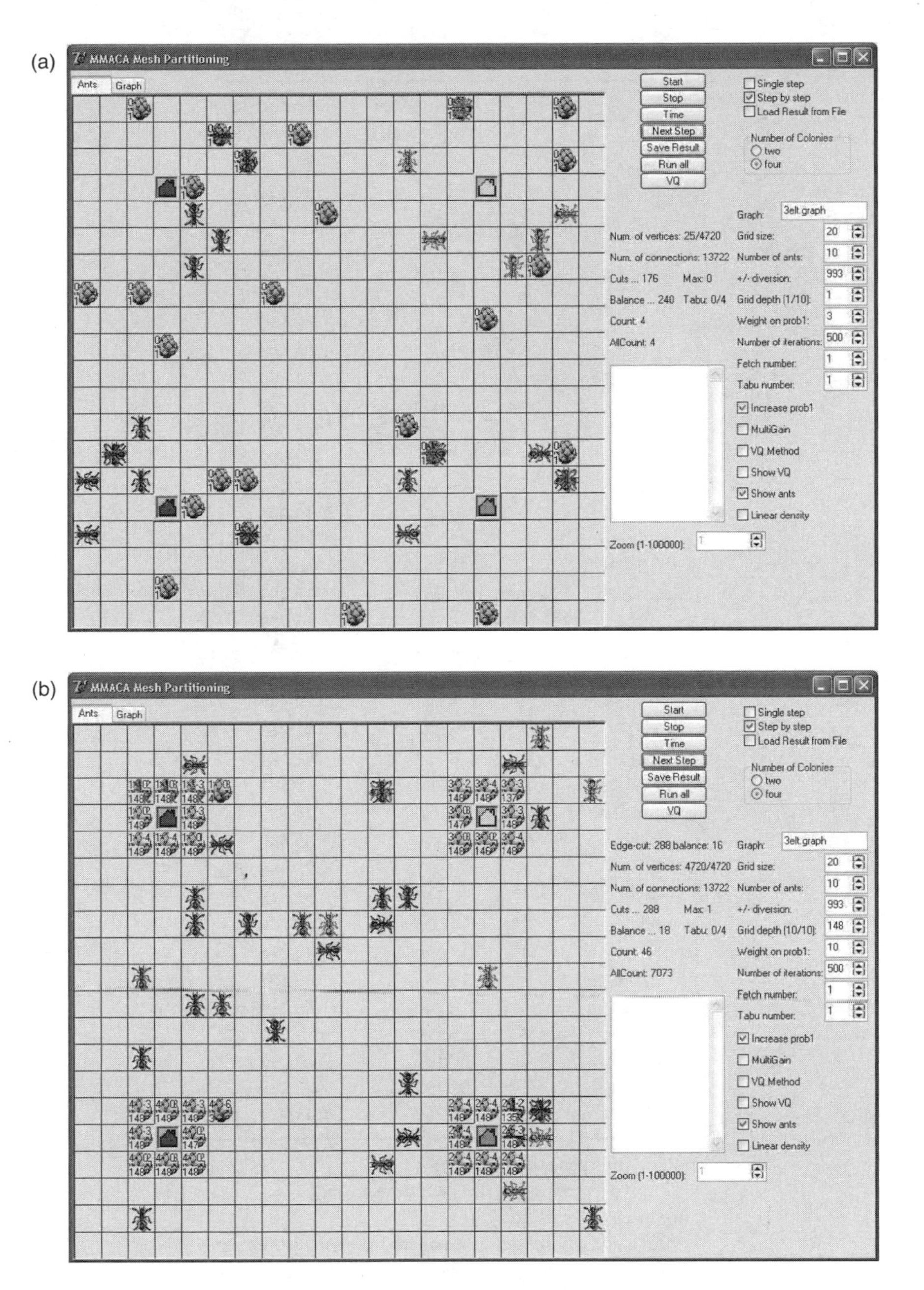

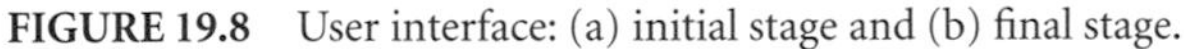

FIGURE 19.8 User interface: (a) initial stage and (b) final stage.

19.3.2.3 Basic Algorithm

Several experiments were made to evaluate the algorithm B-MACA [20]. We compared B-MACA with the well-known partitioning programs k-Metis 4.0 [34], Randomized Tabu Search (RTS) [35] and Hybrid Genetic Algorithm (HGA) [35]. Each test graph was partitioned into two and four domains ($k = 2$ and $k = 4$). The results, which are given in terms of edge-cut $\xi(\boldsymbol{D})$ and balance $\beta(\boldsymbol{D})$, are shown in Table 19.3.

Table 19.3 shows that B-MACA performs well on small graphs ($|\boldsymbol{V}| < 500$). We did additional experiments [20] on larger graphs ($|\boldsymbol{V}| > 1000$) and found that B-MACA performs much worse than RTS or HGA. To remedy this we enhanced the algorithm B-MACA with a multilevel technique.

TABLE 19.2 Benchmark Graphs

Graph	Number of vertices $\|V\|$	Number of edges $\|E\|$
graph1[a]	50	86
graph2[a]	50	143
grid1[c]	252	476
grid1_dual[c]	224	420
graph3[a]	400	546
graph4[a]	400	1,006
netz4504_dual[c]	615	1,171
U1000.5[c]	1,000	2,394
U1000.10[c]	1,000	4,696
U1000.20[c]	1,000	9,339
ukerbel_dual[c]	1,866	3,538
netz4504[c]	1,961	2,578
add20[b]	2,395	7,462
data[b]	2,851	15,093
grid2_dual[c]	3,136	6,112
grid2[c]	3,296	6,432
airfoil[c]	4,253	12,289
3elt[b,c]	4,720	13,722
uk[b]	4,824	6,837
add32[b]	4,960	9,462
ukerbel[c]	5,981	7,852
airfoil_dual[c]	8,034	11,813
bcsstk33[b]	8,738	291,583
3elt_dual[c]	9,000	13,287
whitaker3[b,c]	9,800	28,989
crack[b,c]	10,240	30,380
wing_nodal[b]	10,937	75,488
fe_4elt2[b]	11,143	32,818
4elt[b]	15,606	45,878
big[c]	15,606	45,878
fe_sphere[b]	16,386	49,152
cti[b]	16,840	48,232
whitaker3_dual[c]	19,190	8,581
crack_dual[c]	20,141	30,043
cs4[b]	22,499	43,858
big_dual[c]	30,269	44,929

[a] Randomly generated.
[b] Graph partitioning archive: www.gre.ac.uk/~c.walshaw/partition/.
[c] Graph collection: www.uni-paderborn.de/cs/ag-monien/RESEARCH/PART/graphs.html.

19.3.2.4 Multilevel Algorithm

We again partitioned each of the test graphs into two and four domains ($k = 2$ and $k = 4$). The results are again pairs, $\xi(\boldsymbol{D})$ and $\beta(\boldsymbol{D})$. We allowed a 0.2% unbalance, that is, $\beta(\boldsymbol{D}) \leq 0.002|\boldsymbol{V}|$. The algorithm M-MACA was run 30 times on each graph; and from the solutions obtained we chose the one with minimum $\xi(\boldsymbol{D})$ as the final result for that graph.

The results of our experiments are shown in Table 19.4, where M-MACA is compared with the k-Metis 4.0 [34], the Chaco 2.0 [31] with multilevel Kernighan–Lin global partitioning method, and the new mixed simulated annealing and tabu search algorithm MLSATS [36].

TABLE 19.3 Experimental Results (B-MACA, $k = 2, 4$)

Algorithm		k-Metis 4.0		RTS		HGA		B-MACA	
Graph	k	$\xi(\boldsymbol{D})$	$\beta(\boldsymbol{D})$	$\xi(\boldsymbol{D})$	$\beta(\boldsymbol{D})$	$\xi(\boldsymbol{D})$	$\beta(\boldsymbol{D})$	$\xi(\boldsymbol{D})$	$\beta(\boldsymbol{D})$
graph1	2	15	0	14	0	14	0	13	2
	4	32	12	25	1	25	1	25	1
graph2	2	35	0	33	0	33	0	33	0
	4	77	9	57	1	57	1	57	1
graph3	2	45	8	41	0	40	0	41	0
	4	85	4	82	0	76	0	79	0
graph4	2	208	10	181	0	187	0	182	0
	4	331	2	309	0	310	0	316	2

TABLE 19.4 Experimental Results (M-MACA, $k = 2, 4$)

Algorithm		k-Metis 4.0		Chaco 2.0		MLSATS		M-MACA	
Graph	k	$\xi(\boldsymbol{D})$	$\beta(\boldsymbol{D})$	$\xi(\boldsymbol{D})$	$\beta(\boldsymbol{D})$	$\xi(\boldsymbol{D})$	$\beta(\boldsymbol{D})$	$\xi(\boldsymbol{D})$	$\beta(\boldsymbol{D})$
add20	2	773	23	630	1	696	119	601	5
	4	1,214	45	1,242	1	1,193	57	1,196	3
data	2	253	67	210	1	196	131	199	1
	4	486	42	444	1	395	67	424	1
3elt	2	148	72	124	0	87	108	90	0
	4	250	63	258	0	199	45	225	2
uk	2	33	126	23	0	54	186	20	2
	4	50	47	60	0	261	74	50	2
add32	2	12	62	11	0	10	2	10	2
	4	38	59	53	0	35	66	40	5
bcsstk33	2	13,393	139	10,199	0	10,064	100	10,222	12
	4	22,909	129	25,529	1	22,442	219	22,632	14
whitaker3	2	135	96	135	0	130	0	126	16
	4	407	97	439	0	448	198	386	6
crack	2	221	248	209	0	184	62	184	0
	4	478	106	457	0	401	216	381	8
wing_nodal	2	1,892	315	1,747	1	1,670	543	1,711	23
	4	3,898	163	3,817	1	3,596	273	3,619	11
fe_4elt2	2	132	297	130	1	130	1	130	1
	4	400	87	378	1	350	113	350	2
4elt	2	149	104	242	0	130	1	130	1
	4	385	197	416	1	367	369	389	13
fe_sphere	2	444	418	422	0	384	576	402	4
	4	828	46	835	1	786	329	810	15
cti	2	401	300	369	0	318	120	332	24
	4	1,104	156	1,000	0	966	360	1,045	14
cs4	2	397	653	418	1	418	907	397	1
	4	1,102	312	1,135	1	1,103	403	1,038	21

TABLE 19.5 Results from Graph Partitioning Archive

Graph	k	Algorithm(s)	$\xi(\boldsymbol{D})$ with 1% unbalance
add20	2	M-MACA	601
	4	M-MACA	1,186
data	2	GrPart	196
	4	M-MACA	410
3elt	2	GrPart, M-MACA	89
	4	JE, MLSATS	199
uk	2	M-MACA	19
	4	JE	44
add32	2	J2.2, MLSATS, M-MACA	10
	4	JE	33
bcssth33	2	GrPart	10,109
	4	iJ	21,685
whitaker3	2	JE, M-MACA	126
	4	JE	380
crack	2	M-MACA	183
	4	JE	368
wing_nodal	2	M-MACA	1,697
	4	JE	3,572
fe_4elt2	2	MRSB, MLSATS, M-MACA	130
	4	JE	349
4elt	2	GrPart	138
	4	JE	327
fe_sphere	2	JE	386
	4	N/A	768
cti	2	JE, MLSATS, M-MACA	318
	4	M-MACA	963
cs4	2	JE	367
	4	JE	940

GrPart: A. Kozhushkin's implementation of iterative multilevel Kernighan–Lin; iJ: Iterated JOSTLE — iterated multilevel Kernighan–Lin (k-way) [38]; J2.2: JOSTLE — multilevel Kernighan–Lin (k-way); version 2.2 [38]; JE: JOSTLE Evolutionary — combined evolutionary/multilevel scheme [37]; MLSATS: MultiLevel refinated mixed Simulated Anealing and Tabu Search [36]; MRSB: Multilevel Recursive Spectral Bisection [30]; and M-MACA: Multilevel Multiple Ant-Colony Algorithm [21].

Clearly, M-MACA performed very well. Notice that M-MACA is superior to the classical k-Metis and Chaco algorithms. Notice also that MLSATS produced some results that have better $\xi(\boldsymbol{D})$ but with much higher $\beta(\boldsymbol{D})$ than M-MACA.

The algorithm M-MACA also returned some solutions that are better than currently available (Winter 2003–2004) solutions in the Graph Partitioning Archive (Table 19.5). Furthermore, M-MACA is even comparable with the combined evolutionary/multilevel scheme used in the JOSTLE Evolutionary algorithm [37], which is currently the most promising mesh-partitioning algorithm.

19.3.2.5 Hybrid Algorithm

Now we present and discuss the results of the experimental evaluation of the algorithm H-MACA in comparison with the well-known partitioning programs p-Metis 4.0 [34], Chaco 2.0 [31] with the multilevel Kernighan–Lin global partitioning method and the algorithm M-MACA [21].

TABLE 19.6 Experimental Results (H-MACA, $k = 2, 4$)

Algorithm		Chaco 2.0		p-Metis 4.0		M-MACA		H-MACA	
Graph	k	$\xi(D)$	$\beta(D)$	$\xi(D)$	$\beta(D)$	$\xi(D)$	$\beta(D)$	$\xi(D)$	$\beta(D)$
3elt	2	124	0	98	0	90	0	90	0
	4	258	0	252	0	225	2	212	4
3elt_dual	2	70	0	70	0	44	6	45	8
	4	130	0	120	0	112	4	154	7
airfoil	2	82	1	85	1	74	1	81	1
	4	182	1	179	1	176	2	190	3
airfoil_dual	2	60	0	40	0	37	0	40	14
	4	111	1	84	1	80	7	110	7
big	2	242	0	165	0	141	0	139	0
	4	416	1	405	1	354	11	382	14
big_dual	2	92	1	92	1	78	1	77	11
	4	219	1	196	1	215	18	222	25
crack	2	209	0	206	0	184	0	196	2
	4	457	0	458	0	377	6	371	9
crack_dual	2	130	1	101	1	80	25	87	1
	4	228	1	201	2	169	3	164	9
grid1	2	26	0	20	0	18	0	18	0
	4	48	0	40	0	38	0	38	0
grid1_dual	2	16	0	16	0	16	0	16	0
	4	37	0	35	0	35	0	35	0
grid2	2	38	0	37	0	34	0	34	2
	4	106	0	121	0	94	2	92	2
grid2_dual	2	35	0	32	0	32	0	32	0
	4	99	0	91	0	90	2	96	4
netz4504	2	25	1	26	1	22	1	24	1
	4	66	1	62	1	50	1	49	3
netz4504_dual	2	21	1	21	1	19	1	19	1
	4	54	1	49	1	44	2	44	2
U1000.5	2	10	0	1	0	1	0	2	0
	4	20	0	6	0	7	2	12	2
U1000.10	2	115	0	56	0	39	0	39	0
	4	200	0	108	2	99	2	107	2
U1000.20	2	294	0	253	0	220	4	221	2
	4	554	0	515	2	546	2	497	2
ukerbe1	2	30	1	28	1	27	1	28	1
	4	82	1	64	1	63	2	61	1
ukerbe1_dual	2	25	0	25	0	22	0	22	0
	4	56	1	51	1	52	3	48	1
whitaker3	2	135	0	128	0	126	16	127	0
	4	439	0	424	0	383	0	383	2
whitaker3_dual	2	82	0	74	0	64	18	65	0
	4	251	1	210	1	200	6	195	12

We partitioned each of the graphs into two and four domains ($k = 2$ and $k = 4$). Each score is described with the best obtained edge-cut $\xi(D)$. It is important to mention that the balance $\beta(D)$ was kept inside 0.2% of the $|V|$. The results of our experiment are shown in Table 19.6.

Table 19.6 shows that in most cases the best partition was obtained with the H-MACA algorithm.

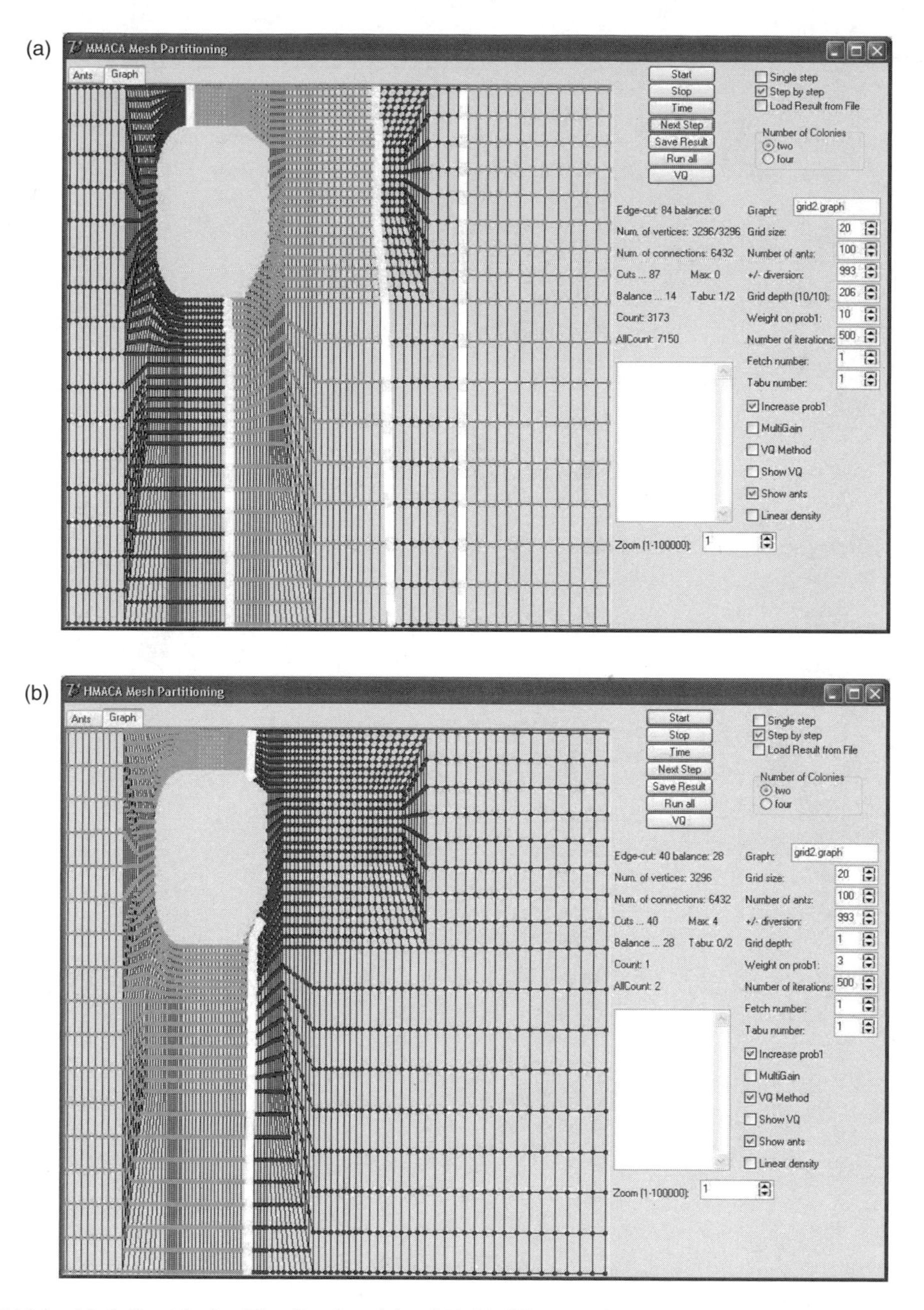

FIGURE 19.9 Mesh-Partitioning Visualization: (a) M-MACA; (b) H-MACA.

Figure 19.9 shows the main drawback of the M-MACA. Here we can see vertices represented as lighter and darker dots, where lighter belongs to the one domain, and darker to the other. With the solid white line we have emphasized the border between these two domains. The edges are hidden again. The final solution obtained by the M-MACA includes "islands" (i.e., a set of connected vertices) that belong to different domains. The islands are due to a bad initial partition and the inability of the M-MACA to merge the islands into homogeneous regions (each having only one border). This drawback is eliminated by using the VQ to obtain the initial partitions. In Figure 19.9(a) (M-MACA) one can see four such islands, whereas there are only two in Figure 19.9(b) (H-MACA).

19.4 Conclusions

The graph-partitioning problem is an important component for mesh partitioning in the domain-decomposition method. The ACO uses a metaheuristic approach for solving hard combinatorial optimization problems. The purpose of this chapter was to give the reader a basic knowledge of ACO, to investigate variants of the MACA for mesh partitioning, to suggest modifications to improve this algorithm, and to evaluate them experimentally.

The B-MACA performed very well on small- or medium-sized graphs ($n < 500$). With larger graphs, which are often encountered in mesh partitioning, we had to use a multilevel or hybrid method to produce results that were competitive with the results given by other algorithms. Both multilevel and hybrid algorithms performed very well on almost all graphs. Both of them were quite similar in producing the best results. The only difference was in the standard deviation of the results [39], which was in favor of the hybrid method. The M-MACA and the H-MACA are very promising algorithms that need to be thoroughly investigated.

An obvious improvement of our algorithm would be to merge these two methods into one. To do this one could apply VQ to produce a starting partition, then coarsen the graph to some extent (to a much smaller extent than in the original multilevel method), and then use the multilevel method to refine the previously (with VQ) obtained partition.

On the other hand, there are many possibilities for improving multilevel and, to some extent, hybrid algorithms. One possibility is in the mapping of the graph onto the grid: with a proper mapping convergence the results can be improved. The use of the load-balancing method between levels would also be a very promising way to go. The next possibility is in determining which and how many vertices from the cell will be picked and with what probability. Here, the Kernighan–Lin gain method [40] might be used. We could also add some daemon actions, like the min-cut algorithm, to improve solutions during the crossing from one level to another. And, finally, we could change the way the pheromone is evaporated, deposited and restored.

There is a wide range of possibilities to be considered in the future. One of the most appealing is a merger of the MACA with some other method through daemon actions and parallel implementation of the MACA.

References

[1] Cook, R.D. et al. *Concepts and Applications of Finite Element Analysis*, 4th ed. John Wiley & Sons, New York, 2001.

[2] Farhat, C. and Lesoinne, M. Automatic partitioning of unstructured meshes for the parallel solution of problems in computational mechanics. *International Journal for Numerical Methods in Engineering*, 36, 745, 1993.

[3] Shephard, M.S. et al. Parallel automated adaptive procedures for unstructured meshes. In *Special Course on Parallel Computing in Computational Fluid Dynamics* (AGARD-R-807), AGARD, Neuilly-sur-Seine, France, 1995, p. 6.1.

[4] Pothen, A., Simon, H.D., and Liou, K.P. Partitioning sparse matrices with eigenvectors of graphs. *SIAM Journal on Matrix Analysis and Application*, 11, 430, 1990.

[5] Zomaya, A.Y. et al. Non-conventional computing paradigms in the new millennium. *IEEE/AIP Computing in Science and Engineering*, 3, 82, 2001.

[6] Kadłuczka, P. and Wala, K. Tabu search and genetic algorithms for the generalized graph partitioning problem. *Control and Cybernetics*, 24, 459, 1995.

[7] Tao, L. et al. Simulated annealing and tabu search algorithms for multiway graph partition, *Journal of Circuits, Systems and Computers*, 2, 159, 1992.

[8] Bahreininejad, A., Topping, B.H.V., and Khan, A.I. Finite element mesh partitioning using neural networks. *Advances in Engineering Software*, 27, 103, 1996.

[9] Żola, J. and Wyrzykowski, R. Application of genetic algorithm for mesh partitioning. In *Proceedings of the Workshop Parallel Numerics*, Bratislava, Slovakia, 2000, p. 209.
[10] Blum, C. and Roli, A. Metaheuristics in combinatorial optimization: Overview and conceptual comparison. *ACM Computing Surveys*, 35, 268, 2003.
[11] Colorni, A., Dorigo, M., and Maniezzo, V. Distributed optimization by ant colonies. In *Proceedings of the 1st European Conference on Artificial Life*, Paris, France, 1991, p. 134.
[12] Dorigo, M., Maniezzo, V., and Colorni, A. The ant system: Optimization by a colony of cooperating agents. *IEEE Transactions on Systems Man and Cybernetics Part B*, 26, 29, 1996.
[13] Dorigo, M. and Gambardella, L.M. Ant colony system: A cooperative learning approach to the traveling salesman problem. *IEEE Transactions on Evolutionary Computation*, 1, 53, 1997.
[14] Gambardella, L.M., Taillard, E., and Dorigo, M. Ant colonies for the quadratic assignment problem. *Journal of Operational Research Society*, 50, 167, 1999.
[15] Teich, T. et al. A new ant colony algorithm for the job shop scheduling problem. In Beyer, H., Canta-Paz, E., GoldBerg, D., Parmee, Spector, L., and Whitley, D., Eds., *Proceedings of Genetic Evolutionary Computation Conference*, Morgan Kaufmann Publishers, San Francisco, CA, 2001, p. 803.
[16] Montemanni, R. et al. A new algorithm for a dynamic vehicle routing problem based on ant colony system. In *Proceedings of the 34th Annual Conference of the Italian Operations Research Society*, Venice, Italy, 2003, p. 140.
[17] Sim, K.M. and Sun, W.H. Multiple ant-colony optimization for network routing. In *Proceedings of the 1st International Symposium Cyber Worlds*, Tokyo, Japan, 2002, p. 277.
[18] Dorigo, M. and Stützle, T. *Ant Colony Optimization*, MIT Press, Cambridge, CA, 2004.
[19] Dorigo, M., Bonabeau, E., and Theraulaz, G. Ant algorithms and stigmergy. *Future Generation Computer Systems*, 16, 851, 2000.
[20] Korošec, P., Šilc, J., and Robič, B. An ant-colony-optimization approach to the mesh-partitioning problem. In *Parallel Numerics '02*, Trobec, R. et al., Eds., University of Salzburg and Jožef Stefan Institute, 2002, p. 123.
[21] Korošec, P., Šilc, J., and Robič, B. A multilevel ant-colony-optimization algorithm for mesh partitioning. *International Journal of Pure and Applied Mathematics*, 5, 143, 2003.
[22] Langham, A.E. and Grant, P.W. Using competing ant colonies to solve k-way partitioning problems with foraging and raiding strategies, *Lecture Notes in Computer Science*, 1674, 621, 1999.
[23] Šilc, J., Korošec, P., and Robič, B. Combining vector quantization and ant-colony algorithm for mesh-partitioning. *Lecture Notes in Computer Science*, 3019, 113, 2004.
[24] Deneubourg, J.-L. et al. The self-organizing exploratory pattern of the argentine ant. *Journal of Insect Behavior*, 3, 159, 1990.
[25] Stützle, T. and Hoos, H.H. MAX–MIN ant system. *Future Generation Computer Systems*, 16, 889, 2000.
[26] Pelikan, M., Goldberg, D.E., and Lobo, F.G. A survey of optimization by building and using probabilistic models. *Computational Optimization and Applications*, 21, 5, 2002.
[27] Zlochin, M. et al. Model-based search for combinatorial optimization: A critical survey. *Annuals of Operations Research*, 131, 373, 2004.
[28] Meuleau, N. and Dorigo, M. Ant colony optimization and stochastic gradient descent. *Artificial Life*, 8, 103, 2002.
[29] Handl, J. and Meyer, B. Improved ant-based clustering and sorting in a document retrival interface. *Lecture Notes in Computer Science*, 2439, 913, 2002.
[30] Barnard, S.T. and Simon, H.D. A fast multilevel implementation of recursive spectral bisection for partitioning unstructured problems. *Concurrency — Practice and Experience*, 6, 101, 1994.
[31] Hendrickson, B. and Leland, R. A multilevel algorithm for partitioning graphs. In *Proceedings of the ACM/IEEE Conference on Supercomputing*, San Diego, CA, 1995, p. 28.
[32] Fiduccia, C.M. and Mattheyses, R.M. A linear time heuristic for improving network partitions, In *Proceedings of 19th IEEE Design Automation Conference*, Las Vegas, NV, 1982, p. 175.

[33] Linde, Y., Buzo, A., and Gray, R.M. An algorithm for vector quantizer design. *IEEE Transactions on Communications*, 28, 84, 1980.
[34] Karypis, G. and Kumar, V. Multilevel k-way partioning scheme for irregular graphs. *Journal of Parallel and Distributed Computing*, 48, 96, 1998.
[35] Šilc, J., Korošec, P., and Robič, B. An experimental evaluation of modified algorithms for the graph partitioning problem. In *Proceedings of 17th International Symposium on Computer and Information Science*, Orlando, FL, October 28–30, CRC Press, Boca Raton, FL, 2002, p. 120.
[36] Banos, R. et al. Multilevel heuristic algorithm for graph partitioning. *Lecture Notes in Computer Science*, 2611, 143, 2003.
[37] Soper, A.J., Walshaw, C., and Cross, M. A combined evolutionary search and multilevel approach to graph partitioning. In *Proceedings of Genetic Evolutonary Computation Conference*, Las Vegas, NV, 2000, p. 674.
[38] Walshaw, C. and Cross, M. Mesh partitioning: A multilevel balancing and refinement algorithm. *SIAM Journal of Scientific Computing*, 22, 63, 2000.
[39] Korošec, P., Šilc, J., and Robič, B. Solving the mesh-partitioning problem with an ant-colony algorithm. *Parallel Computing*, 30, 785, 2004.
[40] Kernighan, B.W. and Lin, S. An efficient heuristic procedure for partitioning graph. *Bell System Technical Journal*, 49, 291, 1970.

20

Simulating the Strategic Adaptation of Organizations Using OrgSwarm

Anthony Brabazon
Arlindo Silva
Ernesto Costa
Tiago Ferra de Sousa
Michael O'Neill

20.1 Introduction

In an organizational setting, a strategy consists of a choice of what activities the organization will perform, and choices as to how these activities will be performed [1]. These choices define the strategic configuration of the organization. Recent work by Levinthal [2] and Rivkin [3] has recognized that strategic configurations consist of interlinked individual elements (decisions), and have applied general models of interconnected systems such as Kauffman's NK model to examine the implications of this for processes of organizational adaptation.

Following a long-established metaphor of adaptation as search [4], strategic adaptation is considered in this study as an attempt to uncover peaks on a high-dimensional strategic landscape. Some strategic configurations produce high profits, others produce poor results. The search for good strategic configurations is difficult due to the vast number of strategic configurations possible, uncertainty as to the nature of topology of the strategic landscape faced by an organization, and changes in the topology of this landscape over time. Despite these uncertainties, the search process for good strategies is not blind. Decision-makers receive feedback on the success of their current and historic strategies, and can assess the payoffs received by the strategies of their competitors [5]. Hence, certain areas of the strategic landscape are illuminated.

Organizations do not exist in isolation but interact with, and receive feedback from their environment. Their efforts at strategic adaptation are guided by *social* as well as *individual* learning. Good ideas discovered by one organization disseminate over time. Particle swarm algorithms (PSAs) also emphasize the importance of individual and social learning processes. Surprisingly, despite the parallels between the learning processes in particle swarm algorithms and those in populations of organizations, as yet the particle swarm metaphor has not been applied to the domain of organizational science. This chapter describes a novel simulation model based on the particle swarm metaphor, and applies this to examine the process of organizational adaptation. This study, therefore, extends the particle swarm metaphor into the domain of organization science.

20.2 Strategic Adaptation

Strategic adaptation and strategic inertia are closely linked. If strategic adaptation is problematic, inertia is a likely contributing cause. Broadly speaking, the strategic inertia of organizations stems from two sources, *imprinting forces*, and as a *consequence of market selection forces.*

Imprinting forces [6] combine to define and solidify the strategic configuration of a newly formed organization. These forces include the dominant initial strategy pursued by the organization, the skills/prior experience of the management team, and the distribution of decision-making influence in the organization at time of founding [6]. All of these influence the initial choice of organizational strategy. As consensus concerning the strategy emerges, it is imprinted on the organization through resource allocation decisions [7]. The imprinting leads to inertia by creating sunk costs, internal political constraints, and a rigid organizational structure. Over time this inertia intensifies due to the formation of an organizational history, which creates barriers to industry exit, and legitimacy issues if adaptation is suggested [8]. The resulting inertia serves to circumscribe the organization's ability to adapt its strategy in the future. Imprinting also occurs as relationships are built up with suppliers and customers. The creation of a web of these relationships can serve to constrain the range of strategic alternatives in the future, as strategic moves that dramatically disrupt the web are less likely to be considered.

The discussion of strategic inertia was extended by Hannan and Freeman [9] who posited that inertia is also created as a natural *consequence* of the market selection process, claiming that "selection processes tend to favor organizations whose structures are difficult to change" (p. 149). The basis of this claim is that organizations that can produce a good or service reliably (consistently of a minimum quality standard) are favored for trading purposes by other organizations, and therefore by market selection processes. The routines required to produce a product or service reliably tend to lead to structural inertia, as the construction of standardized routines leads to an increase in the complexity of the patterns of links between organizational subunits [9,10]. It can, therefore, be posited that efficient organizations are likely to exhibit inertia. As organizations seek better environment-structure congruence, their systems become increasingly specialized and interlinked, making changes to their activities become costly and difficult. Tushman and O'Reilly [11] note that structural inertia is rooted in the size, complexity, and interdependence of the firm's structures, systems, procedures, and processes. Theoretical support for these assertions, that increasing organizational complexity can make adaptation difficult, is found in [3] and [12], as the heightened degree of interconnections between activities within the organization will increase the "ruggedness" of the strategic landscape on which they are adapting.

20.3 Particle Swarm Algorithm

This section provides an introduction to the canonical Particle Swarm Algorithm (PSA). The term PSA is used in place of the commonly used PSO (Particle Swarm Optimization) in this chapter, as the objective is not to develop a tool for "optimizing," but to adapt and apply the swarm metaphor as a model of organizational adaptation. The PSA [13,14] has been widely used for function optimization, and is based on a metaphor of human social interaction [15].

Under the swarm metaphor, a swarm of particles (entities) are assumed to move (fly) through an n-dimensional space, typically looking for a function optimum. Each particle is assumed to have two associated properties, a current position and a velocity. Each particle also has a memory of the best location in the search space that it has found so far (**pbest**), and knows the best location found to date by all the particles in the population (**gbest**). At each step of the algorithm, particles are displaced from their current position by applying a velocity vector to them. The size and direction of this velocity is influenced by the velocity in the previous iteration of the algorithm (simulates "momentum"), and the current location of a particle relative to its **pbest** and **gbest**. Therefore, at each step, the size and direction of each particle's move is a function of its own history (experience), and the social influence of its peer group. A number of variants of the PSA exist.

20.3.1 Description of PSA

The following paragraphs provide a description of the continuous version of the PSA. The algorithm is initially described narratively. This is followed by a description of the particle position-update equations.

1. Initialize each particle in the population by randomly selecting values for its location and velocity vectors.
2. Calculate the fitness value of each particle. If the current fitness value for a particle is greater than the best fitness value found for the particle so far, then revise **pbest**.
3. Determine the location of the particle with the highest fitness and revise **gbest** if necessary.
4. For each particle, calculate its velocity according to Equation (20.1).
5. Update the location of each particle.
6. Repeat steps 2 to 5 until stopping criteria are met.

Each particle i has an associated current position in d-dimensional space $\mathbf{x_i}$, a current velocity $\mathbf{v_i}$, and a personal best position $\mathbf{y_i}$. During each iteration of the algorithm, the location and velocity of each particle is updated using Equations (20.1) to (20.4). Assuming a function f is to be maximized, that the swarm consists of n particles, and that r_1, r_2 are drawn from a uniform distribution in the range (0, 1), the velocity update is described as follows:

$$\mathbf{v_i}(t+1) = W\mathbf{v_i}(t) + c_1 r_1(\mathbf{y_i} - \mathbf{x_i}(t)) + c_2 r_2(\hat{\mathbf{y}} - \mathbf{x_i}(t)), \tag{20.1}$$

where $\hat{\mathbf{y}}$ is the location of the global-best solution found by all the particles. A variant on the basic algorithm is to use a local rather than a global version of **gbest**, and the term **gbest** is replaced by **lbest**. In the local version, **lbest** is set independently for each particle, based on the best point found thus far within a *neighborhood* of that particle's current location.

In every iteration of the algorithm, each particle's velocity is stochastically accelerated toward its previous best position and toward **gbest** (or **lbest**). The weight-coefficients c_1 and c_2 control the relative impact of **pbest** and **gbest** locations on the velocity of a particle. The parameters r_1 and r_2 ensure that the algorithm is stochastic. A practical effect of the random coefficients r_1 and r_2 is that neither the individual nor the social learning terms are always dominant.

Although the velocity update has a stochastic component, the search process is not random. It is guided by the memory of past "good" solutions (corresponding to a psychological tendency for individuals to repeat strategies that have worked for them in the past [15], and by the global best solution found by all particles thus far. W represents a momentum coefficient that controls the impact of a particle's prior-period velocity on its current-period velocity. Each component (dimension) of the velocity vector $\mathbf{v_i}$ is restricted to a range $[-v_{max}, v_{max}]$ to ensure that individual particles do not leave the search space. The implementation of a v_{max} parameter can also be interpreted as simulating the incremental nature of most learning processes [15]. The value of v_{max} is usually chosen to be $k \times x_{max}$, where $0 < k < 1$. Once the velocity update for particle i is determined, its position is updated and **pbest** is updated if necessary, as described in Equations (20.2) to (20.4).

$$\mathbf{x_i}(t+1) = \mathbf{x_i}(t) + \mathbf{v_i}(t+1), \tag{20.2}$$

$$\mathbf{y_i}(t+1) = \mathbf{y_i}(t) \quad \text{if } f(\mathbf{x_i}(t)) \leq f(\mathbf{y_i}(t)), \tag{20.3}$$

$$\mathbf{y_i}(t+1) = \mathbf{x_i}(t) \quad \text{if } f(\mathbf{x_i}(t)) > f(\mathbf{y_i}(t)) \tag{20.4}$$

After all particles have been updated, a check is made to determine whether **gbest** needs to be updated.

$$\hat{\mathbf{y}} \in (\mathbf{y_0}, \mathbf{y_1}, \ldots, \mathbf{y_n}) | f(\hat{\mathbf{y}}) = \max(f(\mathbf{y_0}), f(\mathbf{y_1}), \ldots, f(\mathbf{y_n})). \tag{20.5}$$

20.3.2 Particle Swarm as a Metaphor for Organizational Adaptation

Although particle swarm algorithms have been used extensively in function optimization (PSO), the original inspiration for PSAs arose from observations of animal and human social behavior [13]. Kennedy has published a series of papers that emphasize the social aspects of particle swarm [15–17] and this work was given prominence in the first major book on particle swarm [14].

The velocity update formula (Equation [20.1]) can be divided into *cognitive* and *social* components [15], with the former relating to the adaptive history of a particle, an individual, or an organization. The cognitive term can be considered as an interpretation of Thorndike's *Law of Effect* [18], which states that a behavior that is followed by a (positive) reinforcement becomes more likely in the future. The individual learning component in the velocity update formula $(\mathbf{y_i(t)} - \mathbf{x_i(t)})$ introduces a stochastic tendency to return to previously rewarded strategies, mimicking a psychological tendency for managers to repeat strategies that have worked for them in the past [15]. The social learning component of the formula $(\hat{\mathbf{y}}_\mathbf{i}\mathbf{(t)} - \mathbf{x_i(t)})$ can be compared with social *no-trial learning* [19], in which the observation of a peer being rewarded for a behavior will increase the probability of the observer engaging in the same behavior.

The mechanisms of the canonical PSA bear prima facie similarities to those of the domain of interest, organizational adaptation. It adopts a populational perspective, and learning in the algorithm, just as in populations of organizations, is both distributed and parallel. Organizations persist in employing already-discovered good strategies, and are attracted to, and frequently imitate, good product ideas and business practices discovered by other organizations. However, the canonical PSA requires modification before it can employed as a component of a plausible simulation model of organizational adaptation. These modifications are discussed in the next section.

20.4 Simulation Model

The two key components of the simulation model, the landscape generator (environment), and the adaptation of the basic PSA to incorporate the activities and interactions of the agents (organizations) are described next.

20.4.1 Strategic Landscape

In this chapter, the strategic landscape is defined using the NK model [12,20]. It is noted ab initio that application of the NK model to define a strategic landscape is not atypical and has support from prior literature in organizational science that has adopted this approach [2,3,21,22], and related work on technological innovation [23,24]. The NK model considers the behavior of systems that are comprised of a configuration (string) of N individual elements. Each of these elements are in turn interconnected to K other of the N elements ($K < N$). In a general description of such systems, each of the N elements can assume a finite number of states. If the number of states for each element is constant (S), the space of all possible configurations has N dimensions, and contains a total of $\prod_{i=1}^{N} S_i$ possible configurations.

In Kauffman's operationalization of this general framework [12], the number of states for each element is restricted to two (0 or 1). Therefore the configuration of N elements can be represented as a binary string. The parameter K, determines the degree of fitness interconnectedness of each of the N elements and can vary in value from 0 to $N - 1$. In one limiting case where $K = 0$, the contribution of each of the N elements to the overall fitness value (or worth) of the configuration are independent of each other. As K increases, this mapping becomes more complex, until at the upper limit when $K = N - 1$, the fitness contribution of any of the N elements depends both on its own state, and the simultaneous states of all the other $N - 1$ elements, describing a fully connected graph.

If we let s_i represent the state of an individual element i, the contribution of this element (f_i) to the overall fitness (F) of the entire configuration is given by $f_i(s_i)$ when $K = 0$. When $K > 0$, the contribution of an individual element to overall fitness depends both on its state and the states of K other elements to which it is linked ($f_i(s_i{:}s_{i1}, \ldots, s_{ik})$). A random fitness function ($U(0, 1)$) is adopted, and the overall fitness of each configuration is calculated as the average of the fitness values of each of its individual elements.

Altering the value of K effects the ruggedness of the described landscape, and consequently impacts on the difficulty of search on this landscape [12,20]. The strength of the NK model in the context of this study is that by tuning the value of K it can be used to generate strategic landscapes (graphs) of differing degrees of local-fitness correlation (ruggedness).

The strategy of an organization is characterized as consisting of N attributes [2]. Each of these attributes represents a strategic decision or policy choice, which an organization faces. Hence a specific strategic configuration $\mathbf{s}$, is represented as a vector $s_1, \ldots, s_N$ where each attribute can assume a value of 0 or 1 [3]. The vector of attributes represents an entire organizational form, hence it embeds a choice of markets, products, method of competing in a chosen market, and method of internally structuring the organization [3]. Good consistent sets of strategic decisions — configurations, correspond to peaks on the strategic landscape.

The definition of an organization as a vector of strategic attributes finds resonance in the work of Porter [1,25], where organizations are conceptualized as a series of activities forming a value-chain. The choice of what activities to perform, and subsequent decisions as to how to perform these activities, defines the strategy of the organization. The individual attributes of an organization's strategy interact. For example, the value of an efficient manufacturing process is enhanced when combined with a high-quality sales force. Differing values for K correspond to varying degrees of payoff-interaction among elements of the organization's strategy [3]. As K increases, the difficulty of the task facing strategic decision-makers is magnified. Local-search attempts to improve an organization's position on the strategic landscape become ensnared in a web of conflicting constraints.

20.4.2 Simulation Model

Five characteristics of the problem domain that impact on the design of a simulation model are:

1. The environment is dynamic
2. Organizations are prone to strategic inertia. Their adaptive efforts are anchored by their past
3. Organizations do not knowingly select poorer strategies than the one they already have (election operator)
4. Organizations make errorful *ex-ante* and assessments of fitness
5. Organizations coevolve

Although our simulator embeds all of the above, in this chapter we report results that consider the first three of these factors. We note that this model bears passing resemblance to the *eleMentals* model of [16], which combined a swarm algorithm and an NK landscape to investigate the development of culture and intelligence in a population of hypothetical beings called eleMentals. However, the *OrgSwarm* simulator is differentiated from the eleMental model on grounds of application domain, and because it incorporates the above five characteristics of strategic adaptation.

20.4.2.1 Dynamic Environment

Organizations do not compete in a static environment. The environment may alter as a result of exogenous events, for example a *regime change* such as the emergence of a new technology or a change in customer preferences. This can be mimicked in the simulation by stochastically respecifying the strategic landscape during the course of a simulation run. These respecifications simulate a dynamic environment, and a change in the environment may at least partially negate the value of past learning (adaptation) by organizations. Minor respecifications are simulated by altering the fitness values associated with one of the N dimensions of the strategic landscape, whereas in major changes, the fitness of the entire strategic landscape is redefined. The environment faced by organizations can also change as a result of competition among the population of organizations. The effect of interfirm competition is left for future work.

20.4.2.2 Strategic Anchor

Organizations do not have complete freedom to alter their current strategy. Their adaptive processes are subject to strategic inertia. This inertia springs from the organization's culture, history, and the mental models of its management [6]. In the simulation, strategic inertia is mimicked by implementing a *strategic anchor*. The degree of inertia can be varied from zero to high. In the latter case, the organization is highly constrained from altering its strategic stance. By allowing the weight of this anchor to vary, adaptation processes corresponding to different industries, each with different levels of inertia, can be simulated. Inertia could be incorporated into the PSA in a variety of ways. We have chosen to incorporate it into the velocity update equation, so that the velocity and direction of the particle at each iteration is also a function of the location of its strategic anchor. Therefore for the simulations, Equation (20.1) is altered by adding an additional "anchor" term

$$\mathbf{v_i}(t+1) = \mathbf{v_i}(t) + R_1(\mathbf{y_i} - \mathbf{x_i}(t)) + R_2(\hat{\mathbf{y}} - \mathbf{x_i}(t)) + R_3(\mathbf{a_i} - \mathbf{x_i}(t)), \tag{20.6}$$

where $\mathbf{a_i}$ represents the position of the anchor for organization i (a full description of the other terms such as R_1 is provided in the pseudo-code below). The weight attached to the anchor parameter (R_3) (relative to those attached to pbest and gbest), can be altered by the modeler. The position of the anchor can be fixed at the initial position of the particle at the start of the simulation, or it can be allowed to "drag," thereby being responsive to the adaptive history of the particle. In the latter case, the position of the anchor for each particle corresponds to the position of that particle "x" iterations ago.

20.4.2.3 Election operator

Real-world organizations do not usually intentionally move to strategies that are poorer (i.e produce a lower payoff) than the one they already have. Hence, an election operator (also referred to as a *conditional update* or *ratchet operator*) is implemented, which when turned on ensures that position updates that would worsen an organization's strategic fitness are discarded. In these cases, an organization remains at its current location.

20.4.3 Outline of Swarm Algorithm

As the strategic landscape is described using a binary representation (the NK model), the canonical PSA is adapted for the binary case using the *BinPSO* version of the algorithm [26]. The binary version of the PSA is inspired by the idea that an agent's probability of making a binary decision (yes/no, true/false) is a function of both personal history and social factors. The probability that an agent chooses a value of for example 1, for a particular decision in the next time period, is a function of the agent's history ($\mathbf{x_i(t)}, \mathbf{v_i(t)}$ & **pbest**) and social factors (**lbest**) (see Equation [20.7])

$$\text{Prob}(\mathbf{x_i(t+1)} = 1) = f(\mathbf{x_i(t)}, \mathbf{v_i(t)}, \mathbf{pbest}, \mathbf{lbest}). \tag{20.7}$$

The vector $\mathbf{v_i}$ is interpreted as organization i's predisposition to set each of the N binary strategic choices that it faces to 1. The higher the value of v_i^j for an individual decision j, the more likely that organization i will choose to set decision $j = 1$, with lower values of v_i^j favoring the choice of decision $j = 0$.

In order to model the tendency of managers to repeat historically good strategies, values for each dimension of $\mathbf{x_i}$ that match those of **pbest**, should become more probable in the future. Adding the difference between $pbest_i^j$ and x_i^j for organization i to v_i^j will increase the likelihood that organization i will choose to set decision $j = 1$ if the difference is positive (when $pbest_i^j = 1$ and $x_i^j = 0$). If the difference between $pbest_i^j$ and x_i^j for organization i is negative (when $pbest_i^j = 0$, and $x_i^j = 1$), adding the difference to v_i^j will decrease v_i^j.[1]

In each iteration of the algorithm, the agent adjusts his decision-vector ($\mathbf{x_i(t)}$), taking account of his historical experience (**pbest**), and the best strategy found by his peer group (**lbest**). Hence, the velocity update equation used in the continuous version of the PSA (see Equation [20.6]) can still be used, although now, $\mathbf{v_i(t+1)}$ is interpreted as the updated vector of an agent's predisposition (or probability thresholds) to set each of the N binary strategic choices that it faces to one

$$\mathbf{v_i(t+1)} = \mathbf{v_i(t)} + R_1(\mathbf{pbest_i} - \mathbf{x_i(t)}) + R_2(\mathbf{lbest_i} - \mathbf{x_i(t)}) + R_3(\mathbf{anchor_i} - \mathbf{x_i(t)}). \tag{20.8}$$

To ensure that each element of the vector $\mathbf{v_i(t+1)}$ is mapped into $(0, 1)$, a sigmoid transformation is performed on each element j of $\mathbf{v_i(t+1)}$ (see Equation [20.9])

$$\mathrm{Sig}(v_i^j(t+1)) = \frac{1}{1 + \exp(-v_i^j(t+1))}. \tag{20.9}$$

Finally, the transformed vector of probability thresholds is used to determine the values of each element of $x_i(t+1)$, by comparing each element of $\mathrm{Sig}(\mathbf{v_i(t)})$ with a random number drawn from $U(0, 1)$ (see Equation [20.10]).

$$\text{If } U(0,1) < \mathrm{Sig}(v_i^j(t+1)), \quad \text{then } x_i^j(t+1) = 1; \quad \text{else } x_i^j(t+1) = 0 \tag{20.10}$$

In the binary version of the algorithm, trajectories/velocities are changes in the probability that a coordinate will take on a 0 or 1 value. $\mathrm{Sig}(v_i^j)$ represents the probability of bit x_i^j taking the value 1 [26]. Therefore, if $\mathrm{Sig}(v_i^j) = 0.3$ there is a 30% chance that $x_i^j = 1$, and a 70% chance it is zero.

20.4.3.1 Pseudocode for Algorithm

The pseudocode for the swarm algorithm in the simulator is as follows:

```
    For each entity in turn

    For each dimension (strategic decision) n

    v[n] = v[n] + R1*(pbest[n] - x[n]) + R2*(lbest[n]-x[n]) + R3*(a[n]-x[n])
        If(v[n] > Max) v[n] = Vmax
        If(v[n]<-Vmax) v[n] = -Vmax
```

[1]The difference in each case is weighted by a random number drawn from $U(0, 1)$. Therefore, if $pbest_i^j = 1$, $(pbest_i^j - x_i^j) \times U(0, 1)$ will be nonnegative. Adding this to v_i^j will increase v_i^j, and therefore also increase the probability that $x_i^j = 1$. On the other hand if $pbest_i^j = 0$, v_i^j will tend to decrease, and $\mathrm{Prob}(x_i^j) = 1$ becomes smaller.

```
        If(Pr<Sig(v[n]))t[n] = 1 Else t[n] = 0
    If(fitness(t)*(1 + e))>fitness(x)) //ratchet operator
        For each dimension n x[n] = t[n]
    UpdateAnchor(a) //if iteratively update anchor
                                        //option is selected
```

R_1, R_2, and R_3 are random weights drawn from a uniform distribution ranging from 0 to $R_{1\max}$, $R_{2\max}$, and $R_{3\max}$, respectively, and they weight the importance attached to pbest, lbest and anchor in each iteration of the algorithm. $R_{1\max}$, $R_{2\max}$, and $R_{3\max}$ are constrained to sum up to 4.0 in line with the *BinPSO* algorithm of [26]. x is the particle's actual position, *pbest* is its past best position, *lbest* its local best and a is the position of its anchor. $V_{\max}$ is set to 4.0 to ensure that $\text{Sig}(v[n])$ does not get too close to either 0 or 1, therefore ensuring that there is a nonzero possibility that a bit will flip state during each iteration. Pr is a random value drawn from $U(0, 1)$, and Sig is the sigmoid function: $\text{Sig}(x) = 1/(1 + \exp(-x))$, which squashes v into the range $0 \to 1$ range. t is a temporary record that is used in order to implement the ratchet operator. If the new strategy is considered better than the organization's existing strategy, it is accepted and t is copied into x. Otherwise t is discarded and x remains unchanged. e is the error or noise, injected in the fitness evaluation, in order to mimic an errorful forecast of strategy fitness.

20.4.4 Simulator

Although the underlying code for the *OrgSwarm* simulator is written in C++, the user interacts with the simulator through a series of easy-to-use screens (Figure 20.1 shows one of the screens in the main control menu for the simulator). These screens allow the user to select and alter a wide variety of parameters that determine the nature of the simulation run. In essence, the simulator allows the user to select choices for four items:

1. The form of NK landscape generated
2. The nature of the search heuristics to be employed by inventors
3. The number of simulations to be run
4. The form of output generated during the simulation run

During the simulation run, a series of graphics (see Figure 20.2 for an example of a graphic that shows the status of each particle in the population during the simulation run), and a run report (see Figure 20.3) can be displayed. The report display records the full list of simulation parameters chosen by the modeler, as well as providing a running record of the best design in the population at the end of each iteration. The simulator also facilitates the recording of comprehensive run-data to disk during the simulation.

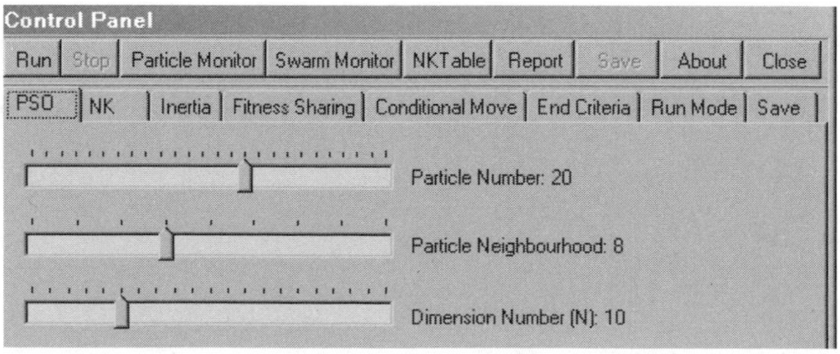

FIGURE 20.1 Main control screen for OrgSwarm.

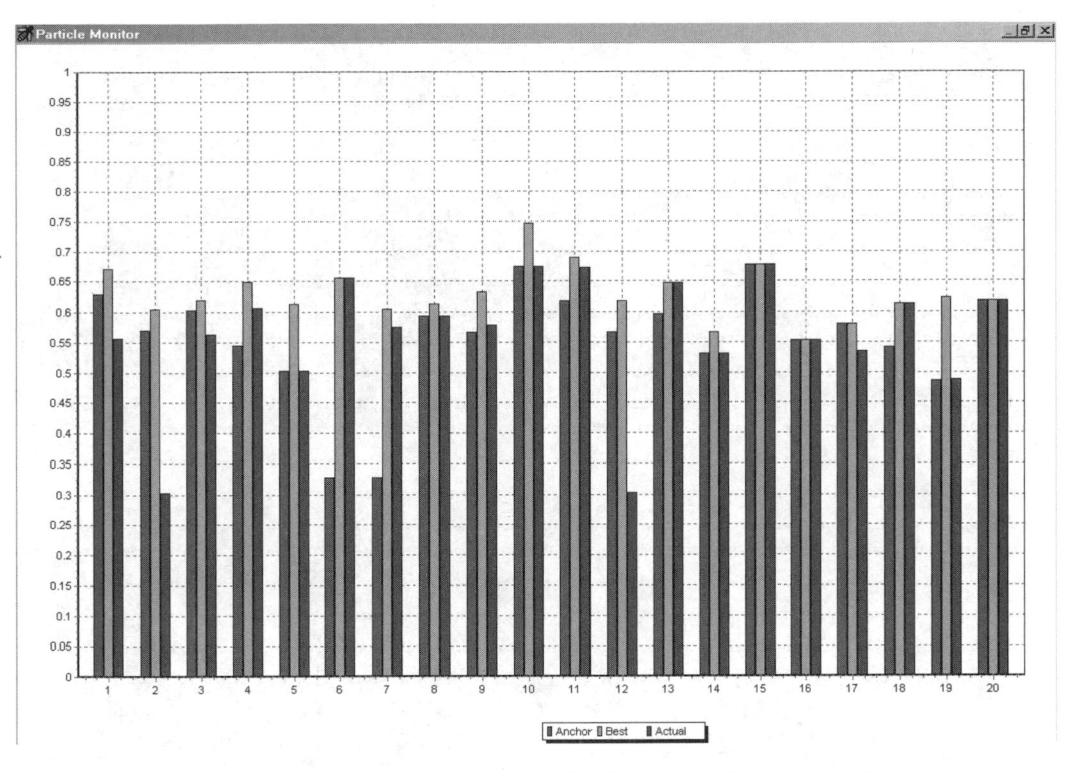

FIGURE 20.2 OrgSwarm screendump showing the status of each particle in the population during the simulation run. Three bars are shown for each of the 20 particles in the population, and these bars represent the fitness of the anchor location, the fitness of the **pbest** location, and the fitness of the current location of each particle.

20.5 Results

All simulations were run for 5000 iterations, and all reported fitnesses are the average population fitnesses, and average environment best fitnesses, across 30 separate simulation runs. On each of the simulation runs, the NK landscape is specified anew, and the positions and velocities of particles are randomly initialized at the start of each run. A population of 20 particles is employed, with a neighborhood of size 18. The choice of a high value for the neighborhood, relative to the size of the population, arises from the observation that real-world organizations know the profitability of their competitors.

Tables 20.1 and 20.2 provide the results for each of 14 distinct PSA variants, at the end of 5000 iterations, across a number of static and dynamic NK landscape scenarios. In each scenario, the same series of simulations are undertaken. Initially, a basic PSA is employed, without an anchor or a ratchet (conditional move) heuristic. This simulates a population of organizations searching a strategic landscape, where the population has no strategic inertia, and where organizations do not utilize a ratchet operator in deciding whether to alter their position on the strategic landscape.

The basic PSA is then supplemented by inclusion of a series of strategic anchor formulations, ranging from an anchor that does not change position during the simulation (initial position anchor) to one that can adapt after a time-lag (moving anchor). Two lag periods are examined, 20 and 50 iterations. Differing weights can be attached to the anchor term in the velocity Equation (20.6), ranging from 0 (anchor is "turned off") to a maximum of 4. To determine whether the weight factor for the anchor term has a critical impact on the results, results are reported for weight values of both 1 and 3, corresponding to low and high inertia weights. Next, to isolate the effect of the ratchet, the conditional move operator is implemented, and the anchor term is dropped. Finally, to ascertain the combined effect of both ratchet and anchor, the anchor simulations outlined above are repeated with the ratchet operator "turned on."

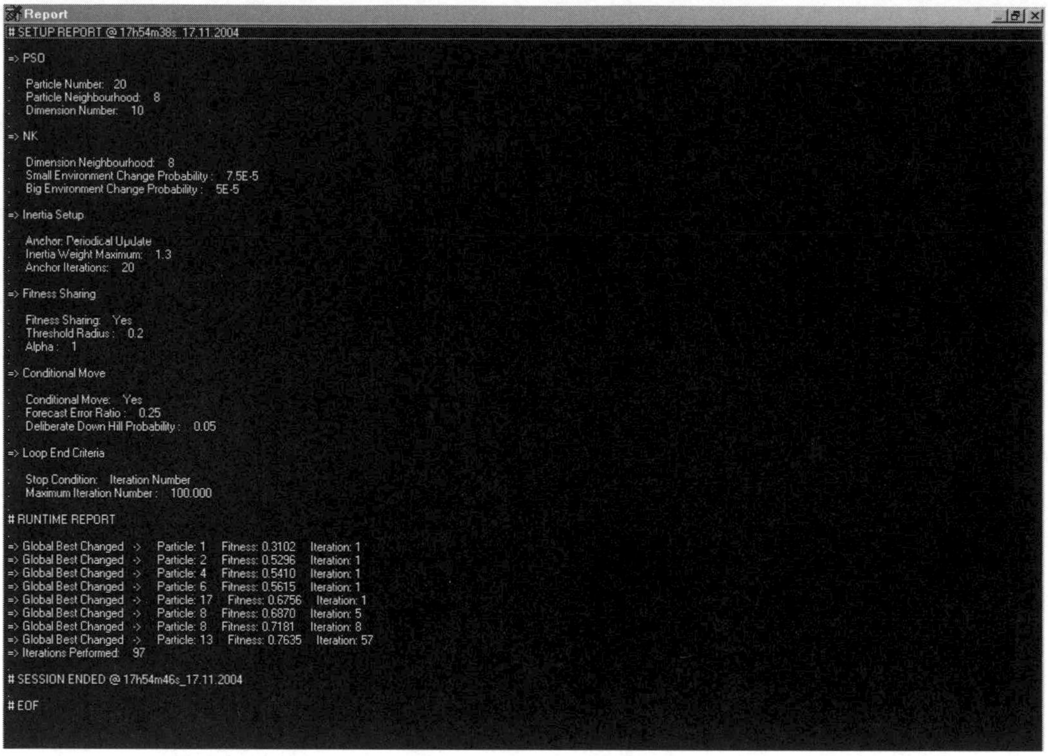

```
Report
# SETUP REPORT @ 17h54m38s_17.11.2004

=> PSO

. Particle Number: 20
. Particle Neighbourhood: 8
. Dimension Number: 10

=> NK

. Dimension Neighbourhood: 8
. Small Environment Change Probability : 7.5E-5
. Big Environment Change Probability : 5E-5

=> Inertia Setup

. Anchor: Periodical Update
. Inertia Weight Maximum: 1.3
. Anchor Iterations: 20

=> Fitness Sharing

. Fitness Sharing: Yes
. Threshold Radius : 0.2
. Alpha : 1

=> Conditional Move

. Conditional Move: Yes
. Forecast Error Ratio : 0.25
. Deliberate Down Hill Probability : 0.05

=> Loop End Criteria

. Stop Condition: Iteration Number
. Maximum Iteration Number : 100.000

# RUNTIME REPORT

=> Global Best Changed -> Particle: 1 Fitness: 0.3102 Iteration: 1
=> Global Best Changed -> Particle: 2 Fitness: 0.5296 Iteration: 1
=> Global Best Changed -> Particle: 4 Fitness: 0.5410 Iteration: 1
=> Global Best Changed -> Particle: 6 Fitness: 0.5615 Iteration: 1
=> Global Best Changed -> Particle: 17 Fitness: 0.6756 Iteration: 1
=> Global Best Changed -> Particle: 8 Fitness: 0.6870 Iteration: 5
=> Global Best Changed -> Particle: 8 Fitness: 0.7181 Iteration: 8
=> Global Best Changed -> Particle: 13 Fitness: 0.7635 Iteration: 57
=> Iterations Performed: 97

# SESSION ENDED @ 17h54m46s_17.11.2004

# EOF
```

FIGURE 20.3 A typical run report generated by OrgSwarm.

TABLE 20.1 Average (Environment Best) Fitness after 5000 Iterations, Static Landscape

Algorithm	Fitness		
	$(N = 96, K = 0)$	$(N = 96, K = 4)$	$(N = 96, K = 10)$
Basic PSA	0.4641 (0.5457)	0.5002 (0.6000)	0.4991 (0.6143)
Initial anchor, $w = 1$	0.4699 (0.5484)	0.4921 (0.5967)	0.4956 (0.6102)
Initial anchor, $w = 3$	0.4943 (0.5591)	0.4994 (0.5979)	0.4991 (0.6103)
Mov. anchor (50,1)	0.4688 (0.5500)	0.4960 (0.6003)	0.4983 (0.6145)
Mov. anchor (50,3)	0.4750 (0.5631)	0.4962 (0.6122)	0.5003 (0.6215)
Mov. anchor (20,1)	0.4644 (0.5475)	0.4986 (0.6018)	0.5001 (0.6120)
Mov. anchor (20,3)	0.4677 (0.5492)	0.4994 (0.6156)	0.4994 (0.6229)
Ratchet PSA	0.5756 (0.6021)	0.6896 (0.7143)	0.6789 (0.7035)
Rach-Initial anchor, $w = 1$	0.6067 (0.6416)	0.6991 (0.7261)	0.6884 (0.7167)
Rach-Initial anchor, $w = 3$	0.5993 (0.6361)	0.6910 (0.7213)	0.6844 (0.7099)
Rach-Mov. anchor (50,1)	0.6659 (0.6659)	0.7213 (0.7456)	0.6990 (0.7256)
Rach-Mov. anchor (50,3)	0.6586 (0.6601)	0.7211 (0.7469)	0.6992 (0.7270)
Rach-Mov. anchor (20,1)	0.6692 (0.6695)	0.7211 (0.7441)	0.6976 (0.7243)
Rach-Mov. anchor (20,3)	0.6612 (0.6627)	0.7228 (0.7462)	0.6984 (0.7251)

"Real world" strategy vectors consist of a large array of strategic decisions. A value of $N = 96$ was chosen in defining the landscapes in this simulation. It is noted that there is no unique value of N that could have been selected, but the selection of very large values are not feasible due to computational limitations. However, a binary string of 96 bits provides 2^{96}, or $\sim 10^{28}$, distinct choices of strategy. It is also noted

TABLE 20.2 Average (Environment Best) Fitness after 5000 Iterations, Entire Landscape Respecified Stochastically

	Fitness		
Algorithm	$(N = 96, K = 0)$	$(N = 96, K = 4)$	$(N = 96, K = 10)$
Basic PSA	0.4761 (0.5428)	0.4886 (0.5891)	0.4961 (0.6019)
Initial anchor, $w = 1$	0.4819 (0.5524)	0.4883 (0.5822)	0.4982 (0.6075)
Initial anchor, $w = 3$	0.5021 (0.5623)	0.4967 (0.5931)	0.4998 (0.6047)
Mov. anchor (50,1)	0.4705 (0.5450)	0.4894 (0.5863)	0.4974 (0.6008)
Mov. anchor (50,3)	0.4800 (0.5612)	0.4966 (0.6053)	0.5010 (0.6187)
Mov. anchor (20,1)	0.4757 (0.5520)	0.4926 (0.5867)	0.4985 (0.6097)
Mov. anchor (20,3)	0.4824 (0.5632)	0.4986 (0.6041)	0.5004 (0.6163)
Ratchet PSA	0.5877 (0.6131)	0.6802 (0.7092)	0.6754 (0.7015)
Rach-Initial anchor, $w = 1$	0.6187 (0.6508)	0.6874 (0.7180)	0.6764 (0.7070)
Rach-Initial anchor, $w = 3$	0.6075 (0.6377)	0.6841 (0.7130)	0.6738 (0.7017)
Rach-Mov. anchor (50,1)	0.6517 (0.6561)	0.7134 (0.7387)	0.6840 (0.7141)
Rach-Mov. anchor (50,3)	0.6597 (0.6637)	0.7049 (0.7304)	0.6925 (0.7225)
Rach-Mov. anchor (20,1)	0.6575 (0.6593)	0.7152 (0.7419)	0.6819 (0.7094)
Rach-Mov. anchor (20,3)	0.6689 (0.6700)	0.7158 (0.7429)	0.6860 (0.7147)

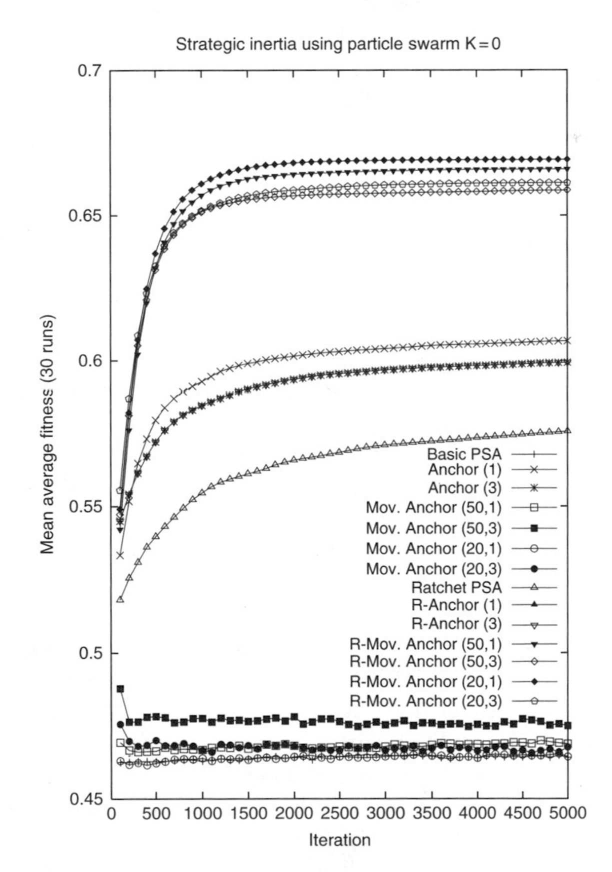

FIGURE 20.4 Plot of the mean average fitness on the static landscape where $K = 0$.

that we would expect the dimensionality of the strategy vector to exceed the number of organizations in the population, hence the size of the population is kept below 96, and a value of 20 is chosen. A series of landscapes of differing K values (0, 4, and 10), representing differing degrees of fitness interconnectivity, were used in the simulations.

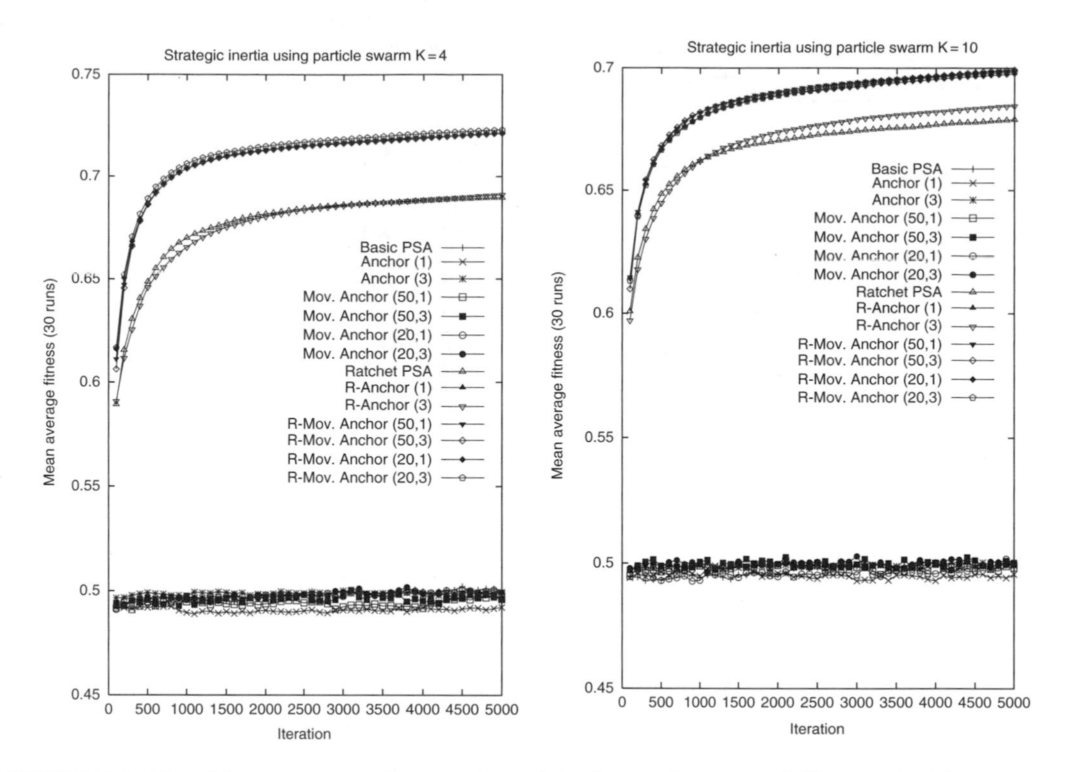

FIGURE 20.5 Plot of the mean average fitness on the static landscape where $K = 4$ (left) and 10 (right).

20.5.1 Static Landscape

Table 20.1 and Figure 20.4 and Figure 20.5 provide the results for a static NK landscape. Examining these results suggests that the basic PSA, without anchor or ratchet heuristics, performs poorly, even on a static landscape. The average populational fitnesses obtained after 5000 iterations (averaged over all 30 runs) is no better than random search, suggesting that unfettered adaptive efforts, based on "social communication" between organizations (gbest), and a memory of good past strategies (pbest) is not sufficient to achieve high levels of populational fitness. When various anchor term mechanisms, simulating strategic inertia, are added to the basic PSA, the results are not qualitatively altered from those of the basic PSA. This suggests that social communication and inertia are not sufficient for the attainment of high levels of populational strategic fitness.

When a ratchet heuristic is added to the basic PSA, a significant improvement (statistically significant at the 5% level) in both average populational, and average environment best fitness is obtained across landscapes of all K values, suggesting that the simple decision heuristic of "only abandon your current strategy for a better one" can lead to notable increases in populational fitness.

Finally, the results of a series of simulations that combine anchor and ratchet mechanisms are reported. Virtually all of these combinations lead to significantly (at the 5% level) enhanced levels of populational fitness against the ratchet-only PSA, *suggesting that strategic inertia can be beneficial, when organizations employ a conditional move test before adopting new strategies.* Examining the combined ratchet and anchor results in more detail, the best results are obtained when the anchor is not fixed at the initial location of each particle on the landscape, but when it is allowed to "drag" or adapt, over time. The results are not qualitatively sensitive to the weight value (1 or 3).

20.5.2 Dynamic Landscape

The real world is rarely static, and changes in the environment can trigger adaptive behavior by agents in a system [27]. Table 20.2 and Figure 20.6 and Figure 20.7 provide results for the case where the entire

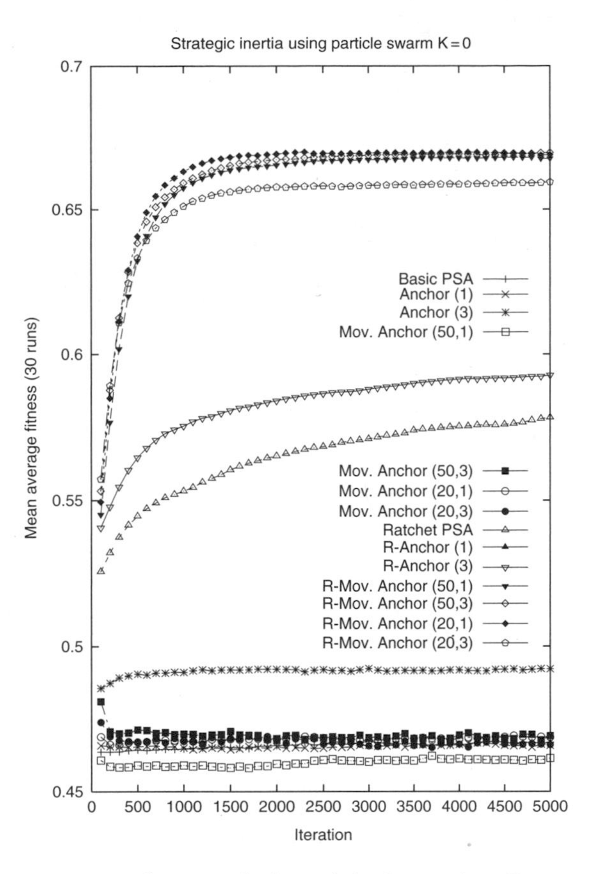

FIGURE 20.6 Plot of the mean average fitness on the dynamic landscape where $K = 0$.

NK landscape is respecified in any iteration with a $p = 0.00025$. When the landscape is wholly or partially respecified, the benefits of past strategic learning by organizations is eroded (see [27–29] for a detailed discussion of the utility of the PSO in tracking dynamic environments).

Qualitatively, the results in both scenarios are similar to those obtained on the static landscape. The basic PSA, even if supplemented by an anchor mechanism, does not perform any better than random search. Supplementing the basic PSA with the ratchet mechanism leads to a significant improvement in populational fitness, with a further improvement in fitness occurring when the ratchet is combined with an anchor mechanism. In the latter case, an adaptive or dragging anchor gives better results than a fixed anchor, but the results between differing forms of dragging anchor do not show a clear dominance for any particular form. As for the static landscape case, the results for the combined ratchet/anchor, are relatively insensitive to the choice of weight value (1 or 3).

20.6 Conclusions

In this chapter, a synthesis of a strategic landscape defined using the NK model, and a Particle Swarm metaphor is used to create a novel simulation model of the process of strategic adaptation of organizations. The results suggest that a degree of strategic inertia, in the presence of an election operator, can assist rather than hamper the adaptive efforts of populations of organizations in static and slowly changing strategic environments. The results also suggest that despite the claim for the importance of social learning in populations, social learning alone is not always enough, unless learnt lessons can be maintained by means of an election mechanism.

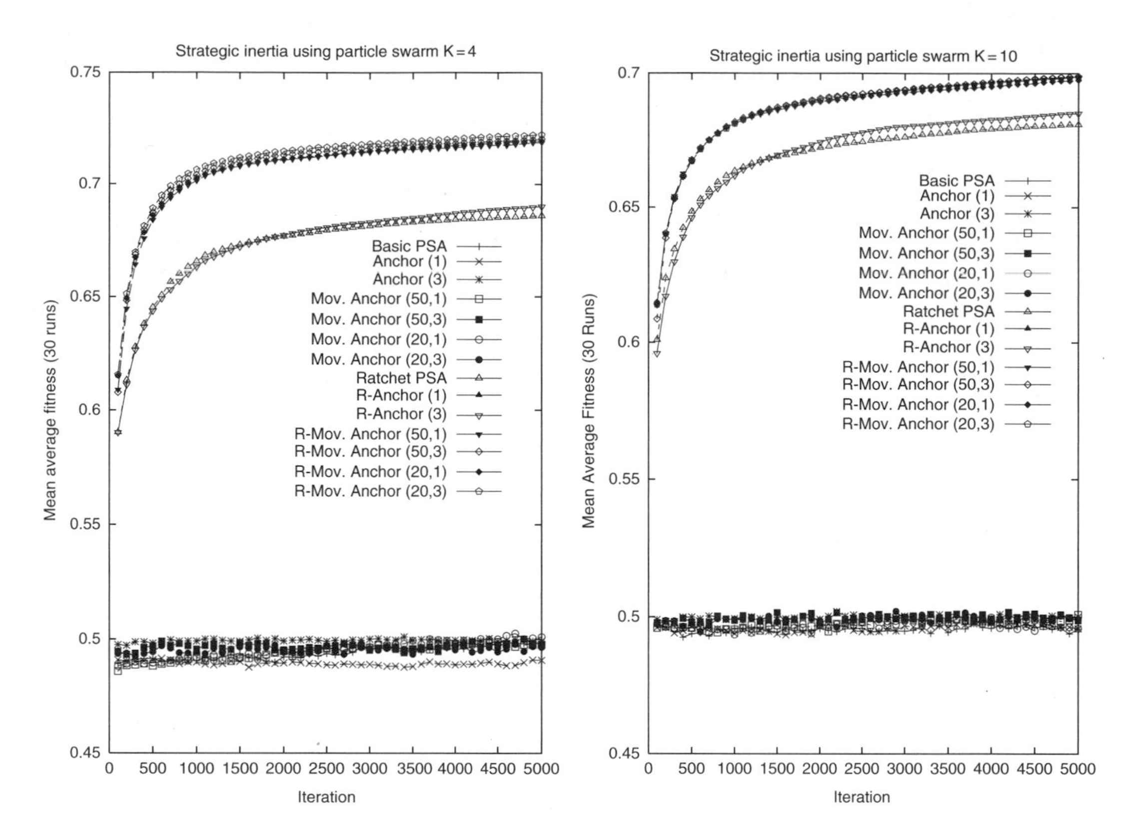

FIGURE 20.7 Plot of the mean average fitness on the dynamic landscape where $K = 4$ (left) and 10 (right).

It is not possible in a single set of simulation experiments to exhaustively examine every possible combination of settings for each parameter in the simulation model. Future work will extend the range of settings examined. However, the initial results cast an interesting light on the role of anchoring in organizational adaptation, and the development of the swarm-landscape simulator extends the methodologies available to researchers to conceptualize and examine organizational adaptation.

Finally, it is noted that the concept of anchoring developed in this chapter is not limited to organizations, but is plausibly a general feature of social systems. Hence, the extension of the social swarm model to incorporate inertia may prove useful beyond this study.

References

[1] Porter, M. (1996). What is strategy? *Harvard Business Review*, Nov.–Dec., pp. 61–78.

[2] Levinthal, D. (1997). Adaptation on rugged landscapes. *Management Science*, 43: 934–950.

[3] Rivkin, J. (2000). Imitation of complex strategies. *Management Science*, 46: 824–844.

[4] Wright, S. (1932). The roles of mutation, inbreeding, crossbreeding and selection in evolution. In *Proceedings of the Sixth International Congress on Genetics*, 1: 356–366.

[5] Kitts, B., Edvinsson, L., and Beding, T. (2001). Intellectual capital: from intangible assets to fitness landscapes. *Expert Systems with Applications*, 20: 35–50.

[6] Boeker, W. (1989). Strategic change: The effects of founding and history. *Academy of Management Journal*, 32: 489–515.

[7] Stuart, T. and Podolny, J. (1996). Local search and the evolution of technological capabilities, *Strategic Management Journal*, 17: 21–38.

[8] Hannan, M. and Freeman, J. (1977). The populational ecology of organizations. *American Journal of Sociology*, 82: 929–964.

[9] Hannan, M. and Freeman, J. (1984). Structural inertia and organizational change. *American Sociological Review*, 49: 149–164.
[10] Levinthal, D. (1991). Random walks and organisational mortality. *Administrative Science Quarterly*, 36: 397–420.
[11] Tushman, M. and O'Reilly, C. (1996). Ambidextrous organizations: Managing evolutionary and revolutionary change. *California Management Review*, 38: 8–30.
[12] Kauffman, S. (1993). *The Origins of Order.* Oxford University Press, Oxford, England.
[13] Kennedy, J. and Eberhart, R. (1995). Particle swarm optimization. In *Proceedings of the IEEE International Conference on Neural Networks*, December 1995, pp. 1942–1948.
[14] Kennedy, J., Eberhart, R., and Shi, Y. (2001). *Swarm Intelligence.* Morgan Kauffman, San Mateo, CA.
[15] Kennedy, J. (1997). The particle swarm: Social adaptation of knowledge. In *Proceedings of the International Conference on Evolutionary Computation.* IEEE Press, Washington, pp. 303–308.
[16] Kennedy, J. (1999). Minds and cultures: Particle swarm implications for beings in sociocognitive space. *Adaptive Behavior*, 7: 269–288.
[17] Kennedy, J. (1999). Small worlds and mega-minds: effects of neighbourhood topology on particle swarm performance. In *Proceedings of the International Conference on Evolutionary Computation.* IEEE Press, Washington, pp. 1931–1938.
[18] Thorndike, E. (1911). *Animal Intelligence*, Macmillan, New York.
[19] Bandura, A. (1986). *Social Foundations of Thought and Action: A Social Cognitive Theory.* Prentice Hall, Englewood Cliffs, NJ.
[20] Kauffman, S. and Levin, S. (1987). Towards a general theory of adaptive walks on rugged landscapes. *Journal of Theoretical Biology*, 128: 11–45.
[21] Gavetti, G. and Levinthal, D. (2000). Looking forward and looking backward: Cognitive and experiential search, *Administrative Science Quarterly*, 45: 113–137.
[22] Porter, M. and Siggelkow, N. (2001). Contextuality within activity systems. Harvard Business School Working paper series, No. 01-053.
[23] Lobo, J. and MacReady, W. (1999). Landscapes: A Natural Extension of Search Theory. Santa Fe Institute Working paper 99-05-037.
[24] Kauffman, S., Lobo, J., and MacReady, W. (1998). Optimal Search on a Technology Landscape. Santa Fe Institute Working paper 98-10-091.
[25] Porter, M. (1985). *Competitive Advantage:Creating and Sustaining Superior Performance.* The Free Press, New York.
[26] Kennedy, J. and Eberhart, R. (1997). A discrete binary version of the particle swarm algorithm. In *Proceedings of the Conference on Systems, Man, and Cybernetics.* IEEE Press, Washington, pp. 4104–4109.
[27] Blackwell, T. (2003). Swarms in dynamic environments. In *Proceedings of GECCO 2003*, Vol. 2723 of *Lecture Notes in Computer Science.* Springer-Verlag, Berlin, pp. 1–12.
[28] Eberhart, R. and Shi, Y. (2001). Tracking and optimizing dynamic systems with particle swarms. In *Proceedings of the CEC 2001.* IEEE Press, Washington, pp. 94–97.
[29] Hu, X. and Eberhart, R. (2002). Adaptive particle swarm optimization: detection and response to dynamic systems. In *Proceedings of CEC 2002.* IEEE Press, Washington, pp. 1666–1670.

21

BeeHive: New Ideas for Developing Routing Algorithms Inspired by Honey Bee Behavior

Horst F. Wedde
Muddassar Farooq

Some parts of the chapter have been taken from H.F. Wedde, M. Farooq, and Y. Zhang. BeeHive: An efficient fault-tolerant routing algorithm inspired by honey bee behavior. Ant Colony Optimization and swarm intelligence, Lecture Notes in Computer Science 3172, by kind permission of Springer-Verlag, Heidelberg, Germany. The paper also won the *best paper award* at ANTS2004 conference.

21.1 Introduction

A honey bee colony manages to react to countless changes in the forage pattern outside the hive and internal changes inside the hive through a decentralized and sophisticated communication and control system. According to Seeley, a honey bee colony can thoroughly monitor a vast region around the hive for rich food sources, nimbly redistribute its foragers within an afternoon, fine-tune its nectar processing to match its nectar collecting, effect cross inhibition between different forager groups to boost its response differential between food sources, precisely regulate its pollen intake in relation to its ratio of internal supply and demand, and limit the expensive process of comb building to times of critical need for additional storage space [1]. A bee colony demonstrates this flexible and adaptive response because it is organized with morphologically uniform individuals but with different temporary specializations. A bee takes up four roles during her life span — cleaner, nurse, food-storer, and forager. The foragers could be further recognized as nectar, pollen, and water collectors [1]. They have two functional roles within each sub specialty: scouts, who discover new food sources around the hive, and foragers, who transport nectar from an already discovered flower site by following the dances of other scouts or foragers. The colony brilliantly allocates, through its communication and control system, its labor force among these individuals to maintain a balance between collection and processing rate of each commodity, as a result, an optimum stock of nectar, pollen, and water is piled inside the colony. On an average, a colony extracts from its environment around 20 kg of pollen, 120 kg of nectar, 25 l of water, and 100 g of resin each year [1].

Karl von Frisch in 1944 made a revolutionary discovery about the communication paradigm that foragers use to communicate the information about flower sites around the hive. He experimentally verified that foragers can inform their fellow foragers, inside the hive on a dance floor, of the direction and the distance to a food source by means of a dance. He deciphered these communication signals into a language, in his book *Tanzsprache und Orientierung der Bienen* [2] (the translation was done by Chadwick [3]). The foragers, according to von Frisch, used two type of dances: round dances, which show that a food source is present near the hive (about 100 m), and waggle dances, which further specify the direction and the distance to a food source (up to few kilometers), by the orientation and duration portion of the waggle portion of each dance circuit. The food sources located in the direction of the sun are represented as an upward direction on the vertical comb, and any angle to the right or left of the sun is encoded as any angle to the right or left of the upward direction. The recruited foragers maintain this angle from the sun to reach in a small region around the flower and then use the flower fragrance, which clings to the body of a forager and which they smelt while observing the dances, to identify the food source unambiguously. Recently, the researchers have also indicated that several components of dance are correlated with food-source profitability, such as dance duration and probability of sound production; however, this information is not taken into account by unemployed (dance following) foragers, instead they randomly choose a dancer before leaving the hive. Nevertheless, the recruited foragers arrive in greater numbers at more profitable food sources because the dances for richer sources are more conspicuous and hence likely to be encountered by the unemployed (dance following) foragers [1].

21.1.1 Organization of the Chapter

In Section 21.2 we will briefly provide an overview of the important organizational principles of a honey bee colony that enabled us to develop a robust routing algorithm. An insight into these principles will help the reader in understanding the motivation for following an engineering approach, discussed in Section 21.3, for a systematic transformation of concepts from *Nature* to *Networks*, described in Section 21.4. In Section 21.5 we will first develop the key ideas of the bee agent model underlying the *BeeHive* algorithm. Based on this we will present our *BeeHive* algorithm in Section 21.5.1. In Section 21.6, we will briefly introduce the important features of three state-of-the-art algorithms, with which we compared the performance of *BeeHive*. Section 21.7 will describe the simulation and the network environment that were used to compare the performance of *BeeHive* with other algorithms. Section 21.8 discusses the results obtained from the extensive simulations. Finally, we conclude our findings, and provide an outlook to future research.

21.2 An Agent-Based Investigation of Honey Bee Colony

In this section we briefly outline the organizational principles of a honey bee colony that will enable the computer scientists to develop agent-based algorithms for different optimization and real world problems. A honey bee colony solves the most interesting multi-objective optimization problem: how to allocate resources to different tasks under constantly changing operating environment so that the colony maximizes its profit gains. This is the same problem that many researchers try to solve in the field of multi-objective optimization under dynamic and time-varying environments. The reader can find details of these principles in Reference 1.

21.2.1 Labor Management

As discussed in Section 21.1 that a honey bee colony is organized with morphologically uniform individuals but with different temporary specializations. The benefit of this approach is that it enhances a colony's flexibility to adapt its response according to the ever changing environment while at the same time doing the tasks with acceptable level of efficiency. For example, a nectar forager is able to extensively forage at a discovered flower site as he/she does not waste time in storing the nectar inside the hive. Moreover, a bee colony is able to adapt the activity level of its specialists according to the group's needs. For example, nectar foragers may become pollen foragers if the amount of protein that they receive from nurse bees falls below a threshold level, or nurse bees might take the role of food-storer bees if the rate of processing nectar is slower than the rate of collecting nectar (foragers indicate this by a tremble dance).

21.2.2 The Communication Network of a Honey Bee Colony

A honey bee colony utilizes a hybrid communication network that consists of *signals* and *cues* for information exchange among its members. Signals are information-bearing actions or structures that have been shaped by the natural selection, specifically to convey information. Cues are variables that likewise convey information about the state of the colony but have not been modeled by the natural selection to convey that information [4]. Signals enable direct communication among the members of a honey bee colony via waggle dance, tremble dance, and shaking signal while cues enable indirect communication among the members through the environment shared by them. Both mechanisms provide an efficient information exchange protocol that empowers the members to, mostly, communicate indirectly (group-to-one paradigm), and when required, directly using one-to-one paradigm.

A good example of cue is the search time for finding a food-storer bee that a forager experiences once she wants to unload her nectar. A nectar forager uses this cue to get an estimate of both rates, nectar collecting and nectar processing, of the colony; higher search time to find a food-storer bee is an indicator to the forager that the rate of processing nectar is slower than the rate of collecting nectar. Consequently, she will decide to perform a tremble dance instead of a waggle dance. The forager by doing a tremble dance will achieve two objectives: one, she will recruit more food-storer bees to increase the rate of processing nectar and two, she will make other foragers, on the dance floor, to stop performing waggle dances, as a result, the rate of collecting nectar will be decreased.

21.2.3 Reinforcement Learning

A colony experiences a strong fluctuation in the external supply or internal demand (or both) for its commodities: nectar, pollen, and water. The feedback signals, both negative and positive, are important to regulate their amount so that the colony has sufficient stockpile of each of these commodities. This is achieved by recruiting unemployed forgers for good supply sites through waggle dances. A forager decides to dance only if the quality of the food site, visited by her, is above a certain threshold or it experiences a very small search time to locate a food-storer bee for its commodity (a cue that the colony needs the commodity). By maintaining the search time within some thresholds, the honey bee colony reinforces the foraging labor at a site in times of need and vice versa. A stochastic model for the foraging behavior is presented in Reference 5. Sumpter used this basic model to come up with an agent-based model in Reference 6. The Sumpter's model provides a solid foundation for developing an agent-based reinforcement learning algorithm [7].

21.2.4 Distributed Coordination and Planning

A honey bee colony achieves coordination among its thousands of members without any central authority. The colony does not have a hierarchy where some individuals require information and then allocate tasks to different members and monitor them. Each individual decides to do a job depending upon the need of the colony that it estimates using the above-mentioned communication network.

21.2.5 Energy Efficient Foraging

The foragers tend to optimize the *energetic efficiency* of foraging at a flower site. For example, if during an average foraging trip a forager collects G units of energy, expends C units of energy, and spends time T, then *energetic efficiency* could be defined as $(G - C)/C$. Consequently, the neighborhood site of the hive get preference than the sites far away from it. That is why von Frisch believed that those foragers (short distance) which return from nearby sites perform round dances while other (long distance) foragers perform waggle dances. This principle enables a colony to collect the commodities at an optimum expenditure of energy.

21.2.6 Stochastic Selection of Flower Sites

The unemployed foragers on the dance floor observe, at the maximum, two or three dancers and then decide to choose one among them at random. They do not broadly survey the dance floor to identify the best flower site. This concept is contradictory to human society where well informed customers are crucial to proper functioning of competitive markets. According to Seeley, this stems from the fact that the individual forager seeks to maximize her colony's profits, whereas the individual human tries to maximize her or his own profit gains. This "sacrifice for group principle" enables a colony to distribute its forager force over different flower sites rather than allocating it to the best site only. Such a policy results in quick reallocation of foragers, which were foraging at the best site, to other discovered sites, once the best flower site is about to fade away.

21.2.7 Group Organization

The employed foragers that collect nectar from the same type of flowers recognize one another in the hive through the flower fragrance that clings to their body. Only group companions respond to the dances and they show no interest in the dances performed by the foragers of other groups, foraging at other types of flowers. However, the employed foragers might switch their group if the quality of their flower site degrades to an extent that is no longer profitable [3].

Section 21.3 introduces the engineering approach that was developed to transform the above-mentioned principles into a routing algorithm for packet switching networks. The engineering approach, however, provides a general set of guidelines for developing algorithms/products inspired from natural systems.

21.3 An Engineering Approach to Nature Centered Routing Protocols

In this section, we will introduce an engineering approach that we followed in the design and development of a routing protocol inspired from a natural system (a honey bee colony).

Definition 21.1 (Natural Engineering). *Natural Engineering is an emerging new engineering discipline that enables scientists/engineers to utilize inspirations and observations from organizational principles of natural systems, and transform them into structural principles of software organization (algorithms) or industrial products, in search of efficient/optimal solutions for real world problems under constraints of the resources.*

The above-mentioned concept emphasizes six aspects:

1. Understanding the working principles of natural systems.
2. Understanding the operational environment of the target system.
3. Developing an algorithmic model of the organizational principles of natural systems.
4. Mapping the concepts from a natural system to a technical system.
5. Adapting the algorithmic model according to the operational environment of a technical system.
6. Follow test, evaluate, improve feedback loop in search of optimum solutions under the resource constraints (time, space, computation, money, labor, etc.).

There is, of course, no clear-cut way to achieve one-to-one match of structures/principles in Nature life organizations with working principles in technical systems. The most important challenge, therefore, is to identify a natural system whose working principles could be easily abstracted to the one in the technical system. If one has to add many nonbiological features into the natural system then, we believe, it is more advisable to look at other natural systems for inspiration. Consequently, we chose honey bees because their foraging behavior could be easily abstracted into a routing problem in telecommunication networks. Both systems have to maximize the amount of a commodity (nectar/data delivered to the hive/nodes) as quickly as possible, under a continuously changing operating environment.

The major focus of our research, however, is to take bio/nature inspired solutions into business and therefore we decided to follow, for developing the *BeeHive* algorithm, a feedback oriented engineering approach as displayed schematically in Figure 21.1 that incorporates most of the features discussed here.

First, we considered the ensemble of constraints under which the envisioned routing protocol is supposed to operate:

- Nonavailability of a global clock for trip time calculation.
- Routers and links could crash.
- Routers have limited queue capacity.
- Links have a BER (Bit Error Rate) associated with them.
- The requirements from the Linux kernel routing framework needed to support the protocol.
- The requirements of the IP protocol which is currently used at the network layer in Internet.

At the same time we decided that the *bee agents* should explore the network, collect important parameters, and make the routing decisions in a decentralized fashion (in the style as real scouts/foragers do decision making during collecting nectar from flowers). *Bee agents* should measure the quality of a route and then communicate to other *bee agents* like foragers do that in Nature. The structure of the routing table should provide the functionality of a dance floor for exchange of information among *bee agents*, and among *bee agents* and data packets. Moreover, later on we should be able to utilize it in a real kernel of the Linux operating system.

We implemented our ideas in a simulation environment and then refined our algorithmic mapping through the feedback channel 1 (see Figure 21.1). During this phase we did not use any simulation specific features that were not available inside the Linux kernel, for example, using vector, stack, or similar data structures. Once we reached a relative optimum of the *BeeHive* concept, we started to develop an engineering model of the algorithm. The engineering model could be easily transported to the Linux kernel routing framework. We tested it on the real network of Linux machines and refined our engineering model through the feedback channel 2 (see Figure 21.1). At the moment we are evaluating our conceptual approach in two prototype projects: BeeHive [8], which deals with the design and development of a routing algorithm for fixed networks, and BeeAdHoc, whose goal is to design and develop an energy efficient routing algorithm for mobile ad hoc networks (MANETS) [9].

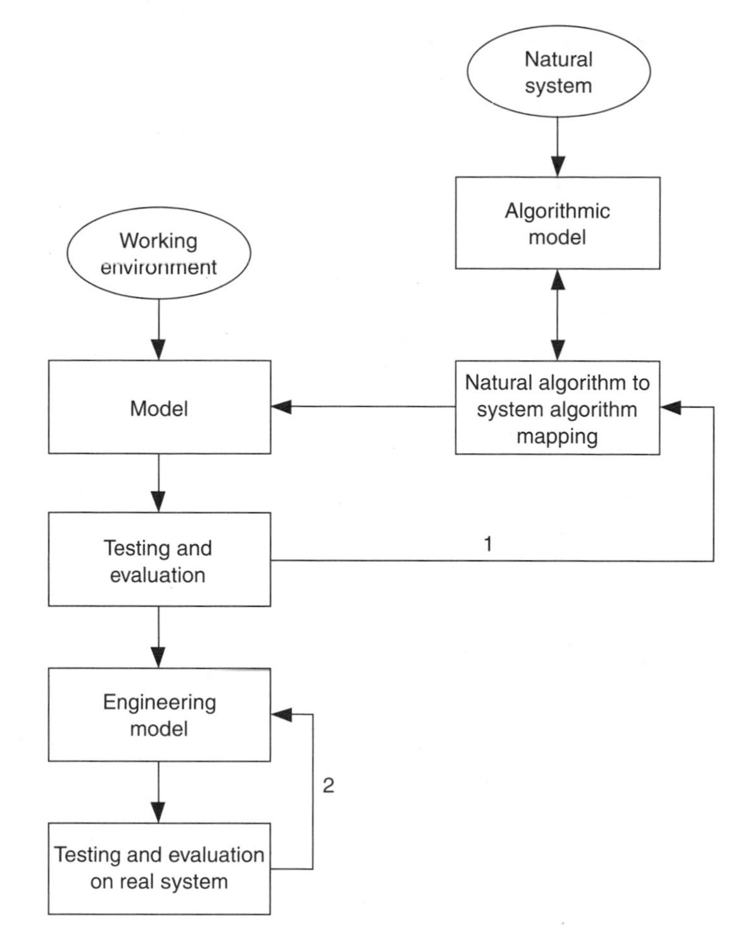

FIGURE 21.1 Nature centered protocol engineering.

In Section 21.4 we will elaborate the most important step of our engineering approach: the mapping of concepts in a honey bee colony, discussed in Section 21.2, to operating environment of real packet switching networks. This step will help the reader in understanding our algorithm described in Section 21.5. We do not discuss the implementation of the *BeeHive* inside the network stack of Linux kernel because it is beyond the scope of the chapter, however, interested reader will find the technical details in Reference 10.

21.4 BeeHive: The Mapping of Concepts from Nature to Networks

This section will briefly illustrate the mapping of concepts from nature to networks, one of the most important step in *Natural Engineering*, that will simplify for the reader to trace back the origin of important features of our *BeeHive* algorithm into the principles of a honey bee colony:

1. We could consider each node in the network as a hive that consists of *bee agents*. Each node periodically launches the *bee agents* that explore the network and collect the routing information that provides the nodes, which they visit, the partial information on the state of the network. These *bee agents* could be considered as scouts that explore and evaluate the quality of multiple paths between their launching node and the nodes that they visit.
2. *Bee agents* provide to the nodes, which they visit, the information on the propagation and queuing delay of the paths, which they explored, leading to their launching nodes from the visited nodes.

One could consider propagation delay as a distance information and queuing delay as a direction information (please remember bee scouts also provide these parameters in their dances).

3. A *bee agent* decides to provide its path information only, if the quality of the path it traversed, is above a threshold. The threshold is controlled by the number of hops that a *bee agent* is allowed to take. Moreover, the agents model the quality of a path as a function of the propagation and the queuing delay of the path; lower values of the parameters result in higher values for the quality parameter.
4. Majority of *bee agents* in *BeeHive* algorithm explore the network near their launching node and very few explore the complete network. The idea is borrowed from a honey bee colony as this reduces not only the overhead of collecting the routing information but also helps in maintaining smaller/local routing tables.
5. We consider routing table as a dance floor where the *bee agents* provide the information about the quality of the paths they traversed. The routing table is used for information exchange among *bee agents*, launched from the same node but arrived at a node via its different neighbors. This information exchange helps in evaluating the overall quality of a node (that has multiple pathways to a destination) for reaching a certain destination.
6. The nectar foragers exploit the flower sites according to their quality while the distance and direction to the sites is communicated through dances made by their fellow foragers, on the dance floor. In our algorithm, we have to map the quality of paths onto the quality of nodes to utilize the bee principle. Consequently, we formulate the quality of a node, for reaching a destination, as a function of proportional quality of only those neighbors, which possibly lie in the path toward the destination.
7. We consider data packets as foragers. Once they arrive at a node, they access the information in the routing tables, stored by *bee agents*, about the quality of different neighbors of the node for reaching their destinations. They select the next neighbor toward the destination in a stochastic manner depending upon its goodness, as a result, not all packets follow the best paths. This will help in *maximizing the system performance though a data packet may not follow the best path*, a concept directly borrowed from a principle of bee behavior: *a bee could only maximize her colony's profit if she refrains from broadly monitoring the dance floor to identify the single most desirable food* [1] (see Section 21.2).

Now we are in a position to introduce our *bee agent model* and *BeeHive* algorithm in the following section.

21.5 The *Bee Agent* Model

Our *bee agent* model consists of two types of agents: *short distance bee agents* and *long distance bee agents.* Both agents undertake the same responsibility: exploring the network and evaluating the quality of the paths that they traverse. However, they only differ in the distance (hops) that they are allowed to take starting from their launching node. *Short distance bee agents* collect and disseminate routing information in the neighborhood (up to a specific number of hops) of their source node while *long distance bee agents* collect and disseminate routing information to all nodes of a network. This helps in collecting the routing information with minimum overhead, both processing and the bandwidth, and as quickly as possible.

The *bee agents* that are launched from the same node form an affinity group in which they show interest in each others information. Once the *bee gents* of the same group arrive at the same node, but via different neighbors of the node, they access the routing information, collected by their fellow *bee agents* in the group, in the routing table. They will decide to discontinue their exploration of the network after storing their information in the routing table, if one of their members has already arrived at the node. The communication model among *bee agents* is known as a *blackboard system* [11]. Note that, in comparison, Ant Colony Optimization (ACO) [12] algorithms utilize the *principle of stigmergy* [13] for communication among agents. The agents, using this principle, communicate indirectly through the shared environment.

Informally, the *BeeHive* algorithm and its main characteristics could be summarized as follows:

1. The network is organized into fixed partitions called *foraging regions.* A partition results from particularities of the network topology. Each *foraging region* has one representative node. Currently the lowest IP address node in a *foraging region* is elected as the representative node. If this node crashes then the next higher IP address node takes over the job.
2. Each node also has a node specific *foraging zone* that consists of all nodes from whom *short distance bee agents* can reach this node.
3. Each nonrepresentative node periodically sends a *short distance bee agent,* by broadcasting replicas of it to each neighbor site.
4. When a replica of a particular *bee agent* arrives at a site it updates the routing information there, and will be flooded again; however, it will not be sent to the neighbor from where it arrived. This process continues until the life span of the agent has expired, or if a replica of this *bee agent* had been received already at a site, the new replica will be killed there.
5. Representative nodes only launch *long distance bee agents* that would be received by the neighbors and propagated as in 4. However, their life span (number of hops) is limited by the *long distance limit.*
6. The idea is that each agent while traveling, collects and carries path information, and it leaves, at each node visited, the trip time estimate for reaching its source node from this node over the incoming link. *Bee agents* use priority queues for quick dissemination of routing information.
7. Thus each node maintains current routing information for reaching nodes within its *foraging zone* and the *representative nodes* of *foraging regions.* This mechanism enables a node to route a data packet (whose destination is beyond the *foraging zone* of the given node) along a path toward the *representative node* of the *foraging region* containing the destination node.
8. The next hop for a data packet is selected in a stochastic fashion according to the quality measure of the neighbors. The motivation for this routing policy is explained in Section 21.4. Note that the currently employed routing algorithms in Internet always choose a next hop on the shortest path [14].

Figure 21.2 provides an exemplary working of the flooding algorithm. *Short distance bee agents* can travel up to 3 hops in this example. Each replica of the shown *bee agent* (launched by node 10) is specified with a different trail to identify its path unambiguously. The numbers on the paths show their costs. The flooding algorithm is a variant of breadth first search algorithm. Nodes 2, 3, 4, 5, 6, 7, 8, 9, 11 constitute the *foraging zone* of node 10.

Now we will briefly discuss the estimation model that *bee agents* utilize to approximate the trip time t_{is} that a packet will take in reaching its source node s from current node i (ignoring the protocol processing delays for a packet at node i and s).

$$t_{is} \approx \frac{ql_{in}}{b_{in}} + tx_{in} + pd_{in} + t_{ns}, \tag{21.1}$$

where ql_{in} is the size of the queue (in bits) for neighbor n at node i, b_{in} is the bandwidth of the link between node i and neighbor n, tx_{in} and pd_{in} are transmission and propagation delays, respectively, of the link between node i and neighbor n, and t_{ns} is trip time from n to s. Bandwidth and propagation delays of all links of a node are approximated by transmitting *hello packets.*

21.5.1 Algorithm Design of BeeHive

In *BeeHive,* each node i maintains three types of routing tables: *Intra-Foraging Zone* (IFZ), *Inter-Foraging Region* (IFR), and *Foraging Region Membership* (FRM). IFZ routing table R_i is organized as a vector of size $|D(i)| \times (|N(i)|)$, where $D(i)$ is the set of destinations in *foraging zone* of node i and $N(i)$ is the set of neighbors of i. Each entry P_{jk} is a pair of queuing delay and propagation delay (q_{jk}, p_{jk}) that a packet will

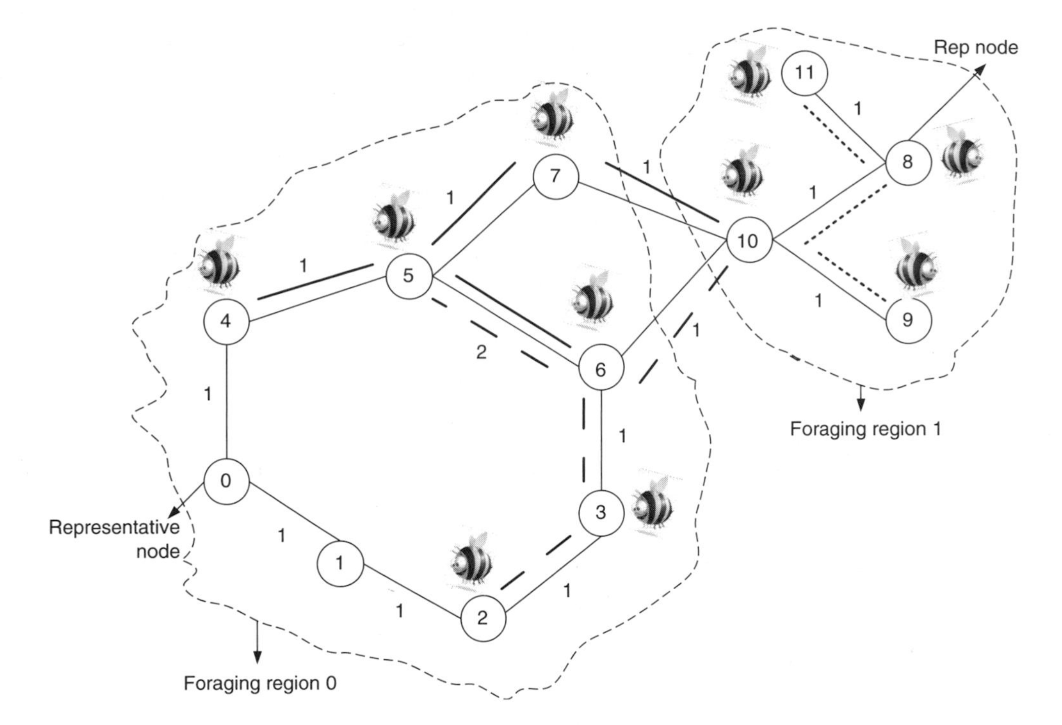

FIGURE 21.2 Bee agents flooding algorithm.

TABLE 21.1 IFZ Routing Table

R_i	$D_1(i)$	$D_2(i)$	$\cdots$	$D_d(i)$
$N_1(i)$	(p_{11}, q_{11})	(p_{12}, q_{12})	$\cdots$	(p_{1d}, q_{1d})
$\vdots$	$\vdots$	$\vdots$	$\ddots$	$\vdots$
$N_n(i)$	(p_{n1}, q_{n1})	(p_{n2}, q_{n2})	$\cdots$	(p_{nd}, q_{nd})

experience in reaching destination k via neighbor j. Table 21.1 shows an example of R_i. In the IFR routing table, the queuing delay and propagation delay values for reaching the *representative node* of each *foraging region* through the neighbors of a node are stored. The structure of the IFR routing table is similar to the one shown in Table 21.1 where destination is replaced by a pair of (representative, region). The FRM routing table provides the mapping of known destinations onto a *foraging region*. In this way we eliminate the need to maintain $O(N \times D)$ (where D is total number of nodes in a network) entries in a routing table as done by *AntNet* and save a considerable amount of router memory needed to store this routing table.

Goodness of a Neighbor: The goodness of a neighbor j of node l (l has n neighbors) for reaching a destination d is g_{jd} and defined as follows:

$$g_{jd} = \frac{(1/p_{jd})(e^{-q_{jd}/p_{jd}}) + (1/q_{jd})(1 - e^{-q_{jd}/p_{jd}})}{\sum_{k=1}^{n}((1/p_{kd})(e^{-q_{kd}/p_{kd}}) + (1/q_{kd})(1 - e^{q_{kd}/p_{kd}}))}. \tag{21.2}$$

The fundamental motivation behind Definition 21.2 is to approximate the behavior of the real network. When the network is experiencing a heavy traffic load then the queuing delay plays the primary role in the delay of a link. In this case it is trivial to say that $q_{jd} \gg p_{jd}$ and we could see from Equation (21.2) that $g_{jd} \approx (1/q_{jd})/\sum_{k=1}^{n} 1/q_{kd}$. When the network is experiencing low traffic then the propagation delay plays an important role in defining the delay of a link. As $q_{jd} \ll p_{jd}$, from Equation (21.2) we get

$g_{jd} \approx (1/p_{jd})/\sum_{k=1}^{n} 1/p_{kd}$. We use *stochastic sampling with replacement* [15] for selecting a neighbor. This principle ensures that a neighbor j with goodness g_{jd} will be selected as the next hop with at least the probability $g_{jd}/\sum_{k=1}^{n} g_{kd}$. Algorithm 21.1 provides the pseudo-code of *BeeHive*.

Algorithm 21.1 (BeeHive)

```
t := current_time, t_end := time to end simulation
// Short_Limit := 7, Long_Limit := 40, Bee_Generation_Interval := 1
// i = current node, d=destination node, s=source node
// n = successor node of i, p=predecessor node of i
// z = Representative node of the foraging region containing s
// w = Representative node of the foraging region containing d
// q is queuing delay estimate of a bee agent from node p to s
// p is propagation delay estimate of a bee agent from node p to s
Δt := Bee_Generation_Interval, Δh := hello packet generation interval
b_ip := estimated_link_band_width_to_neighbor_p
p_ip := estimated_propagation_delay_to_neighbor_p
h_i := hop limit for bees of i, l_ip := size_normal_queue_i_to_p (bits)
foreach Node // concurrent activity over the network
   while (t ≤ t_end)
      if (t mod Δt = 0)
         if (i is representative node of the foraging region)
            set h_i := Long_Limit, b_i is long distance bee agent
         else
            set h_i := Short_Limit, b_i is short distance bee agent
         endif
         launch a bee b_i to all neighbors of i
      endif
      foreach bee b_s received at i from p
         if (b_s was launched by i or its hop limit reached)
            kill bee b_s
         elseif (b_s is inside foraging zone of node s)
            q := q + l_ip/b_ip and p := p + p_ip
            update IFZ routing table entries q_ps = q and p_ps = p
            update q (q := Σ_{k∈N(i)} (q_ks × g_ks)) and p (p := Σ_{k∈N(i)} (p_ks × g_ks))
         else
            q := q + l_ip/b_ip and p := p + p_ip
            update IFR routing table entries q_pz = q and p_pz = p
            update q (q := Σ_{k∈N(i)} (q_kz × g_kz)) and p (p := Σ_{k∈N(i)} (p_kz × g_kz))
         endif
         if (b_s already visited node i)
            kill bee b_s
         else
            use priority queues to forward b_s to all neighbors of i except p
         endif
      endfor
      foreach data packet d_sd received at i from p
         if (node d is within foraging zone of node i)
            consult IFZ routing table of node i to find delays to node d
            calculate goodness of all neighbors for reaching d using equation 21.2
```

```
            else
                consult FRM routing table of node i to find node w
                consult IFR routing table of node i to find delays to node w
                calculate goodness of all neighbors for reaching w using equation 21.2
            endif
            probabilistically select a neighbor n (n ≠ p) as per goodness
            enqueue data packet d_sd in normal queue for neighbor n
        endfor
        if (t mod Δh = 0)
            send a hello packet to all neighbors
            if (time out before a response from neighbor) (4th time)
                neighbor is down
                update the routing table and launch bees to inform other nodes
            endif
        endif
    endwhile
endfor
```

21.6 Routing Algorithms Used for Comparison

The focus of our research is on dynamic algorithms, therefore we now provide an overview of two such algorithms, *AntNet*, which is inspired from the pheromone laying behavior of ants, and *Distributed Genetic Algorithm* (DGA), which uses an evolutionary algorithm for routing. However, we will also briefly introduce Open Shortest Path First (OSPF), a classic static routing algorithm. The reader can find a detailed description of the algorithm in Reference 16 and its complete implementation in Reference 17. We use the algorithm in our comparative simulation for the sake of comprehensiveness.

21.6.1 AntNet: An Algorithm Inspired from Ant Behavior

AntNet was proposed by Di Caro and Dorigo [18]. In *AntNet* the network state is monitored through two ant agents: *Forward_Ant* and *Backward_Ant*. A Forward_Ant agent is launched at regular intervals from a source to a certain destination. The authors proposed a model in Reference 19 that enables a Forward_Ant agent to estimate queuing delay without waiting inside data packet queues. Forward_Ant agent is equipped with a stack memory on which the address and entrance time of each node on its path are pushed. Once the Forward_Ant agent reaches its destination it creates a Backward_Ant agent and transfers all information to it. Backward_Ant visits the same nodes as Forward_Ant in reverse order and modifies the entries in the routing tables based on the trip time from the nodes to the destination. At each node the average trip time, the best trip time, and the variance of the trip times for each destination are saved. The trip time values are calculated by taking the difference of entrance times of two subsequent nodes pushed onto the stack. Backward_Ant agent uses the system priority queues so that it quickly disseminates the information to the nodes. The interested reader may find more details in References 18 and 19. Later on the authors of Reference 20 made significant improvements in the routing table initialization algorithm of *AntNet*, bounded the number of Forward_Ant agents during congestion, and proposed a mechanism to handle routing table entries at the neighbors of crashed routers.

21.6.2 Distributed Genetic Algorithm

The authors of Reference 21 showed that the information needed by *AntNet* for each destination is difficult to obtain in real networks. Their idea of *global information* is that there is an entry in the routing table for each destination. This shortcoming motivated the authors to propose in Reference 22, an evolutionary routing algorithm, DGA, that eliminates the need for having an entry for each destination node in the routing table. In this algorithm ants are asked to traverse a set of n nodes in a particular order, known as a *chromosome*. Once an agent visits the nth node it is then converted into a backward agent that returns to its source node. In contrast to *AntNet*, the backward agents only modify the routing tables at the source node. The source node also measures the fitness of this agent based on the trip time value, and then it generates a new population using single point cross over. New agents enter the network and evaluate the assigned paths. The routing table stores the agents' IDs, their fitness values and trip times to the visited nodes. Routing of a data packet is done through the path that has the shortest trip time to the destination. If no entry for a particular destination is found then a data packet is routed with the help of an agent that has the maximum fitness value. DGA was designed assuming that the routers could crash during network operations. The interested reader will find more details in Reference 22.

Note that in contrast to above-mentioned algorithms, *bee agents* need not be equipped with a stack to perform their duties. Moreover, our agent model requires only forward moving agents and they utilize an estimation model to calculate the trip time from their source to a given node. This model eliminates the need for global clock synchronization among routers, and it is expected that for very large networks routing information could be disseminated quickly with a small overhead as compared with *AntNet*. Our agent model does not require to store the average trip time, the variance of trip times, and the best trip time for each destination at a node to determine the goodness of a neighbor for a particular destination. Last but not the least, *BeeHive* works with a significantly smaller routing table as compared with *AntNet*.

21.6.3 Open Shortest Path First

Open Shortest Path First is currently the state-of-the-art routing algorithm employed as an Interior Gateway Protocol (IGP) in Internet [16]. OSPF stores the entire topology of a network in a weighted directed graph, in which each edge corresponds to a link and each node corresponds to a router. The cost of a link is a function of the propagation delay of the link. However, the network administrators are also allowed to change these costs on the basis of their on-field knowledge about network traffic loads. In our implementation we simply take the fixed propagation costs and do not allow the costs to be changed by the network administrators. Such approach was also taken by the authors of *AntNet*. Each node, in OSPF algorithm, measures the costs of the links to its neighbors. It then encapsulates the addresses of its neighbors and the costs of the links to the neighbors in a link state packet and then broadcasts it to all the neighbors. The neighbors in-turn send the link state packet to their neighbors and so on until all the nodes in the network get the packet. Each node builds the network topology in a distributed fashion. Finally, each node builds a shortest path tree to all destinations by considering itself as a root. The shortest paths from the root to all nodes are calculated using the deterministic Dijkstra Algorithm [23]. The next hop on the shortest path to each destination is stored in a routing table.

21.7 Simulation Environment for BeeHive

In order to evaluate our algorithm *BeeHive* in comparison with *AntNet*, DGA, and OSPF, we implemented all of them in the OMNeT++ simulator [24]. For *AntNet* and DGA we used the same parameters that were reported by the authors in References 18 and 22, respectively. The network instance that we used in our simulation framework is the Japanese Internet Backbone (NTTNet) (see Figure 21.3). It is a 57 node, 162 bidirectional links network. The link bandwidth is 6 Mbits/sec and propagation delay is from 2 to 5 msec. Traffic is defined in terms of open sessions between two different nodes. Each session is characterized completely by sessionSize (2 Mbits), inter-arrival time between two sessions, source, destination, and packet inter-arrival time during a session. The size of data packet is 512 bytes, the size of a *bee agent* is 48 bytes. The queue size is limited to 500 Kbytes in all experiments. To inject dynamically changing data traffic patterns we have defined two states: uniform and weighted. Each state lasts 10 sec and then a state transition to another state occurs. In *Uniform* state (U) a destination is selected from a

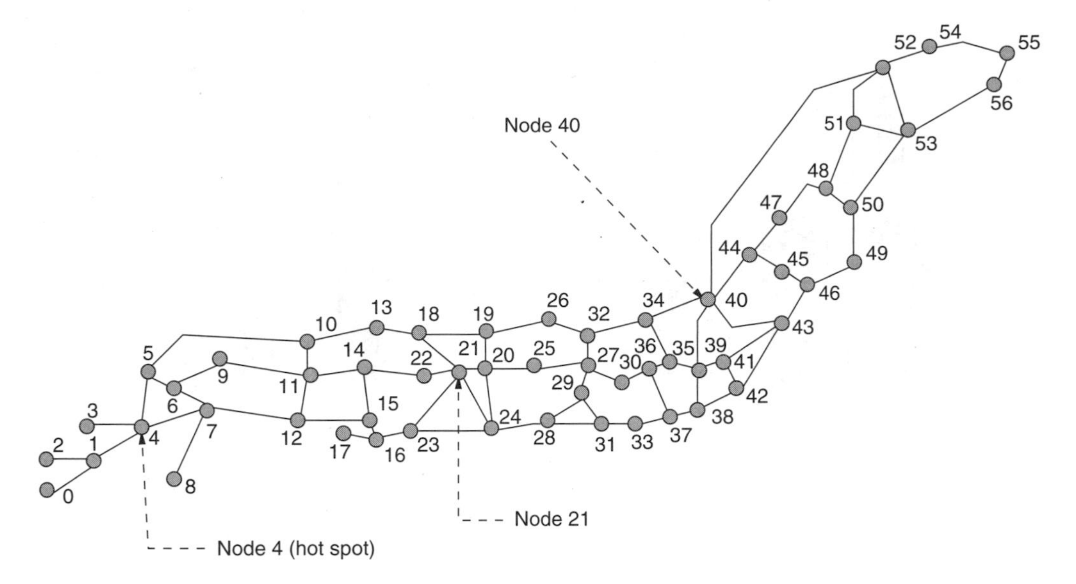

FIGURE 21.3 Japanese Backbone NTTNet.

uniform distribution. While, in *Weighted* state (W), a destination selected in the previous *Uniform* state is favored over other destinations. This approach provides a more challenging experimental environment than the one in which *AntNet* was evaluated.

21.8 Experimental Findings

Now we will report our results obtained from the extensive simulations. MSIA is the mean of session inter-arrival times and MPIA is the mean of packet inter-arrival times during a session. The session inter-arrival and packet inter-arrival times are taken from negative exponential distributions with MSIA and MPIA, respectively. All reported values are an average of the values obtained from ten independent runs. The percentage of deliverable packets that were actually delivered is reported by % Packets Delivered (on top of bars). By deliverable packet we mean a packet whose destination router is up.

Saturating loads: The purpose of the experiments was to study the behavior of the algorithms by gradually increasing the traffic load, through decreasing MSIA from 4.7 to 1.7 sec. MPIA is 0.005 sec during these experiments. Figure 21.4 shows the average throughput and 90th percentile of the packet delay distribution. It is obvious from Figure 21.4 that *BeeHive* delivered approximately the same number of data packets as that of *AntNet* but with lesser packet delay. Both OSFP and DGA are unable to cope with a saturated load yet the performance of DGA is the poorest.

Size of foraging zones: Next, we analyzed the effect of the size of a *foraging zone* in which a *bee agent* updates the routing table. We report the results for sizes of 7, 10, 15, and 20 hops in Figure 21.5. Figure 21.5 shows that increasing the size of *foraging zone* after 10 does not bring significant performance gains. This shows the power of *BeeHive* that it converges to an optimum solution with a size of just 7 hops.

Size of the routing table: The fundamental motivation of the *foraging zone* concept was not only to eliminate the requirement for global knowledge but also to reduce memory needed to store a routing table. *BeeHive* requires 88, 94, and 104 entries, on the average, in the routing tables for *foraging zone* sizes of 7, 10, and 20 hops, respectively. OSFP needs just 57 entries while *AntNet* needs 162 entries on the average. Hence *BeeHive* achieves similar performance as that of *AntNet* but size of the routing table is of the order of OSPF.

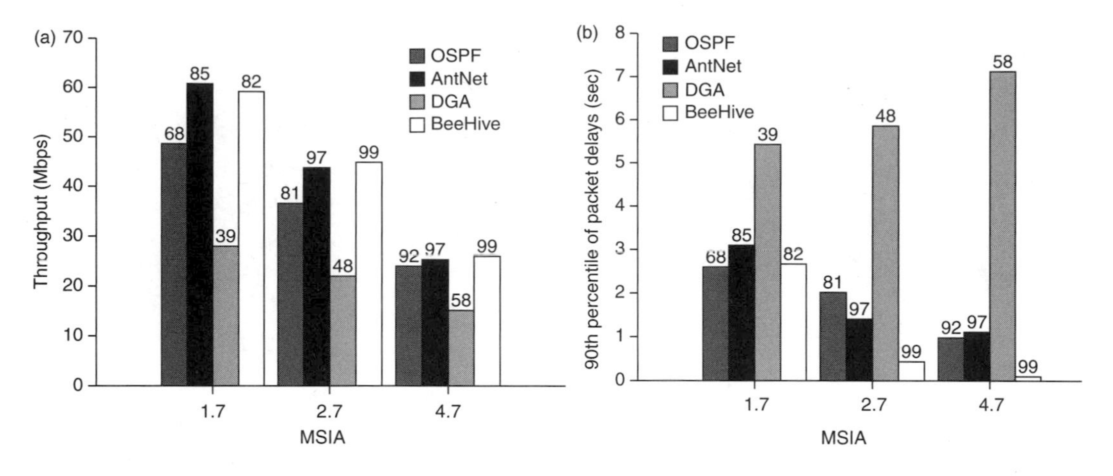

FIGURE 21.4 Behavior under saturating traffic loads.

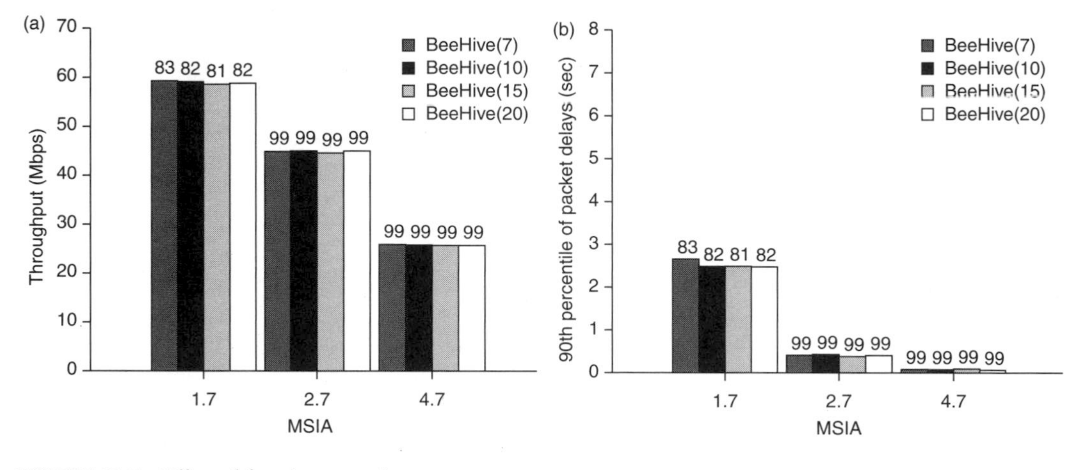

FIGURE 21.5 Effect of *foraging zones* sizes.

Hot spot: The purpose of this experiment is to study the effect of transient overloads in a network. We selected node 4 (see Figure 21.3) as a hot spot. The hot spot was active from 500 to 1000 sec and all nodes sent data to node 4 with MPIA = 0.05 sec. This transient overload was superimposed on a normal load of MSIA = 2.7 sec and MPIA = 0.3 sec. Figure 21.6 shows that both *BeeHive* and *AntNet* are able to cope with the transient overload, however the average packet delay for *BeeHive* is less than 100 msec as compared with 500 msec for *AntNet*. Again DGA shows the poorest performance.

Router crash: The purpose of this experiment was to analyze the fault-tolerant behavior of *BeeHive* so we took MSIA = 4.7 sec and MPIA = 0.005 sec to ensure that no packets are dropped because of the congestion. We simulated a scenario in which Router 21 crashed at 300 sec, and Router 40 crashed at 500 sec and then both were repaired at 800 sec. Figure 21.7 shows the results. *BeeHive* is able to deliver 97% of deliverable packets as compared with 89% by *AntNet*. Observe that from 300 to 500 sec (just Router 21 is down), *BeeHive* has a superior throughput and lesser packet delay but once Router 40 crashes, the packet delay of *BeeHive* increases because of higher load at Router 43. From Figure 21.3 it is obvious that the only path to the upper part of the network is via Router 43 once Router 40 crashed. Since *BeeHive* is able to deliver more packets the queue length at Router 43 increased and this led to relatively poorer packet delay as compared with *AntNet*.

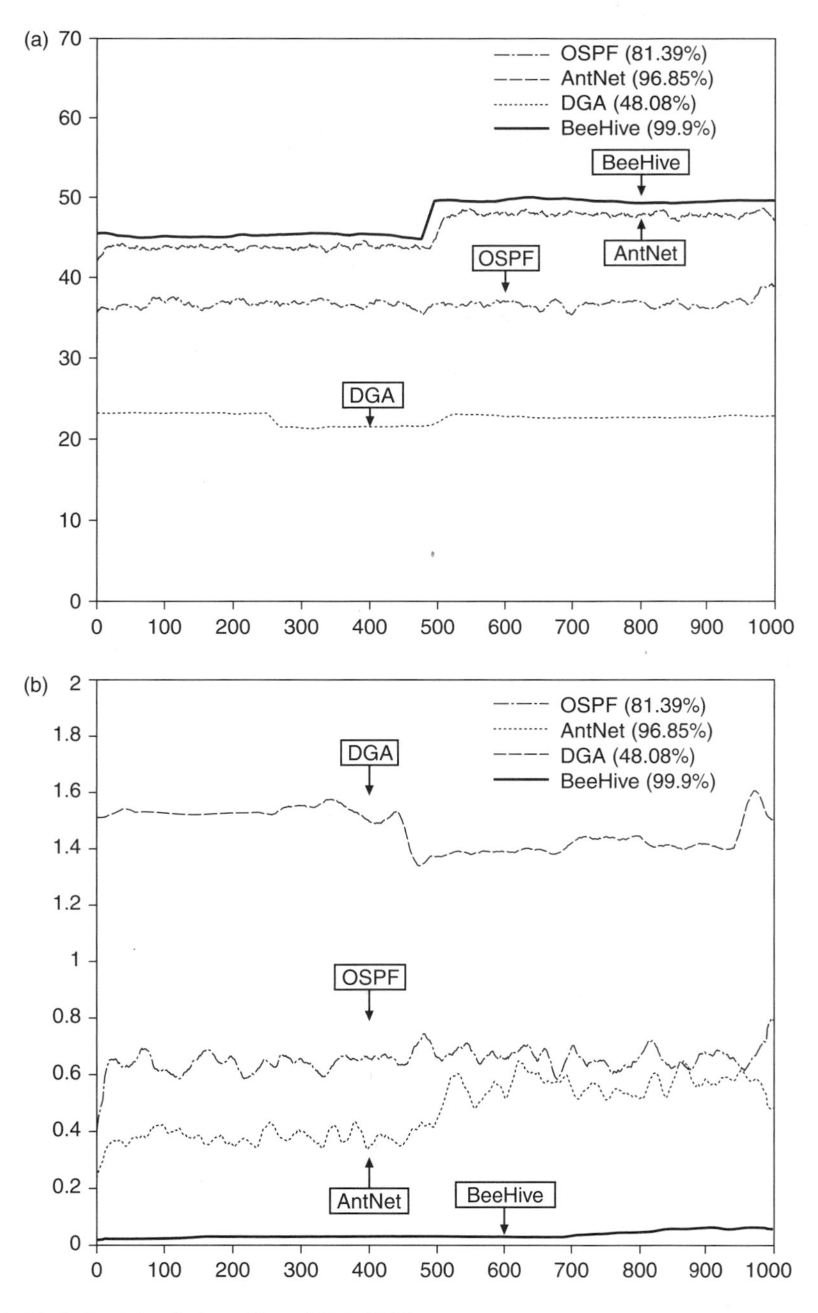

FIGURE 21.6 Node 4 acted as hot spot from 500 to 1000 sec.

In addition, observe that Router 21 is critical but in case of its crash still multiple paths exist to the middle and upper part of the topology via 15, 18, 19, 20, 24.

Overhead of BeeHive: We first define the terms routing overhead and suboptimal overhead.

Definition 21.2 (Routing overhead). *Routing overhead is defined as the ratio between the bandwidth occupied by the routing packets and the total available network bandwidth [18].*

The metric *suboptimal overhead* was introduced in Reference 25 in the context of MANETS but we believe that it is equally relevant in fixed networks as well.

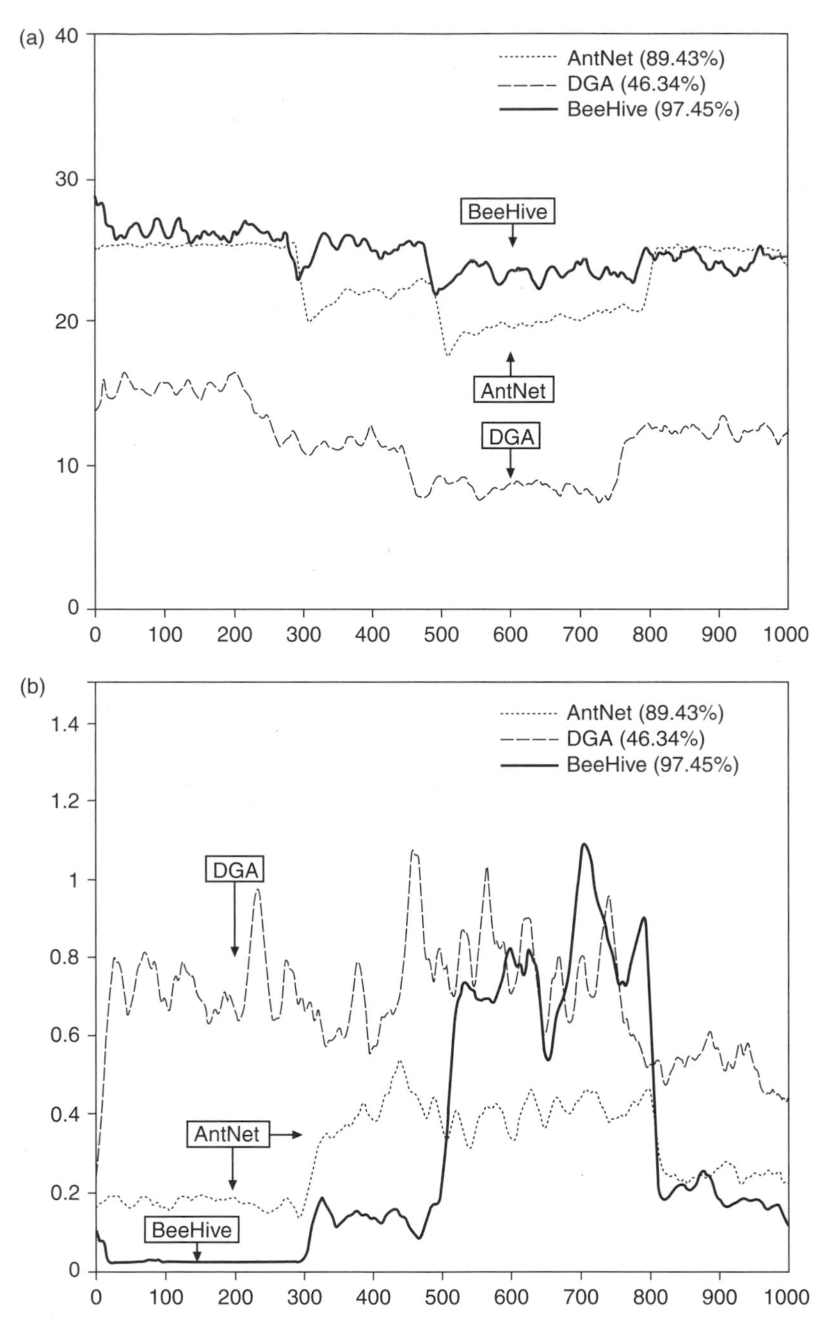

FIGURE 21.7 Router 21 is down at 300 and Router 40 at 500 and both repaired at 800.

Definition 21.3 (Suboptimal overhead). *The difference between the bandwidth consumed when transmitting data packets from all the sources to destinations and the bandwidth that would have been consumed, should the data packets have followed the shortest hop count path. Formally, we could define the parameter as*

$$S_o = \frac{\sum_{d=1}^{n} \sum_{s=1}^{n} \sum_{i=1}^{k} (h_i^{sd} - h_o^{sd}) \times L_i^{sd}}{B_t}, \quad s \neq d, \tag{21.3}$$

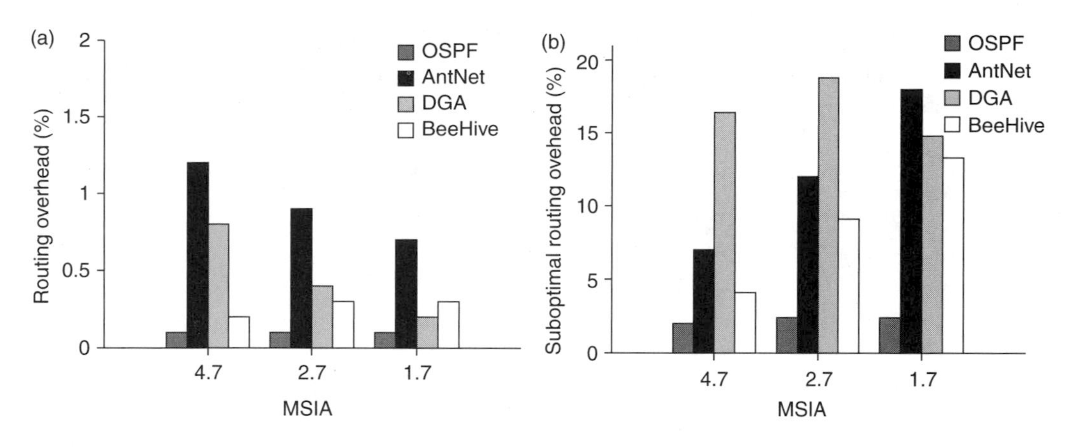

FIGURE 21.8 Saturated loads.

where n is total number of nodes in the network, k is total number of packets generated, h_i^{sd} *number of hops that packet i took to reach its destination d from its source s,* h_0^{sd} *minimum hops between node s and node d,* L_i^{sd} *is length of packet i from source s to destination d, and* B_t *is the total bandwidth of the network.*

We report this parameter because it implicitly includes the overhead of loops.

Figure 21.8 shows the control overhead and suboptimal overhead of the algorithms. It is quite interesting to note that the suboptimal overhead is much higher than the control overhead (note that the x- and y-axes have different scales on the two figures). OSPF has the smallest suboptimal overhead but then it delivers less packets as well. The routing overhead of DGA decreases with an increase in the load and vice versa. This happens due to the genetic algorithm. Remember that the next generation of the agents are launched once four agents from the previous generation are received. Under low load, the return times for the agents are smaller, as a result, the agents are launched at a higher rate and vice versa. Since in *AntNet*, Forward_Ant agents use the same queues that data packets also use, therefore more ants were dropped under increased network load and this explains the decrease in routing overhead behavior of the algorithm. It is evident from Figure 21.8 that *BeeHive* has significantly smaller routing overhead and suboptimal overhead as compared to *AntNet* and DGA.

21.9 Conclusion

A honey bee colony is able to optimize its stockpiles of nectar, pollen, and water through an intelligent allocation of labor among different specialists, which communicate with each other using a sophisticated communication protocol that consists of *signals* and *cues*, in continuously changing internal and external environments. The dance language and foraging behavior of honey bees inspired us to develop a fault-tolerant, adaptive, and robust routing protocol. The algorithm does not need any global information such as the structure of the topology and cost of links among routers, rather it works with the local information that a *short distance bee agent* collects in a *foraging zone*. It works without the need of global clock synchronization, which not only simplifies its installation on real routers but also enhances fault tolerance. In contrast to *AntNet* our algorithm utilizes only forward moving *bee agents* that help in disseminating the state of the network to the routers in real time. The *bee agents* take less than 1% of the available bandwidth but provide significant enhancements in throughput and packet delay.

We implemented two state-of-the-art adaptive algorithms (*AntNet* and DGA) for the OMNeT++ simulator and then compared our *BeeHive* algorithm with them. Through extensive simulations representing dynamically changing operating network environments we have demonstrated that *BeeHive* achieves a better or similar performance as compared with *AntNet*. However, this enhancement in performance is achieved with a routing table whose size is of the order of OSPF.

We should emphasize that the foraging model described in References 6 and 7 could be used for developing any multi-objective optimization algorithm for dynamic environments. The important concepts of *BeeHive*, such as *bee agent propagation* algorithm and *bee agent communication* paradigm, could be applied to any optimization problem that could be represented in a graph. We believe that *BeeHive* concept will motivate the researchers to develop Bee Colony Optimization (BCO) algorithms.

References

[1] T.D. Seeley. *The Wisdom of the Hive*. Harvard University Press, London, 1995.

[2] K. von Frisch. *Tanzsprache und Orientierung der Bienen*. Springer-Verlag, Heidelberg, 1965.

[3] K. von Frisch. *The Dance Language and Orientation of Bees*. Harvard University Press, Cambridge, 1967.

[4] H.A. Simon. *Administrative Behavior: A Study of Decision-Making Processes in Administrative Organization*. Free Press, New York, 1976.

[5] T.D. Seeley and W.F. Towne. Collective decision making in honey bees: How colonies choose among nectar sources. *Behavior Ecology and Sociobiology*, 12: 277–290, 1991.

[6] D.J.T. Sumpter. From bee to society: An agent-based investigation of honey bee colonies. Ph.D. thesis, The University of Manchester, UK, 2000.

[7] L.P. Kaelbling, M.L. Littman, and A.W. Moore. Reinforcement learning: A survey. *Journal of Artificial Intelligence*, 4: 237–285, 1996.

[8] H.F. Wedde, M. Farooq, and Y. Zhang. Beehive: An efficient fault-tolerant routing algorithm inspired by honey bee behavior. In *Proceedings of ANTS Workshop, Lecture Notes in Computer Science 3172*, Springer-Verlag, September 2004.

[9] H.F. Wedde and M. Farooq et al. BeeHive — An Energy-Aware Scheduling and Routing Framework. Technical report, Project Group 439, LSIII, School of Computer Science, University of Dortmund, 2004.

[10] Y. Zhang. Design and implementation of bee agents based algorithm for routing in high speed, adaptive and fault-tolerant networks. Master thesis, LSIII, The University of Dortmund, Germany, 2003.

[11] P. Nii. The blackboard model of problem solving. *AI Magazine*, 7: 38–53, 1986.

[12] E. Bonabeau, M. Dorigo, and G. Theraulaz. *Swarm Intelligence from Natural to Artificial Systems*. Oxford University Press, Oxford, 1999.

[13] P.P. Grassé. La reconstruction du nid et les coordinations interindividuelles chez bellicositermes natalensis et cubitermes sp. la théorie de la stigmergie: essai d'interprétation du comportement des termites constructeurs. *Insectes Sociaux*, 6: 41–81, 1959.

[14] L.L. Peterson and B.S. Davie. *Computer Networks a Systems Approach*. Morgan Kaufmann Publishers, San Francisco, CA, 2000.

[15] D.E. Goldberg. *Genetic Algorithms in Search, Optimization and Machine Learning*. Addison-Wesley, Reading, MA, 1989.

[16] J.T. Moy. *OSPF Anatomy of an Internet Routing Protcol*. Addison-Wesley, Reading, MA, 1998.

[17] J.T. Moy. *OSPF Complete Implementation*. Addison-Wesley, Reading, MA, 2000.

[18] G. Di Caro and M. Dorigo. AntNet: Distributed stigmergetic control for communication networks. *Journal of Artificial Intelligence*, 9: 317–365, 1998.

[19] G. Di Caro and M. Dorigo. Two ant colony algorithms for best-effort routing in datagram networks. In *Proceedings of the Tenth IASTED International Conference on Parallel and Distributed Computing and Systems (PDCS'98)*, IASTED/ACTA Press, 1998, pp. 541–546.

[20] B. Barán and R. Sosa. A new approach for antnet routing. In *Proceedings of the Ninth International Conference on Computer, Communications and Networks*, 2000.

[21] S. Liang, A.N. Zincir-Heywood, and M.I. Heywood. The effect of routing under local information using a social insect metaphor. In *Proceedings of IEEE Congress on Evolutionary Computing*, May 2002.

[22] S. Liang, A.N. Zincir-Heywood, and M.I. Heywood. Intelligent packets for dynamic network routing using distributed genetic algorithm. In *Proceedings of Genetic and Evolutionary Computation Conference*, GECCO, July 2002.

[23] E.W. Dijkstra. A note on two problems in connection with graphs. *Numerical Mathematics*, 1: 269–271, 1959.

[24] A. Varga. OMNeT++: Discrete Event Simulation System: User Manual. http://www.omnetpp.org.

[25] C. Santivanez, B. McDonald, I. Stavrakakis, and R. Ramanathan. On the scalability of ad hoc routing protocols. In *Proceedings of IEEE INFOCOM 2002*, IEEE, June 2002.

22

Swarming Agents for Decentralized Clustering in Spatial Data

Gianluigi Folino
Agostino Forestiero
Giandomenico Spezzano

22.1 Introduction

In recent years several approaches to knowledge discovery and data mining, and in particular, clustering, have been developed, but only a few of them are designed using a decentralized approach. Clustering data is the process of grouping similar objects according to their distance, connectivity, or relative density in space [1]. There are a large number of algorithms for discovering natural clusters, if they exist, in a dataset, but they are usually studied as a centralized problem. These algorithms can be classified into partitioning methods [2], hierarchical methods [3], density-based methods [4], and grid-based methods [5]. Han et al.'s paper [6] is a good introduction to this subject. Many of these algorithms work on data contained in a file or database. In general, clustering algorithms focus on creating good compact representation of clusters and appropriate distance functions between data points. For this purpose, they generally need to be given one or two parameters by a user that indicate the types of clusters expected. Since they use a central representation where each point can be compared with each other point or cluster representation, points are never placed in a cluster with largely different members. Centralized clustering is problematic if we have large data to explore or data is widely distributed. Parallel and distributed computing is expected to relieve current mining methods from the sequential bottleneck, providing the ability to scale to massive datasets, and improving the response time. Achieving good performance on today's high performance systems is a nontrivial task. The main challenges include synchronization and communication minimization, workload balancing, finding good data decomposition, etc. Some existing centralized clustering algorithms have been parallelized and the results have been encouraging. Centralized schemes require high level of connectivity, impose a substantial computational burden, are typically more sensitive to failures than decentralized schemes, and are not scalable, which is a property that distributed computing systems are required to have.

Recently, other algorithms based on biological models [7–10] have been introduced to solve the clustering problem in a decentralized fashion. These algorithms are characterized by the interaction of a large number of simple agents that sense and change their environment locally. Furthermore, they exhibit complex, emergent behavior that is robust with respect to the failure of individual agents. Ant colonies, flocks of birds, termites, swarms of bees, etc., are agent-based insect models that exhibit a collective intelligent behavior (*SWARM intelligence*, SI) [11] that may be used to define new algorithms of clustering. The use of new SI-based techniques for data clustering is an emerging new area of research that has attracted many investigators in the last years. SI models have many features in common with evolutionary algorithms (EAs). Like EA, SI models are population-based. The system is initialized with a population of individuals (i.e., potential solutions). These individuals are then manipulated over many iteration steps by mimicking the social behavior of insects or animals, in an effort to find the optima in the problem space. Unlike EAs, SI models do not explicitly use evolutionary operators, such as crossover and mutation. A potential solution simply "flies" through the search space by modifying itself according to its past experience and its relationship with other individuals in the population and the environment. In these models, the emergent collective behavior is the outcome of a process of self-organization, in which insects are engaged through their repeated actions and interaction with their evolving environment. Intelligent behavior frequently arises through indirect communication between the agents using the principle of *stigmergy* [12]. This mechanism is a powerful principle of cooperation in insect societies. According to this principle, an agent deposits something in the environment that makes no direct contribution to the task being undertaken but is used to influence the subsequent behavior that is task related. The advantages of SI are twofold. First, it offers intrinsically distributed algorithms that can use parallel computation quite easily. Second, these algorithms show a high level of robustness to change by allowing the solution to dynamically adapt itself to global changes by letting the agents self-adapt to the associated local changes. Social insects provide a new paradigm for developing decentralized clustering algorithms.

In this chapter, we present a method for decentralized clustering based on an adaptive flocking algorithm proposed by Macgill [13]. We consider clustering as a search problem in a multiagent system in which individual agents have the goal of finding specific elements in the search space, represented by a large dataset of tuples, by walking efficiently through this space. The method takes advantage of the parallel search mechanism a flock implies, by which if a member of a flock finds an area of interest; the mechanics of the flock will draw other members to scan that area in more detail. The algorithm selects interesting subsets of tuples without inspecting the whole search space guaranteeing a fast placing of points correctly in the clusters. We have applied this strategy as a data reduction technique to perform efficiently *approximate* clustering [14]. In the algorithm, each agent can use hierarchical, partitioned, density-based, and grid-based clustering methods to discover if a tuple belongs to a cluster.

To illustrate this method we present two algorithms: SPARROW (SPAtial clusteRing algoRithm thrOugh sWarm intelligence) and SPARROW-SNN. SPARROW combines the flocking algorithm with the density-based DBSCAN algorithm [15]. SPARROW-SNN combines the flocking algorithm with a shared nearest-neighbor (SNN) cluster algorithm [16] to discover clusters with differing sizes, shapes, and densities in noise, high dimensional data. We have built a SWARM [17] simulation of both algorithms to investigate the interaction of the parameters that characterize them. First, experiments showed encouraging results and a better performance of both algorithms in comparison with the linear randomized search using an entropy-based model.

The rest of this chapter is organized as follows: Section 22.2 presents the SPARROW clustering algorithm, first, introducing the heuristics of the DBSCAN algorithm, used by each agent to discover clusters, and the flocking algorithm for the interaction among the agents. Section 22.3 describes the centralized SNN clustering algorithm and how it is combined with the flocking algorithm to produce the SPARROW-SNN algorithm. Section 22.4 discusses the results obtained. Section 22.5 presents an entropy-based model to theoretically explain the behavior of the algorithm and Section 22.6 draws some conclusions.

22.2 A Multiagent Spatial Clustering Algorithm

In this section, we will present the SPARROW algorithm that combines the stochastic search of an adaptive flocking with the DBSCAN heuristics for discovering clusters in parallel. SPARROW replaces the DBSCAN serial procedure for clusters identification with a multiagent stochastic search that has the advantage of being easily implementable on parallel computers and that is robust compared with the failure of individual agents. We will first introduce the DBSCAN algorithm and then will present the Reynolds' flock model that describes the standard movement rules of birds. Finally, we will illustrate the details of the SPARROW behavioral rules of the Swarming agents (SAs) that move through the spatial data looking for clusters and communicating their findings to each other.

22.2.1 The DBSCAN Algorithm

One of the most popular spatial clustering algorithms is DBSCAN, which is a density-based spatial clustering algorithm. A complete description of the algorithm and its theoretical basis is presented in the paper by Ester et al. [15]. In the following, we briefly present the main principles of DBSCAN. The algorithm is based on the idea that all points of a dataset can be regrouped into two classes: *clusters* and *noise*. Clusters are defined as a set of dense connected regions with a given radius (Eps) and containing at least a minimum number (MinPts) of points. Data are regarded as noise if the number of points contained in a region falls below a specified threshold. The two parameters, Eps and MinPts, must be specified by the user and allowed to control the density of the cluster that must be retrieved. The algorithm defines two different kinds of points in a clustering: *core point*s and *non-core points*. A core point is a point with at least MinPts number of points in an Eps neighborhood of the point. The non-core points in turn are either *border points* if not core points but are density reachable from another core point or *noise points* if not core points and are not density reachable from other points. To find the clusters in a dataset, DBSCAN starts from an arbitrary point and retrieves all points with the same density reachable from that point using Eps and MinPts as controlling parameters. A point p is density reachable from a point q if the two points are connected by a chain of points such that each point has a minimal number of data points, including the next point in the chain, within a fixed radius. If the point is a core point, then the procedure yields a cluster. If the point is on the border, then DBSCAN goes on to the next point in the database and the point is assigned to the noise. DBSCAN builds clusters in sequence (i.e., one at a time), in the order in which they are encountered during space traversal. The retrieval of the density of a cluster is performed by successive spatial queries. Such queries are supported efficiently by spatial access methods such as R*-trees.

22.2.2 The Reynolds' Flock Model

The flocking algorithm was originally proposed by Reynolds [18] as a method for mimicking the flocking behavior of birds on a computer both for animation and as a way to study emergent behavior. Flocking is an example of emergent collective behavior: there is no leader, that is, no global control. Flocking behavior emerges from the local interactions. In the flock algorithm, each agent has direct access to the geometric description of the whole scene, but reacts only to flock mates within a certain small radius. The basic flocking model consists of three simple steering behaviors:

Separation gives an agent the ability to maintain a certain distance from others. This prevents agents from crowding too closely together, allowing them to scan a wider area.

Cohesion gives an agent the ability to cohere (approach and form a group) with other nearby agents. Steering for cohesion can be computed by finding all agents in the local neighborhood and computing the *average position* of the nearby agents. The steering force is then applied in the direction of that *average position.*

Alignment gives an agent the ability to align with other nearby characters. Steering for alignment can be computed by finding all agents in the local neighborhood and averaging together the "heading" vectors of the nearby agents.

22.2.3 SPARROW: A Flocking Algorithm for Spatial Clustering

SPARROW is a multiagent adaptive algorithm able to discover clusters in parallel. It uses a modified version of standard flocking algorithm that incorporates the capacity for learning that can be found in many social insects. In the algorithm, the agents are transformed into hunters with a foraging behavior that allow them to explore the spatial data while searching for clusters.

SPARROW starts with a fixed number of agents that occupy a randomly generated position. Each agent moves around the spatial data testing the neighborhood of each location in order to verify if the point can be identified as a *core point.* In such a case, a temporary label is assigned to all the points of the neighborhood of the core point. The labels are updated concurrently as multiple clusters take shape. Contiguous points belonging to the same cluster take the label corresponding to the smallest label in the group of contiguous points. Each agent follows the rules of movement described in Reynolds' model. In addition, our model considers four different kinds of agents, classified based on the density of data in their neighborhood. These different kinds are characterized by a different color: *red,* revealing a high density of interesting patterns in the data, *green,* a medium one, *yellow,* a low one, and *white,* indicating a total absence of patterns. The main idea behind our approach is to take advantage of the colored agent in order to explore more accurately the most interesting regions (signaled by the red agents) and avoid the ones without clusters (signaled by the white agents). Red and white agents stop moving in order to signal this type of regions to the others, while green and yellow ones fly to find more dense clusters. Indeed, each flying agent computes its heading by taking the weighted average of alignment, separation, and cohesion (as illustrated in Figure 22.1). The following are the main features that make our model different from Reynolds' model:

- *Alignment* and *cohesion* do not consider yellow agents, since they move in a not very attractive zone.
- *Cohesion* is the resultant of the heading toward the average position of the green flockmates (centroid), of the attraction toward reds, and of the repulsion from whites, as illustrated in Figure 22.1.
- A *separation* distance is maintained from all the boids, apart from their color.

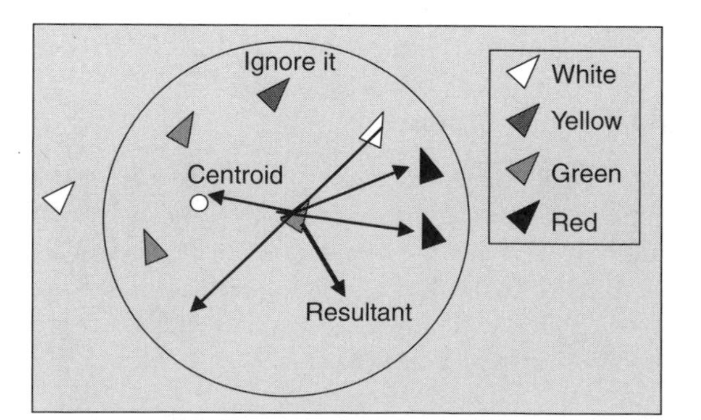

FIGURE 22.1 Computing the direction of an agent in SPARROW.

```
for i = 1..MaxGenerations
    foreach agent (yellow, green)
        age=age+1;
        if (age>Max_Life)
            generate_new_agent(); die();
        endif
        if (not visited (current_point))
            density = compute_local_density();
            mycolor= color_agent();
        endif
    end foreach

    foreach agent (yellow, green)
        dir=compute_dir();
    end foreach
    foreach agent (all)
            switch (mycolor){
                case yellow, green:
                        move(dir, speed(mycolor)); break;
                case white:
                        stop();generate_new_agent(); break;
                case red:
                    stop(); merge(); generate_new_close_agent();
                    break;
                }
    end foreach
```

FIGURE 22.2 Pseudo code of the SPARROW algorithm.

The SPARROW algorithm consists of a setup phase and a running phase shown in Figure 22.2. During the setup phase agents are created, data are loaded, and some general settings are made. In the running phase each agent repeats four distinct procedures for a fixed number of times (*MaxGenerations*). We use the *foreach* statement to indicate that the rules are executed in parallel by the agents whose color is specified in the argument. The *mycolor* procedure chooses the color and the speed of the agents with regard to the local density of the points of clusters in the data. It is based on the same parameters used in the DBSCAN algorithm: *MinPts*, the minimum number of points to form a cluster and *Eps*, the maximum distance that the agents can look at. In practice, the agent computes the local density (*density*) in a circular neighborhood (with a radius determined by its limited sight, i.e., *Eps*) and then it chooses the color according to the following simple rules:

$density > MinPts \rightarrow mycolor = \textbf{red}\ (speed=0)$

$MinPts/4 < density <= MinPts \rightarrow mycolor = \textbf{green}\ (speed=1)$

$0 < density <= MinPts/4 \rightarrow mycolor = \textbf{yellow}\ (speed=2)$

$density = 0 \rightarrow mycolor = \textbf{white}\ (speed=0)$

In the running phase, yellow and green agents will compute their direction, according to the rules previously described, and will move following this direction with the speed corresponding to their color.

In case the agent falls in the same position of that of an older one it will die and will be regenerated in another place.

Agents will move toward the computed destination with a speed depending on their color: green agents more slower than yellow agents since they will explore denser zones of clusters. Green and yellow agents have a variable speed, with a common minimum and maximum for all agents. An agent will speedup to leave an empty or uninteresting region whereas will slowdown to investigate an interesting region more carefully.

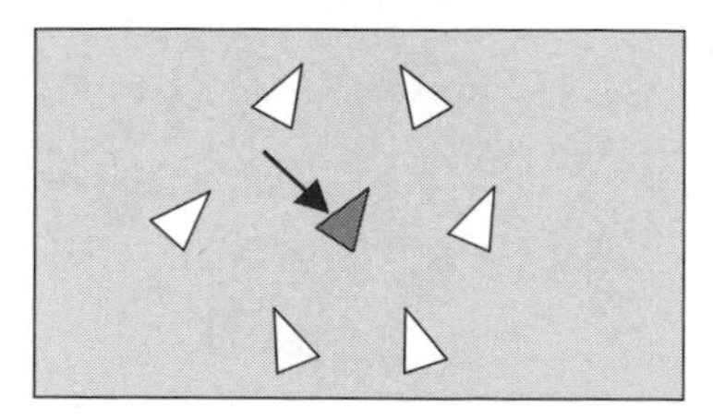

FIGURE 22.3 The cage effect.

The variable speed introduces an adaptive behavior in the algorithm. In fact, agents adapt their movement and change their behavior (speed) based on their previous experience represented from the red and white agents. Red and white agents will stop signaling to the others the interesting and desert regions. Note that for any agent that has become red or white, a new agent will be generated in order to maintain a constant number of agents exploring the data. In the first case, the new agent will be generated in a close random point, since the zone is considered interesting, while in the latter it will be generated in a random point over all the space. Next, red agents will run the *merge* procedure, which will merge the neighboring clusters. The merging phase considers two different cases: when we have never visited points in the circular neighborhood and when we have points belonging to different clusters. In the first case, the points will be labeled and will constitute a new cluster; in the second case, all the points will be merged into the same cluster, that is, they will get the label of the cluster discovered first. The first part of code executed by yellow and green agents was added to avoid a "*cage effect*" (see Figure 22.3), which occurred during the first simulations; in fact, some agents could remain trapped inside regions surrounded by red or white agents and would have no way to go out, wasting useful resources for the exploration. Therefore, a limit was imposed on their life. When their age exceeded a determined value (*Max_Life*) they were made to die and were regenerated in a new randomly chosen position of the space.

SPARROW suffers the same limitation as DBSCAN, that is, cannot cope with clusters of different densities. The new algorithm SPARROW-SNN, introduced in Section 22.3, is more general and overcomes these drawbacks. It can be used to discover clusters with differing sizes, shapes, and densities in noise data.

22.3 The SPARROW-SNN Algorithm

In this section, we present the SPARROW-SNN multiagent clustering algorithm, which combines the stochastic search of the SPARROW algorithm with the SNN heuristics for discovering clusters in spatial data. This approach has a number of nice properties. It can be easily implemented on parallel computers and is robust compared with the failure of individual agents. It can also be applied to perform efficiently *approximate clustering* since the points, that are visited and analyzed by the agents, represent a significant (in *ergodic* sense) subset of the entire dataset. The subset reduces the size of the dataset while keeping the accuracy loss as small as possible. Moreover, SPARROW-SNN can discover clusters with differing sizes, shapes, and densities in noise data in parallel. The behavior requires the individual members to first explore the environment, searching for a goal whose position was not known a priori, and then, after the goal is located, all the flock members should move toward the goal. The same strategy is used by SPARROW to transform the agents into hunters with a foraging behavior in order to explore the spatial data while searching for clusters. Clusters are discovered using the heuristics principles of the SNN clustering algorithm.

22.3.1 The SNN Clustering Algorithm

The SNN is a clustering algorithm developed by Ertöz et al. [16] to discover clusters with differing sizes, shapes, and densities in noise, high dimensional data. The algorithm extends the nearest-neighbor nonhierarchical clustering technique by Jarvis and Patrick [19] redefining the similarity between pairs of points in terms of how many nearest neighbors the two points share. Using this new definition of similarity, the algorithm eliminates noise and outliers, identifies representative points, and then builds clusters around the representative points. These clusters do not contain all the points, rather they represent relatively uniform group of points.

The SNN algorithm starts performing the Jarvis–Patrick scheme. In the Jarvis–Patrick algorithm, a set of objects is partitioned into clusters based on the number of shared nearest neighbors. The standard implementation is constituted by two phases. The first, a preprocessing stage identifies the K nearest neighbors of each object in the dataset. In the subsequent clustering stage, two objects i and j join the same cluster if:

- i is one of the K nearest neighbors of j
- j is one of the K nearest neighbors of i
- i and j have at least $K_{\min}$ of their K nearest neighbors in common

where K and $K_{\min}$ are used-defined parameters. For each pair of points i and j is defined as a link with an associate weight. The strength of the link between i and j is defined as:

$$\text{strength}(i,j) = \Sigma(k+1-m)(k+1-n) \quad \text{where } i_m = j_n.$$

In this equation, k is the nearest-neighbor list size, m and n are the positions of a shared nearest neighbor in i and j's lists. At this point, clusters can be obtained by removing all edges with weights less than a user-specified threshold and taking all the connected components as clusters. A major drawback of the Jarvis–Patrick algorithm is that, the threshold needs to be set high enough since two distinct set of points can be merged into same cluster even if there is only a link across them. On the other hand, if a high threshold is applied, then a natural cluster will be split into many small clusters due to the variations in the similarity in the cluster.

Shared nearest-neighbor addresses these problems introducing the following steps:

1. For every node (data point) calculates the total strength of links coming out of the point.
2. Identify representative points by choosing the point that have high density (>*core_threshold*).
3. Identify noise points by choosing the points that have low density (<*noise_threshold*) and remove them.
4. Remove all links between points that have weight smaller than a threshold (*merge_threshold*).
5. Take connected components of points to form clusters, where every point in a cluster is either a representative point or is connected to a representative point.

The number of clusters is not given to the algorithm as a parameter. Also, note that not all the points are clustered.

22.3.2 The SPARROW-SNN Implementation

As the first step, SPARROW-SNN computes for each data element the nearest-neighbor list using a *similarity threshold* that reduces the number of data elements to take into consideration. The introduction of the similarity threshold produces variable-length nearest-neighbor lists and therefore i and j must have at least $P_{\min}$ of the shorter nearest-neighbor list in common; where $P_{\min}$ is a user-defined percentage.

After the nearest-neighbor list is computed, SPARROW-SNN starts a fixed number of agents that will occupy a randomly generated position. The agents have an attribute that defines their color. From their initial position, each agent moves around the spatial data testing the neighborhood of each location in order to verify if the point can be identified as a *representative* (or *core*) *point*.

Figure 22.4 illustrates the behavioral rules of each agent. All agents execute the same set of rules for a fixed number of times (*MaxGenerations*). When an agent falls on a data point A, not yet visited, it

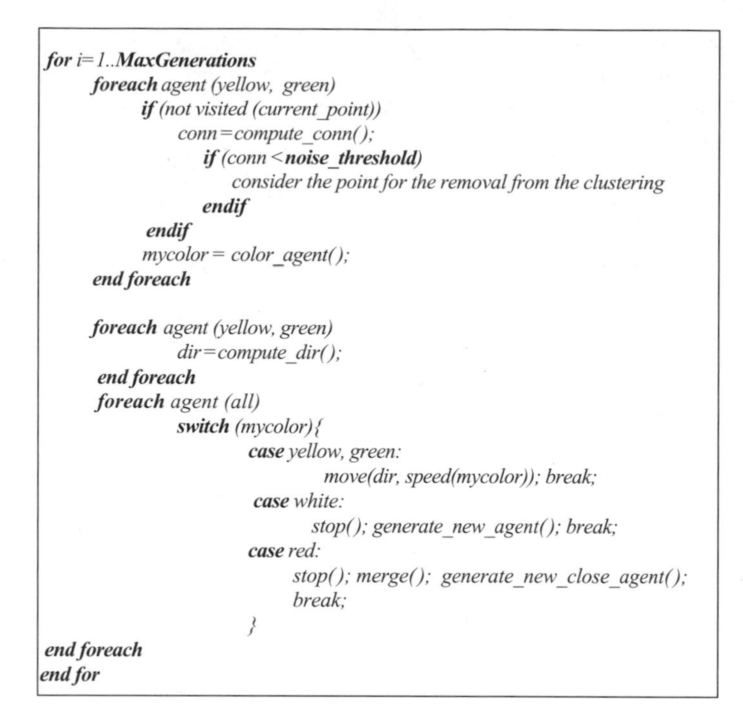

```
for i=1..MaxGenerations
    foreach agent (yellow, green)
        if (not visited (current_point))
            conn=compute_conn();
              if (conn<noise_threshold)
                  consider the point for the removal from the clustering
              endif
        endif
        mycolor= color_agent();
    end foreach

    foreach agent (yellow, green)
            dir=compute_dir();
    end foreach
    foreach agent (all)
            switch (mycolor){
                  case yellow, green:
                          move(dir, speed(mycolor)); break;
                  case white:
                         stop(); generate_new_agent(); break;
                  case red:
                        stop(); merge(); generate_new_close_agent();
                        break;
                  }
end foreach
end for
```

FIGURE 22.4 Pseudo code of the SPARROW-SNN algorithm.

computes the connectivity, conn[A], of the point, that is, computes the total number of strong links the points has, according to the rules of the SNN algorithm. Points having connectivity smaller than a fixed threshold (*noise_threshold*) are classified as noise and are considered for removal from clustering. A color is then assigned to each agent, based on the value of the connectivity computed in the visited point, using the following procedure (called *color_agent()* in the pseudo code):

conn > core_threshold → *mycolor* = ***red***

noise_threshold < conn <= core_threshold → *mycolor* = ***green***

0 < strength < noise_threshold → *mycolor* = ***yellow***

strength = 0 → *mycolor* = ***white***

The colors assigned to the agents are: *red*, revealing representative points, *green*, border points, *yellow*, noise points, and *white*, indicating an obstacle (*uninteresting region*). After the coloration step, the green and yellow agents, compute their movement observing the positions of all other agents that are at some fixed distance (*dist_max*) from their and applying the same rules of SPARROW.

In any case, each red agent (placed on a representative point) will run the merge procedure so that it will include, in the final cluster, the representative point discovered together with the points that share a significant (greater than $P_{\min}$) number of neighbors and that are not noise points. The merging phase considers two different cases: when we have visited none of these points in the neighborhood and when we have points belonging to different clusters. In the first case, the points will be assigned the same temporary label and will constitute a new cluster; in the second case, all the points will be merged into the same cluster, that is, they will get the label corresponding to the smallest one. Therefore, clusters will be built incrementally.

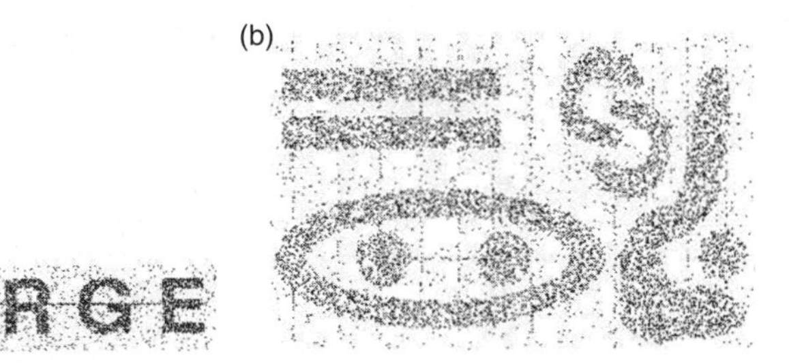

FIGURE 22.5 The two datasets used in our experiments.

In addition, in this case, during simulations a "*cage effect*," was observed. This part is not reported in the pseudo code, as it is the same used in the SPARROW algorithm.

22.4 Experimental Results

In Sections 22.4.1 and 22.4.2, we present the experimental results obtained for the two algorithms. Both algorithms have been implemented using SWARM, a multiagent software platform for the simulation of complex adaptive systems. In the SWARM system, the basic unit of simulation is the *SWARM*, a collection of agents executing a schedule of actions. SWARM provides object-oriented libraries of reusable components for building models and analyzing, displaying, and controlling experiments on those models. More information about SWARM can be obtained from Reference 17.

22.4.1 SPARROW Results

We evaluated the accuracy of the solution supplied by SPARROW in comparison with the one of DBSCAN and the performance of the search strategy of SPARROW in comparison with the standard flocking search strategy and with the linear randomized search. Furthermore, we evaluated the impact of the number of agents on foraging for clusters performance. Results are compared with the ones obtained using a publicly available version of DBSCAN.

For the experiments, we used two synthetic datasets, shown in Figure 22.5(a) and (b). The first dataset, called GEORGE, consists of 5463 points. The second dataset, called DS4, contains 8843 points. Each point of the two datasets has two attributes that define the x and y coordinates. Furthermore, both datasets have a considerable quantity of noise.

Although DBSCAN and SPARROW produce the same results if we examine all points of the dataset, our experiments show that SPARROW can obtain, with an average accuracy about 93% on GEORGE dataset and about 78% on DS4, the same number of clusters with a slightly smaller percentage of points for each cluster using only 22% of the spatial queries used by DBSCAN. The same results cannot be obtained by DBSCAN because of the different strategy of attribution of the points to the clusters. In fact, if we stop DBSCAN before which it has performed the spatial queries on all the points, we should obtain a correct number of points for the clusters already individuated and probably a smaller number of points for the cluster that we were building but of course, we will not discover all the clusters.

Table 22.1 and Table 22.2 show, for the two datasets, the number of clusters and the percentage of points for each cluster found by DBSCAN and SPARROW.

To verify the effectiveness of the search strategy we have compared SPARROW with the random walk search (RWS) strategy of the Reynolds' flock algorithm and with the linear randomized search (LRS) strategy.

Figure 22.6 and Figure 22.7 give the number of clusters found through the three different strategies versus number of visited points for the DS4 and GEORGE dataset.

TABLE 22.1 Number of Clusters and Number of Points for Clusters for GEORGE Dataset (Percentage in Comparison to the Total Point for Cluster found by DBSCAN) when SPARROW Analyzes 7, 12, and 22% Points

Clustering using the GEORGE dataset	Percentage of data points for cluster found by SPARROW		
	7%	12%	22%
G	57.6	83.2	92.4
E	58.2	71.1	91.3
O	61.3	85.8	94.2
R	50.6	72.7	93.3
G	48.8	77.2	89.7
E	61.1	81.2	94.5

TABLE 22.2 Number of Clusters and Number of Points for Clusters for DS4 Dataset (Percentage in Comparison to the Total Point for Cluster found by DBSCAN) when SPARROW Analyzes 7, 12, and 22% Points

Clustering using the DS4 dataset	Percentage of data points for cluster found by SPARROW		
	7%	12%	22%
1	51.16	70.99	78.86
2	45.91	64.74	74.40
3	40.68	59.36	81.95
4	44.21	60.66	81.67
5	54.65	58.72	71.54
6	48.77	59.91	78.10
7	54.29	66.43	79.18
8	51.16	70.99	78.86
9	45.91	64.74	74.40

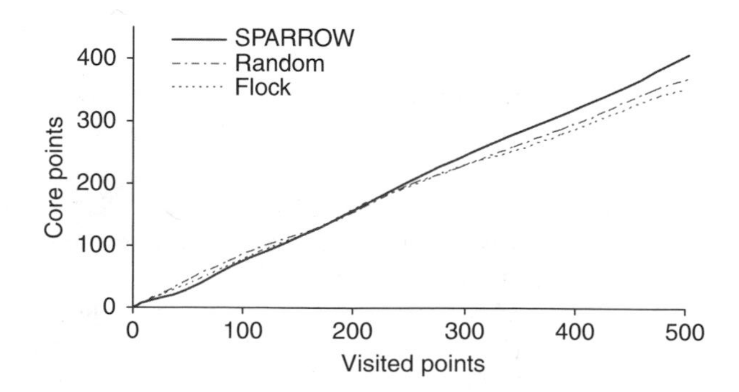

FIGURE 22.6 Number of core points found for SPARROW, random, and flock strategy versus total number of visited points for the DS4 dataset.

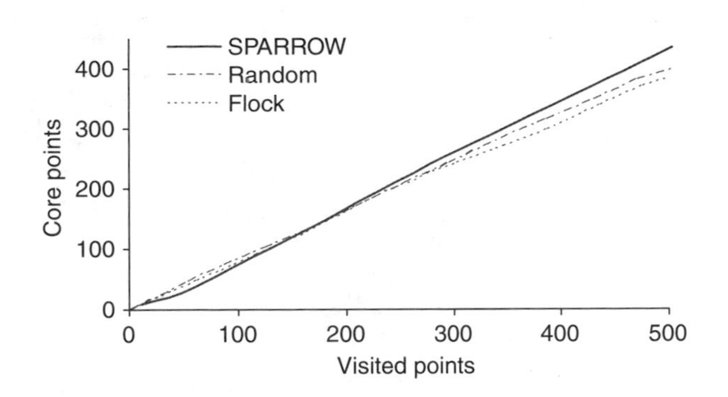

FIGURE 22.7 Number of core points found for SPARROW, random, and flock strategy versus total number of visited points for the GEORGE dataset.

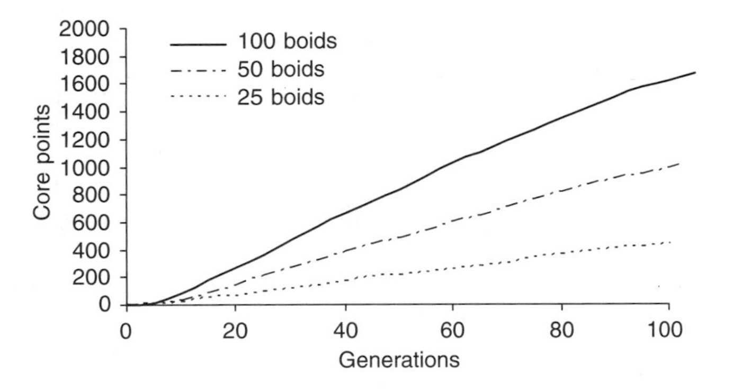

FIGURE 22.8 The impact of the number of agents on the foraging for clusters strategy (DS4).

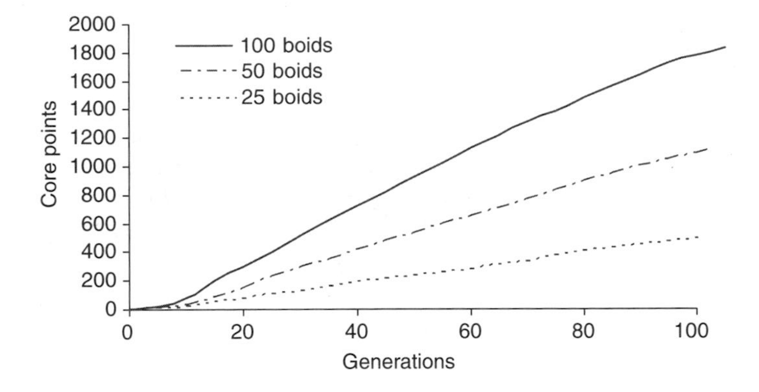

FIGURE 22.9 The impact of the number of agents on the foraging for clusters strategy (GEORGE).

At the beginning, the random strategy, and also (to a minor extent) the flock, overcomes SPARROW; however, after 200 to 250 visited points SPARROW presents a superior behavior on both the search strategies because of the adaptive behavior of the algorithm that allows agents to learn on their previous experience. Finally, we present the impact of the number of agents on the foraging for clusters performance. Figure 22.8 and Figure 22.9 give the number of clusters found in 100 time steps (generations) for 25, 50, and 100 agents. A comparative analysis reveals that a 100-agents population discovers a larger number of clusters than the other two populations with a smaller number of agents (the scalability is almost linear). This scalable behavior of the algorithm determines a faster completion time because a smaller number of iterations are necessary to produce the solution.

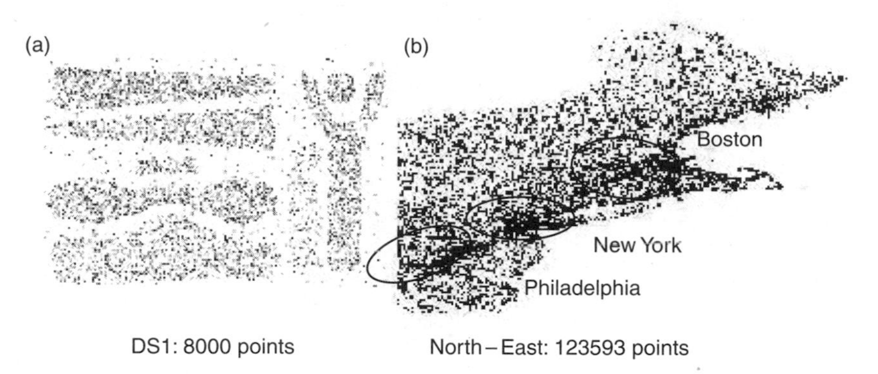

FIGURE 22.10 The two datasets used in our experiments (circles surround the three towns of North–East dataset).

22.4.2 SPARROW-SNN Results

In this section, we present an experimental evaluation of SPARROW-SNN algorithm. We intent to explore the following issues:

1. Determine the accuracy of the approximate solution that we obtain if we run our cluster algorithm on only a percent of points, as opposed to running the SNN clustering algorithm on the entire dataset.
2. Determine the effects of using SPARROW-SNN searching strategy as opposed to random-walk searching strategy in order to identify clusters.
3. Determine the impact of the number agents on the foraging for clusters performance.

For the experiments, we used synthetic datasets and a real life dataset extracted from a spatial database. The structures of these datasets are shown in Figure 22.10(a) and 22.10(b).

The first dataset, called DS1, contains 8,000 points and presents different densities in the clusters. The second dataset, called North–East, contains 123,593 points representing postal addresses of three metropolitan areas (New York, Boston, and Philadelphia) in the North East States.

We first illustrate the accuracy loss of our SPARROW-SNN algorithm in comparison with SNN algorithm when SPARROW-SNN is used as a technique for approximate clustering. For this purpose, we implemented a version of SNN and we computed the number of clusters and the number of points for cluster for the two datasets. Table 22.3 presents a comparison of these results with respect to ones obtained from SPARROW-SNN when a population of 50 agents visited, respectively, 7, 12, and 22% of the entire dataset.

Note that with only 7% of points we can have a clear vision of the found clusters and with a few more points we can obtain the near totality of the points. This trend is well marked in the North–East dataset. For the DS1 dataset, the results are not so well defined because the clusters called numbers 3 and 4 have very few points, so they are very hard to discover. For the real dataset, we only reported the results for the three main clusters representing the towns of Boston, New York, and Philadelphia.

We can explain the good results through the adaptive search strategy of SPARROW-SNN that requires to the individual agents to first explore the data searching for representative points whose position is not known a priori, and then, after the representative points are located, all the flock member are steered to move toward the representative points, that represent the interesting regions, in order to help them, avoiding the uninteresting areas that are instead marked as obstacles and adaptively changing their speed.

To verify the effectiveness of the search strategy we have compared SPARROW-SNN with the RWS strategy and with the standard flocking search strategy. Figure 22.11 and Figure 22.12 show the number of representative points found with SPARROW-SNN and those found with the random search and with the standard flock versus the total number of visited points. Both figures reveal that the number of representative points discovered at the beginning (and until about 170 visited points for DS1 and 150 for

TABLE 22.3 Number of Clusters and Number of Points for Clusters for DS1 and North–East (Percentage in Comparison to the Total Point for Cluster Found by SNN) when SPARROW-SNN Analyzes 7, 12, and 22% Points

Clustering using North–East dataset	Percentage of data points for cluster		
	7%	12%	22%
Philadelphia	42.5	65.2	79.4
New York	38.7	52.3	67.6
Boston	46.5	68.6	82.3
Clustering using the DSI dataset	**Percentage of data points for cluster found by SPARROW-SNN**		
	7%	12%	22%
1	41.35	58.31	70.37
2	30.08	53.58	60.72
3	29.28	40.99	53.02
4	20.9	30.5	51.41
5	53.38	65.36	76.56
6	56.89	69.87	73.63
7	33.89	43.5	61.58

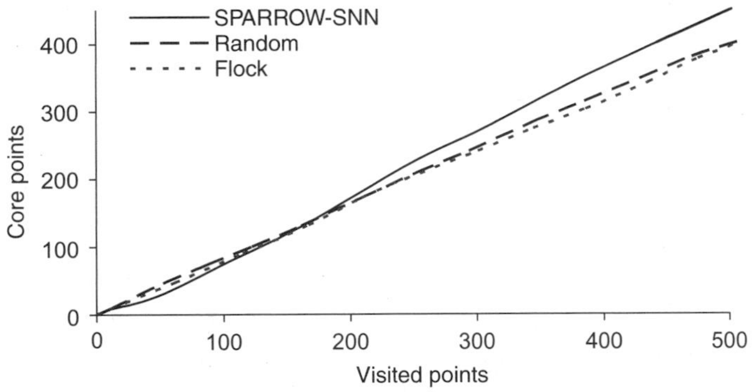

FIGURE 22.11 Number of core points found for SPARROW-SNN, random, and flock versus total number of visited points for DS1 dataset.

North–East dataset) from the random strategy, and (to a minor extent) for the flock, is slightly greater than of SPARROW-SNN.

Next, our strategy presents a superior behavior on both the strategies because of the adaptive behavior of the algorithm that allows to the agents to learn on their previous experience.

Finally, in order to study the scalability of our approach, we present the impact of the number of agents on the foraging for clusters performance. Figure 22.13 and Figure 22.14 give, respectively, the DS1 and North–East dataset, the number of clusters found in 100 time steps (generations) for 25, 50, and 100 agents. A comparative analysis reveals that a 100-agents population discovers a larger number of clusters (almost linear) than the other two populations with a smaller number of agents. This scalable behavior of the algorithm should determine a faster completion time because a smaller number of iterations should be necessary to produce the solution.

We try to partially give a theoretical explanation of the behavior observed in Figure 22.11 and Figure 22.12, that is, the improvement in accuracy of the SPARROW-SNN strategy, introducing an entropy-based model in Section 22.5.

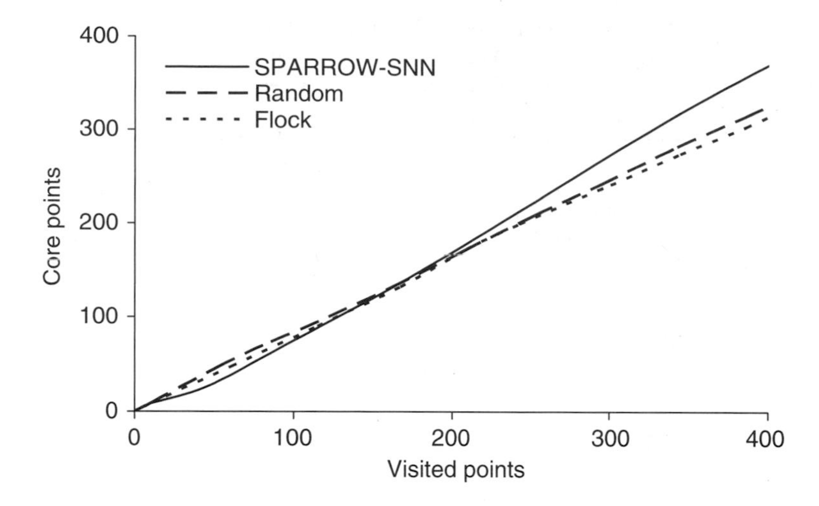

FIGURE 22.12 Number of core points found for SPARROW-SNN, random, and flock versus total number of visited points for North–East dataset.

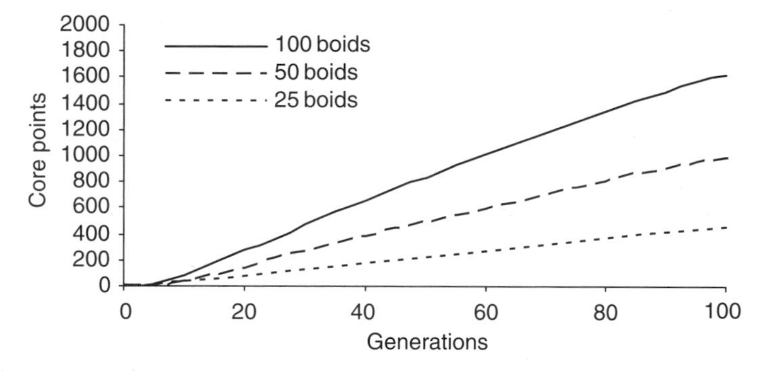

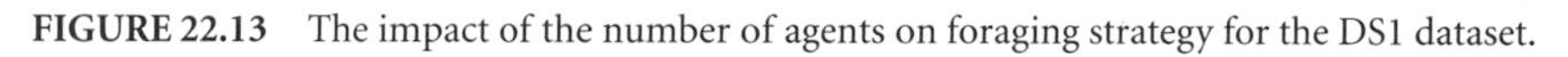
FIGURE 22.13 The impact of the number of agents on foraging strategy for the DS1 dataset.

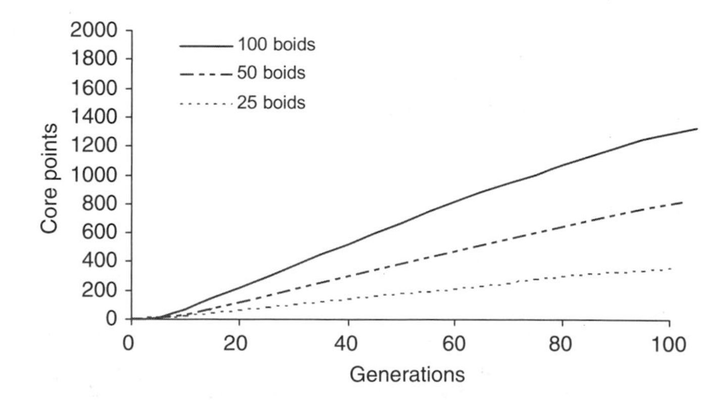

FIGURE 22.14 The impact of the number of agents on foraging strategy for the North–East dataset.

22.5 Entropy Model

In order to verify and explain the behavior of our system, we used a model based on the entropy as introduced in Reference 20. The authors use a measure of entropy to analyze emergence in multiagent systems. The fundamental claim of this chapter is that the relation between self-organization based on emergence in multiagent systems and concepts as entropy is not just a loose metaphor, but can provide quantitative, analytical guidelines for designing and operating agent systems. These concepts can be applied in measuring the behavior of multiagent systems. The main result suggested here concerns the principle that the key to reducing disorder in a multiagent system to achieve a coherent global behavior is coupling that system to another in which disorder increases. This corresponds to a macrolevel where the order increases, that is, a coherent behavior arises, and the microlevel where an increase in disorder is the cause for this coherent behavior at the macrolevel.

A multiagent system should follow the second law of thermodynamics "energy spontaneously disperses from being localized to becoming spread out if it is not hindered," if agents move without any constriction. However, if we add information in an intelligent way the agents natural tendency to maximum entropy will be contrasted and system goes toward self-organization.

Here, we will apply this model to the SPARROW-SNN algorithm. As stated in Reference 20, we can observe two levels of entropy: a macrolevel in which organization takes place, balanced by the micro in that we have an increase of entropy. For the sake of clarity, in SPARROW-SNN, microlevel is represented by red and white agents' positions, signaling, respectively, interesting and desert zones, and the macrolevel is computed considering all the agents positions. Therefore, we expect to observe an increase of micro entropy by the birth of new red and white agents and, on the contrary, a decrease in macro entropy indicating organization in the coordination model of the agents.

In the following, we give a more exact description of the entropy-based model. In information theory, entropy can be defined as

$$S = -\sum_i p_i \log p_i.$$

Now, to adapt this formula to our need, we use the location-based entropy. Consider an agent moving in the space of data divided in a grid $N \times M = K$, where each cell has the same dimension. Hence, if N and M are quite large, each random agent has the same probability to be in one of the K cells of the grid. We measured the entropy running a simulation for 100 times, for 2000 time steps using 50 agents and counting how many times the agent falls in the same cell i for each time step. Dividing this number by T we obtain the probability p_i that the agent will be in this cell.

Then, the locational entropy will be:

$$S = -\frac{\sum_{i=1}^{a} p_i \log p_i}{\log a}. \tag{22.1}$$

In case of random distribution, every state has probability $1/a$, so the overall entropy will be $(\log a)/(\log a) = 1$; this explains the factor of normalization a.

This equation can be generalized for P agents, summing the over all agents and dividing the average by P. Equation (22.1) represents the macro entropy; if we consider only red and white points, it represents the micro entropy.

We computed the micro and macro locational entropy for the real dataset North–East with the SPARROW-SNN algorithm using the parameters cited earlier. The results are showed in Figure 22.15 and Figure 22.16.

As expected, we can observe an increase in micro entropy and a decrease in macro entropy of SPARROW-SNN due to the organization introduced in the coordination model of the agents by the attraction toward red agents and the repulsion of white agents. On the contrary, in random and standard flock model, the curve of macro entropy is almost constant, confirming the absence of organization.

These experiments partially explain the validity of our approach, but further studies are necessary and will be conducted to better demonstrate the positive effects of organization and the correlation between macro entropy and validity of searching strategy.

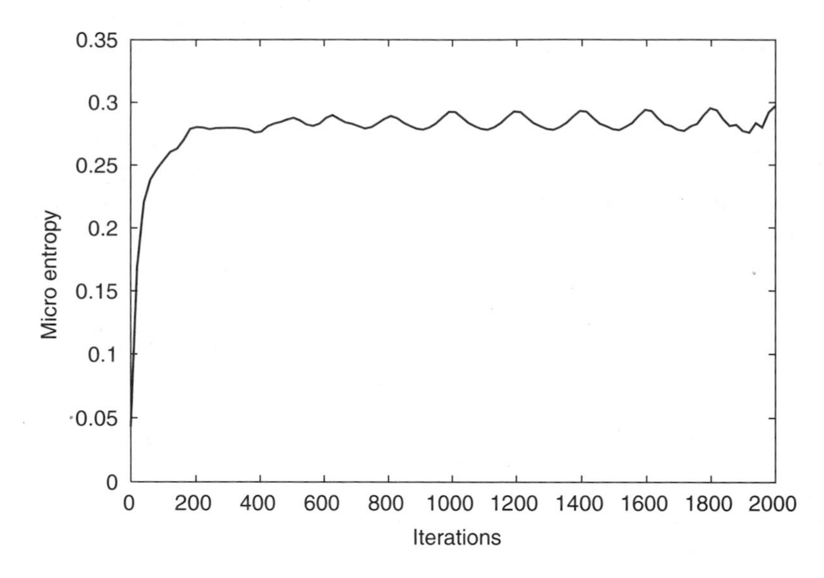

FIGURE 22.15 Micro entropy (red and white agents) for the North–East dataset using SPARROW-SNN.

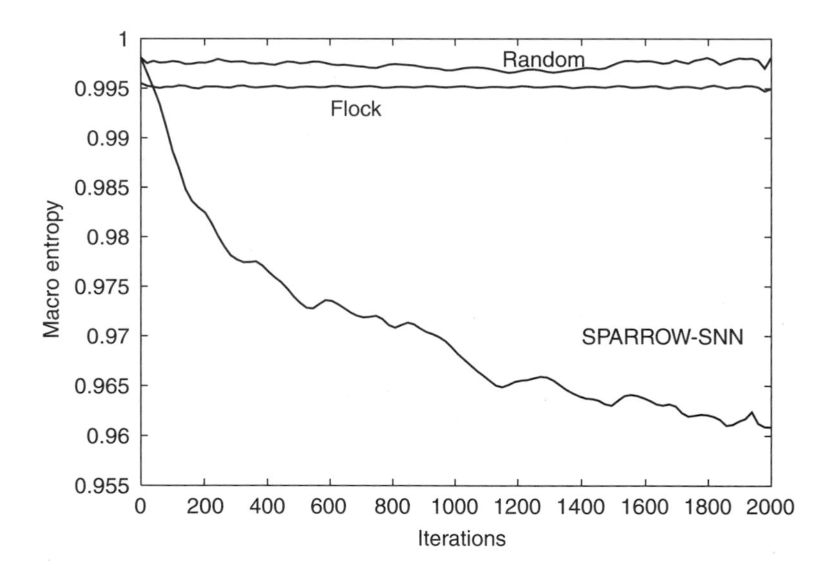

FIGURE 22.16 Macro entropy (all the agents) for the North–East dataset using SPARROW-SNN.

22.6 Conclusions

In this chapter, we have described a fully decentralized approach to clustering spatial data by a multi-agent system. The approach is based on the use of SI techniques. Two novel algorithms that combine density-based and shared nearest-neighbor clustering strategy with a flocking algorithm have been presented. The algorithms have been implemented in SWARM and evaluated using synthetic datasets and one real word dataset. Measures of accuracy of the results show that the flocking algorithm can be efficiently applied as a data reduction strategy to perform approximate clustering. Moreover, by an entropy-based model we have theoretically demonstrated that the adaptive search strategy of SPARROW is more efficient than that of the RWS strategy. Finally, the algorithms show a good scalable behavior.

Acknowledgment

This work was supported by the CNR/MIUR project — legge 449/97-DM 30/10/2000 and by Project "FIRB Grid.it" (RBNE01KNFP).

References

[1] Han, J. and Kamber, M. *Data Mining: Concepts and Techniques.* Morgan Kaufmann, San Mateo, CA, 2000.

[2] Kaufman, L. and Rousseeuw, P.J. *Finding Groups in Data: An Introduction to Cluster Analysis.* John Wiley & Sons, New York, 1990.

[3] Karypis, G., Han, E., and Kumar, V. CHAMELEON: A hierarchical clustering algorithm using dynamic modeling. *IEEE Computer*, 32, 68–75, 1999.

[4] Sander, J., Ester, M., Kriegel, H.-P., and Xu, X. Density-based clustering in spatial databases: The algorithm GDBSCAN and its applications, *Data Mining and Knowledge Discovery*, 2, 169–194, 1998.

[5] Wang, W., Yang, J., and Muntz, R. STING: A statistical information grid approach to spatial data mining. In *Proceedings of International Conference on Very Large Data Bases (VLDB'97)*, 1997, pp. 186–195.

[6] Han, J., Kamber, M., and Tung, A.K.H. Spatial clustering methods in data mining: A survey. In *Geographic Data Mining and Knowledge Discovery*, H. Miller and J. Han (Eds.), Taylor & Francis, London, 2001.

[7] Deneubourg, J.L., Goss, S., Franks, N., Sendova-Franks, A., Detrain, C., and Chretien, L. The dynamic of collective sorting robot-like ants and ant-like robots. In *Proceedings of the first Conference on Simulation of Adaptive Behavior*, J.A. Meyer and S.W. Wilson (Eds.), MIT Press/Bradford Books, Cambridge, MA, 1990, pp. 356–363.

[8] Lumer, E.D. and Faieta, B. Diversity and adaptation in populations of clustering ants. In *Proceedings of the Third International Conference on Simulation of Adaptive Behavior: From Animals to Animats (SAB94)*, D. Cliff, P. Husbands, J.A. Meyer, and S.W. Wilson (Eds.), MIT Press, Cambridge, MA, 1994, pp. 501–508.

[9] Kuntz, P. and Snyers, D., Emergent colonization and graph partitioning, In *Proceedings of the Third International Conference on Simulation of Adaptive Behavior: From Animals to Animats (SAB94)*, D. Cliff, P. Husbands, J.A. Meyer, and S.W. Wilson (Eds.), MIT Press, Cambridge, MA, 1994, pp. 494–500.

[10] Monmarché, N., Slimane, M., and Venturini, G. On improving clustering in numerical databases with artificial ants. In *Advances in Artificial Life: 5th European Conference*, ECAL 99, Vol. 1674 of *Lecture Notes in Computer Science*, Springer-Verlag, Berlin, 1999, pp. 626–635.

[11] Bonabeau, E., Dorigo, M., and Theraulaz, G. *Swarm Intelligence: From Natural to Artificial Systems.* Oxford University Press, Oxford, 1999.

[12] Grassé, P.P., La Reconstruction du nid et les Coordinations Inter-Individuelles chez *Beellicositermes Natalensis et Cubitermes sp.* La Théorie de la Stigmergie: Essai d'interprétation du Comportement des Termites Constructeurs, Insect. Soc. 6, pp. 41–80, 1959.

[13] Macgill, J. Using flocks to drive a geographical analysis engine. In *Artificial Life VII: Proceedings of the Seventh International Conference on Artificial Life.* MIT Press, Reed College, Portland, Oregon, 2000, pp. 1–6.

[14] Kollios, G., Gunopoulos, D., Koudas, N., and Berchtold, S. Efficient biased sampling for approximate clustering and outlier detection in large datasets. *IEEE Transactions on Knowledge and Data Engineering*, 15, 1170–1187, 2003.

[15] Ester, M., Kriegel, H.-P., Sander, J., and Xu, X. A density-based algorithm for discovering clusters in large spatial databases with noise. In *Proceedings of the 2nd International Conference on Knowledge Discovery and Data Mining (KDD-96)*, Portland, OR, 1996, pp. 226–231.

[16] Ertöz, L., Steinbach, M., and Kumar, V. A new shared nearest neighbor clustering algorithm and its applications. In *Workshop on Clustering High Dimensional Data and its Applications at 2nd SIAM International Conference on Data Mining*, Washington, USA, April 2002, pp. 105–115.

[17] Minar, N., Burkhart, R., Langton, C., and Askenazi, M. *The Swarm Development Group*, 1996, http://www.santafe.edu/projects/swarm.

[18] Reynolds, C.W. Flocks, herds, and schools: A distributed behavioral model. *Computer Graphics*, 21, 25–34, 1987.

[19] Jarvis, R.A. and Patrick, E.A. Clustering using a similarity measure based on shared nearest neighbors. *IEEE Transactions on Computers*, C-22(11), 1025–1034, 1973.

[20] Van Dyke Parunak, H. and Brueckner, S. Entropy and self-organization in multi-agent systems. In *Proceedings of the Fifth International Conference on Autonomous Agents*, ACM Press, 2001, pp. 124–130.

23

Biological Inspired Based Intrusion Detection Models for Mobile Telecommunication Systems

Azzedine Boukerche
Kathia Regina Lemos Jucá
João Bosco Mangueira Sobral
Mirela Sechi Moretti
Annoni Notare

23.1 Introduction

Security management against malicious intrusions and frauds in mobile telecommunication systems represents one of the most challenging problems for the wireless networking and mobile phone communities. Currently, mobile telephone carriers had been plagued by fraudulent use of cloned and stolen phone numbers. These fraudulent acts are costing them dearly in many countries, especially in countries where mobile phones can be rented freely and exchanged freely from some telecom carriers [1]. They are also witnessing, every year, a significant increase of their profit losses that are passed, unfortunately, to the mobile phone users. As a consequence, many of these mobile phone carriers are investing a large amount of money in the R&D to develop efficient security management solutions to deal with their security concerns. In this chapter, we focus mainly on the cloning and subscription frauds. In a *subscription fraud*, an impostor may subscribe to the service provided by the mobile phone telecommunication company, with a prior intention of not paying for the service. The impostor can also steal, subscribe, or use, each month, a different mobile phone (e.g., using a different name/number). However, in all of these cases, these mobile phones will most likely have a common usage pattern. The *cloning fraud* is characterized by calls that were made by cloned phones where the calls will appear in a monthly billing statement of a legitimate phone. The cloning may occur via a simple radio that can capture the two identification numbers: ESN, Electronic Serial Number and MIN, Mobile Identification Number) of a legitimate phone. Here again, calls from cloned phones and calls from legitimate phones will most likely have different usage patterns, as usage pattern is specific to each mobile phone user. While it is true that cloning frauds can be significantly reduced using new hardware technologies, subscription frauds are hardware independent. Therefore, new software-based technologies must be developed and deployed to identify these malicious intruders. To the best of our knowledge little work has been done at the software level to deal with the cloning and subscription frauds problem [2–4].

23.1.1 Related Works

Several strategies might be envisaged to treat the cloning and fraud subscriptions problem in mobile phone operation systems. In what follows, we briefly describe each of the available strategies:

1. *Cryptography* schemes will increase the difficulty for malicious intruders to capture the ESN/MIN numbers and perform clandestine hearings. While one might argue that it is simple to use such schemes in digital phones, it is not quite obvious as to how they can be used in analog phones. Thereby, new and efficient schemes are needed to deal with the security problem in mobile phone operations. Furthermore, GSM digital phones using encryption model have already been cloned and broken recently by malicious intruders.
2. *Denial of service* schemes are basically based upon denying international calls, for instance. While this approach might help the mobile phone carriers to reduce their profit losses, it is clear that this solution is an alternative solution and not an efficient security management solution.
3. *User verification* schemes are based upon the use of passwords by a mobile phone user before using his phone set. Note that this approach is widely used by many mobile phone carriers. While this approach provides a certain level of security, it is our belief that a good security solution should not complicate the mobile phone usage by requiring the users to memorize all of these passwords in order to use their mobile phones.
4. *Traffic analysis schemes* are based upon the profile and the behavior of the mobile phone users. Nowadays, the main problem of this approach is its' simplistic and rule based analysis where only the monthly statement bill and the price of the calls are being considered.

In this chapter, we focus upon the last approach. We show how nature and biological inspired techniques can be used efficiently for traffic analysis purposes and the identification of malicious intruders in mobile phone operation systems.

23.1.2 Distributed Management Security System

This chapter proposes a security management system for intrusion detection in wireless telecommunication networks through traffic analysis, where the users are classified into clusters according to their behavior patterns and phone usages — through neural networks and immune human-based system employment. The classification of the users helps the security system to identify when the phone calls do not correspond to their users' usage pattern, which may constitute a possible fraudulent use of the mobile phone. The fraud could take the form of a cloned phone, a fraud subscription, or a frequent delinquent mobile phone user, known through his past, to subscribe to a service with a prior intention of not paying for the service (e.g., user has history of a bad debt). An immediate and automatic message (thereby, minimizing the size of the personal staff to do that) is sent to the mobile phone user in order to confirm the cloning or fraud subscription. As opposed to waiting until the end of the monthly billing statement cycle, this immediate notification will help the telecom carriers to reduce their profit losses, thereby reducing the damage that would have been passed on, somehow, to the mobile phone users, such as increased phone call fees, surcharges, and so on. In the case of detection of fraudulent subscription, personalized reports are quickly generated and used by the system and the telecom carrier to investigate who actually made the suspected calls.

In addition to the security management, Web/WAP application tools could be developed where users can observe online their phone bills, which will allow them to track down their mobile phone statement bill and identify malicious intruders using their phone number, thereby reducing the fraudulent usage of the mobile phone. The system may encompass services of access control, authentication, confidentiality, integrity, availability and nonrepudiation of communication. In our previous work [2,4,5], we have incorporated and implemented all of these security services using Java with the support of CORBA. Moreover, as an important characteristic, our system was formally validated according to ISO 8807 standard, see References 3 and 6.

This chapter is organized as follows: Section 23.2 describes the use of ISO 8807 standard that we have used to specify and validate our security management system. Section 23.3 demonstrates how neural network techniques can be used to detect intrusion by pattern recognition of telecom users. Section 23.4 comments about how the immune human system can be used as an alternative solution to identify malicious intruders and fraudulent usage of mobile phone users. Section 23.5 concludes the chapter.

23.2 Specification and Validation

This section presents the formal specification and validation of the security management system according to ISO 8807 (International Organization for Standardization), FDT (Formal Description Technique) and LOTOS (Language of Temporal Ordering Specifications) [7]. One of the major issues of the use of FDTs, is to demonstrate the correctness of the system. In this work, the LOTOS specifications are validated through the use of the Eucalyptus ToolSet/CADP (Caesar/Aldebaran Development Package) [8], on Sun Solaris environment, that includes the following two main components: (1) Caesar and (2) Aldebaran.

Caesar is a compiler that translates a LOTOS specification into a C program (to be executed or simulated) or into an LTS, Labeled Transition Systems (to be verified using bi-simulation and temporal logic tools). As will be demonstrated in Section 23.2.3, it is possible to compare the protocol LTS with the related LTS' service. Both LTSs are generated using the Caesar compiler and are compared using the Aldebaran tool.

Aldebaran is a tool for communication systems verification. It is represented by LTSs, where the state transitions are labeled by action names. Aldebaran allows the LTSs reduction using several equivalence relations (e.g., strong bi-simulation, observational equivalence, delay bi-simulation and secure equivalence).

The stepwise refinements approach used in our work to build the LOTOS specifications, allows the system to be validated along the system development.

23.2.1 SSTCC Service Specification

At the most abstract level, the SSTCC (Security System for Telecommunications against Cellular Cloning) and Impostors [2,3,9], can be seen as a black box, containing four communication gates (`mail_alarm`, `phone_alarm`, `online_bill`, and `check_owner`), to send/receive messages to and from the telecommunication users. As illustrated in Figure 23.1.

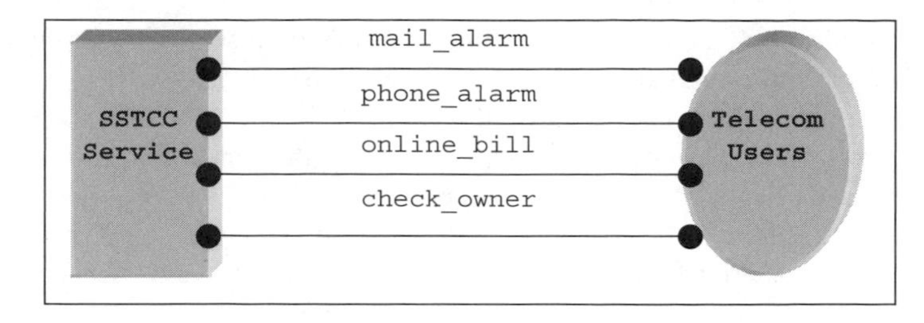

FIGURE 23.1 SSTCCService graphical representation.

```
specification SstccService[mail_alarm,phone_alarm,online_bill,check_owner]:noexit
behaviour SstccService[mail_alarm,phone_alarm,online_bill,check_owner]  where
process SstccService[mail_alarm,phone_alarm,online_bill,check_owner]:noexit:=
(i;mail_alarm;
        (phone_alarm;SstccService[mail_alarm,phone_alarm,online_bill,check_owner]
     [] SstccService[mail_alarm,phone_alarm,online_bill,check_owner]))
[](online_bill;SstccService[mail_alarm,phone_alarm,online_bill,check_owner])
[](check_owner;SstccService[mail_alarm,phone_alarm,online_bill,check_owner])
endproc
endspec
```

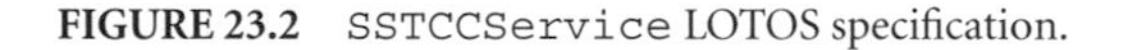
FIGURE 23.2 SSTCCService LOTOS specification.

The mail_alarm gate is used by SSTCC to send alarm signals by snail mail to the users about a malicious intruder. The phone_alarm gate allows the SSTCC system to send these alarms using the mobile phone (e.g., short message or interactive voice message). The advantage of using snail mail is its delivery guarantee, whereas short time has the benefit of using mobile phone. The online_bill gate is used by the users to observe their online phone bill via the Web/WAP tool. Finally, the check_owner gate is used by the telecom carrier to investigate probable impostors (with potential future bad debits). The SSTCC system remains permanently active, thereby characterizing an infinite behavior of this system — This is illustrated in Figure 23.2 using the noexit functionality.

The behavior of our SSTCC system is defined by the process SSCCService, that can execute an action in the gate mail in order to send an alarm by a regular surface mail (this action is always possible); the sequence is followed by a nondeterministic choice with two options. The first option is related to the alarm sent by mobile phone, in the gate phone (this action is not always possible) and, following, the process SSTCCService is called recursively in order to deal with another case. The first option may not be successful, since the phones may not work properly (e.g., the mobile phone users may be out of their area after a certain period of time). Thus, an internal action i occurs (which has not been observed) and the process SSTCCService is executed recursively. Similarity, the online_bill and check_owner are the two gates used to observe the phone bills and verify/confirm the user identity, respectively.

This abstract level of SSTCC specification corresponds to a formal specification of the user requirements of the system, which will represent the basis for future refinements (i.e., for the protocol specification).

23.2.2 SSTCC Protocol Specification

In a more detailed level, the behavior of our SSTCC system is defined by the SstccProtocol specification, where it is possible to observe its three main components: (1) SSCC, Security System against Cellular Cloning, represented by the SsccClone process; (2) SETWeb/Wap, Online Phone Bill by Web/WAP System, represented by the SetwebBill process; and (3) SIPI, System to Identify Probable Impostors, represented by the SipiImpostor process. Figure 23.3 illustrates the general architecture of the refined SSTCC system, including the three main processes that compose it.

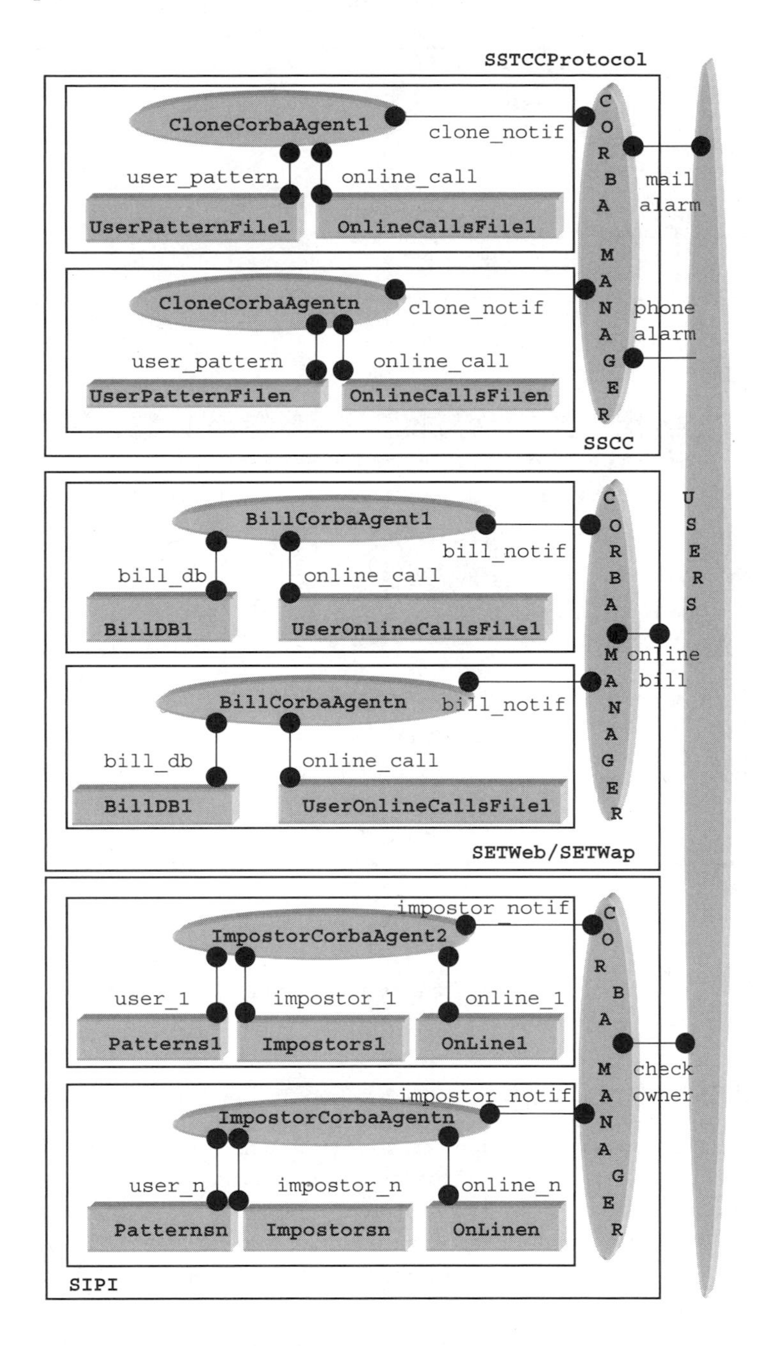

FIGURE 23.3 `SSTCCProtocol` graphical representation.

Figure 23.3 displays each of the three components that are identified by a manager and several distributed agents. The SSCC/SIPI agents compare the online calls database with the users' pattern database identified by the neural networks technology. The SETWeb/Wap agents access both the online and the existing databases to build the phone bills. The procedural definition of the `SSTCCProtocol` behavior is shown in Figure 23.4.

Figure 23.4 presents the three components composed in parallel using the parallel independent composition operator (|||). Based upon the use of the two specifications (`SstccService.lotos` and

```
specification SstccProtocol[mail_alarm,phone_alarm,online_bill,check_owner]:noexit
behaviour
   hide clone_notif,web_notif,impostor_notif in
   SsccClone[mail_alarm,phone_alarm]
|||SetwebBill[online_bill]
|||SipiImpostor[check_owner]
   where
   process SsccClone[mail_alarm,phone_alarm]:noexit:= ...          endproc
   process SetwebBill[online_bill]:noexit:= ...                    endproc
   process SipiImpostor[check_owner]:noexit:= ...                  endproc
endspec
```

FIGURE 23.4 `SSTCCProtocol` LOTOS specification.

BCG monitor
Graph Construction Statistics for SstccProtocol.bcg
Current Number of Explored States 194401 (100%)
Current Number of Remaining States 0 (0%)
Current Variation of Remaining States (completed)
Current Number of Transitions 1645926
ONLINE_BILL
MAIL_ALARM
PHONE_ALARM
CHECK_OWNER
done

FIGURE 23.5 `SSTCCProtocol` LTS.

`SstccProtocol.lotos`, as presented in Sections 23.2.1 and 23.2.2, respectively) it is possible to perform the validation of the security system, as presented in Section 23.2.3.

23.2.3 Formal SSTCC Validation

System's validation can be done via simulation, test and verification. While simulation and testing are used (only) to find errors, verification provides the formal (mathematical) proof and the correctness of the system. In order to obtain the system's correctness proof, initially, the LTS (`SstccProtocol.bcg`) file is generated. This file is related to the protocol specification (file `SstccProtocol.lotos`) using the Caesar tool. Figure 23.5 shows the `SSTCCProtocol` LTS containing 194401 states and 1645926 transitions.

Using a similar approach, the LTS (`SstccService.bcg`) file is generated. This file is related to the service specification (file `SstccService.lotos`) using the Caesar tool. Figure 23.6 shows the `SSTCCService` LTS containing three states and six transitions.

Then, it is possible to verify, by observational equivalence, the protocol LTS with the service LTS using the Aldebaran tool. The obtained result, `TRUE`, signifies that the protocol executes the desired service, that is, the formal (mathematical) proof of correctness of the SSTCC system has been obtained. See Figure 23.7.

This proof aims at satisfying the requirements in order to obtain the ClassA1 of the Security Level Class_A1. Table 23.1 provides the security levels approved by the U.S. – Department of Defense (DoD) [10].

The formal analysis and the mathematical proof presented in this section, aims at guaranteeing the verification of the design based upon the ClassA1 security level.

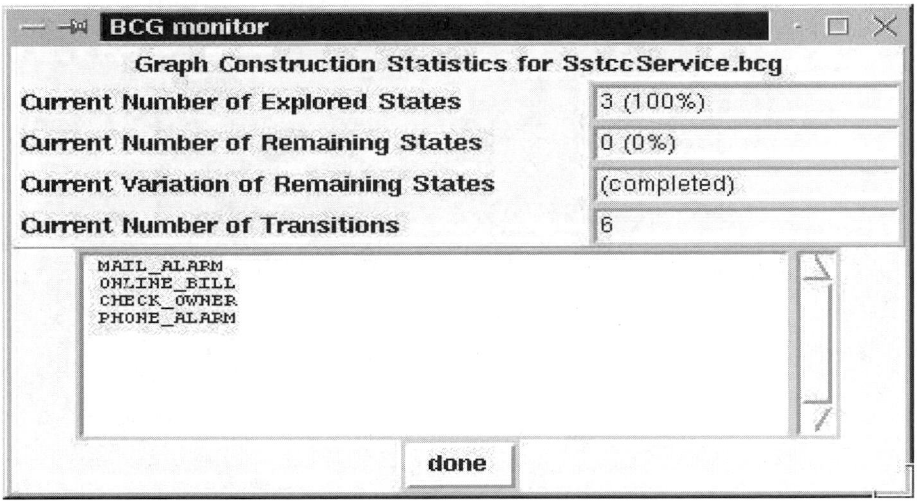

FIGURE 23.6 `SSTCCService` LTS.

```
----------------------------------------------------------
aldebaran -bddsize 4 -oequ -std SstccProtocol_bmin.bcg ./SstccService.bcg | tee aldebaran.seq
TRUE
----------------------------------------------------------
```

FIGURE 23.7 `SstccProtocol` and `SstccService` verification.

TABLE 23.1 Table of Security Levels (DoD of USA)

Security level	Major feature introduced
Division D	Minimal protection
Division C	Discretionary protection
Division B	Mandatory protection
Division A	Verified protection
Class A1	Verified design. Formal analysis and mathematical proof that the computer system matches the system's security policy and its design specifications

23.3 Intrusion Detection by Neural Networks

In this section, we will show how neural networks technology can be used to design intrusion detection systems based upon the traffic analysis and the classification [11] of the mobile phone users. The main reasons behind using neural networks techniques, among others, are: (1) they have the intrinsic capacity of learning input data and capturing the user's behavioral patterns, (2) they are based upon nonparametric-based models and have the potential to provide better understanding of the distribution of input data than the traditional statistical methods (e.g., Bayesian methods), and (3) they have the capabilities of creating decision boundaries that are highly nonlinear in the space of characteristics.

Neural networks and neural brain share certain similarities, among others: (1) the knowledge is acquired via a learning process; and (2) the weights of the connections between neurons, known as synapses, are used to store the knowledge. The procedure used to represent the learning process, commonly called learning algorithm, has the capabilities of modifying the weights of the connections of the network in order to get the targeted design. The algorithms could easily be implemented using the MatLab tool [2].

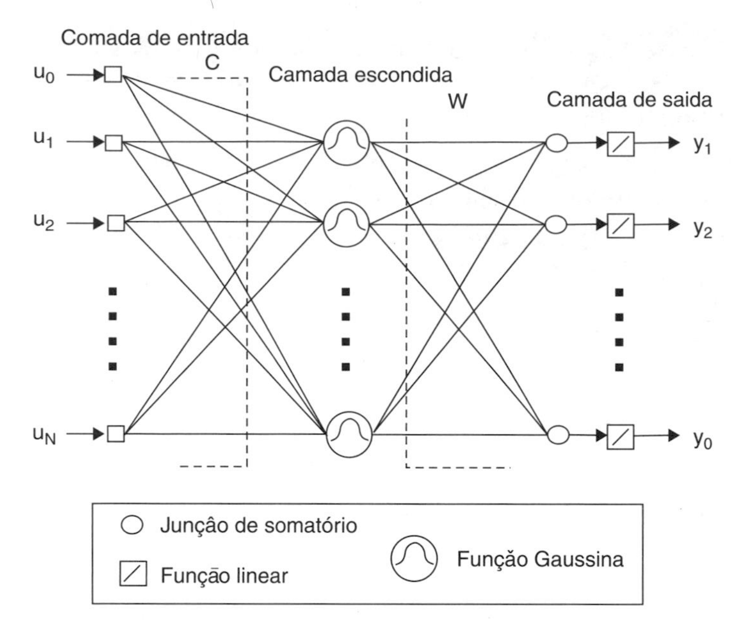

FIGURE 23.8 RBF architecture.

23.3.1 Neural Network Construction

In order to conduct the classification of mobile phone users based upon their phone usage patterns, an artificial neural network using the a Radial Base Function (RBF) can be used [2]. The architecture of the RBF network consists mainly of an input layer, a hidden layer, and an output layer. As illustrated in Figure 23.8.

The input layer is composed of the source nodes (i.e., sensorial units) and the output nodes that constitute a linear combination of the RBF calculated by the nodes of the hidden layer. The RBF in the hidden layer produces a response for the input (stimulus pattern), that is, it produces responses that are different from zero when the input pattern is within a small region located in the input space. The learning component in the hidden (intermediate) layer is executed using the "non-supervised" method, such as a simple heuristic clustering algorithm, or a "supervised" algorithm to find the centers (i.e., the C nodes in the hidden layer). In our work [2], we made use of the K-means algorithm in order to determine the centers that represent the connections between the input layer and the intermediate layer. Gauss function was used in order to determine which of the various centers are similar to the input vector. The Gauss function used to obtaining the output of a hidden layer is defined as follows:

$$f = (x - c) = \frac{1}{(2\pi)^{n/2}\sigma_1, \sigma_2, \ldots, \sigma_n} \exp\left\{-\frac{1}{2}\sum_{j-1}^{n}\left(\frac{x_j - c_j}{\sigma_j}\right)^2\right\}. \quad (23.1)$$

To increase the RBF's functionality, the Mahalanobis distance could be used within the Gauss function, as follows:

$$f = (x - c) = \frac{1}{(2\pi)^{n/2}|K|^{1/2}} \exp\left\{-\frac{1}{2}(x - c)^{\mathrm{T}}K^{-1}(x - c)\right\}, \quad (23.2)$$

where K^{-1} is the inverse of the X covariance matrix, associated with the node of the hidden C layer.

Given n-vectors (input data) of p-samples, representing p-classes, the network can be initiated with the knowledge of the centers (i.e., locations of the samples). If the J-th vector sample is represented, then the weight matrix C can be defined as: $C = [c_1, c_2, \ldots, c_3]^T$ so that the weights of the hidden layer in the j node are composed of the center vector. The output layer is a weight sum of the outputs of the hidden layer. When presenting an input vector for the network, the network implements Equation (23.3)

$$y = W \cdot f(\|x - c\|), \tag{23.3}$$

where f represents the functional output vector of the hidden layer, and C the corresponding center vector. After supplying some data with the desired results, the weights (W) can be determined using the training algorithm either interactively or noninteractively, based upon the descendant or pseudo-inverse gradient scheme, respectively. In order to determine the σ^2 variation parameter for the Gauss function, one has, first, to compute the median distance between all the training data, see Equation (23.4).

$$\sigma_j^2 = \frac{1}{M_j} \sum_{x \in \Theta_j} (x - c)^{\mathrm{T}} (x - c), \tag{23.4}$$

where Θ_j is the group of training patterns grouped in the center of the cluster C_j, and M_j is the number of patterns in Θ.

Another way of choosing the parameter σ^2 is to calculate the distances between the centers in each dimension and use a scaling factor. This approach will allow the p-nearest neighboring algorithm to be used and obtain the variance related to each center.

23.3.2 Neural Network Implementation

In order to implement the neural network algorithm that is used to classify the mobile phone users into groups based upon their phone usage pattern, the following steps must be performed. *Step 1*: identify the number of neurons of the hidden layer. *Step 2*: find the centers c_j $(j = 1, \ldots, M)$ which makeup the basis of an M-dimensional space. *Step 3*: for each input pattern, the $\|x_i - c_j\|$ is sent as a parameter to the RBF, which describes the level of classification of the input patterns. The output of the hidden layer makes up a G matrix, which serves as a basis for the calculation of the weigh W (i.e., connections for the output layer), following the formula: $W = G^{+}t$, where G^{+} is the pseudo-inverse of the G matrix given by $G^{+} = (G^{\mathrm{T}}G + \lambda G_0)^{-1} \cdot G^{\mathrm{T}}$; and t is the matrix that contains the group of training data. In the last step, the output is calculated as a sum of the activated neurons of a hidden layer.

The obtained output, that is, the classified users based upon their phone usage (see Figure 23.9) makes up the database used in the implementation of the SSTCC (see Section 23.4). In our implementation, the online calls are compared to the classified data in order to identify calls that do not match with that of the behavioral pattern of the client, which imply fraud detection.

The results we have obtained are summarized in Figure 23.9. They illustrate the efficiency of neural network and pattern recognition techniques to design efficient nature based intrusion detection system that can be used to identify the cloning and subscription frauds in mobile phone operations [12]. Considering the error rate obtained in the classification (2.5% using 80 neurons in the hidden layer), the currently losses of US$ 500,000 daily (data from IDC, Sept 99) could be reduced to US$ 12.500 daily. This error

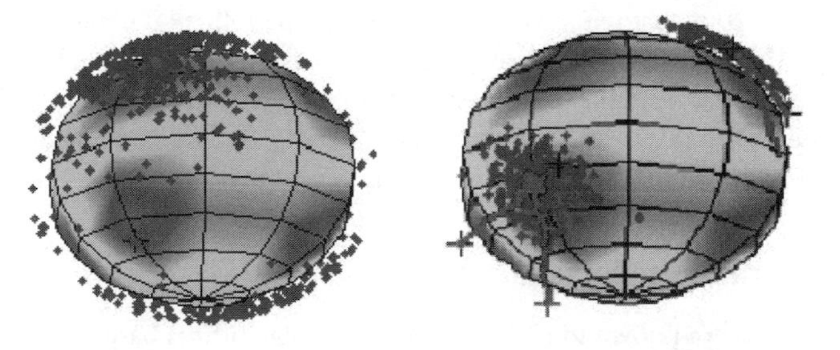

FIGURE 23.9 Users (*before* and *after* classification).

rate is quite encouraging, when compared to previous results obtained by other researchers [13], where a 5.4% error rate was obtained using the Back Propagation algorithm.

23.4 Toward an Artificial Immune Human System for Mobile Phone Intrusion Detection

In this section, we show how an immune human based model can be used to design mobile phone intrusion detection system. First, let us review the basic understanding of the immune human systems. The Natural (Human) Immune System [14] protects our bodies from infected agents such as viruses, bacteria, and fungi. It can distinguish between self and nonself genes. It is able to remember previous experiences and react accordingly. Therefore, once a human being has a flu, the immune human system will prevent him from getting it again. The immune system displays both enormous diversity and extraordinary specificity. Indeed, it is not only able to recognize the many millions of distinctive nonself genes, but it can also produce molecules and cells to match up with nonself genes, and have them counteract each other. An antigen is defined as a substance that reacts with antibody molecules and antigen receptors on lymphocytes. An immunogen is an antigen that is recognized by the body as a nonself gene and stimulates an adaptive immune response [15].

There are two types of immunity, innate and adaptive. Innate immunity primarily consists of the endocytic and phagocytic systems, which involve motile scavenger cells such as macrophages that ingest extracellular molecules and materials, clearing the system of both remaining items and pathogens.

An Adaptive (or Acquired) immunity, allows the immune system to launch an attack against any foreign intruders that the innate system cannot remove. The adaptive system is directed against specific intruders, it learns how to recognize specific kinds of pathogens, and keep track of them, so that it can identify them later if they reappear again, thereby speeding up the response to deal with them. The learning process occurs during a primary response to a kind of pathogen that was not encountered before by the immune system. The primary response is slow, often only becoming apparent, at first, many hours after the initial infection. However, the immune system will retain a memory of the kind of pathogens that has caused the infection for future intrusions [16].

23.4.1 Anamnestic Response (Memory)

As a result of the T-dependent B-lymphocytes being sensitized to a specific antigen, numerous circulating B-memory cells developed primary anamnestic response or memory. A subsequent exposure to the same antigen may result not only in a more rapid production but also of a higher number of antibodies and for a longer period of time. The primary response to a new antigen generally peaks at 10 to 17 days; however, because of the numerous circulating B-memory cells, the secondary anamnestic response peaks in just 2 to 7 days. Due to the clonal expansion and affinity maturation, there is now a pool of B-memory cells having the "fine-tuned" B-cell receptors on their surface. The pool of B-memory cells circulates throughout the body waiting to encounter the original antigen when it again enters the human body. B-memory cells have high longevity. They replicate and produce antibodies periodically when they are exposed to a persisting epitope that remains on the surface of the follicular dendritic cells in the lymphoid organs. This secondary immune response is said to be specific to the antigen that has initiated the immune response the first time and has been the factor behind the faster response that is attributed to the memory cells remaining in the immune system, so that when an antigen, or a similar antigen, is encountered, a new immunity does not need to be built up, since it is already there [17,18].

23.4.2 Artificial Immune System

Artificial Immune Systems are known to use adaptive search algorithms based on the biological immune system with the central task of pattern matching between antigens and antibodies. Two distinct groups of algorithms have emerged as successful implementations of artificial immune systems: (1) the negative selection algorithm developed by Forrest et al. [19] is used to differentiate between normal system operation (i.e., self) and abnormal operation (i.e., nonself), and (2) the immune network based model as discussed by Jerne [20], then formally described by Perelson [21] and Farmer et al. [16]. Our intrusion detection model for mobile phone operations is based upon the Jerne model. Both approaches are particularly well suited to data-mining related tasks that involve searching through large databases and finding the matching patterns [22–25].

The Idiotypic network theory, introduced by Jerne, maintains that interactions in the immune system do not occur only among antibodies and antigens since antibodies may interact with each others as well. Hence, an antibody may be matched to other antibodies, which in turn may be matched to yet other antibodies, and this process can spread throughout the entire population. According to the immune network theory, immunological memory is achieved by the B cells supporting each other through an Idiotypic network. B cells are not only stimulated by antigens, but also stimulated and, to a degree, suppressed by neighboring B cells. The interaction between B cells takes place via the idiotopes on each B cell which acts as a recognizer for other B cells. Therefore, the more neighboring cells a particular B cell has that are similar to itself, the more stimulated that particular B cell becomes. This allows the clustering of a population of identical B cells, and produces a self-supporting structure to aid the immune responses. The network self-organizes and stabilizes within a reasonable time since its survival is achieved by mutual reinforcement among the B cells via a feedback mechanism. Survival of a new B cell produced, as part of the immune response, either by the bone marrow or by hypermutation, depends on its affinity to the antigens that are present, and to its neighbors in the immune network. The new B cells may have an improved match for an antigen and, thus, will proliferate and survive longer than the existing B cells. By repeating the mutation and the selection procedures several times, the immune system learns to produce better antigen matches. This theory could help in explaining how the memory of past infections is retained. Furthermore, it could result in the suppression of similar antibodies, thereby, encouraging a good diversity in the antibody pool [26]. See Figure 23.10.

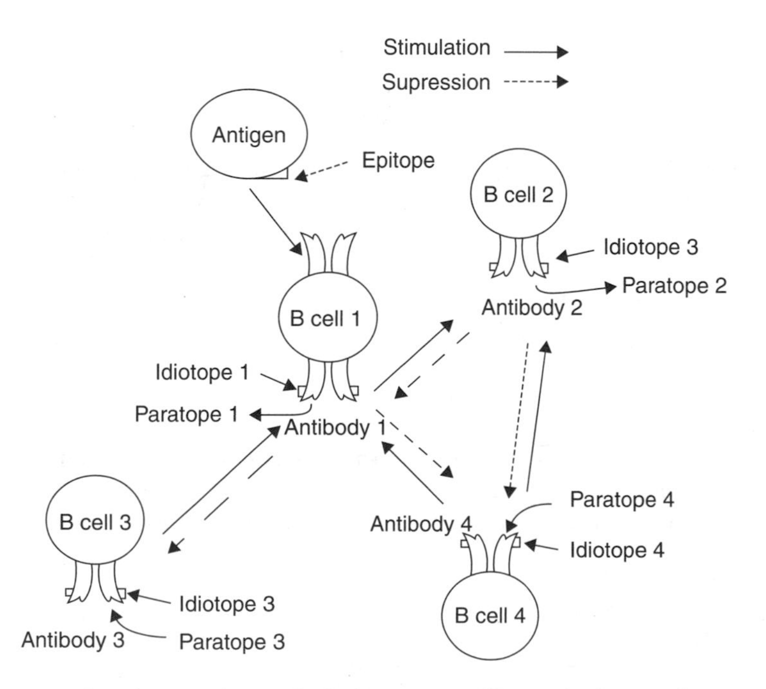

FIGURE 23.10 Jernes Idiotypic Network Hypothesis. (From Jerne, N.K. *Annals of Immunology*, 125, 373–389, 1973. With permission.)

Figure 23.10 shows the basic idea of the immune network hypothesis. There are many antibodies on the surface of the B cells that act as antigen detectors. The specialized portion of the antibody used for identifying other molecules (of antigens or antibodies) is called paratopes. The region on any molecules that can be recognized by the paratopes is called epitope. The binding between idiotopes and paratopes has the effect of stimulating the B cells. This is mainly due to the fact that the paratopes on B cell2 reacts to the idiotope on B cell1 as it would on an antigen. However, to counter the reaction, there is a certain amount of pressure among the B cells to act as a regulatory mechanism. The authors [16,21] have formulized the theory, and they have suggested that the B cells are stimulated by an interaction not only between B cells and antigens, but also the surrounding of the B cells that are connected to form a network of cells. As illustrated in Figure 23.10, B cell1 stimulates three cells, B cell2, B cell3, and B cell4, and it also receives a certain amount of pressure from each one. This creates a network type structure that provides a regulatory effect on the neighboring B cells. The immune network acts as a self-organizing and self-regulatory system that captures antigen information, and is ready to launch an attack against any similar antigens.

In the immune network, a node represents a type of antigens or antibodies, and the links between nodes represent the affinity between them. The affinity is determined by the matching degree between the paratope and the epitope, and the population of each type of antibodies (or antigens). An antibody type is thought to be stimulated when its paratope recognizes the epitope of antigens or other types of antibodies, so that the corresponding lymphocyte is stimulated to reproduce more lymphocytes, and the lymphocytes secrete more antibodies (with a certain mutation rate). This process is called clonal selection. On the contrary, an antibody or antigen may be suppressed if its epitopes are recognized by others [25]. The essential aspect of the immune network model is that the list of antigens and antibody types is dynamically changing as some types that are added or removed, change with time.

23.4.3 Artificial Telecom Immune System

Using similar analogies between the immune human system and the mobile telecommunication systems, the artificial Telecom immune system will consist of a set of B cells and the links between the B cells and their cloned ones, and the data items represent the antigens, as well as the pathogens. These data are, first, converted from their raw state (numerical data) to a normalized form into values within the interval $[0, 1]$. In what follows, we discuss briefly the main components that shall be performed by the immune human-based system for mobile phone operations, as outlined in Figure 23.11.

Figure 23.11 shows a simple diagram of the learning algorithm that is described hereafter:

1. *Data training*: The initial network B cell population is made up by sampling the raw data set and creating the B cell objects (for instance, phone calls on a Sunday between 12 and 14 h).
2. *Data items*: The remainder of the data items are taken to create the antigens set. These data are introduced during the training process and will be repeatedly presented to the network in an attempt to capture the trends of the mobile phone users using the data set.
3. *Expose*: This level computes the B cell stimulation level, which takes into account the match scores with the antigens.
4. *Cloning*: New clones may be produced when the cells are sufficiently stimulated.
5. *Feedback*: New items are introduced in the system.
6. *Results*: Adopting the primary immune response's analogy, the components of the network repeatedly present these new items for training purposes to the network, in an attempt to provoke a reply from B cell, clone and mutate, thereby creating a diverse set of the raw data to keep a network of B cells.

23.5 Conclusion

Biological inspired techniques have received a great deal of interests due to their promises to resolve challenging combinatorial and real-world problems. In this chapter, we have reviewed two promising

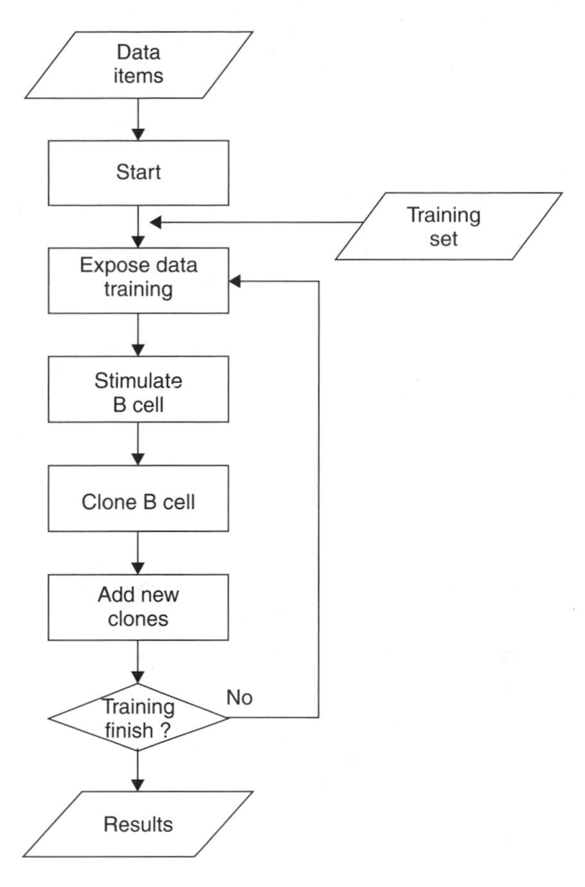

FIGURE 23.11 Immune human-based intrusion detection model for mobile phone operations.

nature-based models, neural networks and immune human-based systems, and show how these two approaches could be used to design efficient distributed security management systems for mobile phone operations.

In the future, we plan to investigate further the immune human-based intrusion detection system and study its performance evaluation for mobile phone operations. Next, we wish to investigate further the design of efficient security models using biological inspired techniques for mobile ad hoc networks.

References

[1] *Ottawa Citizen*, October 11, 2004.

[2] Boukerche, A. and Notare, M.S.M.A. Behavior-based intrusion detection in obile phone systems. *Journal of Parallel and Distributed Computing*, 62, 1476–1490, 2002.

[3] Notare, M.S.M.A., Boukerche, A., Cruz, F.A., Riso, B.G., and Westphall, C.B. Security management against cloning mobile telephones. In *IEEE GLOBECOM'99*, 1999, Rio de Janeiro, Brazil. 1999, pp. 1969–1973.

[4] Boukerche, A., Sobral, J.B.M., Juca, K., and Notare, M.S.M.A. Anomaly and misuse intrusion detection based in the immune human system. In *Proceedings of the 17th IEEE IPDPS/NIDISC'2003 — International Parallel and Distributed Processing Symposium, 2003*, Nice, France. IEEE Press, 2003, p. 146.

[5] Boukerche, A., Juca, K., Notare, M.S.M.A., and Sobral, J.B.M. Intrusion detection based on the immune human system. In *IEEE 16th IPDPS'2002/BioSP3 — International Parallel and Distributed*

Processing Symposium/Workshop on Bio-Inspired Solutions to Parallel Processing Problems, Fort Lauderdale, FL, IEEE, 2002.

[6] Riso, B., Maciel, C., Souza, I., Rossi, L., Notare, M.S.M.A., and Tonin, N.A. *Protocols Engineering with LOTOS/ISO* (in Portuguese). Florianópolis : EdUFSC, 2004. p. 215.

[7] Brinksma, E. *ISO 8807: International Organization for Standardization*, LOTOS — Language of Temporal Ordering Specifications, 1988.

[8] Garavel, H. *CADP: Eucalyptus Manual*. Grenoble, France: INRIA/VASY, 1997. http://www.inrialpes.fr/vasy/cadp/

[9] Notare, M.S.M.A. *Conception, analysis and development of a security management system for telecommunication networks* (in Portuguese). Florianopolis, Brazil, 2000. Thesis (Computer Science Program) — Department of Informatic and Statistic, Technological Center, Federal University of Santa Catarina.

[10] Simon, E. *Distributed Informations Systems: From Client/Server to Distributed Multimedia*. McGraw-Hill, Maidenhead, Berkshire, England, 1996, p. 414.

[11] Duda, R.O. and Hart, P.E. *Pattern Classification and Scene Analysis*. John Wiley and Sons, New York, 1973.

[12] Boukerche, A. and Notare, M.S.M.A. (Wiley and Sons) Applications of neural networks to mobile communication systems. In *Solutions to Parallel and Distributed Computing Problems: Lessons from Biological Sciences*. John Wiley and Sons, New York, 2001, pp. 255–268.

[13] Todesco, J.L. *Pattern Recognition using Artificial Neuronal Networks with a Radial Basis Function: an Application for a Human Chromosome Classification* (in Portuguese). Florianopolis, Brazil, 1995. Thesis (Production Eng. and Systems Program) — Department of Production Engineering, Technological Center, Federal University of Santa Catarina.

[14] NIH/NCI. *Understanding the Immune System*. National Institutes of Health. National Cancer Institute. U.S. Department of Health and Human Services. Public Health Service. http://press2.nci.nih.gov/sciencebehind/immune/immune00.htm. access 08-02-2003.

[15] Alberts, B. *Molecular Biology of the Cell*, 4th ed. Garland Science, Taylor and Francis Group, New York, 2002.

[16] Farmer, J.D., Packard, N.H., and Perelson, A.S. The immune system, adaptation, and machine learning. *Physica* D, 22, 187–204, 1986.

[17] Male, David. *Immunology — An Illustrated Outline*. Gower Medical Publishing Ltd., 1986, p. 45.

[18] Roitt, I.M. *Essential Immunology*, 5th ed. Blackwell Scientific Publications, 1995.

[19] Forrest, S., Perelson, A.S., Allen, L., and Cherukuri, R. Self–nonself discrimination in a computer. In *Proceedings of the 1994 IEEE Symposium on Research in Security and Privacy*, 1994, pp. 202–212.

[20] Jerne, N.K. Towards a network theory of the immune system. *Annals of Immunology*, 125, 373–389, 1973.

[21] Perelson, A.S. *Immune Network Theory*. Immunological. Theoretical Division, Los Alamos National Laboratory, NM 87545.1989. Review 110, pp. 5–36.

[22] Cayzer, S. and Aickelin, U. A recommended system based on the immune network. In *Proceedings of CEC 2002*.

[23] Hofmeyer, S. An Immunological Model of Distributed Detection and Its Applications to Computer Security. Albuquerque, 1999. Ph.D. thesis. Computer Science, University of New Mexico, p. 23

[24] Morrison, T. and Aickelin, U. An artificial immune system as a recommender for web sites. In *Proceedings of the 1st International Conference on Artificial Immune Systems (ICARIS-2002)*, Canterbury, UK, 2002, pp. 161–169.

[25] Yixin, D. and Passino, K.M. Immunity-based hybrid learning methods for approximate structure and parameter adjustment. *Engineering Applications of Artificial Intelligence*, 15, 587–600, 2002.

[26] Timmis, J.I. *Artificial Immune Systems:* A Novel Data Analysis Technique Inspired by the Immune Network Theory. Aberystwyth. Ph.D. thesis. Department of Computer Science, University of Wales, 2000.

24

Synthesis of Multiple-Valued Circuits by Neural Networks

Alioune Ngom
Ivan Stojmenović

24.1 Introduction

24.1.1 Artificial Neural Networks

Several novel methods of computation that are collectively known as soft computing have recently emerged. The raison d'être of these modes is to exploit the tolerance for imprecision and uncertainty in real-world problems to achieve tractability, robustness, and low cost. Soft computing is usually used to find an approximate solution to a precisely or imprecisely formulated problem. Neural computing, fuzzy computing, and evolutionary computing are the major components of this approach.

Artificial neural networks are an attempt to mimic some or all of the characteristics of biological neural networks. This soft computational paradigm differs from a programmed instruction sequence, in that information is stored in form of weights. Each neuron is an elementary processor with primitive operations, such as summing the weighted inputs coming to it and then amplifying or thresholding the sum. Assembly of such neurons can, in principle, perform universal computations for suitably chosen weights.

A well-known model of neuron studied extensively in Reference 1 is called the *perceptron* (see Figure 24.1). The perceptron computes a weighted sum of its input signals and generates an output of 1 if this sum is above a certain threshold $t \in R$, otherwise, an output of 0 results. In General, given a weight vector $\vec{w} = (w_1, \ldots, w_n) \in R^n$ and an input vector $\vec{x} = (x_1, \ldots, x_n) \in R^n$, such neuron computes a simple function of the form $f_{\vec{w}} : R^n \mapsto S$, where $n \geq 1, S \subseteq R$ and

$$f_{\vec{w}}(\vec{x}) = g(\vec{w}\vec{x}) \tag{24.1}$$

for some *transfer function* g: $R \mapsto S$. There are two choices for the set S currently popular in literature. The first is the *discrete model* with $S = \{0, 1\}$. In this case, if a neuron's threshold is t, then its transfer function g is a *linear threshold function* (see Figure 24.2[a]) defined by

$$g(y) = \begin{cases} 0 & \text{if } y < t \\ 1 & \text{if } y \geq t \end{cases} \tag{24.2}$$

and f is called a *weighted linear threshold function.* The second is the *continuous model* with $S = [0, 1]$. In this case, g is typically a monotone increasing function such as the *sigmoid function* (see Figure 24.2[c]) given by

$$g(y) = \frac{1}{1 + a^{-(by)}} \tag{24.3}$$

for $a, b \in R^+$. The continuous model is popular because it is easier to construct. The discrete model is popular because its behavior is easier to analyze (however, it uses more hardware). A neural network is characterized by the network topology, the connection strength (i.e., weights) between pairs of neurons, the neurons properties (i.e., transfer functions), and the learning algorithms.

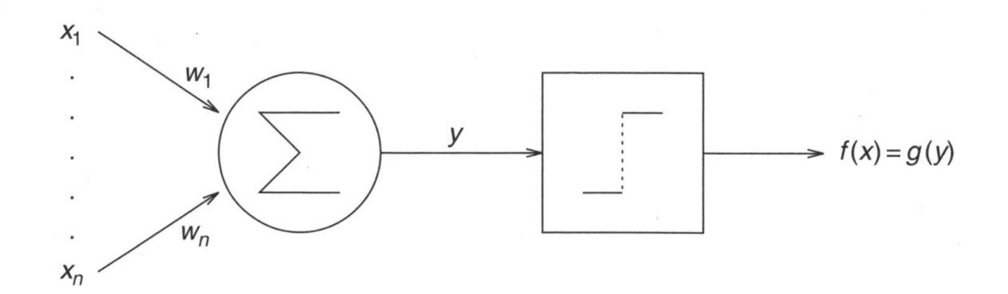

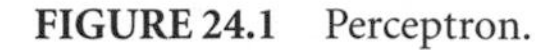
FIGURE 24.1 Perceptron.

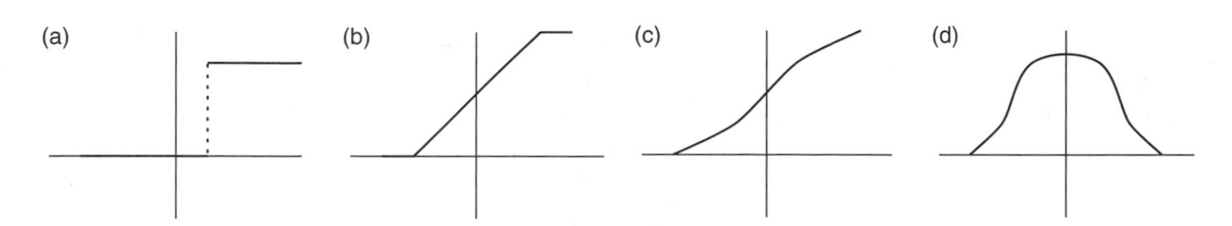

FIGURE 24.2 Examples of transfer functions: (a) linear threshold function, (b) piecewise linear function, (c) sigmoid function, and (d) Gaussian function.

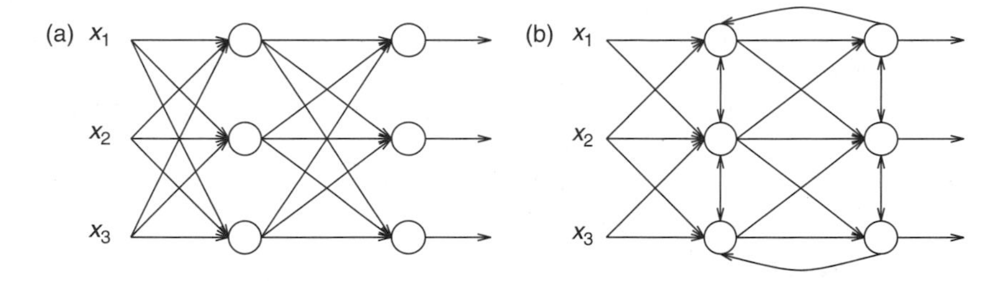

FIGURE 24.3 Examples of neural networks architectures: (a) feed-forward network and (b) feed-back network.

Artificial neural networks can be viewed as weighted directed graphs in which artificial neurons are nodes and directed edges (with weights) are connections between the neuron's outputs and inputs. Based on the interconnection pattern (architecture), artificial neural networks are grouped in two categories: *feed-forward networks* (in which graphs have no loops and cycles) and *feed-back* (or *recurrent*) *networks* (in which loops or cycles occur because of feed-back connections). Different network topologies yield different network behaviors and require appropriate learning algorithms. Figure 24.3 illustrates the two types of network topologies.

A learning process in artificial neural networks is the problem of updating the connection weights so that the network can efficiently perform a specific task. Learning can also be viewed as the problem of minimizing an objective function which is the error between the network output and the desired output. Efficient learning algorithms for specific topologies have been proposed in the literature.

24.1.2 Multiple-Valued Logic

Let k be a fixed positive integer and $K = \{0, \ldots, k-1\}$. A *k-valued logic function* f maps the Cartesian power K^n (of all ordered n-tuples of elements of K) onto K. Denote by P_k^n the set of all such functions $f : K^n \mapsto K$. Thus, P_k^n consists of k^{k^n} multiple-valued logic functions. The set P_k^n is called the set of n-ary operations on K in universal algebras and the set of n-ary functions of k-valued logic in multiple-valued logic algebras. The set P_k defined by $P_k = \cup_{n \geq 1} P_k^n$ is the set of all k-valued logic functions.

Multiple-valued logic has been the subject of research over many years [2]. Besides reduction in chip area and interconnections, multiple-valued logic offers other benefits such as potential for image processing and speech recognition. The development of multiple-valued logic algebras has also proceeded since the pioneering works of Lukasiewicz [3] and Post [4]. Reference 4 presented the first functionally complete set as basic operations in multiple-valued logic. Since then, there have been many multiple-valued logic algebras developed [5] and, applications for image processing network and knowledge information processing systems presented.

24.1.3 Multiple-Valued Logic Neural Networks

A *discrete neuron* is a processing unit whose transfer function outputs a discrete value. An example of such transfer function is the linear threshold function. A *discrete n-input multiple-valued neuron* has a discrete transfer function and realizes a function of n variables ranging in the set $S \subseteq R$ with values in K, that is, computes a function $f : S^n \subseteq R^n \mapsto K$. For $S = K$ we refer to the processing unit as a *multiple-valued logic neuron* since it simulates a multiple-valued logic function $f : K^n \mapsto K$. Multiple-valued logic neural networks are thus neural networks composed of multiple-valued logic neurons as processing units. The first model of multiple-valued logic neural networks were introduced in Reference 6 and since then various other models have been described [7–11].

24.1.4 Problem Statement

The problem we address is that of implementing multiple-valued logic functions by neural networks. In particular, we discuss interesting models of multiple-valued logic neurons (and neural networks) and how they might be used to learn and compute multiple-valued logic functions. In most models, we will be interested in optimizing some identified resource of space or time such as number of thresholds, number of neurons or size of strip. We will also be interested to study the learning abilities of our main model of neuron: the *multiple-valued multiple-threshold perceptron.*

24.1.5 Related Studies

Chan et al. [6] first developed a model of multiple-valued logic neural network. In that model, neurons simulate *Kleenean functions* — a special class of three-valued logic functions — using the well-known back-propagation learning algorithm. The model can be extended to make probabilistic predictions. It can also be used to make decisions with fuzzy input.

A design of multiple-valued logic adaptive linear neuron — adaline — is described in Reference 12. Such adaline contains multiple-valued logic operations in place of the usual linear combiner and, thus, has a faster operation speed and a more robust separation ability. The model uses a modified adaline learning algorithm to learn two-valued logic functions with noisy inputs.

References 13 and 14 defined a continuous multiple-valued neural network containing piecewise linear units as processing elements. A multiple-valued logic algebra with only three basic arithmetic operations is introduced and is proven to be functionally complete. Several logic functions can thus be synthesized using the algebraic system. The authors also described a learning technique, which is largely borrowed from back-propagation training algorithm, to learn any multiple-valued logic function.

Tang et al. [10] described a learning multiple-valued logic network based on back-propagation. The learning network is derived directly from a canonical realization of multiple-valued logic functions and therefore its functional completeness is guaranteed.

It has been shown in Reference 15 that continuous neurons with limited precision are essentially equivalent to discrete multiple-valued neurons. Such model of neurons uses a class of multiple-valued multiple-threshold functions as transfer functions and hence generalizes the discrete binary perceptron model which uses the linear threshold function. The perceptrons discussed in this chapter belong to the more general class of multiple-valued multiple-threshold perceptrons and in Reference 16 it is given the first instance of learning algorithm for multiple-valued multiple-threshold perceptrons.

Some results and methods are known in literature concerning the synthesis of two-valued logic functions by feed-forward multilayer neural networks composed of discrete binary perceptrons [1,17–19]. Minsky and Papert [1] are the first who gave necessary and sufficient conditions for two-valued logic functions to be linearly separable, that is, to be learned and hence simulated by discrete two-valued perceptrons.

Two-valued logic perceptrons are very closely related to *threshold logic elements* which have long been investigated by many authors [20]. Perceptrons can be thought of as threshold logic elements with learning capability. A threshold logic element forms a functionally complete set by itself and thus is used as a building block to construct complex logic circuits implementing logic functions. Therefore, comes the important problem of minimizing the size or depth of such threshold circuits.

Extensions of the two-valued threshold logic elements to the *two-valued multiple-threshold logic elements* were also introduced and analyzed in the literature [21–23]. It is known that any two-valued logic function can be implemented by such a single element. More general circuit elements called *multiple-valued multiple-threshold elements* have also been studied [24–26].

Interesting methods for the synthesis of multiple-valued logic functions have been described in the literature. These new approaches to synthesis, based on natural or physical laws, were introduced mostly to search for a minimal circuit representation (or equivalently, a minimal logic expression) of a given multiple-valued logic function. The only known algorithm for finding minimal multiple-valued logic expressions is exhaustive search. The excessive computation time makes this approach impractical. Especially, multiple-valued sum-of-products expressions are interesting because of the ease with which they can be implemented by programmable logic arrays [26,27]. Because of the computational complexity associated with minimal sum-of-products solutions, there is considerable interest in heuristics. Typical heuristics is that, first a minterm is selected and then an implicant is chosen that covers the minterm [26,28]. This process is repeated until the given expression is covered. Yildirim et al. [29] proposed multiple-valued logic design methods which employ *simulated annealing* [30]. Kaczmarek et al. [31] proposed neural network techniques. Hata et al. [32] proposed solutions using *genetic algorithms.* Lloris-Ruiz et al. [33] used *information theoretic* (entropy) approaches for the minimization of logic expressions.

24.1.6 Motivations

References 6, 10, 12, and 14 do not discuss how to construct minimal neural networks from their respective models. In all of these papers, the size and depth of the networks is fixed in advance. This is a major drawback since many logic functions can be synthesized with minimal size or minimal depth networks.

The homogeneous multiple-valued perceptron learning algorithm presented in Reference 15 has the weakness of being able to learn only a very tiny portion of the set of separable functions, so it has a low capacity. In addtion, no learning algorithm is known, in general, for the multiple-valued multiple-threshold perceptrons.

Multiple-valued multiple-threshold perceptrons are multiple-valued multiple-threshold logic elements with learning abilities. A minimal multiple-valued logic perceptron which computes a given logic function is a perceptron containing the least number of thresholds. The problem of finding such minimal perceptron for a function is difficult and is still left open.

Finding a minimal logic expression for a given function is a very difficult problem. The simulated annealing [29], genetic algorithms [32], and neural networks [31] based-approaches to express minimization introduced in the literature have the tendency to produce *local optimum* solutions. These are still very good solutions; however, it seems to us that better solutions can be obtained by improving these methods.

24.1.7 Organization of this Chapter

In Section 24.2, we introduce the *multiple-valued multiple-threshold perceptron* (or (n, k, s)*-perceptrons*, for short), and define many (new) concepts (such as multilinear partition and multilinear separability, for instance) that are used throughout this chapter. All subsequent discussions will be about this model. The rest of the chapter is subdivided into four main sections. In Section 24.4, the learning ability of (n, k, s)-perceptrons is examined. The previously studied homogeneous $(n, k, k-1)$-perceptron learning algorithm is generalized to the permutably homogeneous (n, k, s)-perceptron learning algorithm with guaranteed convergence property. We also introduce a high capacity learning method that learns any permutably homogeneously separable k-valued function given as input. In Section 24.4, we consider the problem of synthesizing multiple-valued logic functions by neural networks. A genetic algorithm (GA) that finds the longest strip in $V \subseteq K^n$ is described. A strip contains points located between two parallel hyperplanes. Repeated application of GA partitions the space V into certain number of strips, each of them corresponding to a hidden unit. We construct two neural networks based on these hidden units and show that they correctly compute the given but arbitrary multiple-valued function. Preliminary experimental results are presented and discussed. In Section 24.5, an evolutionary strategy (ES) that finds the longest strip in $V \subseteq K^n$ is described and the results obtained are compared with those of Section 24.4. In Section 24.6, we address the problem of minimizing the size of single multiple-valued multiple-threshold perceptrons. Every n-input k-valued logic function can be implemented using a (n, k, s)-perceptron, for some number of thresholds s. We propose a GA to search for an optimal (n, k, s)-perceptron that efficiently realizes a given multiple-valued logic function, that is, to minimize the number of thresholds. Experimental results show that the GA finds optimal solutions in most cases.

24.2 Multiple-Valued Multiple-Threshold Perceptrons

Our main problem is to synthesize multiple-valued logic functions by neural networks. Most of our research focuses on the study of the computational and learning abilities of multiple-valued multiple-threshold perceptrons. As far as we know, this is a new model for neural networks. Some special cases of this model are known in the literature References [9,15,16]describe learning algorithms for these cases. Our main contribution in the domain is the development of learning algorithms for multiple-valued multiple-threshold perceptrons and the constructions of neural networks composed of such perceptrons. Concepts that are known for the binary perceptron such as linear partition, linear separability, and capacity are extended to the multiple-valued perceptron.

24.2.1 The (n, k, s)-Perceptrons

Let R be the set of real numbers. Let $V \subseteq R^n$ and $K = \{0, \ldots, k-1\}$, with $k \geq 2$. An *n-input k-valued real function* $f : V \mapsto K$ is a function with real-valued inputs and k-valued outputs. When $V \subseteq K^n$, we will refer to f as an *n-input k-valued logic function.* We denote by P_k^n the set of all n-input k-valued real (respective logic) functions $f : V \mapsto K$ and by $P_k = \cup_{n \geq 1} P_k^n$ the set of all k-valued real (logic) functions. Thus, P_k^n consists of k^{k^n} k-valued logic functions. The set P_k^n is called the set of n-ary operations on K in universal algebras and the set of n-ary functions of k-valued logic in multiple-valued logic algebras. For instance, for $k = 2$, P_2 is the set of all two-valued logic functions.

A *discrete neuron* is a processing unit whose transfer function outputs a discrete value. An example of such a transfer function is the linear-threshold function. A *discrete n-input multiple-valued neuron* has a discrete transfer function and realizes a function of n variables ranging in a set $V \subseteq R^n$ with values in K, that is, computes a function $f : V \mapsto K$. For $V \subseteq K^n$ we refer to the processing unit as a *multiple-valued logic neuron* since it simulates a multiple-valued logic function $f : K^n \mapsto K$. Our model of multiple-valued logic neuron is the *n-input k-valued s-threshold perceptron* defined in the following paragraph.

In the theory of multiple-valued logic functions there is an important class of functions called *multiple-valued multiple-threshold functions* [25]. Such functions are used in the design of classes of multiple-valued logic circuits called *programmable logic arrays* [26]. A *k-valued s-threshold function* of one variable $y \in R$ is defined as

$$g_{k,s}^{\vec{t},\vec{o}}(y) = \begin{cases} o_0 & \text{if } y < t_1 \\ o_i & \text{if } t_i \leq y < t_{i+1} \text{ for } 1 \leq i \leq s-1 \\ o_s & \text{if } t_s \leq y, \end{cases} \tag{24.4}$$

where $\vec{o} = (o_0, \ldots, o_s) \in K^{s+1}$ is the output vector, $\vec{t} = (t_1, \ldots, t_s) \in R^s$ is the threshold vector — with $t_i \leq t_{i+1}$ $(1 \leq i \leq s-1)$ — and s $(1 \leq s \leq k^n - 1)$ is the number of threshold values.

Multiple-threshold devices [21] are threshold elements containing multiple levels of excitation (thresholds). Among their qualities, given enough thresholds, a single multiple-threshold element can realize any given function operating on a finite domain [25].

An n-input k-valued s-threshold perceptron [8,34,35], abbreviated as (n, k, s)*-perceptron*, computes an *n-input k-valued weighted s-threshold function* $F^n_{k,s}(\vec{w}, \vec{t}, \vec{o})$ given by

$$F^n_{k,s}(\vec{w}, \vec{t}, \vec{o})(\vec{x}) = g^{\vec{t},\vec{o}}_{k,s}(\vec{w}\vec{x}) = \begin{cases} o_0 & \text{if } \vec{w}\vec{x} < t_1, \\ o_i & \text{if } t_i \leq \vec{w}\vec{x} < t_{i+1}, \\ o_s & \text{if } t_s \leq \vec{w}\vec{x}, \end{cases} \tag{24.5}$$

where input vector $\vec{x} = (x_1, \ldots, x_n) \in V$, weight vector $\vec{w} = (w_1, \ldots, w_n) \in V$, threshold vector $\vec{t} = (t_1, \ldots, t_s) \in R^s$, and $t_i \leq t_{i+1}$ $(1 \leq i \leq s - 1$ and $1 \leq s \leq k^n - 1$, the number of threshold values), and output vector $\vec{o} = (o_0, \ldots, o_s) \in K^{s+1}$.

The perceptron's *transfer function* is a k-valued s-threshold function $g^{\vec{t},\vec{o}}_{k,s} : R \mapsto K$. An (n, k, s)-perceptron simulates a k-valued function $f : V \mapsto K$. Depending on V we will refer to either real or logic (n, k, s)-perceptrons. It is well known that any n-input k-valued logic function can be transformed into a k-valued s-threshold function [25,36] for some s.

An (n, k, s)-perceptron is, *monotone* if $\vec{o}$ is monotone, that is, $o_0 \leq \cdots \leq o_s$ or $o_0 \geq \cdots \geq o_s$, otherwise it is *nonmonotone.* For example, the well-known *binary perceptron* (which is an $(n, 2, 1)$-perceptron) studied in Reference 1 is monotone. An (n, k, s)-perceptron is *homogeneous* if $s = k - 1$ and $\vec{o}$ is the identity permutation of K (i.e., $o_i = i$ for $0 \leq i \leq s$), it is *permutably homogeneous* if $s \leq k - 1$ and $\vec{o}$ is any permutation of $s + 1$ elements out of K. Functions computed by such single neurons are called *permutably homogeneous functions.*

The $(n, k, k - 1)$-perceptrons were introduced by Obradović and Parberry [15] as a model for simulating continuous perceptrons (i.e., perceptrons containing analog transfer functions) with limited precision on their inputs. Their computational abilities were extensively studied in Reference 9 whereas Reference 16 examined their learning power. Learning algorithms for some classes of (n, k, s)-perceptrons are investigated in References 7 and 34. Takiyama [23] introduced that the $(n, 2, s)$-perceptrons and their computational power have been intensively studied in References 22 and 37. The neulonets' units discussed in Reference 6 are a class of $(n, 3, 2)$-perceptrons with $\vec{t} = (-d, +d), d \in R$, and $\vec{o} = (1, 0, 2)$.

Multiple-valued logic neural networks, in our definition, are thus neural networks composed of (n, k, s)-perceptrons as processing units. It should be noted that the units are not necessarily the same. For example, one layer of the network may contain $(2, 4, 8)$-perceptrons while another layer may contain $(5, 4, 3)$-perceptrons.

Analog computers are inherently inaccurate due to imperfections in fabrication and fluctuations in operating temperatures. The classical solution to this problem uses extra hardware to enforce discrete behavior. However, the brain appears to compute reliably well with inaccurate components without necessarily resorting to discrete techniques. The *continuous neural network* is a computational model based on certain observed features of the brain. Experimental evidence has shown continuous neural network to be extremely fault tolerant; in particular, their performance does not appear to be significantly impaired when precision is limited. It has been shown by Obradović and Parberry [15] that analog neurons of limited precision are essentially multiple-valued logic neurons. The first model of multiple-valued logic neuron (and neural network) was introduced by Chan et al. [6] and since then various other models have been described (see, for instance, References 11, 15, and 38).

24.2.2 Decomposition of (n, k, s)-Perceptrons

An (n, k, s)-perceptron can be decomposed into a two-hidden-layer network composed of n-input nodes, that is, one linear combiner in the first hidden layer, s $(n, 2, 1)$-perceptrons (usual linear threshold units) in the second hidden layer and one linear combiner in the output layer [25,36]. The transfer function $g^{\vec{t},\vec{o}}_{k,s}(\vec{w}\vec{x})$ of the (n, k, s)-perceptron is expressed as the linear summation of s linear threshold functions

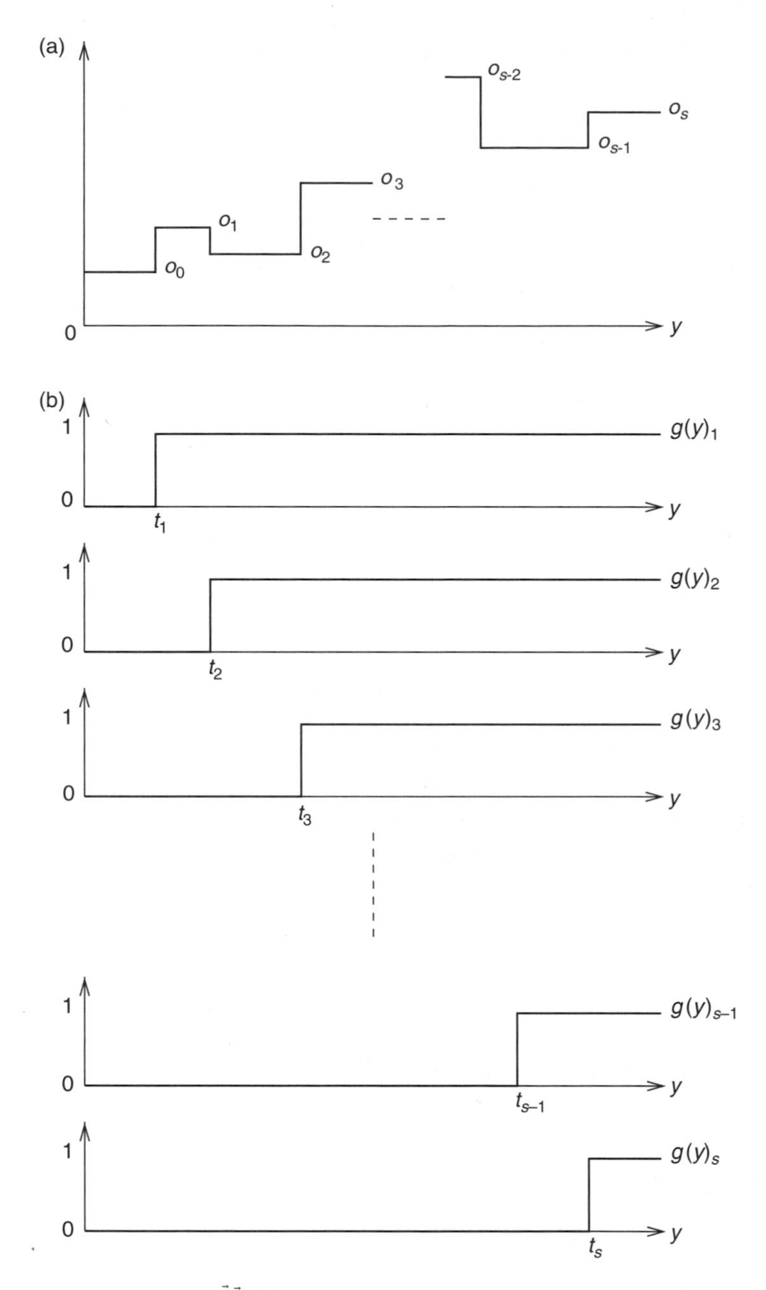

FIGURE 24.4 Decomposition of (a) $g_{k,s}^{\vec{t},\vec{o}}(\vec{w}\vec{x})$ into (b) s linear threshold functions. (From A. Ngom, I. Stojmenović, and V. Milutinović. *IEEE Transactions on Neural Networks*, 12, 212–227, 2001. With permission.)

$g(\vec{w}\vec{x})_1, \ldots, g(\vec{w}\vec{x})_s$, that is,

$$g_s^{\vec{t},\vec{o}}(\vec{w}\vec{x}) = o_0 + \sum_{i=1}^{s} a_i g(\vec{w}\vec{x})_i \quad \text{and} \quad g(\vec{w}\vec{x})_i = \begin{cases} 0 & \text{if } \vec{w}\vec{x} < t_i, \\ 1 & \text{if } \vec{w}\vec{x} \geq t_i, \end{cases} \tag{24.6}$$

where $a_i = o_i - o_{i-1}$ is the weight associated with $g(\vec{w}\vec{x})_i$ and $1 \leq i \leq s$. Figure 24.4 shows an example of such decomposition. Figure 24.5 shows the depth-two network corresponding to Equation (24.6).

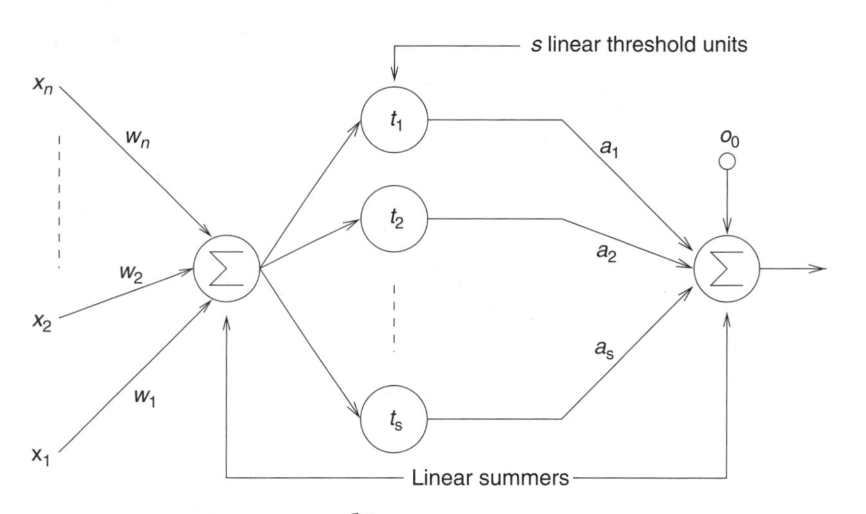

FIGURE 24.5 Two-hidden-layers network for $g_{k,s}^{\vec{t},\vec{o}}(\vec{w}\vec{x})$. (From A. Ngom, I. Stojmenović, and V. Milutinović. *IEEE Transactions on Neural Networks*, 12, 212–227, 2002. With permission.)

24.2.3 Multilinear Separability and Multilinear Partition

The problem of computing (or simulating) a given function f by a (n, k, s)-perceptron is to determine s and a vector $\vec{r} = (\vec{w}, \vec{t}, \vec{o}) \in R^{n+s} \times K^{s+1}$, such that $F_{k,s}^n(\vec{r})(\vec{x}) = f(\vec{x})$ $(\forall \vec{x} \in V)$, that is, $f = F_{k,s}^n(\vec{r})$. We will refer to $\vec{r}$ as an *s-representation* of $F_{k,s}^n$ for f.

Let $V = \{\vec{x}_1, \ldots, \vec{x}_v\} \subseteq R^n$ be a set of v vectors $(v \geq 1)$. A k-valued function f with domain V and specified by the input–output pairs $\{(\vec{x}_1, f(\vec{x}_1)), \ldots, (\vec{x}_v, f(\vec{x}_v))\}$, where $\vec{x}_i \in R^n, f(\vec{x}_i) \in K$, is said to be *s-separable* if there exist vectors $\vec{w} \in R^n, \vec{t} \in R^s$, and $\vec{o} \in K^{s+1}$, such that

$$f(\vec{x}_i) = \begin{cases} o_0 & \text{if } \vec{w}\vec{x}_i < t_1 \\ o_j & \text{if } t_j \leq \vec{w}\vec{x}_i < t_{j+1} \quad \text{for } 1 \leq j \leq s-1 \\ o_s & \text{if } t_s \leq \vec{w}\vec{x}_i \end{cases} \tag{24.7}$$

for $1 \leq i \leq v$. Equivalently, f is s-separable if and only if it has a s-representation defined by $(\vec{w}, \vec{t}, \vec{o})$. A k-valued function over V is said to be *s-nonseparable* if it is not s-separable.

In other words, an (n, k, s)-perceptron partitions the space $V \subset R^n$ into $s + 1$ distinct classes $H_0^{[o_0]}, \ldots, H_s^{[o_s]}$, using s parallel hyperplanes, where $H_j^{[o_j]} = \{\vec{x} \in V | f(\vec{x}) = o_j \text{ and } t_j \leq \vec{w}\vec{x} < t_{j+1}\}$. We assume that, $t_0 = -\infty$ and $t_{s+1} = +\infty$. Each hyperplane equation denoted by H_j $(1 \leq j \leq s)$ is of the form

$$H_j : \vec{w}\vec{x} = t_j. \tag{24.8}$$

Multilinear separability (s-separability) extends the concept of *linear separability* (1-separability of the common binary one-threshold perceptron) to the (n, k, s)-perceptron. Linear separability in two-valued case tells us that an $(n, 2, 1)$-perceptron can only learn from a space $V \subseteq [0, 1]^n$ in which there is a single hyperplane which separates it into two disjoint halfspaces: $H_0^{[0]} = \{\vec{x} | f(\vec{x}) = 0\}$ and $H_1^{[1]} = \{\vec{x} | f(\vec{x}) = 1\}$. From the $(n, 2, 1)$-*perceptron convergence theorem* [1], concepts which are linearly nonseparable cannot be learned by an $(n, 2, 1)$-perceptron. One example of linearly nonseparable two-valued logic function is the n-input parity function. Likewise, the (n, k, s)-*perceptron convergence theorems* [16,34] state that an (n, k, s)-perceptron computes a given function $f \in P_k^n$ if and only if f is s-separable. Figure 24.6 shows an example of two-separable four-valued logic function of P_5^2.

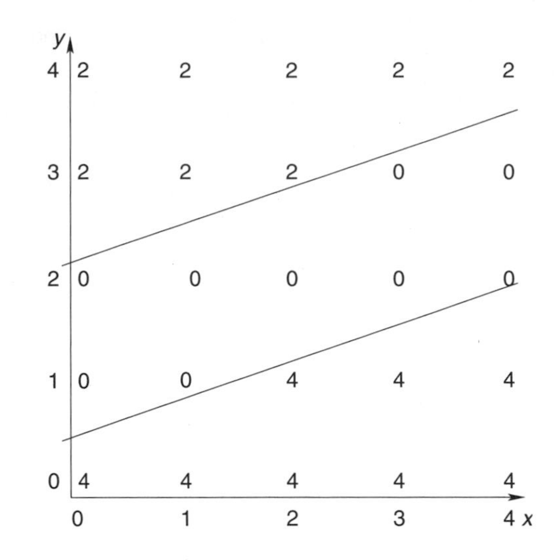

FIGURE 24.6 A two-separable function of P_5^2. (From A. Ngom, I. Stojmenović, and J. Žunić. *IEEE Transactions on Neural Networks.* 14, 469–477, 2004, *Proceedings of the 29th IEEE International Symposium on Multiple-Valued Logic.* IEEE Computer Society Technical Committee on Multiple-Valued Logic, May 1999, pp. 208–213, IEEE Computer Society. With permission.)

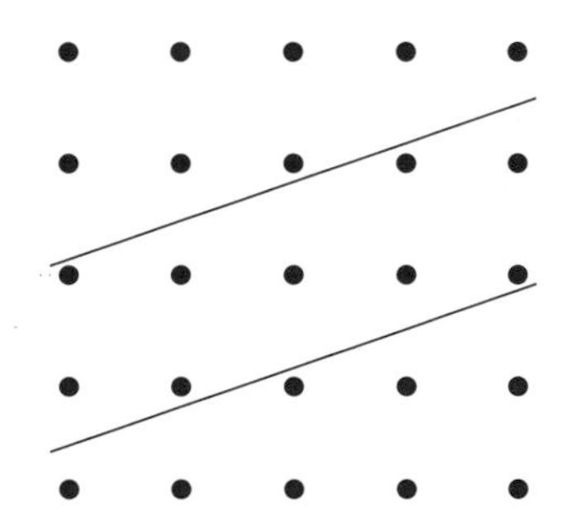

FIGURE 24.7 A $(2, 5^2, 2)$-partition. (From A. Ngom, I. Stojmenović, and J. Žunić. *IEEE Transactions on Neural Networks,* 14, 469–477, 2004, *Proceedings of the 29th IEEE International Symposium on Multiple-Valued Logic,* IEEE Computer Society Technical Committee on Multiple-Valued Logic, May 1999, pp. 208–213, IEEE Computer Society. With permission.)

Let $V \subset R^n$ with $|V| = v \geq 2$. An (n, v, s)*-partition* is a partition of V by $s \leq v-1$ parallel hyperplanes (namely $(n-1)$-planes), which do not pass through any of the v points. For instance, Figure 24.7 shows an example of $(2, 5^2, 2)$-partition.

An (n, v, s)-partition determines $s+1$ distinct classes $S_0, \ldots, S_s \subset V$ separated by s parallel $(n-1)$ planes such that $\bigcup_{i=0}^{s} S_i = V$ and $\bigcap_{i=0}^{s} S_i = \emptyset$. An (n, v, s)-partition corresponds to an s-separable k-valued function $f \in P_k^n$ if and only if all points in the set S_i, for $0 \leq i \leq s+1$, have the same value taken out of K. In addition, we assume that any two neighboring classes have distinct values. If for a given (n, v, s)-partition we have $S_i \neq \emptyset$ $(0 \leq i \leq s+1)$ then, clearly, the number of associated functions is $k(k-1)^s$.

A *linear partition* of a point set V is an $(n, v, 1)$-partition; therefore, only a single $(n-1)$ plane is required to separate an n-dimensional space $V \subseteq R^n$ into two half spaces. The enumeration problem for linear partitions is closely related to the efficiency measurement problem for linear discriminant functions in pattern recognition [17] and to many other algorithmic problems [39].

24.3 Learning with Permutably Homogeneous Multiple-Valued Multiple-Threshold Perceptrons

The (n, k, s)-perceptrons partition the input space $V \subset R^n$ into $s+1$ regions using s parallel hyperplanes. Their learning abilities are examined in this section. The previously studied homogeneous $(n, k, k-1)$-perceptron learning algorithm is generalized to the permutably homogeneous (n, k, s)-perceptron learning algorithm with guaranteed convergence property. We also introduce a high capacity learning method that learns any permutably homogeneously separable k-valued function given as input.

A function implementable by a homogeneous (n, k, s)-perceptron is said to be *homogeneously separable* (or homogeneous, for short). A function computable by an (n, k, s)-perceptron with given output vector $\vec{o}$ is said to be *$\vec{o}$-separable.* A function implementable by an (n, k, s)-perceptron whose output vector is monotone is said to be *monotoneously separable.* A function computable by a permutably homogeneous (n, k, s)-perceptron is said to be *permutably homogeneously separable* (or simply, permutably homogeneous). For instance, the functions f_1, f_2, and f_3 shown in Figure 24.8 are all three-separable and, moreover, are respectively $(0, 2, 1, 3)$-separable, $(0, 1, 2, 3)$-separable, and $(3, 2, 1, 3)$-separable; f_1 is nonmonotoneously separable and permutably homogeneous; f_2 is (permutably) homogeneous and monotoneoulsy separable; f_3 is nonpermutably homogeneous and nonmonotoneously separable.

Note that since $\vec{o} \in K^{s+1}$ then every $\vec{o}$-separable function (for some $\vec{o}$) is also s-separable. However, the converse is not true, that is s-separability does not implies $\vec{o}$-separability (for some $\vec{o}$). The only case where s-separability is equivalent to $\vec{o}$-separability is the two-valued one-threshold case, that is, when $k = 2$, $s = 1$, and $\vec{o} = (0, 1)$ or $(1, 0)$. Every one-separable two-valued logic function is $(0, 1)$-separable, and also, every $(0, 1)$-separable two-valued logic function is one-separable.

24.3.1 Permutably Homogeneous (n, k, s)-Perceptron Learning Algorithm

In this section we propose a learning algorithm for permutably homogeneous (n, k, s)-perceptrons. An (n, k, s)-perceptron is *permutably homogeneous* if its output vector is a $(s+1, k)$-permutation. To the best of our knowledge there are no known learning algorithms for (n, k, s)-perceptrons in general. Our algorithms, when applied to the special case $s = k-1$, that is, the $(n, k, k-1)$-perceptrons, are more powerful than the homogeneous $(n, k, k-1)$-perceptron learning algorithm described in Reference 16, in that they can learn a larger class of multiple-valued logic functions.

A permutation does not need to contain all values of K. There are k-valued logic functions whose set of output values is a subset $S \subseteq K$, that is, functions of the form $f : K^n \mapsto S \subseteq K$. Examples of such functions are, for instance, functions of the form $f : K^n \mapsto \{0, 1\}$, that is, functions with k-valued inputs and two-valued outputs.

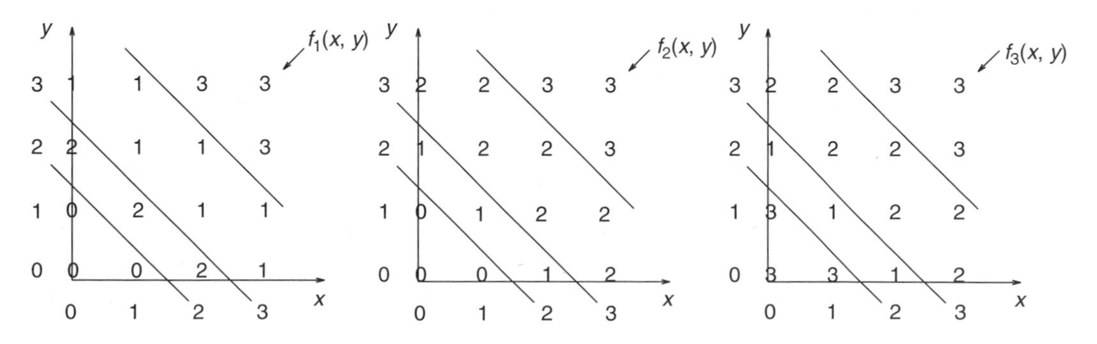

FIGURE 24.8 Examples of three-separable two-input four-valued logic functions. (From A. Ngom, C. Reischer, D.A. Simovici, and I. Stojmenović. *Neural Processing Letters, 12, 2000, Proceedings of the 28th IEEE International Symposium on Multiple-Valued Logic,* May 1998, pp. 161–166. With permission.)

```
Procedure MultiPerceptron(f, n, k, s, o⃗);
  w⃗ := 0⃗;
  t⃗ := 0⃗;
  Repeat
    for each x⃗ ∈ K^n do
      v := F^n_{k,s}(w⃗, t⃗, o⃗)(x⃗);
      if f(x⃗) ≠ v then
        MultiPerceptronUpdate(x⃗, f(x⃗), v, o⃗);
      Output (w⃗, t⃗);
  Until F^n_{k,s}(w⃗, t⃗, o⃗) = f;

Procedure MultiPerceptronUpdate(x⃗, f(x⃗), v, o⃗);
  δ := Pos_o⃗[f(x⃗)] − Pos_o⃗[v];
  if δ < 0 then
    t_{Pos_o⃗[v]} := t_{Pos_o⃗[v]} − ηδ;
  if δ > 0 then
    t_{Pos_o⃗[v]+1} := t_{Pos_o⃗[v]+1} − ηδ;
  for 1 ≤ i ≤ n do
    w_i := w_i + ηδx_i;
```

FIGURE 24.9 Permutably homogeneous (n, k, s)-perceptron learning algorithm. (From A. Ngom, C. Reischer, D.A. Simovici, and I. Stojmenović. *Neural Processing Letters, 12, 2000, Proceedings of the 28th IEEE International Symposium on Multiple-Valued Logic*, May 1998, pp. 161–166. With permission.)

Let $\vec{o} \in K^{s+1}$ be the output vector of an (n, k, s)-perceptron. When $\vec{o}$ is a $(s+1, k)$-permutation, that is, there are no i and j $(i \neq j)$ such that $o_i = o_j$, we propose the *permutably homogeneous* (n, k, s)*-perceptron learning algorithm* (for a fixed $(s+1, k)$-permutation $\vec{o}$) as shown in Figure 24.9.

In Figure 24.9, the constant $0 < \eta \leq 1$ is the learning rate. The initial weights can be set to any (random) values. The initial thresholds can also be set to any (random) values, however, empirical tests show that the algorithm converges faster when the initial thresholds are set in such a way that $t_{i+1} - t_i = c$ (e.g., $c = kn$) and that $t_{v-1} \leq t_v \leq t_{v+1}$ each time we update t_v or t_{v+1}. We can also generate a new random η before each call to *MultiPerceptronUpdate*. $Pos_{\vec{o}}[z]$ is the position (or the index) of z in $\vec{o}$. For example, if $\vec{o} = (3, 0, 2, 1)$ then $Pos_{\vec{o}}[3] = 0, Pos_{\vec{o}}[0] = 1, Pos_{\vec{o}}[2] = 2$, and $Pos_{\vec{o}}[1] = 3$.

In the algorithm, the weight and the threshold vectors are always updated in opposite directions using the error value $\delta = Pos_{\vec{o}}[f(\vec{x})] - Pos_{\vec{o}}[v]$. If $Pos_{\vec{o}}[f(\vec{x})] < Pos_{\vec{o}}[v]$ then $\delta < 0$ means that the weights are too large or $t_{Pos_{\vec{o}}[v]}$ is too small. Therefore, we decrease the weights and increase $t_{Pos_{\vec{o}}[v]}$. If $Pos_{\vec{o}}[f(\vec{x})] > Pos_{\vec{o}}[v]$ then $\delta > 0$ means that the weights are too small or $t_{Pos_{\vec{o}}[v]+1}$ is too large. Thus, we increase the weights and decrease $t_{Pos_{\vec{o}}[v]+1}$. When $Pos_{\vec{o}}[f(\vec{x})] = Pos_{\vec{o}}[v]$ no modification is done and the algorithm goes to the next step. So, $\vec{w}$ and $\vec{t}$ are always updated in opposite directions given by the position of $f(\vec{x})$ relative to that of v in $\vec{o}$. Note that $\vec{o}$ is known and given as input to the algorithm.

Our algorithm can learn any $\vec{o}$-separable logic function as long as $\vec{o}$ is a fixed $(s+1, k)$-permutation. In other words, the algorithm can learn any function whose input vectors can be separated by a set of parallel hyperplanes (i.e., s-separable, for some s) and whose classes — separated by these hyperplanes — have distinct values (i.e., $\vec{o}$-separable, for some $(s+1, k)$-permutation $\vec{o}$). In fact, the permutably homogeneous algorithm generalizes the homogeneous algorithm in which $\vec{o}$ is any permutation (such permutation need to be the identity permutation or a k-permutation).

When $\vec{o}$ is not a permutation, that is, there are i and j $(i \neq j)$ such that $o_i = o_j$, then it becomes difficult to obtain a learning algorithm with guaranteed convergence. This problem is left open for further research.

24.3.2 Permutably Homogeneous (n, k, s)-Perceptrons Convergence Properties

Given a $(s+1, k)$-permutation $\vec{o}$, experiments show that the permutably homogeneous (k, s)-perceptron learning algorithm always converges for $\vec{o}$-separable functions. That is, given the appropriate output vector any permutably homogeneous k-valued logic function will be learned. In this section we give a formal proof of convergence of the algorithm given in Figure 24.9.

The *latency* (or *delay*) of a learning algorithm is the worst case running time between the output of one set of assignments and the next. We will assume *unit-cost* latency; that is, we will assume that the algorithm is implemented on a digital computer with word-size large enough that each elementary arithmetic and logic operation can be implemented in constant time. The *mistake bound* is the worst case total number of distinct assignments outputs. The latency and the mistake bound of the permutably homogeneous (n, k, s)-perceptron learning algorithm are, respectively, $O(n)$ and $\Omega(k^n)$.

Let $\vec{\pi} = (\pi(0), \ldots, \pi(k-1))$ be a k-permutation. The identity permutation is denoted by $\vec{\sigma}$. Thus, a homogeneous (n, k, s)-perceptron is a neuron whose output vector is $\vec{\sigma}$ and whose $s = k - 1$. It has been proven in Reference 16 that the homogeneous $(n, k, k-1)$-perceptron learning algorithm always terminates in learning homogeneous k-valued logic functions.

Let $f \in P_k^n$ be a homogeneous function, then we will denote by $f_\pi \in P_k^n$ the function obtained by permuting the output values of f with respect to π. That is, we define the homogeneous transformation $f_\pi(\vec{x}) = \pi(f(\vec{x}))$ for all $\vec{x} \in K^n$. Clearly, $f_\sigma = f$, and the function f_π is a permutably homogeneous function. In fact, the set of all permutably homogeneous $(k-1)$-separable functions can be constructed in this way from the set of all homogeneous functions.

Lemma 24.1 *Let f be homogeneous and $\vec{\pi}$ be a k-permutation. Then $\vec{r}_\sigma = (\vec{w}, \vec{t}, \vec{\sigma})$ is a $(k-1)$-representation for f if and only if $\vec{r}_\pi = (\vec{w}, \vec{t}, \vec{\pi})$ is a $(k-1)$-representation for f_π.*

Proof: $\Rightarrow$) If $\vec{r}_\sigma$ is a $(k-1)$-representation for f then, for $\vec{x} \in K^n$, we have $f_\pi(\vec{x}) = \pi(f(\vec{x})) = \pi(F_{k,k-1}^n(\vec{r}_\sigma)(\vec{x})) = \pi(g_{k,k-1}^{\vec{t},\vec{\sigma}}(\vec{w}\vec{x})) = \pi(\sigma(i)) = \pi(i) = g_{k,k-1}^{\vec{t},\vec{\pi}}(\vec{w}\vec{x}) = F_{k,k-1}^n(\vec{r}_\pi)(\vec{x})$ (if $t_i \le \vec{w}\vec{x} \le t_{i+1}$ with $t_0 = -\infty$ and $t_k = +\infty$). So $\vec{r}_\pi$ is a $(k-1)$-representation for f_π. $\Leftarrow$) Similar proof using the fact that, for $z \in K$, we have $\sigma(z) = \pi^{-1}(\pi(z)) = z$ where the permutation π^{-1} is the inverse of π (π^{-1} is guaranteed to exist and is unique since the set of all permutations on K forms a group). ■

Lemma 24.1 tells us that, given a $(k-1)$-representation $(\vec{w}, \vec{t}, \vec{\sigma})$ for a homogeneous function f, the homogeneous transformation of f into a permutably homogeneous function f_π leaves the weights and the thresholds invariant. Therefore, the positions of the $k-1$ separating parallel hyperplanes do not change after transformation of f. Clearly, in Figure 24.8, the three hyperplanes remain invariant even after transformation of f_2 into $f_1 = f_{2(0,2,1,3)}$ and vice versa.

Theorem 24.1 *Given the output vector $\vec{\pi}$, the permutably homogeneous $(n, k, k-1)$-perceptron learning algorithm for learning a function $f \in P_k^n$ terminates if and only if f is $\vec{\pi}$-separable.*

Proof: $\Rightarrow$) If the permutably homogeneous $(n, k, k-1)$-perceptron algorithm with output vector $\vec{\pi}$ terminates on learning f then a $(k-1)$-representation $(\vec{w}, \vec{t}, \vec{\pi})$ exists for f. Therefore, f is $\vec{\pi}$-separable and thus permutably homogeneous. $\Leftarrow$) Let f be $\vec{\pi}$-separable, we want to show that the algorithm terminates for f. The algorithm, instead of learning f, learns $f_{\pi^{-1}}$ using the homogeneous $(n, k, k-1)$-perceptron learning algorithm with output vector $\vec{\pi}^{-1}$. Since f is permutably homogeneous and $\vec{\pi}$-separable then from the $\Leftarrow$ part of Lemma 24.1 we have that $f_{\pi^{-1}}$ is homogeneous and thus $\vec{\sigma}$-separable. Once $f_{\pi^{-1}}$ has been learned, f can be reconstructed using the $\Rightarrow$ part of Lemma 24.1. The algorithm is guaranteed to terminate by the homogeneous $(n, k, k-1)$-perceptron convergence theorem of Reference 16. ■

If in Figure 24.9 we replace $\vec{o}$ by $\vec{\pi}$, then we clearly have $Pos_{\vec{o}} = \pi^{-1}$ and therefore the algorithm truly learns $f_{\pi^{-1}}$ and reconstruct f using the parameters found for $f_{\pi^{-1}}$.

These results concern output vectors which are k-permutations of K, that is, full permutations. Next, we consider the case of $(s+1, k)$-permutations for fixed $s \le k-1$. In the sequel we let $\vec{\nu} = (\nu(0), \ldots, \nu(s))$

be a $(s+1,k)$-permutation and $\vec{v}^c = (v^c(0) = v(0), \ldots, v^c(s) = v(s), v^c(s+1), \ldots, v^c(k-1))$ be the k-permutation (full permutation) obtained by adding $k-s$ coordinates to $\vec{v}$.

Lemma 24.2 *f is $\vec{v}$-separable if and only if it is $\vec{v}^c$-separable.*

Proof: Clearly, if $\vec{r} = (\vec{w}, \vec{t} = (t_1, \ldots, t_s), \vec{v})$ is a s-representation for f, then it is easy to see that $\vec{r}^c = (\vec{w}, \vec{t}^c = (t_1^c = t_1, \ldots, t_s^c = t_s, t_{s+1}^c, \ldots, t_{k-1}^c), \vec{v}^c)$ is a $(k-1)$-representation for f and vice versa (where $t_{s+1}^c \leq \cdots \leq t_{k-1}^c$ are arbitrary and their corresponding hyperplanes contain no points between them). ■

Theorem 24.2 (Permutably homogeneous perceptron convergence theorem) *Given the output vector $\vec{v}$, the permutably homogeneous (n,k,s)-perceptron learning algorithm for learning a function $f \in P_k^n$ terminates if and only if f is $\vec{v}$-separable.*

Proof: $\Rightarrow$) Same as in Theorem 24.2 $\Leftarrow$) The algorithm learns f using the output vector $\vec{v}^c$ instead of $\vec{v}$. Since from Lemma 24.2 f is $\vec{v}^c$-separable, then by Theorem 24.2 the algorithm is guaranteed to terminate. ■

24.3.3 Determining the Output Vectors of Permutably Homogeneous (n,k,s)-Perceptrons

A more difficult learning problem is when the output vector $\vec{o}$ is not known and that it should be determined along with $\vec{w}$ and $\vec{t}$. In this section we let $\vec{o}$ be a $(s+1,k)$-permutation. An obvious solution to this problem is to generate each $(s+1,k)$-permutation $\vec{p}$ and apply the learning algorithm with $\vec{o} = \vec{p}$. This method takes $O(enk^n(k!/(k-s-1)!))$ time complexity (where e is the number of learning epochs) and thus is nonrealistic for even small values of k. A better method is, for a given function $f \in P_k^n$, to search for a partial order relation defined over f's values and, if such relation exists then we search for a good linear extension of the corresponding partially ordered set and use it as the output vector of an (n,k,s)-perceptron. Before we describe the algorithm we need a few definitions.

A *partially ordered set* $(P, <_P)$ or (*poset*) is a set $P = \{a_0, \ldots, a_{p-1}\}$ equipped with an irreflexive, antisymmetric, and transitive relation $<_P$. A *linear extension* L of P is a linear (or total) ordering $<_L$ of the elements of P such that $a_i <_L a_j$ whenever $a_i <_P a_j$. For example, $L = \{1 <_L 0 <_L 3 <_L 2 <_L 4\}$ and $L = \{4 <_L 1 <_L 3 <_L 0 <_L 2\}$ are two linear extensions of the poset shown in Figure 24.10. We also define a *covering relation* $\prec_P$ between two elements of P: $x \prec_P y$ if and only if $x <_P y$ and there is no z such that $x <_P z <_P y$.

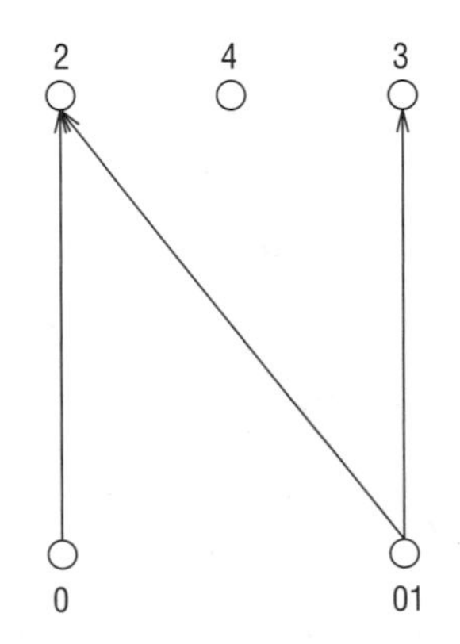

FIGURE 24.10 Example of partially ordered set. (From A. Ngom, C. Reischer, D.A. Simovici, and I. Stojmenović. *Neural Processing Letters, 12, 2000, Proceedings of the 28th IEEE International Symposium on Multiple-Valued Logic,* May 1998, pp. 161–166. With permission.)

Procedure ExtendedLearning(f, n, k, s);
  $K_f \subseteq K :=$ set of output values of f;
  $Contradiction :=$ False;
  $Unique :=$ False;
  $d := 0$;
  Repeat
    $d := d + 1$;
    ConstructPartialOrder$(Contradiction, (K_f, <^d_{K_f}), Unique)$;
    If $Unique =$ True Then
      $\vec{e} := (e_0, \ldots, e_{|K_f|-1})$ the unique linear extension of $(K_f, <^d_{K_f})$;
      $\vec{p} := (p_0, \ldots, p_{s-|K_f|})$ a $(s - |K_f| + 1, |K - K_f|)$-permutation of $K - K_f$;
      $\vec{o} := (e_0, \ldots, e_{|K_f|-1}, p_0, \ldots, p_{s-|K_f|})$;
      MultiPerceptron$(f, n, k, s, \vec{o})$;
  Until $Contradiction =$ True or $Unique =$ True or $d = n$;
  If $Contradiction =$ False and $Unique =$ False then
    CombinePartialOrders$((K_f, <^1_{K_f}), \ldots, (K_f, <^n_{K_f}))$;

FIGURE 24.11 Extended (n, k, s)-perceptron learning algorithm. (From A. Ngom, C. Reischer, D.A. Simovici, and I. Stojmenović. *Neural Processing Letters, 12, 2000, Proceedings of the 28th IEEE International Symposium on Multiple-Valued Logic*, May 1998, pp. 161–166. With permission.)

Our extended permutably homogeneous (n, k, s)-perceptron learning algorithm, for searching an output vector $\vec{o} \in K^{s+1}$ and then learning a given function f using $\vec{o}$, is shown in Figure 24.11. For $f \in P^n_k$, let K_f be its set of values. Denote by $<^d_{K_f}$ an order relation over K_f with respect to the d-th variable, that is x_d, where $1 \leq d \leq n$. We will refer to d as *direction* since, as we will see later, it selects the dimension of the n-cube K^n along which we construct a poset.

The *extended learning algorithm* goes as follows. Using the *partial order construction algorithm* of Figure 24.12 we attempt to construct a poset $(K_f, <^d_{K_f})$ with respect to some variable x_d. If such $(K_f, <^d_{K_f})$ exists and is a chain then $\vec{o}$ is the concatenation of the unique linear extension of $(K_f, <^d_{K_f})$ and a $(s - |K_f| + 1, |K - K_f|)$-permutation of $K - K_f$ and it will be used to learn f. If such $(K_f, <^d_{K_f})$ exists but is not a chain then we attempt to obtain $(K_f, <^{d+1}_{K_f})$ and so on until either we construct a chain poset $(K_f, <^d_{K_f})$ for some d, or there is some d such that a poset $(K_f, <^d_{K_f})$ cannot be obtained, or $d = n$. When n nonchain posets $(K_f, <^1_{K_f}), \ldots, (K_f, <^n_{K_f})$ are constructed then, using the *partial orders combination algorithm* of Figure 24.13, we attempt to combine these n nonchain posets into a chain poset $(K_f, <_{K_f})$.

The *partial order construction algorithm* goes as follows. For a given direction $1 \leq d \leq n$, we start with an antichain $(K_f, <^d_{K_f})$. Then, for every $\vec{x} = (x_1, \ldots, x_n)$, we construct poset $(K_f, <^d_{K_f})$ by adding new comparable pairs $f_1 = f(x_1, \ldots, x_d, \ldots, x_n) <^d_{K_f} f(x_1, \ldots, x_d + 1, \ldots, x_n) = f_2$ whenever $f_1 \not\leq^d_{K_f} f_2$, and also, we add new comparable pairs $y <^d_{K_f} f_2$ whenever $y <^d_{K_f} f_1$ and comparable pairs $f_1 <^d_{K_f} y$ whenever $f_2 <^d_{K_f} y$, for some y. We exit the loops as soon as there is some new comparable pair $y <^d_{K_f} z$ (for some y and z) that cannot be added to $(K_f, <^d_{K_f})$. That is, $z <^d_{K_f} y$ is already in $(K_f, <^d_{K_f})$ and, therefore, adding its inverse leads to inconsistency. In this case, the construction of $(K_f, <^d_{K_f})$ cannot be completed along the direction d (meaning that $(K_f, <^d_{K_f})$ simply do not exist). In case $(K_f, <^d_{K_f})$ exists — its construction can be completed along d — then it has a unique linear extension if and only if it is a chain. In other words, $(K_f, <^d_{K_f})$ is always constructed in the positive direction along the d-th dimension of the n-cube K^n, or equivalently, starting from any point $\vec{y}$ in the hyperplane $x_d = 0$ we move toward the hyperplane $x_d = k - 1$ by following the line segment orthogonal to both hyperplanes and whose origin is $\vec{y}$. Figure 24.14 shows examples of constructed posets. For illustration purpose, we have also shown the graphs obtained from function $g(x, y)$ (Figure 24.6[b]) when attempting to complete the construction

```
Procedure ConstructPartialOrder(Contradiction, (K_f, <^d_{K_f}), Unique);
  (K_f, <^d_{K_f}) := antichain;
  While 0 ≤ x_1 ≤ k − 1 and Contradiction = False do
    ⋱
      While 0 ≤ x_d ≤ k − 2 and Contradiction = False do
        ⋱
          While 0 ≤ x_n ≤ k − 1 and Contradiction = False do
            f_1 := f(x_1, ..., x_d, ..., x_n);
            f_2 := f(x_1, ..., x_d + 1, ..., x_n);
            If f_1 ≮^d_{K_f} f_2 then
              If f_2 <^d_{K_f} f_1 then
                Contradiction := True;
            Else
              Add comparable pair f_1 <^d_{K_f} f_2 in (K_f, <^d_{K_f});
              For every value y such that y <^d_{K_f} f_1 do
                If f_2 <^d_{K_f} y then
                  Contradiction := True;
                  Exit the for loop;
                Else
                  Add comparable pair y <^d_{K_f} f_2 in (K_f, <^d_{K_f});
              For every value y such that f_2 <^d_{K_f} y do
                If y <^d_{K_f} f_1 then
                  Contradiction := True;
                  Exit the for loop;
                Else
                  Add comparable pair f_1 <^d_{K_f} y in (K_f, <^d_{K_f});
  If Contradiction = False and (K_f, <^d_{K_f}) is a chain then
    Unique := True;
```

FIGURE 24.12 Partial order construction algorithm.

with possible contradictory pairs. As one can see, such graph cannot be embedded into a poset. So posets $(K_g, <^1_{K_g})$ and $(K_g, <^2_{K_g})$ do not exist.

For given permutably homogeneous and s-separable function $f \in P^n_k$, not any linear extension $\vec{e}$ of a nonchain $(K_f, <^d_{K_f})$ is good for learning. The (n, k, s)-perceptron learning algorithm may not terminate when $\vec{e}$ is used. For instance, from Figure 24.6(c) the linear extension $(3, 0, 1, 2)$ of $(K_h, <^1_{K_h})$ is not good for learning h since h is $(2, 0, 1, 3)$-separable (assuming $s = 3$). Some directions may give more information on order than others. For example, $f(x_1, x_2) = (x_1 + 1) \bmod k$ is irrelevant on x_2 (or direction 2) and so $(K_f, <^2_{K_f})$ is an antichain whereas $(K_f, <^1_{K_f})$ is a chain. In general, given a direction d there are two possibilities for failure. Either $(K_f, <^d_{K_f})$ cannot be constructed, or $(K_f, <^d_{K_f})$ exists but is a nonchain poset, such that a selected linear extension (among its many linear extensions) does not yield a convergence of the (n, k, s)-perceptron learning algorithm (but some other will do so). Because of this fact, when a nonchain poset $(K_f, <^d_{K_f})$ is obtained for any direction $1 \leq d \leq n$, we must combine these n posets in some way in order to obtain a unique linear extension. Next we describe how to combine them.

The *partial orders combination algorithm* goes as follows. Let $(K_f, <_{K_f})$ be a combination poset of d consistent nonchains posets $(K_f, <^1_{K_f}), \ldots, (K_f, <^d_{K_f})$. Initially $(K_f, <_{K_f})$ is set to $(K_f, <^1_{K_f})$ and is constructed according to some binary string $c = [c_1, \ldots, c_d] \in \{0, 1\}^d$, where $1 \leq d \leq n$. When $c_i = 1$ then the inverse of $(K_f, <^i_{K_f})$, that is poset $(K_f, >^i_{K_f})$, is in $(K_f, <_{K_f})$, otherwise $(K_f, <^i_{K_f})$ itself

```
Procedure CombinePartialOrders((K_f, <^1_{K_f}), ..., (K_f, <^n_{K_f}));
   Contradiction := False;
   Unique := False;
   (K_f, <_{K_f}) := (K_f, <^1_{K_f});
   d := 1;
   c_d := 0;
   Repeat
      If Contradiction = True then
         CutPoset(Contradiction, (K_f, <_{K_f}), [c_1 ... c_d], Unique);
      Else
         If d < n then
            ExtendPoset(Contradiction, (K_f, <_{K_f}), [c_1 ... c_d], Unique);
   Until Unique = True or c_1 = 1 or (Contradiction = False and d = n);
   If c_1 = 1 or (Contradiction = False and d = n) then
      Unique := True; {we force unicity if we did not find a consistent chain}
      (K_f, <_{K_f}) := (K_f, <^1_{K_f}); {or any poset above with the smallest width}
   If Unique = True then
      Repeat
         e⃗ := (e_0, ..., e_{|K_f|-1}) a linear extension of (K_f, <_{K_f});
         p⃗ := (p_0, ..., p_{s-|K_f|}) a (s - |K_f| + 1, |K - K_f|)-permutation of K - K_f;
         o⃗ := (e_0, ..., e_{|K_f|-1}, p_0, ..., p_{s-|K_f|});
      Until MultiPerceptron(f, n, k, s, o⃗) terminates;
```

FIGURE 24.13 Partial orders combination algorithm.

```
Procedure ExtendPoset(Contradiction, (K_f, <_{K_f}), [c_1 ... c_d], Unique);
   d := d + 1;
   c_d := 0;
   (K_f, <_{K_f}) := (K_f, <_{K_f}) ⊕ (K_f, <^d_{K_f});
   {Check for contradiction and unicity during ⊕ operation};
```

FIGURE 24.14 Constructed partial orders for some $f \in P_4^2$.

is in $(K_f, <_{K_f})$ (obviously, $(K_f, <^i_{K_f})$ and $(K_f, >^i_{K_f})$ cannot both be in $(K_f, <_{K_f})$ at the same time). The combination poset $(K_f, <_{K_f})$ is constructed using an algorithm for generating binary strings of lengths $\leq n$ in lexicographic order. For example, for $n = 4$, the lexicographic generation of binary strings of lengths ≤ 4 goes in the following manner:

```
0    00    000    0000
                  0001
           001    0010
                  0011
     01    010    0100
                  0101
           011    0110
                  0111
1    ...
⋮
```

```
Procedure ExtendPoset(Contradiction, (K_f, <_{K_f}), [c_1 ... c_d], Unique);
  d := d + 1;
  c_d := 0;
  (K_f, <_{K_f}) := (K_f, <_{K_f}) ⊕ (K_f, <^d_{K_f});
  {Check for contradiction and unicity during ⊕ operation};
```

FIGURE 24.15 Extension algorithm.

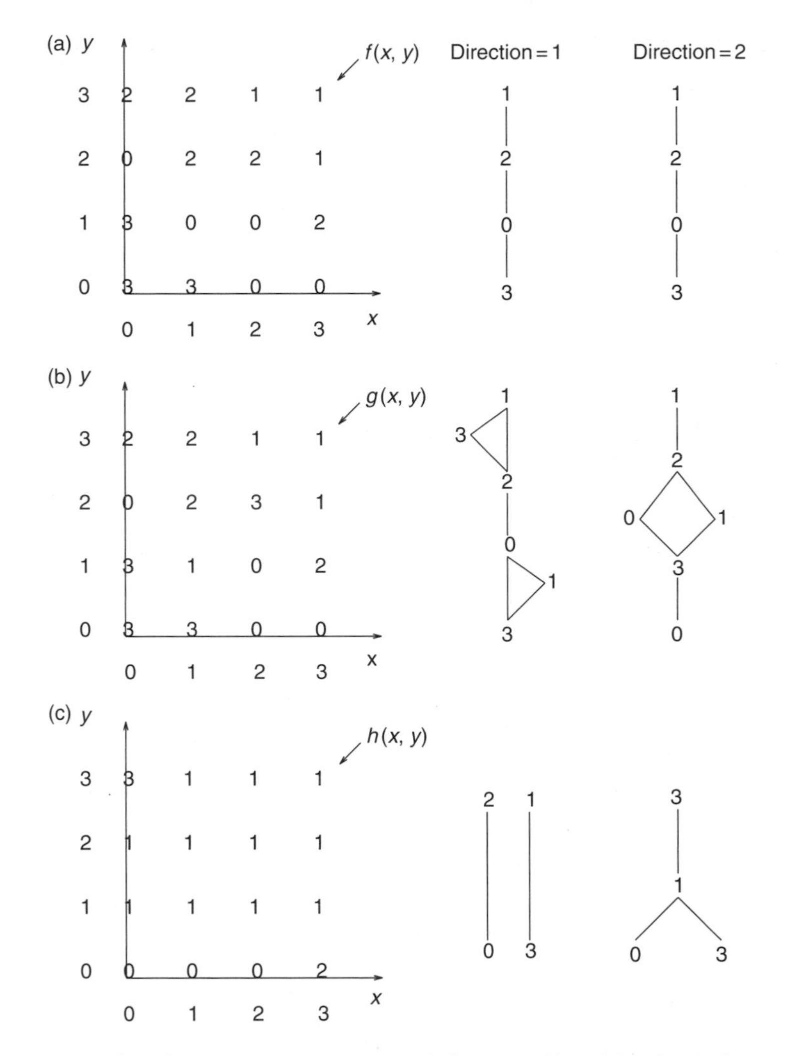

FIGURE 24.16 Cutting algorithm. (From A. Ngom, C. Reischer, D.A. Simovici and I. Stojmenović. *Neural Processing Letters, 12, 2000, Proceedings of the 28th IEEE International Symposium on Multiple-Valued Logic*, May 1998, pp. 161–166. With permission.)

The algorithm is in *ExtendPoset* phase when it goes from left to right staying in a row (Figure 24.15). It is in *CutPoset* phase when the algorithm shifts to some row (possibly far) below (Figure 24.16). The algorithm is used to construct poset $(K_f, <_{K_f})$ as follows. If with the string $[c_1, \ldots, c_d]$ poset $(K_f, <_{K_f})$ exists but is a nonchain for $d < n$ then we extend to next string $[c_1, \ldots, c_d c_{d+1} = 0]$ in the row and add poset $(K_f, <^{d+1}_{K_f})$ into $(K_f, <_{K_f})$. If with the string $[c_1, \ldots, c_d]$ poset $(K_f, <_{K_f})$ cannot be constructed then we do not need to extend since poset $(K_f, <_{K_f})$ simply do not exist and that we cannot add a poset

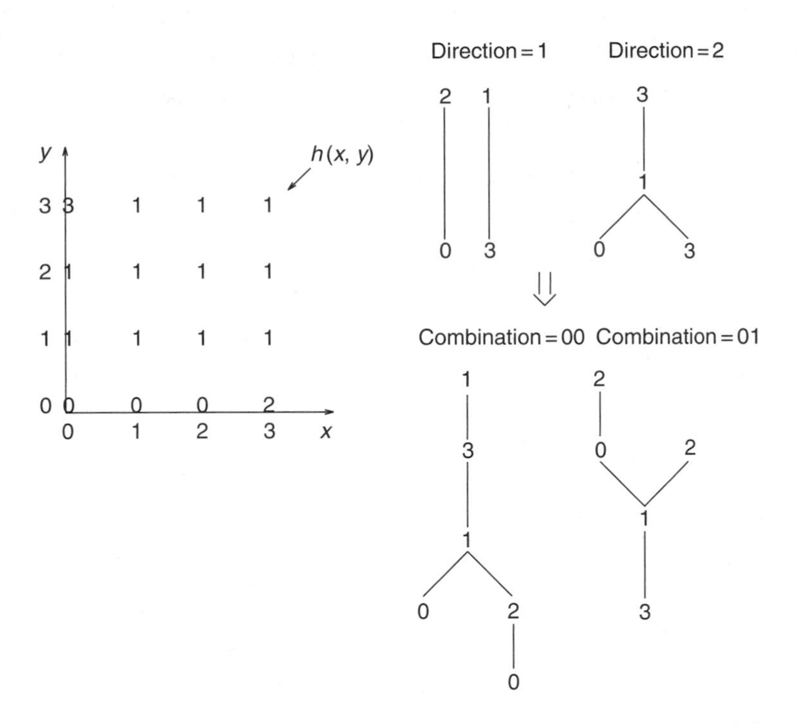

FIGURE 24.17 Examples of combinations posets for Figure 24.14c. (From A. Ngom, C. Reischer, D.A. Simovici, and I. Stojmenović. *Neural Processing Letters, 12, 2000, Proceedings of the 28th IEEE International Symposium on Multiple-Valued Logic,* May 1998, pp. 161–166. With permission.)

to an undefined poset. Hence, in this case we can bypass the lexicographic generation of binary strings to an appropriate point: we say that the algorithm is in *CutPoset* phase. We cut in the following manner. Starting from position d of c we search for the first position $r \leq d$ such that $c_r = 0$. We remove posets $(K_f, <^{r}_{K_f})$ and $(K_f, >^{r<i\leq d}_{K_f})$ from $(K_f, <_{K_f})$ and add poset $(K_f, >^{r}_{K_f})$ into $(K_f, <_{K_f})$, then finally, we set d to r and c_d to 1.

To summarize, with a given string $[c_1, \ldots, c_d]$ we have three possibilities. We generate the next string whenever $d < n$ and $(K_f, <_{K_f})$ is a nonchain poset (the extension phase) and update $(K_f, <_{K_f})$ accordingly. We bypass the lexicographic generation to some row below whenever $(K_f, <_{K_f})$ cannot be constructed (the cutting phase) and compute $(K_f, <_{K_f})$ appropriately. We exit the algorithm as soon as $(K_f, <_{K_f})$ is a chain or $c_1 = 1$ or, $(K_f, <_{K_f})$ is a nonchain and $d = n$. In the first case, we learn f using the unique linear extension of $(K_f, <_{K_f})$. In the second case, when no chain poset $(K_f, <_{K_f})$ is found, we may either randomly select one poset $(K_f, <^{i}_{K_f})$ and look for a good linear extension of it to learn f, or we select among all posets constructed so far the one which has the smallest width (since it will have the smallest number of linear extensions to search); the selection can be done by computing the width of the currently constructed consistent poset and keeping track of the smallest width (and storing the associated poset). The *width* of a poset is the size of its longest antichain.

In addition, we do not need to continue generating new binary strings when c_1 becomes 1. Because they are symmetric to (i.e., complement of) those generated already (the poset constructed according to a string c is dual to the poset constructed according to the complement of c, and hence, both posets behave exactly the same way). See Figure 24.17 for examples of combination posets. Next, we explain how to add or remove from a combination poset $(K_f, <_{K_f})$.

Given $(K_f, <^{d}_{K_f})$ to be added to $(K_f, <_{K_f})$ the addition $(K_f, <_{K_f}) \oplus (K_f, <^{d}_{K_f})$ is defined by $(K_f, <_{K_f}) \oplus (K_f, <^{d}_{K_f}) = (K_f, <_{K_f} \cup \gamma(<^{d}_{K_f}))$, where $\gamma(<^{d}_{K_f})$ is the transitive closure of every comparable pair of relation $<^{d}_{K_f}$ in relation $<_{K_f}$. That is, $O(|K_f|^2)$ comparabilities from $(K_f, <^{d}_{K_f})$ are added to

$(K_f, <_{K_f})$ during addition. In addition, for every such comparability from $(K_f, <^d_{K_f})$ its transitive closure in $(K_f, <_{K_f})$ is also added, that is, $O(|K_f|)$ more comparabilities. In sum, operation $\oplus$ take $O(|K_f|^3)$ steps.

Given a poset $(K_f, <^d_{K_f})$ to be removed from poset $(K_f, <_{K_f})$, the substraction $(K_f, <_{K_f}) \ominus (K_f, <^d_{K_f})$ is defined by $(K_f, <_{K_f}) \ominus (K_f, <^d_{K_f}) = \bigcup_{[c_1,\ldots,c_i,\ldots,c_{d-1}]}(K_f, <^i_{K_f})$ (where $(K_f, <^i_{K_f})$ is reversed when necessary). That is, to remove $(K_f, <^d_{K_f})$ from $(K_f, <_{K_f})$ is equivalent to restore $(K_f, <_{K_f})$ in the state it was before $(K_f, <^d_{K_f})$ was added to it. To achieve efficiency, substraction operation is done in the following way. Whenever we add a poset $(K_f, <^d_{K_f})$ in $(K_f, <_{K_f})$ we store into a separate data structure R_d all comparabilities of $(K_f, <^d_{K_f})$ that are not in $(K_f, <_{K_f})$. So that when we later remove $(K_f, <^d_{K_f})$ from $(K_f, <_{K_f})$ we will eliminate only comparabilities of R_d from $(K_f, <_{K_f})$. The $\oplus$ operation modified in this way still operates in $O(|K_f|^3)$. $O(|K_f|^2)$ comparabilities of R_d are removed from $(K_f, <_{K_f})$. Therefore, substraction takes $O(|K_f|^2)$ steps hence faster than addition. Inconsistency and unicity can be tested, respectively, in $O(1)$ and $O(|K_f|)$ during addition.

In Figure 24.14 we show examples of constructed posets $(K_f, <^1_{K_f})$ and $(K_f, <^2_{K_f})$ for some $f \in P^2_4$. Suppose $s = 3$. Poset $(K_f, <^1_{K_f})$ and $(K_f, <^2_{K_f})$ are chains (Figure 24.14[a]), so they have unique linear extensions and thus f is permutably homogeneous and three-separable and can be learned. In an attempt to construct $(K_g, <^1_{K_g})$ and $(K_g, <^2_{K_g})$ we obtain graphs (Figure 24.14[b]) that cannot be embedded into posets because of inconsistencies, so g is not permutably homogeneous and three-separable and thus g cannot be learned. Poset $(K_h, <^1_{K_h})$ and $(K_h, <^2_{K_h})$ are both nonchains (Figure 24.14[c]), so they have many linear extensions and h is permutably homogeneous and three-separable; however, we do not know which linear extensions are good for learning f, so we must combine $(K_h, <^1_{K_h})$ and $(K_h, <^2_{K_h})$ to search for a unique linear extension. In Figure 24.17 we show two combination posets $(K_h, <_{K_h})$ according to binary strings 00 and 01. As we can see, with string 00 poset $(K_h, <_{K_h})$ cannot be obtained because of inconsistencies whereas with string 01 it is a chain.

A *thick s-separable function* is a function $f \in P^n_k$ for which the distance between any two neighboring separating hyperplanes, in any direction, is strictly greater than one.

Theorem 24.3 *If a permutably homogeneous function $f \in P^n_k$ is thick s-separable then $(K_f, <^d_{K_f})$ is a chain for any $1 \le d \le n$.*

Proof: Let a and b be two neighboring distinct values connected by an edge. There is at least one separating hyperplane between them. However, there is at most one separating hyperplane since, otherwise, two such separating hyperplanes will be at distance strictly less than one along the dimension of that edge. Thus, $a \prec^d_{K_f} b$ or $b \prec^d_{K_f} a$, that is a and b are neighbors in the poset. All such neighboring pairs are detected in at least one dimension. Therefore $(K_f, <^d_{K_f})$ has a unique linear extension. ■

For some nonthick s-separable functions, all combination posets $(K_f, <_{K_f})$ may have many linear extensions and the last *repeat* loop of the algorithm in Figure 24.13 can be modified as follows to make it more efficient. In parallel using several processors, we generate each linear extension of $(K_f, <^d_{K_f})$ and test it for learning, until one processor succeeds. This can be simulated on one processor by time sharing, that is, generate linear extensions and test each of them for the same time in succession, until one successfully terminates. Next, we discuss the time complexity of the extended learning algorithm.

The worst case scenario, in terms of time complexity, for the partial order construction algorithm is when there is no contradiction for a given direction d. Therefore, the *while* loop associated with the selected variable x_d will be iterated $k - 1$ times and the $n - 1$ remaining *for* loops associated with nonselected variables will be iterated each k times. Also, each of the two inner *for* loops will be iterated $|K_f|$ times and it takes $O(|K_f|)$ steps to test whether $(K_f, <^d_{K_f})$ is a chain. Therefore the partial order construction algorithm has a time complexity of $O(k^n|K_f|)$.

The worst case scenario for the partial orders combination algorithm is when there is no contradiction for $d < n$ but always contradiction for $d = n$. Therefore, $2^n - 2$ combination posets are constructed each by either extension or cutting and then $O(|K_f|!)$ linear extensions are checked for learning f. Extension and cutting involve $\oplus$ operations and tests for unicity and inconsistency. Cutting is slower than extension since it also involves $\ominus$ operations and a search for the first bit equals to 0 (starting from the end of the current string). Therefore, extension and cutting take, respectively, $O(|K_f|^3)$ and $O(n|K_f|^2)$ steps. The (n, k, s)-perceptron learning algorithm takes $O(enk^n)$ steps (e is the number of learning epochs) and thus the partial orders combination algorithm has $O(2^n n|K_f|^2 + (s + enk^n)|K_f|!)$ time complexity. Since in practice e is large and that $2^n \leq k^n$ and $|K_f|^2 \leq |K_f|!$ then the complexity becomes $O(enk^n|K_f|!)$.

The worst case scenario for the extended learning algorithm is when poset $(K_f, <^d_{K_f})$ is a nonchain for any direction d. So, n posets are constructed and combined. Consequently, the extended learning algorithm has $O(nk^n|K_f| + enk^n|K_f|!)$, that is, $O(enk^n|K_f|!)$ time complexity.

Recall the first method: generate each $(s+1, k)$-permutation $\vec{p}$ and apply the (k, s)-perceptron learning algorithm with output vector $\vec{o} = \vec{p}$ for learning f until the learning terminates for some permutation $\vec{p}$. This method takes $O(enk^n(k!/(k-s-1)!))$ time complexity. Let us refer to it as the *permutation generating learning algorithm*. Next, we compare the extended algorithm to the permutation algorithm.

First, note that the time complexity of the permutation generating algorithm is always the same for any function. That is not true for the extended algorithm. For instance, for any nonpermutably homogeneous function $f \in P^n_k$ the extended algorithm takes $O(nk^n|K_f|)$ steps; the algorithm takes $O(enk^n)$ steps for any permutably homogeneous thick s-separable function. The worst time complexity is achieved only for permutably homogeneous nonthick s-separable functions f whose any combination poset $(K_f, <_{K_f})$ is a nonchain or cannot be constructed. We believe that the probability to obtain such function f is very close to zero (if not equal to zero), so that in practice, the extended learning algorithm runs in $O(enk^n)$ for permutably homogeneous s-separable functions. This proves its superiority over the permutation generating learning algorithm.

24.3.4 Experiments

We tested our extended learning algorithm on nonpermutably homogeneous functions and on permutably homogeneous thick or nonthick functions. However, we could not obtain nonthick functions whose combination posets are all nonchains. This suggests that such functions are very rare if not inexistent. The nonthick functions we used have at least one chain combination poset. In our test we set the learning rate η to 0.5 and the maximum number of learning epochs e to 5000. We experimented with different values of n and k. Also, the number of threshold s was not given to the learning algorithm, it was to be found by the algorithm itself. The initial weigth vector is set to $\vec{0}$ and the initial threshold vector is set to $(k^n, 2k^n, \ldots, sk^n)$ after s was found.

For nonpermutably homogeneous functions, the algorithm behaved as expected, that is, no learning is effected on these functions. For permutably homogeneous (thick or nonthick) functions the algorithm always terminated after learning the function with its unique linear extensions.

Next, we discuss an example of nonthick function which we have used in our experiment for $k = 4$ and $n = 3$. Consider the two-place function h shown in Figure 24.17. To obtain a three-place function f we project the values h (which will correspond to points in plane $x_3 = 0$) in the three planes $x_3 = 1, x_3 = 2$, and $x_3 = 3$. So, f is also a nonthick function as h. Now, to make it more difficult, to find one of its *good* linear extensions, we replace the value 3 that lies in plane $x_3 = 0$ by value 1, and also change the value 2 that lies in plane $x_3 = 3$ to 0. Here, the function f has no unique linear extension at all in any direction and thus the extended learning algorithm must combine the three constructed posets to search for a good linear extension. The algorithm did indeed, as we expected, find a chain poset which has the unique linear extension $(2, 0, 1, 3)$. The function has been learned successfully in 61 learning epochs. We also obtain same results when extending f to an $(n \geq 3)$-place functions.

Examples of permutably homogeneous thick functions are given by the following formula:

$$f(\vec{x}) = \left\lfloor \left(\sum_{i=1}^{n} \frac{1}{a_i} x_i \right) + n \right\rfloor \bmod k,$$

where $a_i = 2i + 1$. For example, we tested with the four-place four-valued logic function $f(\vec{x}) = \lfloor x_1/3 + x_2/5 + x_3/7 + x_4/9 + 4 \rfloor \bmod f4$. Clearly, such function is permutably homogeneous since it defines itself its separating hyperplanes and their number. It is easy to see that the function has three possible values, namely 0, 1, and 2 and thus there must be 2 separating hyperplanes, also, the three classes of input are separated in the order $(0, 1, 2)$. Therefore, we expect that our extended learning algorithm will find two separating hyperplanes and the output vector $(0, 1, 2)$. Indeed, the function was learned successfully in 946 learning epochs after the algorithm has found the output vector.

24.4 A Strip-Based Neural Network Growth Algorithm for Learning Multiple-Valued Functions

We consider the problem of synthesizing multiple-valued logic functions by neural networks. A GA which finds the longest strip in $V \subseteq K^n$ is described. A strip contains points located between two parallel hyperplanes. Repeated application of GA partitions the space V into certain number of strips, each of them corresponding to a hidden unit. We construct two neural networks based on these hidden units and show that they correctly compute the given but arbitrary multiple-valued function. Preliminary experimental results are presented and discussed.

24.4.1 Problem Statement

The problem we address in this section is that of learning multiple-valued logic functions using minimal neural networks composed of (n, k, s)-perceptrons. Multilayer feed-forward neural networks are in principle able to learn any arbitrary mapping, provided that enough hidden units are present [1]. For these networks, learning algorithms such as back-propagation [40] have been found to be computationally prohibitive. Also, the topology of the networks must be fixed before learning.

Till recently, architecture design is still very much a human expert's job. It depends heavily on the expert experience and a tedious trial and error process. There is no systematic way to design a near optimal architecture for a given task automatically. Research on destructive and constructive algorithms represents an effort toward the automatic design of architectures [41]. Design of the optimal architecture for a neural network can be formulated as a search problem in the architecture space where each point represents an architecture. Given some performance criteria about architectures, the performance level of all architectures forms a discrete surface in the space. The optimal architecture design is equivalent to finding the highest point on this surface.

A way to improve the performance of a neural network is to match its topology to a specific task (i.e., a set of input–output pairs) as closely as possible. However, the problem of deciding whether or not a given task can be performed by a given architecture is known to be NP-complete [42]. In addition, it has been shown in Reference 27 that the problem of finding the absolute minimal architecture for a given task is NP-hard. A third problem known to be NP-complete [19] and which also concerns us here is the following. Given two subsets $S_1, S_2 \subseteq R^n$ such that $|S_1 \cup S_2| = c$, determine subsets $E_1 \subseteq S_1$ and $E_2 \subseteq S_2$ such that $E_1 \cup E_2$ is of maximum cardinality, and for some $\vec{w} \in R^n$ and $t \in R$, $\vec{w}\vec{x} - t > 0$ for all $\vec{x} \in E_1$ and $\vec{w}\vec{x} - t \leq 0$ for all $\vec{x} \in E_2$. This problem is a generalized version of the *maximum cardinality problem* (i.e., find a linearly separable subset of maximum cardinality). When applied to multiple-valued logic, these three problems are also NP-complete since they are known to be NP-complete in the synthesis of two-valued logic functions, which are special case of multiple-valued logic functions.

Our approach to the problem's solution is discussed in Section 24.4.4. The learning method is based on the *general principle of partitioning algorithms* discussed in Section 24.4.3. A partitioning algorithm seeks to construct a minimal network by partitioning the input space into classes that are as large as possible. Each class of partition is then assigned to a new hidden unit. The connections and weights of the new units are determined appropriately in such a way that the constructed network will always give the correct answer for any input. Distinct partitioning algorithms differ in the way the input space is partitioned. Also, network topologies obtained from different partitioning algorithms may differ in the way new hidden units are connected.

24.4.2 Background and Motivations

Chan et al. [6] described three-valued logic networks called *neural logic networks,* which combine the strengths of neural networks and expert systems. Such networks are used to represent a set of nonrecursive propositional rules, and to infer the truth values of the unknown propositions by applying the common back-propagation training method. Tang et al. [10] proposed a multiple-valued logic network with functional completeness properties and learning capabilities. The multiple-valued logic network consists of layered arithmetic piecewise linear units. Since the arithmetic operations of the network are basically wired sums and piecewise linear operations, their implementation should be rather simple and straightforward. In Wang and shi [11], neural networks composed of a novel model of multiple-valued logic neuron suitable for representing arbitrary multiple-valued logic expressions are described. Obradović and Parberry [16] described algorithms for learning multiple-valued logic functions on either a single homogeneous $(n, k, k-1)$-perceptron or a depth-two network composed of k $(n, 2, 1)$-perceptrons in the hidden layer and one homogeneous $(k, k, k-1)$-perceptron in the output layer. Ngom et al. [34] introduced learning algorithms for permutably homogeneous (n, k, s)-perceptrons and proved them to be more powerful than the algorithms mentioned in Reference 16. These are a few mentioned methods for learning multiple-valued logic functions; the reader can refer to Reference 8 for a survey on multiple-valued logic neural networks. The minimal size of a neural network for learning a given but arbitrary multiple-valued logic function has not been studied in the literature. That is, the network is fixed in advance before learning and its size does not change during and after learning. As stated earlier, such networks may not be powerful enough for learning or may overfit the training data. In particular, there is no evidence of any function for which the size of the learning network is significantly smaller than the size of its sum-of-products or other standard gates implementations.

Techniques are known in the literature for learning multiclass functions (multiple-valued logic functions are special cases of multiclass functions). The most powerful one is the *error-correcting output codes* (or ECOC) approach which is a robust method for solving multiclass learning problems. Dieterich and Bakiri [43] has shown that ECOC learning provides a general purpose method for improving the performance of inductive learning programs on multiclass problems. Other simpler and less powerful methods use back-propagation (or any appropriate learning algorithm such as madaline or radial basis function, etc.) networks with multiple output units assigned to distinct classes. All these approaches use fixed-architecture networks for learning arbitrary multiclass functions and thus a given network may not be the optimal one for such functions.

24.4.3 Known Partitioning Methods

In Marchand et al. [44], given a function $f \in P_2^n$, the cube $\{0, 1\}^n$ is partitioned into regions by hyperplanes in such a way that f is constant in each region containing input vectors. The halfspaces defined by these hyperplanes correspond to threshold units in the hidden layer. The construction of the halfspaces implies that one can add an output unit in such a way that the resulting neural network indeed computes f. The hyperplanes are determined sequentially. First, an hyperplane is found such that f is constant on one of its sides. The points in the corresponding halfspace are removed and they continue with the remaining points in a similar manner, until the set of remaining points becomes empty. Thus, each halfspace is used to cut off a set of points with identical function values from the remaining set of points. Let the halfspaces constructed by their algorithm be $H_1, \ldots, H_r$ and define u_i to be 0 (or 1) if f is 0 (or 1) on the region cut off by H_i for $1 \leq i \leq r$. Then adding an output unit with threshold 0 and weight $u_i 2^{r-i}$ for the edge leaving the hidden unit corresponding to H_i, they get a neural network that computes f. They assume that linear threshold units produce an output 0 or 1. In a restricted version of their approach called *regular partitioning*, Ruján and Marchand [45], the hyperplanes do not intersect. This description gives the general principle of partitioning algorithms. Particular implementations depend on the way the next hyperplane is selected and the way new hidden units are connected (in term of network weights and topology). Experiments indicate that partitioning algorithms are successful in the sense that they efficiently construct (near) minimal neural networks [44–47].

Another way to view this process, without any reference to neural networks, is to consider a sequence of halfspaces $H_1, \ldots, H_r$. For each halfspace H_i we specify the value u_i of f for those points in H_i that are not contained in any of the *previous* halfspaces. Thus, for every input vector $\vec{x} \in \{0,1\}^n$ it holds that $f(\vec{x}) = u_i$, where i is the *smallest* index such that $\vec{x} \in H_i$. It is assumed that H_r contains $\{0,1\}^n$, thus i is always defined. This model is called a *linear decision list* [47,48].

More formally, Rivest [48] defines a linear decision list in the following way. A linear test L over the variable $\vec{x} = (x_1, \ldots, x_n) \in \{0,1\}^n$ is of the form $\sum_{i=1}^{n} w_i x_i \geq t$, where $w_1, \ldots, w_n \in R^n$ are the weights and $t \in R$ is a threshold. A linear decision list D over $\vec{x}$ is a sequence $(L_1, u_1), \ldots, (L_r, u_r)$ where L_i is a linear test and u_i is 0 or 1 for $1 \leq i \leq r$. It is assumed that L_r, the last linear test is true for all input vectors $\vec{x}$. The length of D is r. The two-valued function f_D computed by D assigns to every $\vec{x}$ the value u_i, where L_i is the *first* linear test in the list that is satisfied by $\vec{x}$. Every two-valued function $f \in P_2^n$ can be computed by some linear decision list. For instance, a disjunctive normal form with m terms can be represented by a linear decision list of length $m + 1$.

Many heuristics such as Fahlman's *cascade correlation algorithm* [49], Frean's *upstart algorithm* [50], Sethi's *entropy nets* [51], Sirat's *neural trees* [52], Mezard's *tiling algorithm* [53], Barkema's *patch algorithm* [54], Frattale-Mascioli's *oil-spot algorithm* [55], Young's *carve algorithm* [56], *regular partitioning* [45], and other partitioning algorithms [44,47,57–60], are proposed as approaches for building networks which are (near) minimal for a given arbitrary task. These heuristics are known as *constructive* or *growth* algorithms since they all construct a network starting from a fixed small number of units.

Partitioning techniques such as cascade correlation [49], upstart algorithm [50], and entropy nets [51] apply only for functions with Boolean-valued outputs. Moreover, most of the growth algoritms described in literature (see, for instance, References 44, 45, 53–55, 58 and 59) are only applicable to problems with Boolean-valued inputs. Very few constructive methods deal with multiclass problems (k-valued functions are multiclass functions). For instance, the carve algorithm [56] and Marchand's neural decision list Reference [47] both apply to functions $f : R^n \mapsto K$.

The techniques in References 47 and 56 seek to identify the largest subset of input vectors of same class that is separable from vectors of other classes. The hyperplane determined by the maximum separable subset is then assigned to a newly created $(n, 2, 1)$-perceptron. A drawback of Marchand's neural decision list (NDL) [47] is that, since it employs linear programming to determine whether specific subsets of the input vectors are linearly separable, the number of possible subsets of points that could be considered for linear separability is exponential in the size of the inputs set. In order to circumvent this problem, Marchand and Golea [47] worked only with a specific class of functions called *halfspace intersections.* Young's CARVE [56], which is an extension of Marchand et al. [44] to multiclass functions with real-valued inputs, avoids the issue of testing for linear separability of subsets of points by directly searching for hyperplanes that separate sets of points of one class only. Let S_i be a set of points of class i and H_i (the hull set for S_i) be the set of all points in the training set except those in S_i, $i \in K$. Clearly, points outside the convex hull formed by H_i are all of the same class i and thus, only points outside the convex hull can actually be separated by a hyperplane boundary from the hull set. So to find the maximum separable subset, CARVE considers only hyperplanes that touch the boundary (i.e., $(n-1)$-dimensional faces or vertices) of the convex hull. The hyperplane with the largest set of points in its open halfspace outside the convex hull is the maximum separable subset. The algorithm is a simple hill-climbing technique that searches for a (near) optimal hyperplane by partially traversing (or covering) the convex hull (i.e., the boundary of the hull set) starting from a randomly selected convex hull vertex and a hyperplane that passes through that vertex. This is repeated a fixed number of times n_v and each time, the initial hyperplane is randomly rotated a fixed number of times n_r around the boundary of the hull set. The set of points encountered during each rotation is determined; these are potential solutions. The main difficulty with CARVE is that the likelihood to find the maximum separable subset depends on the parameters n_v and n_r. For larger values of the product $n_v n_r$, the more likely is one to obtain a (near) optimal hyperplane since large section of the hull set boundary will be traversed (it should be noted that the number of facets and vertices of the convex hull may be exponential in the dimension of the input space). The optimal values for n_v and n_r depend on the actual problem and the user must find such values by trial and error. In the end, the size of the network obtained by CARVE depends on n_v and n_r.

In this section we introduce a method of partitioning a set $V \subseteq K^n$ using GA [61] to grow a multiple-valued logic neural network for learning a function. In our own approach, the maximum separable subset is treated as a special case of the longest strip (as will be discussed later). We introduce $(n, k+1, 2)$-perceptrons which will freely reduce the network size for given arbitrary k-valued functions.

Both CARVE and NDL use local search to search for good solutions. They extensively search some parts of the space (around faces of a convex hull as in CARVE, or, around neighborhood of separating hyperplanes as in NDL), while leaving some other parts of the search space untouched. On the other hand, GA (in our implementation) performs a global search; it treats all parts of the space equally, due to its implicit parallelism.

24.4.4 Longest Strip-Based Growth Algorithm

In this section, we describe a novel neural network growth algorithm based on the search for the longest strip rather than the maximum separable subset. Our technique can be applied to various kinds of functions. That is, it is capable of handling (1) real-valued and k-valued inputs ($k \geq 2$) and (2) two-class problems as well as multiclass problems. In particular, we apply our growth algorithm to the synthesis of multiple-valued logic functions, that is, functions of the form $f\colon K^n \mapsto K$.

A *strip* is a set of points between two parallel hyperplanes which have the same value. The *longest strip* is the strip with the maximum possible cardinality. A *maximum separable subset* is a set of points having equal values with the maximum possible cardinality that can be separated from all other points by exactly one hyperplane. Examples of longest strip and maximum separable subset are shown, respectively, in Figure 24.18 and Figure 24.19.

We describe a search method for finding the longest strip as well as the maximum separable subset of a currently given set of training examples, namely GA. The maximum separable subset problem is a special case of the longest strip problem. Indeed, they both share the same problem representation used by GA, but differ only by their respective objective functions and constructed network architectures. In the next paragraph we briefly describe our main growth algorithm which constructs a network by

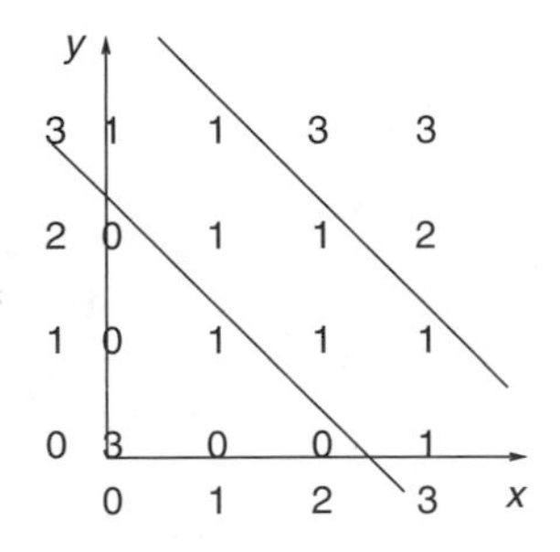

FIGURE 24.18 Example of longest strip for $k = 4$ and $n = 2$. (From A. Ngom, I. Stojmenović, and V. Milutinović. *IEEE Transactions on Neural Networks*, 12, 212–227, 2001. With permission.)

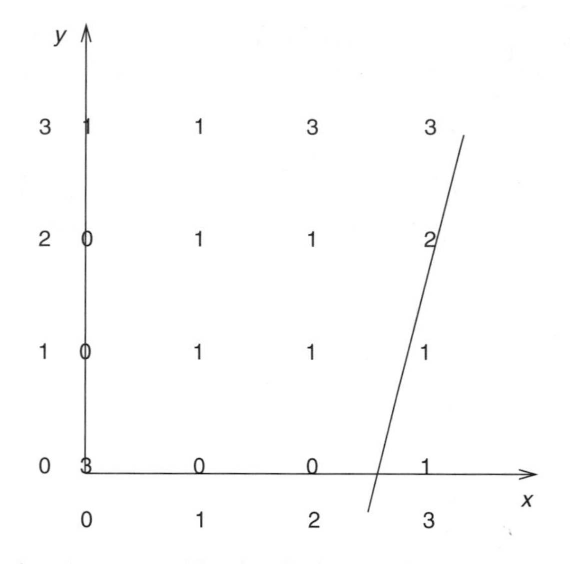

FIGURE 24.19 Example of maximum separable subset for $k = 4$ and $n = 2$. (From A. Ngom, I. Stojmenović, and V. Milutinović. *IEEE Transactions on Neural Networks*, 12, 212–227, 2001. With permission.)

```
Procedure A-BasedSynthesis(n, k, f);
    r := 1;
    S_r := V;
    Repeat
        Apply GA to find a subset G_r such that |G_r| ≈ |A| of S_r;
        Create a new hidden unit U_r with respect to A;
        S_{r+1} := S_r − G_r;
        r := r + 1;
    Until S_r = ∅;
    Construct a network with the r hidden units on the first layer;
```

FIGURE 24.20 A-based synthesis algorithm.

removal of points in a predefined objective subset $A \subseteq V$. A is either the longest strip or the maximum separable subset in V. Our main objective is, using GA, to obtain a subset $G \subseteq V$ such that $|G|$ is as close as possible to $|A|$ if not equal to $|A|$ (of course we have $|G| \leq |A|$). Our maximum separable subset problem approach is an extension of the sequential learning algorithm described in Reference 44 to multiclass functions f: $R^n \mapsto K$. The growth algorithm of Reference 44 is based on the perceptron learning algorithm (more specifically, on the *pocket* algorithm) and its performance was hampered by the fact that the pocket algorithm does not converge in the case where the points are not linearly separable. In our own approach, however, we use an evolutionary approach to find optimal subsets.

In our particular implementation of partitioning algorithm (see Figure 24.20), GA is used to obtain subsequent halfspaces delimited by either one or two hyperplanes (depending on the predefined objective subset A). To each halfspace we assign a hidden unit that correctly classifies all elements of it. Let the objective subset A be the longest strip in the given training set. Our growth algorithm begins with an empty first hidden layer into which new $(n, k+1, 2)$-perceptrons are inserted one after another until no more insertion is possible. An $(n, k+1, 2)$-perceptron implements two parallel hyperplanes in the input domain and the aim is to find two parallel hyperplanes that define a strip G such that $|G|$ is as close as possible to $|A|$. The strip G is then removed from the training set. The next $(n, k+1, 2)$-perceptron to be added to the network aims to separate another (near) longest strip G, but now only from the reduced training set. Once a (near) longest strip for this unit is found, the unit is added to the layer and the strip is removed from the training set. The construction of the first hidden layer continues with each subsequent unit separating a strip from the remaining training examples. The first hidden layer is complete when only points of one class remains in the training set. Once the first hidden layer is complete, the remaining weights, layers and units of the networks are determined to complete the network construction (the details of the network architecture are described in Section 24.4.6).

Our principal objective in this section is to synthesize any arbitrary k-valued logic function by a neural network constructed by our growth algorithms when (portion of) the function is given. We obtain networks that either exactly or approximately implement k-valued logic functions.

24.4.5 Determining Longest Strips by GA

We use an evolutionary method to find the set of partitioning hyperplanes. Holland [62] first proposed GAs in the early 1970s as computer program to mimic the evolutionary processes in nature. GAs manipulate a population of potential solutions to an optimization (or search) problem. Specifically, they operate on encoded representations of the solutions, equivalent to the genetic material of individuals in nature, and not directly on the solutions themselves. Holland's GA encodes the solutions as binary *chromosome* (strings of bits). As in nature, *selection* provides the necessary driving mechanism for better solutions to survive. Each solution is associated with a *fitness value* that reflects how good or bad it is, compared with other solutions in the population. The higher the fitness value of an individual, the higher its chances of survival and reproduction and the larger its representation in the subsequent generations. Recombination of genetic material in GAs is simulated through a *crossover* mechanism that exchanges portions between two chromosomes. Another operation, *mutation*, causes sporadic and random alterations of the chromosomes. Mutation too has a direct analogy from nature and plays the role of regenerating lost genetic material and thus reopening the search.

24.4.5.1 Problem Representation

Fundamental to the GA structure is the encoding mechanism for representing the problem's variables. For the problem of determining A, the search space is the *power set* of the training set (i.e., the set of all subsets of K^n). Each element of the search space is a potential solution and must be represented in such a way that meaningful and suitable genetic operators can be designed for (and applied to) it.

More formally, our population consists of chromosomes which encode the potential solutions of the problem. A potential solution is a subset $G \subseteq K^n$ and the best solution is one whose size is closest to $|A|$. Given a weight vector $\vec{w} = (w_1, \ldots, w_n) \in R^n$ we can find the strip (or separable subset) of maximum cardinality that is associated with $\vec{w}$ (see Section 24.4.5.2). For instance, the strip shown in Figure 24.18 can be obtained given the vector $\vec{w} = (1, 1)$ and the function's table (moreover in this example, the obtained strip is the absolute solution). Given a training set, then searching for a subset whose size is as close to $|A|$ as possible is equivalent to searching the weight vectors space (i.e., R^n) for a vector that generates such subset. In our problem representation, a chromosome will be a weight vector $\vec{w} \in R^n$ (as in Reference 63) since such a vector clearly encodes (i.e., represents) a potential solution to the problem. Each chromosome will uniquely determine a partition of $V \subseteq K^n$ into $s + 1$ classes with s parallel hyperplanes (for some s) and the best chromosome is the one that maximizes the number of points between a pair of parallel hyperplanes. To determine how good a solution the GA needs is an objective function to evaluate each chromosome $\vec{w}$.

We initialize the population with random real-coded chromosomes $\vec{w} \in R^n$ whose coordinates are random real numbers taken from the interval $[-1, 1]$. Each initial chromosome is then normalized to a unit vector. Another method we used for the initialization of the population is to set $w_i = \cos \alpha_i$ (for $1 \leq i \leq n$) for each vector $\vec{w}$, where α_i is a random number in the interval $[-\pi/2, \pi/2]$. Initial population should consist of random unit hyperplanes $\vec{w}$. Starting from the initial population, we apply genetic operators such as selection, crossover, and mutation to create the next generation of chromosomes.

This generational cycle is repeated until a predefined maximum number of generations is reached. Subset G will be the best solution generated so far in the population.

Our main objective here is, for a given function f, to obtain a chromosome $\vec{w}$ which generates G such that $|G| \approx |A|$. Once such $\vec{w}$ is found we create a hidden unit to be inserted in the neural network. We then eliminate all points $\vec{x}$ in G and again apply the GA on the remaining points. The algorithm terminates as soon as there are no points left. The created hidden units will then be collected to construct a feed-forward network. The parameters (weight, threshold, and output vectors) of the hidden units and the topology of the network will be discussed later.

24.4.5.2 Fitness Function

The objective function, the function to optimize, provides the mechanism for evaluating each chromosome.

Let $S \subseteq K^n$ be the set of remaining points. Initially, $S = K^n$. To compute the longest strip generated by $\vec{w}$, we calculate for every $\vec{x} \in S$ the value $\vec{w}\vec{x}$ and construct a sorted list of records of the form $(\vec{w}\vec{x}, f(\vec{x}))$. The list is sorted using $\vec{w}\vec{x}$ as primary key and $f(\vec{x})$ as secondary key. Let these records be sorted as follows: $\vec{x}_1, \ldots, \vec{x}_{|S|}$, or more precisely, $P_i = (\vec{w}\vec{x}_i, f(\vec{x}_i)), 1 \leq i \leq |S|$, where $\vec{w}\vec{x}_i \leq \cdots \leq \vec{w}\vec{x}_{|S|}$. A strip in S is a sequence $T_{\vec{w}}^{(f(\vec{x}_i))} = P_i P_{i+1} \ldots P_{i+j}$ such that

1. $f(\vec{x}_i) = f(\vec{x}_{i+1}) = \cdots = f(\vec{x}_{i+j})$
2. $\vec{w}\vec{x}_{i-1} \neq \vec{w}\vec{x}_i$ and $\vec{w}\vec{x}_{i+j} \neq \vec{w}\vec{x}_{i+j+1}$

with $1 \leq i \leq |S|$ and $0 \leq j \leq |S| - i$. The length of the strip is $j - i + 1$ and $f(\vec{x}_i)$ is the value of the strip. For example, in Figure 24.18 we have $\vec{w} = (1, 1)$ and

$$
\begin{array}{llll}
P_1 = (0,3) & P_2 = (1,0) & P_3 = (1,0) & P_4 = (2,0) \\
P_5 = (2,0) & P_6 = (2,1) & P_7 = (3,1) & P_8 = (3,1) \\
P_9 = (3,1) & P_{10} = (3,1) & P_{11} = (4,1) & P_{12} = (4,1) \\
P_{13} = (4,1) & P_{14} = (5,2) & P_{15} = (5,3) & P_{16} = (6,3)
\end{array}
$$

which gives the longest strip $T_{(1,1)}^{(1)} = P_7 P_8 P_9 P_{10} P_{11} P_{12} P_{13}$ generated by $\vec{w}$.

Given a set of points $S \subseteq K^n$ and a function f over S, let $P_1 \cdots P_{j_1}$ and $P_{j_2} \cdots P_{|S|}$ $(1 \leq j_1 < j_2 \leq |S|)$ be, respectively, the leftmost and rightmost strips generated by $\vec{w}$, with strip values c_1 and c_2. We denote by $L(S, \vec{w})$ the length of the longest strip generated by $\vec{w}$ and denote by $M(S, \vec{w})$ the length of the maximum between the leftmost and rightmost strips, on set S and function f. To evaluate how good is $\vec{w}$ we propose the following fitness function with respect to the definition of A:

- $A =$ longest strip

$$
\text{Fitness1}_L(\vec{w}) = \frac{L(S, \vec{w})}{|S|} \tag{24.9}
$$

- $A =$ maximum separable subset

$$
\text{Fitness1}_M(\vec{w}) = \frac{M(S, \vec{w})}{|S|} \tag{24.10}
$$

Let $S_c = \{\vec{x} \in S | f(\vec{x}) = c, c \in K\}$, that is, the set of points of value c. An alternative objective is to select a strip of value c, that is, $T_{\vec{w}}^{(c)}$, which maximizes $|T_{\vec{w}}^{(c)}|/|S_c|$, where $|T_{\vec{w}}^{(c)}|$ denotes the length of $T_{\vec{w}}^{(c)}$. That is, as in References 44 and 56, the selection criterion chooses the strip that constitutes the largest proportion of a class of points that can be separated. We denote by $L(S_c, \vec{w})$ the length of the largest strip proportion of value c and denote by $L(S_{c_1}, \vec{w})$ and $L(S_{c_2}, \vec{w})$ the lengths of the leftmost and rightmost strips, respectively, on set S and function f. Our alternative fitness function with respect to A is

- $A =$ largest strip proportion

$$\text{Fitness2}_L(\vec{w}) = \max\left(\frac{L(S_c, \vec{w})}{|S_c|}\right) \tag{24.11}$$

- $A =$ largest separable proportion

$$\text{Fitness2}_M(\vec{w}) = \max\left(\frac{L(S_{c_1}, \vec{w})}{|S_{c_1}|}, \frac{L(S_{c_2}, \vec{w})}{|S_{c_2}|}\right), \tag{24.12}$$

where the maximum is over S for every class c presents in the training set. As it was stated in Reference 56, choosing the largest proportion rather than the largest set do have some advantage for some functions $f: S \mapsto K$. For, if the number of points of a class v_1 is small and a strip $T_{\vec{w}_1}^{(v_1)}$ constituting the whole class S_{c_1} is found, it may be that a longer strip $T_{\vec{w}_2}^{(v_2)}$ with a smaller proportion $L(S_{v_2}, \vec{w})/|S_{v_2}|$ can be found. However, it is preferable to select $T_{\vec{w}_1}^{(v_1)}$ because this removes the entire set S_{v_1} from S and brings the neural network construction closer to the hidden layer termination criteria of having only points of one class remaining in the training set. This is illustrated in Figure 24.21 where $f \in P_3^2$ is a random function to which both fitness selection criteria were applied.

As seen in Figure 24.21, it is impossible to obtain a network with exactly three hidden units (which is the absolute minimum here) when using Fitness1_L. The number in circles indicates the order in which a generated strip is assigned to hidden units. So in Figure 24.21[b]) a fourth hidden unit is needed for the last remaining point of value 2. In Figure 24.21(a), Fitness2_L removes the set S_1 immediately after removing S_0 (unlike in Figure 24.2[b] where a proper subset of S_2 is removed after S_0), hence one more unit is needed to remove S_2.

A note on the time complexity of the evaluation function. For a given $\vec{w}$, both fitness functions take $n|S|$ steps to compute the $\vec{w}\vec{x}$'s, $n|S| \log |S|$ steps to sort them and at most $|S|$ steps to compute $L(S, \vec{w})$ or $M(S, \vec{w})$. Therefore, the evaluation of Fitness(1 or 2)$_{(L \text{ or } M)}(\vec{w})$ has a time complexity of $O(n|S| \log |S|)$.

In addition, crossover and mutation operations below take $O(n)$ steps each and the initialization of the population takes $O(pnk^n \log k^n)$ steps (p is the number of chromosomes and all initial chromosomes are evaluated for their fitness). Thus, the evaluation of Fitness$(\vec{w})$ is the most expensive operation in our GA. Let g be the number of generations, then at each new generation $p/2$ new chromosomes are evaluated for their fitnesses and hence, our GA has a time complexity of $O(gpn|S| \log |S|) \approx O(gpn^2 k^n \log k)$.

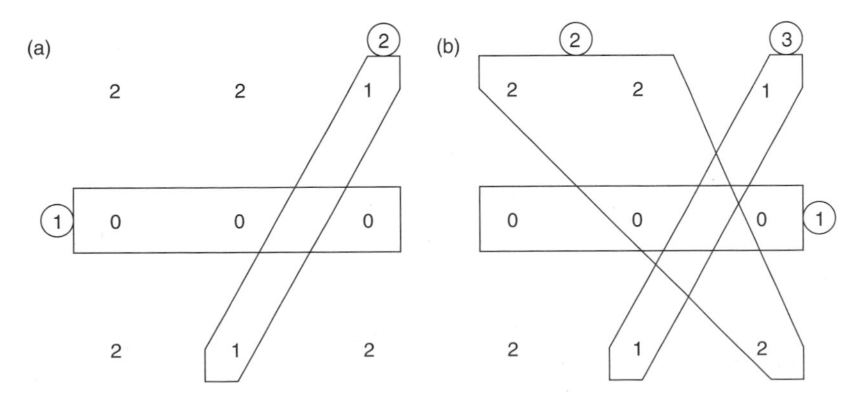

FIGURE 24.21 Behaviors of (a) Fitness2_L and (b) Fitness1_L on some $f \in P_3^2$. (From A. Ngom, I. Stojmenović, and V. Milutinović. *IEEE Transactions on Neural Networks*, 12, 212–227, 2001. With permission.)

24.4.5.3 Crossover

Crossover is the GA's crucial operation. Pairs of randomly selected chromosomes are subjected to crossover. For our problem representation we propose the following mixed crossover method for real-coded chromosomes as described in Reference 64. Let $\vec{p}_1$ and $\vec{p}_2$ be two *unit* vectors to be crossed over and let $\vec{c}_1$ and $\vec{c}_2$ be the result of their crossing. Vectors $\vec{c}_1$ and $\vec{c}_2$ are obtained using, with equal probability, two of the following three crossovers operations:

$$\vec{c}_i = \vec{p}_1 + \vec{p}_2, s \tag{24.13}$$

$$\vec{c}_i = \vec{p}_1 - \vec{p}_2, \tag{24.14}$$

$$c_{i_j} = \begin{cases} p_{1_j} & \text{if random()} \le 0.5, \\ p_{2_j} & \text{otherwise.} \end{cases} \tag{24.15}$$

Crossover in Equation (24.13) is simply the addition of two parents and the child is assured to be their exact middle vector since the parents are unit vectors. Crossover in Equation (24.14) is the substraction of two parents and the child is the vector orthogonal to the sum of its parents. Crossover in Equation (24.15) is a uniform crossover of two parents, that is, at coordinate i each parent has 50% chances to be selected as c_{i_j} $(1 \le j \le n)$. For a more efficient search, $\vec{c}_1$ and $\vec{c}_2$ must not be obtained from the same crossover operator. That is, if $\vec{c}_1$ is obtained using Equation (24.13) then $\vec{c}_2$ must be generated from either Equations (24.14) or (24.15); this helps to maintain a certain level of diversity among chromosomes in the population. Also, crossover is applied only if a randomly generated number in the range 0 to 1 is less than or equal to the crossover probability p_{cros} (in large population, p_{cros} gives the fraction of chromosomes actually crossed).

We must emphasize that each chromosome is a unit vector at any moment in the population. Thus, the initial random vectors are all normalized and the childs are also normalized to unit vectors after any crossover or mutation operation.

24.4.5.4 Mutation

After crossover, chromosomes are subjected to random mutations. We propose three methods of *coordinate-wise* mutations as described in Reference 64. They correspond to bitwise mutation for binary chromosomes. Let $\vec{p}$ be a *unit* vector to be mutated to a child $\vec{c}$.

Random replacement. With some probability of mutation, each coordinate p_i $(1 \le i \le n)$ of a parent $\vec{p}$ may be replaced in the following way:

$$c_i = \text{random}[-1, 1], \tag{24.16}$$

where random$[-1, 1]$ returns a random real number in the interval $[-1, 1]$ with uniform probability.

Orthogonal replacement. With some probability of mutation, each coordinate p_i $(1 \le i \le n)$ of a parent $\vec{p}$ may be replaced in the following way:

$$c_i = \pm\sqrt{1 - p_i^2}. \tag{24.17}$$

Neighborhood replacement. With some probability of mutation, each coordinate p_i $(1 \le i \le n)$ of a parent $\vec{p}$ may be replaced in the following way:

$$c_i = p_i \pm \frac{m}{k^n}, \tag{24.18}$$

where $m \leq k$ is a random constant. Unlike the two previous methods of mutation, this method slightly rotates the current hyperplane $\vec{w}$ to a neighboring one.

Just as p_{cros} controls the probability of crossover, the mutation rate p_{muta} gives the probability for a given coordinate to be mutated. For a vector to be mutated, one of the three mutation operators is selected with probability $\frac{1}{3}$.

Here, we treat mutation only as a secondary operator with the role of restoring lost genetic material or generating completely new genetic material which may be probably (near) optimal. Mutation is not a conservative operator, it is highly disruptive. Therefore, we must set $p_{\text{muta}} \leq 0.1$.

24.4.6 Constructing the Neural Network

In the rth iteration of the A-based synthesis algorithm, GA will find a chromosome $\vec{w}_r$ which generates a subset G_r. Subset G_r is the longest strip (or maximum separable subset) found in the population. The optimization ability of GA makes it possible to attain a solution as close (in size) to $|A|$ as possible, if not equal. The ability of GA to produce $|A|$ depends on its many control parameters and the complexity of the tasks to learn.

1. *Longest strip-based network.* At every iteration r of the A-based synthesis algorithm, GA finds a chromosome $\vec{w}_r$ which produces the longest strip $G_r = P_i \cdots P_{i+j}$, where $1 \leq i \leq |S_r|, 0 \leq j \leq |S_r| - i$ and strip value $v_r = f(\vec{x}_i)$. Let $u_r = v_r + 1$. Then, we create an $(n, k+1, 2)$-perceptron (hidden unit U_r) whose weight vector is $\vec{w}_r$, threshold vector is $\vec{t}_r = (\vec{w}_r\vec{x}_i, \vec{w}_r\vec{x}_{i+j+1})$, and output vector is $\vec{o}_r = (0, u_r, 0)$. In other words, the perceptron has a transfer function of the form $g_{k+1,2}^{(\vec{w}_r\vec{x}_i, \vec{w}_r\vec{x}_{i+j+1}),(0,u_r,0)} : R \mapsto \{0, u_r\}$ (i.e., a $(k+1)$-valued two-threshold function). The $(n, k+1, 2)$-perceptron will output the value u_r for all points $\vec{x} \in G_r$ and will output the value 0 for all points $\vec{x} \in S_r - G_r$.

In order to achieve a good accuracy on the testing set, that is, a good generalization ability of our algorithm when approximating a function, we set the threshold vector to $\vec{t}_r = (\vec{w}_r\vec{x}_i - \tau_1, \vec{w}_r\vec{x}_{i+j+1} + \tau_2)$. Thus, test points of value $v_r = f(\vec{x}_i)$ which are outside but close to the strip — that is, test points that lie between $\vec{w}_r\vec{x}_{i-1}$ and $\vec{w}_r\vec{x}_i$ and between $\vec{w}_r\vec{x}_{i+j+1}$ and $\vec{w}_r\vec{x}_{i+j+2}$ — will be correctly classified by unit U_r, since they are now spanned by U_r. The offsets τ_1 and τ_2 are given by

$$\tau_1 = \frac{|(\vec{w}_r\vec{x}_i - \vec{w}_r\vec{x}_{i-1})|}{2} \quad \text{and} \quad \tau_2 = \frac{|(\vec{w}_r\vec{x}_{i+j+2} - \vec{w}_r\vec{x}_{i+j+1})|}{2}. \tag{24.19}$$

2. *Maximum separable subset-based network.* At every iteration r of the A-based synthesis algorithm, GA finds a chromosome $\vec{w}_r$ which produces a maximum separable subset $G_r = P_1 \cdots P_{j_1}$, or $G_r = P_{j_2} \cdots P_{|S_r|}$ (i.e., the maximum between the leftmost and the rightmost strips), where $1 \leq j_1 < j_2 \leq |S_r|$ and strip value $v_r = f(\vec{x}_1)$ or $f(\vec{x}_{j_2})$. Let $u_r = v_r + 1$. We then create an $(n, k+1, 1)$-perceptron (hidden unit U_r) whose weight vector is $\vec{w}_r$, threshold vector is $\vec{t}_r = (\vec{w}_r\vec{x}_{j_1+1} + \tau_2)$ if G_r is the leftmost strip, or $\vec{t}_r = (\vec{w}_r\vec{x}_{j_2} - \tau_1)$ if G_r is the rightmost strip, and output vector is $\vec{o}_r = (u_r, 0)$ or $\vec{o}_r = (0, u_r)$ depending on G_r. In other words, the perceptron has a transfer function of the form $g_{k+1,1}^{(\vec{w}_r\vec{x}_{j_1+1}+\tau_2),(u_r,0)} : R \mapsto \{0, u_r\}$ or $g_{k+1,1}^{(\vec{w}_r\vec{x}_{j_2}-\tau_1),(0,u_r)} : R \mapsto \{0, u_r\}$ (i.e., a $(k+1)$-valued one-threshold function). The $(n, k+1, 1)$-perceptron will output the value u_r for all points $\vec{x} \in G_r$ and will output the value 0 for all points $\vec{x} \in S_r - G_r$. Offsets τ_1 and τ_2 are determined as earlier.

After defining all units $U_1, \ldots, U_r$ (where r is the number of runs of the A-based synthesis algorithm), the next step is to construct a feed-forward multilayer neural network. We propose two network topologies.

24.4.6.1 Three Hidden Layers and $r + k + 2$ Units Architecture

The network in Figure 24.22 (which shows the case for strip-based method) has three hidden layers and $r + k + 2$ neurons. Hidden layer 1 contains the units (the U_is) obtained by the GA. Each unit is connected to the inputs and their parameters (weight, threshold, and output vectors) are defined as

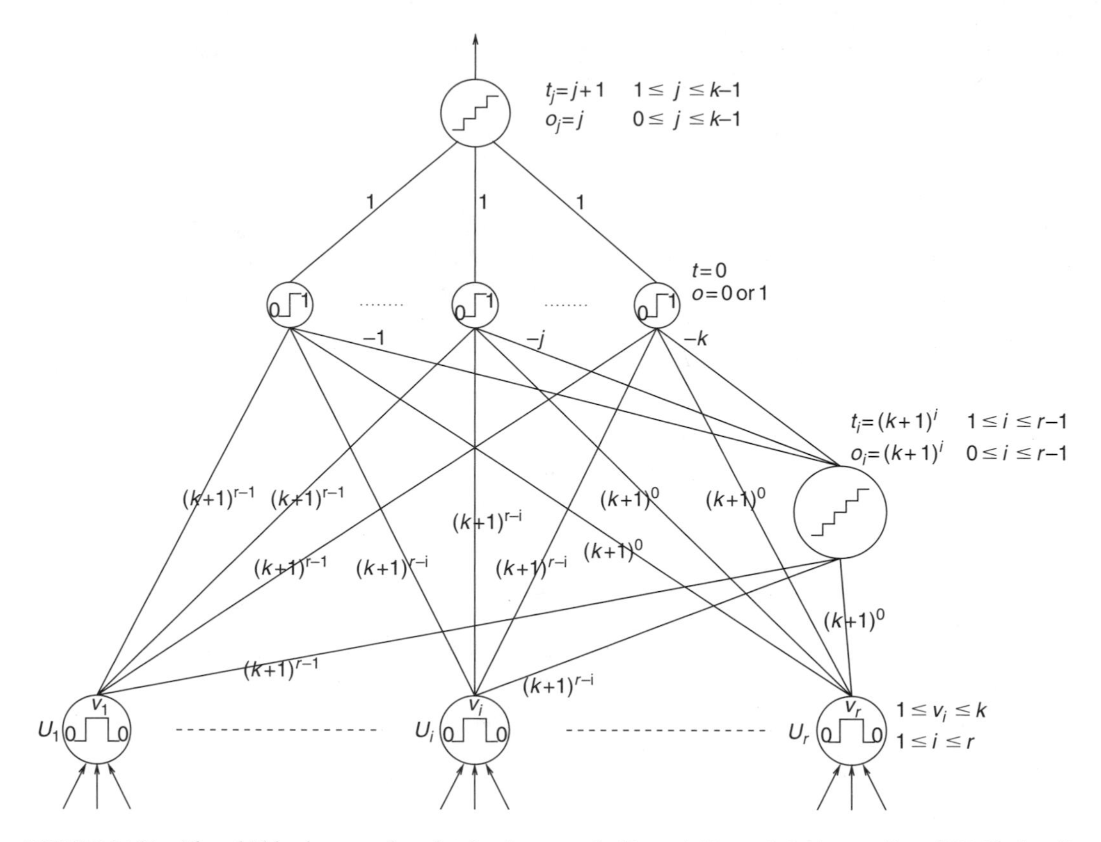

FIGURE 24.22 Three hidden layers and $r+k+2$ units network. (From A. Ngom, I. Stojmenović, and V. Milutinović. *IEEE Transactions on Neural Networks*, 12, 212–227, 2001. With permission.)

described previously. So the units in this layer are all either $(n, k+1, 2)$-perceptrons or $(n, k+1, 1)$-perceptrons, depending on the definition of A, and there are r such units (Figure 24.22 shows the case $A =$ longest strip).

Hidden layer 2 has only one unit which is a $(r, (k+1)^{r-1}+1, r-1)$-perceptron. Its weight vector $\vec{w} = ((k+1)^{r-1}, (k+1)^{r-2}, \dots, (k+1)^0)$, that is, $w_i = (k+1)^{r-i}$ for $1 \le i \le r$; its threshold vector $\vec{t} = ((k+1)^1, (k+1)^2, \dots, (k+1)^{r-1})$, that is, $t_i = (k+1)^i$ for $1 \le i \le r-1$; and its output vector $\vec{o} = ((k+1)^0, (k+1)^1, \dots, (k+1)^{r-1})$, that is, $o_i = (k+1)^i$ for $0 \le i \le r-1$. All units of layer 1 are connected to this unit and the connection weight vector is $\vec{w}$.

Hidden layer 3 contains k units. Each unit of layer 1 and 2 is connected to every unit in this layer. Each unit is an ordinary linear threshold element (thus $\vec{o} = (0, 1)$) and the connection weight vector from layer 1 to that unit is the same as the connection weight vector from layer 1 to the unit at layer 2. The connection weight $w_{i,r+1}$ $(1 \le i \le k)$ from layer 2 to the ith unit in layer 3 is $-i$. The threshold of units in layer 3 are all set to 0.

The output layer has one unit which is a $(k, k, k-1)$-perceptron whose threshold vector $\vec{t} = (2, \dots, k)$, that is, $t_i = i+1$ for $1 \le i \le k-1$, and output vector $\vec{o} = (0, \dots, k-1)$, that is, $o_i = i$ for $0 \le i \le k-1$ (or equivalently, $o_i = t_i - 1$). The connection weight from a unit in layer 3 to the output unit is 1.

24.4.6.2 One Hidden Layer and $r+1$ Units Architecture

The network in Figure 24.23 is equivalent to Marchand's construction [44] but generalized to multiclass functions. It has one hidden layer and $r+1$ neurons. Hidden layer 1 is same as in Figure 24.22 (Figure 24.23 shows the case of maximum separable subsets).

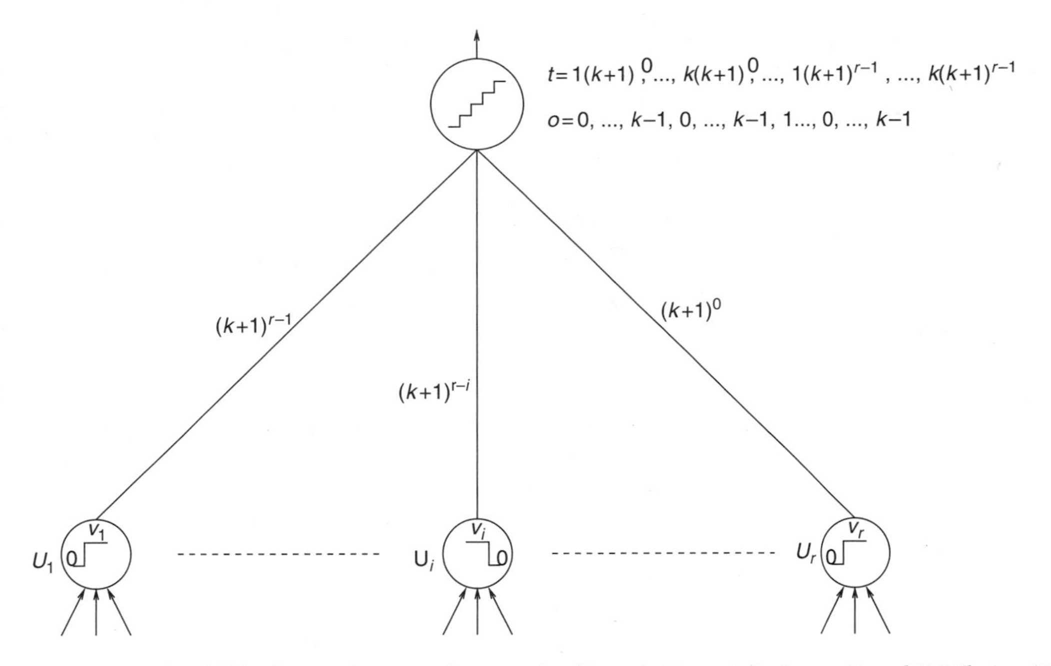

FIGURE 24.23 One hidden layer and $r+1$ units networks. (From A. Ngom, I. Stojmenović, and V. Milutinović. *IEEE Transactions on Neural Networks*, 12, 212–227, 2001. With permission.)

TABLE 24.1 Network Complexities

Networks	#Layers	#Nodes	#Thresholds	#Connections
SEPAR^{d2}	2	$r+1$	$(k+1)r$	$(n+1)r$
STRIP^{d2}	2	$r+1$	$(k+2)r$	$(n+1)r$
CARVE [56]	2	$r+k$	$r+k$	$(n+k)r$
SEPAR^{d4}	4	$r+k+2$	$2r+2k-2$	$(n+k+1)r+2k$
STRIP^{d4}	4	$r+k+2$	$3r+2k-2$	$(n+k+1)r+2k$
NDL [47]	$r+1$	$r+k$	$r+k$	$\frac{1}{2}r^2+(n+k-\frac{3}{2})r$

Source: From A. Ngom, I. Stojmenović, and V. Milutinović. *IEEE Transactions on Neural Networks*, 12, 212–227, 2001. With permission.

The output layer has only one unit which is a $(r, k(k+1)^{r-1}+1, kr)$-perceptron. Its weight vector $\vec{w} = ((k+1)^{r-1}, (k+1)^{r-2}, \ldots, (k+1)^0)$, that is $w_i = (k+1)^{r-i}$ for $1 \leq i \leq r$; its threshold vector $\vec{t} = (1(k+1)^0, \ldots, k(k+1)^0, 1(k+1)^1, \ldots, k(k+1)^1, \ldots, 1(k+1)^{r-1}, \ldots, k(k+1)^{r-1})$; and output vector $\vec{o} = (0, \ldots, k-1, 0, \ldots, k-1, \ldots, 0, \ldots, k-1)$. All units of layer 1 are connected to this unit and the connection weight vector is $\vec{w}$.

24.4.6.3 Networks Complexities

Table 24.1 shows the complexity of our constructed networks (given $f\colon R^n \mapsto K$) along with other networks that deal with multiclass functions, namely the NDL network of Reference [47] and the CARVE network of Reference [56]. In the table, r is the number of created hidden units and #Layers does not include the input layer. Also, STRIP (respectively, SEPAR) are our networks based on strips (respectively, maximum separable subsets) and the superscript $d4$ (or $d2$) specifies the architecture of Figure 24.22 (or Figure 24.23).

As seen from the table, STRIP^{d2}, CARVE, and STRIP^{d4} networks are all within the same order of complexities with respect to r, that is, $O(1)$ number of layers, $O(r)$ number of nodes, thresholds, and weight connections. The total number of parameters, that is, the number of thresholds plus the number of weight connections, is an important complexity measure of neural networks because of their hardware cost and also their influence on generalization performance. With respect to this measure, the NDL network has the worst overall complexity, which is $O(r)$ number of layers and nodes and $O(r^2)$ number of parameters. CARVE network achieves the best overall complexity but is only slightly better than STRIP^{d2} network (which has $2r - k$ more parameters but $k - 1$ less nodes). STRIP^{d4} network is more complex than CARVE and STRIP^{d2} networks because of its asymptotic constant.

An important observation from the experiments is that, as n and k increase in many classes of functions, STRIP^{d2} and STRIP^{d4} networks have significantly much smaller values for r than CARVE, NDL, and SEPAR networks. If GA is used to construct a SEPAR network for a given function, then such SEPAR network should be at least smaller than a CARVE network (for the same function) for reasons explained in the last paragraph of Section 24.4.3. That is, higher values of r in a CARVE network are due to the less efficient search of CARVE algorithm compared with SEPAR algorithm (using GA), given the same task.

In practical applications the relation between the number of new hidden units r and the target function is important. CARVE networks have the simplest operation of the basic units among them, and therefore require more new hidden units than STRIP networks. Thus, we should discuss the overall complexity with the number of new hidden units r for some function realizations. SEPAR algorithm already improves CARVE and NDL for the given reasons. For most classes of functions, STRIP network implementations are significantly smaller in r than maximum separable-based network implementations of the same functions such as CARVE, NDL, SEPAR, and other networks. The reason is that removing a strip generated by a $\vec{w}$ may create a completely new strip which is the union of two strips that enclosed the removed strip. This can happen only when the removed strip is not an end strip. For instance, consider three strips where one has value 1 and the other two have values 0 and suppose strip of value 1 is between strips of values 0, that is, we have the sequence 010. Then removing strip 1 creates a new strip 00, which may be longer than the strips of values 1 and 0 together. If the longer strips are created then the smaller strips will be the number of new added nodes. This situation cannot happen when maximum separable subset techniques are used. That is why the maximum separable method creates more nodes than the longest strip method.

To illustrate this fact, consider the example function of P_9^2 (i.e., $k = 9$ and $n = 2$) in Figure 24.24. This function has a *mirror-symmetric* table, that is, all rows, columns, and the two main diagonals are symmetric about their center; the second half of a row, column, or diagonal is a mirror reflection of the first half. Moreover, at row y $(1 \leq y \leq (k-1)/2)$ the first y entries in that row are equal, and at column x $(0 \leq x \leq (k-1)/2)$ the first $x + 1$ entries in that column are equal. Such two-input mirror-symmetric-table function can be constructed for any odd k and the analysis is similar. This class of functions is very interesting. First, the smallest possible size for a strip-based network or a maximum separable subset-based network that realizes a function in this class can be obtained analytically. Second, like random functions, these functions have small separations between inputs and, therefore, seem least *difficult* to realize. Third, as described later, they clearly demonstrate the power of STRIP compared with

8	6	4	2	0	2	4	6	8
7	6	4	2	0	2	4	6	7
5	5	4	2	0	2	4	5	5
3	3	3	2	0	2	3	3	3
1	1	1	1	0	1	1	1	1
3	3	3	2	0	2	3	3	3
5	5	4	2	0	2	4	5	5
7	6	4	2	0	2	4	6	7
8	6	4	2	0	2	4	6	8

FIGURE 24.24 Mirror-symmetric-table function of P_9^2. (From A. Ngom, I. Stojmenović, and V. Milutinović. *IEEE Transactions on Neural Networks*, 12, 212–227, 2001. With permission.)

CARVE, NDL, and SEPAR algorithms; the difference in size between a smallest STRIP network and a smallest CARVE network for a given function in this class is $O(k^2)$.

Clearly, a function in this class has a minimal representation of exactly k units in the first hidden layer of STRIP. In our example function, STRIP algorithm will extract nine strips out of it and these will have values $0, 1, 2, 3, 4, 5, 6, 7, 8$ in that order. A weight vector that generates the strip of value 0 (the longest strip initially) is $\vec{w} = (0, 1)$ — also note that there are four strips of values 2 (and lengths four) in that direction. After removal of the strip of value 0 the algorithm must change direction, that is $\vec{w} = (1, 0)$, in order to remove the strip of value 1. The next longest strip is the strip of value 2 in direction $\vec{w} = (0, 1)$; the four short strips of length 4 are now joined together into a single strip of length 16, since strip 0 and strips 1 which were between them are removed.

What CARVE, NDL, and SEPAR algorithms can do at best? It can be shown that the minimum number of hyperplanes needed to partition a mirror-symmetric-table function is $(k^2 + 2k - 3)/2$. For instance, the smallest CARVE network associated with our example function contains 48 units in its first hidden layer. Clearly, the ratio between the sizes of a smallest STRIP network and a smallest CARVE network for a function in this class $\rightarrow 0$ as $k \rightarrow +\infty$. A smallest CARVE, NDL, or SEPAR network for such functions is $O(k)$ times larger than a corresponding smallest STRIP network.

Suppose for some function realization, a maximum separable subset-based algorithm, say CARVE, achieves the smallest network size r_m. Suppose, also for the same function realization that STRIP achieves its smallest network size r_s. For STRIP to be *fundamentally better* than CARVE we must have $r_s < \frac{1}{2}r_m$ (this is the case for mirror-symmetric-table functions realizations). Simply put one STRIP neuron (a two-threshold perceptron) is equal in complexity to two CARVE neurons (one-threshold perceptrons). For functions where the inputs are separated by a number of parallel hyperplanes (such as linear, permutably homogeneous, some monotone functions), STRIP is not fundamentally better than CARVE. For these functions, a smallest STRIP network has a value $r_s \geq \frac{1}{2}r_m$. We will discuss more about this fact in Section 24.5.7.

STRIP networks are fundamentally better than CARVE, NDL, SEPAR (and other maximum separable subset based) networks for many classes of functions other than those cited in the previous paragraph. More complex examples than the mirror-symmetric-table functions can be made, for some n and k, where STRIP selects optimal directions for partitioning whereas SEPAR, for instance, cannot *know* such information and therefore will most likely do poor job. Preliminary experiments (see Table 24.2) seem to indicate that such functions (including mirror-symmetric-table functions) are *very* hard to realize by CARVE and similar algorithms; indeed, they look at least harder than random functions. For example, given the function in Figure 24.24, if CARVE algorithm removes a *single* point of value 6 in its first few iterations, then the corresponding CARVE network will never be minimal. CARVE algorithm proceeds by *corner* (and border) separation and it is very likely that one of the points $(1, 0), (1, 9), (8, 0), (8, 9)$ of value 6 will be in a singleton class that may be separated.

We can also compare the complexity and latency of a minimal STRIP network implementation of an arbitrary multiple-valued logic function to the complexity and latency of a minimal direct circuit implementation of the same function. For some functions realizations, STRIP network implementations are sometimes better and sometimes worse (in depth or size) than their corresponding direct circuit implementations. The circuit implementations can be from any basis of gates, such as $\{XOR\}$ or $\{AND, OR, NOT\}$ bases for example. See Section 24.5.7 for a comparison with n-bit parity circuits.

24.4.6.4 Proof of Correctness

In this section we prove that both networks, for both definitions of A, effectively compute any given but arbitrary function $f\colon V \subset R^n \mapsto K$.

Let $\vec{x}$ be an input vector applied to the network (we refer to both networks and both definitions of A). Let $\vec{u} = (u_1, \ldots, u_r)$ be the set of output values of the units of layer 1 when $\vec{x}$ is applied, that is unit U_i outputs the value $u_i \leq k$ on input $\vec{x}$ $(1 \leq i \leq r)$. Let $p \leq (k+1)^{r-1}$ be the output of the unit (we call it P) of layer 2. Let $\vec{q} = (q_1, \ldots, q_k)$ be the outputs of layer 3, that is unit Q_j has value $q_j \leq 1$ $(1 \leq j \leq k)$ on input $\vec{x}$. Let the value of the output unit (we call it Z) be $z \in K$.

TABLE 24.2 Results of 10 runs for $k = 4$ and $n = 4$

		Using 100% of K^n			Using 60% of K^n		
		#Nodes	Min	Time	#Nodes	Time	Accuracy (%)
$p_{muta} = 10\%$ and $p_{cros} = 75\%$							
Random	STRIP^{F2}	28.0 ± 0.89	27 (3)	33	17.6 ± 0.66	16	25.92 ± 2.42
	STRIP^{F1}	28.9 ± 0.94	27 (1)	29	19.2 ± 0.75	16	23.11 ± 1.43
	SEPAR^{F2}	65.8 ± 1.66	62 (1)	58	38.4 ± 1.28	29	25.53 ± 3.39
	SEPAR^{F1}	66.3 ± 2.37	63 (1)	58	40.1 ± 2.12	29	24.95 ± 3.42
Linear	STRIP^{F2}	10.0 ± 0.00	10 (10)	10	09.7 ± 0.46	6	87.28 ± 5.98
	STRIP^{F1}	10.0 ± 0.00	10 (10)	10	09.9 ± 0.30	7	87.09 ± 3.79
	SEPAR^{F2}	20.3 ± 3.58	13 (1)	19	17.8 ± 2.56	11	85.24 ± 6.78
	SEPAR^{F1}	20.2 ± 3.60	13 (1)	19	17.0 ± 2.83	11	83.88 ± 4.21
Monotone	STRIP^{F2}	09.4 ± 0.49	9 (6)	7	08.0 ± 0.77	3	80.68 ± 5.31
	STRIP^{F1}	10.1 ± 0.70	9 (2)	7	08.3 ± 0.46	5	80.97 ± 6.41
	SEPAR^{F2}	10.6 ± 1.28	10 (4)	7	09.0 ± 1.34	4	85.05 ± 6.46
	SEPAR^{F1}	12.7 ± 1.00	11 (2)	7	10.8 ± 1.08	5	80.58 ± 6.22
Permutably	STRIP^{F2}	3.0 ± 0.00	3 (10)	1	3.1 ± 0.30	1	96.99 ± 1.33
	STRIP^{F1}	3.1 ± 0.30	3 (9)	4	3.0 ± 0.00	3	98.16 ± 2.15
	SEPAR^{F2}	3.1 ± 0.30	3 (9)	1	3.1 ± 0.30	1	97.09 ± 2.46
	SEPAR^{F1}	3.3 ± 0.90	3 (9)	5	3.3 ± 0.90	3	96.99 ± 1.76
$p_{muta} = 0\%$ and $p_{cros} = 75\%$							
Random	STRIP^{F2}	27.9 ± 0.94	26 (1)	28	17.8 ± 0.87	11	24.85 ± 3.17
Linear	STRIP^{F2}	10.0 ± 0.00	10 (10)	13	09.9 ± 0.54	9	86.12 ± 3.28
Monotone	STRIP^{F2}	09.8 ± 0.60	9 (3)	4	08.2 ± 0.40	3	82.82 ± 4.64
Permutably	STRIP^{F2}	03.5 ± 0.50	3 (5)	1	03.2 ± 0.40	1	94.85 ± 2.64
$p_{muta} = 75\%$ and $p_{cros} = 0\%$							
Random	STRIP^{F2}	27.5 ± 1.20	26 (3)	36	18.6 ± 0.80	13	24.95 ± 3.16
Linear	STRIP^{F2}	10.0 ± 0.00	10 (10)	15	10.1 ± 0.30	12	87.57 ± 4.27
Monotone	STRIP^{F2}	09.8 ± 0.40	9 (2)	12	07.7 ± 0.64	7	82.14 ± 2.61
Permutably	STRIP^{F2}	03.0 ± 0.00	3 (10)	1	03.0 ± 0.00	1	94.76 ± 2.46

Source: From A. Ngom, I. Stojmenović, and V. Milutinović. *IEEE Transactions on Neural Networks*, 12, 212–227, 2001. With permission.

Clearly, on input $\vec{x}$ each unit U_i has either value $u_i = 0$ or $u_i = v_i \neq 0$, where v_i is the maximum amplitude of the unit (see Figures 24.22 and Figure 24.23). Recall that each U_i corresponds to a subset G_i found and removed by our A-based synthesis algorithm during its ith run. The collection of the G_is is a partition of K^n, that is $\bigcup_{i=1}^{r} G_i = K^n$ and $\bigcap_{i=1}^{r} G_i = \emptyset$, and therefore, the subset G_i such that input $\vec{x} \in G_i$ corresponds to the first unit in layer 1 (starting from the left) which outputs a nonzero value on input $\vec{x}$. That is $\vec{u} = (0, \ldots, 0, u_i = v_i, u_{i+1}, \ldots, u_r)$, where u_j for $i + 1 \leq j \leq r$ is either 0 or v_j. Recall that, according to our definition of U_i, $v_i - 1$ is the function value of all points in G_i, that is $f(\vec{x}) = v_i - 1$.

By definition, unit P always outputs the value $p = (k+1)^{r-i}$ whenever i is the least index in $\vec{u}$ such that $u_i > 0$ on input $\vec{x}$. In the next paragraph we let i be the least such index, that is, such that $u_i = v_i \neq 0$ and $u_1 = \cdots = u_{i-1} = 0$. Let $a = \sum_{l=1}^{r}(k+1)^{r-l}u_l = \sum_{l=i}^{r}(k+1)^{r-l}u_l$ be the dot product of the weights and inputs (i.e., the outputs of layer 1) of P.

Each unit Q_j ($1 \leq j \leq k$) in layer 3 performs the sum $b_j = a - j(k+1)^{r-i}$ (recall that the weight connection from P to Q_j is $-j$ and that the output of P is $(k+1)^{r-i}$). We have $b_j = (k+1)^{r-i}u_i - j(k+1)^{r-i} + \sum_{l=i+1}^{r}(k+1)^{r-l}u_l$. From our definition of the weight connection vector between layer 1 and layer 3 it is easy to see that $0 \leq \sum_{l=i+1}^{r}(k+1)^{r-l}u_l < (k+1)^{r-i}$. Therefore, we obtain $b_j \geq 0$ for $1 \leq j \leq u_i$ and $b_j < 0$ for $u_{i+1} \leq j \leq k$. That is, exactly u_i units in layer 3 will output the value 1.

By definition, the output unit Z which has only unit weights, will perform the sum $c = \sum_{j=1}^{k} 1b_j = \sum_{j=1}^{u_i} 1 = u_i$. Since $1 \leq u_i \leq k$ and the threshold vector of Z is such that $t_{u_i} = u_i + 1$, therefore, we obtain $t_{u_i-1} \leq u_i < t_{u_i}$. From the definition of the output vector of Z we know that the output of the neural network is $z = t_{u_i-1} - 1 = u_i - 1 = v_i - 1 = f(\vec{x})$. Thus, the networks have effectively classified input $\vec{x}$ correctly. This completes our proof for the network of Figure 24.22. The proof for Figure 24.23 is straightforward and therefore omitted.

24.4.7 Experimental Results

In this section we present experimental results from the applications of STRIP and SEPAR algorithms to a number of multiple-valued logic functions, namely: random functions, linear functions, monotone functions, permutably homogeneous functions, n-bit parity functions, mirror-symmetric functions, and mirror-symmetric-table functions. In analyzing the performance of STRIP on these functions we looked at two important aspects, that is, the size of the networks it generates and its generalization ability. We constructed STRIP networks and compared their performances with constructed SEPAR networks along with other well-known construction techniques.

Throughout the experiments, we used the following parameters: 1000 generations for GA, 75% crossover rate, and 10% mutation rate. We also used an elitist strategy in the GA, which is the best individual of the current generation is always reproduced to the next generation. This elitist strategy helps counter balance the disruptive effect of our high mutation rate. On the other hand, we may need this high mutation rate in order to efficiently search the infinite space of weight vectors. These parameters were obtained by trial and error and seemed optimal.

For each given class of functions and method of network construction, we ran our algorithms ten times. All ten runs of the algorithms used the same functions (generated randomly) but with different random seeds (and different random training sets for approximation).

Table 24.2 shows, respectively for both objective functions (the superscripts $F1$ and $F2$ stand for *Fitness*1 and *Fitness*2), the results of ten runs of each method on randomly generated functions from each of the four test classes. We display the average number of created hidden nodes, r (in the first layer of the constructed networks), with its standard deviation, the minimum value found by the method (the number in parenthesis is the number of times it was found), the smallest running time (in minutes) over ten runs, and the average generalization accuracy (with its standard deviation) on the test set over ten runs. One can see from the table that, in general, Fitness2 yields smaller but less accurate networks than Fitness1.

For functions approximations (see column *Using* 60% *of* K^n) we simply used random portions of K^n as training sets. Each of these training sets were generated using a different random seed, also no two elements in the sets are equal. The network obtained with a training set is then tested on the test set, which is the remaining 40% *of* K^n, and its accuracy on the test set is computed.

To see if there is any advantage to set $p_{\text{muta}} > 0$ or $p_{\text{cros}} > 0$ we did two experiments with STRIP using *Fitness*2; however, with either $p_{\text{muta}} = 0\%$ and $p_{\text{cros}} = 75\%$ or $p_{\text{muta}} = 75\%$ and $p_{\text{cros}} = 0\%$ (see Table 24.2). Under the column Using 100% of K^n, results in the third part of the table ($p_{\text{cros}} = 0\%$) are, in general, slightly better than those in the first part of the table ($p_{\text{cros}}, p_{\text{muta}} > 0$) which in turn are slightly better than those in the second part of the table ($p_{\text{muta}} = 0\%$). Thus, there is advantage in setting $p_{\text{muta}} > 0$. Clearly, mutation is the operator that helps GA explore the search space. Crossover helps to focus the search on interesting parts of the space, in parallel. Mutation probability must be set small in order to avoid a *random walk* in space and to better *exploit* good parts of the space. Under the column *Using* 60% of K^n, the situation is somewhat reversed.

Random functions are more difficult to learn because of the smaller separation among their inputs. Therefore, it is not surprising that they give the highest number of nodes in the tables and the lowest generalization accuracy. On the other hand, permutably homogeneous functions have larger separation of inputs which makes them easier and faster to learn. So they naturally produce smaller networks of $O(k)$ nodes on average and also give a very high generalization accuracy. Other classes of functions such as linear, monotone, and other functions lie between these two extremes.

1. *Permutably homogeneous functions.* Functions in this class are partitioned by at most $k-1$ separating parallel hyperplanes such that no two distinct classes have equal values. Therefore, the minimal number of new hidden units for both STRIP and SEPAR is $\leq k$. We have experimented with the following permutably homogeneous functions: $f(\vec{x}) = \lfloor (\sum_{i=1}^{n}(1/a_i)x_i)+n \rfloor \bmod k$, where $a_i = 2i+1$. We randomly generated the four-input four-valued logic function $f(\vec{x}) = \lfloor x_1/3 + x_2/5 + x_3/7 + x_4/9 + 4 \rfloor \bmod 4$ and tested our algorithms with it. The function has three possible values, namely 0, 1, 2. The three classes are separated by two parallel hyperplanes (classes 1 being in the middle) and the distance between two adjacent hyperplanes is ≥ 2.

The results of STRIP and SEPAR are consistent: they gave a value of about the minimal solution $r = 3$. Clearly, there is no benefit of using STRIP for these functions; there is no reduction at all and thus the STRIP network is twice more complex than its corresponding SEPAR network. In general, STRIP is not better than SEPAR for functions where the inputs are separated by *nonintersecting* hyperplanes with distinct values in distinct regions.

2. *Monotone functions.* We experimented with a (random) function of P_4^4 which is monotonic under the *natural* non-decreasing order on K, that is $0 \leq 1 \leq \cdots \leq k-2 \leq k-1$. For such functions, STRIP and SEPAR produced closed results. This suggests that, as for permutably homogeneous functions, SEPAR is better than STRIP for these monotone functions since its achieves smaller network complexity even though STRIP has smaller values for r.

3. *Linear functions.* These functions are partitioned by a number of parallel hyperplanes where, many separated distinct classes of inputs have equal values and any two adjacent classes have distinct values. The minimal STRIP network for such functions will be *exactly* twice smaller than the corresponding minimal SEPAR network, for reasons explained in Section 24.4.6.3. Therefore, even though the STRIP network is twice smaller in r, it will have exactly the same complexity as its corresponding SEPAR network for linear functions realizations. Thus STRIP is no better than SEPAR and CARVE for such functions.

The random four-valued linear function generated was $f(\vec{x}) = (3x_1 + x_2 + 3x_3 + x_4) \bmod k$. Both STRIP and SEPAR algorithms, have difficulties realizing this function. First, the minimum r produced by SEPAR, 13, is very far from the average result, 20.3. Second, STRIP should give a result which is about two times smaller than the minimum obtained by SEPAR. The reason for this is that, for linear functions, the distance between two adjacent separating parallel hyperplanes is very small (≤ 1). Therefore, the algorithms are very sensitive to rotations, that is, small rotations from separating hyperplanes could cause the algorithms to produce large networks.

The n-bit parity functions are linear functions and it is well-known that a single layer minimal solution exists with n hidden $(n, 2, 1)$-perceptrons. We carried out experiments with STRIP and SEPAR algorithms for $n = 0, \ldots, 12$ using 2500 generations of GA to learn such functions. The results obtained were consistently $\lceil n/2 \rceil + 1$ hidden units for STRIP and $n + 1$ hidden units for SEPAR, using both fitness measures, respectively. For SEPAR we cannot obtain n hidden units like the other maximum separable-based methods because the last training set (of points of same class) is always assigned to a new hidden unit. Thus, for instance, we will obtain exactly two hidden units for the binary *AND* function whereas the other methods would obtain one hidden unit. For SEPAR, the results obtained for $n = 10, 11, 12$ were in reality 13, 15, 18 for 2500 generations of GA; however, we obtained the correct results, 11, 12, 13 when we increased the number of generations to 4000. This suggests that our algorithms are able to find the minimal value if they are given enough time.

n-bit parity functions can be implemented by $\{XOR\}$ circuits of depth one and size $O(n)$, $\{XOR\}$ circuits of depth $O(\log_2 n)$ and size n, $\{AND, OR\}$ circuits of depth d and size $O(2^{1/(d-1)} n^{(d-2)/(d-1)})$ or $\{AND, OR, NOT\}$ circuits of depth two and size 2^{n-1}. For such functions, a STRIP^{d2} network has size $\lceil n/2 \rceil + 2$ and a STRIP^{d4} network has size $\lceil n/2 \rceil + k + 3$. As one can see, STRIP networks are not always better than their corresponding direct circuit implementations. The difference in complexity between a STRIP network and a direct circuit depends on factors such as the basis set of gates and the fan-in or fan-out of the gates in the circuit.

4. *Random functions.* The experiments clearly show that STRIP is fundamentally better than SEPAR and CARVE for random functions realizations. The average and the minimum obtained STRIP networks

are at least twice smaller than the corresponding results for SEPAR networks. As already stated, STRIP is able to change (and select good) directions for separation and to create larger classes, while SEPAR cannot do any of these two things. For random functions, the separation between classes is very small. STRIP can *see inside* a random function to decide the best directions to choose and, by doing that, it maximizes the size of all classes (they will get bigger and bigger until they can be removed). SEPAR only maximizes the size of the classes that are at the boundaries of the inputs space.

Table 24.3 contains results reported for the best three constructive algorithms (so far in the literature) that have been applied to learning random two-valued logic functions. The value in parenthesis at the top of each column is the number of trials over which the network is averaged. In each trial of STRIP, we generated a different random function and used 2500 iterations of GA. STRIP produces smaller networks for this classification task than any other growth method. Significantly, STRIP gives much smaller networks than CARVE as n and k increase. Also, SEPAR performs better than Upstart as n and k increases.

5. *Mirror-symmetric functions.* We have carried out experiments with mirror-symmetric functions. A mirror-symmetric function has value 1 if the second half of an input vector is a mirror reflection of the first half, that is the input vector is symmetric about its center. This function is known to have a minimal representation of two hidden units. For $n = 2, \ldots, 7$ we have always found the optimal number of hidden units in STRIP and SEPAR, two, using 1000 generations of GA. For $n = 8, \ldots, 12$ we have always found the optimum when using 3000 generations of GA. Here, as for permutably homogeneous functions, STRIP is no better than SEPAR.

6. *Mirror-symmetric-table (mst) functions.* Table 24.4 shows the comparisons of performances between our example mst function in Figure 24.24 and ten random functions. For each algorithm (STRIP^{F2}, STRIP^{F1}, SEPAR^{F2}, and SEPAR^{F1}), we did ten runs on our mst function and averaged the results. Also, each algorithm was applied on ten distinct random functions (different from the random functions used for the other three algorithms) and the results were then averaged. STRIP^{F2} is a clear winner (except its performance in generalization for random functions); it produced the minimal size for mst. The table also

TABLE 24.3 Comparison of Network Size for Random Functions of P_2^n

n	STRIP^{F2} (100)	CARVE [56] (100)	Sequential [44] (100)	Upstart [50] (100)	SEPAR^{F2} (100)
4	2.81 ± 0.39	2.40 ± 0.69		3	3.62 ± 0.61
5	3.23 ± 0.42	3.73 ± 0.58		4.5	4.93 ± 0.50
6	4.56 ± 0.50	5.88 ± 0.67	7.28 ± 0.82	8	6.99 ± 0.71
7	6.95 ± 0.30	9.47 ± 0.74		16	11.14 ± 0.77
8	11.56 ± 0.50	16.23 ± 0.86	18.3 ± 0.69	33	19.51 ± 1.03

Source: From A. Ngom, I. Stojmenović, and V. Milutinović. *IEEE Transactions on Neural Networks*, 12, 212–227, 2001. With permission.

TABLE 24.4 Performances Comparisons Between Mirror-Symmetric-Table and Random Functions of P_9^2

	STRIP^{F2}	STRIP^{F1}	SEPAR^{F2}	SEPAR^{F1}
Using 100% of K^n				
Mst	09.0 ± 0.00	09.8 ± 0.40	60.2 ± 1.08	69.0 ± 3.69
Random functions	29.2 ± 1.99	31.8 ± 1.60	58.8 ± 2.89	64.7 ± 4.61
Using 60% of K^n				
Mst	09.0 ± 0.00	11.7 ± 1.95	31.0 ± 3.13	36.0 ± 2.49
Random functions	19.1 ± 1.37	21.6 ± 1.91	36.1 ± 2.98	36.7 ± 2.90
Accuracies				
Mst	$65.15\% \pm 12.44$	$57.27\% \pm 09.53$	$18.18\% \pm 06.91$	$13.33\% \pm 05.62$
Random functions	$09.70\% \pm 04.85$	$16.06\% \pm 06.36$	$10.91\% \pm 07.20$	$12.42\% \pm 05.33$

shows that the single mst function is harder to realize by SEPAR than random functions (see the last two entries of the first two rows): the average result of SEPAR for mst is very far from the absolute minimum, 48, and is significantly larger than the averaged result of ten random functions. For reasons explained in Section 24.4.5.2, Fitness2 helps to remove the classes faster than Fitness 1, and therefore, it yields better results in the table for both algorithms (except on generalization of random functions).

24.5 ES for Learning Multiple-Valued Logic Functions

We consider again the problem of synthesizing multiple-valued logic functions by neural networks. However, unlike in Section 24.4, an ES is used (in place of GA) to optimize a network.

In this section we apply STRIP algoritm using evolutionary strategy [64] to grow a multiple-valued logic neural network for learning a function. As in Section 24.4, the maximum separable subset is treated as a special case of the longest strip.

24.5.1 Determining Longest Strips by Evolutionary Strategy

In this section we use *ES* [65], instead of GA, to search for the longest strip or maximum separable subset of training examples. The A-based synthesis algorithm shown in Figure 24.20 is used to grow the network.

The ESs have been proposed by Schwefel [66] as optimization methods for real-valued parameters. ES manipulates a single potential solution to a problem. Specifically, it operates on an encoded representation of the solution, and not directly on the solution itself. Schwefel's ES encodes a solution as a real-valued vector. We will refer to such a vector as *chromosome.* The solution is associated with an *objective value* that reflects how good or bad it is, compared with other potential solutions in the space. ES is a random guided HC technique in which a good candidate solution is obtained by applying a certain number of small mutations to a given parent solution. The best result of mutation is used again to generate the next best solution, and so on until some convergence criterion is satisfied.

Our ES method uses the same solution representation, the same objective functions and the same mutation operators as in Section 24.4. That is, a potential solution G (a subset of K^n) is represented as a weight vector $\vec{w}$ (such a vector can be decoded to obtain G). For the problem of determining A, the search space is the *power set* of the training set (i.e., the set of all subsets of K^n). Each element of the search space is a potential solution and must be represented in such a way that meaningful mutation operators can be designed for (and applied to) it.

We initially start with a unique random unit weight $\vec{w}_0 = (w_{0,1}, \dots, w_{0,n}) \in [-1, 1]^n$ (Another method we used for the initialization is to set $w_i = \cos \alpha_i$ (for $1 \leq i \leq n$) for each vector $\vec{w}$, where α_i is a random number in the interval $[-\pi/2, \pi/2]$. Initial chromosome should consist of random unit hyperplane $\vec{w}$). Then the next weight vector $\vec{w}_1 = (w_{1,1}, \dots, w_{1,n})$ is chosen to be a neighbor of $\vec{w}_0$ which has the best fitness value higher than the fitness of $\vec{w}_0$ and which yields the least change in the ordering of the function values according to $\vec{w}_0$ (see Section 24.4.5.2). Then, we set $\vec{w}_2$ to be the best neighbor of $\vec{w}_1, \dots,$ and continue this process until there is no more best neighbor or a certain number of iterations is reached.

24.5.2 Weight Neighborhood

Recall that to find the subset $T_{\vec{w}}^{(c)}$ for a current $\vec{w}$ in a set of points S with $|S| = v$ we sort the v points with respect to $\vec{w}$ (first key) and $f(\vec{x})$ (second key) and find the longest strip or the maximum separable subset from the sorted list $(\vec{w}\vec{x}_1, f(\vec{x}_1)), \dots, (\vec{w}\vec{x}_v, f(\vec{x}_v))$. The neighbors of $\vec{w}$ are precisely those weight vectors that yield the least change in the ordering of the $f(\vec{x})$'s.

For instance, if $v = 4$ and $\vec{w}$ gives the order $(f(\vec{x}_1), f(\vec{x}_2), f(\vec{x}_3), f(\vec{x}_4))$, then a near neighbor of $\vec{w}$ may produce the order $(f(\vec{x}_2), f(\vec{x}_1), f(\vec{x}_3), f(\vec{x}_4))$ and a far neighbor of $\vec{w}$ may produce the order $(f(\vec{x}_4), f(\vec{x}_3), f(\vec{x}_1), f(\vec{x}_2))$.

Consider the jth coordinate, w_j, of $\vec{w}$, $1 \leq j \leq n$. Let $\vec{w}\vec{x}$ and $\vec{w}\vec{y}$ be two consecutive elements in the order produced by $\vec{w}$ and let $d = \vec{w}\vec{y} - \vec{w}\vec{x} = \vec{w}(\vec{y} - \vec{x}) = w_j(y_j - x_j) + \sum_{i \neq j} w_i(y_i - x_i)$. Likewise, for an arbitrary vector $\vec{\pi}$, we will have $d' = \pi_j(y_j - x_j) + \sum_{i \neq j} \pi_i(y_i - x_i)$.

Vector $\vec{\pi}$ is a neighbor of $\vec{w}$ with respect to coordinate j if $\pi_i = w_i$ for $i \neq j$ and $|\pi_j - w_j|$ is minimal such that the sorted order is changed. Then only w_j is affected and hence $\sum_{i \neq j} w_i(y_i - x_i) = \sum_{i \neq j} \pi_i(y_i - x_i)$. Since $\sum_{i \neq j} w_i(y_i - x_i) = d - w_j(y_j - x_j)$ therefore we have $d' = \pi_j(y_j - x_j) + d - w_j(y_j - x_j) = d + \Delta w_j(y_j - x_j)$, where $\Delta w_j = (\pi_j - w_j)$. Taking $d' = 0$ and solving for Δw_j we obtain

$$\Delta w_j = -\frac{d}{y_j - x_j}. \tag{24.20}$$

Only differences $d_i = \vec{w}(\vec{x}_{i+1} - \vec{x}_i)$, for $1 \leq i \leq v - 1$, between distinct consecutive elements in sorted order are considered. For each coordinate j $(1 \leq j \leq n)$ there should be one neighbor only, chosen such that $|\Delta w_{j_i}| = | - d_i/(x_{i+1,j} - x_{i,j})|$ is minimized and is not equal to zero. Let $\Delta w_j = \min |\Delta w_{j_i}|$ and denote the nearest neighbor of $\vec{w}$ with respect to j by $\vec{\pi}_j$. Then $\vec{\pi}_j$ differs from $\vec{w}$ only in the jth coordinate, by Δw_j. That is $\pi_{j,j} = w_j + \Delta w_j$ and $\pi_{j,c} = w_c$ for $c \neq j$. In other words, the Euclidean distance $E(\vec{w}, \vec{\pi}_j)$ between $\vec{w}$ and $\vec{\pi}_j$ is $|\Delta w_j|$. Thus, there are n neighbors of $\vec{w}$, that is one nearest neigbhor per each coordinate j of $\vec{w}$. We denote the set of all n such neighbors of $\vec{w}$ by $\Pi_{\vec{w}} = \{\vec{\pi}_1, \ldots, \vec{\pi}_n\}$. This set is called the *neighborhood* of $\vec{w}$.

24.5.3 Evolution Strategy

Our ES is shown in Figure 24.25. In the algorithm, $\vec{w}_i$ is the current solution at generation i, $\vec{b}$ is the best solution generated so far, $\vec{\mu}_{\vec{w}_i}$ is the best solution in $\Pi_{\vec{w}_i}$ and $\vec{w}_{i+1}$ is the next solution to generate. We initially start with a random unit vector $\vec{w}_0$ and set $\vec{b}$ to $\vec{w}_0$.

The algorithm works as follows. At generation i, we compute the best neighbor of $\vec{w}_i$, that is, a vector $\vec{\mu}_{\vec{w}_i}$ in $\Pi_{\vec{w}_i}$ that has the highest fitness value. If $\vec{\mu}_{\vec{w}_i}$ is better than $\vec{b}$ then we are done for this generation; $\vec{w}_{i+1}$ and $\vec{b}$ are both set to $\vec{\mu}_{\vec{w}_i}$ and we move to the next generation. If $\vec{\mu}_{\vec{w}_i}$ is worse than $\vec{b}$ then we must decide how to set $\vec{w}_{i+1}$. We first attempt to find a solution better than $\vec{b}$ in the neighborhood of the neighbors of $\vec{w}_i$. That is we search successively in each $\Pi_{\vec{\pi}_j}$ $(1 \leq j \leq n)$ until such solution is found or all n neighborhoods are tried without success. If a solution is found then we set both $\vec{b}$ and $\vec{w}_{i+1}$ to that solution, $\vec{\mu}_{\vec{\pi}_j}$ for some j, and move on to the next generation. If, otherwise, no such solution is found then $\vec{b}$ is possibly a local optimum. We have two choices, with probability $\frac{1}{2}$ we either (i) set $\vec{w}_{i+1}$ to the best between $\vec{\mu}_{\vec{w}_i}$ and $\vec{\mu}_{\vec{\pi}_j}$ (for all j's) and move on to generation $i + 1$, or (ii) to jump out of a possible local optimum, we set $\vec{w}_{i+1}$ to a random unit vector or to a random alteration of $\vec{b}$ or $\vec{\mu}_{\vec{w}_i}$ or $\vec{\mu}_{\vec{\pi}_j}$ for some random j. In the second choice, one of the four possibilities is selected with probability $\frac{1}{4}$ and the alteration of a vector is done by choosing any of our three mutation techniques described in Section 24.4.5.4 (with probability $\frac{1}{3}$).

Let g be the number of generations, then at each new generation n new chromosomes are evaluated for their fitness and hence, our ES has a time complexity of $O(gn^2|S| \log |S|) \approx O(gn^3k^n \log k)$. The time complexity in Section 24.4 is $O(gpn^2k^n \log k)$ where p is the population size of the GA. Since $p \gg n$ then our ES is much faster than the GA in Reference 28.

Using ES in the A-based algorithm, we construct neural networks in the same way as in Section 24.4.

24.5.4 Experimental Results

In this section we present experimental results from the applications of STRIP and SEPAR algorithms, using ES, to the same multiple-valued logic functions as in Section 24.4. We constructed STRIP and SEPAR networks by ES (STRIP_ES and SEPAR_ES) and compared their performances with constructed

```
Procedure EvolutionaryStrategy(k, n, f);
  i := 0;
  w_i := random unit vector;
  b := w_i;
  Repeat
    w_{i+1} := μ_{w_i} = best in Π_{w_i} = {π_1, ..., π_n};
    If Fitness(w_{i+1}) > Fitness(b) then
      b := w_{i+1};
    Else
      j := 0
      Success := false;
      Repeat {we search in the Π_{π_j}'s}
        j := j + 1;
        a := μ_{π_j} = best in Π_{π_j};
        If Fitness(a) > Fitness(b) then
          b := a;
          w_{i+1} := a;
          Success := true;
      Until Success = true or j = n;
      If Success = false then {b may be a local optimum}
        With probability 1/2 do either one of
          1: w_{i+1} := best between μ_{w_i} and all μ_{π_j}'s;
          2: With probability 1/4 do either one of
              2.1: w_{i+1} := random unit vector; {big jump}
              2.2: w_{i+1} := random mutation of b
              2.3: w_{i+1} := random mutation of μ_{w_i}
              2.4: w_{i+1} := random mutation of μ_{π_j} for a random j;
    i := i + 1;
  Until Stopping criteria is true;
```

FIGURE 24.25 Evolution strategy to find G such that $|G| \approx |A|$.

STRIP_GA and SEPAR_GA networks reported in Section 24.4 along with other well-known construction techniques.

Throughout the experiments, we used the following parameters: 2000 generations for ES (since ES is at least twice faster than GA and that in Section 24.4 uses 1000 generations for GA) and 10% mutation rate. Section 24.4 used 10% mutation rate, 75% crossover rate and an elitist strategy in the GA (i.e., the best individual of the current generation is always reproduced to the next generation). In some experiments such as random binary functions (and other functions) learning we increased the number of generation of GA (see Section 24.4), and accordingly use twice as much iterations for the ES.

Tables 24.5 and 24.6 show, respectively, for both objective functions, the results of ten runs of each method on randomly generated functions from each of the four test classes. We display the average number of created hidden nodes, r (in the first layer of the constructed networks), with its standard deviation, the minimum value found by the method (the number in parenthesis is the number of times it was found), the smallest running time (in minutes) over ten runs, and the average generalization accuracy (with its standard deviation) on the test set over ten runs. From these two tables we can see that, in general, Fitness2 yields (slightly) smaller but (slightly) less accurate networks than Fitness1.

TABLE 24.5 Results of 10 Runs for $k = 4$ and $n = 4$ (Using Fitness1)

Fitness1		Using 100% of K^n			Using 60% of K^n		
		#Nodes	Min	Time	#Nodes	Time	Accuracy (%)
Random	STRIP_GA	28.9 ± 0.94	27 (1)	29	19.2 ± 0.75	16	23.11 ± 1.43
	STRIP_ES	32.5 ± 0.67	32 (6)	29	21.4 ± 1.02	14	23.69 ± 3.28
	SEPAR_GA	66.3 ± 2.37	63 (1)	58	40.1 ± 2.12	29	24.95 ± 3.42
	SEPAR_ES	69.2 ± 2.79	64 (1)	53	40.3 ± 1.35	23	25.92 ± 3.86
Linear	STRIP_GA	10.0 ± 0.00	10 (10)	10	09.9 ± 0.30	7	87.09 ± 3.79
	STRIP_ES	10.0 ± 0.00	10 (10)	9	09.8 ± 0.60	6	92.23 ± 4.14
	SEPAR_GA	20.2 ± 3.60	13 (1)	19	17.0 ± 2.83	11	83.88 ± 4.21
	SEPAR_ES	27.9 ± 10.97	13 (2)	17	22.3 ± 7.69	9	73.20 ± 15.84
Monotone	STRIP_GA	10.1 ± 0.70	9 (2)	7	08.3 ± 0.46	5	80.97 ± 6.41
	STRIP_ES	09.6 ± 0.92	9 (6)	6	09.0 ± 0.77	4	83.50 ± 4.62
	SEPAR_GA	12.7 ± 1.00	11 (2)	7	10.8 ± 1.08	5	80.58 ± 6.22
	SEPAR_ES	12.2 ± 0.60	11 (1)	7	10.6 ± 1.11	4	86.89 ± 3.77
Permutably	STRIP_GA	3.1 ± 0.30	3 (9)	4	3.0 ± 0.00	3	98.16 ± 2.15
	STRIP_ES	5.4 ± 0.80	4 (2)	4	4.4 ± 0.66	3	94.08 ± 1.65
	SEPAR_GA	3.3 ± 0.90	3 (9)	5	3.3 ± 0.90	3	96.99 ± 1.76
	SEPAR_ES	7.8 ± 1.17	6 (2)	5	6.9 ± 1.14	3	90.29 ± 3.22

TABLE 24.6 Results of 10 Runs for $k = 4$ and $n = 4$ (Using Fitness2)

Fitness2		Using 100% of K^n			Using 60% of K^n		
		#Nodes	Min	Time	#Nodes	Time	Accuracy (%)
Random	STRIP_GA	28.0 ± 0.89	27 (3)	33	17.6 ± 0.66	16	25.92 ± 2.42
	STRIP_ES	31.6 ± 1.11	30 (2)	34	20.7 ± 1.00	18	24.56 ± 3.88
	SEPAR_GA	65.8 ± 1.66	62 (1)	58	38.4 ± 1.28	29	25.53 ± 3.39
	SEPAR_ES	69.1 ± 3.11	64 (1)	52	42.7 ± 3.13	25	25.63 ± 3.48
Linear	STRIP_GA	10.0 ± 0.00	10 (10)	10	09.7 ± 0.46	6	87.28 ± 5.98
	STRIP_ES	10.0 ± 0.00	10 (10)	11	10.1 ± 0.54	8	88.45 ± 2.55
	SEPAR_GA	20.3 ± 3.58	13 (1)	19	17.8 ± 2.56	11	85.24 ± 6.78
	SEPAR_ES	25.3 ± 7.79	13 (2)	18	23.5 ± 8.27	10	73.88 ± 20.28
Monotone	STRIP_GA	09.4 ± 0.49	9 (6)	7	8.0 ± 0.77	3	80.68 ± 5.31
	STRIP_ES	09.4 ± 0.49	9 (6)	10	8.1 ± 0.54	6	84.37 ± 4.26
	SEPAR_GA	10.6 ± 1.28	10 (4)	7	9.0 ± 1.34	4	85.05 ± 6.46
	SEPAR_ES	10.6 ± 1.43	10 (3)	10	9.1 ± 1.22	6	88.06 ± 5.00
Permutably	STRIP_GA	3.0 ± 0.00	3 (10)	1	3.1 ± 0.30	1	96.99 ± 1.33
	STRIP_ES	5.9 ± 0.94	4 (1)	3	4.4 ± 0.66	2	93.69 ± 2.00
	SEPAR_GA	3.1 ± 0.30	3 (9)	1	3.1 ± 0.30	1	97.09 ± 2.46
	SEPAR_ES	6.1 ± 0.83	5 (2)	4	4.9 ± 0.94	3	93.69 ± 1.58

GA performed slightly better than ES in most experiments, due to the GA manipulation of population and also its implicit parallelism. However, ES is at least twice faster than GA.

Permutably homogeneous functions. GA performed much better than ES.

Monotone functions. GA and ES produced very close results.

Linear functions. GA and ES produced very close results.

Random functions. STRIP_GA and STRIP_ES performed much better than any other constructive technique. Also, ES gave worse results than GA (Table 24.7).

Mirror-symmetric functions. For $n = 2, \ldots, 7$ we have always found the optimal number of hidden units in STRIP and SEPAR, two, using 2000 generations of ES (1000 for GA in Reference 7). For $n = 8, \ldots, 12$ we have always found the optimum when using 6000 generations of ES.

TABLE 24.7 Comparison of Network Size for Random Functions of P_2^n

n	STRIP_{GA}^{F2} [7] (100)	STRIP_{ES}^{F2} [7] (100)	CARVE [56] (100)	Sequential [44] (100)	Upstart [50] (100)	SEPAR_{GA}^{F2} [7] (100)
4	2.81 ± 0.39	3.17 ± 0.49	2.40 ± 0.69		3	3.62 ± 0.61
5	3.23 ± 0.42	3.64 ± 0.52	3.73 ± 0.58		4.5	4.93 ± 0.50
6	4.56 ± 0.50	5.14 ± 0.62	5.88 ± 0.67	7.28 ± 0.82	8	6.99 ± 0.71
7	6.95 ± 0.30	7.84 ± 0.37	9.47 ± 0.74		16	11.14 ± 0.77
8	11.56 ± 0.50	13.04 ± 0.62	16.23 ± 0.86	18.3 ± 0.69	33	19.51 ± 1.03

24.6 Minimization of Multiple-Valued Multiple-Threshold Perceptrons using GAs

We address the problem of computing and learning multiple-valued multiple-threshold perceptrons. Every n-input k-valued logic function can be implemented using an (n, k, s)-perceptron, for some number of thresholds s. We propose a GA to search for an optimal (n, k, s)-perceptron that efficiently realizes a given multiple-valued logic function, that is to minimize the number of thresholds. Experimental results show that the GA find optimal solutions in most cases.

A problem still left open in the domain of multiple-valued multiple-threshold functions is how to minimize the number of thresholds in order to construct the most efficient multiple-valued multiple-threshold networks or units. To minimize the number of thresholds, traditional techniques of multiple-valued multiple-threshold circuit synthesis use either trial and errors, or allow to synthesize only classes of functions for which an optimal number of thresholds can be obtained (multiple-valued multiple-threshold synthesis of *k-valued symmetric functions*, for instance). The multiple-valued multiple-threshold networks considered in literature have no learning capabilities, that is, their parameters are set by the designers once and for all using some traditional techniques of multiple-valued multiple-threshold networks synthesis. In addition only some small classes of k-valued logic functions are considered for multiple-valued multiple-threshold synthesis techniques.

Given $f \in P_k^n$ we are looking for a vector $\vec{r} = (\vec{w}, \vec{t}, \vec{o}) \in R^{n+s} \times K^{s+1}$ such that $F_{k,s}^n(\vec{r})(\vec{x}) = f(\vec{x})$ $(\forall \vec{x} \in K^n)$, that is, $f = F_{k,s}^n(\vec{r})$. We will refer to $\vec{r}$ as a *s-representation* of $F_{k,s}^n$. In this section, we will be mainly interested in finding *minimal* s for which there exist a s-representation for a given $f \in P_k^n$. In other words, given $f \in P_k^n$, we want to find a s-representation $\vec{r}$ with the least possible number of thresholds s such that $F_{k,s}^n(\vec{r}) = f$. We propose GAs as techniques for minimizing multiple-valued multiple-threshold perceptrons.

24.6.1 Computing Optimal s-Representations with GAs

24.6.1.1 Problem Representation

Fundamental to the GA structure is the encoding mechanism for representing the problem's variables. For the s-representation problem, the search space is the space of weight vectors $\vec{w}$ and the representation is more complex. Unfortunately, there is no practical way to encode s-representation problem as a binary chromosome to which the classical genetic operators discussed in Reference 61 can be applied in a meaningful fashion. Therefore it is natural to represent the possible solutions as vectors $\vec{w} \in R^n$ and design appropriate genetic operators which are suitable for the s-representation problem. Each weight vector will uniquely determine a s-representation. To determine how good the solution is, the GA needs a fitness function to evaluate the chromosomes.

A note on the initial population. We initialize the population with random real-coded chromosomes whose coordinates are random real numbers taken from the interval $[-1, 1]$. Each initial chromosome is then normalized to a unit vector. Another method we used for the initialization of the population is to set $w_i = \cos \alpha_i$ (for $1 \leq i \leq n$) for each vector $\vec{w}$, where α_i is a random number in the interval $[-\pi/2, \pi/2]$.

What we are trying to do in both methods of initialization is to generate random hyperplanes (since each $\vec{w}$ represent a hyperplane).

24.6.1.2 Fitness Function

The objective function, the function to be optimized, provides the mechanism for evaluating each chromosome. To describe our fitness function we will need the concept of valid and invalid thresholds (hyperplanes).

To compute the thresholds for a given chromosome $\vec{w}$, we calculate for every $\vec{x} \in K^n$ the value $\vec{w}\vec{x}$ and construct a sorted array (or list) of records of the form $(\vec{w}\vec{x}, f(\vec{x}))$. The array is sorted using $\vec{w}\vec{x}$ as primary key and $f(\vec{x})$ as secondary key. Let these records be sorted as follows: $\vec{x}_1, \ldots, \vec{x}_{k^n}$, or more precisely, $(\vec{w}\vec{x}_i, f(\vec{x}_i)), 1 \leq i \leq k^n$, where $\vec{w}\vec{x}_i \leq \cdots \leq \vec{w}\vec{x}_{k^n}$. Then $\vec{w}\vec{x}_j$ is a threshold if $f(\vec{x}_{j-1}) \neq f(\vec{x}_j)$. We collect all thresholds in a list $\vec{t}$. Some thresholds in $\vec{t}$ may be duplicated (i.e. $t_{i-1} = t_i$ for some i).

Let $T(\vec{w}) = V(\vec{w}) + I(\vec{w})$, where $T(\vec{w})$ is the total number of thresholds generated by $\vec{w}$, $V(\vec{w})$ and $I(\vec{w})$ are, respectively, the number of valid thresholds and invalid thresholds generated by $\vec{w}$. A threshold t_i $(1 \leq i \leq T(\vec{w}) \leq k^n - 1)$ is *valid* if all points $\vec{x} \in K^n$ lying in its corresponding hyperplane H_i (given by $\vec{w}\vec{x} = t_i$) are in the same class (i.e., $f(\vec{x})$ has the same value for all points in hyperplane H_i), otherwise it is *invalid*. In other words, invalid thresholds are those for which there exist at least two points $\vec{x}_1$ and $\vec{x}_2 \in K^n$ such that $\vec{w}\vec{x}_1 = \vec{w}\vec{x}_2$ but $f(\vec{x}_1) \neq f(\vec{x}_2)$. A hyperplane is valid (invalid) if it corresponds to valid (invalid) threshold. With these definitions then duplicated thresholds in $\vec{t}$ are invalid while nonduplicated thresholds are valids.

$T(\vec{w})$ is the total number of thresholds in $\vec{t}$ and can be used to evaluate how good or bad is a chromosome. The best chromosomes are those that have the least $T(\vec{w})$. We can therefore define our fitness function as follows:

$$\text{Fitness1}(\vec{w}) = 1 - \frac{T(\vec{w})}{k^n - 1}. \tag{24.21}$$

Note that a GA always maximizes an objective function and since $1 \leq T(\vec{w}) \leq k^n - 1$, then Fitness1$(\vec{w})$ is maximal when $T(\vec{w})$ is minimal.

However, invalid thresholds must need severe penalty. For instance, assume a n-input k-valued logic function f: $K^n \mapsto \{0, 1\}$ chosen at random. Then one may take hyperplanes $x_1 = 0$, $x_1 = 1, \ldots, x_1 = k - 1$ as invalid thresholds. These k hyperplanes (or k^2 thresholds) will separate in our sense but are not really separating as such random function needs actually an exponential number of thresholds. Because of this fact, instead of using Formula (24.21) we can alternatively use Formula (24.22).

$$\text{Fitness2}(\vec{w}) = \frac{2 - (T\vec{w})/(k^n - 1) - I(\vec{w})/T(\vec{w})}{2} = 1 - \frac{T(\vec{w})}{2 \cdot (k^n - 1)} - \frac{I(\vec{w})}{2 \cdot T(\vec{w})}. \tag{24.22}$$

Here, we not only minimize $T(\vec{w})$ (in second term) but we also punish a chromosome that generates a large number of invalid hyperplanes (in last term). That is we are minimizing $T(\vec{w})$ and $I(\vec{w})$ at the same time. Note that $0 \leq I(\vec{w}) \leq T(\vec{w})$ and thus Fitness2$(\vec{w})$ will be maximal if both $T(\vec{w})$ and $I(\vec{w})$ are minimal.

In all our experiments, both formulae of fitness yield the same results for $I(\vec{w}) = 0$. We do not know for now how they do behave for $I(\vec{w}) \neq 0$ since the $\vec{w}$'s generated valid thresholds only. The probability to generate invalid thresholds seems to be very close to zero.

A note on the time complexity of the evaluation function. For a given $\vec{w}$, it takes $n \cdot k^n$ steps to compute all the $\vec{w} \cdot \vec{x}$'s, $k^n \cdot \log k^n$ steps to sort them and at most k^n steps to compute $T(\vec{w})$. Therefore the evaluation of Fitness$(\vec{w})$ has a time complexity of $O(n \cdot k^n \cdot \log k)$.

Also, crossover and mutation operations take $O(n)$ steps each and the initialization of the population takes $O(n \cdot p \cdot k^n \cdot \log k)$ steps (p is the number of chromosomes and all initial chromosomes are evaluated for their fitness). Thus the evaluation of Fitness$(\vec{w})$ is the most expensive operation in our GA (and is true in general for any GA). Let g be the number of generations, then at each new generation $p/2$ new chromosomes are evaluated for their fitness and hence, our GA has a time complexity of $O(n \cdot g \cdot p \cdot k^n \cdot \log k)$.

24.6.1.3 Crossover and Mutation

They are same as those discussed in Section 24.4.

24.6.2 Experiments and Discussions

In our experiments, the control parameters' setting for the GA were: population size $p = 100$; number of generations $g = 1000$; crossover probability $p_{\text{cros}} = 0.75$; and mutation probability $p_{\text{muta}} = 0.005$. The most important parameters here are p, p_{cros}, and p_{muta} and the values used for them seem to be optimal in that they yield better results (than other possible values) in all experiments we have done. The high crossover rate is necessary to widen the search while the low mutation rate is necessary to avoid too much chromosome disruptions. Because we use an *elitist strategy* some best chromosome in a current generation is always reproduced to the next generation in order to avoid lost of good genetic material (and to reduce the disruptive effect of a high crossover rate). We use a large population size to preserve the diversity of the population, that is to avoid premature convergence. The fact that we used a mixed crossover technique also helps maintain the diversity. In all experiments, we used Fitness2 as our evaluation function (for reasons explained in Section 24.6.1.2). Also, we used *stochastic universal selection shceme* as our reproduction method.

It is interesting that the proposed population (chromosomes) representation does not depend on k. It make us wonder how the number of invalid thresholds vary with k (or n). For a fixed n (or k), larger k (or n) means smaller separation among classes and these problems are typically more difficult to learn. We did some experiments with small k versus large k and small n versus large n in order to see how the number of invalid thresholds changes. Such experiment were performed on random functions and their results are reported in Tables 24.8 and 24.9.

TABLE 24.8 Results for Some Two-Input k-Valued Random Functions

k	#Invalids	#Seconds
2	0	3.25
4	0	8.92
8	0	36.72
16	0	153.76
32	0	700.12
64	0	3,289.59
128	0	14,511.80
256	0	99,434.36

TABLE 24.9 Results for Some n-Input Two-Valued Random Functions

n	#Invalids	#Seconds
2	0	3.25
4	0	10.40
8	0	217.47
9	0	496.61
10	0	1,047.64
11	0	2,155.61
12	0	4,797.34
13	0	9,918.57
14	0	21,959.14

TABLE 24.10 Some Results for 10 Runs

n	Optimal s	Number of runs	Average number of generations
$k = 4$			
2	1	10	0
3	2	10	24.6
4	2	9	430.9
$k = 3$			
5	1	3	669
$k = 2$			
6	1	9	283.78
7	1	3	660.33

As we can see in both tables (and also in Table 24.10), the number of invalid thresholds obtained is always zero. The last column in both tables shows the running time for $s = 100$ and $g = 100$. Although our method is slow, it is no surprise that the algorithm is slower as n grows than as k grows. Such results agree with the complexity analysis given in Section 24.6.1.2. From neural networks applications perspective, results on the number of invalid thresholds for $k \leq 32$ and $k \geq 64$ given in Table 24.8 are interesting, since these values of k correspond to discretizations of *real-valued* neurons by at most 5 bits or at least 6 bits, for fixed n. A more theoretical, not well understood problem would address $n/\log n$, n, or constant number of bits since we know that $n/\log n$ bits is sufficient, n bits is the best known lower bound, while constant number of bits appears to be sufficient in practice for nonmalicious threshold functions.

In Table 24.10 we show results of ten runs of the GA on examples of n-place k-valued logic functions (for $2 \leq k \leq 4$ and $2 \leq n \leq 7$) given by

$$f(\vec{x}) = \left\lfloor \left(\sum_{i=1}^{n} \frac{1}{a_i} x_i \right) + n \right\rfloor \bmod k \tag{24.23}$$

where $a_i = 2i + 1$. We can easily guess the minimum number of thresholds needed for a perceptron to simulate these functions. Indeed, each of these functions defines itself its separating hyperplanes and their number. The number of hyperplanes is simply the number of distinct values of such function minus one, and each hyperplane H_j is defined by the equation $\sum_{i=1}^{n}(1/a_i)x_i = t_j$ for some threshold t_j ($1 \leq t \leq$ number of thresholds). The output vector can also be obtained by computing the value of f for $x_1 = \cdots = x_n = 0$ and $x_1 = \cdots = x_n = k - 1 = 3$ and listing in increasing order modulo k the sequence of other distinct values of f in between. In Table 24.11 we show examples of optimal solutions obtained by the GA.

In Table 24.10, the second column indicates the optimal number of thresholds that the GA must find, the third columns contains the number of runs where the GA reached the optimum, and the fourth column shows the average number of generations over all successful runs needed to obtain the optimal solution. All solutions found by the GA, optimal or not, were valid solutions in that they do not contains invalids thresholds.

As seen from the table, the difficulty for the GA to find an optimal solution within 1000 generations depends mostly on n rather than k. This is not surprising since the search space is exponential on n and thus the GA needs more and more generations (meaning more genetic operations) to successfully obtain an optimum. This is indicated by the fourth column. For $k = 4$ and $n = 5$, for example, the GA could not find an optimum within five runs of 1000 generations each; however, it was successful within one run with 2000 generations. This suggest that given enough time (which depends on n) the GA will always find

TABLE 24.11 Some Optimal Solutions Found

n	$\vec{w}$	$\vec{t}$	$\vec{o}$
$k = 4$			
2	0.953823 0.300371	2.508386	2 3
3	0.830144 0.473697 0.294061	2.303273 4.793706	3 0 1
4	0.785003 0.441867 0.347428 0.260417	2.351256 4.714851	0 1 2
$k = 3$			
5	−0.754707 −0.465486 −0.348600 −0.228958 −0.199491	−2.285579	0 2
$k = 2$			
6	0.487796 0.827506 0.205418 0.075150 0.111406 0.130513	1.595870	0 1
7	0.746777 0.459637 0.299486 0.148247 0.255323 0.191369 0.132577	1.654147	1 0

the minimal s-representation for a logic function f. We do not have rows for higher values of n because of the fact that the algorithm is slow as n grows.

It is interesting to note that the functions we used in our experiments are the most difficult for the GA since the their s-representations are very small (e.g., $s \in \{1, 2\}$). This indicates that for most (random) functions the GA will perform much better than for our test functions because s is larger on average.

We compared our technique with the *extended permutably homogeneous* (n, k, s)*-perceptron learning algorithm* (EPHPLA) described in Section 24.3. A permutably homogeneous perceptron has a $(s + 1, k)$-permutation as its output vector, that is a permutations of $s + 1$ elements out of K with $s \leq k - 1$. The EPHPLA generalizes the homogeneous $(k, k - 1)$-perceptron learning algorithm of Reference 16 and has a time complexity of $O(e \cdot n \cdot k^n)$, where e is the number of learning epochs. The EPHPLA can only learn permutably homogeneous functions and an example of such class of functions are our test functions given by Equation (24.23). It is proven in Reference 34 that the EPHPLA always converges for permutably homogeneous functions, and that also, it always finds a minimal s-representation for a learned function f. The EPHPLA is faster and outperform the GA on learning these same test functions within one run of 1000 learning epochs. The GA converged better only for $n = 2$ and any k. The main advantage of the GA method over the EPHPLA is that it can learn any logic function provided enough time is given.

24.7 Conclusion and Open Problems

Neural net computing is parallel distributed computing with an ensemble of elemental processors interconnected in ways reminiscent of biological neural nets.

Although the idea of computing with arrays of interconnected elemental processors is certainly not new, there is currently a resurgence of interest in this area. This resurgence was initiated through the introduction of a few new algorithms which made it possible, perhaps for the first time, to implement and experiment with some simple realizations of such interconnected nets of elemental processors. Some imaginative demonstrations of such nets helped to activate further interest and fascination with these new developments.

At present there is widespread activity in this area with much interest in the potential use of neural networks in signal and image processing and so on. There is also interest in the possibility of using such nets as computer models for studying the functioning of neuro-biological nets.

This thesis addresses an intellectual issue which is neither of these categories but is more related to the former than to the latter.

The point is that since we can now implement neural nets which can actually perform certain basic information processing functions, it is of interest to see if we can fashion nets or systems of nets which can reproduce (i.e., mimic) certain trains of actions regularly performed by humans. The challenge is to be able to do this without an *oracle* within the net, that is without *devine revelation* or guidance, so to speak, at critical junctures. Positive results gained in activities of such nature would not be interpreted to indicate that that is indeed how human biological nets do function but might serve to suggest preferences among various ways of thinking of processing in biological nets.

The issue we have addressed in this chapter is that of implementing multiple-valued logic systems in neural networks. There are many kinds of multiple-valued logic algebras such as *fuzzy logic, probabilistic logic*, logical calculus with *rough sets*, etc. However, for the present we have concentrated on one such logic system, that is the *classical multiple-valued logic* as defined in this chapter.

In particular, we have discussed original models of multiple-valued neurons and multiple-valued neural networks and studied their learning and computing powers.

24.7.1 Conclusions

We have addressed the issues of synthesizing multiple-valued logic circuits with (n, k, s)-perceptrons. The (n, k, s)-perceptron learning problem is the problem of determining an (optimal) s-representation $\vec{r} = (\vec{w}, \vec{t}, \vec{o})$ required to compute a given function. Another problem related to learning with (n, k, s)-perceptron networks is the search for an optimal size of a network during learning.

1. Learning abilities of (n, k, s)-perceptrons are examined. The previously studied homogeneous $(n, k, k-1)$-perceptron learning algorithm have been generalized to the permutably homogeneous (n, k, s)-perceptron learning algorithm with guaranteed convergence property. A permutably homogeneous perceptron is a neuron whose output vector $\vec{o}$ is a permutation on K. We have obtained a powerful learning method that learns any permutably homogeneous separable k-valued function given as input. When the number of thresholds is not fixed, the algorithm will always find the minimal one to be used for learning a separable function.
2. We have discussed a particular implementation of partitioning algorithm to construct (near) minimal (n, k, s)-perceptron networks for learning given but arbitrary multiple-valued functions. We used GA or ES to find a (near) minimal set of hidden units that partition the space $V \subseteq K^n$ into strips or maximum separable subsets. A strip contains points located between two parallel hyperplanes. We have constructed two neural networks based on these hidden units and show that they correctly compute the given but arbitrary multiple-valued function. STRIP and SEPAR can be used for functions with real-valued inputs, k-valued inputs, two-valued outputs, or k-valued outputs. More research is needed in order to increase the speed of the A-based synthesis algorithm.
3. Every n-input k-valued logic function can be implemented using a (n, k, s)-perceptron, for some number of thresholds s. We proposed a GA to search for a minimal (n, k, s)-perceptron that efficiently realizes a given but arbitray function, that is to minimize its number of thresholds. Experimental evidence show that the genetic search can be very effective however slow it may be.

24.7.2 Open Problems

The following are some interesting open problems related to research in this area:

1. Another complexity measure of neural networks is the *Vapnik-Chervonenkis (VC) dimension.* It is defined as the maximum size of a training set $T \subseteq V$ such that a given network realizes all functions defined on T. If the (n, k, s)-perceptrons have a finite VC-dimension then they may be PAC-learnable. That is, there may exist an efficient (n, k, s)-perceptron learning algorithm which uses only a polynomial number of examples and that, with high probability (which can be made as high as desired), the algorithm outputs a good approximation of a given function within a desired degree of accuracy (which can be made as high as desired). The VC-dimension of (n, k, s)-perceptrons gives a valid lower bound on the VC-dimension of linear decision lists and other related learning architectures.

2. The computing capacity of linear decision lists or any architecture obtained by a partitioning algorithm is not known and is still an open problem. The technique we have applied for deriving capacity results on (n, k, s)-perceptrons may be extended or improved to give results for partitioning architectures. The main question is: In how many ways can we partition a finite set $V \subseteq R^n$ using s hyperplanes? The hyperplanes are not necessarily parallel. The answer to this question gives (bounds on) the capacity of neural networks constructed by partitioning algorithms. Such neural network architectures include neural trees and neural decision lists. This question also motivates for the investigation of the VC-dimension and PAC-learnability of such structures.

3. The combinatorial arguments used to derive our results for the number of linear (or multilinear) partitions and the capacity of (n, k, s)-perceptrons may possibly be extended for the general case of n-dimensional set. For instance, enumerate partitions for higher dimension grids.

4. Suppose the following generalization of a threshold function: $g(P(x)) = 0$ if $P(x) < t$ and, $g(P(x)) = 1$ if $t \leq P(x)$, where $P(x)$ is a polynomial of degree $d \geq 0$ and t is a threshold level. We say that g is a threshold function of order d. For instance, the well-known linear threshold function is a threshold function of order 1 with $P(x) = \vec{w}\vec{x}$. Geometrically, an order d threshold function is a separating hypersurface of degree d. For example, a linear threshold is a separating hyperplane that can be expressed by a polynomial of degree 1. Very little results are known in neural networks literature on using hypersurfaces as discriminant functions, such as quadratic surfaces (parabola, hyperbola, circle, ellipses, spheres, etc.), cubic surfaces, or surfaces of degree $d \geq 4$. It may be that hypersurfaces are better separators than linear surfaces (i.e., hyperplanes). More studies need to be done.

5. An interesting research is to design efficient learning algorithms for multilayer (n, k, s)-perceptron networks. Such network would have the ability to learn any multiple-valued function. Learning and computing abilities of discrete Hopfield networks (or any other type of discrete associative neural network) composed of (n, k, s)-perceptrons can be investigated. Support vector machines (SVMs) with (n, k, s)-perceptrons as processing units can be studied. The capacity of single (n, k, s)-perceptrons may be increased if learning algorithms are designed for output vectors which are $(s+1, k)$-permutations with possible repetitions of their distinct elements.

6. The neural networks we have considered in this thesis are heteregeneous networks, meaning that the (n, k, s)-perceptron elements are not all identical (e.g., the number of thresholds or inputs are not all the same). Designing learning algorithms for homogeneous networks seems (to us) much easier than for heteregeneous networks. However for homogeneous networks, the problem we must solve first is to define a good differentiable error function in order to do learning with gradient descent similar to the back-propagation learning method.

7. The generalization properties of STRIP and SEPAR need to be studied. We believe that our methods generalize better than CARVE since CARVE always overfits the weights. We avoided this overfitting by translating the obtained hyperplanes away from their original positions so to include some test points in the closed space delimited by these hyperplanes. In addition, in higher dimension and for many classes of functions, our STRIP networks are much less complexe (in terms of total number of parameters) than CARVE and others maximum separable based networks, and therefore by the *Occam razor principle* STRIP would generalize better. One can further improve the generalizability by transforming an obtained STRIP network into a *radial basis functions network* and apply back-propagation to the resulting network. Young and Downs [56] have done the same thing with CARVE using sigmoidal units and proved it efficient. In our case, we will use radial basis functions because Gaussian units are continuous version of our $(n, k+1, 2)$-perceptrons.

8. In minimizing the number of thresholds of an (n, k, s)-perceptron, the generalization properties of the GA can be studied when the fitness function is modified to work with small subsets of K^n. Another aspect that would be of interest to neural networks community is related to our proposed GA optimization. If it generates small number of invalid thresholds when working with K^n, it should also be able to discover a near optimal representation for smaller subsets of K^n. If so, and if we believe in *Occam's razor principle* (that the simplest satisfactory explanation of a phenomenon is most likely to be a correct one) then this is interesting from generalization perspective. In particular, instead of calculating $\vec{w}\vec{x}$ for every $\vec{x} \in K^n$, the

same fitness estimation algorithm can be applied to a small enough subset $S \subset K^n$ (e.g., $(k-1)^n$ randomly selected points from K^n). Once a faithful representation for S (called the training set) is learned, it is easy (but interesting) to measure classification error on out of sample data (e.g., total number of mistakes, and maybe sum of squared errors on $K^n - S$). If applied for generalization and if needed, the fitness function can be further modified to allow some error if it results in smaller representation. One major problem with our method is that GA is very slow for even small values of n ($n \geq 10$) and k ($k \geq 8$). Better techniques are needed for efficient computation of (n, k, s)-perceptrons. For example, one possible solution could be to use a population of s-representations $\vec{r}$ (instead of weights $\vec{w}$) and then design a fitness function which minimizes s and minimizes the error between the (n, k, s)-perceptron and a given multiple-valued logic function. Sorting is not needed here since we select only chromosomes whose $\vec{t}$ is sorted. Those with unsorted $\vec{t}$ will have severe penalty.

References

[1] M. Minsky and S. Papert, *Perceptrons: An Introduction to Computational Geometry*, MIT Press, Cambridge, MA, 1969, Expanded edition, 1988.

[2] K.C. Smith, A multiple-valued logic: A tutorial and appreciation. *Computer*, 21, 17–27, 1988.

[3] J. Lukasiewicz, O logice trojwartosciowej. *Ruch Filozoficzny*, 15, 169–171, 1920.

[4] E.L. Post, Introduction to a general theory of elementary propositions. *American Journal of Mathematics*, 43, 163–185, 1921.

[5] A. Ngom, C. Reischer, D.A. Simovici, and I. Stojmenović, Set-valued logic algebra: A carrier computing foundation. *Multiple-Valued Logic — An International Journal*, 2, 183–216, 1997.

[6] S.C. Chan, L.S. Hsu, and H.H. Teh, On neural logic networks. *Neural Networks*, 1 (Suppl. I.), 428, 1988.

[7] V. Milutinović, A. Ngom, and I. Stojmenović, Strip — a strip-based neural network growth algorithm for learning multiple-valued functions. *IEEE Transactions on Neural Networks*, 12, 212–227, 2001.

[8] A. Ngom, Synthesis of Multiple-Valued Logic Functions by Neural Networks, Ph.D. thesis, Computer Science Department, University of Ottawa, Ottawa, Ontario, Canada, October 1998.

[9] Z. Obradović, Computing with nonmonotone multivalued neurons. *Multiple-Valued Logic — An International Journal*, 1, 271–284, 1996.

[10] Z. Tang, Q. Cao, and O. Ishizuka, A learning multiple-valued logic networks: Algebra, algorithm, and application. *IEEE Transactions on Computers*, 47, 247–251, 1998.

[11] G. Wang and H. Shi, Tmlnn: Triple-valued or multiple-valued logic neural network. *IEEE Transactions on Neural Networks*, 9, 1099–1117, 1998.

[12] T. Watanabe, M. Matsumoto, M. Enokida, and T. Hasegawa, A design of multi-valued logic neuron. In *Proceedings of the 20th IEEE International Symposium on Multiple-Valued Logic*, 1990, pp. 418–425.

[13] Q. Cao, O. Ishizuka, Z. Tang, and H. Matsumoto, Algorithm and implementation of a learning multiple-valued logic network. In *Proceedings of the 23rd IEEE International Symposium on Multiple-Valued Logic*, 1993, pp. 202–207.

[14] Z. Tang, O. Ishizuka, Q. Cao, and H. Matsumoto, Algebraic properties of a learning multiple-valued logic network. In *Proceedings of 23rd IEEE International Symposium on Multiple-Valued Logic*, pp. 196–201, 1993.

[15] Z. Obradović and I. Parberry, Computing with discrete multivalued neurons. *Journal of Computer and System Sciences*, 45, 471–492, 1992.

[16] Z. Obradović and I. Parberry, Learning with discrete multivalued neurons. *Journal of Computer and System Sciences*, 49, 375–390, 1994.

[17] R.O. Duda and P.E. Hart, *Pattern Classification and Scene Analysis*, John Wiley & Sons, New York, 1973.

[18] I. Parberry and G. Schnitger, Parallel computation with threshold functions. *Journal of Computing and System Science*, 36, 278–302, 1988.

[19] K.Y. Siu, V. Roychowdhury, and T. Kailath, *Discrete Neural Computation: A Theoretical Foundation*, Information and System Sciences Series. Thomas Kailath, Series Editor. Prentice-Hall, 1995.

[20] S. Muroga, *Threshold Logic and its Applications*, Wiley Interscience, New York, 1971.

[21] D. Haring, Multi-threshold threshold elements. *IEEE Transactions on Electronic and Computer*, 15, 45–65, 1965.

[22] S. Olafsson and Y.A. Abu-Mostafa, The capacity of multilevel threshold functions. *IEEE Transactions on Pattern Analysis and Machine Intelligence*, 10, 277–281, 1988.

[23] R. Takiyama, Multiple threshold perceptron. *Pattern Recognition*, 10, 27–30, 1978.

[24] D. Acketa and J. Žunić, On the number of linear partitions of the (m, n)-grid. *Information Processing Letters*, 38, 163–168, 1991.

[25] O. Ishizuka, Multivalued multithreshold networks. In *Proceedings of the 6th IEEE International Symposium on Multiple-Valued Logic*, 1976, pp. 44–47.

[26] T. Sasao, On the optimal design of multiple-valued plas. *IEEE Computer*, 38, 582–592, 1989.

[27] A. Blum and R.L. Rivest, Training a 3-node neural network is NP-Complete. In *Proceedings of the 1st Workshop on Computational Learning Theory*. Morgan Kaufmann, San Mateo, CA, 1988, p. 9.

[28] J.F. Miller and P. Thomson, Highly efficient exhaustive search algorithm for optimizing canonical ternary Reed-Muller expansions of Boolean functions. *International Journal on Electronics*, 76, 37–56, 1994.

[29] C. Yildirim, J.T. Butler, and C. Yang, Multiple-valued PLA minimization by concurrent multiple and mixed simulated annealing. In *Proceedings of the 23rd IEEE International Symposium on Multiple-Valued Logic*, pp. 17–23, 1993.

[30] S. Kirkpatrick, C.D. Gelatt, and M.P. Vecchi, Optimization by simulated annealing. *Science*, 220, 671–680, 1983.

[31] A. Kaczmarek, V. Antonenko, S. Yanushkevich, and E.N. Zaitseva, Algorithm for network to realize linear MVL functions using arithmetical logic. In *Proceedings of the 12th International Conference on System Science*, pp. 23–30.

[32] Y. Hata, K. Hayase, T. Hozumi, N. Kamiura, and K. Yamato, Multiple-valued logic minimization by genetic algorithms. In *Proceedings of the 27th IEEE International Symposium on Multiple-Valued Logic*, 1997, pp. 97–102.

[33] A. Lloris-Ruiz, J.F. Gomez-Lopera, and R. Roman-Roldan, Entropic minimization of multiple-valued functions. In *Proceedings of the 23rd IEEE International Symposium on Multiple-Valued Logic*, 1993, pp. 24–28.

[34] A. Ngom, C. Reischer, D.A. Simovici, and I. Stojmenović, Learning with permutably homogeneous multiple-valued multiple-threshold perceptrons. *Neural Processing Letters*, 12, 2000, *Proceedings of the 28th IEEE International Symposium on Multiple-Valued Logic*, May 1998, pp. 161–166,

[35] A. Ngom, I. Stojmenović, and R. Tošić, The computing capacity of three-input multiple-valued one-threshold perceptrons. *Neural Processing Letters*, 14, 141–155, 2001.

[36] M.H. Abd-El-Barr, S.G. Zaky, and Z.G. Vranesić, Synthesis of multivalued multithreshold functions for ccd implementation. *IEEE Transactions on Computers*, 35, 124–133, 1986.

[37] R. Takiyama, The separating capacity of a multithreshold threshold element. *IEEE Transactions on Pattern Analysis and Machine Intelligence*, 7, 112–116, 1985.

[38] A. Ngom, I. Stojmenović, and J. Žunić, On the number of multilinear partitions and the computing capacity of multiple-valued multiple-threshold perceptrons. *IEEE Transactions on Neural Networks*, 14, 469–477, 2003, *Proceedings of the 29th IEEE International Symposium on Multiple-Valued Logic*, IEEE Computer Society Technical Committee on Multiple-Valued Logic, IEEE Computer Society, May 1999, pp. 208–213.

[39] H. Edelsbrunner, *Pattern Classification and Scene Analysis*. Springer-Verlag, Heidelberg, 1987.

[40] D.E. Rumelhart, G.E. Hinton, and R.J. Williams, Learning internal representations by error propagation. In *Parallel Distributed Processing: Explorations in the Microstructure of Cognition*,

(J.L. McClelland, D.E. Rumelhart, and the PDP Research Group, Eds.), Vol. I Foundations. MIT Press, Cambridge, MA, 1986.
[41] X. Yao, Evolving artificial neural networks. *Proceedings of the IEEE*, 87, 1423–1447, 1999.
[42] S. Judd, In *Proceedings of the IEEE 1st Conference on Neural Networks*, Vol. 2, San Diego, CA, 1987, p. 685.
[43] T.G. Dieterich and G. Bakiri, Solving multiclass learning problems via error-correcting output codes. *Journal of Artificial Intelligence Research*, 2, 263–386, 1995.
[44] M. Marchand, M. Golea, and P. Ruján, A convergence theorem for sequential learning in two-layer perceptrons. *Europhysics Letters*, 11, 487–492, 1990.
[45] P. Ruján and M. Marchand, A geometric approach to learning in neural networks. *Complex Systems*, 3, 229–242, 1989.
[46] S.A.J. Keibek, H.M.A. Andree, M.H.F. Savenije, G.T. Barkema, and A. Taal, A fast partitioning algorithm and a comparison of feedforward neural networks. *Europhysics Letters*, 18, 555–559, 1992.
[47] M. Marchand and M. Golea, On learning simple neural concepts: From halfspace intersections to neural decisions lists. *Network: Computation in Neural Systems*, 4, 67–85, 1993.
[48] R.L. Rivest, Linear decision list. *Machine Learning*, 2, 229–246, 1987.
[49] S.E. Fahlman and C. Lebière, The cascade-correlation learning architecture. *Advances in Neural Information Processing Systems*, 2, 254, 1990.
[50] M. Frean, The upstart algorithm: A method for constructing and training feedforward neural networks. *Neural Computation*, 2, 198–209, 1990.
[51] I.K. Sethi, Entropy nets: from decision trees to neural networks. In *Proceedings of the IEEE*, 78, 1605–1613, 1990.
[52] J.A. Sirat and J.P. Nadal, Neural trees: A new tool for classification. *Network*, 1, 423–438, 1990.
[53] M. Mezard and J.P. Nadal, Learning in feedforward layered networks: The tiling algorithm. *Journal of Physics A*, 22, 2191–2203, 1989.
[54] G.T. Barkema, H.M.A. Andree, and A. Tall, The patch algorithm: Fast design of binary feedforward neural networks. *Network*, 5, 393–407, 1993.
[55] F.M. Frattale-Mascioli and G. Martinelli, A constructive algorithm for binary neural networks: The oil-spot algorithm. *IEEE Transactions on Neural Networks*, 6, 794–797, 1995.
[56] S. Young and T. Downs, Carve: A constructive algorithm for real-valued examples. *IEEE Transactions on Neural Networks*, 9, 1180–1190, 1998.
[57] S.I. Gallant, Three constructive algorithms for network learning. In *Proceedings of the 8th Annual Conference on Cognitive Science Society*, Amherst, MA, August 15–17, 1986, pp. 652–660.
[58] M. Golea and M. Marchand, A growth algorithm for neural network decision trees. *Europhysics Letters*, 12, 205–210, 1990.
[59] M.M. Muselli, On sequential construction of binary neural networks. *IEEE Transactions on Neural Networks*, 6, 678–690, 1995.
[60] J.P. Nadal, Study of a growth algorithm for neural networks. *International Journal of Neural Systems*, 1, 55–59, 1989.
[61] D.E. Goldberg, *Genetic Algorithms in Search, Optimization, and Machine Learning*. Addison-Wesley, Reading, MA, 1989.
[62] J.H. Holland, *Adaptation in Natural and Artificial Systems*. Michigan University Press, Ann Arbor, MI, 1975.
[63] J.D. Schaffer, L.D. Whitley, and L.J. Eshelman, Combinations of genetic algorithms and neural networks: A survey of the state of the art. In *COGANN-92: International Workshop on Combinations of Genetic Algorithms and Neural Networks*, (L.D. Whitley and J.D. Schaffer, Eds.), IEEE Computer Society Press, Washington, 1992.
[64] A. Ngom, I. Stojmenović, and Z. Obradović, Minimization of multiple-valued multiple-threshold perceptrons by genetic algorithms. In *Proceedings of the 28th IEEE International Symposium on*

Multiple-Valued Logic, Fukuoka, Japan, May 27–29, 1998, IEEE Computer Society Technical Committee on Multiple-Valued Logic, IEEE Computer Society, pp. 209–214.

[65] T. Bäck, *Evolutionary Algorithms in Theory and Practice.* Oxford University Press, Oxford, UK, 1996.

[66] H.-P. Schwefel, *Numerical Optimization of Computer Models.* Wiley, Chichester, New York, 1981.

25

On the Computing Capacity of Multiple-Valued Multiple-Threshold Perceptrons

Alioune Ngom
Ivan Stojmenović
Joviša Žunić

25.1 Introduction

It has been shown in literature that analog neurons of limited precision are essentially discrete multiple-valued neurons. In this chapter, we study the computational abilities of the *Discrete Multiple-Valued Multiple-Threshold Perceptron.* This neuron model extends and generalizes the well-known (two-valued one-threshold) perceptron and other previously studied models. We introduce the concept of multilinear partition of a point set $V \subset R^n$ and the concept of multilinear separability of a function $f: V \mapsto K = \{0, \ldots, k-1\}$. Based on well-known relationships between linear partitions and minimal pairs, an exact and general formula is derived for the number of linear partitions of a given point set V in three-dimensional space, depending on the configuration formed by the points of V. The set V can be a multi-set, that is, it may contain points that coincide. Based on the formula, we obtain an efficient algorithm for counting the number of k-valued logic functions simulated by a three-input k-valued one-threshold perceptron. The n-input k-valued s-threshold perceptrons partition the input space V into $s+1$ regions with s parallel hyperplanes. We obtain results on the capacity of a single such perceptron, respectively, for $V \subset R^n$ in general position and for $V = K^2$. Finally, we describe a fast polynomial-time algorithm for counting the multilinear partitions of K^2.

25.1.1 Multiple-Valued Logic Neural Networks

A *discrete neuron* is a processing unit whose transfer function outputs a discrete value. An example of such transfer function is the linear threshold function. A *discrete n-input multiple-valued neuron* has a discrete transfer function and realizes a function of n variables ranging in the set $S \subseteq R$ with values in $K = \{0, \ldots, k-1\} (k > 0)$, that is, computes a function $f: S^n \subseteq R^n \mapsto K$. For $S = K$ we refer to the processing unit as a *multiple-valued logic neuron* since it simulates a multiple-valued logic function $f: K^n \mapsto K$. Multiple-valued logic neural networks are thus neural networks composed of multiple-valued logic neurons as processing units. The first model of multiple-valued logic neural networks were introduced in Reference 1 and since then various other models have been described References 2 to 6.

25.1.2 The (n, k, s)-Perceptrons

Let R be the set of real numbers. Let $V \subseteq R^n$ and $K = \{0, \ldots, k-1\}$, with $k \geq 2$. An *n-input k-valued real function* $f: V \mapsto K$ is a function with real-valued inputs and k-valued outputs. When $V \subseteq K^n$, we will refer to f as an *n-input k-valued logic function*. We denote by P_k^n the set of all n-input k-valued real (resp. logic) functions $f: V \mapsto K$ and by $P_k = \bigcup_{n \geq 1} P_k^n$ the set of all k-valued real (logic) functions. Thus P_k^n consists of k^{k^n} k-valued logic functions. The set P_k^n is called the set of n-ary operations on K in universal algebras and the set of n-ary functions of k-valued logic in multiple-valued logic algebras. For instance, for $k = 2$, P_2 is the set of all two-valued logic functions.

A *discrete neuron* is a processing unit whose transfer function outputs a discrete value. An example of such transfer function is the linear threshold function. A *discrete n-input multiple-valued neuron* has a discrete transfer function and realizes a function of n variables ranging in a set $V \subseteq R^n$ with values in K, that is, computes a function $f: V \mapsto K$. For $V \subseteq K^n$ we refer to the processing unit as a *multiple-valued logic neuron* since it simulates a multiple-valued logic function $f: K^n \mapsto K$. Our model of multiple-valued logic neuron is the *n-input k-valued s-threshold perceptron* defined below.

In the theory of multiple-valued logic functions there is an important class of functions called *multiple-valued multiple-threshold functions* [7]. Such functions are used in the design of classes of multiple-valued logic circuits called *programmable logic arrays* [8]. A *k-valued s-threshold function* of one variable $y \in R$ is defined as

$$g_{k,s}^{\vec{t},\vec{o}}(y) = \begin{cases} o_0 & \text{if } y < t_1, \\ o_i & \text{if } t_i \leq y < t_{i+1} \quad \text{for } 1 \leq i \leq s-1, \\ o_s & \text{if } t_s \leq y, \end{cases} \tag{25.1}$$

where $\vec{o} = (o_0, \ldots, o_s) \in K^{s+1}$ is the output vector, $\vec{t} = (t_1, \ldots, t_s) \in R^s$ is the threshold vector — with $t_i \leq t_{i+1}$ $(1 \leq i \leq s-1)$ — and s $(1 \leq s \leq k^n - 1)$ is the number of threshold values.

Multiple-threshold devices [9] are threshold elements containing multiple levels of excitation (thresholds). Among their qualities, is that given enough thresholds, a single multiple-threshold element can realize any given function operating on a finite domain [7].

An n-input k-valued s-threshold perceptron [3,10,11], abbreviated as (n, k, s)-*perceptron*, computes an *n-input k-valued weighted s-threshold function* $F^n_{k,s}(\vec{w}, \vec{t}, \vec{o})$ given by

$$F^n_{k,s}(\vec{w}, \vec{t}, \vec{o})(\vec{x}) = g^{\vec{t},\vec{o}}_{k,s}(\vec{w}\vec{x}) = \begin{cases} o_0 & \text{if } \vec{w}\vec{x} < t_1, \\ o_i & \text{if } t_i \leq \vec{w}\vec{x} < t_{i+1}, \\ o_s & \text{if } t_s \leq \vec{w}\vec{x}, \end{cases} \tag{25.2}$$

where input vector $\vec{x} = (x_1, \ldots, x_n) \in V$, weight vector $\vec{w} = (w_1, \ldots, w_n) \in V$, threshold vector $\vec{t} = (t_1, \ldots, t_s) \in R^s$ and $t_i \leq t_{i+1} (1 \leq i \leq s-1$ and $1 \leq s \leq k^n - 1$, the number of threshold values), and output vector $\vec{o} = (o_0, \ldots, o_s) \in K^{s+1}$.

The perceptron's *transfer function* is a k-valued s-threshold function $g^{\vec{t},\vec{o}}_{k,s}: R \mapsto K$. A (n, k, s)-perceptron simulates a k-valued function $f: V \mapsto K$. Depending on V we will either refer to real or logic (n, k, s)-perceptrons. It is well known that any n-input k-valued logic function can be transformed into a k-valued s-threshold function [7,12] for some s.

An (n, k, s)-perceptron is *monotone* if $\vec{o}$ is monotone, that is $o_0 \leq \cdots \leq o_s$ or $o_0 \geq \cdots \geq o_s$, otherwise it is *nonmonotone*. For example, the well-known *binary perceptron* (which is a $[n, 2, 1]$-perceptron) studied in Reference 13 is monotone. A (n, k, s)-perceptron is *homogeneous* if $s = k - 1$ and $\vec{o}$ is the identity permutation of K (i.e., $o_i = i$ for $0 \leq i \leq s$), it is *permutably homogeneous* if $s \leq k - 1$ and $\vec{o}$ is any permutation of $s + 1$ elements out of K. Functions computed by single such neurons are called *permutably homogeneous functions.*

The $(n, k, k-1)$-perceptrons were introduced by Obradović and Parberry [14] as a model for simulating continuous perceptrons (i.e., perceptrons containing analog transfer functions) with limited precision on their inputs. Their computational abilities were extensively studied in Reference 4 whereas Reference 15 examined their learning power. Learning algorithms for some classes of (n, k, s)-perceptrons are investigated in References 2 and 10. Reference 16 introduced the $(n, 2, s)$-perceptrons and their computational power have been intensively studied in References 17 and 18. The neulonets' units discussed in Reference 1 are a class of $(n, 3, 2)$-perceptrons with $\vec{t} = (-d, +d)$, $d \in R$, and $\vec{o} = (1, 0, 2)$.

Multiple-valued logic neural networks, in our definition, are thus neural networks composed of (n, k, s)-perceptrons as processing units. It should be noted that the units are not necessarily the same. For example, one layer of the network may contain $(2, 4, 8)$-perceptrons while another layer may contain $(5, 4, 3)$-perceptrons.

Analog computers are inherently inaccurate due to imperfections in fabrication and fluctuations in operating temperatures. The classical solution to this problem uses extra hardware to enforce discrete behavior. However, the brain appears to compute reliably well with inaccurate components without necessarily resorting to discrete techniques. The *continuous neural network* is a computational model based upon certain observed features of the brain. Experimental evidence has shown continuous neural network to be extremely fault-tolerant; in particular, their performance does not appear to be significantly impaired when precision is limited. It has been shown by Obradović and Parberry [14] that analog neurons of limited precision are essentially multiple-valued logic neurons. The first model of multiple-valued logic neuron (and neural network) was introduced by Chan et al. [1] and since then various other models have been described (see e.g., References 6, 14, and 19).

25.1.3 Problem Statement

The partitioning of points in n-dimensional space has been well studied. The question of how many such partitions are possible arises naturally in the theory of pattern classification and machine learning [13,20–23]. Partitioning by a single hyperplane is fairly limited, and attention has been given to more complex partitioning methods that arise as simple generalization, such as separation by surfaces with polynomial equations [24,25], for instance.

In the context of pattern classification and learning, one proposed way of obtaining more powerful partitioning methods is to use some number s of parallel hyperplanes [17–19,26]. The following question arises: what is the maximum number of ways in which v points in R^n can be partitioned by s parallel hyperplanes (none of which contains any of the points)?

Answering this question, or obtaining good bounds on the answer, has consequences for the *capacity* of generalizations of the threshold functions and threshold elements so central to the theory of artificial neural networks. (These consequences are discussed in Reference 27.)

25.2 Multilinear Separability and Multilinear Partition

The problem of computing (or simulating) a given function f by a (n, k, s)-perceptron is to determine s and a vector $\vec{r} = (\vec{w}, \vec{t}, \vec{o}) \in R^{n+s} \times K^{s+1}$ such that $F^n_{k,s}(\vec{r})(\vec{x}) = f(\vec{x})(\forall \vec{x} \in V)$, that is, $f = F^n_{k,s}(\vec{r})$. We will refer to $\vec{r}$ as an *s-representation* of $F^n_{k,s}$ for f.

Let $V = \{\vec{x}_1, \ldots, \vec{x}_v\} \subseteq R^n$ be a set of v vectors ($v \geq 1$). A k-valued function f with domain V and specified by the input-output pairs $\{(\vec{x}_1, f(\vec{x}_1)), \ldots, (\vec{x}_v, f(\vec{x}_v))\}$, where $\vec{x}_i \in R^n, f(\vec{x}_i) \in K$, is said to be *s-separable* if there exist vectors $\vec{w} \in R^n, \vec{t} \in R^s$ and $\vec{o} \in K^{s+1}$ such that

$$f(\vec{x}_i) = \begin{cases} o_0 & \text{if } \vec{w}\vec{x}_i < t_1, \\ o_j & \text{if } t_j \leq \vec{w}\vec{x}_i < t_{j+1} \quad \text{for } 1 \leq j \leq s-1, \\ o_s & \text{if } t_s \leq \vec{w}\vec{x}_i, \end{cases} \tag{25.3}$$

for $1 \leq i \leq v$. Equivalently, f is s-separable if and only if it has an s-representation defined by $(\vec{w}, \vec{t}, \vec{o})$. A k-valued function over V is said to be *s-nonseparable* if it is not s-separable.

In other words, a (n, k, s)-perceptron partitions the space $V \subset R^n$ into $s+1$ distinct classes $H_0^{[o_0]}, \ldots, H_s^{[o_s]}$, using s parallel hyperplanes, where $H_j^{[o_j]} = \{\vec{x} \in V | f(\vec{x}) = o_j \text{ and } t_j \leq \vec{w}\vec{x} < t_{j+1}\}$. We assume that $t_0 = -\infty$ and $t_{s+1} = +\infty$. Each hyperplane equation denoted by H_j $(1 \leq j \leq s)$ is of the form

$$H_j\colon \vec{w}\vec{x} = t_j. \tag{25.4}$$

Multilinear separability (s-separability) extends the concept of *linear separability* (1-separability of the common binary 1-threshold perceptron) to the (n, k, s)-perceptron. Linear separability in two-valued case tells us that a $(n, 2, 1)$-perceptron can only learn from a space $V \subseteq [0, 1]^n$ in which there is a single hyperplane that separates it into two disjoint halfspaces: $H_0^{[0]} = \{\vec{x} | f(\vec{x}) = 0\}$ and $H_1^{[1]} = \{\vec{x} | f(\vec{x}) = 1\}$. From the *$(n, 2, 1)$-perceptron convergence theorem* [13], concepts that are linearly nonseparable cannot be learned by a $(n, 2, 1)$-perceptron. One example of linearly nonseparable two-valued logic function is the n-input parity function. Likewise, the *(n, k, s)-perceptron convergence theorems* [10,15] state that a (n, k, s)-perceptron computes a given function $f \in P^n_k$ if and only f is s-separable. Figure 25.1 shows an example of 2-separable 4-valued logic function of P^2_5.

Let $V \subset R^n$ with $|V| = v \geq 2$. A *(n, v, s)-partition* is a partition of V by $s \leq v-1$ parallel hyperplanes (namely $(n-1)$-planes), which do not pass through any of the v points. For instance, Figure 25.2 shows an example of $(2, 5^2, 2)$-partition.

A (n, v, s)-partition determines $s+1$ distinct classes $S_0, \ldots, S_s \subset V$ separated by s parallel $(n-1)$-planes such that $\bigcup_{i=0}^{s} S_i = V$ and $\bigcap_{i=0}^{s} S_i = \emptyset$. A (n, v, s)-partition corresponds to an s-separable k-valued function $f \in P^n_k$ if and only if all points in the set S_i, for $0 \leq i \leq s+1$, have the same value taken out of K. Also, we assume that any two neighboring classes have distinct values. If for a given (n, v, s)-partition we have $S_i \neq \emptyset$ $(0 \leq i \leq s+1)$ then, clearly, the number of associated functions is $k(k-1)^s$. In Section 25.6, we consider only partitions where $S_i \neq \emptyset$ for $0 \leq i \leq s$.

A *linear partition* of a point set V is a $(n, v, 1)$-partition, so only a single $(n-1)$-plane is required to separate an n-dimensional space $V \subseteq R^n$ into two halfspaces. The enumeration problem for linear partitions is closely related to the efficiency measurement problem for linear discriminant functions in pattern recognition [22] and to many other algorithmic problems [28].

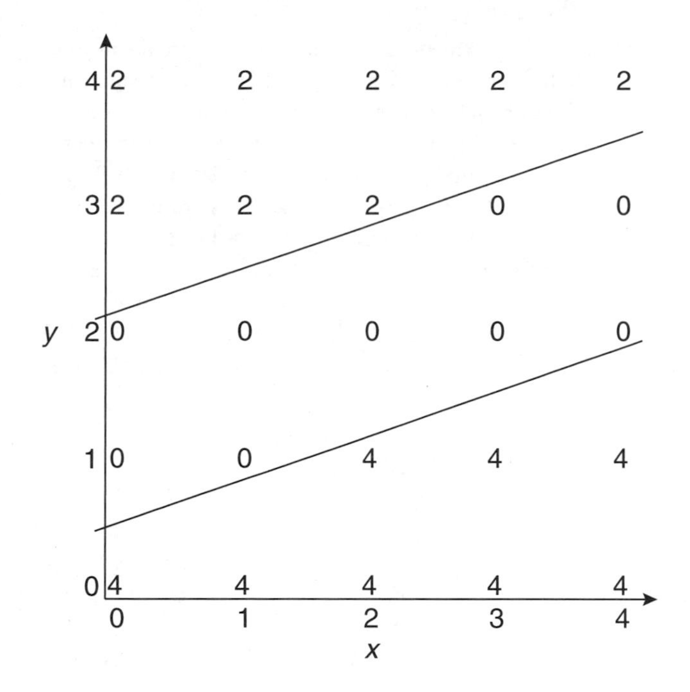

FIGURE 25.1 A 2-separable function of P_5^2.

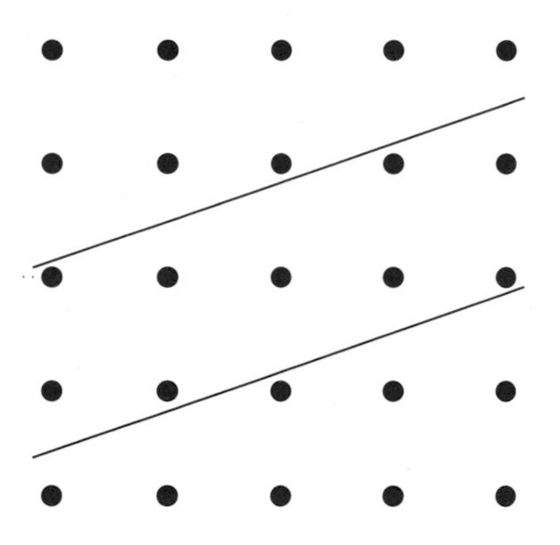

FIGURE 25.2 A $(2, 5^2, 2)$-partition.

25.3 Capacity of Multiple-Valued Multiple-Threshold Perceptrons

In this section we explore the computational power of (n, k, s)-perceptrons. More specifically, the question we will attempt to answer is the following. How many k-valued functions $f\colon V \mapsto K$ can be simulated by a (n, k, s)-perceptron? This will be referred to as the *capacity* of a (n, k, s)-perceptron.

There has been intensive interest in threshold logic as the main component of neural network models. These models provide a direction for pattern recognition systems with distinct natural advantages. The capacity of these models, as well as their computing power, are directly related to the number of threshold functions.

The ability of multiple-threshold devices to simulate a larger number of functions compared to single-threshold devices is vital for the capacity and capabilities of neural network models based on threshold logic. It is therefore of practical as well as theoretical interest to estimate the number of functions that can be modeled as multiple-threshold functions for a given number of inputs and threshold levels.

For given $s \geq 2$, the capacity of a (n, k, s)-perceptron with domain V (i.e., the number of k-valued functions over V that can be simulated by the (n, k, s)-perceptron) is *approximated* by the product of the number of (n, v, s)-partitions and the number of functions associated with each (n, v, s)-partition of V. Thus counting the number of (n, v, s)-partitions of V is a first step toward calculating the capacity of (n, k, s)-perceptrons. We emphasize that the s partitioning $(n-1)$-planes of a (n, v, s)-partition do not pass through any point of V, and therefore we do not obtain the exact number of (n, k, s)-perceptron computable functions.

Let $L_{n,v,s}$ be the number of (n, v, s)-partitions of $V \subset R^n$ and denote by $|F^n_{k,s}|$ the capacity of a (n, k, s)-perceptron. The capacity of $(n, 2, 1)$-perceptrons with domain V [29] is well known and is given by

$$|F^n_{2,s}| = 2\sum_{i=0}^{n}\binom{v-1,}{i} = \begin{cases} 2^v & \text{if } n \geq v-1, \\ 2^v - 2\sum_{i=n+1}^{v-1}\binom{v-1}{i} & \text{otherwise,} \end{cases} \tag{25.5}$$

Reference [17] estimated lower and upper bounds for the capacity of $(n, 2, s)$-perceptrons, using two essentially different enumeration techniques. The paper demonstrated that the exact number of multiple-threshold functions depends strongly on the relative topology of the input set. The results corrected a previously published estimate [18] and indicated that adding threshold levels enhances the capacity more than adding variables.

In order to answer the question, we first describe well-known relationships between linear partitions and minimal pairs in Section 25.4. Based on these relationships, we obtain in Section 25.5 an exact and general formula for the capacity of $(3, k, 1)$-perceptrons. In Section 25.6 we derive results on the capacity of (n, k, s)-perceptrons.

25.4 Linear Partitions and Minimal Pairs

An unordered pair $(\vec{x}, \vec{y})$ of distinct points of a finite set $V \subseteq R^n$ is said to be a *minimal pair* (with respect to V) if there does not exist a third point $\vec{z}$ of V, which belongs to the open line segment $[\vec{x}, \vec{y}]$.

A point set $V \subset R^n$ is said to be in *general position* if and only if no subset of V of $d+1$ points lies on a $(d-1)$-plane (for $1 \leq d \leq n$) and no two $(n-1)$-planes defined from V are parallel (for $2 \leq d \leq n$). The number of linear partitions of a finite planar point set, no three points of which are collinear, is well known [30]. The number of linear partitions of V in general position is a well-known result of pattern recognition [31,32] and is given by

$$L_{n,v,1} = L_{n,v-1,1} + L_{n-1,v-1,1} = \sum_{i=0}^{n}\binom{v-1}{i}. \tag{25.6}$$

It should be noted that Equation (25.6) includes the trivial partition.

A relationship between linear partitions and minimal pairs of a finite point set V in the plane was established in Reference 33. Namely, that the number of linear partitions of V is equal to the number of minimal pairs in V. Moreover, each minimal pair determines two linear partitions of V (see Figure 25.3) and conversely, each linear partition of V has exactly two associated minimal pairs (see Figure 25.4) [34]. These facts provide an easier way of counting linear partitions for arbitrary finite planar point sets: it suffices to count minimal pairs.

In Reference 34 an explicit formula for the number of linear partitions of the (i, j)-grid (a rectangular part of the infinite grid) is stated, using the correspondence with minimal pairs. The same reference includes an efficient algorithm for counting linear partitions of the (i, j)-grid in linear time with respect to the number of its points.

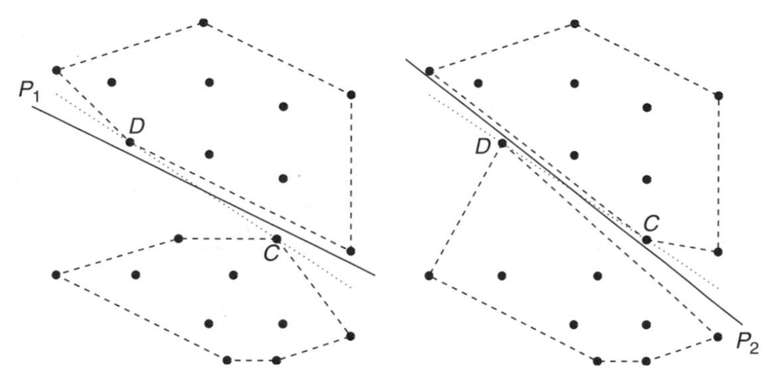

FIGURE 25.3 Minimal pair (C, D) corresponds to separating lines P_1 and P_2.

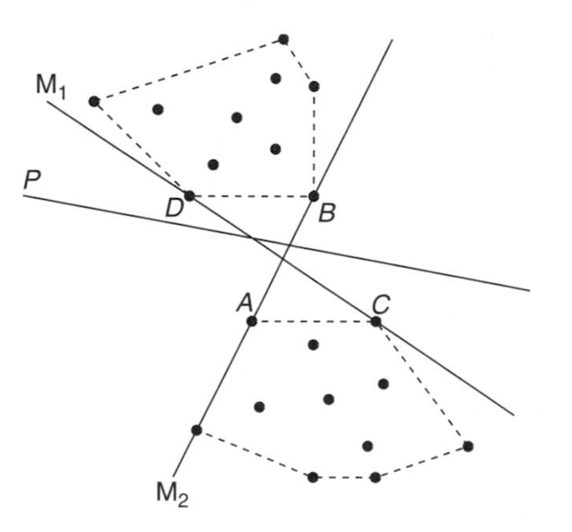

FIGURE 25.4 Separating line P corresponds to minimal pairs (A, B) and (C, D).

Reference 35 derived a formula for the number of linear partitions of a given point set V in two-dimensional and three-dimensional spaces, depending on the configuration formed by the points of V. He considered the case where some points of V may coincide.

25.5 The Computing Capacity of Three-Input Multiple-Valued One-Threshold Perceptron

In this section, an exact and general formula is derived for the number of linear partitions of a given point set V in three-dimensional space, depending on the configuration formed by the points of V. The set V can be a multi-set, that is, it may contain points that coincide. Based on the formula, we obtain an efficient algorithm for counting the number of k-valued logic functions simulated by a three-input k-valued one-threshold perceptron.

A v-set is a set of cardinality v. The cardinality of a set V is denoted by $|V|$. Given a point-set $V \subset R^n$, a linear partition of V is a partition of V into two subsets S_0 and S_1, which is induced by an $(n-1)$-dimensional hyperplane H. It is assumed that the intersection of V and H is empty and that the sets S_0 and S_1 respectively belong to distinct half-spaces with respect to H. In addition, we have $S_0 \cup S_1 = V$ and $S_0 \cap S_1 = \emptyset$. In the sequel we let $|V| = v$.

The capacity of a $(3, k, 1)$-perceptron with domain $V \subset R^3$ is then given by

$$|F^3_{k,1}| = k(k-1)L_{3,v,1} + k, \tag{25.7}$$

where the coefficient $k(k-1)$ is the number of functions associated with each linear partition of V, and the last term k is the number of functions associated with the trivial linear partition $\{V, \emptyset\}$ of V (recall that (n, v, s)-partitions with empty classes are not included in our definition). Determining $L_{3,v,1}$ is the subject of the next sections.

For $V \subset R^3$ in general position, the capacity of $(3, k, 1)$-perceptrons follows directly from Equations (25.6) and (25.7). That is,

Corollary 25.1

$$|F^3_{k,1}| = k(k-1)\sum_{i=1}^{3}\binom{v-1}{i} + k.$$

In this section, however, V is not necessarily in general position and may even contain points that coincide, that is, V can be a multi-set. Thus the formula we obtain will be the most general result for $(3, k, 1)$-perceptrons.

25.5.1 Two-Dimensional Case

From now on, and for ease of notation, we will denote the number of linear partitions of a point set $V \subset R^n$ by $L_{V,n}$ (it includes the trivial partition $\{\emptyset, V\}$). Let V be a planar point set, that is, $V \subset R^2$, and M_V be its number of minimal pairs. The following well-known statement might be very helpful for the enumeration of linear partitions of a planar set.

Lemma 25.2 ([2]) $L_{V,2} = M_V + 1$.

Formula (25.6) follows directly from Lemma 25.2 since every pair of points in V is a minimal. From Lemma 25.2 and Formula (25.6), the following statement can be easily deduced.

Theorem 25.3 *Let $p_1, \ldots, p_d$ be all lines determined by a planar point-set V (each line contains at least two points of V) and let $c_{i,1}$ denote the number of points of V belonging to the line p_i, $1 \le i \le d$. Then*

$$L_{V,2} = 1 + \binom{v-1}{1} + \binom{v-1}{2} - \sum_{i=1}^{d}\binom{c_{i,1}-1}{2}.$$

Proof. If the line p_i contains $c_{i,1}$ points from V, then they determine $\binom{c_{i,1}}{2}$ pairs among which exactly $c_{i,1} - 1$ are minimal pairs; so the number of pairs that are not minimal is $\binom{c_{i,1}}{2} - (c_{i,1} - 1) = \binom{c_{i,1}-1}{2}$. Then the number of nonminimal pairs of points in V is $\sum_{i=1}^{d}\binom{c_{i,1}-1}{2}$. Now, the statement follows from the fact that the total number of pairs in V is $\binom{v-1}{1} + \binom{v-1}{2}$. ■

25.5.2 Generalization of the Two-Dimensional Case

Before proceeding with the two-dimensional case, let us consider the very special case of one dimension. If $V \subset R^1$, then from Formula (25.6), we have

$$L_{V,1} = 1 + \binom{v-1}{1} = v. \tag{25.8}$$

Now, we consider a generalization in R^1, when some points coincide (multiplicity of points). That is there are points in V with multiple occurrence. By A_i^j we denote a point $A_i \in V$ with multiplicity j, that is, j points coincide.

Theorem 25.4 *Let $V = \{A_1^{c_{1,0}}, \ldots, A_d^{c_{d,0}}\}$, where $c_{1,0} + \cdots + c_{d,0} = v$. Then*

$$\overline{L}_{V,1} = 1 + \binom{v-1}{1} - \sum_{i=1}^{d} \binom{c_{i,0}-1}{1}.$$

where $\overline{L}_{V,1}$ is the number of linear partitions of V.

Proof. Obvious. ■

Let $V \subset R^2$ such that $|V| = v$ and $V = \{A_1^{c_{1,0}^V}, \ldots, A_{d_0^V}^{c_{d_0^V,0}^V}\}$, where $c_{i,0}^V$ is the multiplicity of the point A_i in V (for $1 \le i \le d_0^V$), $c_{1,0}^V + \cdots + c_{d_0^V,0}^V = v$ and d_0^V is the number of *distinct* points in V. Let

$$p_1, \ldots, p_{d_1^V} \tag{25.9}$$

be all different lines determined by the points of V (each line contains at least two noncoincident points of V and d_1^V is the number of all such lines in V).

Denote by $c_{i,1}^V$ the number of points of V belonging to the line p_i (with the corresponding multiplicities) for $1 \le i \le d_1^V$. Let $r_{i,0}^V$ be the number of different lines from (25.9) through the point $A_i^{c_{i,0}^V}$, for $1 \le i \le d_0^V$. If we denote by $\overline{L}_{V,2}$ the number of linear partitions of the set V, then the following statement can be proved.

Theorem 25.5

$$\overline{L}_{V,2} = 1 + \binom{v-1}{1} + \binom{v-1}{2} - \sum_{i=1}^{d_0^V} \binom{c_{i,0}^V - 1}{1}$$
$$- \sum_{i=1}^{d_1^V} \binom{c_{i,1}^V - 1}{2} + \sum_{i=1}^{d_0^V} (r_{i,0}^V - 1) \binom{c_{i,0}^V - 1}{2}.$$

Proof. The proof will be by induction on v. Consider a set of points V such that $|V| = v + 1$. Let A be a single point (with multiplicity one) of V, which is a vertex of the convex hull of V. By induction hypothesis, for the set $U = V - \{A\}$ we have

$$L_{U,2} = L_{V-\{A\},2}$$
$$= 1 + \binom{v-1}{1} + \binom{v-1}{2} - \sum_{i=1}^{d_0^U} \binom{c_{i,0}^U - 1}{1}$$
$$- \sum_{i=1}^{d_1^U} \binom{c_{i,1}^U - 1}{2} + \sum_{i=1}^{d_0^U} (r_{i,0}^U - 1) \binom{c_{i,0}^U - 1}{2}. \tag{25.10}$$

Let us determine the number of movable line partitions of U. Project U from the point A into a line a which separates A from U. Let U' be the projection of U. The number of movable line partitions of U is, according to Theorem 25.4

$$\overline{L}_{U',1} = 1 + \binom{v-1}{1} - \sum_{i=1}^{d_0^{U'}} \binom{c_{i,0}^{U'} - 1}{1}. \tag{25.11}$$

From Equations (25.10) and (25.11) it follows that

$$L_{V,2} = L_{U,2} + L_{U',1}$$

$$= 1 + \binom{v}{1} + \binom{v}{2} - \sum_{i=1}^{d_0^V} \binom{c_{i,0}^V - 1}{1} - \sum_{i=1}^{d_1^V} \binom{c_{i,1}^V - 1}{2} + \sum_{i=1}^{d_0^V} (r_{i,0}^V - 1) \binom{c_{i,0}^V - 1}{2}. \tag{25.12}$$

In the case none of the vertices of the convex hull of V is a single point, we shall see what happen when the multiplicity of one point $A \in U(|U| = v)$, increases by one, producing the set V.

Denote, for the sake of simplicity, the sums from the right-hand side of (25.10) by $\sum_0^U, \sum_1^U, \sum_2^U$ respectively and those from (25.12) by $\sum_0^V, \sum_1^V, \sum_2^V$. Since

$$1 + \binom{v}{1} + \binom{v}{2} - \left[1 + \binom{v-1}{1} - \binom{v-1}{2}\right] = v,$$

we need to prove that

$$\left(\sum_0^V + \sum_1^V - \sum_2^V\right) - \left(\sum_0^U + \sum_1^U - \sum_2^U\right) = v. \tag{25.13}$$

Let we have the point $A_i^{c_{i,0}^V+1}$ in V instead of $A_i^{c_{i,0}^U}$ in U. Then

$$\sum_0^V - \sum_0^U = 1. \tag{25.14}$$

Taking into account that $r_{i,0}^V = r_{i,0}^U$ and $c_{i,0}^V = c_{i,0}^U + 1$, we have

$$\sum_2^V - \sum_2^U = (r_{i,0}^U - 1)\binom{c_{i,0}^U}{2} - \binom{c_{i,0}^U - 1}{2},$$

that is,

$$\sum_2^V - \sum_2^U = (r_{i,0}^U - 1)(c_{i,0}^U - 1). \tag{25.15}$$

Let $\{p_1, p_2, \ldots, p_{r_{i,0}^U}\}$ be the set of lines from Equation (25.9) through the point A. Then

$$\sum_2^V - \sum_2^U = \sum_{i=1}^{r_{i,0}^U}\left(\binom{c_{i,1}^V}{2} - \binom{c_{i,1}^V - 1}{2}\right)$$

$$= \sum_{i=1}^{r_{i,0}^U}(c_{i,1}^V - 1) = \sum_{i=1}^{r_{i,0}^U} c_{i,1}^V - \sum_{i=1}^{r_{i,0}^U} 1$$

$$= v - c_{i,0}^U + r_{i,0}^U c_{i,0}^U - r_{i,0}^U$$

$$= v + r_{i,0}^U(c_{i,0}^U - 1) - c_{i,0}^U$$

$$= v + (r_{i,0}^U - 1)(c_{i,0}^U - 1) - 1. \tag{25.16}$$

Now, Equation (25.13) follows from Equations (25.14) to (25.16). ∎

25.5.3 Three-Dimensional Case

Let V be the set of different points in the three-dimensional space such that $|V| = v$. Suppose that $p_1, \ldots, p_{d_1^V}$ are all distinct lines determined by the points of V (each line contains at least two different points of S) and $R_1, \ldots, R_{d_2^V}$ are all distinct planes determined by the points of V (each plane contains at least three different points of V). We introduce the following notations: $c_{i,1}^V$, the number of points of V lying on p_i; $c_{i,2}^V$, the number of points of V lying in R_i; $r_{i,1}^V$, the number of planes from $R_1, \ldots, R_{d_2^V}$ containing the line p_i.

If we denote by $L_{V,3}$ the number of linear partitions of the set V, then the following statement can be proved.

Theorem 25.6

$$L_{V,3} = 1 + \binom{v-1}{1} + \binom{v-1}{2} + \binom{v-1}{3} - \sum_{i=1}^{d_1^V}\binom{c_{i,1}^V - 1}{2} - \sum_{i=1}^{d_2^V}\binom{c_{i,2}^V - 1}{3}$$

$$+ \sum_{i=1}^{d_1^V}(r_{i,1}^V - 1)\binom{c_{i,1}^V - 1}{3}. \tag{25.17}$$

Proof. We use induction (with trivial basis) on $|V|$. Suppose that the statement is valid for $|V| = v$. Consider $|V| = v + 1$.

Take a point A of V. For the sake of simplicity, we may assume (without any loss of generality) that A is a vertex of the convex hull of V. Denote $V - \{A\}$ by U. Consider the projection π of U onto a plane α separating A from U, point A being the center of projection. Those linear partitions of U that can be established by using planes through the point A are said to be *movable* with respect to A.

The number of additional linear partition that are obtained after extension of the set U to V (by adding the point A) is equal to the number of movable (w.r.t. A) linear partitions of U. This last number is equal to $\overline{L}_{Y,2}$, that is, to the number of linear partitions of the planar point set $Y = \pi(U)$. The corresponding bijection is established by the projection π. namely, each movable linear partition (w.r.t. A) may be represented by a plane H through A. The line $h = \pi(H)$ corresponds to a linear partition of $Y = \pi(U)$ in α. Conversely, given a linear partition of Y with the corresponding line h, the plane through h and A determines the associated movable linear partition of V. It follows that

$$L_{V,3} = L_{U,3} + \overline{L}_{Y,2}. \tag{25.18}$$

Now Equation (25.17) can be deduced from Equation (25.18) using induction hypothesis and Theorem 25.5. Namely, by induction hypothesis,

$$L_{U,3} = 1 + \binom{v-1}{1} + \binom{v-1}{2} + \binom{v-1}{3} - \sum_{i=1}^{d_1^U} \binom{c_{i,1}^U - 1}{2} - \sum_{i=1}^{d_2^U} \binom{c_{i,2}^U - 1}{3} + \sum_{i=1}^{d_1^U} (r_{i,1}^U - 1) \binom{c_{i,1}^U - 1}{3}, \tag{25.19}$$

while, according to Theorem 25.5,

$$\overline{L}_{Y,2} = 1 + \binom{v-1}{1} + \binom{v-1}{2} - \sum_{i=1}^{d_0^Y} \binom{c_{i,0}^Y - 1}{1} - \sum_{i=1}^{d_1^Y} \binom{c_{i,1}^Y - 1}{2} + \sum_{i=1}^{d_0^Y} (r_{i,0}^Y - 1) \binom{c_{i,0}^Y - 1}{2}. \tag{25.20}$$

It can be verified that

$$\sum_{i=1}^{d_1^U} \binom{c_{i,1}^U - 1}{2} + \sum_{i=1}^{d_0^Y} \binom{c_{i,0}^Y - 1}{1} = \sum_{i=1}^{d_1^V} \binom{c_{i,1}^V - 1}{2}, \tag{25.21}$$

$$\sum_{i=1}^{d_1^U} \binom{c_{i,2}^U - 1}{3} + \sum_{i=1}^{d_1^Y} \binom{c_{i,1}^Y - 1}{2} = \sum_{i=1}^{d_2^V} \binom{c_{i,2}^V - 1}{3} \tag{25.22}$$

and

$$\sum_{i=1}^{d_1^U} (r_{i,1}^U - 1) \binom{c_{i,1}^U - 1}{3} + \sum_{i=1}^{d_0^Y} (r_{i,0}^Y - 1) \binom{c_{i,0}^Y - 1}{3} = \sum_{i=1}^{d_1^V} (r_{i,1}^V - 1) \binom{c_{i,1}^V - 1}{3}. \tag{25.23}$$

Now, the statement follows from Equations (25.18) to (25.23). ■

25.5.4 Polynomial Time Algorithm for Computing the Capacity of Discrete $(3, k, 1)$-Perceptrons

In this section we describe an algorithm, based on Theorem 25.6, for counting the linear partitions of the three-dimensional square grid K^3, where $K = \{0, \ldots, k-1\}$ and $k \geq 2$.

To efficiently count the linear partitions, we associate each point (x, y, z) of K^3 with an integer $m = xk^2 + yk + z + 1 \in \{1, \ldots, k^3\}$. Also given an integer $m \in \{1, \ldots, k^3\}$ the coordinates of its corresponding point in K^3 are obtained as: $x = ((m-1) \text{ div } k) \text{ div } k$, $y = ((m-1) \text{ div } k) \bmod k$ and $z = (m-1) \bmod k$. Let $V = K^3$. Next we describe the counting algorithm.

Step 1: Generate all d_2^V planes determined by three noncolinear points from V and compute $B = \sum_{i=1}^{d_2^V} \binom{c_{i,2}^V - 1}{3}$ as follows.

1. Initialize B and d_2^V to 0.
2. Generate a candidate plane P as a triple $m_1 < m_2 < m_3$ out of k^3. There are $\binom{k^3}{3}$ candidate planes among which only d_2^V planes are valid.
3. Accept the candidate plane P if m_1 and m_2 are minimal points (i.e., they are the two smallest values) on it and m_3 is minimal point among those noncolinear with m_1 and m_2. That is, P is valid if and only if for each point m in P and $m \notin \{m_1, m_2, m_3\}$ we have $m > m_2$, and, if m is noncolinear with m_1 and m_2 then also $m > m_3$.
4. If P is valid, then
   (a) $d_2^V \leftarrow d_2^V + 1$.
   (b) $c_{d_2^V,2}^V \leftarrow$ number of such points m (described above) $+3$.
   (c) $B \leftarrow B + \binom{c_{d_2^V,2}^V - 1}{3}$.
5. Repeat from Step 2 until no more plane can be generated.

Step 2: Generate all d_1^V lines determined by two points from V and compute the sums $A = \sum_{i=1}^{d_1^V} \binom{c_{i,1}^V - 1}{2}$ and $C = \sum_{i=1}^{d_1^V} (r_{i,1}^V - 1)\binom{c_{i,1}^V - 1}{3}$ as follows.

1. Initialize A, C, and d_1^V to 0.
2. Generate a candidate line L as a pair $m_1 < m_2$ out of k^3. There are $\binom{k^3}{2}$ candidate lines among which only d_1^V lines are valid.
3. Accept the candidate line L if m_1 and m_2 are minimal points on it. That is, L is valid if and only if for each point m in L and $m \notin \{m_1, m_2\}$ we have $m > m_2$.
4. If L is valid, then
   (a) $d_1^V \leftarrow d_1^V + 1$.
   (b) $c_{d_1^V,1}^V \leftarrow$ number of such points m (described above) $+2$.
   (c) $A \leftarrow A + \binom{c_{d_1^V,1}^V - 1}{2}$.
   (d) $r_{d_1^V,1}^V \leftarrow$ number of planes (with third point from V) that pass through L.
   (e) $C \leftarrow C + \left(r_{d_1^V,1}^V - 1\right)\binom{c_{d_1^V,1}^V - 1}{3}$.
5. Repeat from Step 2 until no more line can be generated.

Step 3: Apply Theorem 25.6 and report the results. That is,

1. $L_{V,3} = v + \binom{v}{3} - A - B + C$.
2. $|F_{k,1}^3| = k(k-1)(L_{V,3} - 1) + k = k(k-1)L_{V,3} - k^2 + 2k$.
3. Write d_1^V, d_2^V, $L_{V,3}$, and $|F_{k,1}^3|$.

Figure 25.5 and Figure 25.6 show, respectively, the codes for Step 1 and Step 2.

Remark 1. There is no need to memorize lines and planes, not even to memorize points on them, the algorithms work with one plane (resp. line) at a time. Also, to avoid errors or imprecisions, equations of lines and planes should be made with integer coefficients.

Remark 2. The equation of a plane containing points D, E, F is obtained as follows. Find the normal vector $\vec{n} = \overrightarrow{ED} \times \overrightarrow{EF}$ (cross product), which gives the coefficients a, b, c, then find d from $ax_0 + by_0 + cz_0 + d = 0$, (where (x_0, y_0, z_0) is a point on the plane. Also, three points D, E, F are colinear if and only if $(\vec{D-F}) \times (\vec{E-F}) = 0$ (cross product).

In Step 1, $O(k^9)$ candidate planes are generated and each generated plane is checked for validity at most k^3 times, which gives us a total of $O(k^{12})$ validity tests. Thus Step 1 has a time complexity polynomial

```
B := 0; d_2^V := 0;
For m_1 := 1 to k^3 do
  For m_2 := m_1 + 1 to k^3 do
    For m_3 := m_2 + 1 to k^3 do
      If m_1, m_2, m_3 are non-colinear then
        Good := True;
        Determine equation of plane P : ax + by + cx + d = 0;
          {i.e., find a, b, c, d}
        c^V_{d_2^V,2} := 3;
      m := 1;
      Repeat
        If m ∈ P and m ∉ {m_1, m_2, m_3} then
          If (m < m_2) or (m < m_3 and m, m_1, m_2 not colinear) then
            Good := False;
          Else c^V_{d_2^V,2} := c^V_{d_2^V,2} + 1;
        m := m + 1;
      Until Good = False or m > k^3;
      If Good = True then
        d_2^V := d_2^V + 1;
        B := B + binom(c^V_{d_2^V,2} - 1, 3);
```

FIGURE 25.5 Code for step 1.

TABLE 25.1 Results for $k = 2, \ldots, 8$

k	$d_1^{K^3}$	$d_2^{K^3}$	A	B	C	$L_{K^3,3}$	$\lvert F_{k,1}^3 \rvert$
2	28	20	0	12	0	52	104
3	253	491	49	1,552	0	1,351	8,103
4	1,492	7,502	300	24,422	350	17,356	208,264
5	5,485	52,013	1,338	201,260	4,252	119,529	2,390,565
6	17,092	297,464	3,712	1,031,292	25,852	647,424	19,422,696
7	41,905	1,119,791	10,227	4,322,716	119,598	2,453,869	103,062,463
8	95,140	3,900,890	21,948	14,236,066	418,546	8,399,764	470,386,736

on k. In Step 2, $O(k^6)$ lines are generated and each generation consists of at most k^3 validity checks and $O(k^6)$ time to compute the number of planes that contain the given line. Therefore Step 2 has also a time complexity of $O(k^{12})$ (Table 25.1).

25.6 On the Number of Multilinear Partitions of Multiple-Valued Multiple-Threshold Perceptrons

Based on well-known relationships between linear partitions and minimal pairs, we derive formulae for the number of multilinear partitions of a point set in general position and of the set K^2. The (n, k, s)-perceptrons partition the input space V into $s + 1$ regions with s parallel hyperplanes. We obtain results on the capacity of a single (n, k, s)-perceptron, respectively for $V \subset R^n$ in general position and for $V = K^2$. Finally, we describe a fast polynomial-time algorithm for counting the multilinear partitions of K^2.

```
A := 0; C := 0; d_1^V := 0;
For m_1 := 1 to k^3 do
   For m_2 := m_1 + 1 to k^3 do
      Good := True;
      Determine equation of line L : ax + by + c = 0;
      c^V_{d_1^V,1} := 2;
      m := 1;
      Repeat
         If m ∈ L and m ∉ {m_1, m_2} then
            If (m < m_2) then Good := False;
            Else c^V_{d_1^V,1} := c^V_{d_1^V,1} + 1;
         m := m + 1;
      Until Good = False or m > k^3;
      If Good = True then
         d_1^V := d_1^V + 1;
         A := A + ( (c^V_{d_1^V,1} - 1) choose 2 );
         r^V_{d_1^V,1} := 0;
         For m := 1 to k^3 do
            If m ∉ L then
               First := True;
               m_3 := 1;
               Repeat
                  If m_3 ∈ plane (m_1, m_2, m) and m_3 < m then
                     First := False;
                  m_3 := m_3 + 1;
               Until First = False or m_3 > k^3;
               If First = True then r^V_{d_1^V,1} := r^V_{d_1^V,1} + 1;
         C := C + (r^V_{d_1^V,1} - 1) ( (c^V_{d_1^V,1} - 1) choose 3 );
```

FIGURE 25.6 Code for Step 2.

25.6.1 Counting Multilinear Partitions of Subsets in General Position

In this section, $V \subset R^n$ is in general position and $L_{n,v,s}$ denotes its number of (n, v, s)-partitions. Clearly, every pair of points in V is a minimal pair. We first obtain the number of $(2, v, s)$-partitions, that is, for the two dimensional space, and then we find $L_{n,v,s}$.

We rewrite the number of linear partitions of V in general position [31,32] with a little modification as

$$L_{n,v,1} = L_{n,v-1,1} + L_{n-1,v-1,1} = \sum_{i=1}^{n} \binom{v-1}{i}. \tag{25.24}$$

The difference with the original formula (25.6) is that the index i starts with 1 instead of 0. The reason we start from $i = 1$ is that we do not include partitions containing empty classes.

To count the $(2, v, s)$-partitions, we associate each $(2, v, s)$-partition with a slope $\sigma \in R$ (or equivalently, a minimal pair) as follows:

- For each of the s partitioning lines, choose the corresponding minimal pair which has the smaller slope.
- The associated slope of the given $(2, v, s)$-partition is the maximum among these smaller slopes.

Lemma 25.7 *The number of* $(2, v, s)$*-partitions associated with a given slope* σ *is* $\binom{v-2}{s-1}$.

Proof: For a given minimal pair, rotate its slope σ to increase it a bit in order to obtain the direction of separation P that corresponds to σ. Sort the points of V along this direction, that is, according to their distance to the line P. Consider the selected minimal pair as one point. Then we can choose $s-1$ additional separating lines (parallel to P) for $v-2$ points in $\binom{v-2}{s-1}$ ways. ■

Theorem 25.8

$$L_{2,v,s} \leq \binom{v-2}{s-1}\binom{v}{2}.$$

Proof. Since V is in general position, then there are $\binom{v}{2}$ slopes. The inequality is explained by the fact that two distinct choices of minimal pairs (or slopes) may have sets of associated $(2, v, s)$-partitions that intersect each other. Therefore a given partition may be counted many times, depending on the configuration of V. For instance, consider one of the partitions in Figure 25.7. Clearly, the same partition can be obtained either by selecting the upper minimal pair or by selecting the lower minimal pair (indeed in this example, they both give the same set of associated partitions, even though the points are in general position). We have equality only when $s = 1$. ■

Anthony [36] gave an upper bound on $L_{n,v,s}$, refining an upper bound of Olafsson and Abu-Mostafa [17] which is itself a correction of a claimed upper bound of Takiyama [18]. Their result is as follows:

Theorem 25.9 *(Anthony [36]) The maximum possible number of ways in which v points in* R^n *can be partitioned by s parallel hyperplanes is bounded as follows:*

$$L_{n,v,s} \leq 2\sum_{i=0}^{s}\binom{v-1}{i}\sum_{j=0}^{\lfloor \frac{n-1}{2} \rfloor} S(v-1, n-1-2j),$$

where $S(a, b)$ *is the Stirling number of the first kind, the coefficient of* x^b *in* $\prod_{j=1}^{r}(1 + jx)$

Paper [36] also gave another upper bound on $L_{n,v,s}$, better than that of Theorem 25.9 as follows:

Theorem 25.10 *(Anthony [36]) The maximum possible number of ways in which v points in* R^n *can be partitioned by s parallel hyperplanes is bounded as follows:*

$$L_{n,v,s} \leq \sum_{i=0}^{n+s-1}\binom{vs-1}{i}.$$

Corollary 25.11 *The number of n-input k-valued s-separable functions* $f\colon V \mapsto K$ *is bounded as follows:*

$$|F_{k,s}^{n}| \leq k(k-1)^s \sum_{i=0}^{n+s-1}\binom{vs-1}{i}.$$

Proof. Given a (n, v, s)-partition $S_0, \ldots, S_s$, each class can take one of k values from K such that any two neighboring classes have different values. So there are $k(k-1)^s$ ways to assign values to a (n, v, s)-partition. Each assignment of values to a (n, v, s)-partition defines a unique k-valued s-separable function. The inequality is explained by the fact that some functions can be obtained by at least two different (n, v, s)-partitions (Figure 25.7 shows an example of such function). ■

A *p-permutation* (or permutation of p elements out) of $P = \{a_0, \ldots, a_{p-1}\}$ is an arrangement of $p > 0$ elements into p positions. For example, $a_0a_3a_2a_4a_1$ and $a_3a_0a_1a_4a_2$ are two different five-permutations. The order of the elements is important. There are $p!$ distinct p-permutations. An *(e, p)-permutation* (or permutation of e elements out) of P is an arrangement of e distinct elements of P, with $e \leq p$. For instance, $a_1a_2a_4$ and $a_3a_0a_1$ are two distinct (3, 5)-permutations. The total number of (e, p)-permutations is $p!/(p-e)!$ When $e = p$ we obtain p-permutations. The permutations we consider here are permutations without repetitions (i.e., without repeated elements).

An (n, k, s)-perceptron is *homogeneous* if $\vec{o}$, its output vector, is the identity permutation on K, that is, $o_i = i$ for $0 \leq i \leq s$, otherwise it is *heterogeneous.* An s-separable function $f: V \mapsto K$ is *permutably homogeneous* if and only if $s \leq k-1$ and, for any partitioning of V, no two separate classes have equal value [10]. That is, its output vector is a $(s+1, k)$-permutation. These functions are determined by (n, v, s)-partitions in which each of the $s+1$ classes $S_0, \ldots, S_s$ maps to a distinct value in K. Thus, for permutably homogeneous (n, k, s)-perceptrons we necessarily have $s \leq k-1$.

Theorem 25.12 *A permutably homogeneous s-separable function $f: V \mapsto K$ has a unique (n, v, s)-partition.*

Proof. Suppose f has two distinct (n, v, s)-partitions P_1 and P_2. Consider two points $\vec{x}$ and $\vec{y}$ that are in the same class with respect to P_1 but in different classes with respect to P_2. We have that $f(\vec{x}) = f(\vec{y})$, because $\vec{x}$ and $\vec{y}$ share the same class in P_1. This means that two distinct classes in P_2 have the same value. This contradicts the definition of permutably homogeneous function. ■

Corollary 25.13 *The number of permutably homogeneous n-input k-valued s-separable functions $f: V \mapsto K$ is*

$$|G^n_{k,s}| \leq \frac{k!}{(k-s-1)!} \sum_{i=0}^{n+s-1} \binom{vs-1}{i}.$$

Proof. For a (n, v, s)-partition where $s \leq k-1$, the number of ways to map the $s+1$ classes $S_0, \ldots, S_s$ to distinct values in K equals the number of $(s+1, k)$-permutations, that is, $k!/(k-s-1)!$ Each such $(s+1, k)$-permutation uniquely determines a permutably homogeneous s-separable function. From Theorem 25.12, each permutably homogeneous s-separable function uniquely determines a (n, v, s)-partition. ■

25.6.2 Counting Multilinear Partitions of the (k, k)-Grid

We investigated the computing capacity of the three-input multiple-valued one-threshold perceptron in Reference 11, and obtained a very general formula for the number of linear partitions of a three-dimensional point-set (where the points can be in any possible configuration and the set can be a multi-set). The formula is the most general result for $(3, k, 1)$-perceptrons. In particular, we studied the special case of two-dimensional point-sets and derived a general formula for the number of linear partitions of a planar set. Such planar set can be a multi-set in general or nongeneral position. In Reference 34 an explicit formula for the number of linear partitions of the (i, j)-grid (a rectangular part of the infinite grid) is stated, using the correspondence with minimal pairs. The same reference includes an efficient algorithm for counting linear partitions of the (i, j)-grid in linear time with respect to the number of its points. For a grid configuration, our two-dimension result is a generalization and an extension of Reference 34 to a grid multi-set (i.e., a grid set with multiplicity of points). In this section, we generalize the counting method of Reference 34 to enumerate multilinear partitions of the (k, k)-grid K^2.

An unordered pair $(\vec{x}, \vec{y})$ of distinct points of a finite set $V \subseteq R^n$ is said to be a *minimal pair* (with respect to V) if there does not exist a third point $\vec{z}$ of V which belongs to the open line segment $[\vec{x}, \vec{y}]$.

Let $(\vec{x}, \vec{y})$ be a pair of points from a finite planar grid. Let $a = \max(|x_2 - x_1|, |y_2 - y_1|)$ and $b = \min(|x_2 - x_1|, |y_2 - y_1|)$. One well-known condition for a pair $(\vec{x}, \vec{y})$ to be minimal is that a and b are relatively prime, that is, their greatest common divisor is 1. $a \perp b$ will denote that the integers a and b are mutually simple.

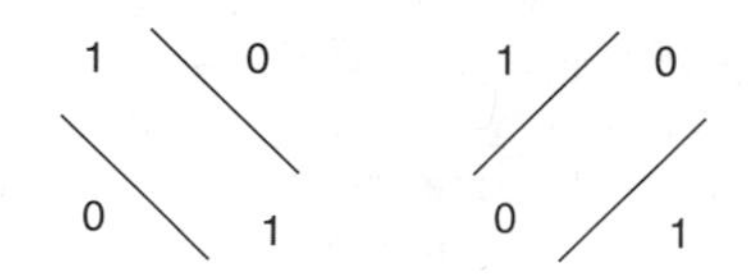

FIGURE 25.7 Two $(2, 4, 2)$-partitions for the same function.

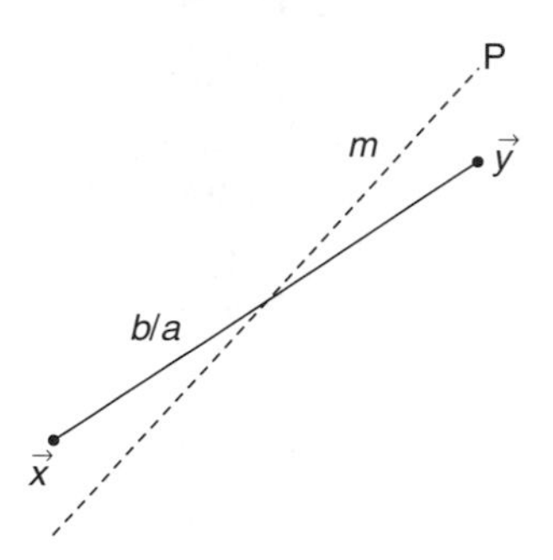

FIGURE 25.8 Rotating slope b/a toward slope m.

Thus the number of minimal pairs corresponds to the number of pairs $(\vec{x}, \vec{y})$ of the (i, j)-grid such that $a \perp b$. Let natural numbers i and j be given so that $i \leq j$. The *generalized Farey* (i, j)*-sequence* $F_{i,j}$ [34] is the strictly increasing sequence of all the fractions of the form b/a, where the integers a and b satisfy: $a \perp b, 0 < b < a \leq j, b \leq i$. Thus the sequence $F_{4,7}$ is as follows:

$$\frac{1}{7}\ \frac{1}{6}\ \frac{1}{5}\ \frac{1}{4}\ \frac{2}{7}\ \frac{1}{3}\ \frac{2}{5}\ \frac{3}{7}\ \frac{1}{2}\ \frac{4}{7}\ \frac{3}{5}\ \frac{2}{3}\ \frac{3}{4}\ \frac{4}{5}.$$

The *Farey i-sequence* F_i for any positive integer i is the set of irreducible rational numbers b/a, with $0 \leq b \leq a \leq i$ and $a \perp b$, arranged in increasing order [37]. So the sequence F_4 is:

$$\frac{0}{1}\ \frac{1}{4}\ \frac{1}{3}\ \frac{1}{2}\ \frac{2}{3}\ \frac{3}{4}\ \frac{1}{1}.$$

The length of the sequence $F_{i,j}$ (resp. F_i) will be denoted by $|F_{i,j}|$ (resp. $|F_i|$). Also, $F_{i,j}^d$ (resp. F_i^d) stands for the d-th fraction in $F_{i,j}$ (resp. F_i), $1 \leq d \leq |F_{i,j}|$ (resp. $|F_i|$).

To count the $(2, k^2, s)$-partitions, we associate to each $(2, k^2, s)$-partition a fraction $(b/a) \in F_{k-1}$ as follows:

- For each of the s partitioning lines, choose one of the two minimal pairs that has the smaller slope.
- The associated slope of the given $(2, k^2, s)$-partition is the maximum among these smaller slopes.

Then to enumerate or generate the $(2, k^2, s)$-partitions associated with a given b/a we will need Lemmas 25.14 and 25.15 below. As in Figure 25.8, rotate a line segment $[\vec{x}, \vec{y}]$ whose slope b/a is irreducible to increase it for a *small* amount, so that we obtain a straight line P with slope $m \notin F_{k-1}$. P is the direction for separation and corresponds to the line segment $[\vec{x}, \vec{y}]$ with slope b/a.

Lemma 25.14 *Slope b/a of the line segment $[\vec{x}, \vec{y}]$ is greater than or equal to the associated slope of any $(2, k^2, s)$-partition in direction parallel to P (where P is the direction of separation that corresponds to the line segment $[\vec{x}, \vec{y}]$).*

Proof. Consider Figure 25.9, where $t \in F_{k-1}$ is the associated slope of a $(2, k^2, s)$-partition parallel to P and $m \notin F_{k-1}$ is the slope of P. From our construction of P we have that $m > (b/a) \in F_{k-1}$ but near b/a.

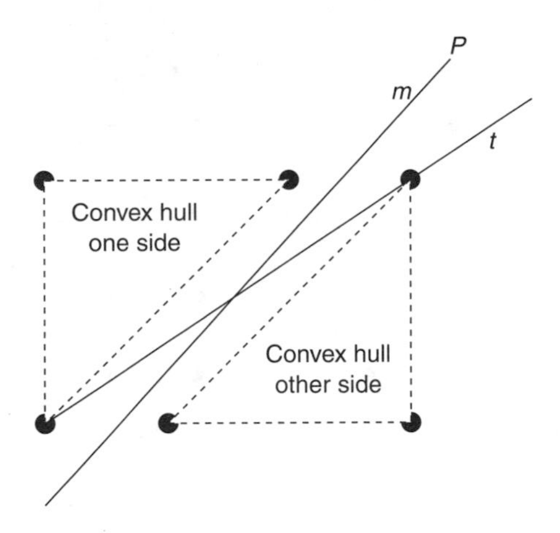

FIGURE 25.9 $t \leq b/a < m$.

Clearly, from Figure 25.9, $t < m$. Let d be the rank of b/a in F_{k-1}. Since $F_{k-1}^{d} = (b/a) < m < F_{k-1}^{d+1}$, $t < m$ and, $t \in F_{k-1}$, therefore we have $t \leq b/a$. ■

Lemma 25.15 *Slope b/a of the line segment $[\vec{x}, \vec{y}]$ is equal to the associated slope of a $(2, k^2, s)$-partition in direction parallel to P if and only if at least one of the s separating lines intersects a minimal pair with slope b/a (where P is the direction of separation that corresponds to the line segment $[\vec{x}, \vec{y}]$).*

Proof. Let $t \in F_{k-1}$ be the associated slope of a $(2, k^2, s)$-partition parallel to P then from Lemma 25.14 we have $t \leq (b/a)$. $\Rightarrow$) Clearly, if none of the separating lines intersects any minimal pair of slope b/a then b/a will not be chosen as the smaller slope for any of the separating lines. Therefore b/a cannot be the associated slope of any $(2, k^2, s)$-partition parallel to P, that is $(b/a) \neq t$. $\Leftarrow$) Since t is the associated slope of a $(2, k^2, s)$-partition parallel to P, then t is greater than or equal to the slopes of the two minimal pairs associated with each separating line. In particular, we have $t \geq (b/a)$ for a separating line that intersects a minimal pair of slope b/a. Since $t \leq (b/a)$ and $t \geq (b/a)$ therefore $t = (b/a)$. ■

Reference 34 obtained an exact formula for the number of linear partitions of the (i, j)-grid. Substituting for the (k, k)-grid we obtain the following corollary.

Corollary 25.16

$$L_{2,k^2,1} = 2k(k-1) + 2(k-1)^2 + 4 \sum_{a \perp b, 0<b<a<k} (k-a)(k-b).$$

Proof. If $b = 0$, then the number of minimal pairs is equal to $2k(k-1)$, which is the number of minimal vertical and horizontal segments of the (k, k)-grid (these are $k(k-1)$ each). This argument explains the first term. One can see from Figure 25.10 that there are $k-1$ segments in each row and column and that there are k rows and columns. If $b > 0$ and $a \perp b$ then $a = b$ implies $a = b = 1$. In that case, the (k, k)-grid contains obviously $2(k-1)^2$ minimal segments with slope $\frac{1}{1}$, which explains the second term. In Figure 25.11, it suffices to count the number of squares of size 1×1. There are $(k-1)^2$ such squares and each square contributes 2 diagonals. Finally, if $a > b > 0$ and $a \perp b$, then there are exactly $2(k-a)(k-b)$ horizontal and vertical rectangles of size $a \times b$ in the (k, k)-grid. The third term comes from the fact that there are two minimal segments per each rectangle (the diagonal segments). In Figure 25.12, the horizontal rectangle in the lower left corner can be translated by one unit length $k-b-1$ times in the horizontal direction to reach the lower right corner, and $k-a-1$ times in the vertical direction to reach the upper left corner (where $k = 5, a = 1, b = 2$). This gives $(k-a)(k-b)$ horizontal rectangles, each contributing two minimal pairs. Similar argument for the vertical rectangle. ■

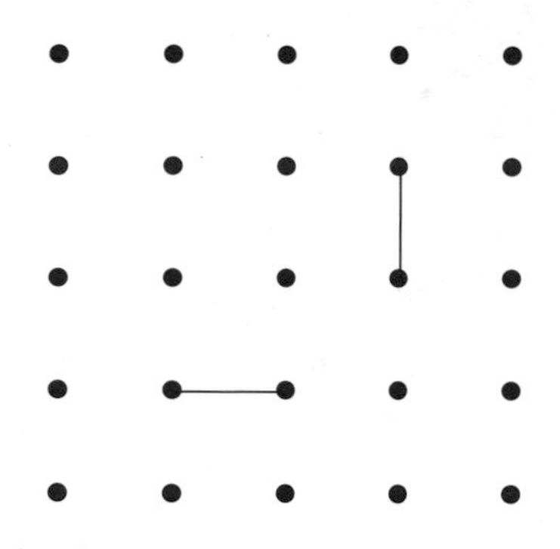

FIGURE 25.10 How many horizontal and vertical segments of length 1?

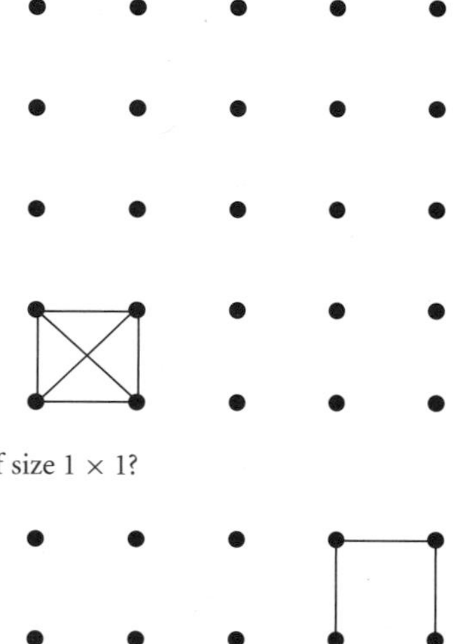

FIGURE 25.11 How many squares of size 1×1?

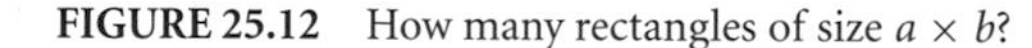

FIGURE 25.12 How many rectangles of size $a \times b$?

Lemma 25.17

$$L_{2,k^2,s} = 4\binom{k^2-1}{s} - 2\binom{k-1}{s} - 2\binom{2k-2}{s} + 4\sum_{a \perp b, 0<b<a<k}\left[\binom{k^2-1}{s} - \binom{ak+bk-ab-1}{s}\right].$$

Proof. Let a slope b/a be such that $a \perp b$ (we consider slopes for directions between 0° and 45°. For each such slope b/a, Lemmas 25.14 and 25.15 give a simple algorithm to construct (i.e., generate) and count the $(2, k^2, s)$-partitions associated with b/a. See Figure 25.13 for an illustration of the proof. ■

- Rotate slope b/a to increase it a bit (this gives the direction of an associated $(2, k^2, s)$-partition P).
- Sort the points along this direction.

- Choose s points $\vec{x}_1, \ldots, \vec{x}_s$ out of $k^2 - 1$ points in $\binom{k^2-1}{s}$ ways (selected points are beginning of classes $S_1, \ldots, S_s$ and the point $\vec{x}_0 = (k-1, 0)$ can always be selected for S_0).
- Eliminate all selections of points where no minimal pair of slope b/a is intersected by a separating line of direction P.

To intersect one of the s separating lines, the lower end of a minimal pair with slope b/a must be selected. From Corollary 25.16 we have:

- If $b = 0$ then there are $k(k-1)$ lower ends (the number of minimal horizontal segments of the (k, k)-grid). None of them can be selected in $\binom{k^2-1-k(k-1)}{s} = \binom{k-1}{s}$ ways. So the number of ways to select a minimal pair with slope $\frac{0}{1}$ to intersect a separating line is $2\binom{k^2-1}{s} - 2\binom{k-1}{s}$.
- If $a = b = 1$ then there are $(k-1)^2$ lower ends (the number of minimal segments with slope 45°). None of them can be selected in $\binom{k^2-1-(k-1)^2}{s} = \binom{2k-2}{s}$ ways. So the number of ways to select a minimal pair with slope $\frac{1}{1}$ to intersect a separating line is $2\binom{k^2-1}{s} - 2\binom{2k-2}{s}$.
- If $a > b > 0$ then there are $(k-a)(k-b)$ lower ends (the number of horizontal rectangles of size $a \times b$ of the (k, k)-grid). None of them can be selected in $\binom{k^2-1-(k-a)(k-b)}{s} = \binom{ak+bk-ab-1}{s}$ ways. So the number of ways to select a minimal pair with slope $0 < (b/a) < 1$ to intersect a separating line is

$$4 \sum_{a \perp b, 0<b<a<k} \left[\binom{k^2-1}{s} - \binom{ak+bk-ab-1}{s} \right].$$

Taking the total sum of all three cases yields the formula. This completes the proof.

Figure 25.13 illustrates the above proof with $a = b = 1, s = 2, k = 3$. The vector $\vec{w}$ gives the direction of separation, which is the dotted line P. The sorted list of points along that direction is $6, 7, 3, 8, 4, 0, 5, 1, 2$. To construct (generate) a $(2, 9, 2)$-partition, we must select three points out of 9 points in the list (these will be beginning of classes S_0, S_1, and S_2, respectively). Point 6 can always be selected as $\vec{x}_0$ for class S_0 and thus we need only to select two points ($\vec{x}_1$ and $\vec{x}_2$) for the remaining classes. A separating line is then placed between $\vec{x}_0$ and $\vec{x}_1$, and another between $\vec{x}_1$ and $\vec{x}_2$ (points $\vec{x}_0, \vec{x}_1, \vec{x}_2$ are assumed sorted along the direction of P). Moreover, if a selected point is a lower end of a minimal pair, we will then place a separating line that intersects the minimal pair. For example, the $(2, 9, 2)$-partition in the figure is generated by selection $x_0 = 6, x_1 = 7, x_2 = 0$. Point 0 is the lower end of minimal pair $(0, 4)$ so we place a separating line between 0 and 4. Among the sorted list of points above, only four points are lower ends of minimal pairs with slope $\frac{1}{1}$, that is, points $0, 1, 3, 4$. To ensure that at least one minimal pair of slope $\frac{1}{1}$ is intersected by a separating line then at least one lower end point must be selected in the construction of a $(2, 9, 2)$-partition. To enumerate the number of $(2, 9, 2)$-partitions parallel to P, it suffices then to count the number of selections of points $(\vec{x}_1, \vec{x}_2)$ that contain at least one lower end point. In Figure 25.13, selections $(6, 7, 8)$, $(6, 8, 5)$, $(6, 8, 2)$, $(6, 7, 5)$, $(6, 7, 2)$, and $(6, 5, 2)$ for instance, will be eliminated since none of them contain lower ends. Also, given selection $(6, 7, 8)$ we can surely have a $(2, 9, 2)$-partition such as in Figure 25.14, even though $(6, 7, 8)$ is an invalid selection. However, such partition will be generated by selection $(6, 3, 0)$ and slope $\frac{1}{2}$, where 0 is the lower end of minimal pair $(0, 7)$ (see dotted separating lines in Figure 25.14).

Theorem 25.18

$$L_{2,k^2,s} = 4\binom{k^2-1}{s}|F_{k-1,k-1}| + 4\binom{k^2-1}{s} - 2\binom{k-1}{s} - 2\binom{2k-2}{s}$$
$$- 4 \sum_{a \perp b, 0<b<a<k} \binom{ak+bk-ab-1}{s}.$$

Proof. Follows from Lemma 25.17 and the fact that the last sum is over $|F_{k-1,k-1}|$. ■

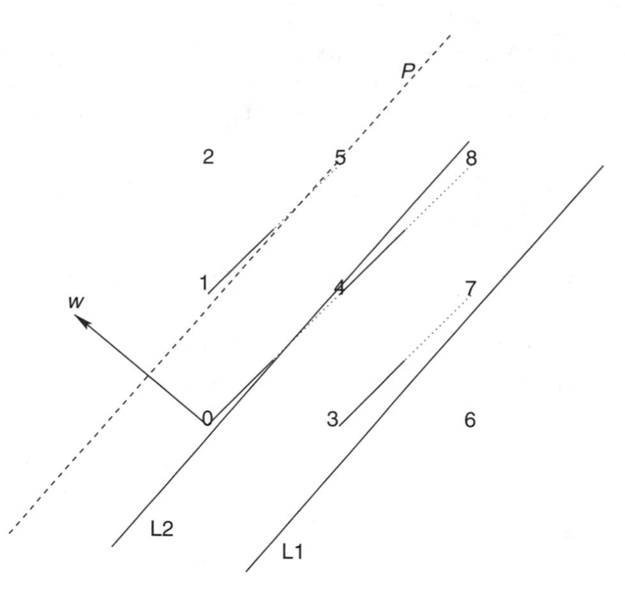

FIGURE 25.13 How many (2, 9, 2)-partitions for slope $\frac{1}{1}$?

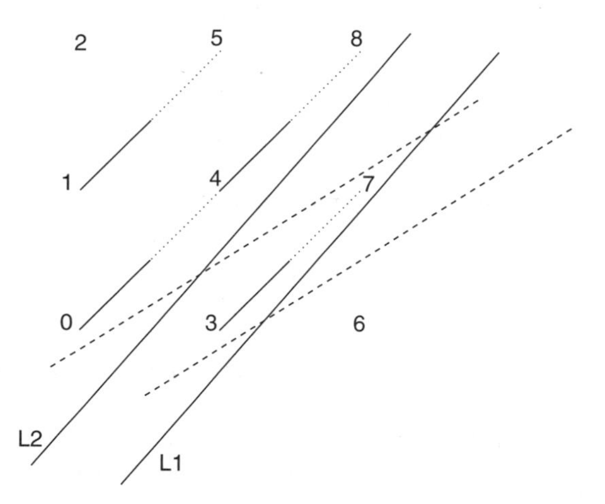

FIGURE 25.14 (2, 9, 2)-partition from slope $\frac{1}{2}$.

Lemma 25.19 $ak + bk - ab - 1 < 2k^2$.

Proof. $ak + bk - ab - 1 < ak + bk < 2k^2$, since $b < a < k$. ■

Corollaries 25.20 and 25.21 below give, respectively, a lower bound and an upper bound on the number of $(2, k^2, s)$-partitions of the (k, k)-grid.

Corollary 25.20

$$L_{2,k^2,s} > 4\binom{k^2-1}{s}(|F_{k-1,k-1}|+1) - 2\binom{k-1}{s} - 2\binom{2k-2}{s} - 4\binom{2k^2}{s}|F_{k-1,k-1}|.$$

Proof. Follows from Lemma 25.19 and Theorem 25.18 because the last sum becomes smaller when substituting $2k^2$ for $ak + bk - ab - 1$. ■

Corollary 25.21

$$L_{2,k^2,s} < 4\binom{k^2-1}{s}(|F_{k-1,k-1}|+1) - 2\binom{k-1}{s} - 2\binom{2k-2}{s}.$$

Proof. Follows from Theorem 25.18 because the last sum is added. ■

The asymptotic formula for the length of the generalized *Farey* (i, j)-sequence (for $i \leq j$) is given in Reference 38 as $|F_{i,j}| = (3i^2j^2/\pi^2) + O(i^2 j \log j) + O(ij^2 \log\log j)$. Hence, substituting for the (k, k)-grid we obtain $|F_{k-1,k-1}| = (3k^4/\pi^2) + O(k^3 \log k) + O(k^3 \log\log k)$.

Theorem 25.22

$$L_{2,k^2,s} \approx \left[4\binom{k^2-1}{s} - 2\binom{2k^2}{s}\right]\frac{3k^4}{\pi^2} + 4\binom{k^2-1}{s} - 2\binom{k-1}{s} - 2\binom{2k-2}{s}.$$

Proof. We take the mean of the lower and upper bounds and replace $|F_{k-1,k-1}|$ by its formula given above. ■

Corollary 25.23 *The number of 2-input k-valued s-separable logic functions is* $|F^2_{k,s}| < k(k-1)^s L_{2,k^2,s}$.

Proof. See Corollary 25.11. ■

The example given in Figure 25.7 explains the inequality in Corollary 25.23. For instance, the XOR function has exactly two $(2, 4, 2)$-partitions. For $s = 1$ we obtain an asymptotic formula.

Corollary 25.24 *The number of permutably homogeneous 2-input k-valued s-separable logic functions is* $|G^2_{k,s}| \approx k!/(k-s-1)! L_{2,k^2,s}$.

Proof. See Corollary 25.13. ■

25.6.3 Complexity of Counting the Number of $(2, k^2, s)$-partitions

Given a point set V in the plane, there are $(v^2 - v)/2$ pairs of points. Generally speaking, each pair requires $v - 2$ tests to check whether it is minimal. Thus the complexity of the general case for the extraction of all minimal pairs is $O(v^3)$. However, if the considered $v = ij$ points are all the points of the (i, j)-grid, then this algorithm turns out to be linear with respect to v, namely $O(v)$, on the basis of Corollary 25.16 and the fact that the successor of a member of $F_{i,j}$ can be calculated in a constant time 34. The simple linear time algorithm for generating $F_{i,j}$ is described in Reference 34.

In Figure 25.15 we show the fast algorithm for counting $(2, k^2, s)$-partitions of the (k, k)-grid. It is a modified version of the 34 algorithm for enumerating the linear partitions of the (i, j)-grid, so, the reader is referred to Reference 34 for a complete proof of correctness of the algorithm. The modifications include replacing the (i, j)-grid by the (k, k)-grid and computing the binomial coefficients present in Lemma 25.17.

Let j/i and d/c be two consecutive elements of $F_{k-1,k-1}$. It has been shown in Reference 34 that the immediate successor b/a of d/c is always determined by the relations $b = rd - j$ and $a = rc - i$, where $r = [(k-1+i)/c]$. This implies that the computation of a consecutive member of $F_{k-1,k-1}$ can be completed in constant time. Another consequence of this fact is that initialization is easy: we can always start with the first two *Farey* numbers $F^1_{k-1,k-1} = 1/k-1$ and $F^2_{k-1,k-1} = (1/k-2)$ and then generate the whole $F_{k-1,k-1}$ sequence in a loop using the above relations.

The last sum of the formula given in Lemma 25.17 necessarily reduces to the sum over the *Farey* $(k-1, k-1)$-sequence. Suppose that k^2 and s require each $O(\log k^2)$ and $O(\log s)$ bits for memory storage, then the complexity of computing the binomial coefficients is $O(s)$. Therefore, for each generation of a number $(b/a) \in F_{k-1,k-1}$, it takes $O(s)$ to compute the number of $(2, k^2, s)$-partitions associated with

Read k and s;
$d := 1; c := k - 1;$
$b := 1; a := k - 2;$
$$L_{2,k^2,s} := 12\binom{k^2-1}{s} - 2\binom{k-1}{s} - 2\binom{2k-2}{s} - 4\binom{k^2-k}{s} - 4\binom{k^2-2k+1}{s};$$
Repeat
  $j := d; i := c;$
  $d := b; c := a;$
  $r := \frac{k-1+i}{c};$
  $b := rd - j;$
  $a := rc - i;$
  $$L_{2,k^2,s} := L_{2,k^2,s} + 4\binom{k^2-1}{s} - 4\binom{ak+bk-ab-1}{s};$$
Until $b = k - 2$ and $a = k - 1$;
Write out $L_{2,k^2,s}$;

FIGURE 25.15 Fast algorithm for counting $(2, k^2, s)$-partitions.

b/a. Since there are k^2 elements in the (k, k)-grid, then it takes $O(sk^2) \leq O(k^4)$ time to calculate $L_{2,k^2,s}$ (recall that $s \leq k^2 - 1$). Clearly, the time complexity is polynomial on k.

25.7 Conclusion and Open Problems

The capacity of a neuron is defined as the total number of functions it can simulate. We introduced, in this chapter, the concept of multilinear partition of a point set $V \subseteq R^n$ and the concept of multilinear separability of a function $f: V \mapsto K = \{0, \ldots, k-1\}$. We obtained a very general formula for the number of linear partitions of a three-dimensional point-set. The points can be in any possible configuration and the set can be a multi-set. Using a polynomial time algorithm, the formula is applied to compute the capacity of discrete $(n, k, 1)$-perceptrons. Based on well-known relationships between linear partitions and minimal pairs, we derived formulae for the number of multilinear partitions of the (k, k)-grids. The counting techniques we used for the (k, k)-grid apply as well for the (i, j)-grid, where $i \leq j$. From the number of multilinear partitions of V, we obtained results on the capacity of a single (n, k, s)-perceptron, respectively for $V \subset R^n$ in general position and for $V = K^2$. Finally, we described a fast polynomial-time algorithm for counting the multilinear partitions of K^2.

The following are some interesting open problems related to research in this area.

Another complexity measure of neural networks is the *Vapnik-Chervonenkis dimension*. It is defined as the maximum size of a training set $T \subseteq V$ such that a given network realizes all functions defined on T. If the (n, k, s)-perceptrons have a finite VC-dimension then they may be PAC-learnable. That is, there may exist an efficient (n, k, s)-perceptron learning algorithm that uses only a polynomial number of examples and that, with high probability (which can be made as high as desired), the algorithm outputs a good approximation of a given function within a desired degree of accuracy (which can be made as high as desired). The VC-dimension of (n, k, s)-perceptrons gives a valid lower bound on the VC-dimension of linear decision lists and other related learning architectures. The computing capacity of linear decision lists or any architecture obtained by a partitioning algorithm is not known and is still an open problem. The technique we have applied for deriving capacity results on (n, k, s)-perceptrons may be extended or improved to give results for partitioning architectures. The main question is: In how many ways can we partition a finite set $V \subseteq R^n$ using s hyperplanes? The hyperplanes are not necessarily parallel.

The answer to this question gives (bounds on) the capacity of neural networks constructed by partitioning algorithms. Such neural network architectures include neural trees and neural decision lists. This question also motivates for the investigation of the VC-dimension and PAC-learnability of such structures. The combinatorial arguments used to derive results for the number of linear (or multilinear) partitions and the capacity of (n, k, s)-perceptrons may possibly be extended for the general case of n-dimensional set. For instance, enumerate partitions for higher dimension grids.

References

[1] S.C. Chan, L.S. Hsu, and H.H. Teh, On neural logic networks. *Neural Networks*, 1 (Suppl. I) 428, 1988.

[2] V. Milutinović, A. Ngom, and I. Stojmenović, Strip — a strip-based neural network growth algorithm for learning multiple-valued functions. *IEEE Transactions on Neural Networks*, 12, 212–227, 2001.

[3] A. Ngom, Synthesis of Multiple-Valued Logic Functions by Neural Networks, Ph.D. thesis, Computer Science Department, University of Ottawa, Ottawa, Ontario, Canada, 1998.

[4] Z. Obradović, Computing with nonmonotone multivalued neurons. *Multiple-Valued Logic — An International Journal*, 1, 271–284, 1996.

[5] Z. Tang, Q. Cao, and O. Ishizuka, A learning multiple-valued logic networks: Algebra, algorithm, and application. *IEEE Transactions on Computers*, 47, 247–251, 1998.

[6] G. Wang and H. Shi, Tmlnn: Triple-valued or multiple-valued logic neural network. *IEEE Transactions on Neural Networks*, 9, 1099–1117, 1998.

[7] O. Ishizuka, Multivalued multithreshold networks. In *Proceedings of the 6th IEEE International Symposium on Multiple-Valued Logic*, 1976, pp. 44–47.

[8] T. Sasao, On the optimal design of multiple-valued plas. *IEEE Computer*, 38, 582–592, 1989.

[9] D. Haring, Multi-threshold threshold elements. *IEEE Transactions on Electronic and Computer*, 15, 45–65, 1965.

[10] A. Ngom, C. Reischer, D.A. Simovici, and I. Stojmenović, Learning with permutably homogeneous multiple-valued multiple-threshold perceptrons. *Neural Processing Letters*, 12, 2000, *Proceedings of the 28th IEEE International Symposium on Multiple-Valued Logic*, May 1998, pp. 161–166.

[11] A. Ngom, I. Stojmenović, and R. Tošić, The computing capacity of three-input multiple-valued one-threshold perceptrons. *Neural Processing Letters*, 14, 141–155, 2001.

[12] M.H. Abd-El-Barr, S.G. Zaky, and Z.G. Vranesić, Synthesis of multivalued multithreshold functions for ccd implementation. *IEEE Transactions on Computers*, 35, 124–133, 1986.

[13] M. Minsky and S. Papert, *Perceptrons: An Introduction to Computational Geometry*, MIT Press, Cambridge, MA, 1969, Expanded edition 1988.

[14] Z. Obradović and I. Parberry, Computing with discrete multivalued neurons. *Journal of Computer and System Sciences*, 45, 471–492, 1992.

[15] Z. Obradović and I. Parberry, Learning with discrete multivalued neurons. *Journal of Computer and System Sciences*, 49, 375–390, 1994.

[16] R. Takiyama, Multiple threshold perceptron. *Pattern Recognition*, 10, 27–30, 1978.

[17] S. Olafsson and Y.A. Abu-Mostafa, The capacity of multilevel threshold functions. *IEEE Transactions on Pattern Analysis and Machine Intelligence*, 10, 277–281, 1988.

[18] R. Takiyama, The separating capacity of a multithreshold threshold element. *IEEE Transactions on Pattern Analysis and Machine Intelligence*, 7, 112–116, 1985.

[19] A. Ngom, I. Stojmenović, and J. Žunić, On the number of multilinear partitions and the computing capacity of multiple-valued multiple-threshold perceptrons. *IEEE Transactions on Neural Networks*, 14, 469–477, 2003, *Proceedings of the 29th IEEE International Symposium on Multiple-Valued Logic, IEEE Computer Society Technical Committee on Multiple-Valued Logic*, May 1999, IEEE Computer Society, pp. 208–213.

[20] M. Anthony and P.L. Bartlett, *Neural Network Learning: Theoretical Foundations*, Cambridge University Press, Cambridge, 1999.

[21] N. Cristiani and J. Shawe-Taylor, *An Introduction to Support Vector Machines*, Cambridge University Press, Cambridge, 2000.

[22] R.O. Dud and P.E. Hart, *Pattern Classification and Scene Analysis*, John Wiley Sons, New York, 1973.

[23] V.N. Vapnik, *Statistical Learning Theory*, John Wiley Sons, New York, 1998.

[24] M. Anthony, Classification by polynomial surfaces. *Discrete Applied Mathematics*, 61, 91–103, 1995.

[25] T.M. Cover, The number of linearly inducible orderings of points in d-space. *SIAM Journal of Applied Mathematics*, 15, 434–439, 1967.

[26] V. Bohossian and J. Bruck, Multiple threshold neural logic. In *Advances in Neural Information Processing* (M. Jordan, M. Kearns, and S. Colla, Eds.), Vol. 10, NIPS'1997, MIT Press, Cambridge, MA, 1998.

[27] M. Anthony, Analysis of data with threshold decision lists. *Center for Discrete and Applied Mathematics Research Report*, CDAM-LSE-2002-12, London School of Economic, December 2002.

[28] H. Edelsbrunner, *Pattern Classification and Scene Analysis*, Springer-Verlag, Heidelberg, 1987.

[29] K.Y. Siu, V. Roychowdhury, and T. Kailath, *Discrete Neural Computation: A Theoretical Foundation*, Information and System Sciences Series (Thomas Kailath, Series Ed.), Prentice Hall, Upper Saddle River, New Jersey, 1995.

[30] J.T. Tou and R.C. Gonzalez, *Pattern Recognition Principles*, Addison-Wesley, Reading, MA, 1974.

[31] T.M. Cover, Geometrical and statistical properties of systems of linear inequalities with applications in pattern recognition. *IEEE Transactions on Electronic and Computer*, 14, 326–334, 1965.

[32] J. Nilsson, *Learning Machines: Foundations of Trainable Pattern Classifying Systems*, McGraw-Hill, New York, 1968.

[33] J. Koplowitz, M. Lindenbaum, and A. Bruckstein, The number of digital straight lines on an $n \times n$ grid. *IEEE Transactions on Information Theory*, 36, 192–197, 1990.

[34] D. Acketa and J. Žunić, On the number of linear partitions of the (m, n)-grid. *Information Processing Letters*, 38, 163–168, 1991.

[35] R. Tošić, On the number of linear partitions. *Review of Research, Mathematical Series*, 22, 141–149, 1992.

[36] M. Anthony, Partitioning points by parallel hyperplanes. *Discrete Mathematics*, to appear.

[37] G.H. Hardy and E.M. Wright, *An Introduction to the Theory of Numbers*, 5th ed., Clarendon Press, Oxford, England,1979.

[38] J. Žunić, On the asymptotic number of linear grid square partitions. *Bild und Ton*, 1991.

26

Advanced Evolutionary Algorithms for Training Neural Networks

Enrique Alba
J. Francisco Chicano
Francisco Luna
Gabriel Luque
Antonio J. Nebro

26.1 Introduction

The interest of making research in Artificial Neural Networks (ANNs) resides in the appealing properties that ANNs exhibit: adaptability, learning capability, and ability to generalize. Nowadays, ANNs are receiving a great attention from the international research community: a large number of studies concerning training, structure design, and real world applications, ranging from classification to robot control or vision [1].

The neural network training task is a capital process in supervised learning, in which a pattern set made up of pairs of inputs plus expected outputs is known beforehand. This set of patterns is used to compute the set of weights that makes the ANN to learn it. To achieve this goal, the algorithm must modify the weights of the neural network in order to get the desired output for a given input, usually in an iterative manner, until a minimum error between the actual and the expected output is attained.

One of the most popular training algorithms in the domain of neural networks is the Backpropagation (BP) technique (generalized delta rule) [2], a gradient-descendent method. Other techniques such as Evolutionary Algorithms (EAs) have been also applied to the training problem in recent years [3,4], trying to avoid the local minima in such a complex problem. Although training is a main issue in ANN's design, many other works exist addressing the evolution of the layered structure of the ANN or even the elementary behavior of the neurons composing the ANN. For example, in Reference 5 a definition of neurons, layers, and the associated training problem is faced by using parallel genetic algorithms (GAs); also, in Reference 6 the architecture of the networks and the weights are evolved by using the EPNet evolutionary system. An exhaustive revision of this topic is really difficult to perform nowadays, however, the work of Yao [7] represents an excellent starting point to get acquired of the research in training ANNs.

The motivation of the present chapter is manyfold. First, we want to perform a standard presentation of results that promotes and facilitates future comparisons. This sounds like common sense, but it is not frequent that authors follow standard rules for comparisons such as the structured Prechelt's set of recommendations [8], a "de facto" standard for many ANN researchers. A second contribution is to include in our study, not only well-known EAs and the BP algorithm, but also the Levenberg–Marquardt (LM) approach [9], and two additional hybrids. The potential advantages coming from an LM utilization merit a detailed study. We have selected a benchmark from the field of Medicine, composed of three classification problems: diagnosis of breast cancer, diagnosis of diabetes in Pima Indians, and diagnosis of heart disease.

The remainder of the chapter is organized as follows. Section 26.2 introduces the ANN computation model. In Section 26.3, we give a brief description of the algorithms under analysis. In Section 26.4, we discuss the mechanisms used for representing solutions and evaluating their quality, a methodological step needed in the application of EAs. The details of the experiments and the analysis of results are shown in Section 26.5. Finally, we summarize our conclusions and future work in Section 26.6.

26.2 Artificial Neural Networks

Artificial Neural Networks are computational models naturally performing a parallel processing of information [10]. Essentially, an ANN can be defined as a pool of simple processing units (*neurons*) which communicate among themselves by means of sending analog signals. These signals travel through weighted connections between neurons. Each of these neurons accumulates the inputs it receives, producing an output according to an internal activation function. This output can serve as an input for other neurons, or it can be a part of the network output. In Figure 26.1 we can see a neuron in detail.

There is a set of important issues involved in the ANN design process. As a first step, the architecture of the network has to be decided. Initially, two major options are usually considered: *feedforward* networks and *recurrent* networks (additional considerations regarding the *order* of the ANN exist, but they are out of our scope). The feedforward model comprises networks in which the connections are strictly feedforward, that is, no neuron receives input from a neuron to which the former sends its output, even indirectly.

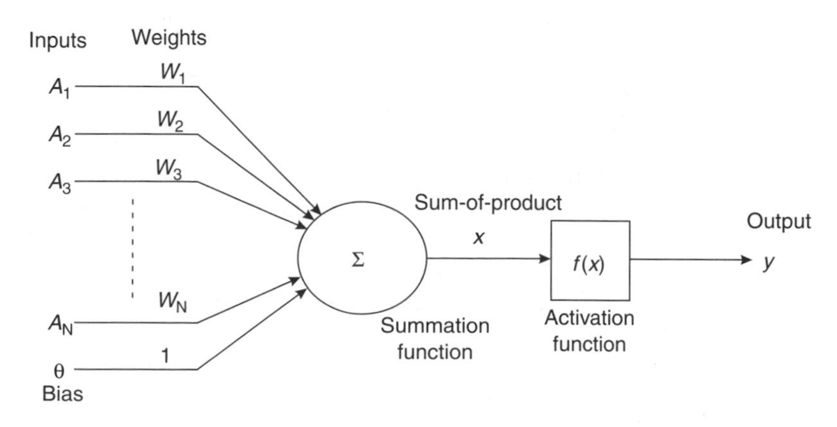

FIGURE 26.1 An artificial neuron.

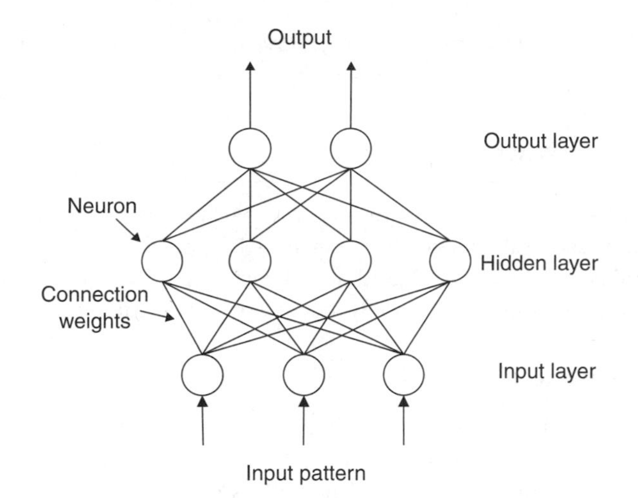

FIGURE 26.2 A MLP with three layers.

The recurrent model defines networks in which feedback connections are allowed, thus inducing complex dynamical properties in the ANN. In this chapter we concentrate on the first and simpler model, the feedforward networks. To be precise, we consider the so-called *multilayer perceptron* (MLP) [11], in which units are structured into ordered layers, and connections are allowed only between adjacent layers in an input-to-output sense (see Figure 26.2).

For any MLP, several parameters such as the number of layers and the number of units per layer must be defined. Then, the last step in the design is to adjust the weights of the network, so that it produces the desired output when its corresponding input is presented. This process is known as *training* the ANN or *learning* the network weights. Network weights comprise both the previously mentioned connection weights as well as *bias* terms for each unit. The latter can be viewed as the weight of a constant saturated input that the corresponding unit always receives. As initially stated, we will focus on the learning situation known as *supervised* training, in which a set of current-input/desired-output patterns is available. Thus, the ANN has to be trained to produce the desired output according to these examples. The input and output of the network are both real vectors in our case.

In order to perform a supervised training we need a way of evaluating the ANN output error with respect to the expected output. A popular measure is the Squared Error Percentage (SEP). This term measures the proximity of the actual output to the desired output. We can compute this error term just for one single pattern or for a set of patterns. In this last case, the SEP is the average value of the separate SEP values. The expression for this global SEP is:

$$\text{SEP} = 100 \cdot \frac{o_{\max} - o_{\min}}{P \cdot S} \sum_{p=1}^{P} \sum_{i=1}^{S} (t_i^p - o_i^p)^2, \tag{26.1}$$

where t_i^p and o_i^p are, respectively, the ith components of the expected vector and the current output vector for the pattern p; $o_{\min}$ and $o_{\max}$ are the minimum and maximum values of the output neurons, S is the number of output neurons, and P is the number of patterns. This SEP value is closely related to the Mean Squared Error (MSE), whose expression is:

$$\text{MSE} = \frac{\sum_{p=1}^{P} \sum_{i=1}^{S} (t_i^p - o_i^p)^2}{P \cdot S}. \tag{26.2}$$

In classification problems, we could use still an additional measure: the Classification Error Percentage (CEP). CEP is the percentage of incorrectly classified patterns, and it is a usual complement to any of the other two (SEP or MSE) crude error values, since CEP reports in a high-level manner the quality of the trained ANN.

26.3 The Algorithms

We use several algorithms to train the networks: the Backpropagation algorithm, the Levenberg–Marquardt algorithm, a canonical Genetic Algorithm, the CHC method, the Hy3 algorithm, an Evolution Strategy (ES), a hybrid between Genetic Algorithm and Backpropagation, and a hybrid between Genetic Algorithm and Levenberg–Marquardt. We briefly describe them in the following paragraphs.

26.3.1 Backpropagation Algorithm

The BP algorithm [2] is a classical domain-dependent technique of supervised training. It works by measuring the output error, calculating the gradient of this error, and adjusting the ANN weights (and biases) in the descending gradient direction. Hence, BP is a gradient-descent local search procedure (foreseeable stagnation in local optima in complex landscapes).

First, we define the squared error of the ANN for a set of patterns:

$$E = \sum_{p=1}^{P} \sum_{i=1}^{S} (t_i^p - o_i^p)^2. \tag{26.3}$$

The actual value of the previous expression depends on the weights of the network. The basic BP algorithm calculates the gradient of E and updates the weights by moving them in the gradient-descendent direction. This can be summarized with the expression:

$$w_{ij}(t+1) = w_{ij}(t) - \eta \frac{\partial E}{\partial w_{ij}}, \tag{26.4}$$

where the parameter $\eta > 0$ is the learning rate that controls the learning speed. A more general BP algorithm adds to the previous expression a momentum term in order to increase the stability of the search process. Then, the final expression for the BP algorithm is:

$$w_{ij}(t+1) = w_{ij}(t) + \alpha \Delta w_{ij}(t) - \eta \frac{\partial E}{\partial w_{ij}}, \tag{26.5}$$

where $\Delta w_{ij}(t)$ is the change in the weight w_{ij} at step t, and α is the momentum constant, whose value must match $0 \leq \alpha < 1$. With this term, the algorithm accelerates the minimization of the error when the error function does not change (in smooth zones of the function). The pseudo-code of the BP algorithm is shown in Figure 26.3.

26.3.2 Levenberg–Marquardt Algorithm

The LM algorithm [9] is an approximation to the Newton method used also for training ANNs. The Newton method approximates the error of the network with a second order expression, which contrasts to the first order information used in the BP algorithm. LM is very popular in the ANN domain (even being adopted as the initially more promising approach for an unseen MLP training task). Curiously, it is not that popular in the metaheuristic field. LM updates the ANN weights as follows:

$$\Delta \mathbf{w} = - \left[\mu I + \sum_{p=1}^{P} J^p(\mathbf{w})^{\mathrm{T}} J^p(\mathbf{w}) \right]^{-1} \nabla E(\mathbf{w}), \tag{26.6}$$

```
InitializeWeights;
for all i,j do
    Δw_ij := 0;
endfor;
while not StopCriterion do
    for all i,j do
        Δw_ij := αΔw_ij − η ∂E/∂w_ij;
        w_ij := w_ij + Δw_ij;
    endfor;
endwhile;
```

FIGURE 26.3 Pseudo-code of the BP algorithm.

```
InitializeWeights;
while not StopCriterion do
    calculate e^p(w) for each pattern;
    e1 := Σ_{p=1}^{P} e^p(w)^t e^p(w); //error
    calculate J^p(w) for each pattern;
    repeat
        calculates Δw;
        e2 := Σ_{p=1}^{P} e^p(w + Δw)^t e^p(w + Δw);
        if (e1 <= e2) then
            μ := μ * β;
        endif;
    until (e2 < e1);
    μ := μ/β;
    w := w + Δw;
endwhile;
```

FIGURE 26.4 Pseudo-code of the LM algorithm.

where $J^p(\mathbf{w})$ is the Jacobian matrix of the vector $\mathbf{e}^p(\mathbf{w})$ evaluated in $\mathbf{w}$, and I is the identity matrix. The vector $\mathbf{e}^p(\mathbf{w})$ is the error of the network for pattern p, that is, $\mathbf{e}^p(\mathbf{w}) = \mathbf{t}^p - \mathbf{o}^p(\mathbf{w})$. The parameter μ is increased or decreased at each step. If the error is reduced then μ is divided by a factor β and it is multiplied by β in other case. LM performs the steps included in Figure 26.4. It calculates the network output, the error vectors, and the Jacobian matrix for each pattern. Then, it computes $\Delta\mathbf{w}$ using Equation (26.6) and recalculates the error with $\mathbf{w} + \Delta\mathbf{w}$ as network weights. If the error has decreased, μ is divided by β, the new weights are maintained, and the process starts again; otherwise, μ is multiplied by β, $\Delta\mathbf{w}$ is calculated with a new value, and it probes again.

26.3.3 Evolutionary Algorithms

Evolutionary Algorithms [12] are heuristic search techniques loosely based on the principles of natural evolution, namely adaptation and survival of the fittest. These techniques have been shown to be very effective in solving hard optimization tasks with similar properties to ANN training, that is, problems in which gradient-descent techniques get trapped into local minima, or are fooled by the complexity and non-differentiability of the search space. The underlying idea in EAs is making individuals represent

```
t := 0;
P(0) = Generate ();
Evaluate (P(0));
while not StopCriterion do
      P'(t) := Select (P(t));
      P''(t) := Recombine (P'(t));
      P'''(t) := Mutate (P''(t));
      Evaluate (P'''(t));
      P(t+1) := Replacement (P(t),P'''(t));
      t := t+1;
endwhile;
```

FIGURE 26.5 Pseudo-code of a GA.

the weights and biases of the ANN, using the network error function to be minimized. Some general considerations must be taken into account when using an evolutionary approach for ANN training. The search features of EAs contrast with those of the BP and LM in that they are not trajectory driven, but population driven. An EA is expected to avoid local optima frequently by promoting exploration of the search space, in opposition to the exploitative trend usually allocated to local search algorithms such as BP or LM. In this chapter, we use four EAs: a canonical GA, the CHC method, the Hy3 algorithm, and an ES.

26.3.3.1 Genetic Algorithms

A GA proceeds in an iterative manner by generating new populations of strings from the old ones (Figure 26.5). Every string is the encoded (binary, real, etc.) version of a tentative solution. The canonical algorithm applies stochastic operators such as selection, crossover, and mutation on an initially random population in order to compute a new population. In *generational* GAs all the populaton is replaced with new individuals. In *steady-state* GAs only one new individual is created and it replaces the worst one in the population if it is better. In our study, the experimental section explains the used parameters and details deeper.

26.3.3.2 CHC

The CHC acronym stands for "Cross generational elitist selection, Heterogeneous recombination, and Cataclysmic mutation" [13]. A CHC is a non-canonical GA that combines a conservative replace strategy (*elitist replace*) with a highly disruptive recombination (HUX). In Figure 26.6 we present the pseudo-code for the CHC method. The main differences between a canonical GA and a CHC are described in the following paragraphs.

First, in CHC, the bias in favor of the best structures occurs in the replacement stage (or survival-selection) rather than in reproduction-selection. More precisely, during the selection for reproduction, each member of the current population is copied to the parent set, that is, the candidates for reproduction are identical to the current population except that the order of the structures has been shuffled. During the replacement, the offspring and the old population are merged and ranked according to the fitness value, the new population is created by selecting the best μ members out of the merged population (where μ is the population size). This mechanism always preserves the best individuals found so far.

Second, a new bias is introduced against mating individuals who are similar (*incest prevention*). The initial threshold (minimum Hamming distance between the parents) for allowing mating is often set to 1/4 of the chromosome length. If no offspring is inserted into the new population during the reproduction, this threshold is reduced by 1.

```
t := 0;
dist := chromosomeLength/4;
P(0) = Generate ();
Evaluate (P(0));
while not StopCriterion do
      P'(t) := Shuffle (P(t));
      P''(t) := HUX (P'(t));
      Evaluate (P''(t));
      P'''(t) := Merge (P''(t), P(t));
      P(t+1) := SelectBest (P'''(t));
      if (P(t+1) == P(t)) then
         dist := dist-1;
      endif;
      if (dist < 0) then //restart
         diverge (P(t+1));
         dist := r * (1.0 - r) * chromosomeLength;
      endif;
      t := t+1;
endwhile;
```

FIGURE 26.6 Pseudo-code of the CHC algorithm.

Third, the recombination operator in CHC is a variant of uniform crossover (HUX), a highly disruptive form of crossover. This operator crosses over exactly half of the bits that differ between the two parent strings. HUX guarantees that the children are always the maximum Hamming distance from their parents.

Finally, a *restart* process reintroduces diversity whenever convergence is detected (i.e., the difference threshold of the incest prevention bias has dropped to zero). The population is restarted by using the best individual found so far as a template for creating the new population.

26.3.3.3 Hy3 Algorithm

In this section we discuss a recent algorithm proposal called Hy3 [14], a kind of distributed technique with heterogeneous components to enhance the search. Distributed GAs (dGAs) are a kind of structured GAs where the population is partitioned into several subpopulations, each one processed by a GA, independently of the others. In addition, sparse migrations of individuals produce an exchange of genetic material between the subpopulations that enhances diversity and usually improves the accuracy and efficiency of the algorithm. Making different decisions on the subalgorithms of a dGA through the application of different search strategies, we obtain the so-called *heterogeneous* dGAs.

The Hy3 algorithm is a heterogeneous dGA that runs eight populations in a cubic topology. This model is suitable for the optimization of continuous functions, because it includes in the basic improvement loop of the algorithm the utilization of crossover operators for real genes (variables), engineered with fuzzy logic technology to deal with the traditional "fuzzy" GA concepts of exploration and exploitation. Hy3 is included into the class of heterogeneous algorithms since it applies a different crossover operator in each of its component subpopulations. There are two important faces in the topology (Figure 26.7) that have to be considered:

- The *front side* is devoted to exploration. It is made up of four subpopulations $E_1, \ldots, E_4$, in which several exploratory crossovers are applied.
- The *rear side* promotes exploitation. It is composed of subpopulations $e_1, \ldots, e_4$, that apply exploitative crossover operators.

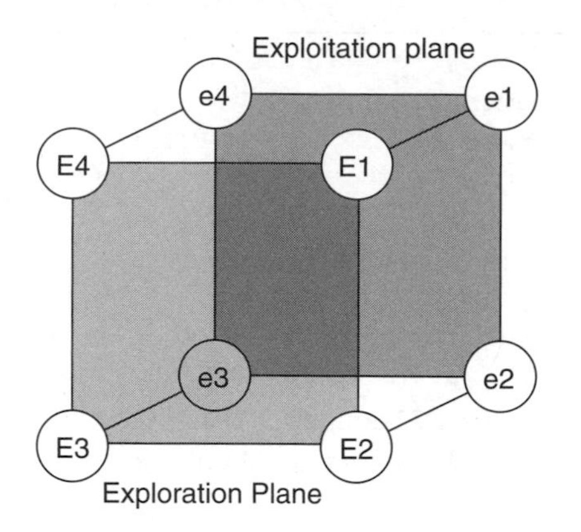

FIGURE 26.7 Structure of Hy3.

```
t := 0;
P(0) = Generate ();
Evaluate (P(0));
while not StopCriterion do
      P'(t) := Select (P(t));
      P''(t) := Recombine (P'(t));
      P'''(t) := GaussianMut (P''(t));
      Evaluate (P'''(t));
      P(t+1) := Replacement (P(t),P'''(t));
      t := t+1;
endwhile;
```

FIGURE 26.8 Pseudo-code of an ES.

The resulting structure is a parallel-suited multi-resolution method using several crossover operators that allow to achieve simultaneously a diversified search (reliability) and an effective local tuning (accuracy).

26.3.3.4 Evolution Strategy

In ES [15] each individual is composed of a vector of real numbers representing the problem variables (**x**), a vector of standard deviations (σ), and, optionally, a vector of angles (ω). These two last vectors are used as parameters for the Gaussian mutation and undergo evolution together with the problem variables, thus allowing the algorithm to self-adapt the way the search is performed.

Figure 26.8 outlines the evolution strategy. First, a random initial population of μ individuals is generated. Then, several individuals are selected from the population for recombination and mutation to obtain λ offsprings. Finally, the new population is formed selecting μ individuals from the offsprings and the current population. If the new population is only formed with the offsprings we have a (μ, λ)-ES, where it must hold $\mu \leq \lambda$. If the current population is taken into account in addition to the offsprings, we have a $(\mu + \lambda)$-ES.

The recombination and mutation affect the three potential vectors contained in the individual, that is, the problem variables and the self-adapted parameters. The mutation of the individual follows the next equations:

$$\sigma_i' = \sigma_i \exp(\tau N(0,1) + \eta N_i(0,1)), \tag{26.7}$$

$$\omega_i' = \omega_i + \varphi N_i(0,1), \tag{26.8}$$

$$\mathbf{x}' = \mathbf{x} + \mathbf{N}(\mathbf{0}, C(\sigma', \omega')), \tag{26.9}$$

where $C(\sigma', \omega')$ is the covariance matrix associated with σ' and ω', $N(0, 1)$ is the standard univariate normal distribution, and $\mathbf{N}(\mathbf{0}, C)$ is the multivariate normal distribution with mean $\mathbf{0}$ and covariance matrix C. The subindex i in the standard normal distribution indicates that a new random number is generated for each component of the vector. The notation $N(0, 1)$ is employed to indicate that the same random number is used for all the components. The parameters τ, η, and φ are set to $(2n)^{-1/2}$, $(4n)^{-1/4}$, and $5\pi/180$, respectively as mentioned in Reference 16.

For the recombination many alternatives may be used. In Reference 17 Fogel summarizes some of them. The three vectors of the individual are recombined in an independent way, that is, for each vector, a different recombination scheme can be chosen.

26.3.4 Hybrid Algorithms

Hybridization is almost "a must" in complex applications if one expects to get efficient and accurate algorithms. In our study, "hybridization" refers to the inclusion of problem-dependent knowledge in a general search template [18,19]. We can distinguish two kinds of hybridization: strong and weak hybridization. In the first one, the knowledge is included using specific operators or representations. In the latter several algorithms are combined somehow. In this last case, an algorithm can be used to improve the results of another one separately or it can be used as an operator of the other.

The hybrid algorithms that we use in this chapter are combinations of two algorithms (weak hybridization), where one of them acts as an operator in the other. We combine a GA with the BP algorithm (GABP), and a GA with LM (GALM). In both cases the problem-specific algorithm (BP and LM) is used as a mutation operation of the general search template (GA). Therefore, we can say that GAxx is a GA (see Figure 26.5) in which the mutation has been replaced by the "xx" algorithm that is applied with probability p_t.

26.4 Representation and Fitness Function

One of the first topics that has to be addressed to solve a problem with an algorithm is the representation of the solutions. This is a very important step in tailoring a general search sheet as an EA for the problem at hands. In our case, a solution is composed of the values of all the weights and biases in the ANN. These values can be included in a real vector where each weight (and bias) is allocated always the same position in it. This k-dimensional real vector could be codified in different ways. In the present analysis we consider a linear encoding of the vector, that is, first, each variable is conveniently encoded in an algorithm-dependent way; subsequently, the codified vector is constructed by concatenating the encoding of each variable into a linear string. This linear encoding of variables raises a question: the best distribution of the variables within the string. This distribution is important in connection with the particular recombination operator used in the EAs. In effect, if this operator breaks the strings into large blocks using them as units for exchange, this distribution might be relevant. On the contrary, using a recombination operator that breaks the strings into very small chunks makes the distribution irrelevant. A good piece of advice is grouping together the input weights and bias for each unit. This way, the probability of transmitting them as a block is increased. Obviously, recombination is not used in many EAs, so this consideration does not apply to all situations.

In BP, LM, ES, and Hy3 each variable is encoded using a machine-dependent codification for real numbers. In the rest, the encoding of solutions is approached via binary strings. More precisely, m bits are used to represent each single variable; subsequently, the k m-bit segments are concatenated into an l-bit binary string, where $l = k \cdot m$. This encoding of the network variables raises a number of issues. Two of them are the choice of m and the encoding mechanism for individual variables, that is, pure binary, Gray-coded numbers, magnitude-sign, etc. In this work we use 16-bit pure binary encoding. The integer value of each variable is mapped linearly into an interval and the result is the weight value.

Now, we discuss the alternatives for the fitness function in order to evaluate the quality of the solutions in EAs. The objective of the network is to classify all the patterns correctly, that is, to obtain 0% of CEP. We can use the CEP as a function to minimize. However, among two networks with the same CEP for the pattern set, the output of one of them can be nearer to the desired output than the other. For this reason, the SEP can better guide the search. In our presentation, the fitness function to be minimized is the SEP for the training pattern set. Conversely, a maximization approach should consider the inverse of the SEP as fitness.

26.5 Computational Experiments

After discussing the algorithms, the representation, and the fitness function, we present in this section the experiments performed and their results. The benchmark for training and the parameters of the algorithms are presented in the next subsection. The analysis of the results is shown in Section 26.5.2.

26.5.1 Benchmark and Parameters

We tackle three classification problems, which consist in determining the class that a certain input vector belongs to. Each pattern from the training set contains an input vector and its desired output vector. These vectors are composed of real numbers. For the classification task we must interpret the network output as a class. The interpretation can be performed in different ways [8]. We adopt here the so called *winner-takes-all* in which an output neuron is associated to each class. When an input vector is presented to the network, the network response is the class associated with the output neuron with the larger value.

The instances solved here belong to the PROBEN1[1] benchmark [8]: Cancer, Diabetes, and Heart. We now briefly detail them:

- *Cancer*: Diagnosis of breast cancer. Classify a tumor as either benign or malignant based on cell descriptions gathered by microscopic examination. There are 699 examples that were obtained by Dr. William H. Wolberg at the University of Wisconsin Hospitals, Madison [20–23].
- *Diabetes*: Diagnose diabetes of Pima Indians. Based on personal data and the results of medical examinations, decide whether a Pima Indian individual is diabetes positive or not. There are 768 examples from the *National Institute of Diabetes and Digestive and Kidney Diseases* by Vincent Sigillito [24].
- *Heart*: Predict heart disease. Decide whether at least one of four major vessels is reduced in diameter by >50%. This decision is made based on personal data and results of medical examinations. There are 920 examples from four different sources: Hungarian Institute of Cardiology in Budapest (Andras Janosi, M.D.), University Hospital of Zurich in Switzerland (William Steinbrunn, M.D.), University Hospital of Basel in Switzerland (Mathhias Pfisterer, M.D.), V.A. Medical Center of Long Beach and Cleveland Clinic Foundation (Robert Detrano, M.D., Ph.D.) [25,26].

[1]Available from `ftp://ftp.ira.uka.de/pub/neuron/proben1.tar.gz`.

The structure of the MLP used for all the instances accounts for three layers (input-hidden-output) having six neurons in the hidden layer. The number of neurons in the input and output layers depends on the concrete instance. The activating function of the neurons is the sigmoid function. Table 26.1 summarizes the network architecture for each instance.

To evaluate an ANN, we split the pattern set into two subsets: the training one and the test one. The ANN is trained with the different algorithms by using the training pattern set, and then it is evaluated on the unseen test pattern set. The training set for each instance is approximately made of the first 75% of the examples, while the last 25% constitutes the test set. The exact number of patterns for each instance is presented in Table 26.1 to ease future comparisons.

After presenting the problems, we now turn to describe the parameters for the algorithms. The parameters for BP, LM, and the hybrid algorithms are shown in Table 26.2. The hybrid algorithms use the same parameters as their elementary components. However, the mutation operator of the GA is not applied but it is replaced by BP or LM, respectively. The BP and LM are applied with an associated probability p_t, just like any other classical genetic operator. When applied, BP/LM only performs one single *epoch*.

The parameters of the rest of the algorithms are shown in Tables 26.3 and 26.4. The ES uses neither vector of angles nor recombination, thus making irrelevant the distribution of variables inside the vector. The ES and Hy3 algorithms employ real representation for the variables of the network, but the GA, the CHC, and the hybrid algorithms use binary vectors. These vectors are 16-bit length and represent a real

TABLE 26.1 MLP Architecture and Patterns Distribution for All Instances

Instance	Architecture	Patterns: Training	Patterns: Test
Cancer	9 — 6 — 2	525	174
Diabetes	8 — 6 — 2	576	192
Heart	35 — 6 — 2	690	230

TABLE 26.2 Parameters for BP, LM, and the Hybrid Algorithms

		BC	DI	HE
BP	Epochs	1000	1000	500
	η	0.01	0.01	0.001
	α	0.0	0.0	0.0
LM	Epochs	1000	1000	500
	μ	0.001	0.001	0.001
	β	10	10	10
GAxx	p_t	1.0	1.0	0.5
	Epochs of BP/LM	1	1	1

TABLE 26.3 Parameters for GA, CHC, Hy3, and ES

	GA	CHC	Hy3	ES
Pop. size	64	64	160 (8 × 20)	1
Selection	Roulette (2 inds.)	CHC selection	LR + SUS	10 inds.
Recombination	SPX ($p_c = 1.0$)	HUX ($p_c = 1.0$)	FCBX ($p_c = 0.6$)	None
Mutation	Bit-Flip ($p_m = 1/length$)		NU ($p_m = 0.125$)	GM ($p_m = 1.0$)
Replacement	Elitist			
Stop criterion	1,064 evals.	10,048 evals.	800,160 evals.	10,001 evals.

TABLE 26.4 Abbreviations Used in Table 26.3

Abbreviation	Meaning
LR	Linear Ranking
SUS	Stochastic Universal Sampling
FCBX	Fuzzy Connective-Based Crossover
SPX	Single Point Crossover
HUX	Highly Disruptive Uniform Crossover
GM	Gaussian Mutation
NU	Non-Uniform Mutation

TABLE 26.5 Results of ANN Training

CEP(%)		BP	LM	GA	CHC	Hy3	ES	GABP	GALM
Cancer	$\bar{x}$	0.98	3.03	16.59	8.56	49.00	0.90	0.75	0.02
	σ_n	0.26	1.34	6.47	0.02	35.40	0.38	3.92	0.10
Diabetes	$\bar{x}$	21.81	25.94	36.46	37.44	53.00	21.30	36.46	28.32
	σ_n	0.37	3.50	0.00	0.01	13.50	1.35	0.00	1.21
Heart	$\bar{x}$	27.13	34.81	39.87	38.06	52.00	31.45	53.83	22.55
	σ_n	1.54	3.77	11.51	0.03	25.30	4.70	21.01	0.73

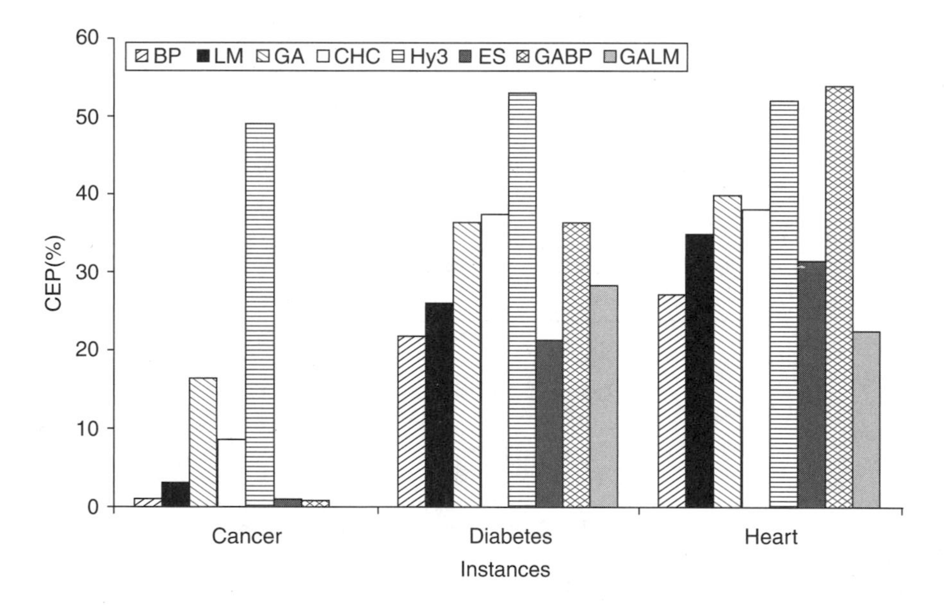

FIGURE 26.9 Comparison among the algorithms.

value in the interval $[-1, +1]$. The weights (variables) associated to input links for a neuron are placed contiguously in the chromosome. The GA uses a steady-state strategy.

26.5.2 Analysis of the Results

In this section we present the results obtained after the application on the three instances of the eight algorithms. We report the mean and the standard deviation of the CEP for the test pattern set after performing 30 independent runs. Table 26.5 and Figure 26.9 show the results.

There are many interesting works related to neural network training that also solve the instances tackled here. But unfortunately, some of the results are not comparable with ours, because they use a different definition of the training and test sets; this is why we consider a capital issue to adhere to a standard way of evaluations like the one proposed by Prechelt [8]. In fact, we did find some works for meaningful comparisons.

For the Cancer instance the best results are obtained by BP, ES, GABP, and GALM. They are followed in a less accurate ranking by LM and, finally, by the GAs. This is not a surprising fact, since the GAs perform a rather explorative search in this kind of problems. For this instance, we found that the best mean CEP in Reference 27 is 1.1%, which represents a lower accuracy compared to our 0.02% obtained with the GALM hybrid, so far the best solution to our knowledge. In Reference 28, a CEP close to 2% for this instance is achieved, and the reader should notice that our GALM is one hundred times more accurate. The mentioned work uses 524 patterns for the training set and the rest for the test set, that is, almost exactly our configuration with only one pattern changed (a minor detail), and therefore the results can be compared. The same happens for the work of Yao and Liu [6], where their EPNet algorithm gets neural networks reaching 1.4% of CEP.

In Diabetes, BP and ES show again the highest accuracy, followed by LM and GALM. This time, GA, CHC, and GABP obtain similar CEP, always below that of Hy3. For this instance, a CEP of 30.11% is reached in Reference 29 (outperformed by many of our algorithms like BP, LM, ES, and GALM) with the same network architecture as here. In Reference 6 the authors report a 22.4% of CEP for this problem, clearly outperformed by our BP and ES techniques.

As to the Heart problem, we found approximately the same behavior as for the two previous instances. However, Hy3 reaches lower CEP than GABP, and GALM gets the lowest CEP. In Reference 29 the authors reported a 45.71% of CEP for the Heart instance using the same architecture. All our algorithms outperform this CEP measure (except Hy3 and GABP).

We observed that BP and ES performed slightly more accurate than LM for all the instances. This is also an unexpected result, since LM is quite accurate in many applications. We conclude that the three selected instances represent problems that do not really need such a complex in-depth local search. In the near future we plan to consider larger and more complex instances to further test LM. With respect to the hybrid algorithms, the results do confirm this hypothesis: GALM is more accurate than GABP.

In summary, we have found some of the more accurate results for the three instances, but there is still need to get ahead in other instances and improve the accuracy on the addressed ones. A learned lesson is to always keep in mind the importance of reporting results in a standard way for meaningful future comparisons.

26.6 Conclusions

In this chapter we have tackled the neural network training problem with eight algorithms: two well-known problem-specific algorithms such as BP and LM, four general metaheuristics such as GA, CHC, Hy3 Algorithm, ES, and two hybrid algorithms combining the GA with the problem-specific techniques. To compare the algorithms we solved three classification problems from the domain of Medicine: the diagnosis of breast cancer, the diagnosis of diabetes in the Pima Indians, and the diagnosis of heart disease.

Our results show that the problem-specific algorithms (BP and LM) and the ES get lower classification error than the other more general search procedures. It is surprising that the ES, also a general search procedure, gets a lower CEP than the LM, a sophisticated problem-specific algorithm.

With respect to hybrids, the algorithm GALM outperforms in two of the three instances the classification error of the problem-specific algorithms. This makes GALM a promising algorithm for neural network training. On the other hand, many of the classification errors obtained in this work are below those found in the literature, what represents a contribution as a reference work for these medical problems. As a future work we plan to add new algorithms to the analysis, and to apply them to more instances, especially in the broader domain of Bioinformatics.

Acknowledgments

This work has been partially funded by the Ministry of Science and Technology (MCYT) and Regional Development European Fund (FEDER) under contract TIC2002-04498-C05-02 (the TRACER project, `http://tracer.lcc.uma.es`).

References

[1] J.T. Alander. Indexed Bibliography of Genetic Algorithms and Neural Networks. Technical report 94-1-1NN, University of Vaasa, Department of Information Technology and Production Economics, 1994.

[2] D. Rumelhart, G. Hinton, and R. Williams. Learning representations by backpropagation errors. *Nature*, 323: 533–536, 1986.

[3] E. Cantú-Paz. Pruning neural networks with distributions estimation algorithms. In Erick Cantú-Paz et al., Eds., Proceedings of *GECCO 2003*, Vol. 2733 of *Lecture Notes in Computer Science*. Springer-Verlag, 2003, pp. 790–800.

[4] C. Cotta, E. Alba, R. Sagarna, and P. Larrañaga. Adjusting weights in articficial neural networks using evolutionary algorithms. In P. Larrañaga and J.A. Lozano, Eds., *Estimation of Distribution Algorithms. A New Tool for Evolutionary Computation*. Kluwer Academic Publishers, 2001, pp. 357–373.

[5] E. Alba, J.F. Aldana, and J.M. Troya. Full automatic ANN design: A genetic approach. In J. Mira, J. Cabestany, and A. Prieto, Eds., *New Trends in Neural Computation*. Springer-Verlag, 1993, pp. 399–404.

[6] X. Yao and Y. Liu. A new evolutionary system evolving artificial neural networks. *IEEE Transactions on Neural Networks*, 8: 694–713, 1997.

[7] X. Yao. Evolving artificial neural networks. *Proceedings of the IEEE*, 87: 1423–1447, 1999.

[8] L. Prechelt. PROBEN1 — A Set of Neural Network Benchmark Problems and Benchmarking Rules. Technical report 21, Fakultät für Informatik Universität Karlsruhe, 76128 Karlsruhe, Germany, September, 1994.

[9] M.T. Hagan and M.B. Menhaj. Training feedforward networks with the Marquardt algorithm. *IEEE Transactions on Neural Networks*, 5: 989–993, 1994.

[10] J.L. McClelland and D.E. Rumelhart. *Parallel Distributed Processing: Explorations in the Microstructure of Cognition*. MIT Press, Cambridge, MA, 1986.

[11] F. Rosenblatt. *Principles of Neurodynamics*. Spartan Books, New York, 1962.

[12] T. Bäck. *Evolutionary Algorithms in Theory and Practice: Evolution Strategies, Evolutionary Programming, Genetic Algorithms*. Oxford University Press, New York, 1996.

[13] L.J. Eshelman. The CHC adaptive search algorithm: How to have safe search when engaging in nontraditional genetic recombination. In *Foundations of Genetic Algorithms*. Morgan Kaufmann, San Mateo, CA 1991, pp. 265–283.

[14] E. Alba, F. Luna, A.J. Nebro, and J.M. Troya. Parallel heterogeneous genetic algorithms for continous optimization. *Parallel Computing*, 2004.

[15] I. Rechenberg. *Evolutionsstrategie: Optimierung technischer Systeme nach Prinzipien der biologischen Evolution*. Fromman-Holzboog Verlag, Stuttgart, 1973.

[16] G. Rudolph. *Evolutionary Computation 1. Basic Algorithms and Operators*, Vol. 1. IOP Publishing Ltd, 2000, chap. 9, pp. 81–88.

[17] D.B. Fogel. *Evolutionary Computation 1. Basic Algorithms and Operators*, Vol. 1. IOP Publishing Ltd, 2000, chap. 33.2, pp. 270–274.

[18] C. Cotta and J.M. Troya. On decision-making in strong hybrid evolutionary algorithms. In *Tasks and Methods in Applied Artificial Intelligence*, Vol. 1415 of Lecture Notes in Artificial Intelligence, 1998, pp. 418–427.

[19] L. Davis, Ed. *Handbook of Genetic Algorithms*. Van Nostrand Reinhold, New York, 1991.

[20] K.P. Bennett and O.L. Mangasarian. Robust linear programming discrimination of two linearly inseparable sets. *Optimization Methods and Software*, 1: 23–34, 1992.

[21] O.L. Mangasarian, R. Setiono, and W.H. Wolberg. Pattern recognition via linear programming: Theory and application to medical diagnosis. In Thomas F. Coleman and Yuying Li, Eds., *Large-Scale Numerical Optimization.* SIAM Publications, Philadelphia, PA, 1990, pp. 22–31.

[22] W.H. Wolberg. Cancer diagnosis via linear programming. *SIAM News*, 23: 1–18, 1990.

[23] W.H. Wolberg and O.L. Mangasarian. Multisurface method of pattern separation for medical diagnosis applied to breast cytology. In *Proceedings of the National Academy of Sciences*, Vol. 87, U.S.A., December 1990, pp. 9193–9196.

[24] J.W. Smith, J.E. Everhart, W.C. Dickson, W.C. Knowler, and R.S. Johannes. Using the ADAP learning algorithm to forecast the onset of diabetes mellitus. In *Proceedings of the Twelfth Symposium on Computer Application in Medical Care.* IEEE Computer Society Press, 1988, pp. 261–265.

[25] R. Detrano, A. Janosi, W. Steinbrunn, M. Pfisterer, J. Schmid, S. Sandhu, K. Guppy, S. Lee, and V. Froelicher. International application of a new probability algorithm for the diagnosis of coronary artery disease. *American Journal of Cardiology*, 64: 304–310, 1989.

[26] J.H. Gennari, P. Langley, and D. Fisher. Models of incremental concept formation. *Artificial Intelligence*, 40: 11–61, 1989.

[27] T. Ragg, S. Gutjahr, and H. Sa. Automatic determination of optimal network topologies based on information theory and evolution. In *Proceedings of the 23rd EUROMICRO Conference*, Budapest, Hungary, September 1997.

[28] W.H. Land and L.E. Albertelli. Breat cancer screening using evolved neural networks. In *IEEE International Conference on Systems, Man, and Cybernetics, 1998*, Vol. 2, IEEE Computer Society Press, October 1988, pp. 1619–1624.

[29] W. Erhard, T. Fink, M.M. Gutzmann, C. Rahn, A. Doering, and M. Galicki. The improvement and comparison of different algorithms for optimizing neural networks on the MasPar MP-2. In M. Heiss, Ed., *Neural Computation — NC'98.* ICSC Academic Press, 1998, pp. 617–623.

27

Bio-Inspired Data Mining

Tiago Sousa
Arlindo Silva
Ana Neves
Ernesto Costa

27.1 Introduction

The dissemination of information systems in organizations and technological progress in terms of computational power, communications technology, and storage capability has the side effect of producing repositories with huge amounts of data. Databases — as these repositories are called — have grown in number and size and it quickly became apparent that all the stored information constituted a valuable resource, especially if reliable methods could be found to recover useful information — knowledge — from the raw data in storage. A completely new interdisciplinary field has grown around this general goal and is now known as Knowledge Discovery in Databases (KDD). As a broadly accepted definition, we can say that KDD is a complex process that aims to extract implicit, previously unknown, and potentially useful information from data, in a nontrivial way.

The central step in the KDD process is usually called data mining and consists of the actual search for interesting regularities or patterns in the data. This step is preceded by a processing stage, where data is prepared for the application of a data mining algorithm. After the data mining algorithm is executed, a postprocessing stage occurs, where the algorithm's results can be refined and simplified.

From the data mining view point, the essential task is to build computer programs capable of searching through the data for nuggets of knowledge. These nuggets are usually represented as data patterns that are expected to have some beneficial characteristics, namely: validity on new data with a high degree of certainty, novelty, potential usefulness measured by some utility function, and "comprehensibility to humans," allowing them a better understanding of the underlying data.

Data used in this task is usually divided into instances described by a set of attributes. The most common problem in data mining is trying to predict the value of a user-defined attribute based on the values of some other attributes. This is called a classification problem. When we want to predict the values of several attributes (instead of just one goal attribute), the problem is now an association one. In some problems we do not really have an attribute that defines the class of a given instance but still would like to know if it were possible to group the data in different groups or classes. This grouping (and class discovery) must then be done by the algorithm and this becomes a clustering problem. Useful overviews of the field can be found in References 1 and 2.

There are several research areas that provide the methods in data mining, the most important being statistics and Artificial Intelligence (AI). Other areas include pattern recognition, databases, data visualization, etc. Within the AI area, the main contributions came from the machine learning community and represent the three main AI paradigms: symbolic, connectionist, and evolutionary. From these, the last two bring us back to the title — and content — of this chapter, since both have a strong biological inspiration. A useful introduction to machine learning approaches to data mining is presented in Reference 3.

Connectionist approaches in data mining are usually synonyms of neural networks. This is one of the most widely known (if not widely understood) technique for data mining. It provides a very flexible way of fitting a model to observed data. This model can then be used to classify new data or make some kind of prediction on it. The major drawback of neural networks is that the resulting models cannot easily fulfill the "comprehensibly to humans" requirement, since their most friendly representation is a network of weighted connections between nodes where some nonlinear operation occurs.

In this chapter, we focus mainly on bio-inspired approaches for data mining that are not neural network related. We start by giving a slight overview of these approaches, especially evolutionary ones. We also describe some novel approaches that are based in new computational models, inspired by biological mechanisms, but not completely accommodated by the connectionist or evolutionary paradigms.

Following this overview, we introduce a new approach to the data mining task of finding classification rules based on the Particle Swarm Optimizer (PSO) algorithm.

The particle swarm algorithm has been originally presented in Reference 4 as a population-based function optimizer in the n-dimensional space of real numbers. Bird flock flight simulations initially inspired the algorithm, and biological inspiration is still present in the current denomination. The swarm or flock metaphor applies to the PSO in the way particles fly in a somewhat coordinated way through an n-dimensional space (the space of parameters of the function being optimized) in search of some desired place (function optimum).

While being considered as a form of evolutionary algorithm, there is no use of genetic operators such as mutation or recombination, as is the case in other evolutionary paradigms such as evolutionary programming, evolutionary strategies, or genetic algorithms. Explicit selection is also not present. Instead, in each iteration, the position of every particle in the search space is updated accordingly to the particle's velocity. The velocity of a particle in a given iteration is a function of the velocity in the previous iteration, its best previous position in the search space, and the best previous position in the search place of all of the particle's neighbors. The behavior of a particle in the swarm is the result of balancing the desire of flying toward the best point in the search place according to its own experience or conforming to the swarm knowledge of where the current best point is. For an extensive discussion of the cultural model behind the PSO, as well as of the PSO itself and several variants see Reference 5.

Our method uses a specialized particle swarm algorithm to build rules for classification tasks. We call the resulting algorithm a Particle Swarm Data Miner (PSDM). We present several variants of the PSDM, more explicitly, a discrete version, two real-valued versions using different stop criteria, and an adaptive version. This last version is based on the Simple Adaptive Predator–Prey Optimizer (SAPPO) algorithm [6,7], initially developed to improve performance on real-valued optimization problems. SAPPO introduces a predator–prey mechanism in the basic swarm algorithm in order to maintain diversity during the search. It also includes a swarm of symbiotic particles, designed to allow the adaptation of the algorithm parameters to the problem being tackled.

These algorithms are compared between themselves and against two industry standard classification algorithms (J48 and PRISM) on a set of benchmark problems. The results are then used to compare and discuss the different characteristics of each variant.

Section 27.2 presents the overview on bio-inspired data mining methodologies. In Section 27.3 we describe the higher level classification algorithm, which can transparently make use of any of the lower level PSO-based algorithms to search for classification rules. The first three variants of the PSO algorithm are discussed in Section 27.4. The SAPPO data miner is presented in Section 27.5. The experimental setup used to test the algorithms is outlined in Section 25.6, where we also report and discuss the results obtained. Finally, we draw some conclusions in Section 27.7.

27.2 A Brief Overview of Bio-Inspired Data Mining

As mentioned previously, we are not going to consider the neural-network-based approaches in this overview. Neural networks were the first bio-inspired computational mechanism to be used in data mining, probably because of being so directly applicable to classification tasks. Its applications are now numerous and well known.

Evolutionary approaches, on the other hand, are relatively new to data mining. We describe the most common ones in the following paragraph. More recently, other biological mechanisms, such as the intelligent behavior of ants and the working of the human immune system have inspired new algorithms and the first attempts of applying these to data mining tasks have been made. We report on these in the following paragraphs. In the rest of this chapter, we introduce and discuss the use of particle swarm-based algorithms for data mining. These algorithms also have a biological base, as their origin is closely related with the study of coordinated movement in "swarms" of living creatures, for example, flock of birds and schools of fish.

The main motivation for using these algorithms in data mining comes from the way in which patterns are searched by more traditional approaches (e.g., tree induction and rule discovery algorithms). These are usually based on some greedy search criterion that is used to fix a part of the pattern, with the search then continuing for the rest. It is well known that greedy search techniques are easily tricked into local optima of the search space. Data mining tasks involving numerous attributes that are strongly related provide search spaces where these sub-optima abound. Biological inspired search procedures take a more global approach to search, usually incorporating stochastic mechanisms that can enable the avoidance of such optima.

27.2.1 Evolutionary Approaches to Data Mining

Evolutionary algorithms are based on the principles behind the theory of evolution, mainly the idea of survival of the fittest. These algorithms act over a group (population) of data structures representing tentative solutions for some problem we want to solve. The quality of these solutions is evaluated (using a fitness function) so that better solutions have a higher probability of being totally or partially copied onto a new group of solutions — the next generation. This mechanism is called selection.

However, selection alone is not enough, since you will only have the individuals in the initial population to work with. We also need variation, so that better individuals can appear, and have their characteristics spread through the population by selection. In evolutionary computation, variation is mainly achieved by randomly changing parts of a solution (mutation) or recombining existing solutions into new ones (recombination).

These basic principles have given origin to many flavors of evolutionary algorithms of which the most widely known (and used) are genetic algorithms [8], evolutionary programming [9], evolutionary strategies [9], and genetic programming (GP) [10]. In spite of their differences, all these algorithms are built around what is usually called the generic evolutionary algorithm loop:

```
P(0) = Generate_initial_population()

Evaluate_population(P(0))

While not Some_stop_criterion() do

      P'(t) = Select_individuals_to_reproduce(P(t))

      P''(t) = Create_offspring_using_recombination(P'(t))

      P(t+1) = Merge_old_population_with_offspring(P''(t))

      Evaluate_population(P(t+1))
```

Of the algorithms given, genetic algorithms and GP are the most widely used for data mining tasks. In Sections 27.2.1.1 and 27.2.1.2 we present, in a generic way, how these algorithms can be used for data mining, more specifically their application to rule discovery. An extended survey on the use of evolutionary algorithms not only for data mining but also for other tasks in the KDD progress can be found in Reference 11.

27.2.1.1 Genetic Algorithms

The classical genetic algorithm acts on populations of fixed length binary strings and uses a one-point crossover recombination operator, as well as some form of bit mutation. Most data mining applications of genetic algorithms focus on using the algorithm to discover prediction (if–then) rules. These applications can be divided accordingly to the way in which the rules (the solution) are encoded. Two main approaches are possible: one individual can represent a single rule — this is sometimes called the Michigan approach, for example, References 12 and 13 or it can represent a whole set of rules — the Pittsburgh approach, for example, References 14 and 15.

The choice of an approach can depend on both the task being tackled and the way the evolutionary algorithm is integrated in the data mining system. As a general guideline, if a rule can be effectively evaluated in an isolated way — such as in association problems — it can be simpler to encode just one rule in each individual. If rules are highly dependent (as is often the case with rule sets for classification tasks), encoding the whole rule set might be more advisable, even if at a cost in terms of representation complexity.

In terms of representation, a rule is usually represented as a bit string. In a typical if-then rule, some attributes are tested in the conditional part of the rule and the value of one or more attributes are predicted in the consequent part. Assuming categorical attributes, each one is represented in the rule by a number of bits equal to the number of its possible values. A test on attribute a of the form `(a = a1 or a = a4)` can then be represented by the substring 1001 (for four possible values). An individual is built by concatenating the substrings for each attribute tested by a rule. If the representation includes several rules then this process is repeated for each one.

Standard operators can be used to create offspring, but specialized ones are often implemented, for example, to specialize or generalize rules.

An essential aspect of every application of a genetic algorithm is the design of the evaluation function. While it is out of the scope of this overview a detailed discussion of evaluation functions, it should be pointed out that not only the accuracy of a rule (or rule set) is important in its evaluation, but also other aspects can be taken into account, for example, its simplicity.

27.2.1.2 Genetic Programming

GP is an evolutionary algorithm designed to evolve computer programs [10]. These are usually represented as parse trees and recombination and mutation involve the exchanging and replacement of sub-trees. Functions, variables, and constants are user chosen, so GP can easily be used to evolve rules for data mining tasks. With some effort these rules can accommodate not only attribute/values comparisons (e.g., `a = a1`), but also complex relations between attributes (e.g., `a < c`). Primary attributes can also be combined into new, more expressive, attributes (e.g., numerical attributes can be combined using arithmetic functions).

Typical recombination (sub-tree exchanging between individuals) and mutation (sub-tree replacement) operators can be used if care is taken to guarantee that correct parameters are given to every function. Especially when mining databases with both numerical and nominal attributes, when functions capable of dealing with both types of attributes must be used, the operators need to guarantee that nominal values are not passed to numerical functions and vice versa (to avoid expressions such as `''high''+''low''`).

Due to its flexibility, as well as its ability to use any user-chosen functions to test and combine attributes, many applications have emerged in data mining using GP, mainly for classification tasks, see, for example, References 16 to 19.

As with genetic algorithm-based data mining applications, the design of the fitness function can incorporate not only an accuracy measure, but also other factors, such as simplicity. With GP this is usually related with the size of the program tree.

27.2.2 Ant Colonies for Data Mining

Another biological mechanism that has served as inspiration for a data mining application is the one imitated by the Ant Colony Optimization algorithm [20]. This algorithm is inspired in the way real ants optimize their tracks. When faced with an obstacle in a food track, an ant initially chooses its way around it in a random fashion. However, while moving, every ant deposits on the ground a chemical trail of pheromone. Ants that choose the shortest path around the obstacle will take less time to reach the food than the ones that follow the longest trajectory, so, with all ants moving at a similar rate and depositing similar quantities of pheromone there will be a tendency to accumulate more pheromone in the shortest path. This tendency will be exacerbated by the fact that all ants have a preference, when in a crossroad, to choose the way with larger deposits of pheromone. This will obviously lead to a rapid increase in the number of ants that choose the shortest path, with a corresponding increase in pheromone depositing. The result is that the final path to the food source, followed by almost all the ants, will emerge as the shortest path between the nest and the food source.

Replacing obstacles by decisions and paths by costs (or gains), this mechanism can be implemented as the Ant Colony Optimization algorithm [21], which has been successful in several combinatorial optimization tasks, as well as in shortest path problems (e.g., in the telecommunications domain). Its use in data mining, for finding if-then rules for classification tasks, was proposed by Parpinelli et al. [22], with the decisions corresponding to the choice of an attribute/value test. A heuristic was used for estimating the cost of a given test. A path represents a rule and pheromone is deposited in the path in proportion to the rule fitness.

27.2.3 Artificial Immune Systems for Data Mining

Artificial immune systems are a new computational paradigm inspired in theoretical immunology and observed immune functions, principles, and models, applied to solve problems [23]. There is not just one artificial immune system inspired algorithm. Instead, we can talk about a framework for the development of artificial immune system algorithms. A specific algorithm includes a representation to create abstract models of immune organs, cells, and molecules; a set of affinity functions, which quantify the interactions of these elements; and a set of algorithms to rule the dynamic behavior of the artificial immune system.

Several mechanisms of the immune system have been used as inspirations for specific algorithms [23,24]. Clonal selection, a mechanism by which immune cells that have a higher affinity to a given pathogen are allowed to reproduce more and with lower mutation rates than cells with lower affinity, has been used in optimization. The "affinity measure" is given by the function being optimized. There is no special reason why this mechanism cannot be used to search for binary represented if–then rules, with affinity being measured by rule fitness.

Another theoretical model of the immune system, immune networks, have been used to develop learning networks, where nodes correspond to immune cells and connection strength represent affinity between the pathogen recognized by each cell. This model has already been used in data analysis, namely for data clustering [25,26].

27.2.4 Particle Swarm Algorithms for Data Mining

In the following sections we present — to the best of our knowledge — the first attempt at the use of particle swarm algorithms for data mining. Being a recent addition to the biological inspired computation area [4,5], particle swarm algorithms have been mostly used for real-valued optimization (e.g., References 27 and 28) but applications to other areas have been appearing quickly, as can be observed in this extensive bibliography of the field [29].

In our work several variants of the PSO algorithm were used in the inner loop of a classification covering algorithm in the search for classification rules. In Section 27.3 we describe the high level covering algorithm and validation procedure. The following sections describe the use of PSO-based algorithms for rule search and present experimental results when compared with industry standard classification techniques.

27.3 Covering Algorithm and Validation Procedure

Our research project has focused in applying, testing, and tuning PSO-based algorithms to classification rule discovery problems.

In a classification problem having a dataset in which each instance is characterized by a set of attributes and belongs to a specific class, the objective is to extract a set of rules that can accurately predict the class of a new instance from its attribute values.

Classification rules are no more than conditional clauses, involving two parts: the antecedent and the consequent. The former is a conjunction of logical tests, and the later gives the class that applies to instances covered by this rule. These rules usually take the following format:

```
IF attribute_a=value_1
       AND attribute_b=value_2
        ...
       AND attribute_n=value_I
THEN class_x
```

A rule set is constructed with rules performing different attributes tests, predicting different classes and a default rule, to capture and classify instances not classified by the previous rules.

A sequential rule set would conform to this structure:

```
Rule #1

Rule #2

...

Rule #n

Default Rule
```

Containing no attribute tests and predicting the same class as the one predominant in the remaining instances, the default rule takes the form:

```
IF true

THEN class_x.
```

As an illustrating example, one of the datasets commonly used as a benchmark, describes animals accordingly to whether they have hair, feathers, or tail, among other attributes, and classifies them as mammals, birds, fish, etc. The resulting set of rules could be:

```
=>RULE #1

       IF milk = true

       THEN type IS mammal

=>RULE #2

       IF feathers = true

       THEN type IS bird

=>RULE #3

       IF fins = true

       THEN type IS fish

=>RULE #4

       IF airborne = false

            AND tail = false

       THEN type IS invertebrate

=>RULE #5

       IF legs=6

       THEN type IS insect
```

```
=>RULE #6
        IF TRUE
        THEN type IS reptile
```

A new instance, the characterization of one animal that we might need to classify, would be iteratively tested against each rule, until its attributes fully matched the attribute tests of one rule. This new instance would be classified with the rule's predicted class.

This type of database that extracted high level knowledge is greatly valued in areas such as loan granting, fraud detection, marketing, etc., where huge databases already exist and predictions are most welcome. Once found, a set of if–then rules is easily comprehended and utilized by humans, which need no special training.

The complexity of this problem — finding the rule set and the set of attribute tests for each rule — lies in the exponentially high number of attribute-value pairs, which generate a wide multidimensional search space.

In this kind of search problem, evolutionary algorithms such as GA and PSO have already proven to be reliable and efficient due to their parallel, population-based search strategies.

Following a typical architecture of the Michigan approach in classification rule discovery, the overall structure of our work was designed contemplating three imbricated procedures, each one fulfilling a specific task.

```
Validation Procedure
|
|      Covering Procedure
|      |
|      |      Classification Rule Discovery Procedure
|      |      |
|      |      A Quality/iteration number criterion is met
|      |
|     A minimum number of uncovered instances is met
|
10 times
```

The classification rule discovery procedure aims to find and return the rule that better classifies the predominant class in a given instance set. One rule is better than another, if it can match a higher number of instances with the specified class and segregate instances with a different class. It is at this level that the PSO-based algorithms are used. The covering procedure, on receiving an instance set (the training set), invokes the classification rule discovery procedure to reduce this set by removing instances correctly classified by the rule returned by the classification rule discovery procedure. This process is repeated until only a predefined number of instances are left to classify in the training set. A sequential rule set is therefore created — hence the Michigan approach. The aim of the validation procedure is, not only to determine the accuracy of a rule set returned by the covering procedure, but also to gauge the liability of the complete classifying procedure — classification rule discovery and covering procedures altogether. This is achieved by iteratively dividing the initial dataset into different test and training sets and computing average indicators, such as accuracy, rule number per set, and attribute tests number per rule.

Having this architecture, clearly segregating tasks to each of the three imbricated procedures, and using alternatively different search strategies in its core procedure — classification rule discovery — we end up having an unbiased platform for testing and benchmarking each of these search strategies.

27.3.1 Validation Procedure

The purpose of the validation procedure is to statistically evaluate the accuracy of the rule set obtained by the covering procedure. This is done using a method known as tenfold cross validation.

The tenfold cross validation consists of dividing the dataset into ten equal partitions and iteratively using one of this sets as a test set and the remaining nine as training sets. In the end, ten different rule sets are obtained and average indicators, such as accuracy, rule number per set, and attribute tests number per rule are computed.

Rule set accuracy is evaluated and presented as the percentage of instances in the test set correctly classified. An instance is considered correctly classified, when the first rule in the rule set whose antecedent matches this instance and the consequent (predicted class) matches this instance's class.

27.3.2 Covering Algorithm

The covering procedure is basically a divide-and-conquer technique. Being given an instance training set, it calls on the rule discovery procedure in order to obtain the highest quality rule for the predominant class in the training set.

Once found, this rule goes through a pruning process where unnecessary attribute tests are removed. This simple process iteratively removes attribute tests if the pruned rule has a better quality value than the original rule.

Correctly classified instances are then removed from the training set and the rule discovery procedure is run once more. A sequential rule set is built iteratively, and the covering procedure runs until only a predefined number of instances are left to classify. This threshold criteria value is user defined as a percentage and it is typically set to 10%.

27.3.3 Postprocessing Routines — Rule Pruning and Rule Set Cleaning

A final aspect of the higher level algorithm worth discussing is the postprocessing routines. Recall that high level knowledge extracted from databases must conform to three main requisites: accuracy, comprehensibility, and interest for the user [1]. In classification rule discovery problems, the number of attribute tests per rule and the number of rules per set is a major contributor for the comprehensibility of the obtained results — fewer attribute tests and rules ease comprehensibility.

After a rule is returned from the classification rule discovery algorithm it goes through a pruning process in order to remove unnecessary attribute tests. This is done by iteratively removing each attribute test whenever the newly obtained rule has the same or higher quality value than the original rule.

Just after the covering algorithm returns a rule set, another postprocessing routine is used: rule set cleaning, where rules that will never be applied are removed from the rule set.

As rules in the rule set are applied sequentially, in this routine, rules are removed from the rule set if:

- There is a previous rule in the rule set that has a subset of the rule's attribute tests.
- If it predicts the same class as the default rule and is located just before it.

Therefore, in the example given, rule number 2 and 3 will be removed and the rule set will be reduced to the first and last rules:

```
Rule #1

        If attribute_a = x_a

        Then class=c_1

Rule #2

        If attribute_a = x_a and attribute_b = x_b

        Then class=c_2

Rule #3

        If attribute_c = x_c

        Then class=c_3

Rule #4 - Default Rule

        If TRUE

        Then class=c_3.
```

27.4 Basic Particle Swarm Data Miners

As already mentioned, this architecture, which clearly segregates tasks to each of the three imbricated procedures, provides an unbiased platform for testing and benchmarking different search strategies in its core procedure — classification rule discovery.

Several variants of PSO algorithm were therefore implemented and tested. Sections 27.4.1 to 27.4.4 give an overview of these algorithms.

27.4.1 The PSO Algorithm

The PSO is inspired in the intelligent behavior of beings as part of an experience sharing community as opposed to an isolated individual reactive response to the environment. The Adaptive Culture Model [5], which is PSO's framing theory, states that the process of cultural adaptation is rooted into three principles: evaluate, compare, and imitate.

In PSO algorithms a particle decides where to move next, considering its own experience, which is the memory of its best past position, and the experience of its most successful neighbor.

There may be different concepts and values for neighborhood; it can be seen as spatial neighborhood where it is determined by the Euclidean distance between the positions of two particles, or as a sociometric neighborhood (e.g., the index position in the storing array). The latter is the most commonly used.

In PSO, a population of point particles "fly" in an n-dimensional real number search space, where each dimension corresponds to a parameter in a function being optimized. A vector X represents the position of the particle in each dimension of the search space, and another vector V represents the velocity of the particle, that is, its change in position. The particle "flies" in the search space by adding the velocity vector to its position vector in order to change its position.

The particle's trajectory is determined by V and depends on two "urges" for each particle i: flying toward its best previous position and flying toward its neighbors' best previous position. Different neighborhood definitions have been tried [30]; here, we assume that every particle is a neighbor to every other particle in the swarm. The general equations for updating the position and velocity for some particle i are the following:

$$\begin{aligned} V_i(t) &= \chi(wV_i(t-1) + \varphi_{1i}(P_i - X_i(t-1)) + \varphi_{2i}(P_g - X_i(t-1))), \\ X_i(t) &= X_i(t-1) + V_i(t). \end{aligned} \tag{27.1}$$

In this formula, χ is the constriction coefficient described in Reference 27, φ_1 and φ_2 are random numbers distributed between 0 and an upper limit and different for each dimension in each individual, P_i is the best position particle i has found in the search space and g is the index of the best individual in the neighborhood. The velocity is usually limited in absolute value to a predefined maximum, $V_{\max}$. The parameter w is a linear decreasing weight. The swarm is usually run for a limit number of iterations or until an error criterion is met.

From Equation (27.1) we can derive the two most usual ways in which convergence and, as a result, the balance between exploration and exploitation is controlled. Reference 28 uses $\chi = 1$ and weight w decreasing linearly from $w_{\max}$ to $w_{\min}$ during the execution of the algorithm. In Reference 27 convergence is guaranteed by choosing appropriated values for χ and $\varphi = \varphi_1 + \varphi_2$. w is fixed and equal to 1 in this approach.

Although some actions differ from one variant of PSO to another, its basic pseudo-code is as follows:

```
Initiate_Swarm()
 Loop
  For p=1 to number of particles
    Evaluate(p)
    Update_past_experience(p)
    Update_neighbourhood_best(p,k)
    For d=1 to number of Dimensions
      Move(p,d)
  Until Criterion
```

The output of this algorithm is the best point in the hyperspace the swarm visited.

There are several variants of PSO, typically differing in the representation: discrete or continuous PSO [5]; in the mechanism used to avoid spatial explosion of the swarm and guaranteeing convergence: linear decreasing weight [28] or constricted PSO [27]; or in the mechanism used to avoid premature convergence to local optima: predator–prey interactions [6] or collision avoiding swarms [31].

27.4.2 Rule Representation

In our particle swarm data miners, each particle encodes a rule, mapping the attribute tests in dimension coordinates. In a preprocessing routine, a data image is created normalizing all attribute values to the range [0.0, t] with $0.0 < t < 1.0$, t being a user predefined value. t stands for the indifference threshold where a higher value will trigger the omission of the corresponding attribute test.

Nominal attributes are normalized assigning to each different attribute value an enumerated index *#idx* and applying Equation (27.2)

$$v_{\text{norm}} = \frac{idx_v \times t}{\#idx}. \tag{27.2}$$

idx_v is the index of the attribute value v and *#idx* the total number of different attribute values.

Both integer and real types are normalized using Equation (27.3)

$$v_{\text{norm}} = \frac{(v - v_{\text{min}}) \times t}{v_{\text{max}} - v_{\text{min}}}. \tag{27.3}$$

v_{min} and v_{max} are the lower and higher attribute values found for this attribute.

Rules/Particles are encoded as a floating-point array and each attribute is represented by either one or two elements on the array, according to its type: nominal attributes are assigned with one element on the array and attribute-matching tests are defined as follows:

$$m(v_r, v_i) = \begin{cases} \text{true} & \text{if } \lfloor v_r \times \#idx \rfloor = \lfloor v_i \times \#idx \rfloor, \\ \text{false} & \text{otherwise.} \end{cases} \tag{27.4}$$

t being the indifference threshold value, v_r the attribute value stored in the rule for testing and v_i the instance value stored in the normalized image of the dataset.

Integer and real attributes are assigned with an extra element in the array in order to implement a value range instead of a single value,

$$m(v_{r1}, v_{r2}, v_i) = \begin{cases} \text{true} & \text{if } v_{r1} \geq t \text{ or } (v_{r1} - v_{r2}) \leq v_i \text{ or } (v_{r1} + v_{r2}) \geq v_i, \\ \text{false} & \text{otherwise.} \end{cases} \tag{27.5}$$

v_{r1} can be seen as the center and v_{r2} as a neighborhood radius, inside which matching will occur.

27.4.3 Rule Evaluation

Rules must be evaluated during the training process in order to establish points of reference for the training algorithm: best particle positioning. The rule evaluation function must not only consider instances correctly classified, but also the ones left to classify as well as the wrongly classified ones.

The formula used to evaluate a rule is expressed in the given Equation [22]:

$$Q(X) = \begin{cases} \dfrac{\text{TP}}{\text{TP} + \text{FN}} \times \dfrac{\text{TN}}{\text{TN} \times \text{FP}} & \text{if } 0.0 \leq x_i \leq 1.0, \quad \forall i \in d, \\ -1.0 & \text{otherwise,} \end{cases} \tag{27.6}$$

where

- TN (True Positives) = number of instances covered by the rule that are correctly classified, that is, its class matches the training target class.
- FP (False Positives) = number of instances covered by the rule that are wrongly classified, that is, its class differs from the training target class.
- TN (True Negatives) = number of instances not covered by the rule, whose class differs from the training target class.
- FN (False Negatives) = number of instances not covered by the rule, whose class matches the training target class.

This formula penalizes a particle, which has moved out of legal values, assigning it with a negative value (−1.0), forcing it to return to the search space.

27.4.4 Data Mining with Particle Swarms

The basic particle swarm data miner simply uses the updated equations presented for the PSO algorithm in Section 27.4.1 to search the rule space defined by the earlier representation. The covering algorithm described in Section 27.4.3 iteratively calls the particle swarm data miner with receeding datasets to build the final rule set.

In our experiments, we implemented and tested three different loop end criteria for the basic PSO data miner:

- *Iteration number*: the loop stops when a predefined number of iterations is achieved (usually 2000).
- *Convergence platform*: the loop stops when the swarm's best particle's quality value is constant for a predefined number of iterations (usually 1000).
- *Convergence radius*: the loop stops when all the swarm's particles are within a predefined normalized Euclidean distance from the swarm's best particle (usually 0.01), as a safeguard against simultaneous multiple optima exploitation, which causes the splitting of the swarm and therefore making this condition unachievable, the loop stops when the number of iterations reaches a predefined value (usually 2000).

The termination criterion is an important factor in the algorithm, since the PSDM is computationally very intensive. An early termination, achieved by radius or platform convergence can help decrease the number of evaluations but it can also result in the algorithm stopping before the best rule is found. Results from early versions of our algorithm can be found in our earlier work [32,33].

27.5 An Adaptive PSDM

The traditional PSO models, such as the ones used for the algorithm described in Section 27.4, while successful in many applications, especially in optimization, suffered from some limitations. One of these was that the basic PSO had no mechanism to introduce diversity after convergence to an optimum had occurred. These meant that if the algorithm converges to a local optimum it cannot escape it. The PSO algorithms presented in References 27 and 28 used carefully chosen parameters to promote a large exploration phase followed by convergence, to increase the probability of finding the global optimum. Another limitation was the algorithm's sensibility to its parameter's values, which can be observed in the results obtained by Clerc and Kennedy [27] for several benchmark problems and parameters sets.

In the previous work [6,7], we introduced a new mechanism in the basic PSO algorithms to help overcome these deficiencies. A predator–prey mechanism was introduced to maintain diversity even after convergence, so that some exploration of the search space was possible in the later stages of the algorithm's execution. A swarm of symbiotic particles was also used to allow the adaptation of the parameters to the problem being solved. The resulting algorithm was called SAPPO. The initial inspiration for the particle swarm optimizer, as described in Reference 4, was the coordinated movement of groups of animals in nature, for example, schools of fish or flocks of birds. The new mechanisms introduced in SAPPO also get their inspiration from nature. The predator-prey interaction is based on the disturbance caused by predators to the groups of animals being hunted. The adaptation scheme used mimes, the symbiotic associations between species so frequently found in nature. Both mechanisms are explained in detail in Sections 27.5.1 and 27.5.2.

27.5.1 The Predator–Prey Interaction

In standard PSO, both the constriction factor version [27] and the linear weight version [28], promote convergence of the swarm to a local optimum by damping the particles' velocity by a deliberate choice of the algorithm's parameters. An undesirable side effect of these strategies is that when the local optimum is not the global optimum being sought, particles cannot gain velocity to jump to another optimum in the search space. This phenomenon is similar to premature convergence in other evolutionary algorithms.

Our motivation for introducing the predator–prey interaction was mainly to introduce a mechanism for creating diversity in the swarm at any moment during the run of the algorithm, by adding velocity to some particles, not depending on the level of convergence already achieved. This would allow the "escape" of particles even when convergence of the swarm around a local sub-optimum had already occurred.

There are other mechanisms used for the same effect in the literature (see Reference 31), but two main reasons led us to prefer the predator–prey scheme. The first reason is its computational simplicity when compared with other approaches. As it will be seen when the mechanism is explained, it only introduces a new particle and little computational effort in the basic algorithm. The simple adjective in SAPPO comes from this fact. A second, and less technical, motive was to maintain the swarm intelligence philosophy behind the algorithm. It seemed more appropriate to introduce a mechanism that could also be implemented as a distributed behavior in the swarm.

The predator–prey model is based on the disturbance caused by a predator to the group of animals being hunted. Animals are driven from their favorite locations, for example, pastures and water sources, by fear of nearby predators. Eventually, this process will result in the finding of even better locations where the arriving animals will also be chased by nearby predators.

It is this predator/prey interaction that we try to reproduce in SAPPO. Here, a new particle is introduced into the swarm to mime the predators' behavior. This particle, called the predator particle, is attracted by the best (fittest) particle in the swarm, according to the following equations:

$$\begin{aligned} V_p(t) &= \varphi_3(X_g(t-1) - X_p(t-1)), \\ X_p(t) &= X_p(t-1) + V_p(t). \end{aligned} \tag{27.7}$$

In Equation (27.7), φ_3 is another random number distributed between 0 and an upper limit, usually 1, and X_g is the present position of the best particle in the swarm.

The predator particle can influence any particle in the swarm by changing its velocity in one or more dimensions. This influence is controlled by a "fear" probability f, which is the probability of a particle changing its velocity in one of the available dimensions due to the presence of the predator. For some particle i, if there is no change in the velocity in a dimension j, the update rules in that dimension still are:

$$\begin{aligned} v_{ij}(t) &= wv_{ij}(t-1) + \varphi_{1ij}(p_{ij} - x_{ij}(t-1)) + \varphi_{2ij}(p_{gj} - x_{ij}(t-1)), \\ x_{ij}(t) &= x_{ij}(t-1) + v_{ij}(t). \end{aligned} \tag{27.8}$$

The only differences from the other approaches are that w is fixed and χ is not explicitly used. However, if the predator "scares" the prey (particle), that is, if there is a change in velocity in dimension j, the rule becomes:

$$\begin{aligned} v_{ij}(t) &= wv_{ij}(t-1) + \varphi_{1ij}(p_{ij} - x_{ij}(t-1)) + \varphi_{2ij}(p_{gj} - x_{ij}(t-1)) + D(d), \\ x_{ij}(t) &= x_{ij}(t-1) + v_{ij}(t). \end{aligned} \tag{27.9}$$

This process is repeated for all dimensions, that is, there can be simultaneous changes in velocity in several dimensions.

The fourth term in the first equation of (27.9) quantifies the repulsive influence of the predator. This term is a function of the difference between the positions of the predator and the particle. The Euclidean distance between predator and prey is d. $D(x)$ is an exponential decreasing distance function:

$$D(x) = ae^{-x/b}. \tag{27.10}$$

$D(x)$ makes the influence of the predator grow exponentially with proximity. The objective of its use is to introduce more perturbation in the swarm when the particles are near the predator, which usually happens when convergence to a local optimum occurs. The a and b parameters define the form of the D function: a represents the maximum amplitude of the predator effect over a prey and b allows to control the distance at which the effect is still significant.

During the initial stages of the algorithm, its behavior is similar to a traditional PSO, since the particles are scattered and the predator's influence is negligible. As convergence occurs and particles start to move to a local optimum, their distance to the best particle diminishes and the predator effect exponentially increases, accelerating some particles in new directions. When the current local optimum is not the same as the global optimum being searched, this will hopefully allow one particle to jump to another near local optima, becoming the new best particle and leading the swarm to a new exploration phase. Several repetitions of this process could lead the swarm through a chain of local optima until the global optimum is found.

27.5.2 The Adaptation Scheme

The analysis of previous work with the PSO has made it clear that a good parameter choice for the PSO, is not only important for the algorithm performance, over a set of problems, but also that the performance of the algorithm, for a specific problem, varies significantly for different set of parameters.

Choosing parameters to generally guarantee convergence to a local optimum can be done using the constriction factor method. Nevertheless, if what we are looking for is not simply convergence to a local optimum, but converge to the global optimum or at least to a "good" local optimum, finding a set of good parameters basically remains a trial and error process.

As in any other optimization or search algorithm, the ideal solution would be to have a built-in procedure to adapt the algorithm itself to the problem that it is trying to solve. Other evolutionary algorithms, for example, evolutionary strategies, tried to do this by encoding some of the algorithm's parameters (e.g., mutation probability, variance, etc.) together with the individuals, thus trying to coevolve solutions and algorithm's parameters.

In spite of the only relative success of past instances of this idea, we believe that PSO-based algorithms, in general, and specially predator–prey based ones, can greatly gain by the inclusion of coevolution-based adaptive schemes. For instance, the a parameter in the $D(x)$ function has its best influence on the algorithm when it is near the average distance between sub-optima in the search space. The algorithm could substantially gain if this parameter could be learned during execution. The same could be said for other parameters, with different roles in the algorithm.

Another important reason is the simple (one almost could say natural) way in which such a mechanism can be included in the PSO framework, as we hope to show in the text that follows.

27.5.2.1 Symbiotic Particles

We introduced a new species of particles in SAPPO as the base for the adaptation mechanism. To the original swarm — constituted by solution particles — we now add a swarm of parameter particles, encoding the algorithm's parameters. These particles interact between themselves as a normal swarm using Equation (27.1) with a constriction factor χ of 0.729 and $\varphi = 4.1$, as recommended in Reference 27, with only the two adjacent particles being considered as neighbors. Each particle encodes seven real-valued parameters corresponding to seven adaptable parameters in SAPPO: a, b, and f for the predator–prey effect, N for the number of neighbors of each particle and c_1, c_2, c_3 for the velocity and position equations, which become:

$$V_i(t) = c_1 V_i(t-1) + c_2(P_i - X_i(t-1)) + c_3(P_g - X_i(t-1)),$$
$$X_i(t) = X_i(t-1) + V_i(t). \tag{27.11}$$

The aim of the adaptation scheme is to find, simultaneously, a solution for some problem and a set of parameters that increase the performance of the algorithm in the search for that solution. To achieve this, we must link in some way the two swarms in the algorithm. We modeled this link on the symbiotic relations so common in nature, where two species live in a close relation from which both get some advantage.

In SAPPO each solution particle lives in symbiosis with a parameter particle, which encodes the parameters used when the algorithm's update equations are applied to that solution particle. The symbiotic relation is implemented through the definition of the parameter particle fitness function.

A parameter particle has a "slower" life cycle than its companion does. While a solution particle is evaluated and has its velocity and position updated every iteration, to a parameter function this only happens every i iterations (usually 10). The parameter particle is then evaluated by comparing the actual fitness of its solution particle companion with the same fitness i iterations ago. If there was an improvement, its value is stored as the parameter particle fitness, unless it is smaller than the improvement value already stored. When a new improvement value replaces an older one, the P vector is also updated with the current parameter particle's position in the search space. Velocity and position updates will then be made using Equation (27.11). As a final element, the fitness of a parameter particle will decay slowly over every iteration of the algorithm (usually being multiplied by a decay factor $\alpha = 0.98$). This decaying ensures that a parameter particle has to keep producing improvements in the associated solution particle to maintain a high fitness.

Our approach to adaptation in SAPPO tries to maintain, as we already did with the predator–prey principle, the biological inspiration of the original PSO algorithm. We also made an effort to develop a mechanism that could be implemented following principles and ideas underlying the paradigm of swarm intelligence. Both the mechanisms introduced are based in the interaction of simple particles, with simple update rules and no centralized control. Both the solution and the behavior of the algorithm emerge from the interactions between the individuals in this ecology of particles, and if there is intelligence in the system it is clearly not preprogrammed but emerges as a property of the system of swarms.

27.5.3 Using SAPPO for Data Mining

We believe that the search space of rules in classification tasks will have irregularities and local optima capable of attracting the basic PSO (and especially traditional, greedier, approaches) to sub-optimal solutions. In Section 27.6 we empirically compare a SAPPO-based data miner to both the other PSO based approach already described and classical data mining algorithms, to investigate if the increase in search space complexity (new dimensions added for the algorithm's parameters) is compensated by an increase in performance.

For these experiments, the SAPPO algorithm was transparently integrated in our PSDM, which, as was explained in the previous sections can accommodate any algorithm for rule search whiteout changes to the higher levels of the PSDM.

27.6 Experimental Results

To evaluate the PSDM introduced in the previous sections we decided to test their performance in terms of both prediction accuracy and comprehensibility in a set of benchmark datasets. These datasets, listed in Table 27.1, were chosen to present different kinds of problems to the algorithms.

In Table 27.1 each dataset is described in terms of number of instances, attributes, classes, possible attribute/value pairs, and, finally, approximate size of the search space for an individual rule. While these values cannot be used directly to evaluate the difficulty of each task, they provide, at least, a coarse grained assessment of that difficulty, specially attending to the size of the search space and the number of instances. Obviously, this can be misleading, since, for example, a large number of examples neatly grouped by its class in the search space will provide a simpler classification task than a smaller number of instances uniformly scattered in the same space.

TABLE 27.1 Dataset Characterization

Dataset	Instances	Attributes	Classes	Attribute/Values	S.S. size
Zoo	101	17	7	43	1.38E+6
Breast cancer	286	10	2	54	1.01E+6
Wisconsin breast cancer	699	10	2	92	2.00E+9
Promoters	106	58	2	230	4.15E+34
Splice	1000	61	3	243	3.99E+36

`Zoo` is a simple dataset that acts as a basic test. It is useful to access the algorithms' ability to deal with multiple classes. `Breast Cancer` is a typical dataset from the field's literature with only two possible classes. `Wisconsin Breast Cancer` presents a similar problem but with a larger search space and number of instances. `Promoters`, with instances being DNA sequences, presents a qualitatively different problem, with much more attributes to be taken into account then the previous datasets. This also results in a much larger search space. `Splice` is again a DNA classification problem, but with the availability of a significantly larger number of examples.

The PSO-based data miners were empirically compared with two standard classification algorithms, namely J48 and PRISM. J48 is a Java implementation of a decision-tree building algorithm (C4.5) and PRISM is a rule discovering algorithm. Both are widely used for classification tasks. J48 was chosen for this comparison since it is probably the best known "industry standard" data mining algorithm. PRISM uses a similar covering strategy to the one used in the higher level of our algorithms, so it is especially useful for accessing the advantages of using swarm particle algorithms for the rule discovering level. Both algorithms can produce rule sets as their final classification model, which provides significant advantages in comparison fairness. Not only do the algorithms produce the same type of knowledge, but also rule sets can be easily compared between themselves in terms of comprehensibility and simplicity.

For each dataset we ran ten tenfold cross-validations for each of the algorithms being compared. This means that 100 classification models where built for each algorithm and dataset. For each model, 90% of the available instances were effectively used in its construction, while the remaining 10% were used for testing. The results are presented in Section 27.6.1.

27.6.1 Results

The tables presented show the results obtained for the described datasets. For each experiment with a given algorithm/dataset combination, we present the average accuracy and variance over the ten tenfold cross-validations. Averages and standard deviations are also presented for the number of rules and attribute test per rule over the same cross-validations. For J48 the number of rules represents the number of tree leafs and the attribute/rule ratio is equal to the number of nodes in the tree divided by the number of leafs. These numbers allow a fair comparison between the complexity of rule set and tree-based models.

Experiments were done with J48, PRISM, three variants of the basic PSDM with different stop criterions and, finally a SAPPO-based data miner. The algorithms with platform and radius stop criterions were investigated in order to find out if temporal complexity — always a major constraint in evolutionary computation — could be reduced without compromising accuracy. The other PSDM algorithms stopped (for each rule) when a 2000 iteration limit was achieved.

On an individual analysis of the results, for the `zoo` dataset, we can observe that, with exception of PRISM, all algorithms performed at similar levels. From these the simplest models were built by the PSO-based data miners, with just around seven rules against an average 10.98 by J48. In the PSO-based approaches, each rule tested just one attribute against an average 1.67 by the J48 algorithm. This makes the models found by the PSDM algorithms significantly simpler for this problem than the ones found by J48.

For the `Wisconsin breast cancer` dataset, the accuracies are again similar, with a slight advantage for PRISM. Regarding simplicity, we find that the tendency observed for the previous dataset is maintained. While J48 (the nonPSDM algorithm with simpler models) uses an average 37.18 rules, all the

PSDM obtained models with a number of rules slightly below or around 10. The number of attributes per rule is similar.

The results for the `breast cancer` dataset show that several PSO data miners are around 2% more accurate than J48, the best standard data miner. The models found are again significantly simpler for the PSDM algorithms.

The experiences with the `promoters` dataset show that the best PSDM algorithm presented an accuracy of 3% above the most accurate nonPSO approach — J48. The other PSDM algorithms all obtained similar accuracy. In terms of simplicity, the PSO-based approaches used, again, a significantly inferior number of rules and attribute tests.

Experimental Results

Algorithm	Particles	Acc	StdDev	#Rules	StdDev	#Att/Rule	StdDev
				Promoters			
J48	—	73.33	11.95	14.60	3.07	1.31	0.28
PRISM	—	72.02	11.43	16.00	0.00	1.94	0.00
PSDM platform	100	76.30	8.60	6.60	0.49	0.87	0.05
PSDM radius	100	75.80	9.00	6.60	0.51	0.88	0.07
PSDM 2000 it.	100	73.60	8.82	6.64	0.50	0.88	0.06
SAPPO	100	74.80	8.70	6.61	0.51	0.92	0.08
				Splice			
J48	—	91.14	3.31	105.21	6.11	1.24	0.07
PRISM	—	81.17	3.71	108.00	0.00	3.03	0.00
PSDM platform	100	76.26	5.71	6.67	1.70	0.88	0.08
PSDM radius	100	78.24	5.46	7.04	1.76	0.92	0.11
PSDM 2000 it.	100	77.56	5.59	6.67	1.62	0.90	0.09
SAPPO	100	83.06	4.62	6.92	1.91	1.17	0.11
				Zoo			
J48	—	92.63	7.36	10.98	2.65	1.67	0.31
PRISM	—	62.43	5.04	40.00	0.00	1.00	0.00
PSDM platform	100	93.20	7.23	7.02	0.14	1.04	0.09
PSDM radius	100	93.90	7.09	7.00	0.00	1.05	0.10
PSDM 2000 it.	100	93.40	7.81	7.01	0.10	1.04	0.09
SAPPO	100	93.30	8.41	7.00	0.00	1.00	0.00
				Wbc			
J48	—	93.24	2.85	37.18	13.90	1.11	0.42
PRISM	—	94.30	2.63	86.00	0.00	1.57	0.00
PSDM platform	100	92.22	5.49	9.98	0.75	0.91	0.03
PSDM radius	100	92.30	5.40	9.92	0.85	0.91	0.03
PSDM 2000 it.	100	92.26	5.46	10.00	0.74	0.91	0.03
SAPPO	100	92.44	4.80	9.87	0.97	0.96	0.04
				Bc			
J48	—	74.91	6.43	11.21	8.84	1.30	0.91
PRISM	—	69.65	7.58	126.00	0.00	2.5	0.00
PSDM platform	100	77.26	8.31	6.51	0.89	1.35	0.18
PSDM radius	100	76.96	8.64	6.52	0.96	1.39	0.18
PSDM 2000 it.	100	77.03	8.55	6.46	0.93	1.38	0.17
SAPPO	100	77.40	8.56	7.00	1.01	1.48	0.15

For the final dataset, `splice`, the accuracy of J48 is almost 8% above the best PSO data miner, which was the SAPPO-based approach. This difference is reflected in terms of simplicity, with the SAPPO models requiring a mere 6.92 average rules, against an average 105.21 from J48. The next best PSDM approach obtained a 5% worse accuracy than the SAPPO-based PSDM.

On a more global analysis, there are also some conclusions to be made. The most relevant one is clearly the competitiveness of the PSO-based data miners — especially SAPPO — with the classical, more established, approaches in terms of accuracy. Inclusively, in two of the datasets (three if we include `zoo` where the results were very similar, but with all PSDM approach getting an accuracy around 1% above the accuracy of J48), the best results were obtained by PSO-based approaches. Only in the `splice` dataset was the best PSO miner clearly outperformed by a classical approach.

When discussing comprehensibility of the results, and assuming simpler models are more comprehensible, results are extremely favorable to the PSO-based approaches. The PSO miners always present the solutions with the least number of rules for each of the tested datasets and the number of attribute tests per rule is also always, at least, slightly lower than the values obtained by the classic approaches. Probably, it is this bias toward simplicity that is accountable for the worse results with the `splice` dataset, this being an issue worthy of further investigation.

Between the PSO-based approaches, in terms of simplicity, all the algorithms performed in a similar way. Regarding accuracy, SAPPO was probably the most robust PSDM, with results near the best PSDM for all datasets, and significantly superior for the last, more complex, problem.

27.7 Conclusions

In this chapter, after briefly discussing other biologically inspired methods for data mining, more specifically for classification problems, we presented a first approach to classification based on particle swarm algorithms. We tested several variants of our algorithm, including an adaptive one, which uses symbiotic particles to adapt its parameters to the problem, and a predator–prey mechanism to maintain diversity in the swarm.

The results obtained until now, for a limited number of datasets, lead us to believe that the PSO-based approaches are clearly competitive with the classical algorithms, used for comparison, in terms of accuracy, and clearly superior in terms of simplicity and, consequently, comprehensibility.

Results with the last dataset seem to indicate that an increase in number of instances can result in a loss of accuracy of the PSDM approaches, probably caused by the identified bias toward simplicity, which can, in these cases become prejudicial by not allowing more complex models to be found by the algorithm. We plan to address this issue in future work.

Especially relevant for us is the fact that, for all but one of the datasets, the PSDM approaches outperformed the PRISM algorithm, which uses a similar covering strategy, but a greedy approach to the search for new rules. The strengths of evolutionary search algorithms are clearly patent in these results.

If a choice was to be made between the PSDM algorithms, if temporal complexity was an issue, the PSDM-radius and PSDM-platform variants are clearly competitive in terms of accuracy — with a possible slight advantage for the platform version — and obviously take less time to find solutions than the remaining approaches. If accuracy was the essential choice criterion, these first results seem to indicate that the SAPPO approach can be more robust over a large set of application domains.

References

[1] U.M. Fayyad, G. Piatetsky-Shapiro, P. Smyth, and R. Uthurusamy. *Advances in Knowledge Discovery and Data Mining*. AAAI/MIT Press, Cambridge, MA, 1995.

[2] S. Weiss and N. Indurkhya. *Predictive Data Mining: A Practical Guide*. Morgan Kaufmann, San Francisco, CA, 1998.

[3] I.H. Witten and E. Frank. *Data Mining: Practical Machine Learning Tools and Techniques with Java Implementations*. Morgan Kaufmann, San Francisco, CA, 1999.

[4] R.C. Eberhart and Y. Shi. Particle swarm optimization. In *Proceedings of the International Conference on Neural Networks and the Brain.* Beijing, China, 1998, pp. PL5–PL13.
[5] J. Kennedy, R.C. Eberhart, and Y. Shi. *Swarm Intelligence.* Morgan Kaufmann, San Francisco, CA, 2001.
[6] A. Silva, A. Neves, and Ernesto Costa. An empirical comparison of particle swarm and predator prey optimisation. In *Proceedings of Artificial Intelligence and Cognitive Science: 13th Irish International Conference, AICS 2002.* Vol. 2464/2002 of *Lecture Notes in Computer Science,* September 12–13, 2002, Springer-Verlag, Limerick, Ireland.
[7] A. Silva, A. Neves, and E. Costa. SAPPO: A simple, adaptive, predator prey optimiser. In *Proceedings of the EPIA'03 — 11th Portuguese Conference on Artificial Intelligence, Workshop on Artificial Life and Evolutionary Algorithms (ALEA),* December 4–7, 2003, Beja, Portugal.
[8] D.E. Goldberg. *Genetic Algorithms in Search, Optimization and Machine Learning.* Addison-Wesley, Reading, MA, 1989.
[9] T. Back. *Evolutionary Algorithms in Theory and Practice: Evolution Strategies, Evolutionary Programming, Genetic Algorithms.* Oxford University Press, Oxford, 1996.
[10] J.R. Koza. *Genetic Programming: On the Programming of Computers by Means of Natural Selection.* MIT Press, Cambridge, MA, 1992.
[11] A.A. Freitas. A survey of evolutionary algorithms for data mining and knowledge discovery. *Advances in Evolutionary Computation* (A. Ghosh and S. Tsutsui, Eds.), Springer-Verlag, 2002.
[12] A. Giordana and F. Neri. Search-intensive concept induction. *Evolutionary Computation,* 3: 375–419, 1995.
[13] D.P. Greene and S.F. Smith. Competition-based induction of decision models from examples. *Machine Learning,* 13: 229–257, 1993.
[14] K.A. DeJong, W.M. Spears, and F.D. Gordon. Using genetic algorithms for concept learning. *Machine Learning,* 13: 161–188, 1993.
[15] C.Z. Janikow. A knowledge-intensive genetic algorithm for supervised learning. *Machine Learning,* 13: 189–228, 1993.
[16] M.L. Wong and K.S. Leung. *Data Mining Using Grammar Based Genetic Programming and Applications.* Kluwer, Dordrecht, 2000.
[17] J. Eggermont, A.E. Eiben, and J.I. van Hemert. A comparison of genetic programming variants for data classification. *Proceedings of the Intelligent Data Analysis (IDA-99),* 1999.
[18] C.C. Bojarczuk, H.S. Lopes, and A.A. Freitas. Discovering comprehensible classification rules by using genetic programming: A case study in a medical domain. *GECCO,* 953–958, 1999.
[19] M.D. Ryan and V.J. Rayward-Smith. The evolution of decision trees. In *Proceedings of the Third Annual Conference on Genetic Programming,* Morgan Kaufmann, San Francisco, CA, 1998.
[20] E. Bonabeau, M. Dorigo, and G. Théraulaz. *Swarm Intelligence: From Natural to Artificial Systems.* Oxford University Press, Oxford, 1999.
[21] M. Dorigo, V. Maniezzo, and A. Colorni. The ant system: optimization by a colony of cooperating agents. *IEEE Transactions on Systems, Man, and Cybernetics — Part B,* 26: 29–41, 1996.
[22] R.S. Parpinelli, H.S. Lopes, and A.A. Freitas. An ant colony algorithm for classification rule discovery. In *Data Mining: A Heuristic Approach* (H. Abbass, R. Sarker, and C. Newton, Eds.), Idea Group Publishing, London, 2002, pp. 191–208.
[23] L.N. de Castro and J. Timmis. *Artificial Immune Systems: A New Computational Intelligence Approach.* Springer-Verlag, 1998.
[24] D. Dasgupta, Z. Ji, and F. González. Artificial immune system (AIS) research in the last five years. In *Proceedings of the International Conference on Evolutionary Computation Conference (CEC),* Canbara, Australia, December 8–12, 2003.
[25] L.N. de Castro and Fernando José Von Zuben. aiNet: An artificial immune network for data analysis. In *Data Mining: A Heuristic Approach* (Hussein A. Abbass, Ruhul A. Sarker, and Charles S. Newton, Eds.), Idea Group Publishing, USA, March, 2001.

[26] O. Nasraoui, F.A. González, C. Cardona, C. Rojas, and D. Dasgupta. A scalable artificial immune system model for dynamic unsupervised learning. *GECCO*, 219–230, 2003.
[27] M. Clerc and J. Kennedy. The particle swarm-explosion, stability, and convergence in a multidimensional complex space. *IEEE Transactions on Evolutionary Computation*, 6: 58–73, 2002.
[28] Y. Shi and R.C. Eberhart. Empirical study of particle swarm optimization. In *Proceedings of the IEEE Congress on Evolutionary Computation (CEC 1999)*, Piscataway, NJ, 1999, pp. 1945–1950.
[29] http://web.ics.purdue.edu/~hux/bibliography.shtml
[30] J. Kennedy. Small worlds and mega-minds: Effects of neighborhood topology on particle swarm performance. In *Proceedings of IEEE Congress on Evolutionary Computation (CEC 1999)*, Piscataway, NJ, 1999, pp. 1931–1938.
[31] T.M. Blackwell and P.J. Bentley. Don't push me! collision-avoiding swarms. In *Proceedings of the IEEE Congress on Evolutionary Computation (CEC 2002)*, Honolulu, Hawaii, USA, 2002.
[32] T. Sousa, A. Neves, and A. Silva. Swarm optimisation as a new tool for data mining. In *Proceedings of the Parallel and Distributed Processing Symposium*, 2003, pp. 144–149.
[33] T. Sousa, A. Silva, and A. Neves. A particle swarm data miner. In *Proceedings of 11th Protuguese Conference on Artificial Intelligence (EPIA 2003)*, Vol. 2902 of *Lecture Notes in Computer Science*, Progress in Artificial Intelligence, Beja, Portugal, 2003, pp. 43–53.

28

A Hybrid Evolutionary Algorithm for Knowledge Discovery in Microarray Experiments

L. Jourdan
M. Khabzaoui
C. Dhaenens
El-Ghazali Talbi

28.1 Introduction

Experiments with DNA microarray technology generate a large amount of data as they are able to measure the expression levels of thousands of genes in a single experiment [1]. To explore these data it is necessary to develop knowledge discovery techniques that can extract biological significance and use the data to assign functions to genes. This is the goal of data mining (also known as Knowledge Discovery in Databases [KDDs]) that has been defined as "The nontrivial extraction of implicit, previously unknown, and potentially useful information from data" [2].

Several kinds of representations to express knowledge that can be extracted from microarray data can be found in the literature [6]. Most current approaches in this area group genes, using clustering algorithms such as hierarchical clustering [3] or Kohonen maps, that is, self-organizing maps [4]. Association rule mining is an advanced data mining technique that is useful in deriving meaningful rules from a given dataset. Rule mining, as well as other datamining approaches, may be seen as a NP-hard combinatorial problem that can be solved with combinatorial optimization methods such as evolutionary algorithms.

In this chapter, we propose a multicriteria hybrid metaheuristic for microarray data analysis. We first introduce the biological context of the work. Then we set the rule mining context. After this positioning, we present a multicriteria hybrid algorithm and describe every feature of the algorithm (representation, operators, etc.). Finally, results on real datasets are presented and analyzed.

28.2 Microarray Analysis

Microarray technology (also called DNA microarrays, DNA arrays, DNA chips, gene chip, etc.) is now widely used in many areas of biomedical research. It provides access to expression levels of thousands of genes at once in order to identify coexpressed genes, relationships between genes, patterns of gene activity, changes in gene activity under some medical treatments, etc. A microarray is typically a glass (or some other material) slide, on to which probes are attached at fixed locations (spots). There may be tens of thousands of spots on a single array. The technology involves hybridizing DNA of a reference sample and a test sample, previously labeled with different fluorescent dyes, with probes that match a single gene. Hence, measuring the fluorescence allows us to calculate the relative abundance of DNA of each gene and evaluate their expression level. Arrays are scanned and images are produced and analyzed to obtain an intensity value for each probe.

There are two variants of the DNA microarray technology: Synteni/Stanford chips and Affymetrix chips. They differ in

- How DNA sequences are put down (spotting/photolithography)
- Length of DNA sequences (complete sequences or a series of fragments)

In the Synteni/Stanford chips, probes of cDNA (500/5000 bases long) are immobilized to a solid surface such as glass using robot spotting and exposed to a set of targets either separately or in a mixture. Affymetrix chips are powerful technology for resequencing DNA and polymorphisms detection. They use photolabile agents and photolithography techniques.

Although fundamental differences exist between those two technologies, their strength lies in a massively parallel analysis of thousands of genes and the generation of a lot of data.

Analyzing DNA microarray data requires a preprocessing phase [1,5,6]. Figure 28.1 recalls the different steps required to produce new biological assumptions from microarray experiments. Major steps are described as follows:

Relative gene expression: The differential gene expression is calculated by dividing the intensity of the gene in the sample under study by its intensity level in the control. This intensity ratio has a highly asymmetric distribution. To avoid this problem, a $\log_2$-transformation is usually used to make a normal-like distribution:

$$\text{Gene expression} = \log_2 \frac{\text{Expression of a gene in the sample}}{\text{Expression of the same gene in the control}}. \qquad (28.1)$$

Normalization: There are many sources of variations of measures in microarray experiments (variations in cells or individuals, mRNA extraction, isolation, hybridization condition, optical measurement, scanner noise, etc.). The purpose of normalization is to adjust (or correct) a signal in order to make the comparison with other signals more meaningful. Many techniques for normalization aim to make the data more normally distributed (log transformation per chip and per gene). This is an important issue for data analysis.

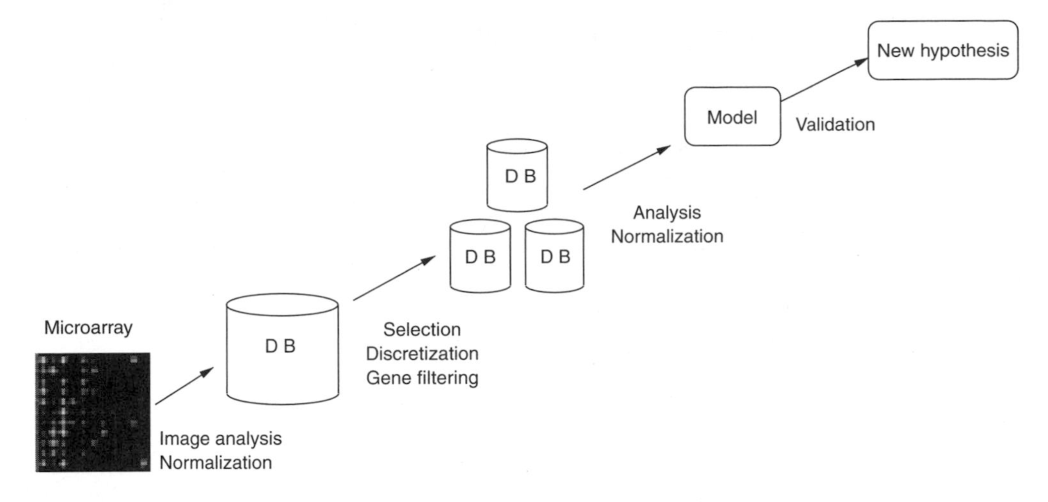

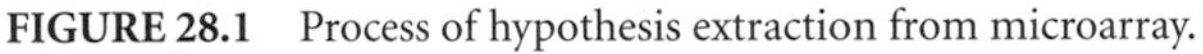

FIGURE 28.1 Process of hypothesis extraction from microarray.

Gene filtering and discretization: We are interested in finding genes that show significant differences between two groups of patients. Hence, the filtering process may remove genes that do not differentiate the sample under study from the test sample (their relative expression is not significant). Most of the time, gene expression data is discretized into under/over expressed genes thanks to cutoffs.

Hence after these different phases, data may be considered as large tables indicating, with discretized values, the relative gene expression for thousands of genes under different experiments.

28.3 Rule Mining

In this chapter, we propose to use a rule mining approach to explore the data provided by microarray experiments. We first introduce what rule mining is, and a classical rule mining approach. Then, we present a state-of-the-art bioinspired methods for rule mining. Finally, we justify the use of rule mining, in analysis of microarray experiments.

28.3.1 Association Rules

The search of association rules is an important task in knowledge discovery from databases [7–9]. Originally proposed for Market Basket data to study consumer-purchasing patterns, it has potential applications in many areas. Given a database describing instances (e.g., transactions) according to different attributes (e.g., items purchased), this problem involves discovering rules in the form: *IF C THEN P* or in a more detailed expression, *IF* $Cond_1$ *AND* $Cond_2$ *AND* ... *AND* $Cond_m$ *THEN P*. The condition *C* is a conjunction of terms. For binary representation, a term indicates whether the corresponding attribute has been chosen or not. In nominal data, where an attribute may take different values, a term may be of the form ⟨*attribute operator value*⟩. The prediction *P* is also represented by a conjunction of terms.

The association rule search problem is a complex combinatorial problem as every attribute may be candidate to participate, with one of its possible values, to the rule. Hence, the number of possible combinations is exponential to the number of attributes. Moreover, this problem has been shown to be NP-hard by Angiulli et al. [10]. The problem has been reduced into a MAX clique problem that is a well-known NP-hard problem [11]. Therefore, algorithms dedicated to this problem have to be carefully designed.

28.3.2 A Constructive Method for Rule Mining

A classical approach for finding association rules is to look for frequent itemsets (set of items — or terms — that often occur simultaneously) as done by the Apriori [12] algorithm. This algorithm relies on a fundamental property of frequent itemsets, called the Apriori property: "Every subset of a frequent itemset is also a frequent itemset." This monotony allows to incrementally construct interesting sets of items in a very efficient way. Then items belonging to a same set are combined in order to obtain rules. There exists numerous variants of Apriori [13–15] that allow to deal with problems of interesting sizes, but each of them requires the monotony of the criterion used.

28.3.3 Bioinspired Methods and Rule Mining

As the number of rules that are candidate for extraction is exponential [16], the use of an evolutionary algorithm such as a genetic algorithm is well adapted to explore the large search space of candidate rules. Those algorithms have already shown their ability to solve large combinatorial optimization problems and have already been applied with success to the rule discovery problem [17].

The paradigm of evolutionary algorithms consists of stochastic search algorithms that are based on an abstraction of the process of Darwinian evolution [18]. An evolutionary algorithm maintains a population of "individuals," each of them representing a potential solution to a given problem. Each individual is evaluated by a fitness function, which measures the quality of the solution (its adaptation to the problem). Individuals evolve toward better and better individuals via a selection procedure, based on the fitness of individuals, thanks to crossover (recombination) and mutation operators, and the replacement procedure.

In previous studies on association rule discovery with genetic algorithms, we can find two major representations: the *Michigan* [19] representation and the *Pittsburg* [20] one. Each representation has its drawbacks and advantages. The *Pittsburg* representation, where an individual is a set of rules, is more expensive in memory but has been used in several genetic algorithms: GABIL [21], GIL [22], and HDPDCS [23]. In the *Michigan* representation each individual of the population of the genetic algorithm represents a candidate rule. The main drawback of the *Michigan* approach is that it could, without the use of specific mechanisms, converge to a single rule and not to a set of rules. For example, in Reference 24 Weiss uses the sharing fitness to avoid the convergence to a single rule in order to predict rare events. Classical genetic algorithms for rule mining using the *Michigan* representation are COGIN [25] and REGAL [26].

28.3.4 Interest of Rule Mining for Microarray Analysis

Microarray data format is very similar to the Market Basket data format. Microarray data may be represented in two different ways [27]:

Gene table: The genes constitute the rows of the data table whereas experiments to which genes were exposed are the columns. The values represent the abundance of transcript for each spot on the microarray. Clustering and classification may be applied to the gene table to divide the dataset into clusters/classes by grouping the rows (genes).

Treatment table: The reverse way to present data is to flip the gene table. Thus, genes are now the columns whereas treatments are the rows. The values are the gene expression level. The objective is then to find associations between columns (genes). Therefore, association rule mining is a great challenge. This data mining model is general and allows to find, for example, associations between subsets of genes. Moreover, thanks to the condition and prediction notions, relations obtained are precise and give more information than only grouping genes together.

There has been some recent works on using association rule mining to analyze gene expression data. Chen et al. [28] apply association rules to mine the transcription factors essential to certain gene expressions using the Apriori algorithm. They manage, with a small value of support to extract a small number of rules they can analyze. Creighton and Hanash [29] propose to also use the Apriori algorithm to reveal biological relevant associations between different genes or between environmental effects and gene expression. They apply it with success to a yeast database. Kotala et al. [27] introduce a new approach to mine association rules from microarray gene expression data using Peano count tree. Icev et al. [30] focus on the combinatorial analysis of motifs involved in transcriptional control and introduce a notion of association rules with distance information.

In our study we will consider data in the treatment table form (genes are the columns, treatments — or comparison of individuals of different status — are the rows. Our goal is to look for rules combining genes where a term can be in the form $\langle gene = value \rangle$. The value belongs to the discretized gene expression level. An example of a rule could be: *IF* $(gene_{12} = over\_expressed)$ *AND* $(gene_{504} = under\_expressed)$ *THEN* $(gene_{8734} = over\_expressed)$.

28.4 A Hybrid Multiobjective Metaheuristic

The algorithm proposed has been developed thanks to the open source framework PARADISEO (PARAllel and DIStributed Evolving Objects) [31] `http://www.lifl.fr/~cahon/logiciels.html` that allows, once you have designed your algorithm, to easily develop multicriteria evolutionary algorithms and to use parallelism and distributed mechanisms. The originality of the proposed approach is to deal with rule mining through a multiobjective point of view. Hence, we first recall some principles of multiobjective combinatorial optimization problem, then we present the modeling of the rule mining problem as a multiobjective problem. Finally, we present the algorithm and its hybridization with an exact enumeration.

28.4.1 A Multicriteria Model for Rule Mining

28.4.1.1 Multicriteria Definitions

We first describe and define Multi-Objective combinatorial Problems (MOPs) in a general case. Assume that a solution to such a problem can be described by a decision vector $(x_1, x_2, \ldots, x_n)$ in the decision space X. A fitness function $f : X \rightarrow Y$ evaluates the quality of each solution by assigning it an objective vector $(y_1, y_2, \ldots, y_p)$ in the objective space Y (Figure 28.2) [32]. So, multiobjective optimization consists in finding the solutions, in the decision space, optimizing (minimizing or maximizing) p objectives.

For the following definitions, we consider the minimization of p objectives. For maximization problems, definitions are similar.

In the case of a single objective optimization, the comparison between two solutions x^1 and x^2 is immediate. If $y^1 < y^2$ then x^1 is better than x^2. For multiobjective optimization, comparing two solutions x^1 and x^2 is more complex. Here there exists only a partial order relation, known as the Pareto dominance concept illustrated in Figure 28.3.

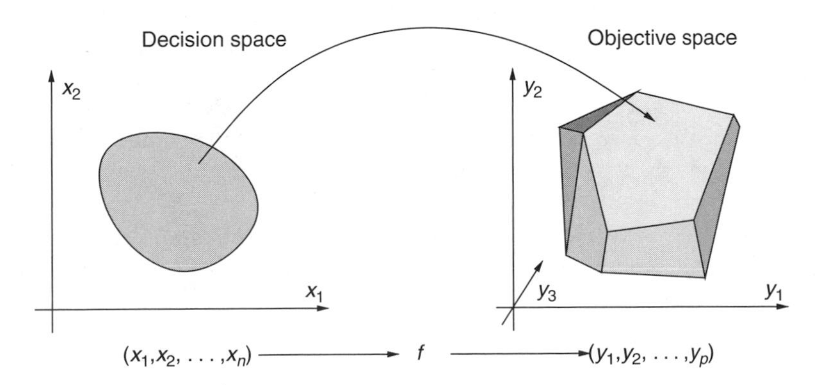

FIGURE 28.2 Example of a MOP.

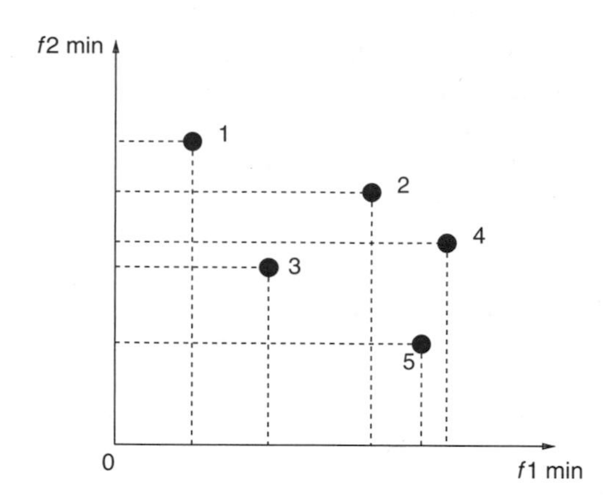

FIGURE 28.3 Example of dominance. In this example, points 1, 3, and 5 are nondominated. Point 2 is dominated by point 3, and point 4 by points 3 and 5.

Definition 28.1. *A solution x^i dominates a solution x^j if and only if:*

$$\begin{cases} \forall k \in [1..p], f_k(x^i) \leq f_k(x^j), \\ \exists k \in [1..p]/f_k(x^i) < f_k(x^j). \end{cases} \tag{28.2}$$

Definition 28.1. *A solution is Pareto optimal if it is not dominated by any other solution of the feasible set.*

The set of optimal solutions in the decision space X is denoted as the Pareto set, and its image in the objective space is the Pareto front. In MOP, we are looking for all the Pareto optimal solutions.

In the Pareto front two types of solutions may be distinguished: the supported solutions (that are on the convex hull of the set of solutions and that may be found by a linear combination of criteria), and nonsupported solutions [33]. These solutions are important, because for some problems only few Pareto solutions are supported (the extremes) and to get a good compromise between the two criteria, it is necessary to choose one of the nonsupported solution.

28.4.1.2 Multicriteria Association Rules

In order to solve association rule discovery problem as a combinatorial optimization problem, the optimization criterion has to be defined. A lot of measures exist for estimating the quality of association rules. For an overview, readers can refer to Freitas [34], Tan et al. [35], or Khabzaoui et al. [36]. In Reference 36, we made a statistical study of different criteria found in the literature. This study lead us to determine five groups where each group represents correlated criteria. We choose to select one criterion of each group and obtain five complementary criteria that allow to evaluate rules in a complete way: Support, Jmeasure, Interest, Surprise, and Confidence. These criteria are described below.

Rules are evaluated for a set of N instances, where $|C|$ (respectively $|P|$) represents the number of instances satisfying the C (respectively P) part of the rule and $|C\&P|$ the number of instances satisfying simultaneously the C and the P parts of the rule.

Support (S): It is the classical measure of association rules. It enables to measure rule frequency in the database. It is the percentage of transactions containing, both the C part and the P part, in the database. It is used to find frequent itemsets in Apriori

$$S = \frac{|C\&P|}{N}. \tag{28.3}$$

Confidence (C_f): The Confidence measures the validity of a rule. It is the conditional probability of P given C. It is used in Apriori to find interesting rules in frequent itemsets

$$C_f = \frac{|C\&P|}{|C|}. \tag{28.4}$$

J-measure (Jm): Smyth and Goodman [37] have proposed the Jmeasure, which estimates the degree of interest of a rule and combines support and confidence. It is used in optimization methods [38,39]

$$Jm = \frac{|P|}{N} \times \frac{|C\&P|}{|P|} \log\left(\frac{N \times |C\&P|}{|C| \times |P|}\right). \tag{28.5}$$

Interest (I): The Interest measures the dependency while privileging rare patterns in the region of weak support

$$I = \frac{N \times |C\&P|}{|C| \times |P|}. \tag{28.6}$$

Surprise (R): It is used to measure the affirmation. It enables the search for surprising rules

$$R = \frac{|C\&P| - |C\&\bar{P}|}{|\bar{P}|}. \tag{28.7}$$

Hence, the use of these five criteria allow an evaluation of a rule in a multicriteria manner. As they are complementary, this model is interesting to select interesting rules and to reduce the numbers of possible rules.

28.4.2 The Genetic Algorithm

We present the general scheme of the algorithm and the genetic operators used. The algorithm presented here is a multicriteria version of ASGARD presented in References 17 and 40.

28.4.2.1 General Scheme

Figure 28.4 presents the scheme of the genetic algorithm with its multicriteria aspects. The algorithm starts with a set of randomly generated solutions. Then, solutions are selected according to their fitness quality to create new solutions (offsprings). Best solutions encountered over generations are archived into a secondary population called the "Pareto archive."

28.4.2.2 Genetic Operators

Genetic operators allow diversification and intensification of the search:

Crossover: The proposed crossover operator has two versions:

- *Exchange crossover*: If two individuals X and Y have one or several common attribute(s) in the C parts, one common attribute is randomly selected. The value of the selected attribute in X is exchanged with its counterpart in Y (see Figure 28.5).
- *Insert crossover*: Conversely, if X and Y have no common attribute, one term is randomly selected in the C part of X and inserted in Y with a probability inversely proportional to the length of Y. The similar operation is performed to insert one term of Y in X (see Figure 28.6).

Mutation: There are four mutation operators designed. The mutation operator called "Value mutation" replaces the value of an attribute by a randomly chosen one (see Figure 28.7). The second one called "Attribute mutation" replaces an attribute by another; the value of this attribute is randomly

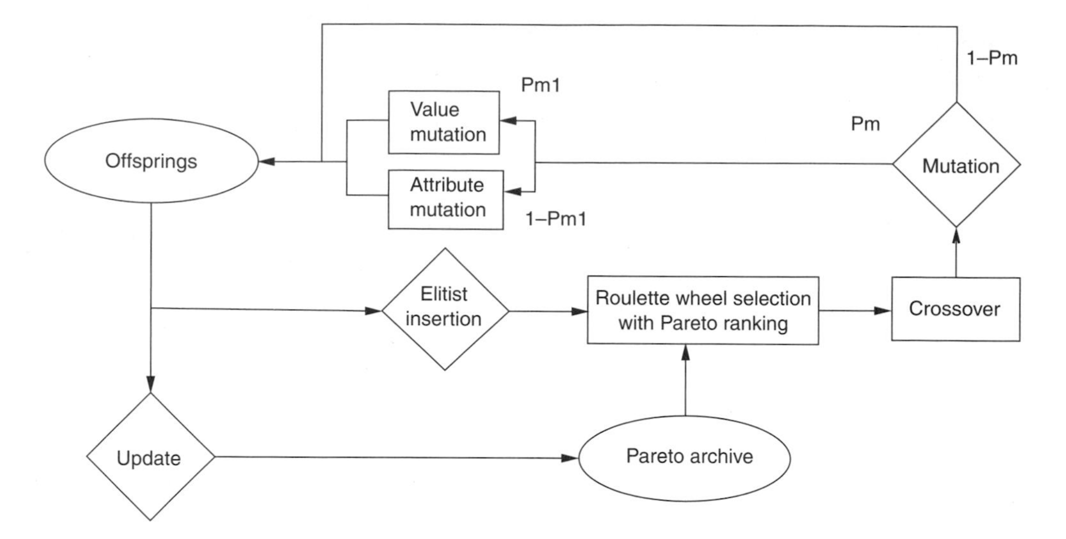

FIGURE 28.4 General scheme of the genetic algorithm.

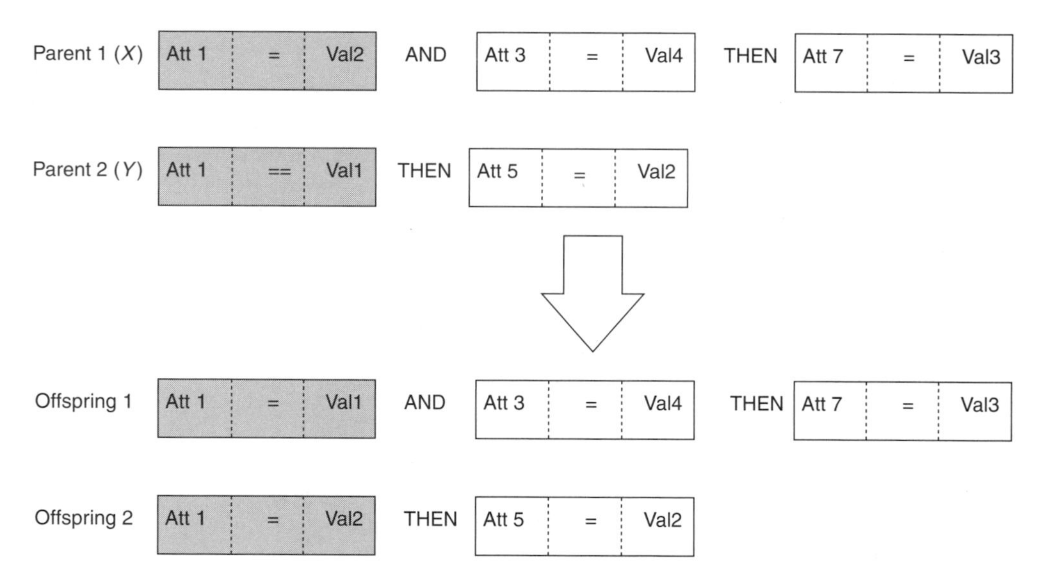

FIGURE 28.5 Exchange crossover.

chosen (see Figure 28.8). The third one is a reduction mutation that randomly removes one term of the rule. The last one is an augmentation mutation that randomly adds a term to the rule.

28.4.3 Multicriteria Mechanisms

In order to deal with multicriteria optimization problems, different mechanisms have to be used. For example, the notion of dominance has to be defined (to be able to compare solutions) and population management has to be carefully studied.

Selection operator: The classical roulette selection based on the ranking notion has been used. The probability of selection of a solution is proportional to its rank. We use two ranking methods:

1. *Pareto ranking*: The rank of a solution corresponds to the number of solutions, in the current population, by which it is dominated (see Figure 28.9 for a minimization problem) [4].

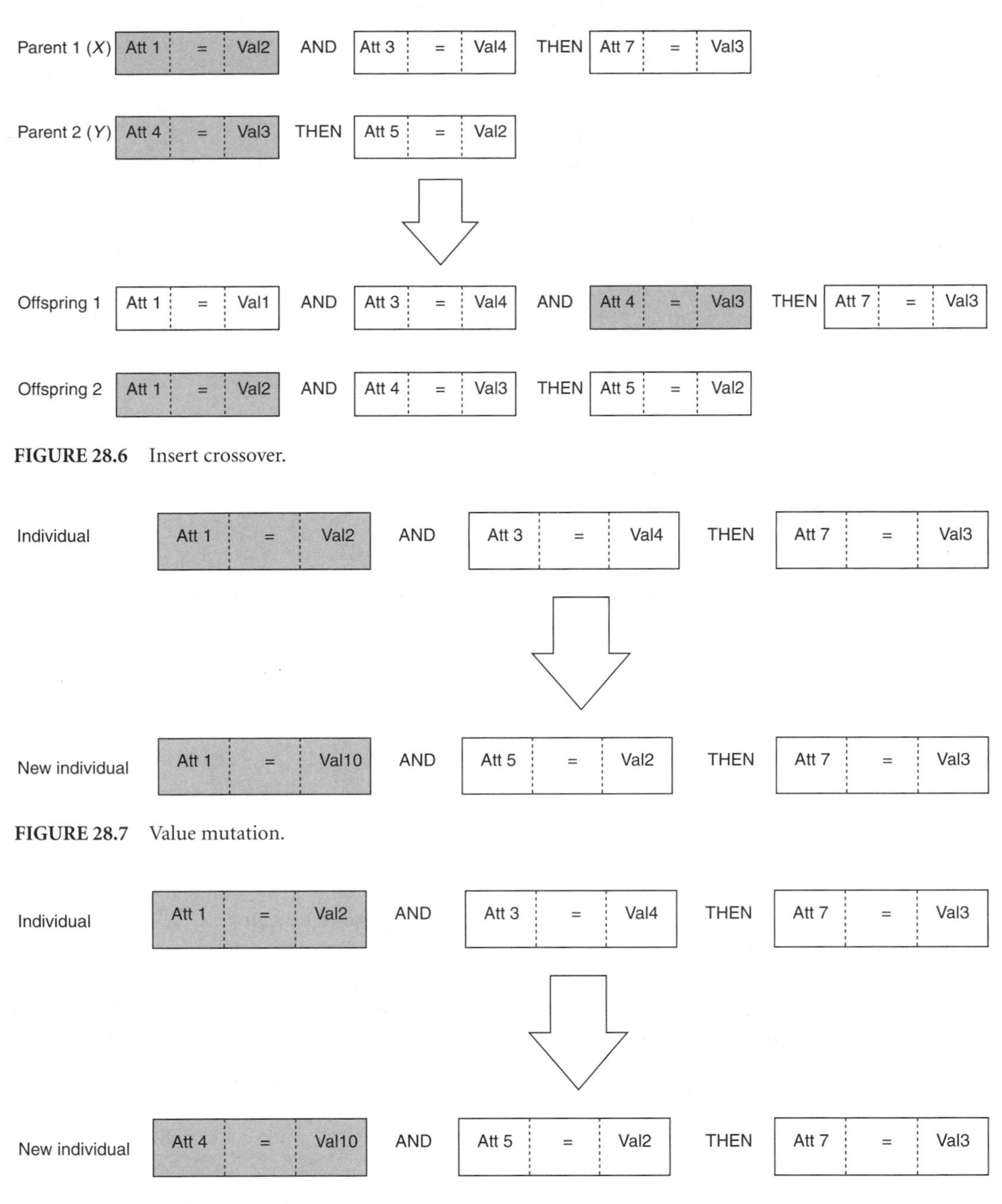

FIGURE 28.6 Insert crossover.

FIGURE 28.7 Value mutation.

FIGURE 28.8 Attribute mutation.

2. *Nondominated sorting GA (NSGA)*: This method assigns ranks to solutions by first finding the set of nondominated solutions in the current population. Those solutions are removed from the population and assigned rank 1. As these solutions are removed, a new so-called front of nondominated solutions is now present in the remainder of the original population. This second front is extracted and assigned rank 2. This procedure is repeated till there is no solution present in the population (see Figure 28.10 for a minimization problem) [42].

Experiments have shown that with the proposed algorithm, both ranking methods give interesting results with a small superiority for the Pareto ranking. Therefore we will use this Pareto ranking for the selection process.

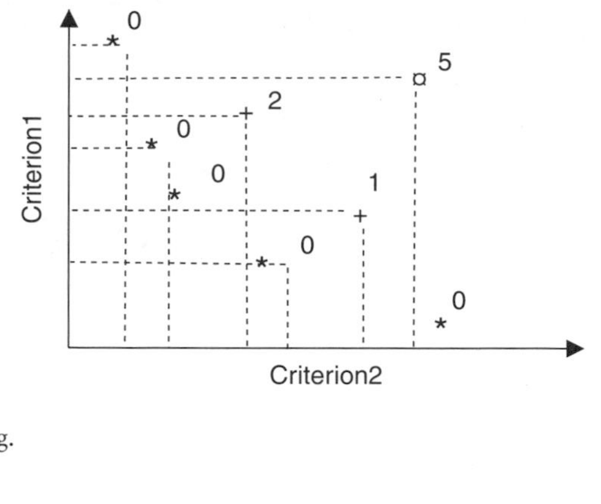

FIGURE 28.9 Pareto ranking.

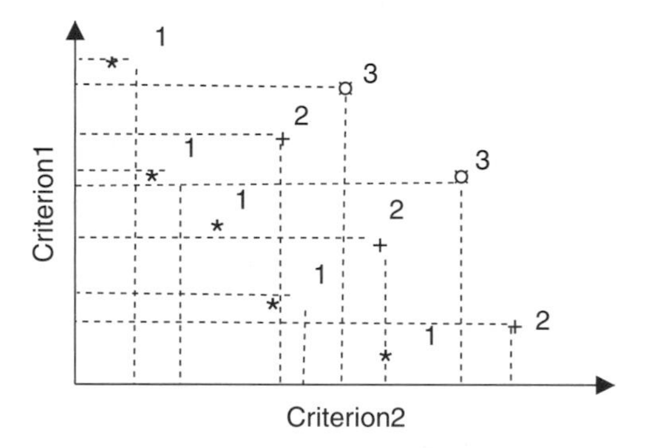

FIGURE 28.10 NSGA ranking.

Replacement operator: We use the elitist nondominated sorting replacement. The worst ranked solutions are replaced by the dominating solutions (if there exists any) generated by mutation and crossover operators (offsprings). The size of the population remains unchanged.

Archive: Nondominated association rules are archived into a secondary population called the "Pareto Archive" in order to keep track of them. It consists of archiving all the Pareto association rules encountered over generations. When a new Pareto solution is added to the archive, an update has to be done (some solutions may become dominated).

Elitism: The Pareto solutions (best solutions) are not only stored permanently, they also take part in the selection and may participate to the reproduction.

28.4.4 Hybridization

In order to increase the robustness of the approach, we hybridize it with an exact enumeration procedure. As the search space is large, this enumeration is realized for a small subspace defined thanks to solutions of the population. Hence, we design an operator which makes an exhaustive search of all the possible rules generated with attributes selected in two rules.

This operator may be seen as a quadratic crossover operator. It takes as input two individuals, each coding a rule. It examines all the possible itemsets that can be derived from the items composing the rules, while taking into account the different possible values of the attributes. All the possible rules that can be constructed from the generated itemsets are evaluated and introduced if necessary in a local Pareto archive. Finally, only the global Pareto solutions of the local Pareto archive are introduced in the global Pareto archive with the replacement operator. Two offsprings are candidates to take part in the population (Figure 28.11).

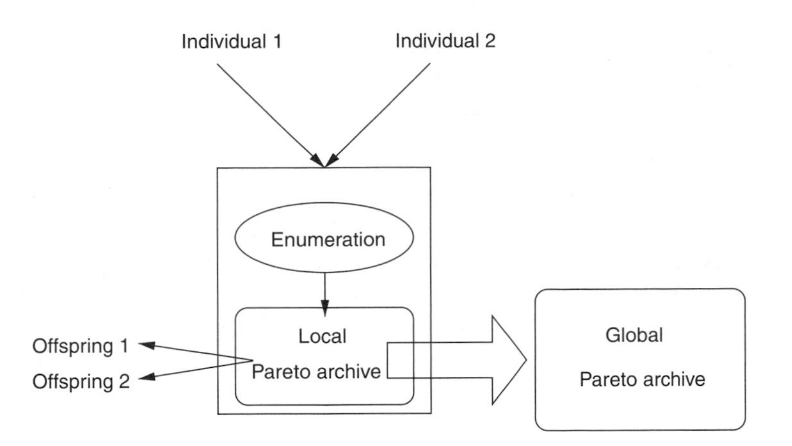

FIGURE 28.11 The enumeration operator.

28.4.5 Adaptive Rate

Probabilities of mutation are hard to set when several mutation operators compete and are often set experimentally. To overcome this problem, we implement an adaptive strategy for calculating the rate of application of each mutation operator. Many authors have worked on setting automatically probabilities of applying operator [43–45]. Hong et al. [46] proposed to compute the new rate of mutation by calculating the progress of the jth application of mutation M_i, for an individual *ind* mutated into an individual *mut* as follows:

$$\text{progress}_j(M_i) = \text{Max}(\text{fitness(ind)}, \text{fitness(mut)}) - \text{fitness(ind)}. \tag{28.8}$$

Then for each mutation operator M_i, assume $Nb\_mut(M_i)$ applications of the mutation are done during a given generation ($j = 1, \ldots, Nb_{\text{mut}}(M_i)$). Then we can compute the profit of a mutation M_k:

$$\text{Profit}(M_k) = \frac{\sum_j \text{progress}_j(M_k)/Nb\_\text{mut}(M_k)}{\sum_i \left(\sum_j \text{progress}_j(M_i)/Nb\_\text{mut}(M_i)\right)}. \tag{28.9}$$

We set a minimum rate δ and a global mutation rate p_{mutation} for N mutation operators. The new mutation ratio for each M_i is calculated using the following formula [46]:

$$p(M_i) = \text{Profit}(M_i) \times (p_{\text{mutation}} - N \times \delta) + \delta. \tag{28.10}$$

The sum of all the mutation rates is equal to the global rate of mutation p_{mutation}. The initial rate of application of each mutation operator is set to p_{mutation}/N.

28.5 Experiments

28.5.1 Data

In order to evaluate the algorithm, we evaluate it on two microarray databases:

- A confidential microarray data containing 22,376 human genes for 45 Affymetrix chips (*DB*1).
- A public database, the "MIPS yeast genome database" containing 2467 genes for 79 chips (*YeastDB*).

Genes expressions have been discretized and may take five values: Increase (I), Marginal Increase (MI), when the gene is over-expressed, Decrease (D) and Marginal Decrease (MD) when it is under-expressed and No Change (NC), when the difference of expression is not significant.

For an initial study, a set of 514 genes (numbered from 1 to 514) that show an interesting differential expression over the set of experiments (filtered on the number of No Changes), have been selected for *DB*1. For the *yeastDB* all the genes (2467) have been considered.

28.5.1.1 Missing Value

Missing values lead to problems in the data analysis, so that they influence the computation of statistical tests and quality criteria of association rules problems. There may have numerous missing values in microarray data due to the empty spots or because the background intensity is higher than the spot intensity. Two methods are commonly used for treatment of missing values: they may be replaced by estimated values (Median for example), or corresponding instances are deleted. In an association rule problem we propose that all genes that have missing values are kept without modification but when these missing values are on the attributes of a rule, these genes are excluded from the computation of the quality of this rule. So $|N|$ may be different for each rule.

28.5.2 Evaluation Protocol and Parameters

28.5.2.1 Configurations and Parameters

Several propositions have been done in this chapter in order to improve the standard genetic algorithm: elitism, adaptive strategy for mutation rates, hybridization. In order to evaluate the contribution of each of these mechanisms, several configurations of the genetic algorithm have been compared:

- *Conf A*: Pareto ranking
- *Conf B*: Pareto ranking + Adaptive strategy
- *Conf C*: Pareto ranking + Elitism
- *Conf D*: Pareto ranking + Adaptive strategy + Elitism
- *Conf E*: Conf D + Hybridization
- *Conf E′*: Conf D + Hybridization 1/10

In the *Conf E* configuration, the crossover operator has been replaced by the enumeration operator presented for the hybridization. Using this operator for every crossover is very time consuming, hence we compared with *Conf E′* algorithm where the hybridization operator replaces the classical crossover by only one iteration over ten.

For all of these configurations, the same parameters have been used:

- *Population size*: around $Nb_{\text{Attributes}}/2$, which gives 250 for *DB*1, 1200 for *YeastDB*
- *Selection in Pareto archive*: 0.5
- *Global mutation rate*: 0.5
- *Crossover rate*: 0.8
- *Number of generations*: 500

28.5.2.2 Evaluation Measure Used

In multicriteria optimization, solutions quality can be assessed in different ways. Some approaches compare the obtained front with the optimal Pareto front [47]. Other approaches evaluate a front with a reference point [48]. Some performance measures do not use any reference point or front to evaluate an algorithm [49,50], especially when the optimal Pareto front is not known at all.

Here, we use the *contribution metric* [51,52] to evaluate the proportion of Pareto solutions given by each front.

The contribution of a set of solutions PO_1 relatively to a set of solutions PO_2 is the ratio of nondominated solutions produced by PO_1 in PO^*, where PO^* is the set of Pareto solutions of $PO_1 \cup PO_2$.

- Let PO be the set of solutions in $PO_1 \cap PO_2$.
- Let W_1 (respectively W_2) be the set of solutions in PO_1 (respectively PO_2) that dominate some solutions of PO_2 (respectively PO_1).

- Let L_1 (respectively L_2) be the set of solutions in PO_1 (respectively PO_2) that are dominated by some solutions of PO_2 (respectively PO_1).
- Let N_1 (respectively N_2) be the other solutions of PO_1 (respectively PO_2): $N_i = PO_i \backslash (PO \cup W_i \cup L_i)$.

$$Cont(PO_1/PO_2) = \frac{(\|PO\|/2) + \|W_1\| + \|N_1\|}{\|PO^*\|}. \tag{28.11}$$

Let us remark that $\|PO^*\| = \|PO\| + \|W_1\| + \|N_1\| + \|W_2\| + \|N_2\|$ and $Cont(PO_1/PO_2) + Cont(PO_2/PO_1) = 1$ (with $Cont(PO_1/PO_2) \in [0, 1]$). Hence a contribution greater than 0.5 indicates that the Pareto front has been improved.

For example, we evaluate the contribution of the two sets of solution PO_1 and PO_2 in Figure 28.12: solutions of PO_1 (respectively PO_2) are represented by circles (respectively crosses). We obtain $Cont(PO_1, PO_2) = 0.7$ and $Cont(PO_2, PO_1) = 0.3$.

28.5.3 Results

Genetic algorithms are stochastic methods. Hence to evaluate the proposed approach, we have executed 10 runs for each configuration. Results are given with different indicators: mean of the value, minimum, maximum, and its standard deviation. This allows a more efficient comparison.

28.5.3.1 Evaluation of the Mechanisms

In order to evaluate the different mechanisms, comparisons of the final Pareto fronts obtained with the different configurations described before ($Conf\ A, B, C, D, E, E'$) are reported in Table 28.1. This table indicates the average of contribution as well as the minimum contribution, the maximum, and the standard deviation.

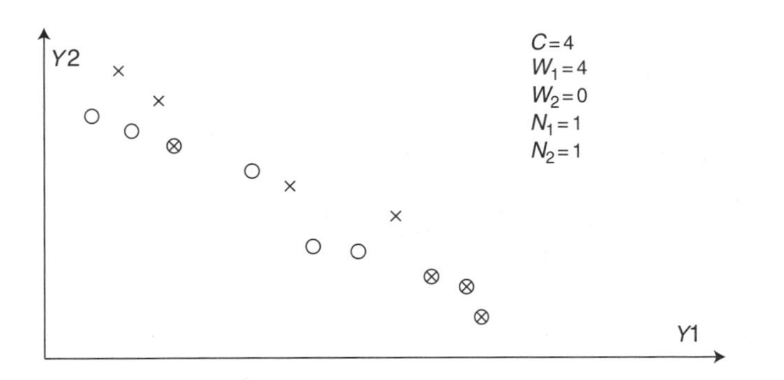

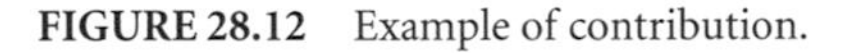
FIGURE 28.12 Example of contribution.

TABLE 28.1 Quality Comparison of the Different Configurations (Contribution Metric)

Contribution	Average	Minimum	Maximum	σ
Conf B/Conf A	0.55	0.39	0.78	0.09
Conf C/Conf A	0.62	0.43	0.90	0.09
Conf D/Conf B	0.72	0.52	0.86	0.07
Conf D/Conf C	0.61	0.43	0.76	0.08
Conf E/Conf D	0.74	0.59	0.87	0.06
Conf E'/Conf D	0.60	0.45	0.81	0.09
Conf E/Conf E'	0.58	0.41	0.74	0.09

28.5.3.1.1 Elitism

The gain of elitism may be evaluated by comparing similar configurations with and without elitism (here, C/A and D/B). Table 28.1 indicates that on an average, the Pareto fronts obtained using elitism are of better quality than those obtained without elitism ($Cont(C/A) = 0.62$ and $Cont(D/B) = 0.72$). Moreover this contribution may reach 0.9 for best improvements. This shows the interest of this mechanism.

28.5.3.1.2 Adaptive Strategy

In a similar way, the gain of using the adaptive strategy may be evaluated by comparing similar configurations with and without it (here, B/A and D/C). Table 28.1 shows that using the adaptive strategy also improves, but in a smaller way, the results obtained ($Cont(B/A) = 0.55$ and $Cont(D/C) = 0.61$). Let us remark that the improvement is greater for more complete configurations (Pareto ranking + elitism with and without adaptive strategy).

28.5.3.1.3 Hybridization

Two questions may arise concerning the use of the hybridization. Is it interesting to use such an operator and is it worth using it at each iteration. Hence we can compare configurations E/D and E/E'. Table 28.1 shows that using the hybridization allows to improve the more complete configuration (Pareto ranking + elitism + adaptive strategy). In this case, the minimum contribution encountered is equal to 0.59, which means that the hybridization always improves the Pareto front. When applying the operator only one generation over ten (*Conf* E'), the results are still interesting ($Cont(E'/D) = 0.60$), but Pareto front obtained are in general dominated by those produced by *Conf* E ($Cont(E/E') = 0.58$), where the operator is applied at each generation. However, the drawback of using *Conf* E is the large amount of time required and the compromise between the quality of the solution and the computing time allowed has to be considered.

28.5.3.2 Example of Rules Obtained

Table 28.2 describes some rules of the Pareto front obtained thanks to *Conf* E for the *YeastDB*, whereas Table 28.3 presents values of these rules for the five criteria considered (Support, Interest, Surprise, Confidence, and Jmeasure). Interest of such rules is difficult to estimate from a biological point of view; however one idea would be to compare with relations already known between genes belonging to a same rule.

Table 28.2 shows that there exists genes that appear in several rules, and we may suppose that those genes are relevant for the problem under study. Those genes should be the first the biologists have to study.

Moreover, Table 28.3 shows the interest of using the proposed multicriteria model as good solutions for one criteria which are not always interesting for another one.

TABLE 28.2 Description of Some Pareto Solutions Obtained with *Conf* E (YeastDB)

Rules	Description
R1	IF ((VPS35 = D) and (PRB1 = I)) then (VPH1 = D)
R2	IF ((DAK2 = I) and (SVS1 = D) and (KTR7 = D) and (RPS27A = D) and (PRO2 = D)) then (ARE2 = I)
R3	IF ((POP3 = D) and (TAF67 = I) and (VPS35 = D) and (SRP14 = D)) then (MF(ALPHA)2 = I)
R4	IF ((RPS6A = D) and (RPO26 = D) and (SVS1 = D) and (KTR7 = D)) then (DAK2 = I)
R5	IF ((RNR4 = I) and (IPT1 = I) and (GPM1 = I) and (YTA12 = I) and (PUP3 = I) and (SAP185 = D) and (ESC2 = D)) then (GAR1 = I)
R6	IF ((PUS1 = D)) then (POP3 = D)

Rules indicate the level of expression of the genes: I, Increase; D, Decrease.

TABLE 28.3 Quality of Pareto Solutions Presented in Table 28.2

Rules	S	C_f	I	R	Jm
R1	0.582	0.978	1.288	2.368	0.147
R2	0.177	0.875	3.638	0.200	0.228
R3	0.278	0.916	2.263	0.425	0.227
R4	0.253	0.909	2.762	0.339	0.257
R5	0.063	1.000	7.900	0.072	0.130
R6	0.835	0.929	1.005	10.16	0.004

S, Support; I, Interest; R, Surprise; C_f, Confidence; Jm, Jmeasure.

28.6 Conclusion

In this chapter we have presented a multiobjective genetic algorithm for rule mining problems. Therefore, a multicriteria model has been proposed for association rules mining. We have proposed a genetic algorithm which helps look for the Pareto solutions regarding the five selected criteria. We have presented its application to analyze microarray experiment data. Through the experiments, the advanced mechanisms proposed have been validated, and in particular the hybridization with an exact enumerative procedure.

In order to improve the use of such an algorithm and to speed up executions that may take several hours, we now work on a parallel implementation of the method. The parallelism should allow for results in a more reasonable time, but should also allow to an execution of more intense searches with the hybridization operator. Hence, we will be able to give biologists different hypothesis for their evaluation.

References

[1] D.P. Berrar, W. Dubitzky, and M. Granzow, Eds. *A Practical Approach to Microarray Data Analysis.* Kluwer Academic Publishers, New York, 2003.

[2] U. Fayyad, G. Piatetsky-Shapiro, and P. Smyth. From data mining to knowledge discovery: An overview. *Advances in Knowledge Discovery.* MIT Press, Cambridge, MA, 1996, pp. 1–34.

[3] P.T. Spellman, G. Sherlock, M.Q. Zhang, V.R. Iyer, K. Anders, M.B. Eisen, P.O. Brown, D. Botstein, and B. Futcher. Comprehensive identification of cell cycle regulated genes of yeast *Saccharomyces cervisiae* by microarray hybridization. *Molecular Biology of the Cell*, 9: 3273–3297, 1998.

[4] P. Tamayo, D. Slonim, J. Mesirov, Q. Zhu, S. Kitareewan, E. Dmitrovsky, E. Lander, and T. Golub. Interpreting patterns of gene expression with self-organizing maps: Methods and application to hematopoietic differentiation. *Proceedings of National Academy of Sciences*, 96: 2907–2912, 1999.

[5] B. Phimister. The chipping forecast. *Nature Genetics*, 21(Suppl.): 1–60, 1990.

[6] Collective works. The human genome project. *Nature*, 409: 813–959, 2001.

[7] R. Agrawal, T. Imielinski, and A. Swami. Mining association rules between sets of items in large databases. In P. Buneman and S. Jajodia, Eds., *Proeceedings of the 1993 ACM SIGMOD International Conference on Management of Data.* ACM Press, Washington, DC, May 1993, pp. 207–216.

[8] S. Morishita and A. Nakaya. Parallel branch-and-bound graph search for correlated association rules. In *Large-Scale Parallel Data Mining*, 1999, pp. 127–144.

[9] T. Scheffer. Finding association rules that trade support optimally against confidence. In *Principles of Data Mining and Knowledge Discovery*, 2001, pp. 424–435.

[10] F. Angiulli, G. Ianni, and L. Palopoli. On the complexity of mining association rules. In *Atti del Nono Convegno su Sistemi Evoluti per Basi di Dati (SEBD)*, Venise, Italie, 2001, p. 8.

[11] M.R. Garey and D.S. Johnson. *Computers and Intractability. The Guide to the Theory of NP-Completeness.* W.H. Freeman and Company, San Francisco, CA, 1979.

[12] R. Agrawal and R. Srikant. Fast algorithms for mining association rules. In J.B. Bocca, M. Jarke, and C. Zaniolo, Eds., *Proceedings of the 20th International Conference on Very Large Data Bases, VLDB*, Morgan Kaufmann, San Francisco, CA, 12–15 1994, pp. 487–499.

[13] S. Brin, R. Motwani, and C. Silverstein. Beyond market baskets: Generalizing association rules to correlations. In *ACM SIGMAD*, 1997, pp. 265–276.

[14] A. Savasere, E. Omiecinski, and S.B. Navathe. An efficient algorithm for mining association rules in large databases. In *The VLDB*, Zurich, Switzerland, 1995, pp. 432–444.

[15] H. Toivonen. Sampling large databases for association rules. In T.M. Vijayaraman, A.P. Buchmann, C. Mohan, and Nandlal L. Sarda, Eds., *Proceedings of the 1996 International Conference on Very Large Data Bases.* Morgan Kaufmann, San Francisco, CA, 1996, pp. 134–145.

[16] M.J. Zaki. Parallel and distributed association mining: A survey. *IEEE Concurrency*, 7:14–25, 1999.

[17] L. Jourdan, C. Dhaenens, and E.-G. Talbi. Rules extraction in linkage disequilibrium mapping with an adaptive genetic algorithm. In *European Conference on Computational Biology (ECCB) 2003*, Paris, France, 2003, pp. 29–32.

[18] C. Darwin. *On the Origin of Species.* John Murray, London, 1859.

[19] J. Holland. *Adaptation in Natural and Artificial Systems.* University of Michigan Press, Ann Arbor, MI, 1975.

[20] S. Smith. Flexible learning of problem solving heuristics through adaptive search. In *Proceedings of the Eighth International Joint Conference on Artificial Intelligence*, Karlsruhe, Germany, 1983, pp. 422–425.

[21] K.A. De Jong, W.M. Spears, and D.F. Gordon. Using genetic algorithms for concept learning. *Machine Learning*, 13: 161–188, 1993.

[22] C.Z. Janikow. A knowledge-intensive genetic algorithm for supervised learning. *Machine Learning*, 13: 189–228, 1993. J.J. Grefenstette, Ed., Kluwer Academic Publishers, Massachusetts.

[23] M. Pei, E.D. Goodman, and W.F. Punch III. Pattern discovery from data using genetic algorithms. In *Proceedings of the First Pacic-Asia Conference on Knowledge Discovery and Data Mining*, February 1997. Available via www URL: http://garage.cps.msu.edu/papers/papers-index.html.

[24] G.M. Weiss. Timeweaver: A genetic algorithm for identifying predictive patterns in sequences of events. In W. Banzhaf, J. Daida, A.E. Eliben, M.H. Garzon, V. Honavar, M. Jakiela, and R.E. Smith, Eds., *Proceedings of the Genetic and Evolutionary Computation Conference*, Vol. 1. Morgan Kaufmann, Orlando, Florida, USA, San Francisco, CA, 1999, pp. 718–725.

[25] D.P. Greene and S.F. Smith. Competition-based induction of decision models from examples. *Machine Learning*, 13: 229–257, 1993. J.J. Grefenstette, Ed., Kluwer Academic Publishers, Massachusetts.

[26] A. Giordana and F. Neri. Search-intensive concept induction. *Evolutionary Computation*, 3: 375–419, 1995.

[27] P. Kotala, P. Zhou, S. Mudivarthy, W. Perrizo, and E. Deckard. Gene expression profiling of dna microarray data using peano count trees (p-trees). In *Online Proceedings on the First Virtual Conference on Genomics and Bioinformatics*, 2001.

[28] R. Chen, Q. Jiang, H. Yuan, and L. Gruenswald. Mining association rules in analysis of transcription factors essential to gene expressions. In *Atlantic Symposium on Computational Biology and Genome Information Systems and Technology*, 2001.

[29] C. Creighton and S. Hanash. Mining gene expression databases for association rules. *Bioinformatics*, 19: 79–86, 2003.

[30] A. Icev, C. Ruiz, and E.F. Ryder. Distance-enhanced association rules for gene expression. In *ACM SIGKDD Workshop on Data Mining in Bioinformatics (BIOKDD03)*, 2003, pp. 34–40.

[31] S. Cahon, N. Melab, E.-G. Talbi, and M. Schoenauer. Paradiseo based design of parallel and distributed evolutionary algorithms. In *Evolutionary Algorithms EA'03*, 2003, pp. 195–207.

[32] E. Zitzler, M. Laumanns, and S. Bleuler. A tutorial on evolutionary multiobjective optimization. In *Metaheuristics for Multiobjective Optimisation, Lecture Notes in Economics and Mathematical Systems*, Vol. 535. Springer-Verlag, Berlin, 2004, pp. 3–37.
[33] E.L. Ulungu, J. Teghem, and Ph. Fortemps. Heuristics for multi-objective combinatorial optimization by simulated annealing. In J. Gu, G. Chen, Q. Wei, and S. Wang, Eds., *Proceedings of the Sixth National Conference on Multiple Criteria Decision Making*. Sci-Tech, 1995, pp. 228–238.
[34] A.A. Freitas. On rule interestingness measures. *Knowledge-Based Systems Jounral*, 12: 309–315, 1999.
[35] P.-N. Tan, V. Kumar, and J. Srivastava. Selecting the right interestingness measure for association patterns. In *Proceedings of the Eighth ACM SIGKDD International Conference on Knowledge Discovery and Data Mining*, 2002. http://www.cse.unsw.edu.au/~qzhang/proceedings/research.html.
[36] M. Khabzaoui, C. Dhaenens, A. N'Guessan, and E.-G. Talbi. Etude exploratoire des critères de qualité des règles d'association en datamining. In *XXXVèmes JOURNEES DE STATISTIQUE*, 2003, pp. 583–586.
[37] P. Smyth and R.M. Goodman. Rule induction using information theory. In G. Piatetsky-Shapiro and J. Frawley, Eds., *Knowledge Discovery in Databases*, MIT Press, Cambridge, MA, 1991, pp. 159–176.
[38] D.L.A. Araujo, H.S. Lopes, and A.A. Freitas. A parallel genetic algorithm for rule discovery in large databases. In K. Ilto, Ed., *Proceedings of the 1000 IEEE Systems, Man and Cybernetics Conference*, Vol. III. IEEE, Tokyo, October 1999, pp. 940–945.
[39] K. Wang, S.H.W. Tay, and B. Liu. Interestingness-based interval merger for numeric association rules. In R. Agrawal, P.E. Stolorz, and G. Piatetsky-Shapiro, Eds., *Proceedings of the Fourth International Conference on Knowledge Discovery and Data Mining, KDD*. AAAI Press, New York, USA, 27–31 1998, pp. 121–128.
[40] L. Jourdan. Métaheuristiques pour l'extraction de connaissances: Application à la génomique. Ph.D. thesis. Université des Sciences et Technologies de Lille, November 2003.
[41] C.M. Fonseca and P.J. Fleming. An overview of evolutionary algorithms in multiobjective optimization. *Evolutionary Computation*, 3: 1:1–16, 1995.
[42] K. Deb. Introduction to selection. *Evolutionary Computation 1: Basic Algorithms and Operators*, Institute of Physic, 2000.
[43] L. Davis. Adapting operator probabilities in genetic algorithms. In J.D. Schaffer, Ed., *Proceedings of the Third International Conference on Genetic Algorithms*. Morgan Kaufmann, San Mateo, CA, 1989, pp. 61–69.
[44] F. Herrera and M. Lozano. Adaptation of genetic algorithm parameters based on fuzzy logic controllers. In F. Herrera and J.L. Verdegay, Eds., *Genetic Algorithms and Soft Computing*. Physica-Verlag, Heidelberg, 1996, pp. 95–125.
[45] B.A. Julstrom. What have you done for me lately? Adapting operator probabilities in a steady-state genetic algorithm. In L.J. Eshelman, Ed., *Proceedings of the Sixth International Conference on Genetic Algorithms*. Morgan Kaufmann, San Francisco, CA, 1995, pp. 81–87.
[46] T.P. Hong, H.S. Wang, and W.C. Chen. Simultaneously applying multiple mutation operators in genetic algorithms. *Journal of Heuristics*, 6: 439–455, 2000.
[47] D.A. Van Veldhuizen and G.B. Lamont. On measuring multiobjective evolutionary algorithm performance. In *2000 Congress on Evolutionary Computation*, Vol. 1. Piscataway, NJ, July 2000, pp. 204–211.
[48] A. Jaszkiewicz. On the Performance of Multiple Objective Genetic Local Search on the 0/1 Knapsack Problem. A Comparative Experiment. Technical report RA-002/2000, Institute of Computing Science, Poznan University of Technology, Poznan, Poland, July 2000.
[49] J.D. Knowles, D.W. Corne, and M.J. Oates. On the assessment of multiobjective approaches to the adaptive distributed database management problem. In *Proceedings of the Sixth International Conference on Parallel Problem Solving from Nature (PPSN VI)*, September 2000, pp. 869–878,

[50] E. Zitzler, L. Thiele, M. Laumanns, M.C. Fonseca, and V.G. da Fonseca. *Performance Assess in IEEE Transactions on Evolutionary Computation*, 7(2): 117–132, 2003.

[51] M. Basseur, F. Seynhaeve, and E.-G. Talbi. Design of multi-objective evolutionary algorithms: Application to the flow-shop scheduling problem. In *Congress on Evolutionary Computation (CEC'02)*, Honolulu, USA, 2002, pp. 1151–1156.

[52] H. Meunier, E.G. Talbi, and P. Reininger. A multiobjective genetic algorithm for radio network optimisation. In *CEC*, Vol. 1, IEEE Service Center, Piscataway, NJ, July 2000, pp. 317–324.

29

An Evolutionary Approach to Problems in Electrical Engineering Design

Gregor Papa
Jurij Šilc
Barbara Koroušić-Seljak

> [Genetic algorithms] have been shown to rapidly converge to near-optimal solutions in a wide variety of application domains. Further, they have been shown to be computationally efficient, and to be well suited for solving problems characterized by local minima.
>
> Karr, Yakushin, and Nicolosi,
> Engineering Applications of Artificial Intelligence, 2000.

This chapter presents two engineering design problems, both of which were solved with biology-inspired algorithms. The evolutionary approach is used in *universal electromotor geometry optimization* (UM design) and *integrated circuits area/time optimization* (IC design).

In the UM design we improve the efficiency of a universal motor; here the goal is to find a new set of independent geometrical parameters for the rotor and the stator with the aim of reducing the motor's power losses, which occur in the iron and the copper. In the IC design we improve some parts of the high-level synthesis process of integrated circuits by considering the concurrency of operation scheduling and resource allocation constraints to ensure a globally optimal solution in a reasonable time.

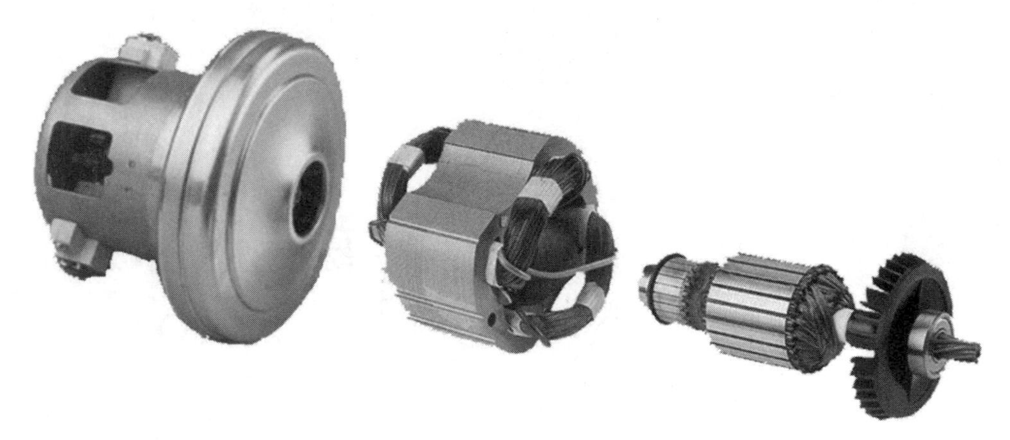

FIGURE 29.1 A UM used in a vacuum cleaner, showing its rotor and stator parts.

29.1 Problems in Engineering Design

29.1.1 Universal Motor Design

Home appliances, such as vacuum cleaners and mixers, as well as power tools, such as drills and saws, are generally powered by a UM [1] (see Figure 29.1). This type of motor has many advantages that make the UM such a popular choice for home appliances and power tools: a large output power in relation to its small size, high starting and running torque, variable speed that can be regulated in a simple way, and low manufacturing costs.

Home appliances and power tools need as low an energy consumption, that is, input power, as possible, while still satisfying the needs of the user by providing sufficient output power. The ratio of the output power to the input power defines the efficiency of the motor, which can be improved by reducing some of the main power losses in the motor, that is, those that originate in the iron and the copper. This can be done by optimizing the geometry of both the rotor and the stator. Because of the high magnetic saturation of the iron in a UM the problem is a highly nonlinear one.

29.1.1.1 The Geometry of the Rotor and the Stator

The rotor-and-stator unit of a UM is constructed by stacking the rotor/stator iron laminations (see Figure 29.2). The shape and the profile of the rotor/stator lamination are described by several two-dimensional geometrical parameters. There are two types of parameter: the invariable and the variable. Invariable parameters are fixed; they cannot be altered, either for technical reasons or because of the physical constraints of the motor. Some of the more important invariable parameters of a UM are: the air gap, the external radius of the stator, the radius of the rotor's shaft, the radius on the stator's side hole, and the width of the rotor's jag.

Variable parameters are those that do not have predefined optimum values. Some of these variable parameters are mutually independent and without any constraints, they include: the thickness of the horizontal of the stator's yoke, the thickness of the vertical of the stator's yoke, the width of the stator, the height of the stator, the radius of the stator's internal edge, the thickness of the stator's yoke at the hole, the length between the bisector and the slot edge, the radius of the stator's teeth, the external radius of the rotor, the width of the rotor's pole, the thickness of the rotor's yoke, and the height of the rotor's teeth. Other variable parameters are dependent, either on some invariable parameters or on mutually independent ones.

In our case we optimize 12 mutually independent variable parameters, while taking into account the following constraints:

- The parameters (both independent and dependent) should be changed simultaneously to achieve proper electromagnetic conditions in the material.

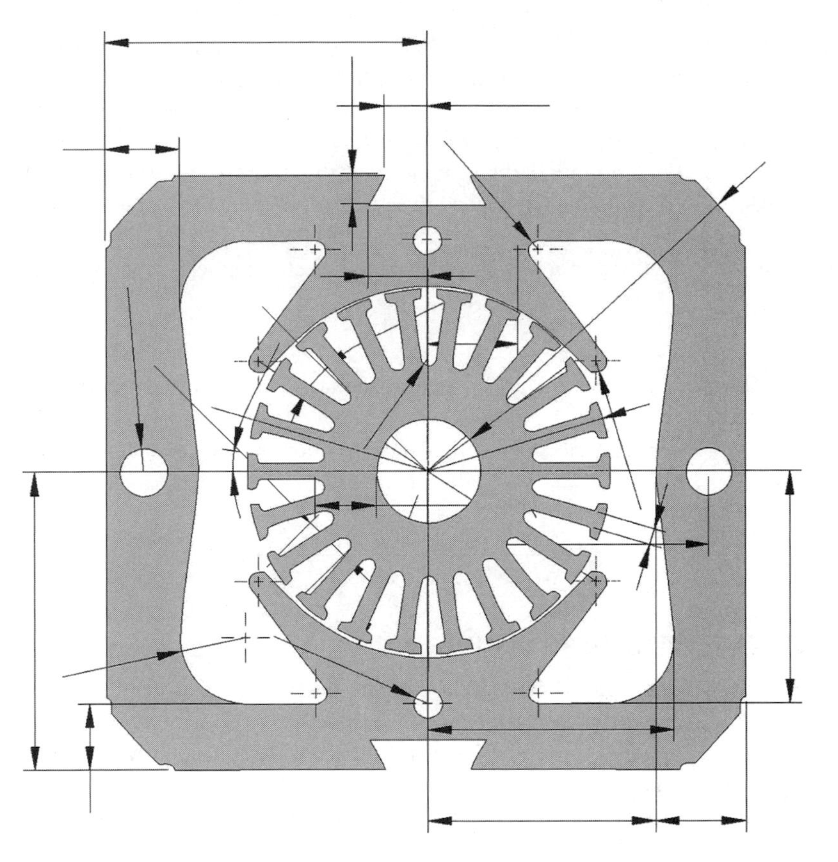

FIGURE 29.2 Geometrical parameters of the stator and rotor of a UM.

- Each parameter's dimension should only be varied within a predefined feasible limit.
- Parameter transformations and their evaluation should be done as quickly as possible.

29.1.1.2 The Efficiency of a UM

The efficiency of a UM is defined as the ratio of the output power to the input power, and it depends on various power losses, which include:

- *Copper losses*: the joule losses in the windings of the stator and the rotor.
- *Iron losses*: including the hysteresis losses and the eddy-current losses, which are primarily in the armature core and in the saturated parts of the stator core.
- *Other losses*: such as brush losses, ventilation losses, and friction losses.

The overall copper losses (in all stator and rotor slots) are as follows:

$$P_{\text{Cu}} = \sum_i (J^2 \cdot A \cdot \rho \cdot l_{\text{turn}})_i, \tag{29.1}$$

where i stands for each slot, J is the current density, A is the slot area, ρ is the copper's specific resistance, and l_{turn} is the length of the winding turn.

Because of the nonlinear magnetic characteristic, the calculation of the iron losses is less exact. The iron losses are separated into two components: the hysteresis losses and the eddy-current losses. Consequently, a motor's iron losses can be expressed by the following equation [1]:

$$P_{\text{Fe}} = k_{\text{e}} \cdot B^2 \cdot f_{\text{rot}}^2 \cdot m_{\text{rot}} + k_{\text{e}} \cdot B^2 \cdot f_{\text{stat}}^2 \cdot m_{\text{stat}} + k_{\text{h}} \cdot B^2 \cdot f_{\text{stat}} \cdot m_{\text{stat}}, \tag{29.2}$$

where k_e is an eddy-current material constant of 50 Hz, k_h is a hysteresis material constant of 50 Hz, B is the maximum magnetic flux density, f is the frequency, and m is the mass.

Besides the iron and the copper losses, three additional types of losses also occur in a UM, that is, brush losses P_{Brush}, ventilation losses P_{Vent}, and friction losses P_{Frict}. All three types of losses mainly depend on the speed of the motor. When optimizing the geometry of the rotor and the stator, the motor's speed is fixed, so P_{Brush}, P_{Vent}, and P_{Frict} have no impact on the efficiency of the motor. Therefore, these parameters are not significantly affected by the geometry of the rotor and the stator.

The output power P_2 of the motor is a product of the electromagnetic torque, T, and the angular velocity, ω,

$$P_2 = T \cdot \omega, \tag{29.3}$$

where ω is set by the motor's speed, and T is a vector product of the distance from the origin, r, and the electromagnetic force, F.

When considering all the mentioned losses and the output power, the overall efficiency of a UM can be defined as follows:

$$\eta = \frac{P_2}{P_2 + P_{Cu} + P_{Fe} + P_{Brush} + P_{Vent} + P_{Frict}}. \tag{29.4}$$

29.1.2 Design of Integrated Circuits

Whenever a new IC is designed, the problem of selecting the best register-transfer level (RTL) specification has to be faced. And as circuits get bigger, so too does the problem: more combinations have to be examined before the optimum combination is found. Therefore, automatic circuit optimization is required, as this speeds up the whole design process and eliminates some of the errors. More specifically, the case is focused on application-specific integrated circuits (ASICs), which need an even more sophisticated design (in terms of size and speed) because of their specific uses.

High-level synthesis [2] is an automatic design process that transforms the initial behavioral description into the final specification of the RTL. The process consists of the following: compilation, transformation, scheduling, allocation, and binding. Of these, the operation scheduling and the resource allocation are the most important tasks of the high-level synthesis because they are at the core of the design and crucially influence both the design and the final layout. Due to the interdependence of these two tasks, the solution of one task depends on an estimation of the solution of the other task, which is not solved yet. The scheduling of the operation into different control steps therefore affects the allocation of operations to different units. The interaction of these two tasks presents formidable obstacles to the goal of optimization [3]. There are, however, some approaches to concurrent solving, but their solutions, to some extent, are less than optimal.

Optimally scheduled operations are not necessarily optimally allocated to units. To enable optimal allocation we need to consider some allocation criteria while the scheduling is being done. Therefore, algorithms with concurrent scheduling and allocation produce the best results. However, these algorithms are also time-consuming. It is obvious that we have to deal with a trade-off between the quality of the solution and its design time: an algorithm might be fast, but its solutions will not be as good, and vice versa.

29.1.2.1 Data-Flow Graphs of ICs

An IC described with a hardware description language, for example, VHDL, can be presented as a data-flow graph (DFG) [2] (see Figure 29.3). Each node i represents the operation to be executed, and the type of node determines the type of operation. The edges e represent dependencies between operations. An edge e_{ij}, between nodes i and j, represents the data produced by the node i and used by the node j. All edges are unidirectional.

In the final design there is a group of resources (functional units, storage units or registers, and connection units or buses) that define the implementation. Functional units (adders, multipliers, etc.) perform transformations on data values; connection units transport values from one unit to another; while storage units preserve those values over time. We have N different functional units $F_i, i = 1, 2, \ldots, N$, n_r registers, and n_b buses, while T represents the execution time of the DFG.

The parameters are calculated as follows:

- The number n_{f_i} is the highest number of the i-th functional unit needed in a separate control step.
- The number n_r is the highest number of variables needed in a separate control step. We consider variables that are needed by the functional unit as input data, variables that are returned as output data, and variables that are not used at the moment but will be used in some of the later control steps or must be available until the end of the execution of all operations.
- The number n_b is the highest number of data transmissions (into or from the functional units) in a separate moment.
- The execution time, T, is the time needed to execute all the operations of the schedule.
- The weights w_{fi}, w_r, w_b, and w_t are the weights of functional units, registers, buses, and time, respectively, to be considered in the IC quality-evaluation cost function. The first three weights are proportional to their silicon area in the IC, while w_t reflects our IC speed constraints.

29.1.2.2 Concurrency of Scheduling and Allocation

Algorithms that perform concurrent scheduling and allocation are better in terms of the optimality of the solutions. These algorithms are, however, very time-consuming. Therefore, we have to deal with a trade-off between the quality of the final solution and the length of the design-time for it. There are some approaches to concurrent solving, but their solutions are, to some extent, less than optimal.

- Landwehr et al. [4] suggest solving with integer linear programming. The procedure ensures optimal solutions, but is computationally expensive and therefore practically useful only for small circuits.
- Zhu and Gajski [5] suggest a "soft scheduling," where operations are temporarily scheduled, but are later adjusted according to some physical characteristics. Operations are finally scheduled after the allocation and binding tasks, when all the parameters that affect the scheduling optimality are known. Actually, there is no concurrency, but there is some iterative refinement of the scheduled operations.
- Mandal and Zimmer [6] suggest concurrent scheduling and allocation using a genetic algorithm, and with a stress on the reduction of the number of connections. In the optimization process, the whole circuit is partitioned into blocks consisting of a functional unit, a storage unit and internal connections. Operations are scheduled according to the units in those blocks, and according to additional global storage units and interconnections, which are to be minimized. Here, the problem of various functional units being present in each block arises, to ensure that all the operations of the block are executed.
- Kim [7] suggests a concurrent procedure where the operations subset is chosen for each control step. These operations are later bound to functional units and the variables are bound to storage units in order to achieve the highest possible connection utilization. The order of operations to be scheduled is based on their readiness for execution and their influence on the number of storage and connection units, while considering the lowest possible number of control steps.
- Grajcar [8] describes the concurrency of scheduling and allocation on a multiprocessor system. The procedure is based on a GA and list scheduling, which considers only some values (critical points) in the chromosome code.

As mentioned earlier, when tasks are performed separately (Figure 29.4[a]) the solution is not necessarily optimal; it is better to use an approach with iterative repetition of the scheduling and allocation (Figure 29.4[b]). Here, the problem of the next operation/unit to be changed appears, since the order of changes can influence the final solution. It is a similar situation with the approach that involves partitioning of the operations into small groups, within which there is an iterative repetition of the scheduling and

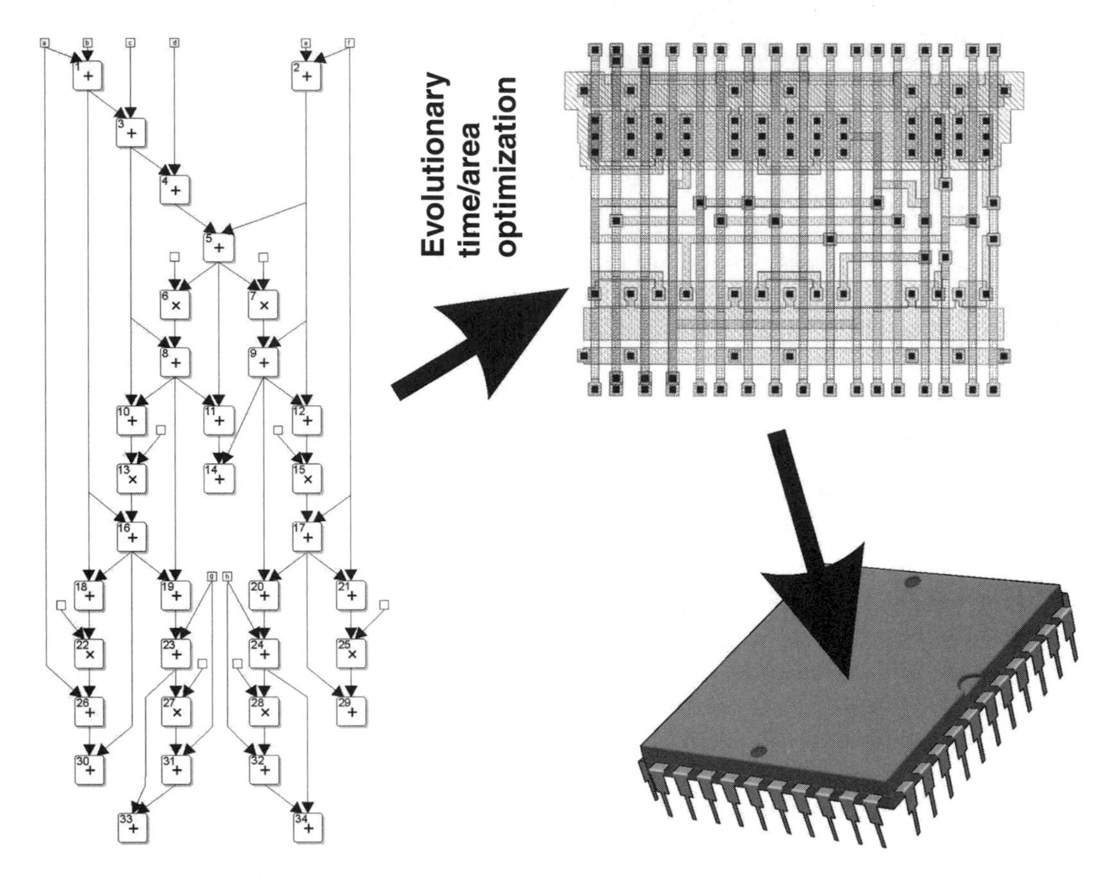

FIGURE 29.3 DFG and its IC.

the allocation (Figure 29.4[c]). Since there are fewer operations in the group there is no problem with the order of changes, but there is a problem with the appropriate partitioning of the operations. Obviously, the best approach is the one with purely concurrent scheduling and allocation (Figure 29.4[d]), where the iterative-refinement order does not influence the quality of the solution [9]. Concurrency is achieved by using algorithms that do not depend on the order of the transformations. Therefore, there is no influence of the altered start time on the allocated unit, nor is there any influence of the allocated unit on the start time. When all the transformations are made, then the appropriateness of the changes is checked.

29.2 Evolutionary Optimization in Problems of Engineering Design

Evolutionary techniques are used in various search methods for a range of different optimization areas. Their undetermined approach gives them an advantage when it comes to multicriteria problems and problems with more local optima [10]. For this reason we adopted an artificial approach to the solving of our design problems using a genetic algorithm (GA) [11,12]. The GA, as a frequent implementation of evolutionary techniques, is an optimization method based on the mechanism of evolution and natural genetics. This algorithm has already proved to be very efficient in a wide range of different optimization procedures where the exact equations are not available or some nonlinearities are present [13]. In spite of its simplicity, the GA has proved to be an efficient method for solving various optimization and classification problems, in areas ranging from economics and game-theory to control-system design [14–18].

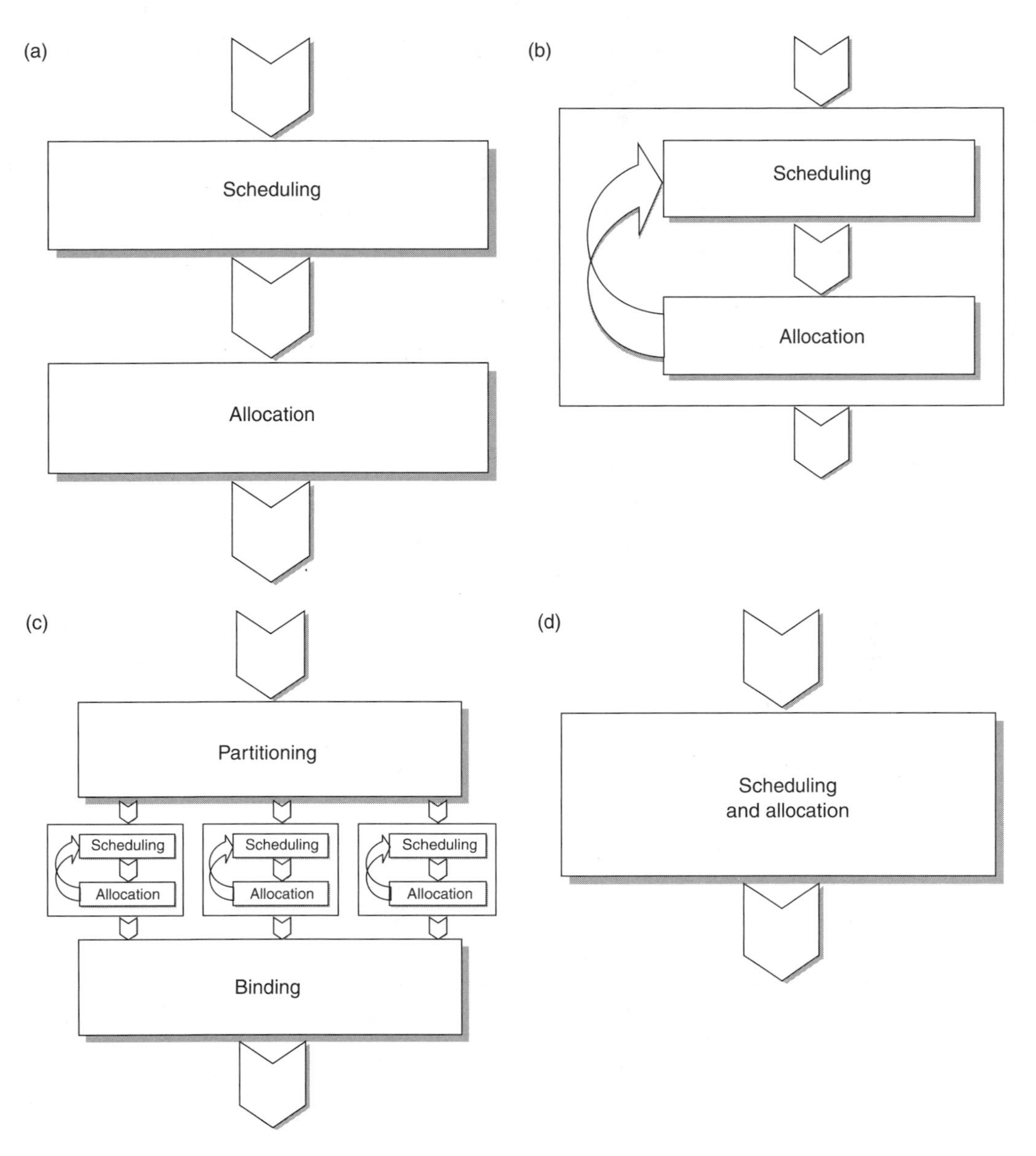

FIGURE 29.4 Scheduling and allocation concurrency.

29.2.1 The Genetic Algorithm in UM and IC Design

When reducing the main power losses in a UM design by optimizing the geometry of the rotor/stator lamination, we have to deal with a complex search space and its nonlinear behavior. In IC design we apply the concurrency of interdependent tasks with their opponent constraints. Because traditional search-and-optimization methods have proved to be inefficient at finding the solution under such conditions, we decided to apply a GA. This heuristic method requires only a little information to provide a robust, yet flexible, search in a wide and complex search space.

29.2.1.1 The Basics of the GA

The GA codes parameters of the problem's search space as finite-length strings over some finite alphabet. It works with a coding of the parameter set, not the parameters themselves. The algorithm employs an initial population of strings, which evolve into the next generation under the control of probabilistic transition rules — known as randomized genetic operators — such as selection, crossover, and mutation. The objective function evaluates the quality (or fitness) of solutions coded as strings. This information is then used to perform an effective search for better solutions. There is no need for any other auxiliary knowledge. The GA tends to take advantage of the fittest solutions by giving them greater weight, and concentrating the search on the regions of the search space that show likely improvement.

29.2.1.2 Encoding

One of the most important parts of the GA is the encoding. By encoding the proper parameters and using the proper encoding type we can significantly influence the efficiency of the algorithm.

In the UM design, the mutually independent variable geometrical parameters of the rotor/stator lamination are coded as strings over the alphabet $\mathbb{R}$ of real values. There is no need to normalize the physical parameters. They do differ in ranges but the crossover operation switches the values of the same parameter, no matter where the crossover point is, since each parameter is always encoded at the same place in the chromosome.

In the IC design the chromosome string consists of the numbers that represent the starting time of each operation and the allocated unit for each operation, where the position in the string depends on the order of the operations in the input IC description. This means that the chromosome consists of pairs of time/unit information for each operation. And the genetic operators can influence both parts of that information, either together or separately. The selected encoding type is chosen because of its convenience. When strings have to be further transformed, checked, and analyzed, there is no need for any additional conversion of their values. In addition, the used implementation of genetic operators can check the changed values (their feasibility) instantly, without any transformation. The correctness of the transformation can therefore be checked within the function itself.

29.2.1.3 Cost Function

In the UM design each string of the population is decoded into a set of rotor and stator parameters. Its fitness is estimated by performing a finite-element numerical simulation to calculate the copper and the iron power losses (using Equations [29.1] and [29.2], respectively). Their sum corresponds to the solution's fitness.

In the IC design a multiobjective [19,20] function is used, which means it takes control over more criteria or objectives. The cost function, represented by Equation (29.5), considers the number of functional units, the number of registers, the number of buses, and the execution time of all the operations in the DFG.

$$
\begin{aligned}
\text{Cost} &= \sqrt{\sum_{i=1}^{N} (\text{cost}_{f_i})^2 + \text{cost}_\text{r}^2 + \text{cost}_\text{b}^2 + \text{cost}_\text{t}^2}, \\
\text{cost}_{f_i} &= w_{f_i} F_i, \\
\text{cost}_\text{r} &= w_\text{r} n_\text{r}, \\
\text{cost}_\text{b} &= w_\text{b} n_\text{b}, \\
\text{cost}_\text{t} &= w_\text{t} T.
\end{aligned}
\qquad (29.5)
$$

To obtain the cost of a certain DFG, the algorithm has to evaluate the required number of resources. In contrast to the other multiobjective functions that give more than one final solution, this one already includes the decision-making part, that is, it chooses one solution from all the solutions on the Pareto front. The chosen solution has the shortest distance to the origin, where the origin represents the ideal, costless, solution and the axis represents the considered objectives.

29.2.1.4 The Genetic Operators

To evolve the best solution candidate, the GA employs the genetic operators for manipulating the strings in a population. The GA uses these operators to combine the strings of the population in different arrangements, seeking a string that optimizes the objective function. In the IC design, operators also consider data dependencies and the given library of available functional units. This combination of strings results in a new population.

29.2.1.4.1 Basic Operators

- The selection operator is used for creating a new generation. We apply the elitism strategy, where a number of least-fit members of the current population are interchanged with an equal number of the best-ranked strings. This approach ensured better starting positions for the best-ranked solutions, as all solutions have equal chances for reproduction.
- The crossover operator is used for exchanging information between the selected strings. In our case it proceeds in two steps. First, the strings are mated randomly to pair off the couples. Second, the mated string couples cross over, using a given probability p_c. Whether the one-point or two-point crossover method is chosen, characteristic values are swapped between the mated couples to create two new strings. Moreover, we might use a constant probability p_r to select a case in which the values of the swapped characteristics are calculated as a mean value of the parent characteristic values. While the first crossover approach ensures that the offspring solutions preserve the "genetic material" from both parents, the second approach helps to seek other solutions close to the solutions that appear to be good. In the IC design, after crossover points are determined, either the unit information is changed between the two chromosomes and the start times are adapted, or the start times are changed and a suitable unit is allocated. So the dominancy is expressed either in functional units or the start times of operations.
- The mutation is a process by which strings resulting from selection and crossover are perturbed; the process serves to create random diversity in the population. We use a mutation approach where each string is subjected to the mutation operator. The values of the chromosome string are either slightly changed in the UM design, or start times are changed with a respective faster/slower unit allocated in the IC design. Mutation is performed on a characteristic-by-characteristic basis, each characteristic mutating with a probability p_{m}. We also apply the possibility of annealing the mutation rate, where p_{m} is the variable mutation probability that decreased linearly with each new population. In other words, we assume that each new population is generally fitter than the previous one. Such an approach is used to overcome a possible disruptive effect of mutation, and to speed up the convergence of the GA to the optimum solution.
- The variation operator swaps some values of the similar chromosome positions. After two operations are selected, and when they are of the same type, for example, additions in the IC design, their functional units are switched. If needed, their start times are also updated.

29.2.1.4.2 Independent Operators

The advantage of the independent GA approach [9] is that there is no need to preset some working parameters, for example, the number of generations, the population size, and the probabilities of crossover, mutation, and variation. These parameters are set automatically during the optimization phase, depending on the progress and the speed of the optimization.

- *Setup.* If the chromosome that represents a solution is large, then the population size also has to be large enough to ensure that many different chromosomes will be involved in a search. The population size therefore depends on the size of the chromosome or the complexity of the problem.
- *Crossover.* Considering four candidates — two parents and their two offspring — only the first and the third, rated according to their fitness, pass to the next generation. This forces, very probably, at least one of the offspring to be passed to the next generation in addition to the best candidate. Otherwise the offspring have only a small influence on new generations, since the crossing of two

good parents probably produces offspring that are not so good. They might, however, be good after a few more transformations.

- *Mutation.* Chromosomes with low fitness are mostly exposed to mutation. Each position in the chromosome string is mutated if that position of the chromosome is of the same value in the majority of "poorly fitted" chromosomes in the population. This is the way to change the bad characteristics in "poorly fitted" chromosomes and to redirect the search to another direction. In the case of "well-fitted" chromosomes, values are mutated if they differ from the majority of values in other good chromosomes at the same position. This ensures faster convergence in the final stages of the optimization.
- *Variation.* The interchange of the values of two positions, as described for the basic operators, is performed if the frequency of the value in that position in the population of one position is high and the frequency of another bit is low.

29.3 Procedures and Evaluations

29.3.1 UM Design Procedure

29.3.1.1 The Conventional versus the Evolutionary Design Procedure

In a conventional design procedure for a motor:

1. The initial estimation for the geometry of the rotor and the stator is made based on experience.
2. The appropriateness of this geometry is then usually analyzed by means of a numerical simulation of the electromagnetic field. In our case, the analysis is performed with commercial ANSYS software [21], which applies a finite-element method (FEM) with an automatic finite-element-mesh generation. The result is a magnetic vector potential on every node of the finite-element mesh.
3. If the results of the numerical simulation show an inconvenient electromagnetic field structure, the direct design procedure is repeated until the motor geometry is optimized.

The advantage of this approach is that engineers can significantly influence the progress of the design process by using their experience, and they can react intelligently to any noticeable electromagnetic response with proper geometry redesign. The drawback of this approach is that an experienced engineer and a large amount of time are needed.

The described conventional motor design can be upgraded with a genetic algorithm. The concept of this evolutionary design can be roughly explained as follows:

1. The GA can start its optimization from any configuration of geometrical parameters while it defines a feasible solution.
2. Each geometrical configuration is analyzed using the ANSYS finite-element program. This step requires a prior decoding of the strings into a set of geometrical parameters for the rotor and the stator.
3. After the calculation of the fitness, the reproduction of the individuals and the application of the genetic operators to a new population are made. The GA repeats this procedure until a predefined number of iterations have been accomplished.

The advantages of this approach are that: there is no need for an experienced engineer to be present during the whole process, only at the beginning to decide on the initial design, and there is no need to know the mechanical and physical details of the problem. The problem can be solved without any knowledge of the problem.

The drawbacks of this approach are that it can lead to the improper use of genetic operators, and an initial solution that is too loosely set, can lead to a longer convergence time.

29.3.1.2 The Optimization Software

In accordance with the proposed evolutionary design approach described earlier we developed software that links together: (a) the genetic algorithm that optimizes the geometry of the rotor and the stator of a UM, and (b) the finite-element program needed to evaluate the optimized geometry.

The DOptiMeL software [22] is a tool for experimenting with different settings of the GA's parameters, which may significantly influence the quantity and the quality of the program solutions. The tool consists of two parts:

- A setup part for setting the GA parameters and the geometry limits of the rotor/stator lamination of an initial UM.
- An optimization part for optimizing the geometry of the rotor/stator lamination.

The program was developed using the Microsoft® Visual C++® programming tool and runs under the Microsoft® Windows® operating systems.

29.3.1.3 Parameters

29.3.1.3.1 Geometrical Parameters

As stated earlier, there are 12 mutually independent geometrical parameters that need to be optimized. These parameters can only be varied within their predefined dimension limits to find an optimum configuration that will increase the motor's efficiency. Solutions in which the parameters exceed the limits are rejected as being inoperable.

There are some invariable parameters that have a strong influence when defining the outline of the lamination:

1. *The external radius of the stator*: which roughly defines the amount of iron and copper, and consequently, the price of the motor. This is held constant during the optimization to ensure cost-comparable solutions.
1. *The radius of the rotor's shaft*: which we fixed at 5.5 mm. From our experience we know that when the rotor-shaft radius is less than 5.5 mm, the rotor's natural frequency can fall below the maximum frequency of the motor and the resulting resonance would cause the rotor's vibrations to exceed allowable limits.
3. *The radius of the stator's side-hole*: which would be 0 mm in the ideal case, is set to a small positive value because holes for the rivets are required in order to bind the stator.
4. *The air gap*: The angles of the symmetrical and tangential parts of the air gap are set to fixed values because the commutation, which is conditioned by the air gap, is not taken into account during the optimization.

29.3.1.3.2 GA Parameters

For the GA to work well, robust parameter settings have to be found for the population size, the number of generations, the selection criteria, and the genetic operator probabilities:

- If the population is too small, the GA converges too quickly to a local optimum solution and can miss the best solution. On the other hand, a large population requires a long time to converge to a region of the search space with significant improvement. The best results are obtained when the population size is between 30 and 50.
- By applying the elitism strategy, fitter solutions have a greater chance of reproducing. But when the ratio of least-fit solutions to be exchanged with best-fit ones is too high, the GA is trapped too quickly into a local optimum solution. However, this number is subject to the population size, and appears to be acceptable within 20 to 30% of the population size.
- As a crossover probability that is too low preserves solutions from being interchanged and a longer time is required for them to converge, the probability is set to at least 60%, so that the algorithm can act satisfactorily.

- A mutation rate that is too high introduces too much diversity and takes a longer time to reach an optimum solution. A mutation rate that is too low tends to miss some near-optimum solutions. Using the annealing strategy — a linearly decreasing mutation probability rate with each new generation — the effects of a too high or too low mutation rate can be overcome. The operator is useful if the value of the probability is 0.1%, while in the annealing strategy it starts with 1% and ends at 0.1%.

29.3.1.4 The UM Evaluation

The proposed evolutionary design approach is evaluated by estimating the actual improvement in the efficiency of an initial UM that is designed using the conventional and evolutionary design approaches.

1. With the ANSYS software we calculated the efficiency of an initial UM. An outline of the rotor/stator lamination of this motor is shown in Figure 29.5(a). The power losses of this motor were calculated to be 313 W and the output power was calculated to be 731 W (Table 29.1). In the outline, the levels of magnetic flux density through the rotor/stator lamination are shown, expressed as T. The darkest gray color indicates the areas with the highest level of magnetic flux density, which results in high iron losses. The copper losses are not shown in this area.
2. After several runs of the DOptiMeL software, a set of promising solution candidates was collected. We applied the following settings for the GA parameters: population size 30, number of generations 100, selection ratio 0.3, crossover probability 0.7, and mutation rate 0.01. For each candidate we make a finite-element numerical simulation followed by the calculation of the objective function value (fitness). Because each design is verified with finite-element numerical calculations, the optimization is a lengthy process. It takes around 3000 runs for the optimization to converge. Most of the solutions that are given by the DOptiMeL program show a significant reduction in the iron and the copper losses in comparison with the losses in the initial motor. The best solution results in a power-loss reduction of 24%, and gives us a motor with iron and copper losses of 239 W (see Figure 29.5[b]).

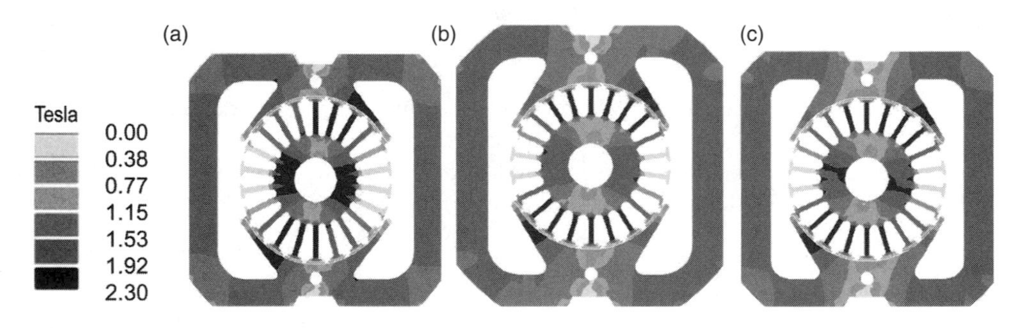

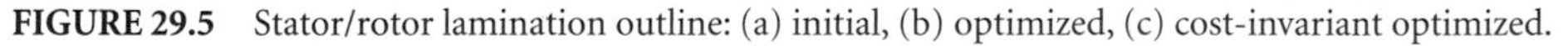
FIGURE 29.5 Stator/rotor lamination outline: (a) initial, (b) optimized, (c) cost-invariant optimized.

TABLE 29.1 UM Evaluation Results

	Analytic calculation			Prototype measurement		
	Initial	Optimized 1	Optimized 2	Initial	Optimized 1	Optimized 2
Input power (W)	1044	970	982	1050	990	1000
Efficiency (%)	70.0	75.8	74.8	69.5	73.7	73.0
Output power (W)	731	731	731	730	730	730
Power losses (W)	313	239	251	320	260	270
Efficiency difference (%)*	—	5.8	4.8	—	4.2	3.5
Power losses difference (W)*	—	74	62	—	60	50

* According to the initial UM.

The main differences between the initial design (Figure 29.5[a]) and the optimized design (Figure 29.5[b]) are: (a) the height of the rotor-and-stator laminations is increased by 13%, (b) the rotor radius is increased by 5%, (c) the slot (copper) areas in the stator and the rotor are larger, and (d) the iron area in the rotor is larger.

A comparison of the magnetic flux densities in the initial and the optimized motor shows a clear reduction of the areas with the highest levels of magnetic flux density in the optimized motor.

29.3.1.4.1 Copper and Iron Losses

In the optimized lamination the copper losses in the rotor and the stator are significantly lower than in the initial lamination. The reason for this effect is that the slot area of the optimized stator lamination is larger. The number of ampere-turns in the optimized lamination that are necessary for obtaining the required torque are lower in comparison with the initial lamination due to the optimized lamination design.

The copper losses increase with the square of the number of ampere-turns and they are inversely proportional to the slot area [1]. In the rotor, the slot area is increased, and this results in a reduction of 52% for the copper losses in the rotor. For the same reason, the stator's copper losses are 52% lower as well.

In contrast, the iron losses are higher than that in the initial lamination. The iron losses are increased in both the rotor and the stator, mainly because of the slightly larger iron area in the optimized design. The magnetic flux density in the stator remains in the same range as in the initial design. In the rotor, the magnetic flux density is decreased, but the iron area of the rotor is larger.

Overall, the optimized design has a much better torque capability, and this results in a total loss reduction of 24%.

After considering these results we might be tempted to think that the next step after a reduction of the copper losses would be a reduction of the iron losses. However, if we design a motor that has lower iron losses, the copper losses would automatically increase — and at an even higher rate. It should be remembered that a lamination is optimized when the sum of the copper and the iron losses is a minimum.

29.3.1.4.2 Material Costs

The results presented earlier consider an optimized UM. During the optimization procedure the material costs of the motor are not taken into account, that is, the external radius of the stator is not considered to be a variable parameter. The solution resulted in much larger stator dimensions than that in the initial lamination. As the amount of iron used for manufacturing the motor depends only on the width and the height of the stator, the 13% increase in the stator's height (as in our optimized solution) would cause the same increase in the amount, and consequently the cost, of the iron required.

The amount of copper that is used in the motor depends on the size of the slot areas; assuming the copper-fill factor is constant during the optimization. In the optimized solution, the slot area is 18% larger than in the initial one. As a consequence, the copper costs increased by the same percentage.

The figures for the estimated material costs suggest that the optimized design would be much more expensive than the initial design. Therefore, we repeated the optimization procedure, this time by considering the material price criterion. We extended the definition of the technical quality of the UM to be a function of both power losses and materials costs. The new criterion is based on the product of efficiency and an estimation of the material costs. A unity factor is set to the material costs of the initial design: an increase in material costs decreased the material costs factor linearly and vice versa. The goal of the optimization is to maximize this new criterion. The outline of the new costs-optimized rotor/stator lamination is shown in Figure 29.5(c).

This lamination has a different profile than the first optimized design. The stator covers the same area as the initial design, so the amount of iron is not increased. The slot areas are slightly larger than in the initial design, so 6.5% more copper would be needed. This would mean a more expensive motor, but the increases in cost would be negligible in comparison with the much better performance. Such a motor would have a power loss of 251 W, which is 20% better than the initial design. Comparing the optimized design that does not consider the material costs criterion with this new costs-optimized design, the first one is better: its power loss is estimated to be 12 W lower. By fixing the stator's outer radius, we mostly lose the gain of the first design in terms of the decrease of the stator's copper losses.

29.3.1.4.3 Prototyping

We made prototypes of both the optimized and the costs-optimized motors and measured the real power losses and the efficiencies of the motors. These values are shown in Table 29.1. The results are only slightly different from those calculated with the ANSYS finite-element program. The main reason for this difference can be explained by the non-exact calculation of the iron losses, due to a variation in the material's properties.

29.3.2 IC Design Procedure

The promising results of different evaluations [14–16] led us to the Evolutionary Concurrent Scheduling and Allocation (ECSA) design approach [9]. This approach considers scheduling and allocation constraints, allows a short design time, and can find globally optimal solutions. The input description of the circuit is transformed into two basic (initial) schedules, obtained with the ASAP and ALAP algorithms. The functional units used in the first case are those that are the fastest for each operation, and in the second case are those that are the slowest for each operation. These two schedules present some kind of boundary solutions, since all the other solutions are executed in between the time limits defined by these two schedules. In other words, no other solution can be faster or slower, irrespective of the combinations of used units.

Each solution has to be properly encoded, that is, each operation's start time and functional unit have to exist in the chromosome. The initial population is built upon the two initial solutions, which are multiplied to form the population with the so-called boundary solutions. The optimal solution has to be somewhere in between the boundaries, therefore genetic operators (crossover, mutation, variation) transform those encoded solutions. With the transformations their start times and allocated functional units are changed.

The appropriateness of the proposed approach is tested by a computer implementation of the ECSA algorithm, which is used with test-bench ICs.

29.3.2.1 Test-Bench Circuits

The ICs used for the evaluation are chosen on the basis of their appearance in the literature and similar studies. They differ in terms of size and the number of operation types.

29.3.2.1.1 Differential Equation

The relatively small circuit of differential equation [23] has only 11 operations, but 4 different operation types (6 multiplications, 2 additions, 2 subtractions, 1 comparison). This circuit is useful when testing libraries with different implementations of the same operation types.

29.3.2.1.2 Elliptic Filter

This filter [24] consists of 34 operations, but only 2 operation types: 26 additions and 8 multiplications. The circuit is suitable for comparison, due to its size and operation dependencies, since they form two independent, similar critical paths, both influencing the circuit delay.

29.3.2.1.3 Bandpass Filter

One of the implementations of the bandpass filter [25] is the circuit used for our evaluation. It consists of 29 operations: 11 multiplications, 10 additions, and 8 subtractions. Due to data dependencies, almost all the operations influence the circuit delay.

29.3.2.1.4 Least-Mean-Square Filter

This filter for signal adaptation (noise reduction) is based on the least-mean-square method [26]. It consists of 47 operations: 24 multiplications and 23 additions. This test-bench circuit is useful because of its size and unique data dependencies.

TABLE 29.2 Technical Characteristics of the Functional Units

Functional unit {operations}	Delay [No. of steps]	No. of gates
FE1 {+, −}	6, 6	370
FE2 {+, −}	1, 1	665
FE3 {<}	6	353
FE4 {+, −, <}	1, 1, 1	696
FE5 {×}	4	3040
FE6 {×}	2	7296
FE7 {÷}	21	3040
FE8 {÷}	9	7296
FE9 {×, ÷}	4, 21	3344
FE10 {×, ÷}	2, 9	8025
FE11 {+, −, <, ×, ÷}	1, 1, 1, 4, 21	3692
FE12 {+, −, <, ×, ÷}	1, 1, 1, 2, 9	8373

29.3.2.2 Functional Units

For an easier and more realistic comparison of different algorithms when testing the size and delay of implemented circuits we made a library of different functional units, which differ in terms of their sizes and delays. Table 29.2 shows the sizes and delays of various implementations of the arithmetic logic operations. Here, different types of logic are used to make the units with different delays and sizes. The values are based on an analysis of data on circuits and their complexities [27–29]. The number of gate transitions defines the delay, whereas the overall number of gates needed to implement the unit defines its size. These values are just for orientation, since the real numbers depend on the chosen technology [27]. The delays presented in Table 29.2 are relative, for example, normalized to the fastest functional unit among all the operations. Most of the units are multifunctional, that is, they can perform different types of operation.

29.3.2.3 Parameters

By considering 18,750 different schedules of each circuit with the ECSA algorithm and 3,125 different combinations of the parameters, we statistically compared (using the procedure described in Reference 30) the results according to their cost function (Equation [29.5]). To ensure that most solutions are time-constrained, that is, executed in the shortest possible time, the weight w_T is set to an extremely high value.

Figure 29.6 presents the results of different parameter-set evaluations for different test circuits. Figure 29.6(a) shows the influence of the number of generations on the percentage of good/bad solutions within the set of solutions with a specified number of generations. Here, and in other figures and subfigures, the column marked as high represents high-quality (good) solutions, while the column marked as low represents the low-quality (bad) solutions. So, from among all the solutions we count the number of solutions with high fitness and the number of solutions with low fitness for each number of generations. Figure 29.6(b) shows the influence of the population size on the percentage of good/bad solutions within the set of solutions with a specified population size. Figure 29.6(c) shows the influence of the crossover probability on the percentage of good/bad solutions within the set of solutions with a specified crossover probability. Figure 29.6(d) shows the influence of the mutation probability on the percentage of good/bad solutions within the set of solutions with a specified mutation probability. Figure 29.6(e) shows the influence of the variation probability on the percentage of good/bad solutions within the set of solutions with a specified variation probability.

As shown in Figure 29.6 and Table 29.3, the high-quality solutions are mostly obtained with the following values of the parameters: probability of crossover equal to 0.7, probability of mutation equal to 0.04, and probability of variation equal to 0.03. In addition, taking into account the sizes of the circuits,

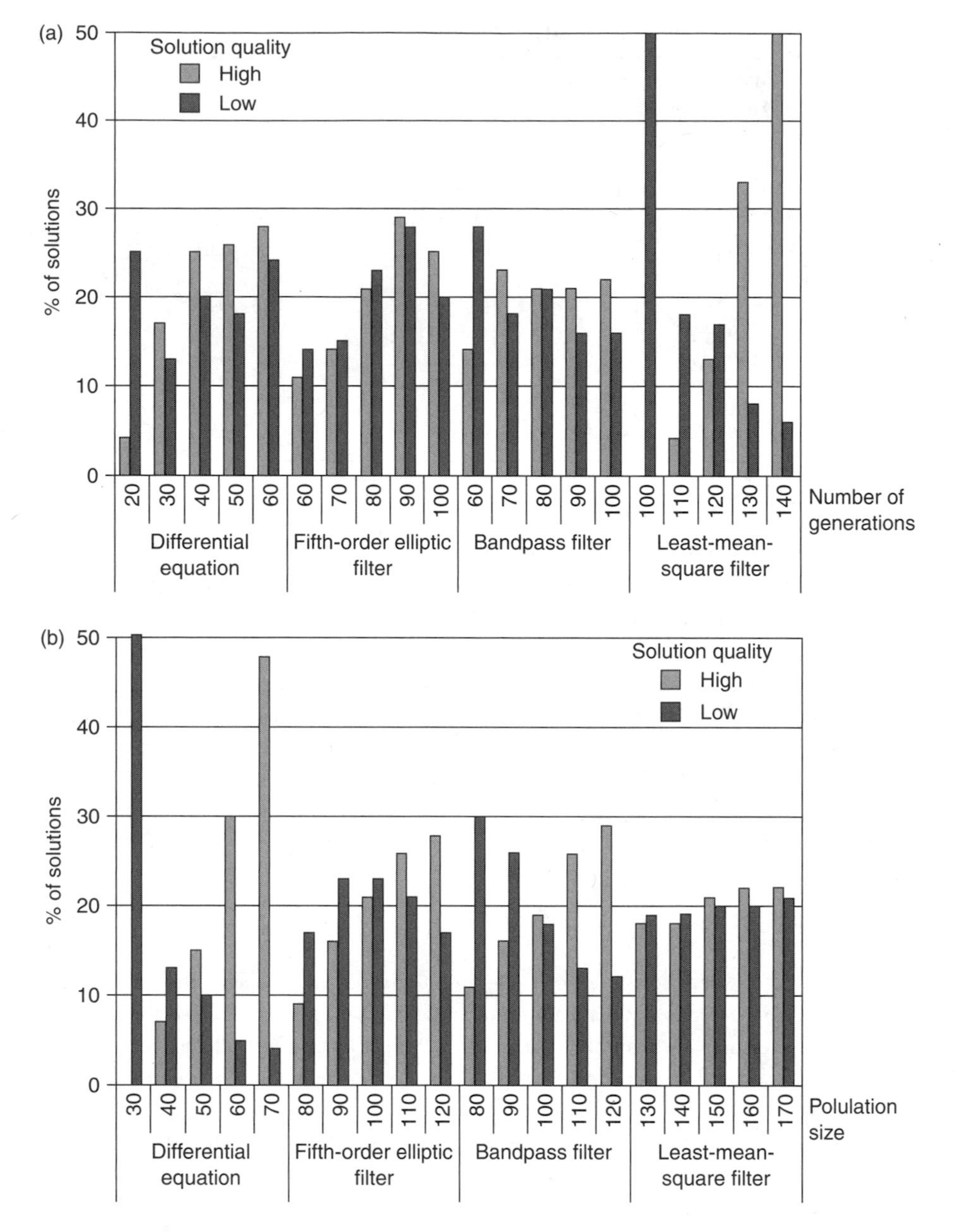

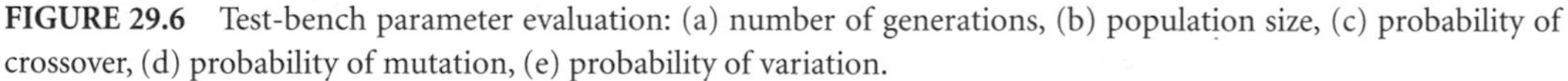

FIGURE 29.6 Test-bench parameter evaluation: (a) number of generations, (b) population size, (c) probability of crossover, (d) probability of mutation, (e) probability of variation.

the number of generations and the population size should be set to 3- and 3.5-times the size of the circuit, respectively.

The values of the parameters in this combination are referred to as the optimal values. These optimal values are determined on the basis of the percentage of solutions with certain parameters from among the good solutions. A parameter value that is to be considered as optimal should have at least a 25% share of the high-quality solutions, as well as having a <10% share of the low-quality solutions.

The ECSA algorithm is used with the values of the parameters as presented in Table 29.3. Other parameters needed to run the force-directed scheduling (FDS) and ECSA algorithms and the cost function depend on the sizes of the functional units.

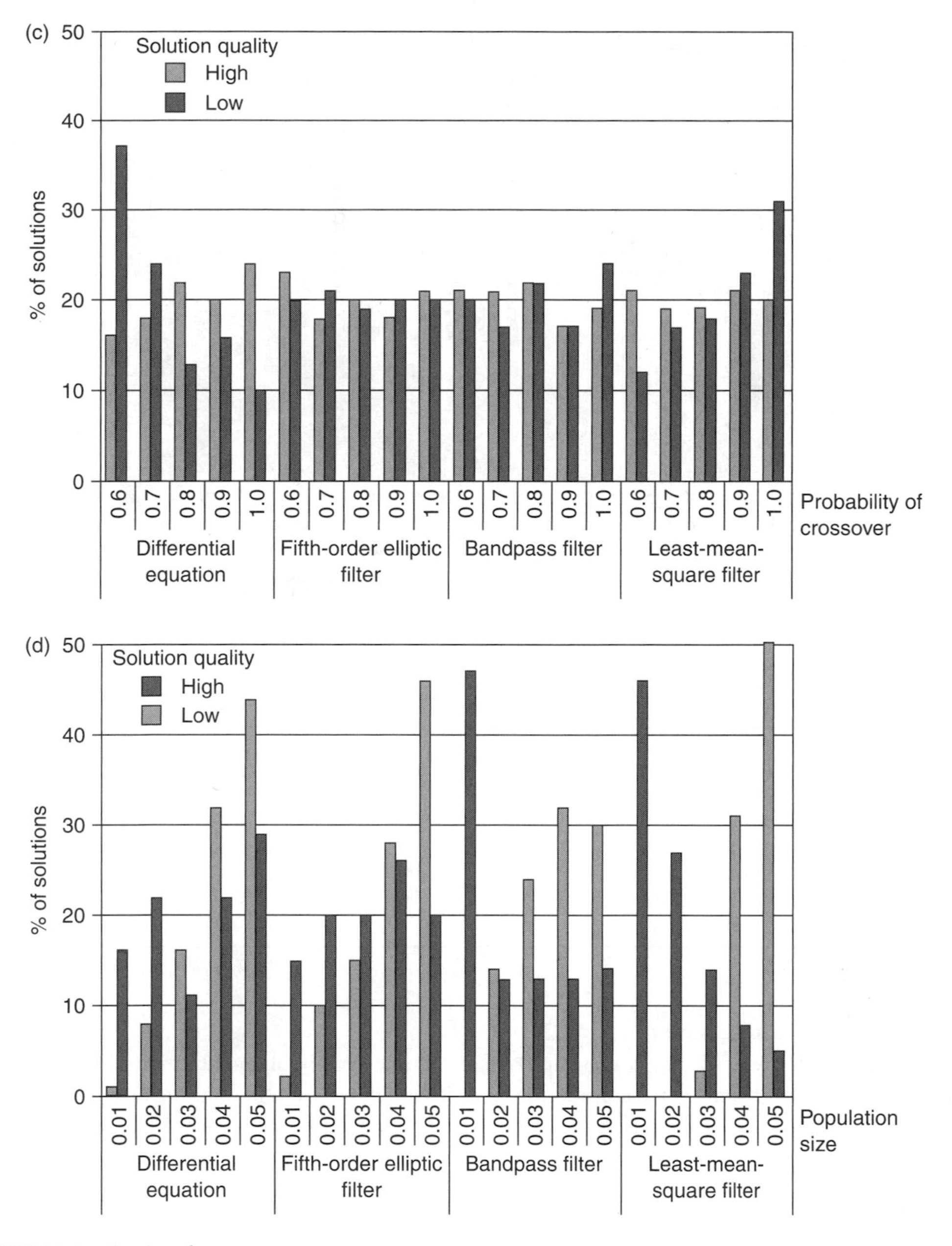

FIGURE 29.6 *Continued.*

29.3.2.4 The IC Evaluation

The ECSA algorithm is evaluated by a comparison with nearly optimal [31] FDS [32]. FDS tries to schedule optimally the DFG considering a uniform distribution of the operations of the same type over the available control steps.

Table 29.4 shows the results of the following evaluations: FDS with fast units, FDS with slow units, and ECSA with basic and independent genetic operators. There are two types of DFGs for each circuit. The first, or plain, is an ordinary data-flow graph with nodes that represent operations, as described in similar studies; and the second, or improved [9], considers the input variables (start registers) via some additional nodes to ensure a more accurate estimation of the registers and the buses needed to implement the circuit.

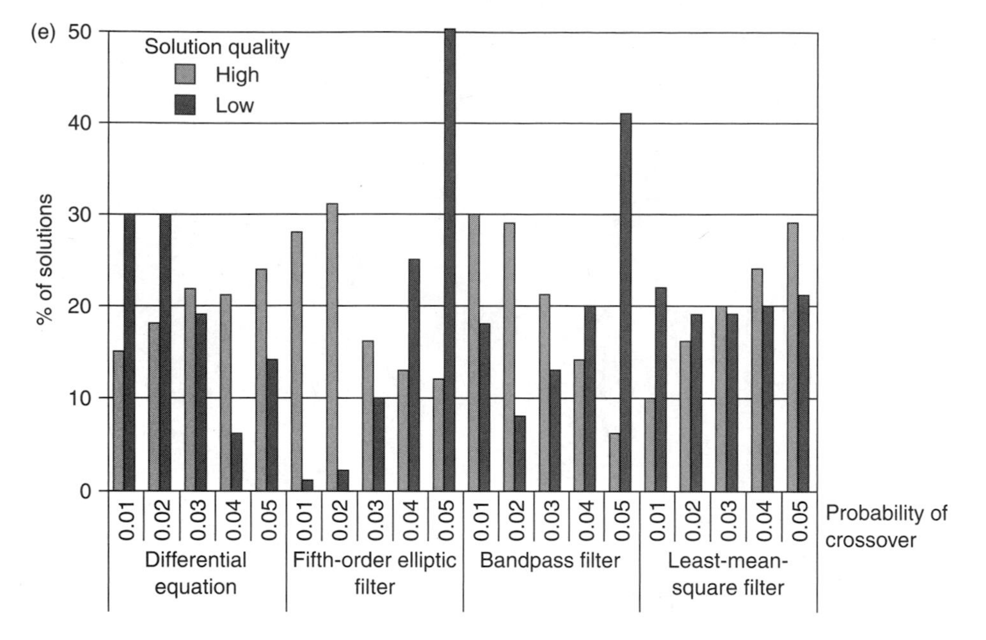

FIGURE 29.6 *Continued.*

TABLE 29.3 Optimal Values of the Parameters for Different Test-Bench Circuits

	Differential equation	Fifth-order elliptic filter	Bandpass filter	Least-mean-square filter	Average optimal values
Number of generations	40	100	90	130	3 × DFG size
Population size	55	120	110	160	3.5 × DFG size
Probability of crossover	0.8	0.6	0.7	0.6	0.7
Probability of mutation	0.04	0.05	0.04	0.05	0.04
Probability of variation	0.04	0.02	0.02	0.05	0.03

29.3.2.4.1 Differential Equation

Because of the small circuit size there is no improvement in the solutions obtained with the ECSA algorithm (either basic or independent) when considering an ordinary DFG — all the solutions are of a larger size. But when we consider the start registers (input variables) there are some ECSA solutions with a slightly larger size and a smaller number of buses.

29.3.2.4.2 Fifth-Order Elliptic Filter

The evolutionary method with a basic approach finds a smaller circuit with a smaller number of buses and a slightly longer execution time for the ordinary DFG, whereas the independent approach could not find any improved solution. When dealing with the improved DFG, both approaches (basic and independent) find considerably smaller circuits with a slight increase in the execution time, while the independent approach also finds the solution with a substantial decrease in the required number of registers and buses.

29.3.2.4.3 Bandpass Filter

Both ECSA methods find, when dealing with the ordinary DFG, the solutions with a smaller number of registers and buses; the basic approach also finds the smaller circuit, but with a slightly longer execution time. When dealing with the improved DFG, both approaches find the solutions with the same circuit size and execution time as the comparable FDS solution, but the required number of registers and buses is considerably smaller for the ECSA solutions.

TABLE 29.4 The Evaluation Results of the ECSA Algorithm with Different Test-Bench ICs

Algorithm	Functional units	Size	Registers	Buses	Delay	Runtime [sec]
Differential equation						
FDS-fast	1 × FE2 + 1 × FE4 + 3 × FE6	23,249	17	6	6	0.01
FDS-slow	2 × FE1 + 1 × FE3 + 2 × FE5	7,173	18	6	20	0.01
ECSA-basic	2 × FE2 + 1 × FE4 + 3 × FE6	23,914	18	8	6	0.11
ECSA-independent	2 × FE2 + 1 × FE4 + 3 × FE6	23,914	17	6	6	0.09
Differential equation with start registers						
FDS-fast	1 × FE2 + 1 × FE4 + 3 × FE6	23,249	10	9	6	0.01
FDS-slow	2 × FE1 + 1 × FE3 + 2 × FE5	7,173	11	9	20	0.01
ECSA-basic	2 × FE2 + 1 × FE4 + 4 × FE6	31,210	10	7	6	0.15
ECSA-independent	2 × FE2 + 1 × FE4 + 3 × FE6	23,914	10	7	6	0.35
Fifth-order elliptic filter						
FDS-fast	3 × FE2 + 3 × FE6	23,883	26	8	17	0.02
FDS-slow	5 × FE1 + 3 × FE5	10,970	30	6	78	0.03
ECSA-basic	2 × FE2 + 1 × FE5 + 2 × FE6	18,962	29	4	21	3.80
ECSA-independent	4 × FE2 + 4 × FE6	31,844	30	8	17	1.60
Fifth-order elliptic filter with start registers						
FDS-fast	3 × FE2 + 3 × FE6	23,883	21	16	17	0.02
FDS-slow	5 × FE1 + 3 × FE5	10,970	25	16	78	0.04
ECSA-basic	2 × FE2 + 1 × FE5 + 2 × FE6	18,962	24	16	19	4.80
ECSA-independent	2 × FE2 + 2 × FE6	15,922	18	9	21	3.40
Bandpass filter						
FDS-fast	3 × FE2 + 4 × FE6	31,179	34	10	10	0.01
FDS-slow	4 × FE1 + 3 × FE5	10,600	35	8	44	0.04
ECSA-basic	3 × FE2 + 3 × FE6	23,883	33	8	11	1.70
ECSA-independent	3 × FE2 + 1 × FE5 + 4 × FE6	34,219	33	8	10	1.30
Bandpass filter with start registers						
FDS-fast	3 × FE2 + 4 × FE6	31,179	25	23	10	0.02
FDS-slow	4 × FE1 + 3 × FE5	10,600	26	23	44	0.04
ECSA-basic	3 × FE2 + 4 × FE6	31,179	23	19	10	2.40
ECSA-independent	3 × FE2 + 4 × FE6	31,179	23	19	10	2.90
Least-mean-square filter						
FDS-fast	3 × FE2 + 6 × FE6	45,771	68	12	13	0.40
FDS-slow	3 × FE1 + 4 × FE5	13,270	72	8	70	2.79
ECSA-basic	3 × FE2 + 6 × FE5 + 3 × FE6	42,123	67	10	14	6.30
ECSA-independent	9 × FE2 + 6 × FE5 + 9 × FE6	89,889	69	30	15	8.05
Least-mean-square filter with start registers						
FDS-fast	3 × FE2 + 6 × FE6	45,771	33	29	13	0.48
FDS-slow	3 × FE1 + 4 × FE5	13,270	37	27	70	3.52
ECSA-basic	4 × FE2 + 2 × FE5 + 5 × FE6	45,220	32	25	13	9.20
ECSA-independent	5 × FE2 + 3 × FE5 + 7 × FE6	63,517	33	25	13	12.30

29.3.2.4.4 Least-Mean-Square Filter

At the expense of a small increase in the delay, the basic ECSA is able to decrease the size and lower the number of register and buses of the ordinary DFG; but the independent ECSA is not able to improve any parameter. When dealing with the improved DFG, the basic ECSA is able to keep the initial delay, to decrease the circuit size and to lower the number of required registers and buses. The independent ECSA is only able to decrease the number of buses while increasing the circuit size.

There are slightly longer runtimes when the ECSA algorithm is used. But considering the speed (a few seconds) and the evaluation presented in Reference 33, where the runtimes for larger circuits increase enormously (exponentially) when the FDS algorithm is used, we can conclude that small and large circuits can be designed and optimized with the use of the proposed evolution-based algorithm, which exhibits a linear increase in the design time with an increase of circuit size.

29.4 Conclusions

In the UM geometry optimization we used an evolutionary approach to improve the efficiency of a UM, the motor that is typically used in home appliances and power tools. The goal of our optimization was to find the new set of independent geometrical parameters of the rotor and the stator with the aim of reducing the motor's power losses, which occur in the iron and the copper. The approach proves to be a simple and efficient search-and-optimization method for solving this day-to-day design problem in industry. It outperforms, by a significant improvement of the motor's efficiency, a conventional design procedure that was used previously. By using the GA we are able to reduce the iron and the copper losses of an initial UM by at least 20%, and increasing the GA running time or setting its parameters more appropriately could improve on this result.

In the IC area/time optimization we used an evolutionary approach to some parts of IC design. The work was focused on ASICs that need an even more sophisticated design due to their specific use. Optimally scheduled operations are not necessarily optimally allocated to functional units. To ensure optimum allocation we need to consider some allocation criteria while the scheduling is being done. The evolutionary approach considers scheduling and allocation constraints and ensures a globally optimal solution in a reasonable time. To evaluate our method we built an algorithm and implemented it with a computer. It is used with a group of test-bench ICs. These circuits are chosen because the same types were used in similar studies. They differ in terms of their size and the number of operation types. The results of the evaluation of a computer-implemented algorithm show that the evolutionary methods are able to find a solution that is more appropriate in terms of all the considered and important objectives than the classical deterministic methods.

References

[1] Sen, P.C. *Principles of Electric Machines and Power Electronics.* John Wiley & Sons, New York, 1996.

[2] Gajski, D. et al. *High-Level Synthesis: Introduction to Chip and System Design.* Kluwer Academic Publishers, Boston, MA, 1992.

[3] Armstrong, J.R. and Gray, F.G. *VHDL Design: Representation and Synthesis.* Prentice Hall PTR, Upper Saddle River, NJ, 2000.

[4] Landwehr, B., Marwedel, P., and Dömer, R. Optimum simultaneous scheduling, allocation and resource binding based on integer programming. In *Proceedings of EuroDAC*, Grenoble, France, 1994, p. 90.

[5] Zhu, J. and Gajski, D. Soft scheduling in high level synthesis. In *Proceedings of the 36th ACM/IEEE Design Automation Conference*, New Orleans, LA, 1999, p. 219.

[6] Mandal, C. and Zimmer, R.M. A genetic algorithm for the synthesis of structured data paths. In *Proceedings of the IEEE VLSI Design*, Calcutta, India, 2000, p. 206.

[7] Kim, T. Scheduling and Allocation Problems in High-Level Synthesis, Ph.D. thesis, University of Illinois at Urbana-Champaign, 1993.

[8] Grajcar, M. Genetic list scheduling algorithm for scheduling and allocation on a loosely coupled heterogeneous multiprocessor system. In *Proceedings of the 36th ACM/IEEE Design Automation Conference*, New Orleans, LA, 1999, p. 280.

[9] Papa, G. Concurrent operation scheduling and unit allocation with an evolutionary technique in the process of integrated-circuit design, Ph.D. thesis, Faculty of Electrical Engineering, University of Ljubljana, 2002.

[10] Karr, C.L., Yakushin, I., and Nicolosi, K. Solving inverse initial-value, boundary-value problems via genetic algorithm. *Engineering Applications of Artificial Intelligence*, 13, 625, 2000.

[11] Bäck, T. *Evolutionary Algorithms in Theory and Practice.* Oxford University Press, New York, 1996.
[12] Goldberg, D.E. *Genetic Algorithms in Search, Optimization, and Machine Learning.* Addison-Wesley, Reading, MA, 1989.
[13] Drechsler, R. *Evolutionary Algorithms for VLSI CAD.* Kluwer Academic Publishers, Boston, MA, 1998.
[14] Papa, G. et al. Universal motor efficiency improvement using evolutionary optimization. *IEEE Transactions on Industrial Electronics*, 50, 602, 2003.
[15] Filipič, B. and Štrancar, J. Tuning EPR spectral parameters with a genetic algorithm. *Applied Soft Computing*, 1, 83, 2001.
[16] Papa, G. and Šilc, J. Automatic large-scale integrated circuit synthesis using allocation-based scheduling algorithm. *Microprocessors and Microsystems*, 26, 139, 2002.
[17] Koroušić-Seljak, B. Heuristic methods for a combinatorial optimization problem — real-time task scheduling problem. In *Smart Engineering System Design: Neural Networks, Fuzzy Logic, Evolutionary Programming, Data Mining, and Complex Systems*, Dagli, C.H. et al., Eds., ASME Press, New York, 1999, p. 1041.
[18] Dasgupta, D. and Michalewicz, Z. *Evolutionary Algorithms in Engineering Applications.* Springer-Verlag, Berlin, 1997.
[19] Coello Coello, C.A. A comprehensive survey of evolutionary-based multiobjective optimization techniques. *Knowledge and Information Systems*, 1, 269, 1999.
[20] Van Veldhuizen, D. and Lamont, G.B. Multiobjective evolutionary algorithms: analyzing the state-of-the-art. *Evolutionary Computation*, 8, 125, 2000.
[21] *ANSYS User's Manual*, ANSYS version 5.6, 2000.
[22] Papa, G. *DOptiMeL — User's Manual.* Technical report IJS DP 8440, Jožef Stefan Institute, Ljubljana, 2001.
[23] Paulin, P.G., Knight, J.P., and Girczyc, E.F. HAL: a multiparadigm approach to automatic data path synthesis. In *Proceedings of the 23rd ACM/IEEE Design Automation Conference*, Las Vegas, NV, 1986, p. 263.
[24] Kung, T., Whitehouse, H.J., and Kailath, T. *VLSI and Modern Signal Processing.* Prentice Hall, New York, 1985.
[25] Grewal, G.W. and Wilson, T.C. An enhanced genetic algorithm for solving the high-level synthesis problems of scheduling, allocation, and binding. *International Journal of Computational Intelligence and Application*, 1, 91, 2001.
[26] Benesty, J. and Duhamel, P. A fast exact least square adaptive algorithm. *IEEE Transactions on Signal Processing*, 40, 2904, 1992.
[27] Parhami, B. *Computer Arithmetic: Algorithms and Hardware Designs.* Oxford University Press, New York, 2000.
[28] Šilc, J., Robič, B., and Ungerer, T. *Processor Architecture: From Dataflow to Superscalar and Beyond.* Springer-Verlag, Berlin, 1999.
[29] Thornton, M.A., Gaiche, J.D., and Lemieux, J.V. Trade-off analysis of integer multiplier circuits implemented in FPGAs. In *Proceedings of the IEEE Pacific Rim Conference on Communications, Computers and Signal Processing*, Victoria, Canada, 1999, p. 301.
[30] Papa, G. and Šilc, J. Evolutionary synthesis algorithm — genetic operators tuning. In *Advances in Intelligent Systems, Fuzzy Systems, Evolutionary Computation*, Grmela, A. and Mastorakis, N., Eds., WSEAS Press, Athens, 2002, p. 256.
[31] Paulin, P.G. and Knight, J.P. Scheduling and binding algorithms for high-level synthesis. In *Proceedings of the 26th ACM/IEEE Design Automation Conference*, Las Vegas, NV, 1989, p. 1.
[32] Paulin, P.G. and Knight, J.P. Force-directed scheduling in automatic data path synthesis. In *Proceedings of the 24th ACM/IEEE Design Automation Conference*, Miami, FL, 1987, p. 195.
[33] Papa, G., Šilc, J., and Wyrzykowski, R. Scheduling algorithms based on genetic approach. In *Proceedings of the 4th Conference on Neural Networks and Their Application*, Zakopane, Poland, 1999, p. 469.

30

Solving the Partitioning Problem in Distributed Virtual Environment Systems Using Evolutive Algorithms

Pedro Morillo
Marcos Fernández
Juan Manuel Orduña

30.1 Introduction

The widespread use of both fast Internet connections and also high-performance graphic cards have made possible the current growth of Distributed Virtual Environment (DVE) systems. These systems allow multiple users, working on different computers that are interconnected through different networks (and even through Internet) to interact in a shared virtual world. This is achieved by rendering images of the environment as the user would perceive them if he was located at that point of the virtual environment. Each user is represented in the shared virtual environment by an entity called *avatar*, whose state is controlled by the user. Since DVE systems support visual interactions between multiple avatars, every change in each avatar must be notified to the neighboring avatars in the shared virtual environment.

DVE systems are currently used in different applications [1], such as collaborative design [2], civil and military distributed training [3], e-learning [4], or multiplayer games [1,5–7].

Designing an efficient DVE system is a complex task, since these system show an inherent heterogeneousness. Such heterogeneousness appears in several elements:

Hardware: Each client computer controlling an avatar may have installed different hardware: a very different range of resources like processor speed, memory size, and graphic card technology can be specified for different client computer.

Connection: Different connections can be found in a single system. From shared medium topologies like Ethernet or Fast-Ethernet to other network connections like ISDN, fiber-optic or ATM can be simultaneously found in some DVEs.

Communication rate of avatars: Depending on the application, different communication rates of avatars can be found. For example, the communication rate of avatars in a collaborative three-dimensional (3D) environment may greatly differ from the communication rate of avatars in a 3D virtual military battle.

Additionally, other factors help to increase the complexity of designing an efficient DVE system. Each of them have now become an open research field:

Data model: This concept describes some conceivable ways of distributing persistent or semipersistent data in a DVE [8]. Data can be managed in a replicated, shared, or distributed methodology.

Communication model: Network bandwidth determines the size and performance of a DVE. The system behavior is related to the way all the scene clients are connected. Broadcast, peer-to-peer, or unicast schemes define different network latency values for exchanging information between avatars.

View consistency: This problem has already been defined in other computer science fields such as database management [9]. In DVE systems, this problem involves ensuring that all avatars sharing a virtual space with common objects have the same local vision of them.

Message traffic reduction: Keeping a low amount of messages allows DVE systems to efficiently scale with the number of avatars in the system. Traditionally, techniques like dead-reckoning described in Reference 1 offered some level of independence to the avatars. With network support, broadcast or multicast solutions [10,11] decrease the number of messages used to keep a consistent state of the system.

Most of the issues described above are related to the *partitioning problem* or *p-problem*. This problem consists of efficiently distributing the workload (avatars) among different servers in the system [12]. The partitioning problem may seriously affect the overall performance of the DVE system, since it determines not only the workload that each server must support, but also the inter-server communication requirements (and therefore the network traffic).

Some methods for solving the partitioning problems have been already proposed [13–15]. These methods provide efficient solutions even for large DVE systems. However, there are still some features in the proposed methods that can still be improved. For example, different heuristic search methods can be used for finding the best assignment of clients to servers, instead of using ad hoc heuristics. In this chapter, we present a comparison study of several evolutive heuristics for solving the partitioning problem in DVE systems. We have implemented five different heuristics, ranging over most of the current taxonomy of heuristics: Genetic Algorithms (GAs) [16], two different implementations of Simulated Annealing [17], Ant Colony Systems (ACSs) [18], and Greedy Randomized Adaptive Search (GRASP) [19]. Performance evaluation results show that the execution cost of the partitioning algorithm (in terms of execution times) can be dramatically reduced, while providing similar or even better solutions than the ones provided by the ad hoc heuristic proposed in [14].

The rest of the chapter is organized as follows: Section 30.2 describes the partitioning problem and the existing proposals for solving it. Section 30.3 describes the proposed implementations of the heuristics considered for this study. Next, Section 30.4 presents the performance evaluation of the proposed heuristics. Finally, Section 30.5 presents some concluding remarks.

30.2 The Partitioning Problem in DVE Systems

30.2.1 Architectures for DVE Systems

Several architectures have been traditionally used for simulating a large set of avatars sharing the same virtual world. Internet multiplayer games as Quake [5] and Kali [7] or educational systems as VES [4] are examples of client–server systems (Figure 30.1[b]). In these applications, each client computer has a single connection to the only existing server in the system. This server maintains the global state of the simulation, but it becomes a single point of failure in the system. Instead of sending messages to a central server, in peer-to-peer architectures (Figure 30.1[a]) avatars exchange messages directly. Several systems have been developed with this architecture, such as NPSNET [10]. Although these systems obtain low latencies, they do not properly scale. When the number of avatars greatly increases, clients are not able to handle the amount of messages from other avatars and simultaneously offering an interactive 3D virtual world to the user. In order to improve scalability, peer–server and server–network architectures group sets of avatars. Following a peer–server scheme (Figure 30.1[d]), systems like ATLAS [11] reduces the volume of information using multicast messaging. However, this architecture will be useful only when multicast protocols are fully available in the Internet [13].

Architectures based on networked servers are becoming a de facto standard for DVE systems [1,12,15]. In these architectures, the control of the simulation relies on several interconnected servers. Figure 30.1(c) depicts how multi-platform client computers are attached to only one of the servers of the simulation.

In this architecture, when a client modifies an avatar, it also sends an updating message to its server, that in turn must propagate this message to other servers and clients. Servers must render different 3D models, perform positional updates of avatars and transfer control information among different clients. Thus, each new avatar represents an increase in both the computational requirements of the application and also in the amount of network traffic. When the number of connected clients increases, the number of updating messages must be limited in order to avoid a message outburst. In this sense, concepts like Areas of Interest (AOI) [1], locales [20], or auras [21] have been proposed for limiting the number of neighboring avatars that a given avatar must communicate with. All these concepts define a neighborhood area for avatars, in such a way that a given avatar must notify his movements (by sending an updating message) only to those avatars located in that neighborhood. Depending on their origin and destination avatars, messages in a DVE system can be intra- or inter-server messages (see Figure 30.2). Inter-server messages are those messages involving two or more servers. On the contrary, intra-server messages are those messages exchanged between avatars whose client computers are attached to the same server. In order to design a scalable DVE systems, the number of intra-server messages must be maximized. Effectively, when clients send intra-server messages they only concern a single server. Therefore, they are minimizing

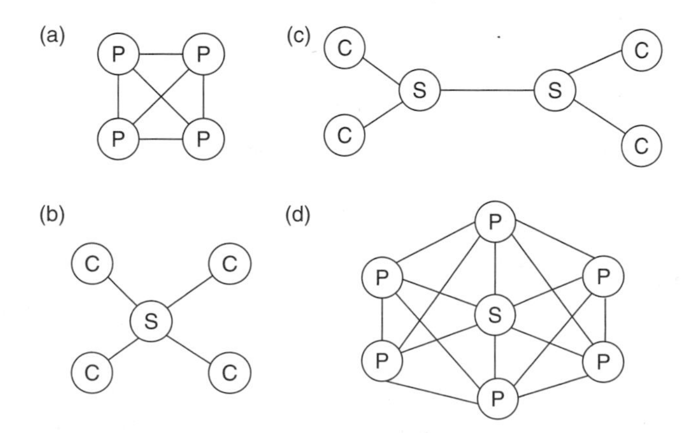

FIGURE 30.1 Architectures: (a) peer-to-peer, (b) server–network, (c) client–server, and (d) peer-to-server.

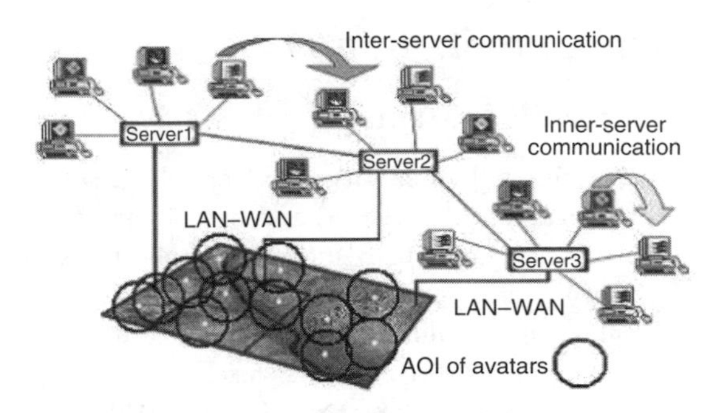

FIGURE 30.2 A multi-server architecture for a basic DVE.

the computing, storage, and communication requirements for maintaining a consistent state of the avatars in a DVE system.

The *partitioning problem* consists of efficiently distributing the workload (assigning avatars) among the different servers in the system. Lui and Chan have shown the key role of finding a good assignment of avatars to servers in order to ensure both a good frame rate and a minimum network traffic in DVE systems [12,14]. They propose a quality function, denoted as C_{p}, for evaluating each partition (assignment of avatars to servers). This quality function takes into account two parameters. One of them consists of the computing workload generated by clients in the DVE system, and is denoted as C_{p}^{W}. In order to minimize this parameter, the computing workload should be proportionally shared among all the servers in the DVE system, according to the computing resources of each server. The other parameter of the quality function consists of the overall number of inter-server messages, and it is denoted as C_{p}^{L}. In order to minimize this parameter, avatars sharing the same AOI should be assigned to the same server. Thus, quality function C_{p} is defined as

$$C_{\mathrm{p}} = W_1 C_{\mathrm{p}}^{W} + W_2 C_{\mathrm{p}}^{L}, \tag{30.1}$$

where $W_1 + W_2 = 1$. W_1 and W_2 are two coefficients that weighs the relative importance of the computational and communication workload, respectively. These coefficients should be tuned according to the specific features of each DVE system. Using this quality function (and assuming $W_1 = W_2 = 0.5$) Lui and Chan propose a partitioning algorithm that reassigns clients to servers [14]. The partitioning algorithm should be periodically executed for adapting the partition to the current state of the DVE system as it evolves (avatars can join or leave the DVE system at any moment, and they can also move everywhere within the simulated virtual world). Lui and Chan also have proposed a testing platform for the performance evaluation of DVE systems, as well as a parallelization of the partitioning algorithm [14].

The partitioning method proposed by Lui and Chan, known as LOT or *Linear Optimization Technique*, currently provides the best results for DVE systems. However, it uses an ad hoc heuristic. We propose a comparative study of several heuristics, ranging over most of the current taxonomy of heuristics, in order to determine which one provides the best performance when applied to the partitioning problem in DVE systems. In this study, we propose the same approach as Lui–Chan: using the same quality function, we will obtain an initial partition (assignment) of avatars to servers, and then we will test the implementation of each heuristic to provide a near optimal assignment.

30.2.2 Linear Optimization Technique (LOT)

Linear Optimization Technique (LOT) was initially published by Lui et al. [12] and revisited in Reference 14 from their ideas published in Reference 22 about graph theory.

This ad hoc approach models the 3D virtual scene as a graph. Each avatar is modeled by a node and two avatars are linked by an edge if their AOI collide. Using C_{p} as the evaluation function, LOT provides an efficient partition of this graph following three steps. The first step is named the *Recursive Bisection Procedure* (RBP), and it consists of using a divide-and-conquer procedure that provides an initial partition.

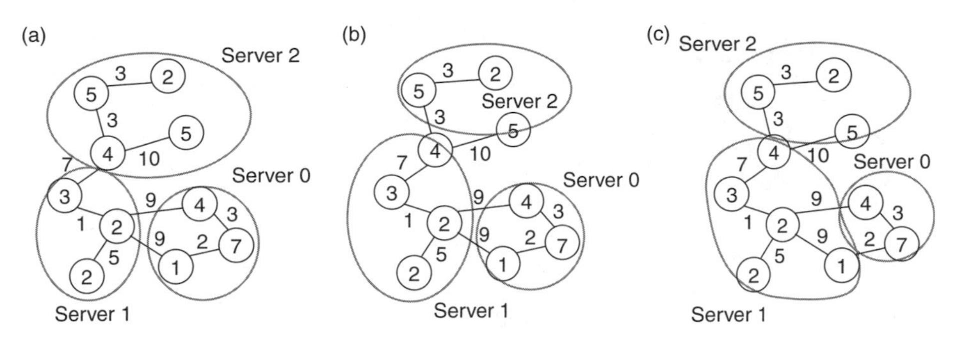

FIGURE 30.3 Partition of a DVE performed by LOT method: (a) RBP, (b) CRP, and (c) LP.

Then, the *layering partitioning procedure* (LP) and the *Communication Refinement Partitioning* (CRP) are applied on that initial partition of the graph. Each of these procedures performs workload balancing and minimizes the number of inter-server messages, respectively.

Figure 30.3 shows the partitions that this method would provide when applied to a small DVE system. In this example, a DVE composed by 10 avatars is simulated with three identical servers. Nodes and edges obtained from the associated graph has been labeled. The label of a node (avatar) represents a estimation of the workload generated by this avatar to the server where it is going to be assigned. Ranging from 1 to 10, the label of each edge represents the nearness of these two avatars. Figure 30.3(a) represents the result obtained by a RBP phase. Although the number of avatars assigned to each server seems to be balanced, the workload that those avatars generate must be also uniform in order to achieve actual workload balancing. By adding the labels of all the nodes assigned to the same server, we obtain that the workload assigned to each server is 16, 7, and 12 units of workload, respectively. Then, CRP balances (Figure 30.3[b]) the existing workload in sets of 12, 11, and 12 units. At that point, the number of inter-server messages generated by these sets of avatars is reduced by LP, as shown in Figure 30.3(c). Since strategies of CRP and LP techniques could be opposed, this couple of steps is repeated three times.

30.2.3 Other Approaches to the Partitioning Problem

In addition to LOT technique, two other relevant approaches have been also proposed for solving the partitioning problem. One of them divides the whole virtual scene into hexagonal cells, and each cell is mapped to a multicast group [13]. With this division, all avatars located in the same cell share the same multicast address. Since the size of AOI is bigger than the area of the cells, avatars can list different multicast addresses but they can only send information to the cell where they are located.

Figure 30.4 shows an example of this technique. In this figure, the represented avatar is assigned to group A and also lists multicast information from groups B, C, E, F, and G. In order to ensure efficient management, this algorithm generates a group leader for each cell. The oldest avatar is selected as the group leader, using a timestamp mechanism. This leader controls how avatars join and leave the multicast group, and he also implements a flow control mechanism for the messages.

Despite its apparent scalability, this approach presents several problems. First, it does not obtain good performance when avatars are located following a nonuniform distribution. Second, system performance is very low when avatars can quickly cross the virtual scene. This is due to the associated cost of leaving and joining different multicast groups. Finally, multicast is not fully implemented within the Internet.

Another different approach was proposed by Tam [15]. In this method, dynamic concepts associated with avatars like AOI, aura, or locale are rejected. This partitioning method divides the scene in square cells. The side length of the cells S is related to the radius of the AOI of avatars, following the next relation:

$$\frac{S}{2} \leq \text{AOI} \leq S. \qquad (30.2)$$

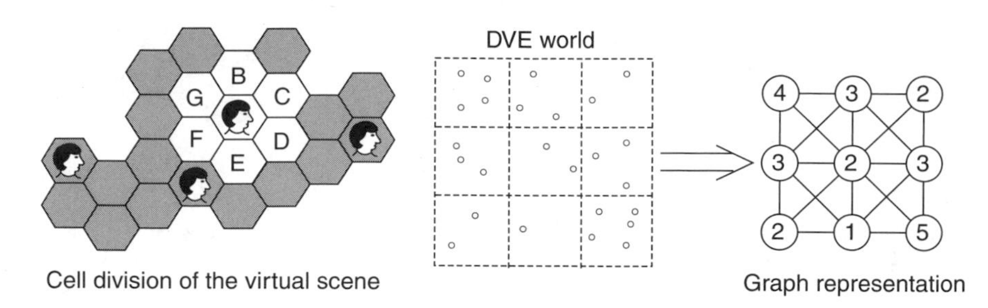

FIGURE 30.4 Other approaches: cell division and graph representation.

A graph representation of the virtual scene is obtained from the volume of avatars contained in each cell, as also shown in Figure 30.4. Next, this graph is divided in partitions and each partition is assigned to a server of the DVE. In order to accomplish this division an exhaustive and greedy algorithms are compared. Using a quality different from C_P, these algorithms take also into account the number of inter-server messages and workload balancing among the servers. Although this approach provides a fast way of solving the partitioning problem, the performance of the static partitioning is quite low when avatars show a clustered distribution. In this case, the servers controlling the crowded areas are overloaded, increasing the overall cost of the quality function.

30.3 Metaheuristic Procedures

Evolutive computing has become a important paradigm in the development of high-quality solution for NP-Complete problems. This paradigm, which models computational problems as natural processes, is used to provide solutions for complex problems that in most cases could not be solved by using deterministic computing techniques. Problems such as multi-objective optimization have been successfully adapted using these strategies [23]. In this section, we present five implementations of different heuristics for solving the partitioning problem in DVE systems. Following the approach presented by Lui and Chan (and using the same quality function C_P), the idea is dynamically applying a heuristic search method that provides a good assignment of clients to servers as the state of the DVE system changes. All the presented solutions have been adapted in the same way. That is, a initial partition obtains good assignments only for several avatars, while the evolutive algorithm itself performs successive refinements of the initial partition that lead to a near optimal partition of the DVE system. Also, the evolutive method is used periodically for updating the obtained partition to the current state of the DVE system (avatars change their locations, new avatars can join the system, and some avatars can leave the system at any time).

In this section, we describe the implementation of each heuristic search method and the tuning of its parameters for solving the partitioning problem. Concretely, this tuning has been performed on the DVE evaluation test described by Lui and Chan in [14]. This test defines two kinds of DVE systems: a SMALL virtual world, composed of 13 avatars and 3 servers and also a LARGE world, composed of 2500 avatars and 8 servers. Effectively, the purpose of solving the partitioning problem is to provide scalable DVE systems. Therefore, the partitioning method must efficiently work in LARGE virtual worlds. On other hand, since the performance of the method may heavily depend on the location of avatars, this evaluation test also considers three different distributions of avatars: uniform, skewed, and clustered distribution.

30.3.1 Obtaining the Initial Population

All of the implemented heuristics start from an initial partition (assignment) of the n avatars in the DVE system. We tested several clustering algorithms for obtaining this initial partition. Although they are not shown here due to space limitations, we obtained the best results for a *Density-Based Algorithm* (DBA) [24].

This algorithm divides the virtual 3D scene in square sections. Each section is labeled with the number of avatars that it contains (*na*), and all the sections are sorted (using Quick-sort algorithm) by their *na* value. The first *S* sections in the sorted list are then selected and assigned to a given server, where *S* is the number of servers in the DVE system. That is, all the avatars in a selected region are assigned to a single server. The next step consists of computing the mass-center (*mc*) of the avatars assigned to each server. Using a round-robin scheme, the algorithm then chooses the closest free avatar to the *mc* of each server, assigning that avatar to that server, until all avatars are assigned.

The proposed implementation of the DBA method consists of the following steps (expressed as pseudo-code statements):

```
program Initial_Partition (avatar, Int n, Int S)

const
   n_sections = 25x25

type Cell
    sum,idx: Int

var
    assigned,represent:Int[]
    pivot,elect,ncentr:Int
    min_dis,dist_tmp   :Real
    na                 :Cell[n_sections]

begin
    DivideSceneInSquareSections(n_sections)
    for i:=0 to n_sections do
        na[i].sum:=CountAvatarsInSection(i)
        na[i].idx:=i
    end_for
    QuickSort (na)
    for i=0 to S do
        representant [i]= ObtainMC (na[i].idx)
    end_for;
    pivot := 0, elect := -1, ncentr:=0
    for i:=0 to n do
        elect:=-1,min_dis := 100000
        for  j:=0 to n do
           if (assigned[j] = NOT_ASSIGNED)
                dist_tmp:= EuclideanDistance(avatar[j],represent([ncentr])
                if (dist_tmp < min_dis)
                    pivot   := j
                    min_dis := dist_tmp
                endif
           endif
        end_for
        avatar[pivot].assignment := assigned[pivot] := ncentr
        ncentr : (ncentr + 1) mod S
    end_for
end
```

Since the assignment of avatars follows a round-robin scheme, this algorithm provides a good balancing of the computing workload (the number of avatars assigned to each server does not differ in more than one). On other hand, avatars that are grouped in a small region and close to the mass-center of a server will be assigned to that server by the DBA. Additionally, since these avatars are located so closely, they will probably will share the same AOI. Therefore, the DBA also provides an initial partition with low inter-server communication requirements for those avatars.

However, the assignment of avatars equidistant (or located far away) from the mass-center of the servers is critical for obtaining a partition with minimum inter-server communication requirements (minimum values of the quality function C_P), particularly for SMALL virtual worlds with only a few servers. DBA inherently provides good assignments for clustered avatars, but it does not properly focus on the assignment of these critical avatars. Each of the following evolutive methods should be used at this point to search a near optimal assignment that properly reassigns these avatars.

30.3.2 Genetic Algorithms (GA)

This heuristic consists of a search method based on the concept of evolution by natural selection [16]. GA starts with an initial population (the initial partition) and then it evolves a certain number of *generations* (iterations), providing an *evolved population* (final solution). The proposed implementation for solving the partitioning problem, proposed initially in Reference 25, starts with a population composed of a set of elements called *genomes* or *chromosomes*. The number of chromosomes is the number of partial solutions that each iteration must maintain. Each chromosome is defined by a descriptor vector containing a given assignment of avatars to servers. The length of this vector is equal to the number of *Border Avatars* (BA) within the DVE system. BA is an avatar that, although it lies within the AOI of another avatar, it is not assigned to the same server. These kinds of avatars offers a higher probability of giving a successful permutation, since they minimize inter-server communication.

Starting from the initial population, this approach applies an auto-fertilization mechanism for generating new chromosomes [26]. This mechanism is based on a single-point random crossover, where each generation is found by exchanging some elements of the restricted population in such a way that in each of the N chromosomes, two border avatars assigned to different servers are randomly chosen and exchanged. Thus, an iteration performed on a population of N chromosomes will produce a new population of $2N$. From these $2N$ chromosomes, the N elements with the lower value of C_P will be chosen.

GA is also capable of escaping from local minima due to the *mutation* process. Once the child-vector has been obtained, mutation involves changing at random the server assigned to one of the elements of the population. Some other mutations such as the modification of several elements or the crossover between the characteristics of pairs of chromosomes have also been tested. But after several tests it has been found that the mutation of just one "bit" is the one offering the best results. It is important to mention that the mutation is a random process controlled by a parameter. This parameter needs to be carefully chosen for each specific experiment in order to achieve solutions with low C_P System Cost.

Next code describes the behavior of the proposed approach based on GA. It has been expressed as pseudo-code statements:

```
program GA (Int chromo, Int iterations, Real mut_rate)

var
   B,temp_cost,Cp_GA :Real
   av_i,av_j,av_k    :Integer

begin
   Initial_Partition (DBA)
   B := ObtainBorderAvatars()
   Cp_GA := Compute_Cp()
   For i:=0 to iterations do
       For j:=0 to chromo do
            SelectAndCopyChromosome(j)
            Choose2DifRamdomAvatars(B,av_i,av_j)
```

```
            ExchangeServerAssignment(av_i,av_j)
            temp_cost := Compute_Cp()
            if (HaveItoMutate(mut_rate))
                Chose1RandomAvatar(av_i)
                ForceServerAssignment(av_i)
            endif
            AddChromosomeToPopulation(this)
        end_for
        CWPSortPopulationByCp(2*chromo)
        Cp_GA := SelectBestIndividuals(chromo)
    end_for
end
```

Figure 30.5 shows a example of generation of new avatars in a SMALL DVE system where six BA have been obtained from the full set of avatars. These avatars define a chromosome, and they can be assigned to any of the three servers in the DVE system. This figure represents a possible crossover and mutation on this chromosome.

Although this basic GA approach performs reasonably well, it is based on the generation of an initial population obtained by deriving a unique solution provided by a clustering algorithm described in Section 30.3.1. Despite the fact that this feature offers an improved initial population of feasible solutions, it does not focus on maximizing the structural diversity of chromosomes. As described in References 27 and 28, this low level of structural diversity can lead the algorithm to reach a local minimum or even a poorer approximation of this value. Additionally, the crossover mechanism used by this algorithm is based on an auto-fertilization technique, where chromosomes are derived following a single-point crossover [26]. This crossover mechanism is excessively generalist, and it is possible to offer new crossover strategies more oriented to problem specifications.

30.3.3 Improving the Performance of the Proposed GA Approach

In order to improve the performance of our current approach for solving the partitioning problem in DVE systems, we propose a refined version of the genetic algorithm. This version focuses on maximizing the structural diversity of initial population and improving the crossover operator. The new algorithm replaces the current generation of the initial population, described in Section 30.3.1 with a new method based on random projections. Additionally, the crossover is performed by randomly choosing a mechanism from a list composed of five different crossover operators. All these five operators are very oriented to the specifications of partitioning problem in DVE systems.

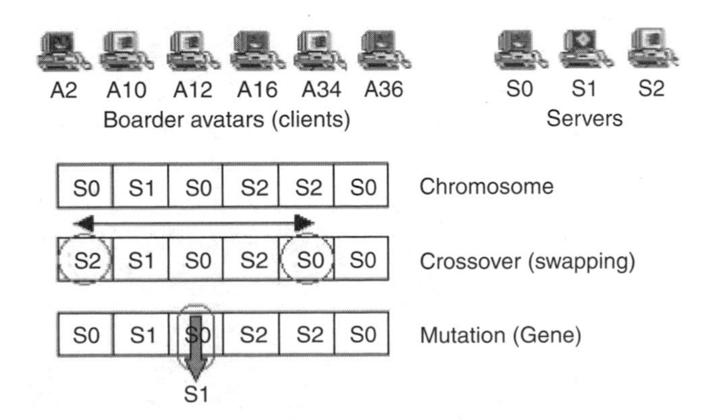

FIGURE 30.5 Generation of new individuals in GA approach.

30.3.3.1 Generation of a Heuristic Initial Population

Most metaheuristics are based on the fast generation of an initial population of elements [27,29]. This initial population usually represents a set of poor solutions to the problem, and it is evolved through a crossover operator until a stopping criterion (for example, a given number of iterations) is reached.

If the initial population has been correctly defined, then the metaheuristic algorithm easily obtains a good approximation to the global optimum. Moreover, as described in Reference 27, if the initial population is not randomly selected then the algorithm should maintain a certain level of structural diversity among all the chromosomes, in order to properly represent the whole set of feasible solutions and to avoid the premature convergence of the search.

Taking into account these considerations, we propose a new mechanism, called *Projections algorithm* (*PA*), in order to generate the new population of initial solutions. This fast algorithm provides a set of independent and well-diversified initial solutions to the problem.

PA consist of a given number of iterations n_c that defines the number of chromosomes in the population (population size). Each one of the n_c chromosomes represents a complete partitioning solution to the problem where all the N avatars $(A_0, \ldots, A_{N-1})$ in the DVE are assigned to the M servers $(S_0, \ldots, S_{M-1})$ in the system. For the sake of clearness, we will consider in the following description that avatars move across a 2D virtual world. The extrapolation to 3D worlds is reasonably trivial. Each one of the n_c iterations consists of four steps illustrated in Figure 30.6.

First, each avatar in the system is assigned to server S_0 (Figure 30.6[a]). Next, a random value θ between 0 and $\pi/2$ is generated. Since all avatars are located on a Cartesian plane, PA draws a straight line which passes through the zero coordinates (0,0) with a slope of θ radians (Figure 30.6[b]). This line and its perpendicular define a new coordinate axis that is rotated θ radians with respect to the original position. Using a simple affine transformation (Figure 30.6[c]), the old coordinates (X_i, Y_i) of each avatar can now be expressed with respect to the rotated axis by the new coordinates (X_i', Y_i'). At this point, PA generates two different *binary search trees* with X_i' and Y_i' search keys of each avatar. Once both trees are created, the N/M different avatars in both trees with the highest and the lowest keys are put into four different sets [30]. In order to assign a set of avatars to a server, the third step of PA evaluates separately the different sets of avatars using the C_p function. Since both sets have the same cardinality N/M, PA algorithm only computes the C_p^L term. The term C_p^W is not computed, since it evaluates the standard deviation of the assigned avatars with respect to the perfect balancing. The last step is to select the set with the lowest C_p^L value, and all avatars contained in this set are assigned to server S_1 (Figure 30.6[d]). At this point, $N(M-1)/M$ avatars are still assigned to server S_0, and N/M avatars are assigned to server S_1. Next iteration allows the server S_0 to lose another N/M avatars, which are assigned to S_2. PA algorithm finishes when the last group of avatars (with a size less or equal to N/M avatars) is assigned to server S_{M-1}. At this point, a number of avatars very close to N/M are assigned to each server in the DVE system.

It is important to mention that the use of four binary search trees allows to improve diversity when selecting the sets of avatars to be assigned to the target server for each iteration. The random generation of rotation angles guarantees inherently independent solutions in the generation of individuals of the initial population. Moreover, this population allows the genetic approach to evolve solutions with good values of quality function C_p. These good C_p values are achieved by balancing the number of avatars among the servers and assigning to the same server the avatars that are closely located in the virtual scene.

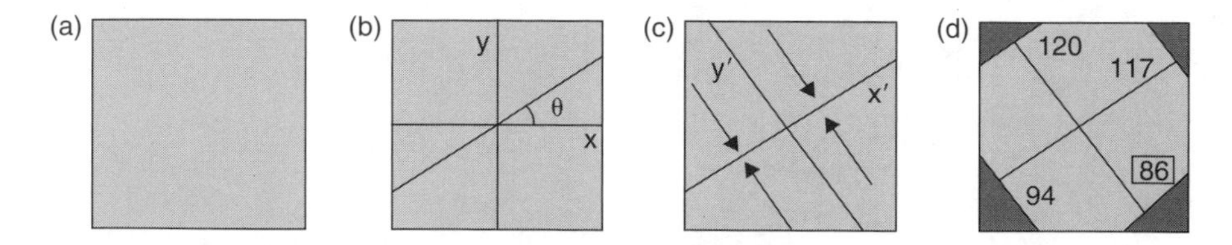

FIGURE 30.6 Generation of the initial population based on *Projection Algorithm*.

30.3.3.2 Providing Randomness to Chromosome Crossover

In order to evolve chromosomes, the first approach for solving the partitioning problem in DVE systems, based on a genetic algorithm uses an auto-fertilization technique [25]. In this technique, each chromosome generates a new chromosome following a single-point crossover with a probability equal to one [26]. When the crossover operator is applied to all individuals of the population (the child population has been completely created) a elitist selection guarantees the survival of the best individuals. This crossover operator is excessively generalist and has been used often for solving different combinatorial problems.

The proposed technique is based on the random selection of crossover operators from a list. This list, which is accessed for each derivation, consists of five operators very oriented to problem specifications. The operator list consists of the following elements:

Operator 1: Random exchange of the current assignment for two *border* avatars. A given avatar A_i is a border avatar if it is assigned to a certain server S_r in the initial partition and any of the avatars in its AOI is assigned to a server different from S_r [31].

Operator 2: Once a border avatar A_i has been randomly selected, it is randomly assigned to one of the servers S_f hosting the border avatars of A_i.

Operator 3: Besides the step described in the previous operator, if there exists an avatar A_j such that A_j is assigned to S_f and it is a neighbor avatar of A_i, then A_j is assigned to S_r.

Operator 4: Since each avatar generates a certain level of workload in the server where it is assigned to [12], then it is possible to sort the servers of a DVE system according to the level of workload they support. If S_m and S_n are the servers with the highest and the lowest level of workload in the system, respectively, then a random avatar A_k assigned to S_m is assigned to S_n.

Operator 5: Besides the step described in the previous operator, a random avatar A_l, initially assigned to S_n, is now assigned to S_m.

The main parameters to be tuned in GA method are the population size P, the number of iterations N and the mutation rate M. Figure 30.7 depicts the values of que quality function C_p (denoted as system cost) as the number of generations and individuals grows. In both cases C_p decreases significantly while the algorithm does not reach a threshold value. From this threshold value (close to 300 for the number of generations and 15 for the population size) the quality of the obtained solutions remains constant, showing the impossibility of finding better solutions in this search domain. Therefore it has no sense in spending more time in searching new partitioning solutions.

Figure 30.8 shows the tuning of the mutation rate in a given DVE system. The behavior of the algorithm is different for this parameter. In this case, the algorithm is able to provide a high-quality solution when mutation rate is close to 1%.

If GA approach uses values much lower than 1% for this parameter, the system can be trapped in a local minimum. In the opposite case, when GA selects rates higher than this threshold value, then the search method spends too much CPU time testing useless solutions.

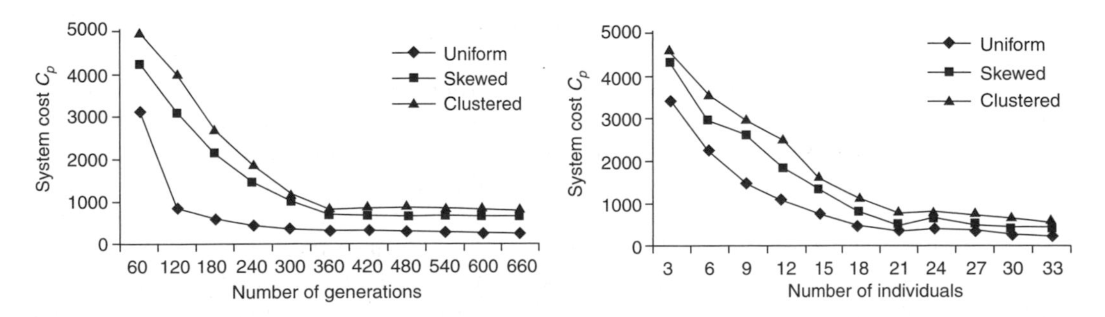

FIGURE 30.7 Values of the quality function C_p for different number of generations and population sizes.

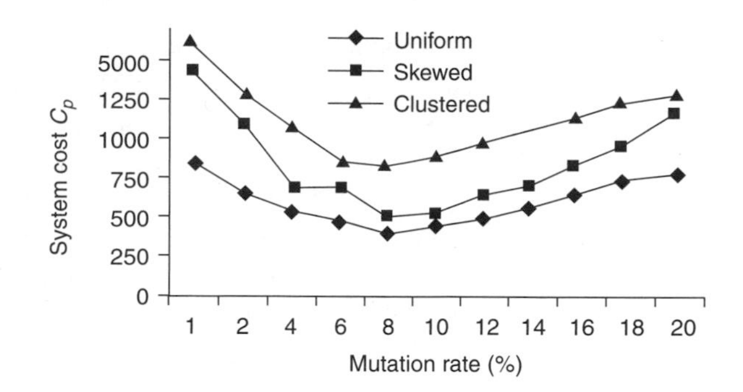

FIGURE 30.8 Values of the quality function C_p for different mutation coefficients.

As a conclusion of this tuning phase, we can state that for this particular application GA method provides good results for a population of 15 individuals, a mutation probability of 1%, and 100 iterations as stop criterion.

30.3.4 Ant Colony System (ACS)

Ant Colony Systems (ACSs) search method is based on the behavior shown by ant colonies when searching possible paths to their food [18]. They use a hormone called *pheromone* to communicate among them. The path that a given ant follows when searching food depends on the amount of pheromone each possible path contains. Additionally, when a given ant chooses a path to the food, she adds pheromone on that path, thus increasing the probability for the ants behind her to choose the same path. This system makes the food search to be initially random. Nevertheless, since the ants that choose the shortest path will add pheromone more often, the probability of choosing the shortest path increases with time (positive feedback). On the other hand, pheromone evaporates at a given rate. Therefore, the associated pheromone for each path decreases if that path is not used during certain period of time (negative feedback). Evaporation rate determines the ability of the system for escaping from local minima.

ACS search method has been implemented in different ways as it has been used for solving several discrete optimization problems [32]. Concretely, we propose a new implementation of the ACS search method, to be used for solving the partitioning problem in DVE systems. This implementation was originally proposed in Reference 31. As with the rest of evolutive approaches, this implementation follows two steps. First, the DBA initial partition (described in Section 30.3.1), obtains good assignments only for several avatars. At that point, the ACS algorithm performs successive refinements of the initial partition that lead to a near optimal partition.

The first step in the ACS method is to select the subset of border avatars from the set of all the avatars in the system. A given avatar is selected as a border avatar if it is assigned to a certain server S in the initial partition and any of the avatars in its AOI is assigned to a server different from S. For each of the border avatars, a list of candidate servers is constructed, and a level of pheromone is associated to each element of the list. This list contains all the different servers that the avatars in the same AOI are assigned to (including the server that the avatar is currently assigned). Initially, all the elements in the list of candidate servers have associated the same pheromone level.

ACS method consists of a population of *ants*. Each ant consists of performing a search through the solution space, providing a given assignment of the B border avatars to servers. The number of ants N is a key parameter of the ACS method that should be tuned for a good performance of the algorithm. Each iteration of the ACS method consists of computing N different ants (assignments of the B border avatars). When each ant is completed, if the resulting assignment of the B border avatars produces a lower value of the quality function C_p, then this assignment is considered as a partial solution, and a certain amount of pheromone is added to the servers that the border avatars are assigned to in this assignment (just the same way that each ant adds pheromone to the search path that she follows). Otherwise, the ant (assignment) is discarded. When each iteration finishes (the N ants have been computed), the pheromone level is then equally decreased in all the candidate servers of all of the border avatars, according to the evaporation rate (the pheromone evaporates at a given rate). The ACS method ends when all the iterations have been performed.

In the process described earlier, each ant must assign each border avatar to one of the candidate servers for that avatar. Thus, a *selection value* is computed for each of the candidate servers. The selection value S_v is defined as

$$S_v = \alpha \times \text{pheromone} + \beta \times C_p, \tag{30.3}$$

where pheromone is the current pheromone level associated to that server, C_p is the resulting value of the quality function when the border avatar is assigned to that server instead of the current server, and α and β are weighing coefficients that must be also tuned. The server with the highest selection value will be chosen by that ant for that border avatar.

On other hand, when a partial solution is found then the pheromone level must be increased in those servers where the border avatars are assigned to in that solution. The pheromone level is increased using the following formula:

$$\text{pheromone} = \text{pheromone} + Q \times \frac{1}{C_p}. \tag{30.4}$$

Following this description, the proposed implementation of the ACS search method consists of the following steps:

```
program ACS (Int Ants, Int iterations, Real evap_rate)

const
   alpha  = 1.0
   beta   = 7.0
   Q      = 1000

var
   temp_sol    :Real[Number_of_Avatars]
   L           :Integer[]
   B,Cp_ACS,temp_cost :Real

begin
   Initial_Partition (DBA)
   B := ObtainBorderAvatars()
   Cp_ACS := Compute_Cp()
   For i:=0 to iterations do
       For j:=0 to N do
           For k:=0 to B do
               L:=ChooseServer(alpha,beta,Q)
           end_for
           temp_sol := Compose_Solution(B)
           temp_cost:= Obtain_Cp(temp_sol)
           if (temp_cost < Cp_ACS)
               Cp_ACS := temp_cost
               IncreasePheromone (B,Q)
           endif
       end_for
       DecreasePheromone(evap_rate)
   end_for
end
```

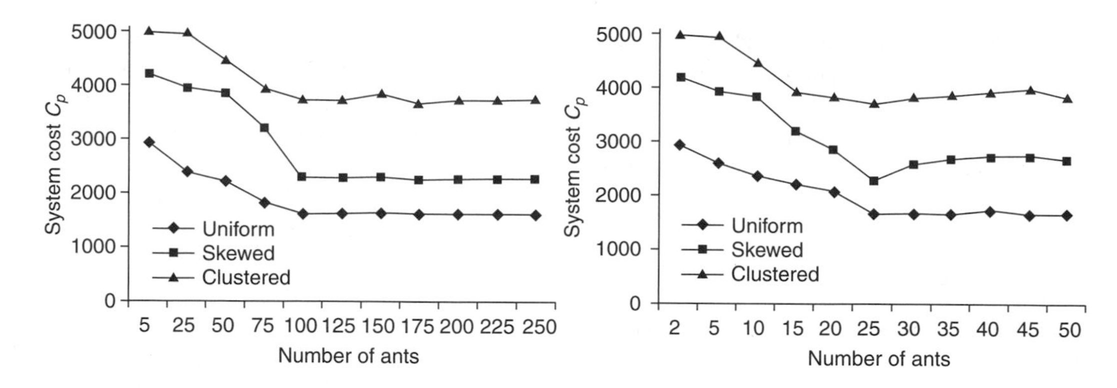

FIGURE 30.9 Values of the quality function C_p for different number of iterations.

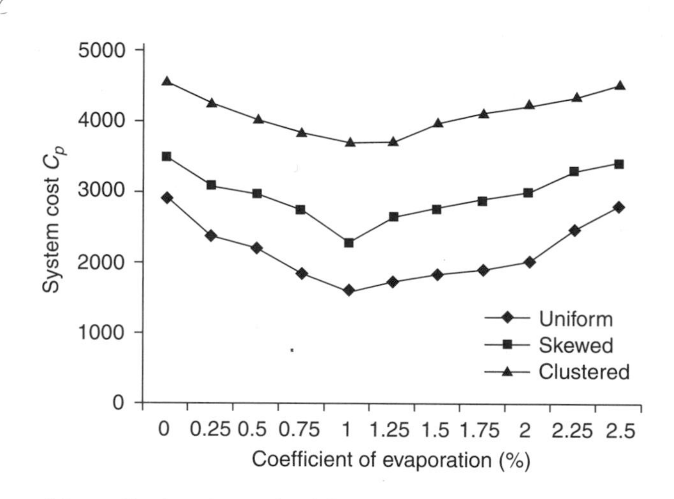

FIGURE 30.10 Values of the quality function C_p for different evaporation rates.

Like in GA approach, there are some parameters in ACS search method that must be properly tuned. In particular, the values for the number of ants N, the pheromone evaporation rate and the number of iterations that ACS method must perform should be tuned.

Figure 30.9 shows the values of the quality function C_p (denoted as *system cost*) reached by the ACS method when different number of ants and iterations are considered. It shows that C_p decreases when the number of iterations increases, until value of 25 iterations is reached. The same behavior manifests when the number of ants is grown. In this case is 100 the top value. From that point, system cost C_p slightly increases or remain constant, depending on the considered distribution of avatars. The reason for this behavior is that the existing pheromone level keeps the search method from finding better search paths even when more iterations are performed. Thus, the number of iterations and ants selected for ACS method has been 25 and 100, respectively.

Finally, Figure 30.10 shows the values of C_p reached by the ACS method when different pheromone evaporation rates are considered. This figure shows on the x-axis the percentage decreasing in pheromone level that all candidate servers suffer after each iteration. It shows that for all the considered distributions C_p decreases when the evaporation rate increases, until a value of 1% is reached. The reason for this behavior is that for evaporation rates lower than 1% the pheromone level keeps the search method from escaping from local minima, thus decreasing performance. From that point, system cost C_p also increases, since pheromone evaporation is too high and the search method cannot properly explore good search paths. Thus a coefficient of 1% has been selected as the optimal value of evaporation rate.

Additionally, we have performed empirical studies in order to obtain the best values for α, β, and Q coefficients. Although the results are not shown here for the sake of shortness, we have obtained the best behavior of the ACS method for $\alpha = 1.0$, $\beta = 7.0$, and $Q = 1000$. These are the values that the pseudo-code shown earlier uses for ACS algorithm.

30.3.5 Simulated Annealing (SA)

SA was initially proposed for solving complex optimization problems [17]. Although some authors do not consider SA as an evolutive technique, it can be modeled as a simplified version of GA approach. SA has shown to be effective in the solution of a large set of applications such as staff scheduling problems, timetabling problems and locomotive allocation problem in railway networks [33].

This heuristic search method is based on a thermodynamic theory establishing that the minimum energy state in a system can be found if the temperature decreasing is performed slowly enough. SA is a heuristic search method that always accepts better solutions than the current solutions, and also accepts worse solutions according to a probability system based on the system temperature. SA starts with a high temperature (a high probability of accepting a worsening movement), and in each iteration system the temperature is decreased. Thus, SA can leave local minima by accepting worsening movements at intermediate stages. The search method ends when system temperature is so low that worsening movements are practically impossible. Since the method cannot leave local minima, it cannot find better solutions neither (the algorithm ends when certain amount of iterations N are performed without finding better solutions).

The proposed partitioning method based on SA was previously presented in Reference 25. As with the rest of evolutive procedures, this approach starts from an initial population. This initial population has been obtained with DBA algorithm. In SA method each iteration consists of randomly selecting two different border avatars assigned to different servers. Then, the servers where this two border avatars are assigned to are exchanged. If the resulting value of the quality function C_p is higher than the previous one plus a threshold T, that change is rejected. Otherwise, it is accepted (the search method must decrease the value of the quality function C_p associated with each assignment). The threshold value T used in each iteration i of the search depends on the rate of temperature decreasing R, and it is defined in this implementation as

$$T = R - \left(\frac{R \times i}{N}\right), \tag{30.5}$$

where N determines the finishing condition of the search. When N iterations are performed without finding a partition that decreases the value of quality function C_p, then the search finishes.

The next code shows the described implementation based on SA:

```
program SA (Int iterations, Real dec_t_rate)

var
   B,temp_cost,Cp_SA :Real
   delta_sup         :Real
   av_i,av_j,        :Integer

begin
   Initial_Partition (DBA)
   B := ObtainBorderAvatars()
   Cp_SA := Compute_Cp()
   For i:=0 to iterations do
      Choose2DifRandomAvatars(B,av_i,av_j)
      ExchangeServerAssignment(av_i,av_j)
      temp_cost := Compute_Cp()
```

```
        delta_sup := dec_t_rate - (dec_t_rate*i/interations)
        if (temp_cost - delta_sup < Cp_SA)
           Cp_SA := temp_cost
        else
           ExchangeServerAssignment(av_i,av_j)
     end_for
  end
```

In order to improve the performance of the SA method, we measured the impact of exchanging groups of avatars, instead of exchanging one avatar in each iteration. Table 30.1 compares the results (in terms of both the value for the quality function C_p and also in terms of execution times) of exchanging groups of two, three, and four avatars. We tested each option under three different distributions of avatars in the virtual world (uniform, skewed, and clustered). These distributions are detailed in Section 30.4. On the other hand, these results have been obtained for a LARGE world composed by 2500 avatars and 8 servers. This table shows that the best option for SA method is to exchange the lowest number of avatars as possible in each permutation. Therefore, we exchanged a single avatar in all the simulations performed in our study.

On other hand, the two key issues for properly tuning this heuristic search method are the number of iterations N and the temperature decreasing rate R [33]. Figure 30.11 shows the performance (in terms of C_p values) obtained with SA algorithm for a LARGE world when the number of iterations increases.

From Figure 30.11 we can conclude that performing more iterations results in providing better values of C_p. However, the slope of this plot decreases from a certain number of iterations. We have considered the value of 3000 iterations at that point, and we have tested GA method with this number of iterations.

System temperature shows a different behavior in terms of C_p. Figure 30.12 shows the values of C_p obtained when the temperature decreasing rate is modified in a LARGE world. It clearly shows that the

TABLE 30.1 Different Types of Exchanges Performed on SA Approach

	Two avatars		Three avatars		Four avatars	
	C_p	Time (sec)	C_p	Time (sec)	C_p	Time (sec)
Uniform	1707.62	6.35	1808.38	6.51	1817.90	7.02
Skewed	2628.46	13.79	2826.44	14.32	2992.30	15.82
Clustered	4697.61	29.62	5046.27	30.25	5283.705	33.98

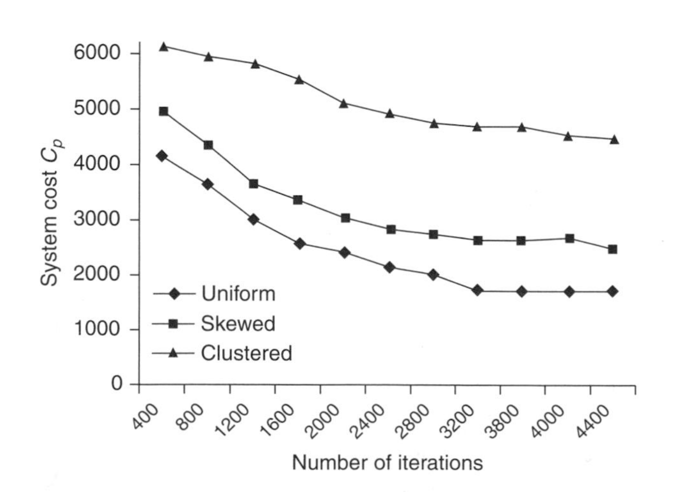

FIGURE 30.11 Values of the quality function C_p for different number of iterations.

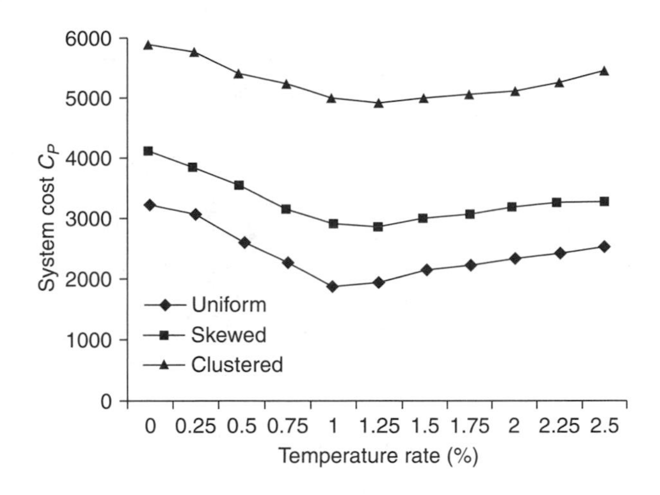

FIGURE 30.12 Values of the quality function C_p for different temperature decreasing rate.

quality of the obtained solutions do not follow a lineal progression. Effectively, since the temperature decreasing rate allows SA approach to escape from local minima, a threshold value appears when this parameter is modified. As this rate comes closer to 1.15, algorithm abandons local minima much more fast, and therefore the quality of the obtained solution increases. Beyond this value, the risk of accepting inefficient exchange of avatars is much too high and thus the algorithm is unable to find the right path.

30.3.6 Greedy Randomized Adaptive Search (GRASP)

Greedy Randomized Adaptive Search (GRASP) is a constructive technique designed as a multi-start heuristic for combinatorial problems [34]. Although GRASP is not an evolutive technique, it has been elected as a metaheuristic method to be compared with LOT and the set of evolutive approaches. Several problems such as circuit partitioning, set covering, or graph planarization have been successfully addressed by GRASP algorithms [35]. GRASP implementations are very robust and it is unusual to find examples where this method performs badly.

GRASP implementation starts with a initial partition of the DVE. Unlike the evolutive techniques described above, the initial partition used by GRASP does not provide any assignment for the border avatars. In this case, GRASP exploits an important property of the DBA method presented in Section 30.3.1. In DBA algorithm the assignment of avatars to servers is calculated iteratively. The avatar with the lowest distance to a given server of the DVE (following a round-robin scheme) is assigned in each iteration. In order to measure these distances, servers are represented by mass-centers (see DBA algorithm). Therefore, as avatars are assigned they are clustered around DVE servers.

Because DBA obtains a perfect balancing of avatars (C_p^W), system cost (C_p) depends only (see Equation 30.1) on the evaluation of inter-server communications (C_p^L). Thus, the best partitioning solutions are obtained when neighboring avatars are assigned to the same server. This property makes avatars that are located far away or equidistant of different mass-centers to be critical. These avatars are denoted as *critical avatars.* In order to properly assign critical avatars it is necessary to spend more resources. In order to avoid such spending, a different version of DBA algorithm is used. This version, denoted as DBA-R, does not assign critical avatars. These avatars are the inputs to GRASP algorithm.

Figure 30.13 shows an example of a DBA execution. In this example 66 avatars are simulated with three servers in a DVE system. Each server received a balanced group of 18 elements each located in a round pattern. The remaining 12 avatars have been intentionally nonassigned.

The GRASP-based method for solving the partitioning problem in DVE systems consists of several iterations [36]. Each iteration assigns a critical avatar and consists of two steps: construction and local search. The construction phase builds a feasible solution choosing one border avatar by iteration, and the

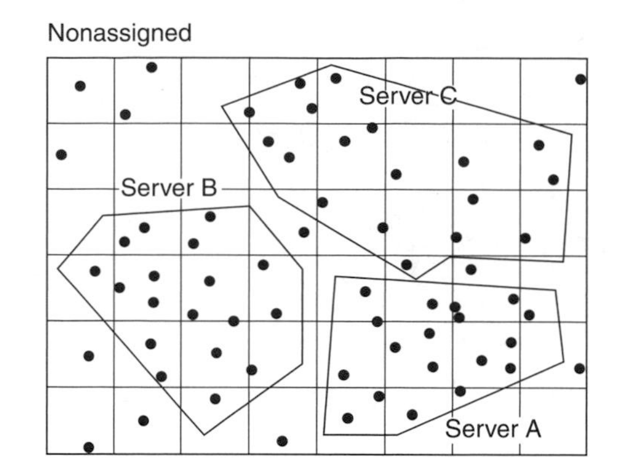

FIGURE 30.13 Result from a initial solution obtained by DBA-R algorithm.

local search also provides a server assignment of that border avatar in the same iteration, following the next procedure: first, the resulting cost C_p of adding each nonassigned critical avatar to the current (initial) partition is computed. Since each border avatar can be assigned to different servers, the cost for assigning each border avatar to each server is computed, forming the *List of Candidates (LCs)* (each element in this list has the form (*nonassigned border avatar, server, resulting cost*). This list is sorted (using Quick-sort algorithm) by the resulting cost C_p in descendent order, and then is reduced to its top quartile. One element of this reduced list of candidates (RLCs) is then randomly chosen (construction phase). Next, an extensive search is performed in the AOI of that selected avatar. That is, all the possible assignments of the avatars in the AOI of the selected avatars are computed, and the assignment with the lowest C_p is kept.

Next code describes how the GRASP approach works when n avatars are assigned to S servers in a DVE system:

```
program GRASP (Int threshold)

type New_sol
     idx_av,idx_ser: Int
     new_cost       : Real
     avatar,server : Int

var
   tmp_cost,Cp_GRASP :Real
   Non_assig          :Integer
   list               :New_sol[]

begin
   Initial_Partition (DBA-R,threshold)
   non_assig := n - threshold
   For i:=0 to non_assig do
      for j:=0 to n do
        for k:=0 to S do
             tmp_cost=TestSolution(j,k)
             AddToList(list,tmp_cost,j,k)
        end_for
      end_for
      QuickSort(list)
      ReduceToFirsQuartile(list)
      ChooseRandomElement(list,avatar,server)
```

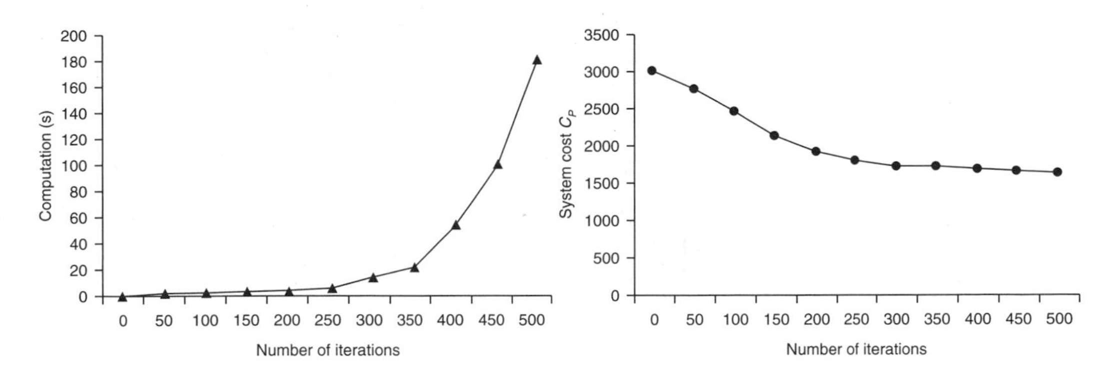

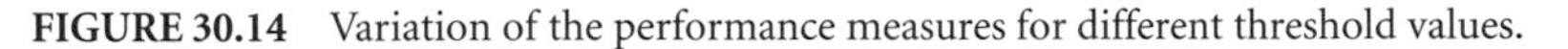

FIGURE 30.14 Variation of the performance measures for different threshold values.

```
        Cp_GRASP = tmp_cost = 10000
        for j:=0 To AvatarsInAOI(avatar) do
            for k:=0 to S do
              tmp_cost=TestSolution(j,k)
              AddToList(list,tmp_cost,j,k)
              if (tmp_cost < Cp_GRASP)
                 Cp_GRASP := tmp_cost
                 savej := j
                 savek := k
              end_for
            end_for
        end_for
        AssignSolutionServer(savej,savek)
    end_for
  end
```

The quality of the solutions provided by GRASP search method depends on the quality of the elements in the RLC, and the range of solutions depends on the length of the RLC. Thus, the main parameter to be tuned in this case is the number of *nonassigned N* or *critical* avatars that the initial partition must leave.

Figure 30.14 shows the results in this tuning phase in order to compose an intermediate solution. In this example a LARGE world composed of 2500 avatars are assigned to eight servers. The avatars are located following a uniform distribution. This figure represents the variations of two performance measures as the number of critical avatars (iterations) is increased. The quality of the obtained solutions C_p and the execution time for GRASP algorithm have been elected as performance measures.

Figure 30.14 shows that as the number of critical avatars increases the quality of the provided solutions also increases (C_p values decreases), but the execution time for GRASP algorithm (labeled as computations) also increases. We chose in this case a compromise solution of 250 iterations. It is worth mentioning that a larger number of iterations result in higher execution times and do not reach significant solutions.

30.4 Performance Evaluation

In this section, we present the performance evaluation of the heuristics described in the previous section when they are used for solving the partitioning problem in DVE systems. Following the evaluation methodology shown in Reference 14, we have empirically tested these heuristics in two examples of a DVE system: a SMALL world, composed of 13 avatars and 3 servers, and a LARGE world, composed of 2500 avatars and 8 servers. We have considered two parameters: the value of the quality function C_p for the partition provided by the search method and also the computational cost, in terms of execution time,

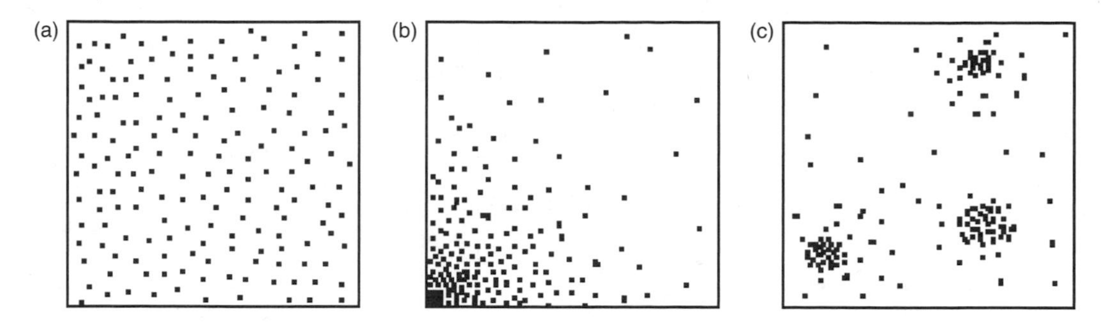

FIGURE 30.15 Distributions of avatars: (a) uniform, (b) skewed, and (c) clustered.

TABLE 30.2 Results for a SMALL DVE System

	Uniform distribution		Skewed distribution		Clustered distribution	
	Time (sec)	C_p	Time (sec)	C_p	Time (sec)	C_p
Exhaustive	3.4110	6.54	3.8430	7.04	4.7830	7.91
LOT	0.0009	6.56	0.0010	8.41	0.0011	8.89
SA	0.0040	6.82	0.0050	7.46	0.0050	7.91
ACS	0.0007	6.59	0.0030	7.61	0.0024	8.76
GA-basic	0.0020	6.54	0.0030	7.04	0.0050	7.91
GA-improved	0.0030	6.54	0.0030	7.04	0.0060	7.91
GRASP	0.0002	7.42	0.0002	8.63	0.0003	11.88

required by the search method in order to provide that partition. For comparison purposes, we have also implemented the LOT [14]. This method currently provides the best results for the partitioning problem in DVE systems. In the case of SMALL worlds we have also performed an exhaustive search through the solution space, obtaining the best partition possible. The hardware platform used for the evaluation has been a 1.7 GHz Pentium IV with 256 Mbytes of RAM.

Since the performance of the heuristic search methods may heavily depend on the location of avatars in the virtual world, we have considered three different distributions of avatars: uniform, skewed, and clustered distribution. Figure 30.15 shows an example of how avatars would be located in a 2D world when following each one of these distributions.

Table 30.2 shows the C_p values corresponding to the final partitions provided by each heuristic search method for a SMALL virtual world, as well as the execution times required for each method in order to obtain that final partition. For this proposed approach based on genetic algorithms, GA-B represents the basic approximation described in Section 30.3.2 and GA-I incorporates both the PA method and the proposed crossover operator detailed in Section 30.3.3. It can be seen that all of the heuristics provide better (lower) C_p values than the LOT search method for a uniform distribution of avatars. For the skewed and clustered distributions, most of the heuristics also provides better C_p values than the LOT search method, and some of them (GA and SA methods) even provide the minimum value. However, the execution times required by most of the heuristics are longer than the ones required by the LOT method. Only GRASP method provides worse C_p values than the LOT method, but it requires much shorter execution times. Although these results do not clearly show which heuristic provides the best performance, they validate any of the proposed heuristics as an alternative to the LOT search method.

However, in order to design a scalable DVE system, the partitioning method must provide good performance when the number of avatars in the system increases. That is, it must provide a good performance specially for LARGE virtual worlds. Table 30.3 shows the required execution times and the C_p values obtained by each heuristic search method for a LARGE virtual world.

When uniform distributions of avatars are considered, GA-I not only obtains the best partitioning solutions (in terms of C_p) but also achieved them by spending the minimum execution time. The rest

TABLE 30.3 Results for a LARGE DVE System

	Uniform distribution		Skewed distribution		Clustered distribution	
	Time (sec)	C_p	Time (sec)	C_p	Time (sec)	C_p
LOT	30.939	1637.04	32.176	3460.52	43.314	5903.80
SA	6.350	1707.62	13.789	2628.46	29.620	4697.61
ACS	5.484	1674.08	14.050	2286.16	23.213	3736.69
GA-B	6.598	1832.21	14.593	2825.64	29.198	4905.93
GA-I	6.410	321.31	14.590	450.80	28.740	791.94
GRASP	6.622	1879.76	13.535	2883.84	26.704	5306.24

of heuristics provides similar values of C_p than the one provided by LOT method, while requiring much shorter execution times. When nonuniform distributions of avatars are considered, then all the heuristics provide much better C_p values than the LOT method and they also require much shorter execution times than the LOT method. In particular, GA-I method provides the best C_p values for nonuniform distributions, requiring also the shortest execution time in the case of a skewed and clustered distribution of avatars.

These results show that the performance of the partitioning algorithm can be significantly improved by simply using any of the proposed heuristics instead of the LOT method, thus increasing the scalability of DVE systems. In particular, GA-I method provides the best performance as a partitioning algorithm for LARGE worlds.

30.5 Conclusions

In this chapter, we have proposed a comparison study of modern heuristics for solving the partitioning problem in DVE systems. This problem is the key issue that allows to design scalable and efficient DVE systems. We have evaluated the implementation of different metaheuristics, ranging over most of the current taxonomy of modern heuristics. We have tested the proposed heuristics when applied for both SMALL and LARGE DVE systems, with different distributions of the existing avatars in the system. We have compared these results with the ones provided by the LOT, the partitioning method that currently provides the best solutions for DVE systems. For SMALL virtual worlds, we can conclude that in general terms any of the implemented heuristics provides a partition with similar values of the quality function C_p, but the execution times required by the implemented heuristics are longer than the time required by the LOT search method. Although SA and GA methods provide the minimum value of the quality function, only GRASP method provides execution times shorter than the ones required by the LOT method for all the tested distributions of avatars. These results validate any of the proposed heuristics as an alternative to the LOT search method when considering SMALL DVE systems. However, for LARGE virtual worlds any of the proposed heuristics provides better C_p values and requires shorter execution times than the LOT method for nonuniform distributions of avatars. In particular, GA-I method provides the best results. Since a scalable DVE system must be able to manage large amounts of avatars, we can conclude that these results validates GA-I search method as the best heuristic method for solving the partitioning problem in DVE systems.

References

[1] S. Singhal and M. Zyda. *Networked Virtual Environments.* ACM Press, 1999.

[2] J.M. Salles, R. Galli, A.C. Almeida, C.A.C. Belo, and J.M. Rebordão. mworld: A multiuser 3d virtual environment. *IEEE Computer Graphics,* 17(2): 55–65, 1997.

[3] D.C. Miller and J.A. Thorpe. Simnet: The advent of simulator networking. *Proceedings of the IEEE,* 83: 1114–1123, 1995.

[4] C. Bouras, D. Fotakis, and A. Philopoulos. A distributed virtual learning centre in cyberspace. In *Proceedings of the Fourth International Conference on Virtual Systems and Multimedia (VSMM'98)*, November 1998.

[5] M. Abrash. Quake's game engine. *Dr. Dobb's Journal*, 51–63, Spring 1997.

[6] M. Lewis and J. Jacboson. Game engines in scientific research. *Communications of the ACM*, 45: 17–31, 2002.

[7] Kali networked game support software. http://www.kali.net.

[8] M. Macedonia. A taxonomy for networked virtual environments. *IEEE Multimedia*, 4(1): 48–56, 1997.

[9] P.A. Berstein, V. Hadzilacos, and N. Goodman. *Concurrency, Control and Recovery in Database Systems*. Addison-Wesley, Reading, MA, 1997.

[10] J. Falby, M. Zyda, D. Pratt, and R. Mackey. Npsnet: Hierarchical data structures for real-time three-dimensional visual simulation. *Computers & Graphics*, 17(1): 65–69, 1993.

[11] D. Lee, M. Lim, and S. Han. Atlas — A scalable network framework for distributed virtual environments. In *Proceedings of ACM Collaborative Virtual Environments (CVE 2002)*, September 2002, pp. 47–54.

[12] J.C.S. Lui, M.F. Chan, and K.Y. Oldfield. Dynamic Partitioning for a Distributed Virtual Environment. Department of Computer Science, The Chinese University of Hong Kong, 1998.

[13] P. Barham and T. Paul. Exploiting reality with multicast groups. *IEEE Computer Graphics and Applications*, 15: 38–45, 1995.

[14] J.C.S. Lui and M.F. Chan. An efficient partitioning algorithm for distributed virtual environment systems. *IEEE Transaction on Parallel and Distributed Systems*, 13: 193–211, 2002.

[15] P.T. Tam. Communication Cost Optimization and Analysis in Distributed Virtual Environment. Technical report RM1026-TR98-0412, Department of Computer Science and Engineering. The Chinese University of Hong Kong, 1998.

[16] R.L. Haupt and S.E. Haupt. *Practical Genetic Algorithms*. John Wiley & Sons, New York, 1997.

[17] S. Kirkpatrick, C.D. Gelatt, and M.P. Vecchi. Optimization by simulated annealing. *Science*, 220: 671–679, 1983.

[18] M. Dorigo, V. Maniezzo, and A. Coloni. The ant system: Optimization by a colony of operation agent. *IEEE Transactions on Systems, Man and Cybernetics*, 96: 1–13, 1996.

[19] H. Delmaire, J.A. Díaz, E.M. Fernández, and M. Ortega. Comparing New Heuristics for the Pure Integer Capacitated Plant Location Problem. Technical report DR97/10, Department of Statistics and Operations Research, Universitat Politecnica de Catalunya (Spain), 1997.

[20] D.B. Anderson, J.W. Barrus, and J.H. Howard. Building multi-user interactive multimedia environments at MERL. *IEEE Multimedia*, 2(4): 77–82, 1995.

[21] F.C. Greenhalgh. Analysing movement and world transitions in virtual reality tele-conferencing. In *Proceedings of Fifth European Conference on Computer Supported Cooperative Work (ECSCW'97)*, 1997, pp. 313–328.

[22] J.C.S. Lui and W.K. Lam. General methodology in analysing the performance of parallel/distributed simulation under general computational graphs. In *Proceedings of Third International Conference on the numerical Solution of Markov Chain*, September 1996.

[23] C. Coello, G. Lamont, and D. Van Veldhuizen. *Evolutionary Algorithms for Solving Multi-Objective Problems*. Kluwer Academic Publishers, 2002.

[24] R. Duda, P. Hart, and D. Stork. *Pattern Classification*. Wiley Interscience, 2000.

[25] P. Morillo, M. Fernández, and N. Pelechano. A grid representation for distributed virtual environments, acrossgrid'2003. In *Proceedings of the First European Across Grids Conference*, February 2003.

[26] K.E. Kinnear. Alternatives in automatic function definition: A comparison of performance. In K.E. Kinnear, Ed., *Advances in Genetic Programming*. MIT Press, Cambridge, MA, 1994, pp. 119–141.

[27] Z. Michalewicz. *Genetic Algorithms + Data Structures = Evolution Programs,* 2nd ed. Springer-Verlag, Heidelberg, 1992.
[28] J.H. Holland and D.E. Goldberg. Genetic algorithms and machine learning: Introduction to the special issue on genetic algorithms. *Machine Learning,* 3, 1988.
[29] P. Morillo, M. Fernández, and J.M. Orduña. A comparison study of modern heuristics for solving the partitioning problem in distributed virtual environment systems. In *International Conference in Computational Science and Its Applications (ICCSA'2003) Vol. 2669 of Lecture Notes in Computer Science,* Springer-Verlag, Heidelberg, 2003, pp. 458–467.
[30] R. Sedgewick. *Algorithms in C,* 3rd ed. Addison-Wesley, Reading, MA, 1998.
[31] P. Morillo, M. Fernández, and J.M. Orduña. An ACS-based partitioning method for distributed virtual environment systems. In *Proceedings of 2003 IEEE International Parallel and Distributed Processing Symposium, NIDISC-IPDPS'2003,* April 2003, p. 148.
[32] M. Dorigo, G. Di Caro, and M. Sampels. *Ant Algorithms: Third International Workshop, Ants 2002.* Springer-Verlag, Heidelberg, 2002.
[33] C. Koulamas, R. Jaen, and S.R. Antony. A survey of simulated annealing applications to operations research problems. *International Journal of Management Science,* 22: 41–56, 1994.
[34] T.A. Feo and M.G.C. Resende. Greedy randomized adaptive search procedures. *Journal of Global Optimization,* 6: 109–133, 1995.
[35] M. Resende and C. Ribeiro. Handbook of Metaheuristics. Kluwer Academic Publishers.
[36] P. Morillo and M. Fernández. A grasp-based algorithm for solving DVE partitioning problem. In *Proceedings of 2003 IEEE International Parallel and Distributed Processing Symposium, IPDPS'2003,* April 2003, p. 60.

31

Population Learning Algorithm and Its Applications

Piotr Jedrzejowicz

31.1 Population-Based Algorithms

The techniques used to solve difficult combinatorial optimization problems have evolved from constructive algorithms to local search techniques, and finally to population-based algorithms. Population-based methods have become very popular. They provide good solutions since any constructive method can be used to generate the initial population, and any local search technique can be used to improve each solution in the population. But population-based methods have the additional advantage of being able to combine good solutions in order to get possibly better ones. The basic idea behind this way of doing is that good solutions often share parts with optimal solutions (Hertz and Kobler, 1998).

Population based methods are optimization techniques inspired by natural evolution processes. They handle a population of individuals that evolves with the help of information exchange procedures. Each individual may also evolve independently. Periods of cooperation alternate with periods of self-adaptation.

Among best-known population-based method are *evolutionary algorithms*. These can be perceived as a generalization of *genetic algorithms*, which operate on fixed-length binary strings, which need not be the case for evolutionary algorithms (Michalewicz, 1994). Evolutionary algorithms are stochastic search methods that mimic the metaphor of natural biological evolution. Evolutionary algorithms operate on a population of potential solutions applying the principle of survival of the fittest to produce better and better approximations to a solution.

Evolutionary algorithms model natural processes, such as selection, recombination, mutation, migration, locality, and neighborhood. Evolutionary algorithms work on populations of individuals instead of single solutions. At the beginning of the computation, a number of individuals (the population) are randomly initialized. The objective function is then evaluated for these individuals. The first/initial generation is produced. If the optimization criteria are not met the creation of a new generation starts. Individuals are selected according to their fitness for the production of offspring. Parents are recombined to produce offspring. All offspring will be mutated with a certain probability. The fitness of the offspring is then computed. The offspring are inserted into the population replacing the parents, producing a new generation. This cycle is performed until the optimization criteria are reached.

Such a single population evolutionary algorithm is powerful and performs well on a broad class of problems. However, better results can be obtained by introducing many populations, called islands. Every island evolves for a number of generations isolated (like the single population evolutionary algorithm) before one or more individuals are exchanged between the islands. The *island-based evolutionary algorithm* models the evolution of a species in a way more similar to nature than the single population evolutionary algorithm (Gordon and Whitley, 1993).

Development and proliferation of evolutionary algorithms have been initiated by and advanced in seminal works of Holland (1975), Goldberg (1989), Davis (1991), and Michalewicz (1994). Despite the fact that evolutionary algorithms, in general, lack a strong theoretical background, the application results were more than encouraging. This success has led to emergence of numerous techniques, algorithms, and their respective clones that are now considered as belonging to the population-based class, and which differ in some respects from "classic" evolutionary algorithms.

Among well known population-based methods are *scatter search techniques* (Glover, 1977) maintaining a population of reference points and generating offspring by weighted linear combinations. Different type of population-based techniques are adaptive memory algorithms (Golden et al., 1997). Here a central memory is used to keep track of the best components of the solutions visited during the search. These components are combined to create new solutions, which are subject to repair to assure feasibility. Finally, the components of the new solutions are considered as candidates that may be selected for replacing old components in the central memory.

Another population based approach — *ant colony optimization* (ACO) studies artificial systems that take inspiration from the behavior of real ant colonies and are used to solve discrete optimization problems. Recently, the ant colony optimization metaheuristic was defined in Dorigo and Di Caro (1999). Ant colony algorithms are inspired by biological observations on the behavior of ant colonies. Simple agents called ants explore a region of the search space and share information gathered throughout their individual searches. In a basic ant system (Colorni et al., 1992), each ant is a constructive procedure generating solution to the considered problem. At each step of the construction, each ant has to decide how to make one more step towards the completion of a partial solution. Such choice is based on two factors. The trace factor guides the ants to choices that gave good results in earlier constructions. The intensity of this factor (an analogue of the pheromone trail) informs about the quality of the solutions generated by making the corresponding choice. The desirability factor guides the ants to choices that induce the best value of a partial solution. Once all ants have completed their construction, the trace factors are updated. A choice that led to good solutions increases the corresponding trace factor. The solutions constructed by ants are further submitted to an improvement algorithm.

Particle swarm optimization (PSO) is a population based stochastic optimization technique developed by Eberhart and Kennedy (1995), inspired by social behavior of bird flocking or fish schooling. PSO shares many similarities with evolutionary computation techniques such as genetic algorithms (GAs). The system is initialized with a population of random solutions and searches for optima by updating generations. However, unlike GA, PSO has no evolution operators such as crossover and mutation.

An idea that culture might be symbolically encoded and transmitted within and between populations as another inheritance mechanism have triggered the development of *cultural algorithms* (CAs). Reynolds (1994) developed a computational model in which cultural evolution is seen as an inheritance process that operates at two levels: the microevolutionary and the macroevolutionary. Genetic algorithms are used to model the microevolutionary process, and so-called Version Spaces are used to model the macroevolutionary process of a cultural algorithm. A cultural algorithm models the evolution of the culture component of an evolutionary computational system over time. The culture component provides an explicit mechanism for acquisition, storage, and integration of individual and group's problem solving experience, and behavior. In a cultural algorithm, there are two main spaces: the normal population of the evolutionary algorithm and the belief space, which is the place where the shared acquired knowledge is stored during the evolution of the population (Chung and Reynolds, 1998).

Moscato and Norman (1989) have introduced the term *memetic algorithm* (MA) to describe evolutionary algorithms in which local search plays a significant part. This term is motivated by Richard Dawkins's notion of a *meme* as a unit of information that reproduces itself as people exchange ideas (Dawkins, 1976). Moscato and Norman liken this thinking to local refinement, and therefore promote the term "memetic algorithm" to describe genetic algorithms that use local search heavily. While genetic algorithms have been inspired by a biological evolution, (MAs) would try to mimic cultural evolution. MA is a marriage between a population-based global search and the heuristic local search made by each of the individuals.

Given a representation of an optimization problem, a certain number of individuals are created. The state of these individuals can be randomly chosen or set according to a certain initialization procedure. A heuristic can be chosen to initialize the population. After that, each individual makes local search. After that, when the individual has reached a certain development, it interacts with the other members of the population. The interaction can be a competitive or a cooperative one. The cooperative behavior can be understood as the mechanisms of crossover in GA or other types of breeding that result in the creation of a new individual. More generally, cooperation is understood as an interchange of information. The local search and cooperation (mating, interchange of information) or competition (selection of better individuals) is repeated until a stopping criterion is satisfied. Usually, it should involve a measure of diversity within the population.

An important addition to the family of the population-based method is GRASP. A greedy randomized adaptive search procedure is a metaheuristic for combinatorial optimization. It is a multi-start or iterative process in which each iteration consists of two phases, a construction one, in which a feasible solution is produced and a local search one, in which a local optimum in the neighborhood of the constructed solution is sought (Feo and Rosende, 1995). In the construction phase a feasible solution is iteratively constructed, one element at a time. At each construction step, the choice of the next element to be added is determined by ordering all candidate elements in a candidate list C with respect to a greedy function measuring the benefit of selecting each element. The heuristic is adaptive because the benefits associated with every element are updated at each iteration of the construction phase to reflect the changes brought on by the selection of the previous element. The probabilistic component of a GRASP is characterized by randomly choosing one of the best candidates in the list, but not necessarily the top one. The list of best candidates is called the restricted candidate list (RCL).

The solutions generated by a GRASP construction are not guaranteed to be locally optimal. Hence, it is almost always beneficial to apply a local search in attempt to improve each constructed solution. It terminates when no better solution is found in the neighborhood. While such local optimization procedures can require exponential time from an arbitrary starting point, empirically their efficiency significantly improves as the initial solution improves. The result is that often many GRASP solutions are generated in the same amount of time required for the local optimization procedure to converge from a single random start. Furthermore, the best of these GRASP solutions is generally significantly better than the single solution obtained from a random starting point. GRASP can be easily implemented in parallel. Each processor can be initialized with its own copy of the procedure, the instance data, and an independent random number sequence. The GRASP iterations are then performed in parallel with only a single global variable required to store the best solutions found over all processors.

The above review, by no means exhaustive, allows to draw the following general conclusions:

- Population-based algorithms and techniques have become a standard tool to deal effectively with computationally complex problems.
- There are several ingredients common to all population-based approaches. Among them, the most characteristic include diversification by means of generating a population of individuals that are solutions or parts of solutions and introducing some random noises at various stages of searching for the solution. All population-based algorithms are equipped with some tools allowing to exploit information gathered during computation with a view to drop less promising direction of search. Finally, all population-based approaches mimic some natural biological or social processes.
- A common framework of properties characterizing population-based methods still does allow for design flexibility and development of new population-based algorithms, each having its own distinctive features and strengths, which may prove effective in solving target problem types.

31.2 Population Learning Algorithm

Population learning algorithm (PLA), was proposed in Jedrzejowicz (1999) as another population-based method, which can be applied to solve combinatorial optimization problems. PLA has been inspired by analogies to a social phenomenon rather than to evolutionary processes. Whereas *evolutionary algorithms* emulate basic features of natural evolution including natural selection, hereditary variations, the survival of the fittest, and production of far more offspring than are necessary to replace current generation, PLAs take advantage of features that are common to social education systems:

- A generation of individuals enters the system.
- Individuals learn through organized tuition, interaction, self-study, and self-improvement.
- Learning process is inherently parallel with different schools, curricula, teachers, etc.
- Learning process is divided into stages.
- More advanced and more demanding stages are entered by a diminishing number of individuals from the initial population (generation).
- At higher stages, more advanced education techniques are used.
- The final stage can be reached by only a fraction of the initial population.

In PLA, an individual represents a coded solution of the considered problem. Initially, a number of individuals, known as the initial population is randomly generated. Once the initial population has been generated, individuals enter the first learning stage. It involves applying some, possibly basic and elementary, improvement schemes. These can be based, for example, on some simple local search procedures. The improved individuals are then evaluated and better ones pass to a subsequent stage. A strategy of selecting better or more promising individuals must be defined and duly applied. In the following stages the whole cycle is repeated. Individuals are subject to improvement and learning, either individually or through information exchange, and the selected ones are again promoted to a higher stage with the remaining ones dropped-out from the process. At the final stage, the remaining individuals are reviewed and the best represents a solution to the problem at hand. Basic idea of the population learning algorithm is shown in Figure 31.1.

Figure 31.1. covers a simple, nonparallel version of the algorithm. For a parallel PLA the following features need to be defined:

- A set of rules controlling how individuals are grouped into concurrent populations at various stages.
- The respective rules for running learning and improvement algorithms in parallel at various stages.
- Rules for information exchange and coordination between concurrent processes.

Three example types of the PLA structure are shown in Figure 31.2. Scheme (a) is a simple sequential, nonparallel, implementation of the algorithm. Schemes (b) and (c) depict parallel implementations.

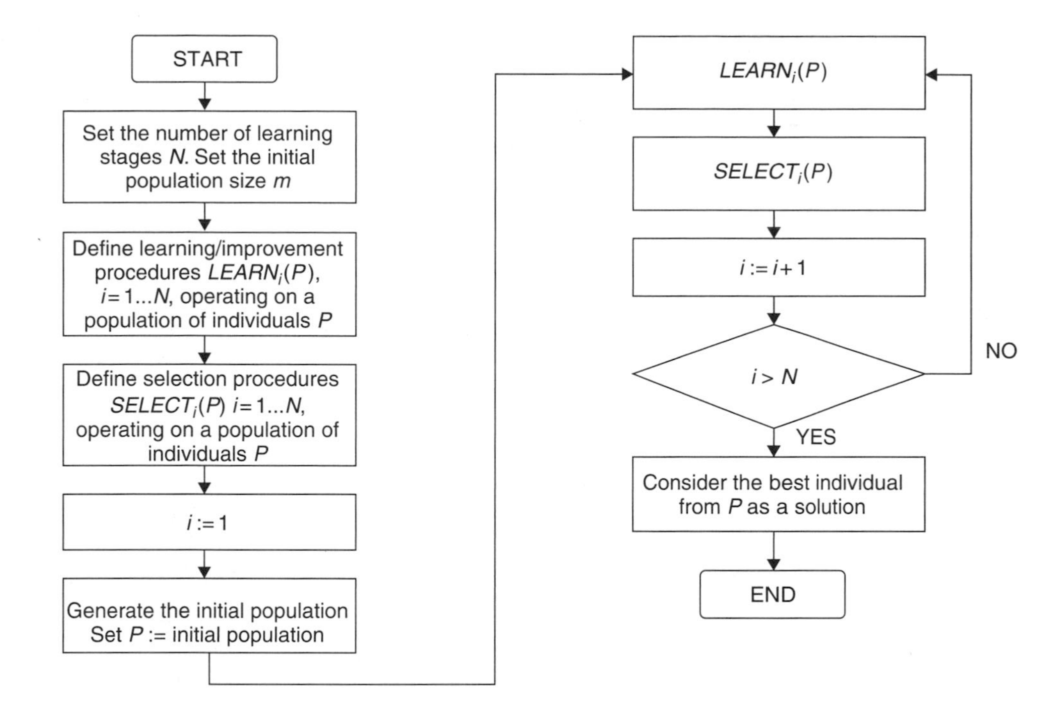

FIGURE 31.1 General idea of the population learning algorithm.

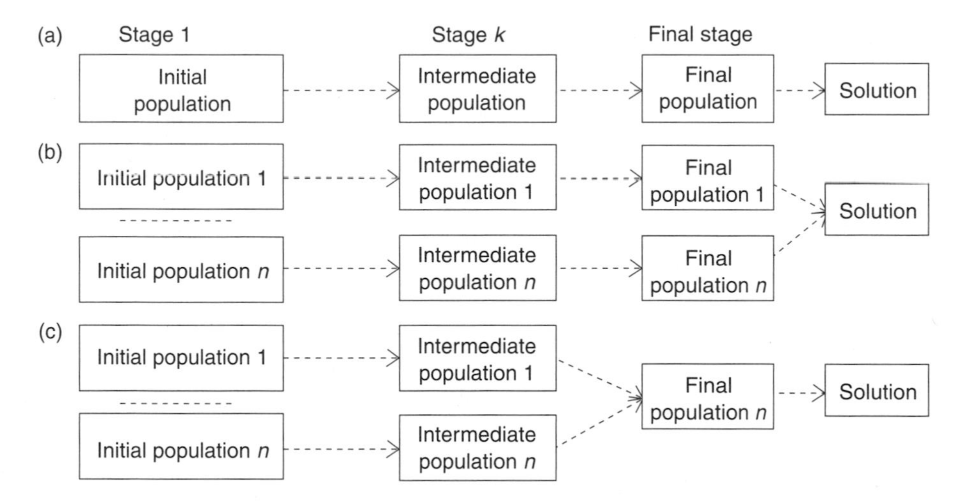

FIGURE 31.2 Three example types of the PLA structure.

Scheme (b) is a simple parallel implementation, assuming that the learning and improvement procedures used at various stages do not require any information exchange between the concurrent populations. Finally, Scheme (c) is a parallel implementation with information exchange. Such scheme (or its variants) should be used in case some information is exchanged between the concurrent populations during the learning and improvement process. Choice of the appropriate PLA scheme depends on both — the computational resources available and characteristics of learning and improvement procedures used.

TABLE 31.1 Population Learning Algorithm Design Elements

Design element	Comment
Definition of an individual	An individual is a coded feasible solution or part of a solution. The designer should aim at achieving ease of manipulation and storage, ease of transformation into a solution and ease of fitness evaluation
Procedure for generating the initial population of individuals	The designer should aim at achieving unbiased individuals, if possible representing all regions of the feasible solution space. A good construction heuristics can be used to produce a seed from which the initial population can be generated
Size of the initial population	Must be set at the PLA fine-tuning stage. Should be chosen as a compromise between the requirement of the sufficient and adequate representation assuring good quality of results and the available computational resources
Number of learning and improvement stages	Depends on availability of learning and improvement algorithms. Should be chosen with a view of finding a satisfactory compromise between quality of solutions and computation time
Number and size of parallel groups of individuals at each stage	Should be chosen considering the available computational resources, complexity of the problem at hand, and complexity of the learning and improvement algorithms used. Another compromise between quality of solutions and computation time is involved
Learning and improvement algorithms for each stage/each group	Basic and simple procedures at earlier stages should be replaced by more complex and sophisticated at later ones. The designer should try to assure adequate diversity of learning and improvement algorithms used
Fitness function of an individual	Should be simple (easily computable) and directly related to the quality of the solution represented by an individual
Selection strategy	Must be set at the PLA fine-tuning stage. Rules for rejection and promotion of individuals may differ at various stages. The designer should aim at achieving good efficiency of the algorithm not losing, too early, representation sufficiency and adequacy. Another compromise between quality of solutions and computation time is involved

Designing population learning algorithm intended for solving a particular problem type allows the designer a lot of freedom (as, in fact, happens in case of majority of other population-based algorithms). Moreover, an effective PLA would certainly require a lot of fine-tuning and experimenting. This could be considered as a disadvantage, at least as long as the process of setting different parameters of the population learning algorithm is rather based on heuristics instead of some theoretical or statistical rules, which, unfortunately, are not yet appropriately developed. Main PLA design elements are summarized in Table 31.1.

Although PLA shares many features with other population-based approaches, it clearly has its own distinctive characteristics. Brief comparison of several example algorithms belonging to the discussed class is shown in Table 31.2.

In the following sections several example implementations of the PLA applied to solving different computationally difficult problems are discussed. The PLA is seen here as a general framework for constructing hybrid solutions to difficult computational problems. From such a perspective the PLA role is to structure, organize, sequence, and eventually help to apply in a parallel environment variety of techniques. Strength of the PLA stems from combining in an "intelligent" manner the power of population-based algorithms using some random mechanism for diversity assurance, with efficiency of various local search algorithms. The later may include, for example, reactive search, tabu search, simulated annealing as well as the described earlier population based approaches.

Computational intelligence embedded into the population learning algorithm scheme is based on using the following heuristic rules:

- To solve difficult computational problems apply a cocktail of methods and techniques including random and local search techniques, greedy and construction algorithms, etc., building upon their strengths and masking weaknesses.

TABLE 31.2 Comparison of Example Population-based Algorithms

Algorithm	Processing object	Iterative search mode	Intensification	Diversification	Information exchange
PLA	Population of potential solutions	Two level iterative search (stages with a decreasing population size and iterative local search procedures)	A cocktail of local search and random search procedures	Random initial population, plus random local search at some stages	Different schemes used by different learning and improvement algorithms
EA	Population of potential solutions	Generation replacement	None	Random initial population, plus mutation operator	Crossover operator
ACO	Population of construction processes	Search at each construction step	Improvement algorithm	Random initial population	Trace factor mechanism
PSO	Population of potential solutions	Generation replacement	None	Random initial population	Through *pbest*, *lbest*, and *gbest* values
CA	Population of potential solutions	Generation replacement	None	Random initial population, plus mutation	Knowledge sharing through version spaces
MA	Population of potential solutions	Two level iterative search (generation replacement through competitive interactions plus local search)	Local search procedure	Random initial population, plus mutation operator	Cooperative interactions mechanism
GRASP	Population of construction processes	Two level iterative search (iterative local search procedure plus multiple construction processes)	Local search procedure	Multiple start construction processes	None

- To escape getting trapped into a local optimum generate or construct an initial population of solutions called individuals, which in the following stages will be improved, thus increasing chances for reaching a global optimum.
- Another means of avoiding getting trapped into local optima is to apply at various stages of search for a global optimum some random diversification algorithms.
- To increase effectiveness of searching for a global optimum divide the process into stages retaining after each stage only a part of the population consisting of "better" or "more promising" individuals.
- Another means of increasing effectiveness is to use at early stages of search, improvement algorithms with lower computational complexity as compared to those used at final stages.

To conclude this section it should be noted that the PLA is an addition to the family of the population-based techniques, which can be seen as a hybridization framework allowing for an effective integration of different deterministic and random local search techniques. Figure 31.3 summarizes main features of the population learning algorithm.

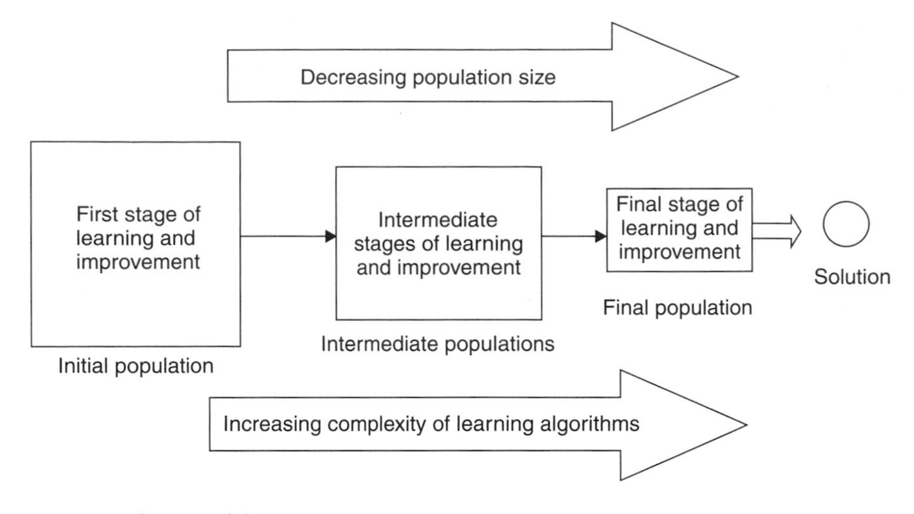

FIGURE 31.3 Main features of the PLA.

31.3 Example Applications of the PLA to Solving Combinatorial Optimization Problems

This section contains a review of several example PLA implementations applied to solving combinatorial optimization problems. In all cases the problem description is followed by a presentation of some main features of the respective implementation and a discussion of the computational experiments carried out.

31.3.1 Permutation Scheduling

Permutation scheduling problems are defined as scheduling problems where the solution is a permutation of tasks or jobs to be processed on one or more machines (processors). In the review, four permutation scheduling problems are considered — common due date, permutation flow shop, a single machine weighted tardiness, and multiprocessor task scheduling in multistage hybrid flowshops.

In a *common due date* problem there is a set of n non-preemptable tasks available at time zero. Tasks are to be processed on a single machine. Processing times of the tasks p_i, $i = 1, \ldots, n$ are known. There is also a common due date d. Earliness and tardiness of tasks are defined as $E_i = \max\{d - C_i, 0\}$ and $T_i = \max\{C_i - d, 0\}$, respectively ($i = 1, \ldots, n$) where C_i is the completion time of job i. Penalties per time unit of the task i being early or tardy are a_i and b_i, respectively. The objective is to minimize the sum of earliness and tardiness penalties:

$$\Sigma_i a_i E_i + \Sigma_i b_i T_i \rightarrow \min, \quad i = 1, \ldots, n.$$

It should be also noted that the due date is called unrestrictive if the optimal sequence of tasks can be constructed without considering the value of the due date. Otherwise the common due date is called restrictive. Obviously a common due date for which $d \geq \Sigma_i p_i$ holds is unrestrictive.

In a *permutation flow shop* there is a set of n jobs. Each of n jobs has to be processed on m machines $1, \ldots, m$ in this order. The processing time of job i on machine j is p_{ij}, where p_{ij} is fixed and nonnegative. At any time, each job can be processed on at most one machine, and each machine can process at most one job. The jobs are available at time 0 and the processing of a job may not be interrupted. In permutation flow shop problem (PFSP) the job order is the same on every machine. The objective is to find a job sequence minimizing schedule makespan (i.e., completion time of the last job).

TABLE 31.3 Neighborhood Structures used in the PLA Implementations

Notation	Move	Neighbourhood pace
$\mathcal{N}_1(x)$	Exchange of two consecutive tasks in x	All possible exchanges
$\mathcal{N}_2(x)$	Exchange of two nonconsecutive tasks in x	All possible exchanges
$\mathcal{N}_3(x)$	Finding the order of four consecutive tasks in x by enumeration	All quadruplets of consecutive tasks
$\mathcal{N}_4(x)$	A single step rotation of three random tasks in x	All possible triples
$\mathcal{N}_5(x)$	Moving a random task in x to another location and adjusting other tasks	All possible moves

In a *single machine weighted tardiness* problem there is a set of n jobs. Each of n jobs (numbered $1, \ldots, n$) is to be processed without interruption on a single machine that can handle no more than one job at a time. Job $j(j = 1, \ldots, n)$ becomes available for processing at time zero, requires an uninterrupted positive processing time p_j on the machine, has a positive weight w_j, and has a due date d_j by which it should ideally be finished. For a given processing order of the jobs, the earliest completion time C_j and the tardiness $T_j = \max\{C_j - d_j, 0\}$ of job $j(j = 1, \ldots, n)$ can readily be computed. The problem is to find a processing order of the jobs with minimum total weighted tardiness $\Sigma_j w_j T_j$.

Finally, a *multiprocessor task scheduling in multistage hybrid flow shops* is considered. The discussed problem involves scheduling of n jobs composed of tasks in a hybrid flow shop with m stages. All jobs have the same processing order through the machines, that is, a job is composed of an ordered list of multiprocessor tasks where the i-th task of each job is processed at the i-th flow shop stage (the number of tasks within a job corresponds exactly to the number of flow shop stages). Processing order of tasks flowing through stages is the same for all jobs. At each stage $i, i = 1, \ldots, m$, there are available m_i identical parallel processors. For the processing at stage i, task i being a part of the job $j, j = 1, \ldots, n$, requires $size_{i,j}$ processors simultaneously. That is, $size_{i,j}$ processors assigned to task i at stage i start processing the task simultaneously and continue doing so for a period of time equal to the processing time requirement of this task, denoted $p_{i,j}$. Each subset of available processors can process only the task assigned to them at a time. The processors do not break down. All jobs are ready at the beginning of the scheduling period. Preemption of tasks is not allowed. The objective is to minimize makespan, that is the completion of the lastly scheduled task in the last stage.

To solve permutation scheduling problems three versions of the population learning algorithm, denoted respectively as PLA1, PLA2, and PLA3 have been designed and implemented. PLA1 has been used to solve the restricted instances of the common due date scheduling problem, as well as instances of flow shop and total tardiness problems. PLA2 has been used to solve the unrestricted instances of the common due date problem. PLA3 has been designed to solve instances of permutation flow shop, total weighted tardiness problem, as well as instances of multiprocessor task scheduling in multistage hybrid flow shops. The proposed algorithms make use of different learning and improvement procedures, which, in turn, are based on five neighborhood structures shown in Table 31.3 In what follows x denotes an individual encoded as a set of natural numbers (a permutation of tasks) and $g(x)$ its fitness function.

All learning and improvement (l & p) procedures operate on the population of individuals P, and perform local search algorithms based on the above defined neighborhood structures. General structure of a l & p procedure is shown in the following pseudo-code:

```
Procedure LEARN(i,P):
begin
  for each individual x in P do
    Local_ search(i,x);
  end for
end
```

Local search algorithms used are shown in Table 31.4.

TABLE 31.4 Local Search Algorithms used in the PLA Implementations

i	Idea of the local search algorithm
1	Perform all moves from the neighborhood structure $\mathcal{N}_1(x)$; accept moves improving $g(x)$; stop when no further improvements of $g(x)$ are possible
2	Mutate x producing x' {mutation procedure is selected randomly from the two available ones — the two point random exchange or the rotation of all tasks between two random points}; perform all moves from the neighborhood structure $\mathcal{N}_1(x')$; accept moves improving $g(x')$; stop when no further improvements of $g(x')$ are possible
3	Repeat k' times {k' is a parameter set at the fine-tuning phase; in the reported experiment $k' = 3^*$ initial population size}; generate offspring y and y' by a single point crossover of x and a random individual x'; perform all moves from the neighborhood structure $\mathcal{N}_1(y)$ and $\mathcal{N}_1(y')$; accept moves improving $g(y)$ and $g(y')$; stop when no further improvements of $g(y)$ and $g(y')$ are possible; adjust P by replacing x and x' with the two best individuals from $\{x, x', y, y'\}$
4	Perform all moves from the neighborhood structure $\mathcal{N}_2(x)$; accept moves improving $g(x)$; stop when no further improvements of $g(x)$ are possible
5	Perfor Perform all moves from the neighborhood structure $\mathcal{N}_3(x)$; accept moves improving $g(x)$; stop when no further improvements of $g(x)$ are possible
6	Perform *SIMULATED ANNEALING* (x) based on $\mathcal{N}_4(x)$ neighborhood structure
7	Perform *TABU SEARCH* (x) based on $\mathcal{N}_5(x)$ neighborhood structure

Simulated annealing is a metaheuristic introduced in Kirkpatrick et al. (1983). Implementation of simulated annealing used within PLA as the sixth local search algorithm is based on random moves from the $\mathcal{N}_4(x)$ neighborhood structure.

Tabu search is yet another metaheuristic (see Glover, 1990). General idea of the present implementation using a short term memory (STM) and a long term memory (LTM), is shown in the following pseudo code (STM and LTM contain parameters of the moves and are managed according to the reverse elimination method):

```
Procedure TABU SEARCH (x):
begin
xbest := x;
Gbest := g(x);
STM, LTM := Ø;
  for j = 1 to it do {where it is the number of iterations}
    Do := - ∞;
    for i = 1 to lo do {where lo is the number of iterations}
      Perform current_move, which is a random move from N5(x) and not
      in STM, producing y from x;
      D := g(x) -- g(y);
      Update STM;
      if D > Do then Do = D; x' := y; best_move := current_move;
    end for
    if g(x') < Gbest then Gbest := g(x'); xbest := x'; x := x';
      else if best_move ∉ LTM then x := x';
    end if
    Update LTM;
    Update STM;
    end for
  x := xbest;
  end
```

In the PLA implementations a simple procedure *SELECT*(P, s) is used, where s is a percentage of best individuals from a current population promoted to higher stages. For any population P, let *LEARNING*(P) stand for the following:

```
Procedure LEARNING(P):
begin
  for i = 1 to  5 do
    LEARN(i,P);
    SELECT(P, s);
  end for
end
```

The last learning procedure *LEARN_REC*(x) is recursive and operates on overlapping partitions of the individual x. For example, for $x = (5, 2, 4, 6, 1, 3, 7)$ the individuals $s = (5, 2, 4)$, $t = (4, 6, 1)$, $u = (1, 3, 7)$ make an overlapping partition of x, and the merge of (s, t, u) is x.

```
Procedure LEARN_REC(x):
begin
  let (y1,..., ys) be an overlapping partition of x
  for i = 1 to s do
    generate randomly a ''small'' population Pi of individuals for yi;
    LEARNING(Pi);
    yi':=the best individual from Pi;
  end for
  output the merge of (y1',...,ys');
end
```

Now, the structure of the implemented population learning algorithms can be shown as:

```
Procedure PLA1:
begin
  generate randomly initial_population;
  P := initial_population;
  LEARNING (P);
  for each individual x in P do
    LEARN_REC (x);
  end for
  output the best individual from P;
end
```

For the unrestrictive common due date scheduling not only the sequence of tasks but also the starting time of the first task need to be found. To achieve this PLA2 is based on the following assumptions:

- In the course of computation it is assumed that the deadline equals 0 and each task can be scheduled prior to and after time 0.
- Each individual $x = (x_0, x_1, \ldots, x_n)$ in the population consists of a sequence of n tasks, $1 \leq x_j \leq n$ for $j = 1, \ldots, n$ and $x_0 = i \leq n$, which is the number of the last task completed before deadline ($x_1, \ldots, x_i$ are completed before time 0 and $x_{i+1}, \ldots, x_n$ after time 0).
- starting time of the schedule is $(-\sum_{j=1}^{x_0} p_{x_j})$. In what follows x_0 is called the boundary task.
- In the final step of the computation the real starting time of the first task is computed from: $t_0 = d - \sum_{j=1}^{x_0} p_{x_j}$.

PLA2 uses slightly modified learning and improvement procedures as compared with PLA1. For instance, *LEARN* (3, *P*) produces an offspring using the following rule: If two individuals x and x' produce y, then the boundary task of y equals that of x if the crossover point is greater than the boundary task of x, or the value: (the crossover point + boundary task of x–1) otherwise. Same applies to y'.

Further modifications include two new learning and improvement procedures using the following local search procedures:

```
Procedure LEARN (8, P):
begin
  for each individual x in P do
    Local_ search (8); {exchanges two tasks i, j, not necessarily

    consecutive, where i (j) is less (greater) than the boundary task}
  end for
end

Procedure LEARN (9, P):
begin
  for each individual x in P do
    Local_ search (9); {if the boundary task is less than the number
    of tasks then it is incremented by one}
  end for
end
```

Now PLA2 is defined as follows:

```
Procedure PLA2:
begin
  generate randomly initial_population;
  P := initial_population;
  for i = 1,2,3,4,5,8  do
    LEARN (i, P);
    SELECT (P, s);
  end for
    LEARN (9, P);
    output the best individual from P;
end
```

Finally, PLA3 has the following structure:

```
Procedure PLA3:
begin
  generate randomly initial_population;
  P := initial_population;
  for i = 1, 7  do
    LEARN (i, P);
    SELECT (P, s);
  end for
    LEARN (6, P);
    output the best individual from P;
end
```

TABLE 31.5 Benchmark Data Sets

Problem	Data set description	Source
Common due date	240 problem instances with 20, 50, 100, 200, 500, and 1000 tasks (40 instances for each problem size). Due dates for instances in each problem size group are calculated as: $d = h\Sigma_i pi$ with $h = 0.2, 0.4, 0.6,$ and 0.8, respectively. Problems with $h = 0.6$ and 0.8 are considered as unrestrictive. Upper bounds are provided	www.wiwi.uni-bielefeld.de/~kistner/Bounds.html
Flow shop	120 problem instances; 10 instances for each combination of number of tasks and number of machines (20-5, 20-10, 20-20, 50-5, 50-10, 50-20, 100-5, 100-10, 100-20, 200-10, 200-20 and 500-20 tasks-machines). Upper bounds are provided	OR-LIBRARY, http://people.brunel.ac.uk/~mastjb/jeb/info.html (see also [1993])
Weighted tardiness	375 problem instances; 125 instances for each problem size (40, 50, and 100 tasks). Optimal solutions are provided for instances with 40 and 50 tasks. Upper bounds are provided for instances with 100 tasks	OR-LIBRARY, http://people.brunel.ac.uk/~mastjb/jeb/info.html
Multiprocessor task scheduling in a multistage hybrid flow shops	160 problem instances. The dataset includes 80 instances with 50 jobs each and 80 instances with 100 jobs each. Each group of 80 problem instances is further partitioned into four subgroups of 20 instances each with 2, 5, 8, and 10 stages, respectively. Processor availability at various stages and processor requirements per task varied between 1 and 10	OR-LIBRARY, http://people.brunel.ac.uk/~mastjb/jeb/info.html

In case of the multiprocessor task scheduling in a multistage hybrid flow shop problem, the PLA3 uses the following fitness function:

```
Procedure FITNESS:
begin
  for i = 1 to m do
    for j = 1 to n do
      allocate task π(j) at stage i to the required number of
      processors, scheduling it as early as feasible;
    end for
  end for
  g(π) := finishing time of task number π(n) at stage m;
end
```

To evaluate the proposed algorithms several computational experiments have been carried out using a PC with Pentium III, 850 Mhz processor (experiment results were originally reported in Jedrzejowicz & Jedrzejowicz, (2003a, 2003b). The results have been compared with upper bounds or optimal solutions (if known) of several sets of benchmark problems. Table 31.5 contains short descriptions of benchmark data sets.

Table 31.6 shows mean relative errors (MREs) and mean computation times (MCT) for the experiment with common due date scheduling instances. These have been calculated by comparing the best result out of the three experimental runs with the upper bounds (i.e., best up-to-now known results for benchmark problems). MRE and MCT values in Table 31.6 have been calculated taking the average from ten instances available for each problem size and due date class. Negative value of the MRE shows, in fact, the relative improvement in comparison to previously established upper bounds. In the experiment the initial population size has been set to 2000 and the selection coefficient value s has been set to 0.5.

TABLE 31.6 MRE and MCT for the Common due Date Scheduling Instances

n	MRE (PLA1)		MRE (PLA2)			
	$h = 0.2$ (%)	$h = 0.4$ (%)	$h = 0.6$ (%)	$h = 0.8$ (%)	Overall (%)	MCT (%)
20	0.00	0.01	−0.63	−0.41	−0.26	6.3s.
50	−0.05	−0.01	−0.33	−0.24	−0.16	28.9s.
100	−0.01	−0.01	−0.17	−0.17	−0.09	91.5s.
200	−0.51	−0.45	−0.11	−0.11	−0.29%	205.2s.
500	−5.84	−5.91	−2.34	−2.34	−4.11	21.2m.
1000	−7.42	−4.39	−0.03	−0.04	−2.97	43.3m.
Overall	−2.31	−1.80	−0.60	−0.55	−1.31	—

TABLE 31.7 MRE, SDE and MCT for the Flow Shop Scheduling

Measure	Processors	$n = 20$	$n = 50$	$n = 100$	$n = 200$	$n = 500$
	5	0.0000%	0.0000%	−0.0097%	—	—
MRE	10	−0.0073%	0.0566%	0.1456%	0.1471%	—
	20	0.0084%	0.4885%	0.2944%	0.3405%	0.3818%
	5	0	0	0.000442	—	—
SDE	10	0	0.001897	0.001234	0.003874	—
	20	0.000279	0.000957	0.002313	0.001238	0.001631
	5	2.7m.	9.75m.	28.4m.	—	—
MCT	10	8.5m.	23.0m.	29.1m.	81.0m.	—
	20	15.2m.	33.4m.	38.8m.	140.0m.	295m.

Experiment results prove that PLA1 and PLA2, as applied to the common due date scheduling, perform well. Mean relative error for the whole considered population of problem instances is negative and equals −1.31% , which means that the total amount of penalties to be paid as a result of scheduling all 240 problem instances has decreased by 1.31% . Out of 240 problem instances involved in the experiment in 77.5% of cases, that is, in 186 instances, it has been possible to find a better upper bound than previously known. Only in case of 8 problem instances out of 240, results obtained by applying PLA were worse than respective benchmark values. The discussed results have been obtained with a reasonable computational effort.

Table 31.7 shows mean relative errors, standard deviation of errors (SDEs), and mean computation times for the experiment with flow shop scheduling instances. These have been calculated by comparing results of a single experimental run of the PLA3 with the upper bounds, that is best up-to-now known result for benchmark problems. MRE and MCT values in Table 31.7 have been calculated taking an average from 10 benchmark instances available for each problem size and number of processors class. The initial population size has been set to 100 and the selection coefficient value s has been set to 0.5.

Experiment results prove that PLA3 as applied to permutation flow shop scheduling perform quite well. Mean relative error for the whole considered population of benchmark problem instances equals only 0.1776% and it has been obtained in a single run. Out of 120 problem instances involved in the experiment in 10% of cases, that is, in 12 instances, it has been possible to find a better upper bound than previously known. Considering the research effort invested during the last 20 years into finding algorithms solving flow shop problems the quality of PLA3 is more than satisfactory. The discussed results, however, required a substantial computational effort.

Table 31.8 shows mean relative errors and standard deviation of errors for the total tardiness scheduling. These have been calculated by comparing results of a single experimental run of PLA3 with the optimal solutions in case of 40 and 50 tasks and upper bounds in case of 100 tasks instances (strongly suspected to be optimal solutions). MRE values in Table 31.8 have been calculated for different initial population sizes

TABLE 31.8 MRE and SDE for the Total Tardiness Scheduling Experiment

Initial population size	$n = 40$ MRE	$n = 40$ SDE	$n = 50$ MRE(%)	$n = 50$ SDE	$n = 100$ MRE(%)	$n = 100$ SDE
25	0.0000	0	0.0107	0.000526	0.0870	0.005079
50	0.0000	0	0.00134	0.000144	0.0632	0.004727
100	0.0000	0	0.00116	0.000130	0.0227	0.001916
200	0.0000	0	0.00116	0.000130	0.0164	0.001453
400	0.0000	0	0.00000	0	0.00074	0.000045
800	0.0000	0	0.00000	0	0.00030	0.000027

TABLE 31.9 MRE, SDE, and MTC for the Multiprocessor Task Scheduling in a m-h Flow Shop Experiment

Measure	Number of jobs	Number of stages			
		2	5	8	10
MRE	50	−0.9814%	−0.4036%	−0.3367%	0.3712%
	100	−0.6127%	−0.3180%	−0.2535%	−0.4540%
SDE	50	0.0174%	0.2971%	0.4104%	0.5992%
	100	0.0116%	0.2253%	0.3042%	0.2122%
MCT(m.)	50	1.20	3.35	5.76	10.40
	100	3.55	8.80	14.15	27.42

taking the average from 125 instances available for each problem size. The initial population sizes have been set to 25, 50, 100, 200, 400, and 800 individuals with the selection factor set to 0.5.

Experiment results prove that PLA3 applied to total tardiness scheduling perform very well. However, the computational effort involved is rather high (from an average 72s. per instance of 100 tasks and the initial population size set to 25, up to about 3000s. with the initial population size set to 800).

Finally, MRE, SDE, and MCT calculated after a single run of the PLA3 applied to solving benchmark instances of the multiprocessor task scheduling in a multistage hybrid (m-h) flow shop problem are shown in Table 31.9. Negative values of MRE show a percentage of improvement achieved over known upper bounds within the respective clusters of benchmark dataset.

Application of the population learning algorithm has resulted in improving the total of upper bounds of the considered cases by 0.34% . Out of 160 instances solved it has been possible to improve currently known upper bounds in 73 instances, which is more than 45% of all considered instances.

31.3.2 Other Combinatorial Optimization Problems

PLA has been used to solve a number of instances of the following combinatorial optimization problems belonging to the NP-hard class:

- Scheduling multiple variant programs (Jedrzejowicz and Wierzbowska, 2000).
- Scheduling multiple processor tasks with correlated failures, variants, ready times, deadlines, and variant reliabilities on identical multiple processors to maximize schedule reliability (Jedrzejowicz and Ratajczak, 2001; Czarnowski et al., 2003).
- Generalized segregated storage problem (GSSP) (Barbucha and Jedrzejowicz, 2000).
- Resource constrained project-scheduling problem with makespan as an objective function (Jedrzejowicz and Ratajczak, 2003).
- Optimization of system reliability structure under multiple criteria (Jedrzejowicz et al., 2002).

In all of the above listed implementations, PLA proved to be an effective framework for designing a hybrid algorithms producing competitive solutions. In this section, because of the length constraints, only two example implementations out of the above listed ones are briefly described.

A GSSP is a combinatorial optimization problem that involves allocation of certain number of goods to the available compartments subject to segregation (physical separation) constraints. Let m be a number of divisible consignments consisting of the homogenous goods, n be a number of available (internal) compartments. The size of each consignment to be stored — a_i ($i = 1, \ldots, m$) and the capacity of each compartment, $b_j(j = 1, \ldots, n)$ are known. Let c_{ij} be the cost of storing a unit of goods from consignment i in the compartment $j(i = 1, \ldots, m, j = 1, \ldots, n)$. For a given set of goods a *segregation matrix, S* is introduced. Each element $s_{ij} \in Z^+$ of the matrix $S(i, j = 1, \ldots, m)$ defines the required segregation distance between goods. Element s_{ij} is equal to 0, if and only if, good i can be stored together with j (one without any restrictions) and element s_{ij} is greater than 0, if some segregation type between cargos $i, j(i, j = 1, \ldots, m)$ is required (for example $s_{ij} = 2$ means that goods i and j must be stored in two separated compartments). On the other hand, for a given set of compartments an additional *compartment segregation matrix, CS* is defined. Element $cs_{ij} \in Z^+$ of CS matrix $(i, j = 1, \ldots, n)$ specifies segregation type guaranteed when storing goods in two compartments i and $j(i, j = 1, \ldots, n)$. It is also required that all consignments will be stored. To meet this constraint, it is assumed that an external storage space, denoted as $(n+1)^{\text{st}}$ compartment, is also available. It can accommodate any consignment at a higher unit cost $c_{i,n+1}(i = 1, \ldots, m)$. In this compartment segregation requirements do not have to be satisfied. Let x_{ij} denote the amount of cargo i stored in the compartment $j(i = 1, \ldots, m, j = 1, \ldots, n+1)$. Generalized Segregated Storage Problem can be then formulated as follows:

$$Z = \min \sum_{i=1}^{m} \sum_{j=1}^{n+1} c_{ij} x_{ij}$$

subject to:

$$\sum_{j=1}^{n+1} x_{ij} = a_i \qquad i = 1, \ldots, m,$$

$$\sum_{i=1}^{m} x_{ij} \leq b_j \qquad j = 1, \ldots, n$$

$$\sum_{i=1}^{m} \sum_{j=1}^{m} x_{ik} x_{jl} h_{ijkl} = 0 \qquad k, l = 1, \ldots, n,$$

$$x_{ij} \geq 0 \qquad i = 1, \ldots, m, j = 1, \ldots, n+1,$$

where

$$h_{ijkl} = \begin{cases} 0 & \text{if } s_{ij} \leq cs_{kl}, \\ 1 & \text{if } s_{ij} \geq cs_{kl}, \end{cases}$$

Elements h_{ijkl} form a binary matrix H. Each element of H is defined for two pairs: segregation requirement for goods and segregation type for compartments, (i, k) and (j, l), where $i, j = 1, \ldots, m$, and $k, l = 1, \ldots, n$.

Let $I = \{1, 2, \ldots, m\}$ be a set of consignments, each consisting of a certain number of units and $H = \{1, 2, \ldots, n\}$ be a set of internal compartments. Let g be a function $g : \{1, \ldots, mn\} \to I \times H$ defined as follows:

$$g(v) = (\lfloor (v-1)/n \rfloor + 1, (v-1) \bmod n + 1).$$

An ordered pair (p,q) is called *allocation (distribution)* of some units of goods p to compartment $q(p \in I, q \in H)$.

PLA for GSSP is based on the following assumptions:

- Individuals are represented by permutations of numbers from the $\{1, \ldots, mn\}$ set where each number can be transformed to an allocation of goods to a compartment. For example, the chromosome $ch_k = [2\,5\,3\,8\,1\,7\,4\,6]$ with $m = 2, n = 4$ is transformed into an allocation $(1, 2)$ $(2, 1)$ $(1, 3)$ $(2, 4)$ $(1, 1)$ $(2, 3)$ $(1, 4)$ $(2, 2)$.
- Each solution represented by an individual is directly evaluated in terms of its fitness by the decoding procedure *(DECODE)*. The fitness of the individual, f, reflects the cost of allocating all consignments to available compartments:

$$f(ch_k) = \sum_{i=1}^{m} \sum_{j=1}^{n+1} c_{ij} x_y^k$$

where $x_{ij}^k DECODE(ch)_k$.

- Initial population J consists of randomly generated permutations from the $\{1, \ldots, mn\}$ set.
- There are three learning stages.
- Selection criterion requires that all individuals with fitness below an average be rejected.

The respective learning-improvement procedures are shown in the following pseudo code:

```
Procedure L1:
begin
  for each individual in J do
    Create string Zj containing allocations (p,q) for which x^k_pq>0
    and Nj containing allocations for which x^k_pq=0. Create new
    individual by substituting the most expensive allocation i.e.
    allocation (p,q) for which the expression x^k_pq(c_p,n+1-c_pq) has
    the greatest value with element randomly chosen from Nj string.
    Accept exchange if the new individual is an improvement,
    otherwise recover an old individual;
  end for
end
Procedure L2:
begin
  for each individual in J do
    Choose randomly the position tj (tj∈ {1,...,mn}). Create a new
    individual from the old one cyclically shifting its elements
    to the left by tj positions. Accept if the new individual is an
    improvement otherwise recover an old individual;
  end for
end
Procedure L3:
begin
  for each individual in J do
    Create two strings as in L1. A new individual is created from
    the old one by randomly changing position of each element of
    the Zj string. Accept if the new individual is an improvement
    otherwise recover an old individual;
  end for
end
Procedure DECODE(ch):
```

```
begin
  ai' ← ai;
  bj' ← bj;
  for k=1 to mn do
  (p,q) ← g(chk)
  if allocation (p,q) is feasible then
    xpq ← min(ap', bq');
    ap' ← ap' - xpq;
    bq' ← bq' - xpq;
  end if
  end for
Store non-allocated goods into external ((n+1)-th) compartment;
end
```

To validate the approach, the performance of the proposed PLA implementation has been compared with that of the "classic" evolutionary algorithm. The respective experiment involved two randomly generated data sets of 10 instances each with number of goods and a number of compartments from $U[5,20]$ denoting the discrete uniform distribution between 5 and 20 inclusively. In data set 1, the internal storage costs c_{ij} were randomly drawn from $U[10,19]$, the external storage costs $c_{i,n+1}$ from $U[20,24]$. In data set 2, the internal storage costs c_{ij} were randomly drawn from $U[100,199]$, the external storage costs $c_{i,n+1}$ from $U[200,249]$. Quantities of cargos, a_i and capacities of compartments, b_j were randomly generated from $U[1,9]$ in both data sets. Elements of segregation matrices S and CS were generated from $U[0,5]$ and $U[0,4]$, respectively.

All generated problem instances have been solved by EA and PLA. Optimum solutions have been also obtained using a CPLEX solver. Mean relative error from optimal solution in case of the PLA is 2.17% (with 2.61% in case of EA). Mean relative computational effort is 91 and 100 for the PLA and EA, respectively.

The next example involves applying the population learning algorithm to solve the *resource constrained project scheduling* problem (RCPSP) with makespan minimization as the objective function. A single-mode RCPSP is considered. In a single-mode case a project consists of a set of activities, where each activity has to be processed in a single, prescribed way (mode). Each activity requires some resources, availability of which is constrained. The discussed problem is computationally difficult and belongs to the NP-hard class. Because of its practical importance RCPSP has attracted a lot of attention and many exact and heuristic methods have been proposed for solving it (see e.g., Christofides et al. [1987]). Exact algorithms seem suitable for solving smaller instances of RCPSP. On the other hand, heuristic approaches, used for solving its larger instances, can only be evaluated experimentally using benchmark datasets with known optimal solutions or upper bounds.

A project consists of a set of n activities, where each activity has to be processed without interruption to complete the project. The dummy activities 1 and n represent the beginning and end of the project. The duration of an activity j is denoted by d_j, where $d_1 = d_n = 0$. There are R renewable resource types. The availability of each resource type k in each time period is R_k units, $k = 1, \ldots, r$. Each activity j requires r_{jk} units of resource k during each period of its duration where $r_{1k} = r_{nk} = 0, k = 1, \ldots, r$. All parameters are nonnegative integers. There are precedence relations of the finish-start type with a zero parameter value (i.e., $FS = 0$) defined between the activities. In other words activity i precedes activity j if j cannot start until i has been completed. The structure of a project can be represented as an activity-on-node network $G = (V, A)$, where V is the set of activities and A is the set of precedence relationships. $S_j(P_j)$ is the set of successors (predecessors) of activity j. It is further assumed that $1 \in P_j, j = 2, \ldots, n$, and $n \in S_j, j = 1, \ldots, n-1$. The objective is to find a schedule S of activities, that is, a set of starting times $(s_1, \ldots, s_n)$ where $s_1 = 0$ and resource constraints are satisfied, such that the schedule duration $T(S) = s_n$ is minimized.

The respective PLA implementation involves three learning and improvement stages. The value of the goal function is directly used as a measure of quality of individuals and, hence, as a selection criterion. An individual is represented as a vector of activities (schedule) $S = [a_1, \ldots, a_n]$, each activity a_j is an object consisting of: starting time, s_j, duration, d_j, set of required units of resources, set of predecessors, SP_j, and set of successors, SS_j.

The algorithm requires also that values of the following parameters are set:

- p — a multiplier used to calculate the size of the initial population.
- $xi1$ — a coefficient used to calculate the number of iterations at the first learning and improvement stage.
- $xm1$ — a coefficient used to calculate a frequency of calls to a selection procedure within the first learning and improvement stage.
- $xi2$ — a coefficient used to define a number of iterations at the second learning and improvement stage.

Values of the above control parameters are set at the algorithm fine-tuning phase. All random moves within the algorithm are drawn from the uniform distribution. All parameter values used in the following pseudo-code have been set by trials and errors during the fine-tuning phase. In the pseudo-code shown below P denotes the population, and $|P|$ its size.

```
Procedure PLA:
begin
  Set size of P, |P|= p*n;
  Set values of xi1, xm1, xi2;
  Generate the initial population;
  {-- the first learning stage}
  for it=1 to xi1*n do
    for i=1 to 0.4*|P| do
      Crossover(Random(P), Random(P));
    end for
    for i=1 to 0.05*|P|do
      Mutation(Random(P));
    end for
    for i=1 to 0.05*|P| do
      LSA(Random(P),2,6,10);
    end for
    if it mod (xm1*n)=0 then
      Selection(medium makespan);
    end if
  end for
  Selection(medium makespan);
  {-- the second learning stage}
  for it=1 to xi2*n do
    for i=1 to 0.4*|P|do
      Crossover(Random(P), Random(P));
    end for
    for 2 best solutions S∈P do
      EPTA(S,6,4);
    end for
    for i=1 to 0.2*|P|do
      LSA(Random(P),2,6,10);
    end for
  end for
  Selection(medium makespan);
```

```
  {-- the third learning stage}
  for every solution S∈P do
    EPTA(S,6,2);
    LSA(S,10,2,10_);
  end for
end
```

The algorithm creates an initial population by producing four individuals using simple construction heuristics and generating randomly the remaining ones. Heuristics are based on the following rules: shortest duration first, shortest duration last, longest duration first, and longest duration last. The first learning stage uses evolutionary operators and a simple local search algorithm (LSA). Three procedures, such as Crossover, Mutation, and LSA are repeated $xi1 * n$ times. The Random(P) function chooses randomly an individual from the population P.

```
Procedure Crossover(S1,S2):
begin
  One-point crossover on two randomly chosen individuals S1,S2∈P;
end
Procedure Mutation(S):
begin
  Move an activity from randomly chosen position in the schedule S
  to another randomly chosen position such that no successor of this
  activity can be found before it;
end
Procedure LSA(S, itNumber, iStep, fStep):
begin
  for it=1 to itNumber do
    for s= iStep to fStep do
      for j=1 to scheduleLength-s do
        Exchange activities at positions j and j+s;
        if not the new solution is better
        then recall the exchange;
      end for
    end for
  end for
end
```

The LSA procedure requires four variables. The first (S) denotes an individual, the second (*itNumber*) defines a number of iterations within the procedure. The last two *(iStep, fStep)* indicate the distance between activities under exchange.

At the second learning stage a crossover and two heuristics, EPTA (exact precedence tree algorithm) and LSA are used. EPTA is based on the precedence tree approach. It finds an optimum solution by enumeration for a part of the schedule consisting of a sequence of activities, with the number of activities in such sequence denoted as *partExtent.* In the following pseudo-code the variable S denotes an individual, and $|S|$ its current size (a number of activities). The *step* variable denotes a distance between starting points of the considered partitions.

```
Procedure EPTA(S, partExtent, step):
begin
  i=1;
  while i*step + partExtent < |S|do
    Find an optimal solution for a part of the schedule beginning
```

```
    from activity at position i*step and ending with activity at
    position i*step + partExtent;
  end while
end
```

Finally, at the third learning stage two heuristics EPTA and LSA are again used with settings resulting in more iterations and higher granularity of the neighborhood explored as compared with the previous stages.

To validate the proposed approach a computational experiment has been carried based on 1440 benchmark instances of the single-mode RCPSP. The benchmark data set used in the reported experiments includes 480 instances for each out of the three problem sizes (30, 60, 90, and 120 activities, respectively). The benchmark data set together with known upper bounds can be found at http://www.wior.uni-karlsruhe.de/rcpsp

The fine-tuning phase of the experiment has been devoted to finding values of the PLA parameters assuring acceptable compromise between computation time and quality of solutions. This search has been carried using the subset of available benchmark instances consisting of all RCPSP with 30 activities. The resulting setting is shown in Table 31.10.

The experiment involved solving all benchmark instances twice. The results were evaluated in terms of mean and maximum relative errors (MRE, max RE), percent of solutions equal to respective upper bounds (eqUB) as well as mean computation time (MCT) needed to solve a single instance. Relative errors have been calculated considering the available upper bounds. Experiment results are shown in Tables 31.11 and 31.12. The proposed PLA implementation has contributed to finding an improved upper bound for one benchmark instance with 90 activities. The experiment was carried out on a PC computer with the AMD XP 1600+ processor and 256 MB RAM.

Application of the population learning algorithm to solving several difficult combinatorial problems proves that it is a useful tool extending the range of available techniques. Although the PLA does not seem to be able to produce very good solution quickly it is a technique providing consistently a good to very good solution in reasonable time, leading in many instances to improving upper bounds on most difficult combinatorial problems.

TABLE 31.10 Six Variants of the Parameter Settings for the Computational Experiment

Variant	*p*	*xi*1	*xm*1	*xi*2
a	3	0.5	0.333	0.333
b	3	0.5	0.333	0.5
c	3	0.5	0.5	0.5
d	4	0.5	0.333	0.333
e	4	0.5	0.333	0.5
f	4	0.5	0.5	0.5

TABLE 31.11 The Experiment Results (30 and 60 Activities)

Variant	30 activities				60 activities			
	MRE%	max RE%	eqUB%	MTC [s.]	MRE%	max RE%	eqUB%	MTC [s.]
a	0.11	4.35	94.17	1.32	0.53	6.06	76.25	14.75
b	0.11	3.45	93.96	1.57	0.47	4.55	75.28	18.25
c	0.13	5.17	93.54	1.55	0.48	5.05	77.08	18.28
d	0.10	3.95	94.38	1.66	0.51	5.05	76.25	18.71
e	0.10	3.95	94.38	1.97	0.43	5.05	78.13	23.07
f	0.11	3.95	93.96	1.97	0.46	5.05	77.08	23.14

TABLE 31.12 The experiment results (90 and 120 activities)

Variant	90 activities				120 activities			
	MRE%	max RE%	eqUB%	MTC [s.]	MRE%	max RE%	eqUB%	MTC [s.]
a	0.64	6.42	75.83	73.46	1.05	8.09	74.92	217
b	0.58	5.65	76.04	91.93	0.93	6.85	74.60	242
c	0.58	5.50	76.04	91.02	0.92	8.22	75.42	462
d	0.63	6.42	76.04	93.15	0.91	6.86	77.08	384
e	0.54	5.50	76.46	116.92	0.87	6.92	78.04	487
f	0.52	5.45	76.88	115.05	0.86	6.69	78.02	437

31.4 Example Applications of the PLA to Machine Learning

31.4.1 Application of the Population Learning Algorithm to Training Feed-Forward Artificial Neural Networks (ANN)

Artificial neural networks (ANN) are, nowadays, being used for solving a wide variety of real-life problems like, for example, pattern recognition, prediction, control, combinatorial optimization, or classification. Main advantages of the approach include the ability to tolerate imprecision and uncertainty, while still retaining the possibility of achieving tractability, robustness, and low cost in practical applications. Since training a neural network for practical application is often very time-consuming, an extensive research work is being carried out in order to accelerate this process. Another problem with ANN training methods is danger of being caught in a local optimum. Hence, researchers look not only for algorithms that train neural networks quickly but rather for quick algorithms that are not likely, or less likely, to get trapped in a local optima.

In this section an application of the population learning algorithm to ANN training is investigated. The idea of applying various implementations of the population learning algorithm to ANN training has been suggested in Czarnowski and Jedrzejowicz (2002a, 2002b). Several implementations of the PLA have been proposed and applied to solving variety of benchmark problems. Initial results were promising showing a good performance of the PLA as a tool for ANN training.

The proposed implementations are based on the following assumptions:

- An individual is a vector of real numbers from the predefined interval, each representing a value of weight of the respective link between neurons in the considered ANN.
- The initial population of individuals is generated randomly.
- There are five learning/improvement procedures used — standard mutation, local search, nonuniform mutation, gradient mutation, and gradient adjustment.
- There is a common selection criterion for all stages. At each stage, individuals with fitness below the current average are rejected.

All five learning and improvement procedure $L(1)$ to $L(5)$ used for ANN training are shown in the following pseudo-code:

```
Procedure L(1): { standard mutation}
begin
 for i=1 to J do
  for i=1 to k do {k -- number of iteration}
   Select randomly two different elements x1 and x2 within individual i;
   Generate new random value of x1 and x2 producing new individual;
   Calculate the fitness of a new individual;
   if new fitness < old fitness then accept changes;
  end for
```

```
 end for
end
Procedure L(2): {local search}
begin
 for i=1 to J do
  for i=1 to k do {k -- number of iterations}
   Select randomly two different elements x1 and x2 within individual i;
   Exchange values between x1 and x2 producing a new individual;
   Calculate the fitness of a new individual;
    if new fitness < old fitness then accept changes;
  end for
 end for
end
Procedure L(3): {non-uniform mutation}
begin
 for i=1 to J do
  for t=1 to T do
   Select random point within the individual i;
   Apply non-uniform mutation;
   Calculate the fitness of a new individual;
   if new fitness < old fitness then accept change;
  end for
 end for
end
Procedure L(4): {gradient mutation}
begin
 for i=1 to J do
  for i=1 to k do {k -- number of iteration}
   Select randomly two different elements x1 and x2 within
   individual i;
   Generate a random binary digit;
   Change values x1 and x2 to x1 + ξ and x2 + ξ if a random digit
   is 0 and to x1 -- ξ and x2 -- ξ otherwise, producing a new
   individual;
   Calculate the fitness of a new individual;
   if new fitness < old fitness then accept changes;
  end for
 end for
end
Procedure L(5): {gradient adjustment operator}
begin
 for i=1 to J do
  α =1;
  while α > 0 do
   Apply gradient adjustment operator;
   Calculate fitness of the new individual;
   if new fitness < old fitness then accept changes and break;
   α = α − 0.02;
  end while
 end for
end
```

The above-proposed procedures require some additional comments. The *first* procedure, standard mutation, modifies an individual by generating new values of its two randomly selected elements. If the operation improves the fitness function value then the change will be accepted. The *second* learning and improvement procedure involves exchange of values between the two randomly selected elements within an individual. If the operation improves the fitness function value then the change will be accepted. The *third* procedure, a nonuniform mutation, involves modifying an individual by repeatedly adjusting value of the randomly selected element (in this case a real number) until the fitness function value has improved or until a number of the consecutive improvements have been attempted unsuccessfully. The value of the adjustment is calculated as:

$$\Delta(t, y) = y\left(1 - r^{(1-(t/T))\cdot r}\right),$$

where r is the uniformly distributed real number from (0, 1], T is equal to the length of the vector representing an individual and t is a current number of adjustment. The *fourth* learning and improvement procedure, a gradient mutation, changes two randomly selected elements within an individual by incrementing or decrementing their values. Direction of change (increment/decrement) is random and has identical probabilities equal to 0.5. The value of change is proportional to the gradient of an individual. If the fitness function value of an individual has improved then the change is accepted. Number of iterations for learning and improvement procedures 1, 2, and 4 has to be set at the fine-tuning phase. Finally, the *fifth* procedure adjusts the value of each element of the individual by a constant value delta (Δ), proportional to its gradient. Delta is calculated as $\Delta = \alpha \cdot \xi$, where α is the factor determining a size of the step in direction of ξ, known as a momentum. Factor α takes values from (0, 1]. In the proposed algorithm its value iterates starting from 1 with the step equal to 0.02. Here, ξ is a vector determining the direction of search and is equal to the gradient of an individual.

Population learning algorithm for ANN training has been implemented in two versions — sequential and parallel. A general structure of the sequential implementation is shown in the following pseudo-code:

```
Procedure Sequential_PLA
begin
  Generate initial population;
  P =: initial population;
  for i=1 to 4 do
    Procedure L(i);
    P = Select(P);
  end for
  Procedure L(5);
  Consider best individual in P as a solution;
end
```

The parallel PLA implementation is based on the cooperation between the master worker (server) whose task is to manage computations and a number of slave workers, who act in parallel, performing computations as requested by the master. The approach allows a lot of freedom in designing population-learning process. The master worker is managing communication flow during the population learning. It allocates computational tasks in terms of the required population size and the number of iterations and also controls information exchange between slaves. The latter task involves upgrading, at various learning stages, current populations maintained by all slaves with globally best individuals. Communication flow between the master and slave workers is shown in Figure 31.4.

The parallel PLA implementation for ANN training denoted Parallel_PLA is based on the following rules:

- Master worker defines the number of slave workers and the size of the initial population for each of them.

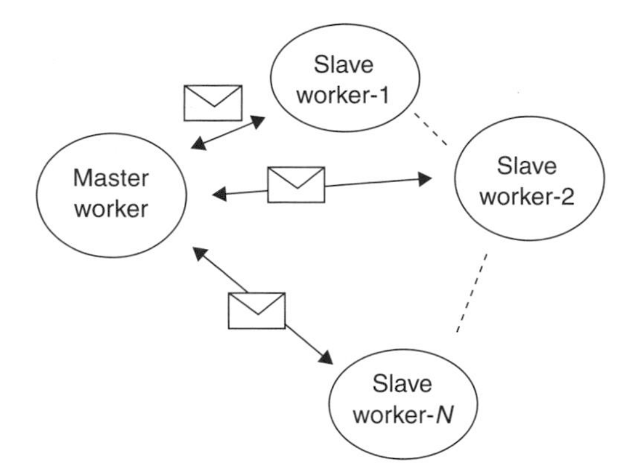

FIGURE 31.4 Communication flow between the master and slave workers.

- Each slave worker uses the same, described earlier, learning and improvement procedures.
- Master worker activates parallel processing.
- After completing each stage, workers inform master about the best solution found so far.
- Master worker compares the received values and sends out the best solution to all the workers replacing their current worst individual.
- Master worker can stop computations if the desired quality level of the objective function has been achieved. This level is defined at the beginning of computations through setting the desired value of the mean squared error on a given set of training patterns. Alternatively, computations are stopped after the predefined number of iterations at each stage has been executed.
- Slave workers can also stop computations if the above condition has been met.

A variant of the above approach, denoted Parallel-PLAs, differs with respect to the strategy of using the best solution sent to slaves from the master. Now, the best solution is used to produce offspring by applying two types of the crossover operators — a single point crossover and a position crossover. Both are applied in turn to generate an offspring from all of the current individuals for each slave. Each individual in each slave is coupled with the best current solution forwarded from the master to produce an offspring. This takes place after completing each stage.

The proposed implementations of the PLA have been used to train several neural networks applied to solving popular benchmarking classification problems — two spirals, 10-parity, Wisconsin breast cancer, Cleveland heart disease, and credit approval.

Artificial neural network algorithms solving the two spirals problem has two real-valued inputs corresponding to the x and y coordinates of the point, and one binary target output that classifies the point as belonging to either of the two spirals coiling three times around the origin. The two spirals are constructed from 97 points given by the respective coordinates from the training set. Solving the two spirals problem appears to be a very difficult task for back-propagation networks (Shang and Wah, 1996). The topology of the network for this problem has been adopted from (Fahlman and Lebiere, 1990), where also an efficient constructive training algorithm called CASCOR was suggested. This so-called "shortcut" topology has 4 hidden units and 25 links with 25 weights.

In 10-parity problem ANN has to be trained to produce the Boolean "Exclusive OR" function of ten variables. A two-layer neural network with 10, 10 and 1 neurons in layers 1, 2, and 3, respectively is used.

Diagnosis of breast cancer involves classifying a tumor as either benign or malignant based on cell descriptions gathered by microscopic examination. Breast cancer databases were obtained from Dr. William H. Wolberg, University of Wisconsin Hospitals, Madison (see [Mangasarian and Wolberg, 1990]). It includes 699 examples, 9 inputs and 2 outputs each. The corresponding ANN has, 9, 9, and 1 neuron in layers 1, 2, and 3, respectively.

TABLE 31.13 Correct Classification Ratios (%), Mean Errors and Mean Training Times for the PLA Implementations

		TTV			10CV					Training
Problem	Approach	mean	max	min	mean	max	min	MRE	MSE	time [m.]
Two spirals	Sequential-PLA	71.6	82.0	58.9	74.7	83.2	65.0	0.29	0.25	45.0
Two spirals	Parallel-PLA	89.2	98.4	70.2	88.4	96.6	74.0	0.07	0.04	20.0
Two spirals	Parallel-PLAs	91.7	98.0	77.0	92.0	98.2	80.0	0.07	0.04	18.0
10 parity	Sequential-PLA	88.1	92.0	72.0	88.0	97.4	70.0	0.12	0.11	3.50
10 parity	Parallel-PLA	93.6	98.0	80.5	92.2	96.0	75.0	0.06	0.04	2.00
Breast	Sequential-PLA	95.4	98.5	82.2	96.4	98.0	78.0	0.02	0.01	2.50
Breast	Parallel-PLAs	96.6	98.0	82.2	96.6	98.0	80.8	0.02	0.01	1.50
Heart	Sequential-PLA	81.0	88.0	65.0	81.3	89.0	62.3	0.21	0.17	2.70
Heart	Parallel-PLAs	85.7	90.0	68.0	86.5	89.2	70.0	0.11	0.10	1.50
Credit	Sequential-PLA	82.0	86.4	63.0	82.6	87.4	63.0	0.20	0.18	3.40
Credit	Parallel-PLAs	88.1	91.9	74.0	86.6	90.4	79.2	0.11	0.11	2.50

Cleveland heart disease problem involves predicting heart disease, that is, deciding whether at least one of four major vessels is reduced in diameter by more than 50% . The binary decision is made based on personal data. The data set includes 303 examples with 13 inputs and 2 outputs and the corresponding ANN has 13, 13, and 1 neuron in layers 1, 2, and 3, respectively.

Credit card approval involves predicting the approval or rejection of a credit card request. The data set consists of 690 examples with 15 inputs and 2 outputs with a good mix of attributes. The corresponding ANN to be trained has 15, 15, and 1 neuron in layers 1, 2, and 3, respectively.

All data sets for the above problems are available at the UCI repository (Merz and Murphy, 1998) and are often used to compare various techniques for ANN training. Data set for the 2-spirals problem is available and has been obtained from the home page http://www.cae.wisc.edu/~ece602/data/CMUbenchmark

Computational experiment designed to validate the proposed PLA implementations have been based on the following quality measures:

- Correct classification ratio for the "10 cross validation" (10CV) approach.
- Correct classification ratio for the training, test, and validation sets constructed using the 50–25–25% (TTV) principle (Prechelt and Proben, 1994).
- Mean squared classification error, MSE.
- Mean relative classification error, MRE.

All artificial neural networks used in the experiment have sigmoid activation function with the sigmoid gain equal to 1. The initial population size in all PLA implementations has been set to 200. A maximum number of repetitions for each learning and improvement procedure has been set to 20 except for the 2-spirals problem in both, sequential and parallel versions, where it has been set to 40. Each benchmarking problem has been solved 50 times and the reported values of the quality measures have been averaged over these 50 runs. All computations have been carried on the Sun Challenge R4400 and, additionally, for the Parallel_PLA and Parallel_PLAs using the PVM environment. Experiment results are shown in Table 31.12. In Table 31.13 the performance of the PLA implementations is compared with some reported results from the literature.

It can be observed that PLA implementations guarantee good or even competitive level of training quality. It seems that the approach is also interesting in terms of computational time requirements. The author has not been able to find comparable data on time performance of the alternative approaches. Some conclusions can be drawn from the information on time performance of the Novel algorithm, which required 900 min to train ANN solving 2-spirals problem and 35 min to train ANN solving 10-parity problem (Shang and Wah, 1996).

31.4.2 An Approach to Instance Reduction in Supervised Learning

As it has been observed by Wilson and Martinez (2000) in the supervised learning, a machine-learning algorithm is shown a training set, which is a collection of training examples called instances. Each instance has an input vector and an output value. After learning from the training set, the learning algorithm is presented with additional input vectors, and the algorithm must generalize, that is, decide what the output value should be. It is well known that in order to avoid excessive storage and time complexity, and possibly, to improve generalization accuracy by avoiding noise and overfitting it is often necessary to reduce original training set by removing some instances before learning phase or to modify the instances using a new representation.

In the instance reduction approach a subset S of instances to keep from the original training set T can be obtained using incremental, decremental, or batch search. An incremental search begins with an empty subset S and adds an instance to S if it fulfills some criteria. The decremental search begins with $S = T$, and then searches for instances to remove from S. Finally, in a batch search mode, all instances are evaluated using some removal criteria. Then, all those that do meet these criteria are removed in a single step.

This section presents a collection of heuristic algorithms *IRA*1 to *IRA*4 used to decide whether an instance should be kept or removed from the training set. The algorithms were originally proposed in Czarnowski and Jedrzejowicz (2003).

Let N denote a number of instances in T and n be a number of attributes. Total length of each instance (i.e., training example) is equal to $n + 1$, where element numbered $n + 1$ contains the output value. Let also $X = \{xij\}(i = 1, \ldots, N,\ j = 1, \ldots, n + 1)$ denote a matrix of $n + 1$ columns and N rows containing values of all instances from T. Each of the proposed algorithms groups instances from input set T into a number of clusters with identical value of identity factor, which is used to indicate similar instances.

*IRA*1 involves the following steps:

Step 1. Transform X normalizing value of each x_{ij} into interval [0, 1] and then rounding it to the nearest integer, that is 0 or 1.

Step 2. Calculate for each instance from the original training set the value of its identity factor I_i:

$$I_i = \sum_{j=1}^{n+1} x_{ij} s_j, \quad i = 1, \ldots, N,$$

where

$$S_j = \sum_{i=1}^{N} x_{ij}, \quad j = 1, \ldots, n + 1.$$

Step 3. Map input vectors (i.e., rows from X) into t clusters denoted as $Y_v, v = 1, \ldots, t$. Each cluster contains input vectors with identical value of the identity factor I_i, where t is a number of different values of I_i.

Step 4. Set value of the representation level K, which denotes the maximum number of input vectors to be retained in each of t clusters defined in step 3. Value of K is set arbitrarily by the user.

Step 5. Select input vectors to be retained in each cluster. Let y_v denote a number of input vectors in cluster $v, v = 1, \ldots, t$. Then the following rules for selecting input vectors apply:

- If $y_v \leq K$ then $S = S \cup Y_v$.
- If $y_v > K$ then the order of input vectors in Y_v is randomized and the cluster is partitioned into $q = y_v/\mathrm{K}$ subsets denoted as $D_{vj}, j = 1, \ldots, q$. Generalization accuracy of each subset denoted as A_{vj} is calculated carrying out the *leave-(q-1)-out* test with $X' = X - Y_v + D_{vj}$ as a training set. Subset of input vectors from cluster v maximizing value of A_{vj} is kept in the reduced training set.

IRA2 uses two additional parameters: pr, precision and ml, multiple. The first one determines a number of digits after the decimal point. The second determines a way the identity factor value is rounded. The algorithm proceeds as follows:

Step 1. Set value of the representation level K, which is the maximum number of input vectors to be retained in each cluster.
Step 2. Set values pr $= 0$ and ml $= 1$.
Step 3. Transform X normalizing value of each x_{ij} into interval $[0, 1]$.
Step 4. Round x_{ij} to the nearest value of 0 or 1 with precision pr digits after the decimal point.
Step 5. Calculate for each instance from the original training set the value of its identity factor I_i.
Step 6. Round I_i to the nearest multiple of ml.
Step 7. Map instances (i.e., rows from X) into t clusters $Y_v, v = 1, \ldots, t$. Each cluster contains instances with identical value of the identity factor I_i, and t is a number of such clusters.
Step 8. Select instances to be retained in each cluster. Let y_v denote a number of input vectors in cluster $v, v = 1, \ldots, t$. Then the following rules for selecting input vectors apply:

- If $y_v \leq K$ then $S = S \cup Y_v$.
- If $y_v > K$ and $K = 1$ then the order of input vectors in Y_v is randomized. Generalization accuracy of each Y_v is calculated carrying out the leave-(q-1)-out test. An instance from cluster v maximizing the generalization accuracy is kept in the reduced training set.
- If $y_v > K$ and $K > 1$ set *pr* to 1 and then for instances in Y_v and for several arbitrary values of ml (i.e., ml $= \{10, 20, 30\}$) repeat steps 4 to 8 calculating new values of identity factor and creating q subsets denoted as $D_{vj}, j = 1, \ldots, q$, until a number of elements in each subset is at least equal to K. Generalization accuracy of each subset A_{vj} is calculated carrying out the leave-(q-1)-out test with $X' = X - Y_v + D_{vj}$ as a training set. The subset of instances from cluster v maximizing the value of A_{vj} is kept in the reduced training set.

IRA3 uses the population-learning algorithm for the selection of instances to be kept. Steps 1 to 4 in *IRA3* are identical as in *IRA1* but step 5 differs:
Step 5. Select instances to be retained in each cluster:

- If $y_v \leq K$ then $S = S \cup Y_v$.
- If $y_v > K$ and $K = 1$ then $S = S \cup \{x_v\}$, where x_v is a selected reference instance from the cluster Y_v, where the distance $d(x_v, \mu_v) = \sqrt{\sum_{i=1}^{n}(x_i^v - \mu_i^v)^2)}$ is minimal and $\mu^v = (1/y_v)\sum_{j=1}^{y_v} x_v$ is the mean vector of the cluster Y_v.
- If $y_v > K$ and $K > 1$ then $S = S \cup \{x_j^v\}$, where $\{x_j^v\}, j = 1, \ldots, K)$ are reference instances from the cluster Y_v selected by applying the PLA.

The PLA implemented for the selection of reference instances maps instances x_v from Y_v into K subsets $D_{vj} (j = 1, \ldots, K)$, such that the sum of the squared Euclidean distances between each instance $x_v \in D_{vj}$ and the mean vector μ^j of the subset D_{vj} is minimal. Vectors with minimal distance to the mean vector in each subset are selected as K reference vectors. This selection method can be associated with one of the clustering technique known as k-means algorithm (Likas et al., 2001).

A potential solution p is coded as the $(K + y_v)$-element vector. Its first K positions inform how many subsequent elements from all y_v elements are contained in each $D_{vj} (j = 1, \ldots, K)$. The next y_v positions of the potential solution represent input vector numbers from Y_v.

The fitness of an individual $p \in P$ where P is a population of individuals can be evaluated as:

$$J(p) = \sum_{j=1}^{K} \sum_{z \in T} \| p[z] - \mu^j \|^2,$$

where

$$T = \begin{cases} (K, K + p[j]) & \text{if } j = 1 \\ (K + \sum_{i=1}^{j-1} p[i], K + \sum_{i=1}^{j} p[i]) & \text{otherwise} \end{cases}$$

and $p[z]$ is an instance number stored at position z of individual p.

The remaining assumptions for the discussed PLA implementation can be summarized as follows:

- The initial population is generated randomly.
- Four learning and improvement procedures are used — random local search, partially-mapped crossover (PMX), steepest ascent local search and tabu search.
- There is a common selection criterion for all stages — individuals with fitness below the current average are rejected.

Random local search involves selecting randomly two elements of an individual each representing a number of the input vector. Both numbers are then exchanged, providing such action improves value of the fitness function. The procedure is run for a predefined number of iterations.

PMX crossover operates on two parents (individuals) and includes the following steps (Michalewicz, 1994) — make a copy of the second parent, choose an arbitrary part from the first parent, and make minimal changes in the offspring necessary to achieve the chosen pattern. A crossover is performed on each individual in turn. A partner is always selected randomly. After the operation is performed the best out of parents and offspring is selected.

The steepest ascent local search procedure involves selecting randomly an element representing a number of the input vector and allocating it to a cluster, minimizing the Euclidean distance to the mean vector, providing such an allocation improves value of the fitness function. The procedure is run for a predefined number of iterations.

The tabu-search procedure is based on a pair-wise exchange of elements representing numbers of the input vectors. A standard tabu-search arrangements including tabu list and a short and long term memories are used (Glover, 1990). The procedure is run for a predefined number of iterations.

IRA *4* is identical to IRA *3* except that for the selection of reference instances the PLA with a real number representation instead of the permutation one is used.

To validate the proposed instance reduction algorithms it has been decided to use generalization accuracy as a criterion. A set of artificial neural networks (ANN) was used as learning machines. The machines were presented with the reduced sets of instances during the supervised learning stage and the results were compared with those obtained without reducing the respective training sets. To train artificial neural networks an implementation of the population learning algorithm, was used. Neural network trained using PLA is further on referred to as the PLAANN. Possibility of applying population learning algorithms to train ANN was discussed in the previous section. In order to increase efficiency of the approach it has been decided to use a PVM parallel computing environment.

The experiment involved five datasets from UCI Machine Learning Repository (Merz and Murphy, 1998), which are Wisconsin breast cancer (699 input vectors), Cleveland heart disease (303 input vectors), credit approval (690 input vectors), and thyroid disease (3772 input vectors). Additional problem is customer intelligence in banking provided under the EUNITE World Competition (12000 input vectors) (EUNITE, 2002). All the above problems require classification based on input vectors with both — continuous and binary attributes.

All ANN used during the experiment have had the MLP structure with 3 layers — input, hidden, and output. The range of weights has been set to $[-10, 10]$ and the sigmoid activation (transfer) function has had the sigmoid gain value set to 1.0. Number of neurons in layers 1, 2, and 3, respectively, has been set to the following values:

- Wisconsin breast cancer — 9, 9, and 1
- Cleveland heart disease — 13, 13, and 1

TABLE 31.14 PLA Versus the Reported Accuracy of Training Algorithms (source for the literature reported accuracy: http://www.phys.uni.torun.pl/kmk/projects/datasets.html)

Problem	Approach	Literature reported accuracy (%)	Best achieved PLA accuracy (%)
Two spirals	Novel	94.0	92.0
Two spirals	Simulated annealing	84.8	92.0
Two spirals	CASCOR	62.0	92.0
Two spirals	Back propagation	52.0	92.0
Breast	Back propagation	96.7	96.6
Breast	C 4.5	94.7	96.6
Heart	Back propagation	81.3	86.5
Credit	Back propagation	91.0	89.0

TABLE 31.15 Mean Percentages of Input Vectors Retained by *IRA1* to *IRA4*

Problem	Representation level (%)		
	$K = 1$	$K = 4$	$K = 10$
Credit	28	55	69
Heart	34	55	67
Breast	37	63	81
Thyroid	9	18	30
CI	1	3	5

- Credit approval — 15, 15, and 1
- Thyroid disease — 21, 21, and 3
- Customer intelligence in banking (CI) — 36, 15, and 1

The above proposed instance reduction algorithms have been used to generate training sets for all considered problems. For each problem five reduced instance sets have been generated with the representation levels varied between 1 and 10. In the computational experiment "10 *cross-validation*" approach was used. Each dataset was divided into 10 partitions and each reduction technique was applied to a training set T consisting of 9 of the partitions, from which it reduced a subset S. The ANN was trained using only the instances in S. Ten such trials were run for each dataset with each reduction algorithm, using a different partitions as the test set in each trial. Mean percentages of input vectors retained are shown in Table 31.14.

The PLAANN classifier has been run 20 times for each representation level and for each training set generated by *IRA1* to *IRA4*. Results of thus obtained classifications, averaged over 20 runs, are shown in Table 31.15. The column "Original set" in Table 31.15 shows results obtained by applying the PLAANN to the original, non-reduced training set using the "10 *cross-validation*" approach. Best results for each of the 5 benchmark problems and for each of the representation levels tried are shown in bold. Best overall results are underlined.

Applying the proposed variants of *IRA* clearly results in a substantial reduction of the training set size as compared with an original data set. Overall performance of the PLAANN classifier seems quite satisfactory. It is also clear that in majority of cases increasing the representation level leads to a better performance in terms of the classifier quality at a cost of higher requirements in terms of the computation time. For the credit, heart and breast problems the PLAANN classifier trained using the reduced training set, performs with the accuracy of classification comparable to the accuracy achieved on the original training set. The respective training time decreases, on average, 4 to 6 times (for $K = 10$). It might be worth noting that in case of the Customer Intelligence problem the PLAANN applied to the original set of instances has not been able to find any satisfactory solution in a reasonable time and in fact classification

TABLE 31.16 Mean Accuracy (%) of PLAANN in the Classification Experiment

Reduction algorithm	Problem	Representation level			
		$K=1$	$K=4$	$K=10$	Original set
IRA1	Credit	74.00	79.87	82.70	85.70
	Heart	**85.34**	78.31	85.10	88.10
	Breast	93.04	93.68	95.84	96.60
	Thyroid	**93.45**	94.71	95.69	93.10
	CI	67.44	69.45	71.80	58.70
IRA2	Credit	74.00	80.04	83.15	85.70
	Heart	**85.34**	83.45	85.87	88.10
	Breast	93.04	94.21	94.81	96.60
	Thyroid	**93.45**	94.50	94.68	93.10
	CI	67.44	75.31	76.48	58.70
IRA3	Credit	**76.00**	81.32	**84.33**	85.70
	Heart	80.10	**87.20**	87.28	88.10
	Breast	**93.80**	94.92	**96.20**	96.60
	Thyroid	**93.45**	94.80	96.60	93.10
	CI	**70.00**	79.50	80.50	58.70
IRA4	Credit	**76.00**	**81.45**	83.73	85.70
	Heart	80.10	87.10	**88.10**	**88.10**
	Breast	**93.80**	**95.10**	96.15	96.60
	Thyroid	**93.45**	**98.21**	**98.43**	93.10
	CI	**70.00**	**80.50**	**83.40**	58.70

TABLE 31.17 Performance Comparison of Different Instance Reduction Algorithm

	Breast		Heart		Credit	
Algorithm	Accuracy	% Retained	Accuracy	% Retained	Accuracy	% Retained
IRA($K=1$)	93.80	37.34	76.00	33.66	85.34	28.32
IRA($K=10$)	96.20	81.00	**84.33**	67.00	**88.10**	69.00
CNN	95.71	7.09	73.95	30.84	77.68	24.22
SNN	93.85	8.35	76.25	33.89	81.31	28.38
IB2	95.71	7.09	73.96	30.29	78.26	24.15
IB3	**96.57**	3.47	81.16	11.11	85.22	4.78
DROP3	96.14	3.58	80.84	12.76	83.91	5.96
*k*NN	96.28	100.00	81.19	100.00	84.78	100.00
PLAANN	96.60	100.00	85.70	100.00	88.10	100.00

process has been stopped. Accuracy of classification obtained with the original set was then only 58% . However the PLAANN trained on the reduced set guarantees accuracy better then 80% already 120 sec of computation time.

For the thyroid problem the PLAANN trained using the original data set produces accuracy of classification at the level of about 92%, which is a standard performance. The accuracy of classification for the discussed problem has grown substantially with the reduction of the original dataset size.

Comparison of the proposed IRAs with other approaches to instance reduction is shown in Table 31.16. The column "Retained" in Table 31.16 shows a percentage of training vectors from the original training set that have been retained by the respective instance reduction algorithm. All the results other than these produced by the proposed *IRA*s were reported in Wilson and Martinez (2000). Acronyms used in Table 31.16 stand for *k*-Nearest-Neighbor (kNN), Condensed Nearest Neighbor (CNN), Selective Nearest Neighbor (SNN), Instance Based (IB), and Decremental Reduction Optimization Procedure (DROP).

Computational experiment was carried on a SGI Challenge R4400 workstation with 12 processors. A number of slave workers used by the master varied in different runs from 5 to 15 and in each run were chosen randomly. A size of the initial population in the *IRA3* implementation of the PLA was set to 100. A maximum number of repetitions for each learning and improvement procedure was set to 500. In the *IRA4* implementation the respective parameters were set to 50 for the initial population size and 500 for the number of iterations.

The proposed and provisionally validated simple, PLA-based heuristic instance reduction algorithms, can be used to increase efficiency of the supervised learning. Computational experiment results support the claim that reducing training set size still preserves basic features of the analyzed data and can be even beneficial to a classifier accuracy. The approach extends a range of available instance reduction algorithms. Moreover, it is shown that the proposed algorithm can be, for some problems, competitive in comparison with the existing techniques. Possible extension of the approach could focus on establishing decision rules for finding a representation level suitable for each cluster thus allowing a variable representation level for different clusters.

References

Barbucha D. and Jędrzejowicz, P. A population learning algorithm for solving the generalized segregated storage problem. In P. Sincak, J. Vascak, V. Kvasnicka, and R. Mesiar (Eds.), The State of the Art. In Computational Intelligence", Advances in Soft Computing, Physica-Verlag, Heidelberg, New York, 2000, pp. 355–360.

Chung, C-J. and Reynolds, R.G. CAEP: An evolution-based tool for real-valued function optimization using cultural algorithms. *Journal of Artificial Intelligence Tools*, 7, 1998, 239–292.

Colorni, A., Dorigo, M., and Maniezzo, V. An investigation of some properties of an ant algorithm. In R. Manner and B. Manderick (Eds.), *Proceedings of the 2nd European Conference on Parallel Problem Solving from Nature*, Elsevier, Amsterdam, 1992, pp. 509–520.

Christofides, N., Alvarez-Valdes, R., and Tamarit, J.M. Project scheduling with resource constraints: A branch and bound approach. *European Journal of Operational Research*, 29, 1987, 262–273.

Czarnowski, I., and Jędrzejowicz, P. Application of the parallel population learning algorithm to training feed-forward ANN. In P. Sincak et al. (Eds.), *Intelligent Technologies–Theory and Applications*, IOS Press, Amsterdam, 2002a, pp. 10–16.

Czarnowski, I. and Jędrzejowicz, P. An approach to artificial neural network training. In M. Bremer, A. Preece, F. Coenen (Eds.), *Research and Development in Intelligent Systems XIX*, Springer-Verlag, Heidelberg, 2002b, pp. 149–160.

Czarnowski, I. and Jędrzejowicz, P. An approach to instance reduction in supervised learning. In F. Coenen, A. Preece, and A.L. Macintosh (Eds.), *Research and Development in Intelligent Systems XX*, Springer, London, 2003, pp. 267–280.

Czarnowski, I., Gutjahr, W.J., Jędrzejowicz, P., Ratajczak, E., Skakowski, A., and Wierzbowska, I. Scheduling multiprocessor tasks in presence of correlated failures. *Central European Journal of Operations Research*, 11, 2003, 163–182.

Davis, L. *Handbook of Genetic Algorithms*, Van Nostrand Reinhold, New York, 1991.

Dawkins, R. *The Selfish Gene*, Oxford University Press, Oxford, 1976.

Dorigo, M., and Di Caro, G. The ant colony optimization meta-heuristic. In D. Corne, M. Dorigo, and F. Glover, (Eds.), *New Ideas in Optimization*, McGraw-Hill, New York, 1999, pp. 11–32.

The European Network of Excellence on Intelligent Technologies for Smart Adaptive Systems (EUNITE) — EUNITE World competition in domain of Intelligent Technologies — 0, 2002.

Fahlman, S.E. and Lebiere, C. The cascade-correlation learning architecture. In E. Touretzky (Ed.), *Advances in Neural Information Processing II*, Morgan Kauffman, San Mateo, CA, 1990, pp. 524–532.

Feo, T.A. and Rosende, M.G.C. Greedy randomized adaptive search procedures. *Journal of Global Optimization*, 6, 1995, 109–133.

Glover, F. Heuristics for integer programming using surrogate constraints. *Decision Sciences*, 8, 1977, 156–166.

Glover, F. Tabu search: A tutorial. *Interfaces*, 20, 1990, 74–94.

Goldberg, D.E. *Genetic Algorithms in Search, Optimization, and Machine Learning.* Addison-Wesley, Reading, MA, 1989.

Golden, B.L., Laporte, G., and Taillard, E.D. An adaptive memory heuristic for a class of vehicle routing problems with minmax objective. *Computers & Operations Research*, 24, 1997, 445–452.

Gordon, V. and Whitley, D. Serial and parallel genetic algorithms as function optimizers. In S. Forrest (Ed.), *Proceedings of the 5th Internation Conference on Genetic Algorithms*, Morgan Kaufman Publishers, San Mateo, CA, 1993, pp. 177–183.

Hertz, A. and Kobler, D. A framework for the description of population based methods, 16 European Conferencee on Operational Research, *Tutorials and Research Reviews*, 1998, 1–21.

Holland, J.H. Adaptation in Natural and Artificial Systems, The University of Michigan Press, Ann Arbor, MI, 1975.

Jędrzejowicz, P. Social learning algorithm as a tool for solving some difficult scheduling problems. *Foundation of Computing and Decision Sciences*, 24, 1999, 51–66.

Jedrzejowicz, J., Jędrzejowicz, P. PLA-based permutation scheduling, *Foundation of Computing and Decision Sciences*, 28, 2003, 159–177.

Jedrzejowicz, J. and Jędrzejowicz, P. Population-based approach to multiprocessor task scheduling in multistage hybrid flowshops. In V. Palade, R.J. Howlett, and L. Jain (Eds.), *Knowledge-Based Intelligent Information and Engineering Systems, Lecture Notes in Artificial Intelligence*, No. 2773, Springer, Berlin, 2003, pp. 279–286.

Jędrzejowicz, P., Rosicka, L., and Wierzbowska, I. Population learning algorithms for optimizing system structure under multiple objectives. In *Proceedings of the XVIII European Conference on System Dependability and Safety – ESREL'02*, Lyon, 2002, pp. 420–424.

Jędrzejowicz, P. and Ratajczak, E. Scheduling multiprocessor tasks with correlated failures using population learning algorithm. V. Kurkova, (Ed.), *Artificial Neural Nets and Genetic Algorithms*, Springer Computer Science, Wien-New York, 2001, pp. 296–300.

Jędrzejowicz, P. and Ratajczak, E. Population learning algorithm for resource-constrained project scheduling. In D.W. Pearson, N.C. Steele, and R.F. Albrecht (Eds.), *Artificial Neural Nets and Genetic Algorithms*, Springer Computer Science, Springer, Wien, 2003, pp. 223-228.

Jędrzejowicz, P. and Wierzbowska, I. Scheduling multiple variant programs under hard real time constraints, *European Journal of Operational Research*, 127, 2000, 458–465.

Kennedy, J. and Eberhart, R.C. Particle swarm optimisation. In *Proceedings of IEEE International Conference on Neural Networks*, Piscataway, NJ, 1995, pp. 1942–1948.

Kirkpatrick, S., Gelatt, C.D., Jr., and Vecchi, M.P. Optimization by simulated annealing, *Science*, 220, 1983, 671–680.

Likas, A., Vlassis, N., and Verbeek, J.J. The Global k-Means Clustering Algorithm, Technical report, Computer Science Institute, University of Amsterdam, IAS-UVA-01-02, 2001.

Mangasarian, O.L. and Wolberg, W.H. Cancer diagnosis via linear programming, *SIAM News*, 23, 1990, 1–18.

Merz, C.J. and Murphy, M. UCI Repository of machine learning databases, (http://www.ics.uci.edu/~mlearn/MLRepository.html), University of California, Department of Information and Computer Science, Irvine CA, 1998.

Michalewicz, Z. *Genetic Algorithms + Data Structures = Evolution Programs*, 2nd Extended edition, Springer-Verlag, Berlin, Heidelberg, New York, 1994.

Norman M.G. and Moscato P. A Competitive and Cooperative Approach to Complex Combinatorial Search, Caltech Concurrent Computation Program, C3P report 790, 1989.

Prechelt, L. and Proben, I. A Set of Benchmark and Benchmarking Rules for Neural Network Training Algorithm, Fakultät für Informatik, Universität Karlsruhe, Anonymous /pub/papers/techraports/1994/1994-21.ps.z. on ftp.ira.uka.de, Technical report no. 21/94, Karlsruhe, 1994.

Reynolds, R.G. An introduction to Cultural Algorithms. In A.V. Sebald, L.J. Fogel (Eds.), *Proceedings of the 3rd Annual Conference on Evolutionary Programming*, World Scientific, River Edge NJ, 1994, pp. 131–139.

Shang, Yi. and Wah, B.W. A global optimization method for neural network training, Conference of Neural Networks. *IEEE Computer*, 29, 1996, 45–54.

Taillard, E. Benchmarks for basic scheduling instances. *European Journal of Operational Research*, 64, 1993, 278–285.

Wilson, D.R. and Martinez, T.R. Reduction techniques for instance-based learning algorithm. *Machine Learning*, 38, 2000, 257–286.

32

Biology-Derived Algorithms in Engineering Optimization

Xin-She Yang

32.1 Introduction

Biology-derived algorithms are an important part of computational sciences, which are essential to many scientific disciplines and engineering applications. Many computational methods are derived from or based on the analogy to natural evolution and biological activities, and these biologically inspired computations include genetic algorithms, neural networks, cellular automata, and other algorithms. However, a substantial amount of computations today are still using conventional methods such as finite difference, finite element, and finite volume methods. New algorithms are often developed in the form of a hybrid combination of biology-derived algorithms and conventional methods, and this is especially true in the field of engineering optimizations. Engineering problems with optimization objectives are often difficult and time consuming, and the application of nature or biology-inspired algorithms in combination with the conventional optimization methods has been very successful in the last several decades.

There are five paradigms of nature-inspired evolutionary computations: genetic algorithms, evolutionary programming, evolutionary strategies, genetic programming, and classifier systems (Holland, 1975; Goldberg, 1989; Mitchell, 1996; Flake, 1998). Genetic algorithm (GA), developed by John Holland and his collaborators in the 1960s and 1970s, is a model or abstraction of biological evolution, which includes the following operators: crossover, mutation, inversion, and selection. This is done by the representation within a computer of a population of individuals corresponding to chromosomes in terms of a set of character strings, and the individuals in the population then evolve through the crossover and mutation of the string from parents, and the selection or survival according to their fitness. Evolutionary programming (EP), first developed by Lawrence J. Fogel in 1960, is a stochastic optimization strategy similar to GAs. But it differs from GAs in that there is no constraint on the representation of solutions in EP and the representation often follows the problem. In addition, the EPs do not attempt to model genetic operations closely in the sense that the crossover operation is not used in EPs. The mutation operation simply changes aspects of the solution according to a statistical distribution, such as multivariate Gaussian perturbations, instead of bit-flopping, which is often done in GAs. As the global optimum is approached, the rate of mutation is often reduced. Evolutionary strategies (ESs) were conceived by Ingo Rechenberg and Hans-Paul Schwefel in 1963, later joined by Peter Bienert, to solve technical optimization problems (Rechenberg, 1973). Although they were developed independently of one another, both ESs and EPs have many similarities in implementations. Typically, they both operate on real-values to solve real-valued function optimization in contrast with the encoding in Gas. Multivariate Gaussian mutation with zero mean are used for each parent population and appropriate selection criteria are used to determine which solution to keep or remove. However, EPs often use stochastic selection via a tournament and the selection eliminates those solutions with the least wins, while the ESs use deterministic selection criterion that removes the worst solutions directly based on the evaluations of certain functions (Heitkotter and Beasley, 2000). In addition, recombination is possible in an ES as it is an abstraction of evolution at the level of individual behavior in contrast to the abstraction of evolution at the level of reproductive populations and no recombination mechanisms in EPs.

The aforementioned three areas have the most impact in the development of evolutionary computations, and, in fact, evolutionary computation has been chosen as the general term that encompasses all these areas and some new areas. In recent years, two more paradigms in evolutionary computation have attracted substantial attention: Genetic programming and classifier systems. Genetic programming (GP) was introduced in the early 1990s by John Koza (1992), and it extends GAs using parse trees to represent functions and programs. The programs in the population consist of elements from the function sets, rather than fixed-length character strings, selected appropriately to be the solutions to the problems. The crossover operation is done through randomly selected subtrees in the individuals according to their fitness; the mutation operator is not used in GP. On the other hand, a classifier system (CFS), another invention by John Holland, is an adaptive system that combines many methods of adaptation with learning and evolution. Such hybrid systems can adapt behaviors toward a changing environment by using GAs with adding capacities such as memory, recursion, or iterations. In fact, we can essentially consider the CFSs as general-purpose computing machines that are modified by both environmental feedback and the underlying GAs (Holland, 1975, 1995; Michaelewicz, 1996; Flake, 1998).

Biology-derived algorithms are applicable to a wide variety of optimization problems. For example, optimization functions can have discrete, continuous, or even mixed parameters without any a priori assumptions about their continuity and differentiability. Thus, evolutionary algorithms are particularly suitable for parameter search and optimization problems. In addition, they are easy for parallel implementation. However, evolutionary algorithms are usually computationally intensive, and there is no absolute guarantee for the quality of the global optimizations. Besides, the tuning of the parameters can be very difficult for any given algorithms. Furthermore, there are many evolutionary algorithms with different suitabilities and the best choice of a particular algorithm depends on the type and characteristics of the problems concerned. However, great progress has been made in the last several decades in the application of evolutionary algorithms in engineering optimizations. In this chapter, we will focus on some of the important areas of the application of GAs in engineering optimizations.

32.2 Biology-Derived Algorithms

There are many biology-derived algorithms that are popular in evolutionary computations. For engineering applications in particular, four types of algorithms are very useful and hence relevant. They are

GAs, photosynthetic algorithms (PAs), neural networks, and cellular automata. We will briefly discuss these algorithms in this section, and we will focus on the application of GAs and PAs in engineering optimizations in Section 32.3.

32.2.1 Genetic Algorithms

The essence of GAs involves the encoding of an optimization function as arrays of bits or character strings to represent the chromosomes, the manipulation operations of strings by genetic operators, and the selection according to their fitness to find a solution to the problem concerned. This is often done by the following procedure: (1) encoding of the objectives or optimization functions; (2) defining a fitness function or selection criterion; (3) creating a population of individuals; (4) evolution cycle or iterations by evaluating the fitness of all the individuals in the population, creating a new population by performing crossover, mutation, and inversion, fitness-proportionate reproduction, etc., and replacing the old population and iterating using the new population; (5) decoding the results to obtain the solution to the problem.

One iteration of creating new populations is called a generation. Fixed-length character strings are used in most GAs during each generation although there is substantial research on variable-length string and coding structures. The coding of objective functions is usually in the form of binary arrays or real-valued arrays in adaptive GAs. For simplicity, we use the binary string for coding for describing genetic operators. Genetic operators include crossover, mutation, inversion, and selection. The crossover of two parent strings is the main operator with highest probability p_c (usually, 0.6 to 1.0) and is carried out by switching one segment of one string with the corresponding segment on another string at a random position (see Figure 32.1). The crossover carried out in this way is a single-point crossover. Crossover at multiple points is also used in many GAs to increase the efficiency of the algorithms. The mutation operation is achieved by the flopping of randomly selected bits, and the mutation probability p_m is usually small (say, 0.001 to 0.05), while the inversion of some part of a string is done by interchanging 0 and 1. The selection of an individual in a population is carried out by the evaluation of its fitness, and it can remain in the new generation if a certain threshold of fitness is reached or the reproduction of a population is fitness-proportionate. One of the key parts is the formulation or choice of fitness functions that determines the selection criterion in a particular problem.

Further, just as crossover can be carried out at multiple points, mutations can also occur at multiple sites. More complex and adaptive GAs are being actively researched and there is a vast literature on this topic (Chipperfield et al. 1994; Pohlheim, 1999). In Section 32.3, we will give examples of GAs and their applications in engineering optimizations.

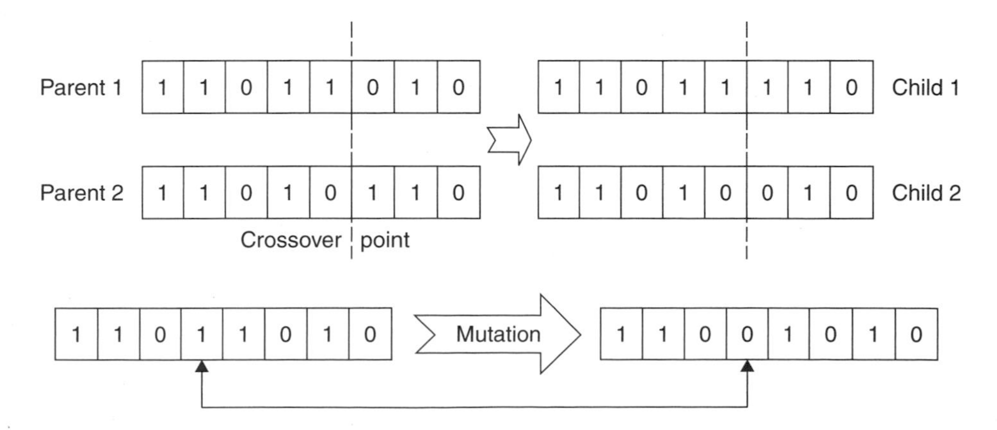

FIGURE 32.1 Diagram of crossover and mutation in a GA.

32.2.2 Photosynthetic Algorithms

The PA was first introduced by Murase (2000) to optimize the parameter estimation in finite element inverse analysis. The PA is a good example of biology-derived algorithms in the sense that its computational procedure corresponds well to the real photosynthesis process in green plants. Photosynthesis uses water and CO_2 to produce glucose and oxygen when there is light and in the presence of chloroplasts. The overall reaction

$$6CO_2 + 12H_2O \xrightarrow{\text{light and green plants}} C_6H_{12}O_6 + 6O_2 + 6H_2O$$

is just a simple version of a complicated process. Other factors, such as temperature, concentration of CO_2, water content, etc., being equal, the reaction efficiency depends largely on light intensity. The important part of photosynthetic reactions is the dark reactions that consist of a biological process including two cycles: the Benson–Calvin cycle and photorespiration cycle. The balance between these two cycles can be considered as a natural optimization procedure that maximizes the efficiency of sugar production under the continuous variations of light energy input (Murase, 2000).

Murase's PA uses the rules governing the conversion of carbon molecules in the Benson–Calvin cycle (with a product or feedback from dihydroxyacetone phosphate or DHAP) and photorespiration reactions. The product DHAP serves as the knowledge strings of the algorithm and optimization is reached when the quality or the fitness of the products no longer improves. An interesting feature of such algorithms is that the stimulation is a function of light intensity that is randomly changing and affects the rate of photorespiration. The ratio of O_2 to CO_2 concentration determines the ratio of the Benson–Calvin and photorespiration cycles. A PA consists of the following steps: (1) the coding of optimization functions in terms of fixed-length DHAP strings (16-bit in Murase's PA) and a random generation of light intensity (L); (2) the CO_2 fixation rate r is then evaluated by the following equation: $r = V_{max}/(1 + A/L)$, where V_{max} is the maximum fixation rate of CO_2 and A is its affinity constant; (3) either the Benson–Calvin cycle or photorespiration cycle is chosen for the next step, depending on the CO_2 fixation rate, and the 16-bit strings are shuffled in both cycles according to the rule of carbon molecule combination in photosynthetic pathways; (4) after some iterations, the fitness of the intermediate strings is evaluated, and the best fit remains as a DHAP, then the results are decoded into the solution of the optimization problem (see Figure 32.2).

In the next section, we will present an example of parametric inversion and optimization using PA in finite element inverse analysis.

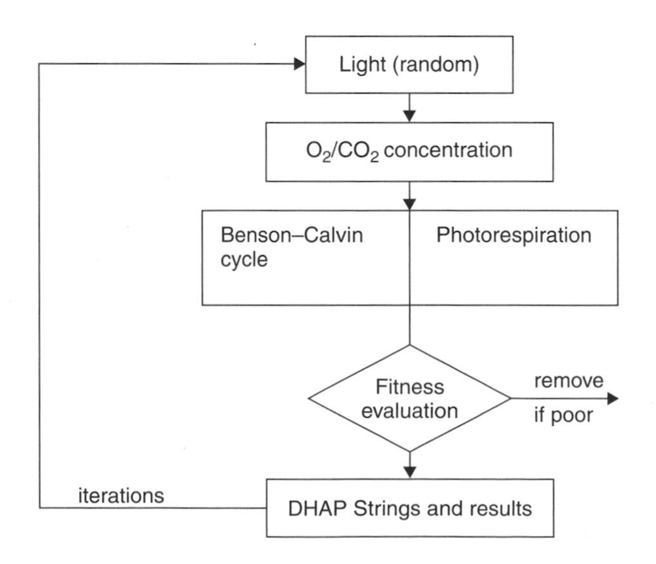

FIGURE 32.2 Scheme of Murase's PAs.

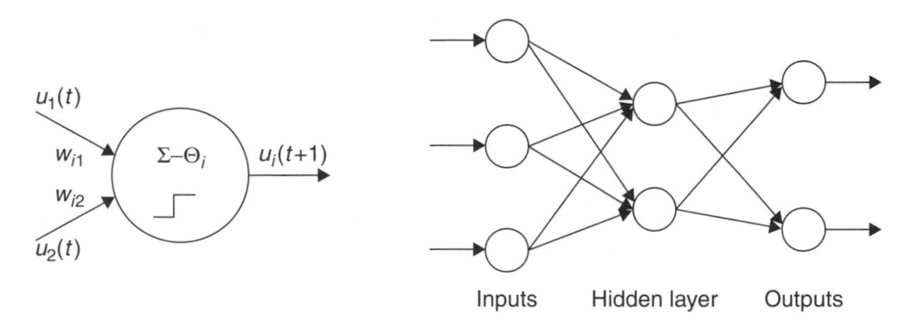

FIGURE 32.3 Diagram of a McCulloch–Pitts neuron (left) and neural networks (right).

32.2.3 Neural Networks

Neural networks, and the associated machine-learning algorithms, is one more type of biology-inspired algorithms, which uses a network of interconnected neurons with the intention of imitating the neural activities in human brains. These neurons have high interconnectivity and feedback, and their connectivity can be weakened or enhanced during the learning process. The simplest model for neurons is the McCulloch–Pitts model (see Figure 32.3) for multiple inputs $u_1(t), \ldots, u_n(t)$, and the output $u_i(t)$. The activation in a neuron or a neuron's state is determined by

$$u_i(t+1) = H\left[\sum_{j=1}^{n} w_{ij} u_i(t) - \Theta_i\right],$$

where $H(x)$ is the Heaviside unit step function that $H(x) = 1$ if $x \geq 0$ and $H(x) = 0$ otherwise. The weight coefficient w_{ij} is considered as the synaptic strength of the connection from neuron j to neuron i. For each neuron, it can be activated only if its threshold Θ_i is reached. One can consider a single neuron as a simple computer that gives the output 1 or yes if the weighted sum of incoming signals is greater than the threshold, otherwise it outputs 0 or no.

Real power comes from the combination of nonlinear activation functions with multiple neurons (McCulloch and Pitts, 1943; Flake, 1998). Figure 32.3 also shows an example of feed-forward neural networks. The key element of an artificial neural network (ANN) is the novel structure of such an information processing system that consists of a large number of interconnected processing neurons. These neurons work together to solve specific problems by adaptive learning through examples and self-organization. A trained neural network in a given category can solve and answer what-if type questions to a particular problem when new situations of interest are given. Due to the real-time capability and parallel architecture as well as adaptive learning, neural networks have been applied to solve many real-world problems such as industrial process control, data validation, pattern recognition, and other systems of artificial intelligence such as drug design and diagnosis of cardiovascular conditions. Optimization is just one possibility of such applications, and often the optimization functions can change with time as is the case in industrial process control, target marketing, and business forecasting (Haykins, 1994). On the other hand, the training of a network may take considerable time and a good training database or examples that are specific to a particular problem are required. However, neural networks will, gradually, come to play an important role in engineering applications because of its flexibility and adaptability in learning.

32.2.4 Cellular Automata

Cellular automata (CA) were also inspired by biological evolution. On a regular grid of cells, each cell has finite number of states. Their states are updated according to certain local rules that are functions of the states of neighbor cells and the current state of the cell concerned. The states of the cells evolve with time in a discrete manner, and complex characteristics can be observed and studied. For more details on this topic, readers can refer to Chapters 1 and 18 in this handbook.

There is some similarity between finite state CA and conventional numerical methods such as finite difference methods. If one considers the finite different method as real-valued CA, and the real-values are always converted to finite discrete values due to the round-off in the implementation on a computer, then there is no substantial difference between a finite difference method and a finite state CA.

However, CA are easier to parallelize and more numerically stable. In addition, finite difference schemes are based on differential equations and it is sometimes straightforward to formulate a CA from the corresponding partial differential equations via appropriate finite differencing procedure; however, it is usually very difficult to conversely obtain a differential equation for a given CA (see Chapter 18 in this handbook).

An optimization problem can be solved using CA if the objective functions can be coded to be associated with the states of the CA and the parameters are properly associated with automaton rules. This is an area under active research. One of the advantages of CA is that it can simulate many processes such as reaction–diffusion, fluid flow, phase transition, percolation, waves, and biological evolution. Artificial intelligence also uses CA intensively.

32.2.5 Optimization

Many problems in engineering and other disciplines involve optimizations that depend on a number of parameters, and the choice of these parameters affects the performance or objectives of the system concerned. The optimization target is often measured in terms of objective or fitness functions in qualitative models. Engineering design and testing often require an iteration process with parameter adjustment. Optimization functions are generally formulated as:

$$\begin{aligned} &\text{Optimize: } f(\mathbf{x}), \\ &\text{Subject to: } g_i(\mathbf{x}) \geq 0, \quad i = 1, 2, \ldots, N; \qquad h_j(x) = 0, \quad j = 1, 2, \ldots, M. \\ &\quad \text{where } \mathbf{x} = (x_1, x_2, \ldots, x_n), \quad \mathbf{x} \in \Omega(\text{parameter space}). \end{aligned}$$

Optimization can be expressed either as maximization or more often as minimization (Deb, 1995, 2000). As parameter variations are usually very large, systematic adaptive searching or optimization procedures are required. In the past several decades, researchers have developed many optimization algorithms. Examples of conventional methods are hill climbing, gradient methods, random walk, simulated annealing, heuristic methods, etc. Examples of evolutionary or biology-inspired algorithms are GAs, photosynthetic methods, neural network, and many others.

The methods used to solve a particular problem depend largely on the type and characteristics of the optimization problem itself. There is no universal method that works for all problems, and there is generally no guarantee to find the optimal solution in global optimizations. In general, we can emphasize on the best estimate or suboptimal solutions under the given conditions. Knowledge about the particular problem concerned is always helpful to make the appropriate choice of the best or most efficient methods for the optimization procedure. In this chapter, however, we focus mainly on biology-inspired algorithms and their applications in engineering optimizations.

32.3 Engineering Optimization and Applications

Biology-derived algorithms such as GAs and PAs have many applications in engineering optimizations. However, as we mentioned earlier, the choice of methods for optimization in a particular problem depends on the nature of the problem and the quality of solutions concerned. We will now discuss optimization problems and related issues in various engineering applications.

32.3.1 Function and Multilevel Optimizations

For optimization of a function using GAs, one way is to use the simplest GA with a fitness function: $F = A - y$ with A being the large constant and $y = f(\mathbf{x})$, thus the objective is to maximize the fitness function and subsequently minimize the objective function $f(\mathbf{x})$. However, there are many different ways of defining a fitness function. For example, we can use the individual fitness assignment relative to the whole population

$$F(x_i) = \frac{f(x_i)}{\sum_{i=1}^{N} f(x_i)},$$

where x_i is the phenotypic value of individual i, and N is the population size. For the generalized De Jong's (1975) test function

$$f(\mathbf{x}) = \sum_{i=1}^{n} x_i^{2\alpha}, \quad |x| \leq r, \quad \alpha = 1, 2, \ldots, m,$$

where α is a positive integer and r is the half-length of the domain. This function has a minimum of $f(\mathbf{x}) = 0$ at $\mathbf{x} = 0$. For the values of $\alpha = 3$, $r = 256$, and $n = 40$, the results of optimization of this test function are shown in Figure 32.4 using GAs.

The function we just discussed is relatively simple in the sense that it is single-peaked. In reality, many functions are multi-peaked and the optimization is thus multileveled. Keane (1995) studied the following bumby function in a multi-peaked and multileveled optimization problem

$$f(x, y) = \frac{\sin^2(x - y)\sin^2(x + y)}{\sqrt{x^2 + y^2}}, \quad 0 < x, \quad y < 10.$$

The optimization problem is to find (x, y) starting $(5, 5)$ to maximize the function $f(x, y)$ subject to: $x + y \leq 15$ and $xy \geq 3/4$. In this problem, optimization is difficult because it is nearly symmetrical about $x = y$, and while the peaks occur in pairs one is bigger than the other. In addition, the true maximum is $f(1.593, 0.471) = 0.365$, which is defined by a constraint boundary. Figure 32.5 shows the surface variation of the multi-peaked bumpy function.

Although the properties of this bumpy function make it difficult for most optimizers and algorithms, GAs and other evolutionary algorithms perform well for this function and it has been widely used as a test

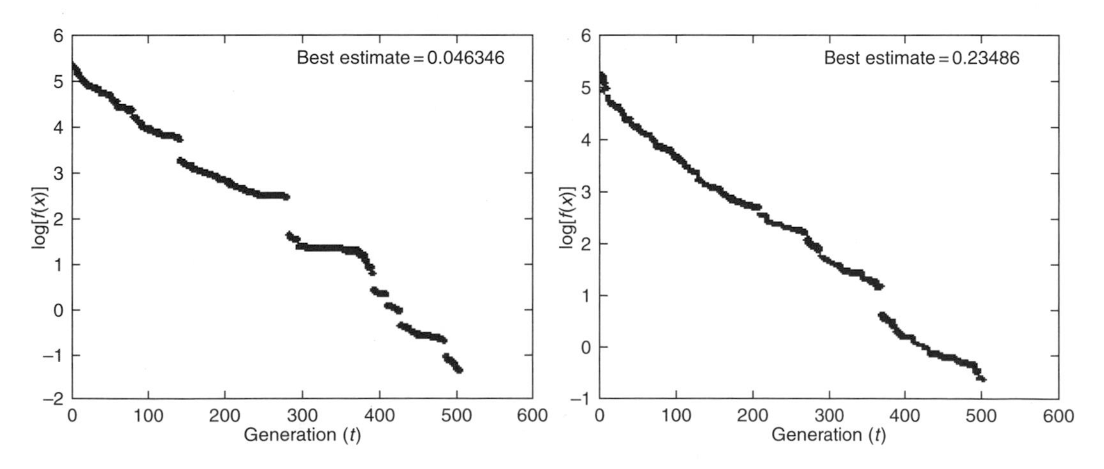

FIGURE 32.4 Function optimization using GAs. Two runs will give slightly different results due to the stochastic nature of GAs, but they produce better estimates: $f(\mathbf{x}) \to 0$ as the generation increases.

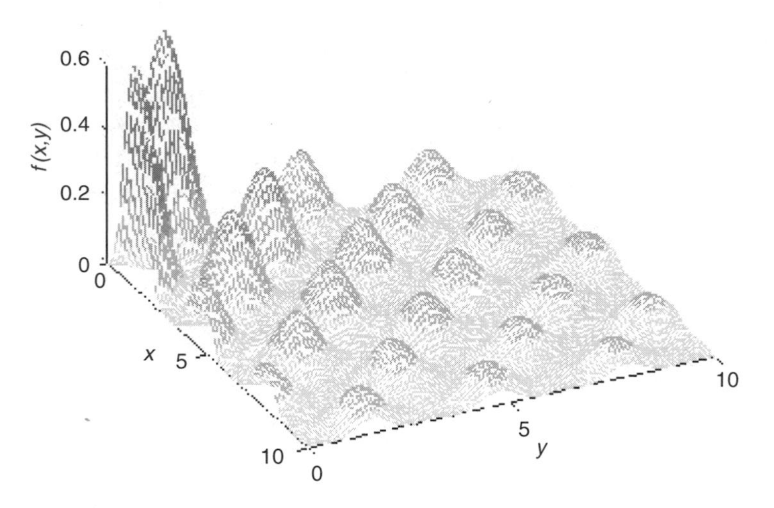

FIGURE 32.5 Surface of the multi-peaked bumpy function.

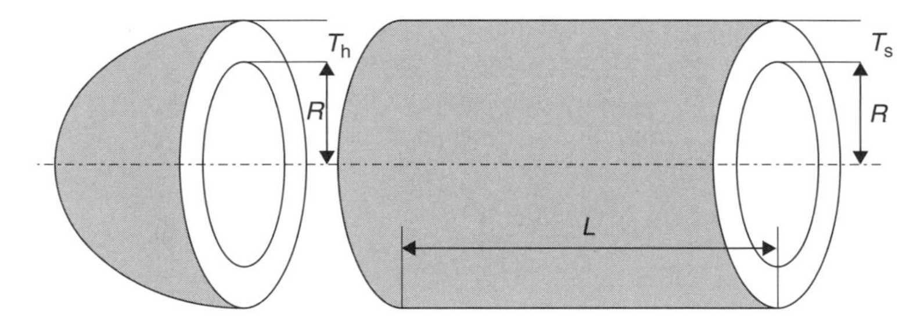

FIGURE 32.6 Diagram of the pressure vessel.

function in GAs for comparative studies of various evolutionary algorithms or in multilevel optimization environments (Jenkins, 1997; El-Beltagy and Keane, 1999).

32.3.2 Shape Design and Optimization

Most engineering design problems, especially in shape design, aim to reduce the cost, weight, and volume and increase the performance and quality of the products. The optimization process starts with the transformation of design specification and descriptions into optimization functions and constraints. The structure and parameters of a product depend on the functionality and manufacturability, and thus considerable effort has been put into the modeling of the design process and search technique to find the optimal solution in the search space, which comprises the set of all designs with all allowable values of design parameters (Renner and Ekart, 2003). Genetic algorithms have been applied in many areas of engineering design such as conceptual design, shape optimization, data fitting, and robot path design.

A well-studied example is the design of a pressure vessel (Kannan and Kramer, 1994; Coello, 2000) using different algorithms such as augmented Lagrangian multiplier and GAs. Figure 32.6 shows the diagram of the parameter notations of the pressure vessel. The vessel is cylindrical and capped at both ends by hemispherical heads with four design variables: thickness of the shell T_s, thickness of the head T_h, inner radius R, and the length of the cylindrical part L. The objective of the design is to minimize the total cost including that of the material, forming, as well as welding. Using the notation given by Kannan and

Kramer, the optimization problem can be expressed as:

$$\text{Minimize: } f(\mathbf{x}) = 0.6224x_1x_2x_3 + 1.7781x_2x_3^2 + 3.1611x_1^2x_4 + 19.84x_1^2x_3.$$

$$\mathbf{x} = (x_1, x_2, x_3, x_4)^T = (T_s, T_h, R, L)^T,$$

$$\text{Subject to: } g_1(\mathbf{x}) = -x_1 + 0.0193x_3 \leq 0, \quad g_2(\mathbf{x}) = -x_2 + 0.00954x_3 \leq 0,$$

$$g_3(\mathbf{x}) = -\pi x_3^2 x_4 - 4\pi x_3^3/3 + 1296000 \leq 0, \quad g_4(\mathbf{x}) = x_4 - 240 \leq 0.$$

The values for x_1, x_2 should be considered as integer multiples of 0.0625. Using the same constraints as given in Coello (2000), the variables are in the ranges: $1 \leq x_1, x_2 \leq 99, 10.0000 \leq x_3 . x_4 \leq 100.0000$ (with a four-decimal precision). By coding the GAs with a population of 44-bit strings for each individual (4-bits for $x_1 \cdot x_2$; 18-bits for $x_3 \cdot x_4$), similar to that by Wu and Chow (1994), we can solve the optimization problem for the pressure vessel. After several runs, the best solution obtained is $x_* = (1.125, 0.625, 58.2906, 43.6926)$ with $f(\mathbf{x}) = 7197.9912\$$, which is compared with the results $x_* = (1.125, 0.625, 28.291, 43.690)$ and $f(\mathbf{x}) = 7198.0428\$$ obtained by Kannan and Kramer (1994).

32.3.3 Finite Element Inverse Analysis

The usage and efficiency of Murase's PA described in Section 32.3.4 can be demonstrated in the application of finite element inverse analysis. Finite element analysis (FEA) in structural engineering is forward modeling as the aims are to calculate the displacements at various positions for given loading conditions and material properties such as Young's modulus (E) and Poisson's ratio (ν). This forward FEA is widely used in engineering design and applications. Sometimes, inverse problems need to be solved. For a given structure with known loading conditions and measured displacement, the objective is to find or invert the material properties E and ν, which may be required for testing new materials and design optimizations. It is well known that inverse problems are usually very difficult, and this is especially true for finite element inverse analysis.

To show how it works, we use a test example similar to that proposed by Murase (2000). A simple beam system of 5 unit length $\times$ 10 unit length (see Figure 32.7) consists of five nodes and four elements whose Young's modulus and Poisson's ratio may be different. Nodes 1 and 2 are fixed, and a unit vertical load is applied at node 4 while the other nodes deform freely. Let us denote the displacement vector

$$\mathbf{U} = (u_1, v_1, u_2, v_2, u_3, v_3, u_4, v_4, u_5, v_5)^T,$$

where $u_1 = v_1 = u_2 = v_2 = 0$ (fixed). Measurements are made for other displacements. By using the PA with the values of the CO_2 affinity $A = 10000$, light intensity $L = 10^4$ to 5×10^4 lx, and maximum CO_2

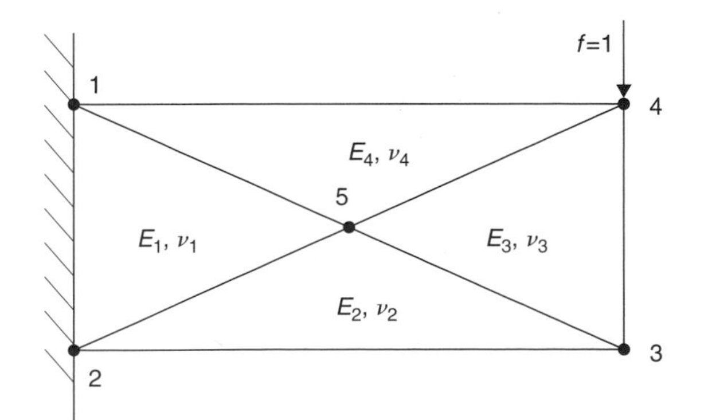

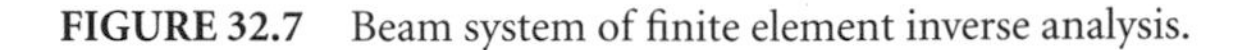
FIGURE 32.7 Beam system of finite element inverse analysis.

fixation speed $V_{max} = 30$, each of eight elastic modulus $(E_i, \nu_i)(i = 1, 2, 3, 4)$ is coded as a 16-bit DHAP molecule string. For a target vector

$$\mathbf{Y} = (E_1, \nu_1, E_2, \nu_2, E_3, \nu_3, E_4, \nu_4) = (600, 0.25, 400, 0.35, 450, 0.30, 350, 0.32),$$

and measured displacements $\mathbf{U} = (0, 0, 0, 0, -0.0066, -0.0246, 0.0828, -0.2606, 0.0002, -0.0110)$, the best estimates after 500 iterations from the optimization by the PA are $\mathbf{Y} = (580, 0.24, 400, 0.31, 460, 0.29, 346, 0.26)$.

32.3.4 Inverse Initial-Value, Boundary-Value Problem Optimization

Inverse initial-value, boundary-value problem (IVBV) is an optimization paradigm in which GAs have been used successfully (Karr et al., 2000). Some conventional algorithms for solving such search optimizations are the trial-and-error iteration methods that usually start with a guessed solution, and the substitution into the partial differential equations and associated boundary conditions to calculate the errors between predicted values and measured or known values at various locations, then the new guessed or improved solutions are obtained by corrections according to the errors. The aim is to minimize the difference or errors, and the procedure stops once the given precision or tolerance criterion is satisfied. In this way, the inverse problem is actually transformed into an optimization problem.

We now use the heat equation and inverse procedure discussed by Karr et al. (2000) as an example to illustrate the IVBV optimization. On a square plate of unit dimensions, the diffusivity $\kappa(x, y)$ varies with locations (x, y). The heat equation and its boundary conditions can be written as:

$$\begin{aligned}&\frac{\partial u}{\partial t} = \nabla \cdot [\kappa(x, y)\nabla u], \quad 0 < x, y < 1, \quad t > 0,\\&u(x, y, 0) = 1, \quad u(x, 0, t) = u(x, 1, t) = u(0, y, t) = u(1, y, t) = 0.\end{aligned}$$

The domain is discretized as an $N \times N$ grid, and the measurements of values at $(x_i, y_j, t_n), (i, j = 1, 2, \ldots, N; n = 1, 2, 3)$. The data set consists of the measured value at N^2 points at three different times t_1, t_2, t_3. The objective is to inverse or estimate the N^2 diffusivity values at the N^2 distinct locations. The Karr's error metrics are defined as

$$E_u = A\frac{\sum_{i=1}^{N}\sum_{j=1}^{N}\left|u_{i,j}^{\text{measured}} - u_{i,j}^{\text{computed}}\right|}{\sum_{i=1}^{N}\sum_{j=1}^{N}\left|u_{i,j}^{\text{measured}}\right|}, \quad E_\kappa = A\frac{\sum_{i=1}^{N}\sum_{j=1}^{N}\left|\kappa_{i,j}^{\text{known}} - \kappa_{i,j}^{\text{predicted}}\right|}{\sum_{i=1}^{N}\sum_{j=1}^{N}\left|\kappa_{i,j}^{\text{known}}\right|},$$

where $A = 100$ is just a constant.

The floating-point GA proposed by Karr et al. for the inverse IVBV optimization can be summarized as the following procedure: (1) Generate randomly a population containing N solutions to the IVBV problem; the potential solutions are represented in a vector due to the variation of diffusivity κ with locations; (2) The error metric is computed for each of the potential solution vectors; (3) N new potential solution vectors are generated by genetic operators such as crossover and mutations in GAs. Selection of solutions depends on the required quality of the solutions with the aim to minimize the error metric and thereby remove solutions with large errors; (4) the iteration continues until the best acceptable solution is found.

On a grid of 16×16 points with a target matrix of diffusivity, after 40,000 random $\kappa(i, j)$ matrices were generated, the best value of the error metrics $E_u = 4.6050$ and $E_\kappa = 1.50 \times 10^{-2}$. Figure 32.8 shows the error metric E_κ associated with the best solution determined by the GA. The small values in the error metric imply that the inverse diffusivity matrix is very close to the true diffusivity matrix.

With some modifications, this type of GA can also be applied to the inverse analysis of other problems such as the inverse parametric estimation of Poisson equation and wave equations. In addition, nonlinear problems can also be studied using GAs.

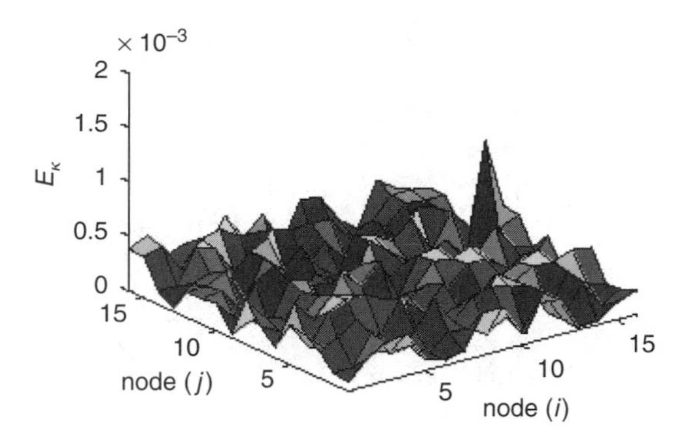

FIGURE 32.8 Error metric E_K associated with the best solution obtained by the GA algorithm.

The optimization methods using biology-derived algorithms and their engineering applications have been summarized. We used four examples to show how GAs and PAs can be applied to solve optimization problems in multilevel function optimization, shape design of pressure vessels, finite element inverse analysis of material properties, and the inversion of diffusivity matrix as an IVBV problem. Biology-inspired algorithms have many advantages over traditional optimization methods such as hill-climbing and calculus-based techniques due to parallelism and the ability to locate the best approximate solutions in very large search spaces. Furthermore, more powerful and flexible new generation algorithms can be formulated by combining existing and new evolutionary algorithms with classical optimization methods.

References

Chipperfield, A.J., Fleming, P.J., and Fonseca, C.M. Genetic algorithm tools for control systems engineering. In *Proceedings of Adaptive Computing in Engineering Design and Control*, Plymouth, pp. 128–133 (1994).

Coello, C.A. Use of a self-adaptive penalty approach for engineering optimization problems. *Computers in Industry*, **41** (2000) 113–127.

De Jong, K. Analysis of the Behaviour of a Class of Genetic Adaptive Systems, Ph.D. thesis, University of Michigan, Ann Arbor, MI (1975).

Deb, K. *Optimization for Engineering Design: Algorithms and Examples*, Prentice-Hall, New Delhi (1995).

Deb, K. An efficient constraint handling method for genetic algorithms. *Computer Methods and Applications of Mechanical Engineering*, **186** (2000) 311–338.

El-Beltagy, M.A. and Keane, A.J. A comparison of various optimization algorithms on a multilevel problem. *Engineering Applications of Artificial Intelligence*, **12** (1999) 639–654.

Flake, G.W. *The Computational Beauty of Nature: Computer Explorations of Fractals, Chaos, Complex Systems, and Adaptation*, MIT Press, Cambridge, MA (1998).

Goldberg, D.E. *Genetic Algorithms in Search, Optimization and Machine Learning*, Addison-Wesley, Reading, MA (1989).

Haykins, S. *Neural Networks: A Comprehensive Foundation*, MacMillan, New York (1994).

Heitkotter, J. and Beasley, D. *The Hitch-Hiker's Guide to Evolutionary Computation: A List of Frequently Asked Questions (FAQ)*, (2000) Available via anonymous FTP from ftp://rtfm.mit.edu/pub/usenet/news.answers/ai-faq/genetic/ About 110 pages.

Holland, J. *Adaptation in Natural and Artificial Systems*, University of Michigan Press, Ann Anbor, (1975).

Holland, J.H. *Hidden Order: How Adaptation Builds Complexity*, Addison-Wesley, Reading, MA (1995).

Jenkins, W.M. On the applications of natural algorithms to structural design optimization. *Engineering Structures*, **19** (1997) 302–308.

Kannan, B.K. and Kramer, S.N. An augmented Lagrange multiplier based method for mixed integer discrete continuous optimization and its application to mechanical design. *Journal Mechanical Design*, Transaction of ASME, **116** (1994) 318–320.

Karr, C.L., Yakushin, I., and Nicolosi, K. Solving inverse initial-value, boundary-valued problems via genetic algorithms. *Engineering Applications of Artificial Intelligence*, **13** (2000) 625–633.

Keane, A.J. Genetic algorithm optimization of multi-peak problems: Studies in convergence and robustness, *Artificial Intelligence in Engineering*, **9** (1995) 75–83.

Koza, J.R. *Genetic Programming: On the Programming of Computers by Natural Selection*, MIT Press, Cambridge, MA (1992).

McCulloch, W.S. and Pitts, W. A logical calculus of the idea immanent in nervous activity, Bulletin of Mathematical Biophysics, **5** (1943) 115–133.

Michaelewicz, Z. *Genetic Algorithm + Data Structure=Evolution Programming*, Springer-Verlag, New York (1996).

Mitchell, M. *An Introduction to Genetic Algorithms*, Cambridge, MIT Press, MA (1996).

Murase, H. Finite element inverse analysis using a photosynthetic algorithm. *Computers and Electronics in Agriculture*, **29** (2000) 115–123.

Pohlheim, H. *Genetic and Evolutionary Algorithm Toolbox for Matlab* (geatbx.com) (1999).

Rechenberg, I. *Evolutionsstrategie: Optimerrung Technischer Systeme nach Prinzipien der Biologischen Evolution*, FrommannHolzboog, Stuttgart (1973).

Renner, G. and Ekart, A. Genetic algorithms in computer aided design. *Computer-Aided Design*, **35** (2003) 709–726.

Wu, S.Y. and Chow, P. T. Genetic algorithms for solving mixed-discrete optimization problems. *Journal of the Franklin Institute*, **331** (1994) 381–401.

33

Biomimetic Models for Wireless Sensor Networks

Kennie H. Jones
Kenneth N. Lodding
Stephan Olariu
Ashraf Wadaa
Larry Wilson
Mohamed Eltoweissy

33.1 Introduction

With the advent of Micro-Electrical-Mechanical Systems (MEMSs) technology, complex and ubiquitous control of the physical environment by machines will facilitate the diversity of mechanization and automation, long promised by visionaries. Wireless sensor networks have appeared as the first generation of this revolutionary technology. The small size and low cost of sensor devices will enable deployment of massive numbers, but initially place severe limitations on the computing, communicating, and power capabilities of these devices. With these constraints, research efforts have concentrated on developing techniques for executing simple tasks with minimal energy expense. But, as MEMS evolve, computing and communicating capabilities are expected to improve at an accelerating rate and new techniques for supplying energy will significantly reduce the low power constraint. Increased capabilities will be possible, and it is predicted that societies of machines will evolve to be autonomous, cooperative, fault-tolerant, self-regulating, and self-healing. Improvements in biomimetic software and evolvable hardware will lead to self-sustaining communities of machines with emergent behavior that autonomously operate and adapt to changes in the environment. The main goal of this chapter is to investigate biomimetic models in relation to their potential application to the evolution of these systems, thus providing a framework that guides the evolution from the current primitive organizations of sensor nodes to pervasive societies of intelligent electromechanical systems.

33.2 Background

The dream of ubiquitous machinery was a result of the Industrial Revolution: that machinery could eventually be developed to service man's needs, particularly replacing man's direct participation in physical work for provision of food, clothing, housing, and other necessities. The realization of this dream has progressed for several centuries but has taken a revolutionary leap through the invention of electronic computers. Computers have vastly improved the performance of mechanical machines by allowing more sophisticated "thought" processes for control of the mechanics (e.g., capabilities of industrial robots used in manufacturing are far more autonomous and complex, automobile engines are much more efficiently controlled, etc.). Futurists have long predicted that machines would eventually communicate and cooperate with each other to accomplish extraordinarily complex tasks without human intervention.

The rapidly accelerating improvement of computing capabilities over the last 50 years has contributed to the realization of this dream. Four major factors are

1. *Increasing computational performance*: Moore's Law has reliably predicted that computing performance doubles every 18 months.
2. *Reduction in physical size*: Computer technology has gone from vacuum tubes to transistors, to integrated circuits. Soon *nano-technology* will create another revolution by further reducing the physical size of circuitry.
3. *Decreasing cost of production*: Improved technology has contributed, but commodity pricing has had a greater affect.
4. *Reduction in power requirements and improvements in power sources*: Advances in technology combined with reductions in size will continue to reduce power requirements. However, the next paradigm shift will come from techniques making it possible to scavenge power from the ambient environment (from various sources including vibration, heat, light, and background radio noise).

While futurists long dreamed of machines working with other machines, a giant step toward the realization of this dream may be credited to a DARPA-sponsored program, SmartDust, originated in 1999 (Kahn et al., 1999). The title of the program creatively described its goal: to make machines, with self-contained sensing, computing, transmitting, and powering capabilities, so small and inexpensive that they could be released into the environment in massive numbers. Whether intended to be mobile (e.g., small enough to be cast into the wind to stay aloft for extended periods) or immediately stationary (e.g., deployed from an airplane over a large geographical area for ground surveillance), the sensors have the formidable task of self-organizing into a network that can transmit information to a user while being severely constrained by the onboard energy supply. As they were funded by the Department of Defense, much of the research concerned surveillance of battlefield scenarios, but it was immediately apparent that many peace-time applications could benefit from wireless sensor networks (WSNs).

Observations of the SmartDust project suggests the following definition of WSNs:

> A wireless sensor network is a deployment of massive numbers of small, inexpensive, self-powered devices that can sense, compute, and communicate with other devices for the purpose of gathering local information to make global decisions about a physical environment.

The National Research Council's (NRC 2001) Committee on Networked Systems of Embedded Computers published a report expanding this definition. They defined the concept of the embedded network, *EmNet*, as a network of heterogeneous computing devices pervasively embedded in the environment of interest. Their stated objective was to "develop a research agenda that could guide federal programs related to computing research and inform the research community (in industry, universities, and government) about the challenging needs" of research in EmNets. They recognized a difference between EmNets and traditional computer networks in that the former will be "more tightly integrated with their physical environment, more autonomous, and more constrained in terms of space, power, and other resources. They will also need to operate, communicate, and adapt in real time, often unattended." The Committee enlarged the scope for SmartDust to paint a picture of the eventual embedding of sensor (and effector) nodes into every aspect of our world: "computing and communications technologies will be embedded into everyday objects of all kinds to allow objects to sense and react to their changing environments. Networks comprising thousands or millions of nodes could monitor the environment, the battlefield, or the factory floor; smart spaces containing hundreds of smart surfaces and intelligent appliances could provide access to computational resources." A subtle but important prediction is made here: these networks *will not only gather information about the environment but will affect environmental conditions or effect new actions.* Thus, a feedback mechanism is instantiated where input to the network may be affected by its own actions. Diverse applications are anticipated:

> EmNets will be implemented as a kind of digital nervous system to enable instrumentation of all sorts of spaces, ranging from *in situ* environmental monitoring to surveillance of battlespace conditions; EmNets will be employed in personal monitoring strategies (both defense related and civilian), combining information from nodes on and within a person with information from laboratory tests and other sources; and EmNets will dramatically affect scientific data collection capabilities, ranging from new techniques for precision agriculture and biotechnological research to detailed environmental and pollution monitoring.

When this point is reached, one could argue that the universe becomes one gigantic EmNet. The Massachusetts Institute of Technology's (MIT) Amorphous Computing group sees nanoscale computers "combining microsensors, actuators and communications devices integrated on the same chip to produce particles that could be mixed with bulk materials, such as paints, gels, and concrete" (Abelson et al., 1999). These groups envision an evolution from SmartDust's passive observation of the environment to active manipulation of the environment, driven by the coordinated action of massive numbers of sensors and actuators, coupled with vast computing resources. Clearly, the term *wireless sensor network* describes only a small portion of this vision.

To accomplish this vision, research challenges abound. Despite constraints in power, memory, bandwidth, etc., these devices will be embedded into systems designed to last for long periods of time. While SmartDust predicted massive numbers of nodes per network, many issues resulting from density and scale remain unsolved. EmNets will be self-configuring upon deployment and adaptive to changes in both the network and environment. Trust and fault tolerance models will have to be developed to solve problems well beyond those posed by conventional networks. EmNets will control real-time processes where great costs will be incurred or life is at risk upon failure making reliability, security, and quality of service the major research issues to be considered. There are also nontechnical issues. Ubiquitous and pervasive devices constantly monitoring and controlling everything present many potential legal, ethical, and policy controversies regarding privacy, security, reliability, intellectual property rights, etc. There are other issues concerning production standards, commercialization, business models for implementation and coordination, etc. For the sake of making some progress, many of these requirements are being ignored for current prototypes and present significant challenges for future research.

The NRC expanded the definition of wireless sensor networks:

> Wireless sensor networks are massive numbers of small, inexpensive, self-powered devices *pervasive throughout electrical and mechanical systems and ubiquitous throughout the environment that monitor (i.e., sense) and control (i.e., effect) most aspects* of our physical world.

While the NRC envisions ubiquitous use of WSNs in the next few years, some visionaries look beyond that horizon to predict even greater developments. Kurzweil (Richards et al., 2002) predicts that within the next 20 years, a $1000 computer (typical price of today's PC) will have computational power that matches the human brain. He states, "This level of processing power is necessary but not sufficient for achieving human-level intelligence in a machine. Organizing these resources — the 'software' of intelligence — will take us to 2029, by which time your average personal computer will be equivalent to a thousand brains."

Unlike humans who must share knowledge through speech or vicariously through print or other media, computers can instantly transmit all of their knowledge to other computers in a relative instant. Science fiction writers and Hollywood often portray such an intelligent machine as a walking, talking robot (e.g., Arnold Schwartzenegger's character in the Terminator), but a more likely scenario is a distributed network of intelligent machines. Kurzweil further predicts that humans will be directly connected to these networks via neural implants. Some of these implants are available today as "standalone" devices to counteract Parkinson's disease, for hearing (cochlear implants), and, soon, visual implants for the blind. Wearable devices currently monitor a person's health and transmit this information to a network of resources for analysis and action. Kurzweil envisions these implants will progress to allow humans to immerse themselves in the network of intelligent machines (e.g., to enter virtual reality environments). While his focus is on the expectation of vast improvements in the intelligence of machines, it is a reasonable assumption that similar scales of improvements will be obtained for the size of components and power requirements.

Moravec (1988) believes that machines will not only exceed the capabilities of humans in a similar time period but will eventually become self-sufficient. He sees a future in which, "the human race has been swept away by the tide of cultural change, usurped by its own artificial progeny." Thus, when machines are more intelligent than humans, autonomous, adaptable, self-sustaining, and self-replicating, they will have no further use for mankind.

These visionaries further extend the definition of WSNs predicting their destiny:

> The ever increasing capabilities of pervasive and ubiquitous wireless sensor networks will improve the *intelligence, autonomy, and adaptability* of electrical and mechanical systems such that they *will soon converge with and surpass the capabilities of humans.*

Kurzweil's and Moravec's predictions do not go unopposed. Their vision of man–machine convergence, titled *Strong AI (Artificial Intelligence)*, is challenged by many (Richards et al., 2002). One would expect theologians to contest these predictions, but many metaphysicians as well argue that there are characteristics of humanity that cannot be created or duplicated through a programmed series of chemical and physical reactions. These include consciousness, curiosity, creativity, compassion, freedom, etc. The debate rages on as to whether machines could ever identify with humans. But, even if they never do, and Kurzweil and Moravec are off by orders of magnitude in their projections, capabilities will improve to the extent that networked machines in the near future will go far beyond those of today.

Throughout this sequence of definitions, it is apparent that the term *wireless sensor network* is inadequate to describe this increasing complexity. *EmNets* is neither very descriptive, nor often used. *Sensor/effector network* may be more complete but is still not comprehensive. As the original term is well understood and accepted by the research community, it will be retained for the remainder of the discussion using the abbreviation, WSN. The individual sensor device is often called a *sensor mote* (or *mote*), but in the context of the network, a mote is referred to as a *sensor node* (or *node*).

The progression from simple machines to complex arrangements of machines that are autonomous, self-regulating, self-healing, and even self-reproducing will require increasing sophistication not only of individual nodes and the network but also of the supporting software. As will be shown, the evolution needed is not unlike that of life from simple unicellular organisms to multicellular societies. Section 33.2 presents a set of defining characteristics of WSNs that differentiate them from conventional computing networks by describing their functionality and organization. Section 33.3 describes the current status of WSN implementations. Section 33.4 describes the evolution of living organisms in terms of their organization and behavior. Of prime importance are

- The diversity of life and how such diversity is required of future WSNs.
- The differences between innate behavior and learned behavior and their effect on emergent behavior from individual organisms acting as a group.

The key to the application of these principles to WSNs is a new discipline of engineering, *engineering for emergent behavior*, which will be defined in later sections of this chapter. Section 33.5 presents our vision that current WSN implementations parallel early life forms and describes what types of biomimetics will apply to future WSNs to build an *ecosystem* of WSNs. This section also identifies research techniques currently used in Artificial Intelligence (AI) and Artificial Life (Alife) that apply to engineering for emergent behavior of WSNs. Section 33.6 presents a philosophical discussion of the importance of engineering for emergent behavior to the future of WSNs. Section 33.7 offers the concluding remarks.

33.3 Characteristics of a WSN

Current WSN research was largely spawned by SmartDust and the challenges it raised. A necessary first step for any implementation is to design a network that can be constructed ad hoc to enable communication amongst randomly distributed sensor nodes. Operating with a tight energy constraint presents particular demands, and much current work focuses on the completion of work in an energy efficient manner. Radio, the most common communication mechanism for WSNs, is known to expend a lot of power. To conserve power, the technique typically used is to restrict the nodes to a relatively short transmission distance. They must then form a multi-hop network where a source node broadcasts a message, which is received by a neighboring node, and rebroadcast by that node. This continues until the message arrives at its destination. Ad hoc networks present new issues for the traditional routing problem and the resource constraints of WSNs add even more issues. For WSNs, research abounds at all levels of the network model along with such related issues as security, quality of service, reliability, etc. Reflecting this concentration, Tilak et al. (2002) proposed a taxonomy for WSNs "from a communication protocol perspective." They first categorize by data the delivery requirements of applications and define the categories of continuous, event driven, observer-initiated, and a hybrid of these. They further categorize by network dynamics or routing in this dynamic environment.

Providing an efficient networking infrastructure is a necessary first step in the evolution of WSNs. However, assuming such low-level problems will be solved soon, this discussion will focus on modeling a WSN in terms of the relative complexity of its organization, its functionality, and the effect of its behavior. The universal network will be compared to an ecosystem, and the subsystem networks will become organisms in that ecosystem. The questions to be addressed in this section are how does one identify a subsystem and to what subsystems of WSNs might biomimetic models apply?

The conundrum of defining a WSN is reminiscent of a famous statement from United States Supreme Court Justice Potter Stewart in 1973 concerning a Supreme Court decision. Placed in the context of WSNs, it would read, "I shall not today attempt further to define [a WSN]; and perhaps I could never succeed in intelligibly doing so. But I know it when I see it ..." Although a precise definition may not be possible, the following characteristics are common to all WSNs:

- *Close ties to the physical world*: The purpose of WSNs is to interact with the physical world. Although their name denotes sensing, actuating may be as important as sensing. The Internet is dissimilar from a WSN as most nodes function either as information repositories (servers) or information users (clients) with virtual locations that can be physically moved with no consequence.
- *Granularity of action*: Small size and low cost allow massive numbers of sensor nodes to be deployed over a geographical region instead of a small number of larger, more costly devices each of which must act upon a larger space.
- *Collection of local action results in global effects*: It is not necessary for any sensor node to know or understand the function of the network, but, like a soldier in a battalion, if each sensor node does its job, the network will succeed. The global view will emerge from an aggregation of the local information.

The following list of characteristics will be useful in categorizing WSNs:

- *Random distribution of sensor nodes*: Sensor nodes are deployed by means that preclude a pre-determined spatial topography (e.g., dumped from an airplane). Distribution density guarantees functionality and fault-tolerance, as failure of some nodes will not cause the network to fail. Because

the WSN is closely tied to the physical environment, node location must be determined. Wadaa et al. (2004a) devised a scalable energy-efficient training protocol to provide locations for nodes that are initially anonymous, asynchronous, and unaware of their locations.
- *Large numbers of sensor nodes*: Many of the challenging and new issues of WSNs are concerned with coordinating massive numbers of sensor nodes into a functioning network.
- *Small physical size of sensor nodes*: Initially, reduction in size is expensive. However, as manufacturing improves, smaller size and increased volume should contribute to reduced prices. Also, smaller sized nodes will widen the range of applications.
- *No pre-assigned network topology*: If the location of the sensor nodes is not predetermined, neither can the network topology. Modest power budgets may prevent nodes from transmitting directly to a common destination, therefore an ad hoc process will be applied to form a multi-hop network.
- *Anonymous network*: Scalability of massive numbers of nodes within a WSN precludes the assignment of unique node identifiers during deployment. Anonymity can be useful in securing a WSN. Wadaa et al. (2004b) describe how anonymity can be used to prevent denial of service attacks.
- *Wireless communication*: Random distribution dictates wireless communication.
- *Limited onboard energy supply*: Current battery technology ties capacity to physical size. Therefore, the need for small sensor nodes competes with the need for more onboard energy. If technology improves sufficiently, this will no longer be a major issue.
- *Node mobility*: Motion of sensor nodes may be:
    - *Static*: Once deployed, the sensor nodes do not move.
    - *Relative*: The sensors move, but as a group (e.g., a group of satellites may be placed in parallel orbits but remain fixed relative to each other; a "satellite constellation").
    - *Fully mobile with independent movement*: For example, a swarm of sensing robots.
- *Application code distributed across nodes*: WSNs range from those containing homogeneous nodes reporting simple observations to heterogeneous nodes with differing (and perhaps changing) functions.
- *Data aggregation*: In addition to individual sensor nodes reporting to a single destination, sensor nodes may form clusters whereby local preprocessing of sensed data improves efficiency by transmitting only the aggregated data. In fact, a hierarchy of clusters and aggregation may be used.
- *Performance and reliability*: Because of the interaction with the physical world and the potential for damage to it in many applications, new demands will be placed on these networks for real-time performance and reliability.
- *Security*: As stated, these networks must perform and be reliable; therefore, nodes must be trusted network members. Jones et al. (2003) proposed a new security paradigm in which security is based upon parameterized frequency hopping and cryptographic keys in a unified framework.

Many of these characteristics may not apply to networks resembling WSNs. Consider for example a "smart skin" where many sensors are attached to a surface (e.g., an airplane wing). Sensor and actuators may be pervasive throughout a larger structure, serving not only as surface sensors to sample environmental input but internally as "muscles" to move an object. Such a network may be massive but fixed such that each sensor's position and identity may be predetermined, its network connectivity may be predetermined and need not be wireless, and its supply of power may not be limited. But it still has large numbers of small, spatially distributed nodes that must communicate efficiently and aggregate large amounts of locally collected information for global decisions. Because of its close ties to the physical world, real-time performance, reliability, and security are extremely important.

WSNs have many close cousins that go by a variety of names such as:

- *Heterogeneous WSNs*: This is the nomenclature of Intel's EcoSense research project, that is, "tackling a difficult challenge: how to network large numbers of inexpensive wireless sensor nodes while maintaining a high level of network performance" (*Intel Research — Exploratory Research — Deep Networking — Heterogeneous Sensor*, 2004).
- *Deep networking*: Intel defines Deep Networking as

> Locally networking billions of embedded nodes, driving computing deeper into the infrastructure that surrounds us. . . .Small, inexpensive, low-powered sensors and actuators, deeply embedded into the physical environment can be [deployed] in large numbers, interacting and forming networks to communicate, adapt, and coordinate high-level tasks. . . .As these micro devices are networked, the Internet will be pushed not just into different locations but deep into the embedded platforms within each location. This will enable a hundredfold increase in the size of the Internet beyond the growth we are already anticipating. New and different methods of networking devices to one another and to the Internet must be developed. (*Intel Research — Exploratory Research — Deep Networking*, 2004)

- *Amorphous computing*: This is MIT's term for "the development of organizational principles and programming languages for obtaining coherent behavior from the cooperation of myriads of unreliable parts that are interconnected in unknown, irregular, and time-varying ways." (Abelson et al., 2000)
- *Sensor webs*: NASA's Jet Propulsion Laboratory (JPL) describes a sensor web as "an independent network of wireless, intra-communicating sensor pods, deployed to monitor and explore a limitless range of environments. This adaptable instrument can be tailored to whatever conditions it is sent to observe." (Delin, 2004)
- *Mesh networks*: This term describes a type of Wi-Fi network where nodes do not communicate through a central controller, or access point, but rather, mobile ad hoc peers in the network form a mesh topology to transmit data from source to destination via multiple hops throughout the network. This technique improves on the reliability and efficiency of the centralized approach. Although designed for connection of conventional computers, this technology shares many issues with WSNs.
- *Sensor constellation*: This term is usually used to describe multiple satellites cooperating in a space science experiment (*Leveraging the Infosphere: Surveillance and Reconnaissance in 2020*, 1995).
- *Pervasive computing*: The Centre for Pervasive Computing (2004) defines this as,

> the next generation computing environments with information & communication technology everywhere, for everyone, at all times. Information and communication technology will be an integrated part of our environments: from toys, milk cartons and desktops to cars, factories and whole city areas — with integrated processors, sensors, and actuators connected via high-speed networks and combined with new visualization devices ranging from projections directly into the eye to large panorama displays. . . .Pervasive computing goes beyond the traditional user interfaces, on the one hand imploding them into small devices and appliances, and on the other hand exploding them onto large scale walls, buildings and furniture.

- *Ubiquitous computing*: This term was coined by Mark Weiser (1996) whose goal is to "Activate the world. Provide hundreds of wireless computing devices per person per office, of all scales (from 1″ displays to wall sized). This has required new work in operating systems, user interfaces, networks, wireless, displays, and many other areas."
- *Invisible computing*: An ACM SIGGRAPH conference in 2000 explored "the tiny, cheap, special-purpose devices that experts expect to diffuse into our lives over the next two decades. Though these devices themselves may be visible, our common goal is to make them so comfortable to use that they seem not to be computers at all — the computing power they use stays invisible" (*Invisible Computing: Scope*, 2000).

33.4 The Current Status of WSNs

Prototypes such as SmartDust have begun the implementation of futurist dreams. In March 2001, a team of researchers led by Pister and Culler of the University of California at Berkeley (UCB) held a demonstration at the Marine Corps Air/Ground Combat Center, Twentynine Palms, CA (*UCB/MLB 29*

Palms UAV-Dropped Sensor Network Demo). They dropped six sensor motes from an Unmanned Aerial Vehicle (UAV) along a road. These motes self-organized by synchronizing their clocks and forming a multi-hop network. They magnetically detected vehicles passing and reported the time of the passing. Using information collected from all motes, velocity of passing vehicles was computed.

On August 27, 2001, researchers from UCB and the Intel Berkeley Research Lab demonstrated a self-organizing WSN to those attending the kickoff keynote of the Intel Developers Forum. Several students each brought on stage a wireless sensor mote and activated it at different times. As the motes were initiated, their icons appeared on a display with lines connecting all motes that could "hear" each other and highlighted lines depicting a multi-hop routing structure over which sensor data could be transmitted to a central collector, a PC. The network grew as the nodes were initiated and adapted to changing conditions. Additionally, color cues on the display indicated changing lighting conditions on the stage as sensors detected these changes. As the students left the stage, the network disintegrated. In a second large-scale demonstration of the networking capability, the quarter-sized motes were hidden under 800 chairs in the presentation hall and simultaneously initiated forming what Culler described as, "the biggest ad hoc network ever to be demonstrated" (Lammers, 2001).

In the spring of 2002, Culler's group (Mainwaring et al., 2002) collaborated with the College of the Atlantic in Bar Harbor to instate a WSN on Great Duck Island, Maine. The initial application was to monitor the microclimates of nesting burrows in the Leach's Storm Petrel, and, by disseminating the data worldwide, to enable researchers anywhere to nonintrusively monitor sensitive wildlife habitats. The sensor motes were placed in the habitat and formed a multi-hop network to pass messages in an energy-efficient manner back to a laptop base station. The data was intermediately stored at the base station and eventually passed by satellite to servers in Berkeley, CA, where it was distributed via the Internet to any interested viewer. The sensors measured temperature, humidity, barometric pressure, and mid-range infrared by periodically sensing and relaying the sensed data to the base station. The largest deployment had 190 nodes with the most distant placement over 1000 feet from the nearest base station.

The FireBug system (Chen et al., 2003) is a network of GPS-enabled, wireless thermal sensors motes, communicating through a control layer for processing sensor data, and a command center for interactive monitoring and control of the WSN. The FireBug network self-organizes into clusters in which cluster leader motes act as base stations, receiving sample data from cluster members and brokering commands to these members. The controller is a personal computer running the Apache web server interfaced with MySQL using PHP. The FireBug Command Center allows user interaction for controlling the FireBug network and displays real-time changes in the network.

33.5 Biological Organizations and Behavior

Life began as an organization of entities performing functions. The entities and their functions survived or became extinct based on the sustainability of those functions. Proteins catalyzed chemical reactions. Aggregates of proteins resulted in more complex reactions eventually evolving RNA and then DNA that direct protein production. Further aggregation eventually produced cells. Cells aggregated in symbiotic relationships to form more complex cells and then to form tissues and organs composing multicellular organisms. Throughout this progression, organisms developed more complex forms and functionality. Successful species adapt to an ecological niche providing sustenance, and live to reproduce at a rate that keeps them in existence. Critical to understanding this process is the realization that evolution occurs not because complexity is *better* but because, while simple, successful organisms remain, random mutations result in increasing complexities that are sustainable in some niche (i.e., one did not develop eyes so one can see; rather one sees because one developed eyes — and this enhances success). Detrimental mutations result in unsuccessful species. Development of WSNs will also follow such an evolution but with two very important distinctions:

- While biological evolution occurs randomly with resulting successes and failures over long periods of time, initially, electromechanical systems can be directly designed and these designs evaluated and improved rapidly. If Kurzweil and Moravec are correct, eventually electromechanical systems may design themselves. Again, random mutation and selection for sustainability may prevail but at a much faster pace than ever occurred in life.

- Functionality, as are all other characteristics of living organisms, is only important as it affects sustainability. While humans are in control, functionality will be most important for a WSN, whereas sustainability only affects issues such as cost, efficiency, longevity, etc. Evolution of WSNs can be directed for increasing functionality with sustainability only a secondary issue. Again, Kurzweil and Moravec predict that this may change back if machines become autonomous giving primary importance to sustainability.

Organization models and behavior influence functionality. In general, with more complexity comes more functionality and more diverse behavior. It is important to note that whereas man continually strives to divide living systems into distinct categories organisms have developed as a continuum of both organizations and behavior in which there are typically exceptions to any rule applied for distinction. Development of machines can be perceived as such a continuum, but it is necessary to adopt a model in which clear distinctions are made. A distinction will be made between the capabilities of *unicellular organisms compared with those of multicellular organisms* and, furthermore between *organisms surviving alone* and those *organized into groups* (colonies, swarms, societies, etc.).

Behaviors will be categorized as those that *result from responses to stimuli* that ultimately are determined genetically (i.e., innate) and those that are *cognitive.* Cognition is defined as the mental process by which *knowledge* is acquired, which is a result of *awareness, perception, intuition, reasoning, memory,* and *judgment.* Thus, a behavior is said to be *cognitive* if its effect is *known* and *understood* by the effector. Cognitive behavior begets new cognitive behavior that is not genetically encoded. For this discussion, new behavior that results from prior experience defines *learning.* A simple response to stimuli is not considered learning. An interesting effect of cognition in living systems is that it may introduce selection factors other than sustainability.

33.5.1 Biological Organizations

Life can be represented hierarchically with the most complexity at the root of the tree and the simplest organization at the leaves. Levels of organization are defined by degree of complexity:

- *Simple organization, cells*: Although cells are composed of identifiable structures, and some independent but subcellular components function in a capacity that arguably demonstrates characteristics of life, cells are the basic unit of life for the purpose of this discussion. All cells are contained by a plasma membrane in which the chemical "soup" of life operates. All cells have a genetic structure, DNA, controlling protein production that, in turn, catalyzes all metabolic processes within the cell. Cells are at least potentially capable of self-replication. A cell may serve many functions and exist independently as a distinct organism or have a very specialized function while cooperating symbiotically with many other adjacent cells forming a multicellular organism.
- *Building blocks of complexity, tissues, and organs*: Tissues are contiguous, homogeneous, highly specialized cells, each serving a function where the collective result of their actions yields a cumulative result. For example, muscle cells contract, applying a pulling force, when stimulated. While the power of individual cells is small, the cumulative power of millions of cells contracting simultaneously produces a significant force. Unlike unicellular organisms that must forage for food, cells of tissues are not self-sustaining as they are provided nourishment by the multicellular organism to which they belong and cannot survive independently. Organs are a heterogeneous organization composed of many kinds of tissues to serve a cumulative function greater than the individual functions of their component tissues (e.g., a stomach has muscle tissues and tissues to secrete acid and mucus; it functions to digest food). Organs may combine as components in an organ system to perform a higher function (e.g., the digestive system has many organs working together to intake, digest, and distribute food while eliminating waste).

- *Complex organization, multicellular organisms*: Multicellular organisms are composed of tissues and, usually, organs. They range from the simplest of these composed of few tissues (e.g., Coelenterates such as jellyfish) to plants with simple organs to mobile animals with complex motor and nervous systems. Of particular interest to this discussion are the latter. Because of these complex systems, they can not only respond to stimuli but can learn from these stimuli to alter their behavior. Multicellular organisms reproduce as a unit.
- *Organizations of organisms*: Colony, population, community, and society are nearly synonymous words describing a group of organisms living or growing together. Although they are often used interchangeably, colony, population, and society usually denote a homogeneous group of the same kind of animals, plants, or unicellular organisms, while community usually indicates all life in a given habitat.
- *Add the environment to comprise the ecosystem*: An ecosystem is the community of living organisms in a habitat together with all nonliving (abiotic) components of the environment with which the community interacts.
- *The root of the tree is the biosphere*: On Earth, the biosphere is all of the area of Earth from the highest altitude in the atmosphere to the lowest depths of the oceans and land where life exists. Thus, life on Earth can be represented as a tree with the biosphere the root, which is composed of ecosystems, which are composed of communities and the abiotic environment, etc. The beauty of the evolutionary process is that the biosphere comprises all levels of organizations and behavior: if an organism fills a niche for sustainability, it survives. Life did not begin with the simple organisms and, as it evolved to more complex organisms, discard the simple organisms. There is a delicate balance within an ecosystem between all of the interdependent organisms together with their environmental resource requirements. With the obvious competition for food, it may appear that life is a constant fight for survival. While this is true for an individual organism, when the biosphere is considered, life is what is sustained. Margulis and Sagan (1986) stated, "Life did not take over the globe by combat, but by networking." The biosphere as a whole may be viewed as a giant symbiotic relationship.

Applying the ecosystem model to future WSNs encourages one to plan a diverse collection of interacting (i.e., networked) sensor/effector systems of all sizes and complexities to fulfil every required niche.

33.5.2 Biological Behavior

Biological behavior ranges along a continuum from responses to stimuli resulting from genetic encoding to complex, learned behavior that build upon experience, to social organizations where the society exhibits emergent behavior as the result of individual actions of its members. Organisms exhibiting cognitive behavior must have the ability to learn and often form societies.

33.5.2.1 Non-Cognitive Behavior of Singletons

What distinguishes unicellular organisms from multicellular organisms is their relative independence from all other cells. Although many may cohabitate spatially, they remain relatively independent having simple, independent functions that result in their sustainability. They reproduce asexually with mainly two functions: eat and reproduce. Although they may exhibit other functionality, these two are necessary for sustainability. While their metabolism may greatly affect the ecosystem (e.g., bacteria metabolize dead organisms returning chemicals to the ecosystem that are used to sustain other organisms), these are *side effects unknown and not understood* by the organism. Dusenbery (1996) describes a diversity of behavior in microbes such as locomotion, navigation, migration, habituation, and communication. He gives examples of how they can anticipate the future (e.g., circadian responses). For habituation, he gives examples in which protozoa appear to have memory as they will "learn" to ignore a stimulus for the duration of a few hours. He concludes that "most microbes are capable of only the most specialized forms of learning: sensory adaptation, setting internal clocks, and crude habituation. Many biologists and most psychologists would not consider any of these to be 'real' learning." Primio, et al. (2000) argue that life and cognition are synonymous, "Bacteria and other unicellular organisms are autonomous and social beings showing (the lowest levels of) cognition. They have the fundamental cognitive abilities to identify elements of the environment and to differentiate between them (and self), to choose among alternatives, to adapt to changes, to coordinate their behavior in groups, to act purposefully." While these cells respond to external stimuli, their response is ultimately preprogrammed within and limited by their DNA. In view of previous definitions of cognition and learning, *these behaviors are not cognitive.* Although improvements occur through generations of randomly mutated cells with mutations selected for sustainability, this is also not considered learning.

Multicellular organisms often act as singletons. Other than to mate, mosquitoes normally function independently. Mosquitoes cohabitate in massive numbers, often appearing to swarm and are described as such. In northern ecosystems, these swarms congregate in such density as to inhibit breathing (Conniff, 1996) and have been known to drain enough blood to cause the death of both cows and caribou (Budiansky, 2002). However, as defined below, this differs from the behavior exhibited by swarms of bees or ants. This action is the result of massive singletons simultaneously attacking the same food source; they are not cooperating but are competing. Again, for this discussion, these organisms do not exhibit cognitive behavior.

33.5.2.2 Non-Cognitive Behavior of Swarms

Swarms are organisms at any level that form a group in which relatively simple and individual behavior combine to result in what is known as *emergent* behavior (i.e., the behavior of the group *emerges* from individual behavior). The emergent behavior of the group is neither individually possible nor understood by the individuals. From a view outside of the colony, insects such as ants, termites, and bees exhibit behavioral characteristics that could appear to be intelligent. Bonabeau et al. (1999) catalogue a list of behaviors often demonstrated by social insects that seem to indicate cognitive thought: optimal route finding, division of labor and task allocation, cemetery organization, and brood sorting. For example, consider that to maintain a healthy beehive specific gas-exchange requirements must be met. When necessary, honeybees will actively ventilate the hive by positioning workers at the hive entrance to fan their wings and control the flow of gases through the hive. This wing fanning event is not a simple operation, as the bees must synchronize their fanning actions to form an *on–off* pattern that allows the hive to inhale and exhale: the hive breathes. In the context of a colony, this behavior has been termed *social homeostasis* (Turner, 2000).

A decade ago, Kevin Kelly (1994) in his landmark book, *Out of Control*, wrote of what he termed the *hive mind*:

> The marvel of the "hive-mind" is that no one is in control, and yet an invisible hand governs, a hand that emerges from very dumb members. The marvel is that more is different. To generate a colony organism from a bug organism requires only that the bugs be multiplied so that there are many, many more of them, and that they communicate with each other. At some stage the level of complexity reaches a point where new categories like "colony" can emerge from simple categories of "bug." Colony is inherent in bugness, implies this marvel. Thus, there is nothing to be found in a beehive that is not submerged in a bee. And yet you can search a bee forever with cyclotron and fluoroscope, and you will never find the hive."

Social homeostasis, hive-minds, swarming, and similar behavior associated with large groups of social insects share a common thread: *the intelligence that these behaviors seem to exhibit is not the result of a cognitive process.* The intelligence of the swarm is *emergent* and results from the interactions of many thousands of basically dumb, autonomous individuals. Each individual follows its own set of rules and reacts to local state information. The rules are primarily a result of a manifestation of genetic encoding in the presence of environmental stimuli. There is no central control command issuing orders. Individuals are highly connected within their immediate neighborhood and can share information, but they do not have a central *server* with which to communicate. Control and management of the swarm is distributed throughout and within the swarm members. It is distributed control with no single point of command and no single point of failure.

What level of intelligence (i.e., the extent of their cognition) is required of bees to perform such behavior? There is evidence that some degree of memory is required. Gould and Gould (1988) believed a bee's ability for spatial recollection not only provided for storage of many mental maps to food locations but also the ability to calculate shortcut routes between visited sites. However, Dyer and Seeley (1991) found that bees are limited to storing route maps comprising of memorized landmarks (i.e., a much condensed representation of spatial topology) and found no evidence that bees can form spatial relationships between different routes. Bees also use visions of solar movement together with some internal circadian indicator to relate landmarks for directional cues.

Gregory (1997) stated,

> One's first thought might be that insects are mere automata, lacking ability to learn, but this is not so. Bees can learn flower patterns and complex routes to food. They can navigate, and communicate in many ways including the famous dance-language discovered by Karl von Frisch. The waggle dance... is related to the position of the Sun, indicating to other workers direction and distance and food quality of flowers. No doubt there is always a danger of reading too much intelligence into innate behavior, but it seems impossible to see this as completely stereotyped, for it is adjusted to particular conditions and needs. ...Insect learning may be limited to immediate adaptive uses; but insects do have associative learning by Pavlovian conditioning with rewards from previously meaningless stimuli. ...They do show latent learning, though probably without "insight." They cannot reorganize their memories for a new situation (at least not in laboratory conditions), and there is little transfer learning from one situation to another. But it remains remarkable that such small brains can do so much.

Reinhard et al. (2004) demonstrated a "Pavlov's dog" response in bees by exposing a hive to two food sources each laced with a different fragrances: rose and lemon. Later, all food and scent was removed from the stations. When exposed to a scent, the bees returned to the empty feeding station that previously contained that scent. While this demonstrates memory and learning, they do not give evidence for long duration memory in bees.

Remarkable as the behavior is, Gregory (1997) sees no evidence of concept formation and insight in bees. Gregory agreed with Gould and Gould (1988) when they said,

> (bee) communication and navigation are the most complex known among invertebrates and, excepting humans, quite likely among vertebrates as well. ...Far and away the most complex animal behaviors we know of usually are innate... Orb-weaving spiders make their characteristic webs in total darkness with no previous experience or learning. ...In our opinion, selection has operated to make complex behavior innate for the simple reason that, if it were not, animals could not hope to discover it by trial and error or to learn it by observation in time to be able to perform it. In the end, complexity of behavior is one of the worst guides of all to intelligence.

While these behaviors demonstrate the innate ability for efficient calculations using simple memories of environmental cues such as landmarks and the movement of the sun, this does not compare with the cognitive abilities of higher animals.

33.5.2.3 Cognitive Behavior of Societies

The term "social" is often applied to colonies of organisms described as swarms: a society of bees, social ants, etc. In this discussion, societies will be limited to those that are cognitive. A society is similar to a swarm in that individuals combine their actions for an aggregate result that could not be accomplished individually. It differs from a swarm in that the behavior of individuals are cognitive: they understand and exploit the cumulative power of the group and depend upon each other for support. A pack of wolves exemplifies a cognitive society. An individual wolf is able to track down and kill prey unassisted. But it understands that the chance of success is greater in a pack and, therefore, prefers to participate as a member of the pack. In this case, individual abilities are similar although differing duties may be assumed. Wolves have behaviors for greeting pack members, keeping the pack together, and identifying social ranking. The entire pack participates in raising the young: they feed the mother and the young, protect them from harm, and nurse the young when the mother is away. Human societies are much more complex with more diverse duties of individuals contributing to the good of the community. In a successful society, a symbiotic relationship develops among individuals, where actions that are good for the society are also good for the individual.

Most will agree that cognitive organisms require a higher level of intelligence. But that does not add to understanding without a definition of intelligence. Pfeifer and Scheier (2000) summarized the opinions of major experts in the field of psychology in 1921 on the definition of intelligence:

- The ability to carry on abstract thinking.
- Having learned or having the ability to learn to adjust oneself to the environment.
- The ability to adapt oneself adequately to relatively new situations in life.
- A biological mechanism by which the effects of a complexity of stimuli are brought together and given a somewhat unified effect in behavior.
- The capacity to acquire capacity.
- The capacity to learn or profit by experience.

They point out interesting consistencies and inconsistencies. Opinions vary at the level in which intelligence operates ("abstract thinking" versus "biological mechanism"). Some mention interaction with the environment; some do not. Most indicate the ability to change behavior ("adjust", "adapt," "effect in behavior," "learn . . . by experience").

For this discussion, both interacting with the environment and adapting to changes in the environment are apropos. It has been stated that even unicellular organisms do both, but lack the ability to creatively develop new behavior for adaptation. Albert Einstein said, "Insanity is doing the same thing over and over again and expecting a different result." In this context, a restatement is, "The unintelligent do the same thing over and over and get the same result." Conversely, "The intelligent create new behavior to improve the result." What do intelligent animals have that lower life forms lack? The ability to learn is paramount.

33.5.2.4 Learning Models

A necessary but not sufficient component for learning is memory. In higher animals, there are different types of memory for different functions (short term, long term, etc.). Learning requires more than just memory: learning requires relationships among the memories so that new behavior can be conceived. Long-term memory is critical for cognitive behavior (i.e., if you cannot collect memories over long periods of time, the development of relationships is limited). As described, paramecia "remember" for minutes or hours, bees perhaps for weeks. However, after memory is lost, the experience must be relearned.

Various learning models have been popularized. In the early years of the 20th century, two models for behavioral modification gained wide acceptance. Pavlov developed the *respondent conditioning* model informally known as "Pavlov's dog" response. In this model, a subject can be trained to associate an unrelated stimulus with a desired result (e.g., when a bell rings, food is presented and the dog salivates). After enough exposure, the dog will salivate if a bell rings and no food is present. Skinner proposed the *operant conditioning* model: behavior can be changed by applying rewards when a desired behavior is observed (reinforcement is a reward to continue a desired behavior; negative reinforcement is a reward when an undesirable behavior ceases) and punishment to extinguish an undesirable behavior. Clark (1997) added an interesting observation concerning these models of learning: "Biological cognition is highly selective and it can sensitize an organism to whatever (often simple) parameters reliably specify states of affairs that matter to the specific life-form."

Cognitive science was influenced greatly by the development of electronic computers. Behaviorists came to believe that all behavior can be understood by the language of information processing. Lachman et al. (1979) stated that behavior can be modeled by "how people take in information, how they recode and remember it, how they make decisions, how they transform their internal knowledge states, and how they translate these states into behavioral outputs." Pfeifer and Scheier (2000) described this as *cognitivistic paradigm* or *functionalism*, which formed the basis of *informational processing psychology*. This conveniently aligned with computer simulation, where thinking could be modeled as a computer program having input, data structures, computation, and output.

While this model served well for simulating problem solving (culminating with the chess championship of IBM's "Deep Blue" computer over the reigning world champion chess player), it proved less than adequate for modeling simple behavior even a child can perform, particularly those requiring interaction with the environment (e.g., identification of colors and shapes, depth perception, navigation, etc.). Pfeifer and Scheier (2000) promote a new model of learning they call *embodied cognitive science* (alias *new artificial intelligence* or *behavior-based artificial intelligence*). Summarizing the view of Rodney Brooks of the MIT Artificial Intelligence Laboratory, one of the founders of this new model, they agree that the *cognitivistic paradigm's* model of thinking, logic, and problem solving is fundamentally flawed by our own introspection of how we see ourselves. Brooks suggested we abandon this approach and focus on interaction with the real world. Intelligence *emerges* from the interaction of an organism with its environment: intelligence must operate within a *body*. This effort has overcome some of the problems encountered by use of the *cognitivistic paradigm*, and developed the approach of *designing for emergent behavior*, producing some unexpected though desirable results.

Pfeifer and Scheier (2000) specify two additional concepts that pertain to this discussion: First, "The essence of learning is that the [organism] can use its own experience to improve its behavior." Because experience is nondeterministic, predicting changes in behavior is difficult. Second, adaptive behavior requires two components, compliance with existing "tried and true" rules and applying diverse, new rules. They call this *diversity–compliance tradeoff*. It is the application of diverse, new behavior related to previously learned behavior (rules) that has the best chance for improvement.

Cognitive behavior described above is defined as intra-generational; each individual of a generation learns based on its own *experience*. Response to stimuli is evaluated by a fitness function (e.g., was lifting my hand quickly from a hot stove good or bad?); behavior is either reinforced or altered based on the result. In higher animals, stimuli and the resulting chosen behavior are remembered, organized, related, and prioritized. Thus, when the same stimuli are encountered again, the best behavior is exhibited. But inter-generational *learning* may be man's greatest achievement. All organisms benefit from information that is inherited (i.e., "learned") from the previous generation through genetic codes. Genetic codes are passed through DNA in reproduction and determine behavior as responses to stimuli (e.g., phototropism in plants — plants "know how" to grow toward a light source). But as we have defined, this is not cognitive behavior. In humans, this inter-generational transfer is augmented as information is acquired from parents and other members of previous generations. In most animals this transfer of knowledge occurs only by direct interaction with parents or contemporaries (e.g., other animals in the pack). Man however has developed the ability to permanently capture knowledge through books and other media and indirectly pass it on to future generations. The power of this technique cannot be overemphasized.

33.5.2.5 Social Models

In lower organisms, behavior is learned either through inheritance or by experience. However, higher animals may control behavior through social means. These means may be represented as a continuum between two extremes, centralized and distributed control:

- *Centralized*: Behavior of the society is dictated by a central authority. This approach has the weakness of a single point of failure.
- *Federated*: Behavior of the society is dictated by a federation of authorities. This approach mitigates the single point of failure but increases the complication of decision making.
- *Distributed*: Behavior of the society can emerge from actions of individuals on the basis of rewards, extinction, and punishment or other learning models.

Imagine a town is newly formed. Who decides how many tailors the town needs? How many merchants? How many carpenters? Obviously, a leader could emerge that would make those decisions for all. But the leader could be wrong. He could fail and chaos would develop until new leadership emerges. He could misappropriate resources for personal gain, but to the detriment of the society. Alternatively, individuals could decide their own role in the society. A balance would naturally evolve: if too many people decide to be a tailor, there will not be enough work and some will starve or be forced to change. It is likely that the better tailors will get what work is available and survive. When individuals are free to make their own choices of behavior, the choices can be detrimental to the society. The most successful societies have been those in which individuals have been free to work for individual rewards within certain bounds (i.e., individuals cannot seek personal rewards to the detriment of the society).

33.6 Application of Biomimetics to WSNs

The current WSN implementations described in Section 33.4 represent pioneering work that demonstrate the enormous promise of the WSN concept. They share a number of similarities. Most of the motes used are similar in capability to those used in the UCB-Intel large-scale demonstration (Lammers, 2001): a 4 MHz microprocessor with 16 KB of flash instruction memory, 512 B of SRAM, a 256 KB EEPROM for secondary storage, and a 10 Kbpsec radio transmitter. Each network has homogeneous motes although the motes may be multifunctional having multiple sensing capabilities and serving other functions such as routers, cluster leaders, etc. PCs serve as base stations for interfacing with the world outside the WSN. However, all are passive, sensing systems (i.e., no effectors) observing the environment and reporting these observations to a central authority where decisions will be made often by human observers. These demonstrations used between 6 and 800 motes. Will the techniques used scale to massive numbers? In these designs, behavior is predetermined, its results collected, and otherwise managed by a central authority. Motes appear to be working together, though only in simple ways: multi-hop networks are formed and clocks are synchronized. An examination of the method used for clock synchronization in the Twentynine Palms demonstration (UCB/MLB 29 Palms UAV-Dropped Sensor Network Demo, 2001) reveals that it closely models singleton behavior. Each mote periodically broadcasts its current time. Each mote that receives the broadcast updates its time if the broadcast value is greater than its own. Though effective, are the motes really working together? Or are they singletons following simple rules: they periodically broadcast their time and they listen for the broadcast of packets containing time values. Do they know or care that other motes exist? From the individual application of these simple rules, a simple behavior for the group *emerges*: their times are synchronized. However, neither the individual nor the group behavior can be described as cognitive.

Thus, it can be argued that much of the behavior of current WSNs parallels the behavior of simple singleton organisms. The singleton model with simple emergent behavior will be insufficient as sensing increases and diversifies, effecting becomes a player, and, thus, environmental influences become nondeterministic. To reach full potential, cognitive WSNs must evolve to apply the *diversity–compliance tradeoff*: deciding when to apply established rules and when to try new behavior based on previous learning from experience. The more new behavior is tried, the more interaction with the environment becomes nondeterministic and, therefore, unpredictable. Cognition becomes a great advantage.

As with biological societies, the WSN is expected to operate for a long time, although it is both acceptable and expected that individual motes will not. For longevity of such a cognitive network, inheritance is required to preserve knowledge between generations of motes. Consider the changing of a guard in a combat scenario. A guard is deployed with the assignment of watching for the appearance of a threat. If the guard was a sensor mote of today, he may be asked to notify the command center if a light, a noise, or a magnetic disturbance is detected. The command center would take the observations and decide whether or not to act. But a human guard would represent a *cognitive* mote. He would apply his cumulative experience and training to his observation of the environment. He would apply known rules (e.g., he would report seeing a tank), but he would exhibit new behavior when presented with new experience.

If he sees a light or hears a noise, he may not report the incident but store that information for further consideration. He may pay more attention to that area in the future. He may also rely on inter-generational learning: when faced with a new situation, he may consult the command center for more information; he may consult a manual. What happens during the changing of the guard (i.e., generational replacement)? The guard could tell the central command all information to be transferred to the new guard, let the central command inform the new guard, and pass the new guard without saying a word. Perhaps it would be more efficient for the current guard to directly discuss the information with the new guard allowing interaction; the new guard could ask questions for clarification. The current guard could do a complete "data dump" to the new guard telling him every incident that happened so that the new guard had all experience that happened during the previous shift. Perhaps it would be more efficient for the current guard to use his awareness, perception, intuition, reasoning, memory, and judgment (i.e., *cognition*) to decide what *knowledge* to impart to the new guard. Similarly, cognitive nodes can significantly improve the efficiency and effectiveness of many WSN implementations.

33.6.1 Potential Applications

Lessons from life that should be applied to future WSNs include aspects of both organizational and behavioral models. Organization characteristics directly affect the sustainability and the functionality of WSNs:

- *Genetic material*: The "genes" of sensor nodes are represented by characteristics endowed in their creation. These can be both software and hardware. Flexibility requires that this genetic material be mutable. Current nodes can change their software "genes" via their wireless connection but hardware is constrained to the genes granted to them at "birth" (i.e., their manufacture). Efforts are currently underway to design hardware that is evolvable such that it can redesign itself for new functions.
- *Metabolism*: The metabolism of a sensor node is represented by electromechanical processes. The source of "food" to energize these processes is the power source. Currently, power in WSNs is provided by an onboard battery. The lifetime of a sensor node can be extended by providing it with more powerful batteries, and technological improvements will surely provide more power in smaller sized batteries. But this approach, no matter how potent, is still a finite power source. As the NRC predicts, WSNs will be deployed requiring a long lifetime. A better approach is to design sensor nodes that "eat": mobile nodes forage for energy as do animals; stationary nodes must acquire energy from a nearby, renewable source as do plants (photosynthesis using sunlight and water). Research in this area is active, discovering techniques to make it possible to scavenge power from the ambient environment. Potential sources include vibration, heat, light, and background radio noise.
- *Self-healing*: Just as a multicellular animal does not fail because of malfunction or the death of individual cells, a WSN must be able to compensate for malfunction or death of nodes.
- *Symbiosis*: Some organisms have developed symbiotic relationships in which each benefit from a direct relationship with others. WSNs must be designed to organize in cooperation; complimenting each other rather than competing.

Behavioral characteristics also directly affect the sustainability and the functionality of WSNs:

- *Adaptability*: The ability to adapt to a changing environment is paramount for any sustainable system. If WSNs are to be autonomous, cooperative, fault-tolerant, self-regulating, and self-healing, they must be able to adapt to changes in their environment. As with living organisms, some adaptability can be innate, but to achieve maximum flexibility, cognition is required.
- *Functional mobility*: Obviously, mobile nodes can carry their functionality to different locations. However, functionality can also migrate using stationary nodes. This migration is facilitated by the reassignment of individual node duties.

- *Sharing knowledge*: Cognitive organisms pass information among their community in order to reduce collectively the cost of learning. By sharing knowledge amongst nodes, WSNs can get universal benefits for the experiences of a single node.
- *Generational learning*: As with communities of organisms, individual nodes will be "born" (i.e., manufactured and deployed) and die (i.e., cease to function). New nodes coming into an existing WSN represent a new generation. WSNs must be designed such that knowledge is preserved and made available to newer generations.
- *Policing the society*: It must be expected that nodes will not only fail, they may "misbehave." In a self-sustaining WSN, such rogue nodes must be discovered and dealt with to prevent damage to the functioning network (i.e., the WSN must police itself). The ideal condition would be that all damage is prevented, but, as with human societies, this may not be possible. However, the policing must keep damage at an acceptable level.
- *Cooperative sustenance*: In animal societies, weak or sick individuals are cared for by the healthy and strong members. In a well-regulated WSN, duties may be assumed by a stronger node, or perhaps they may "nurse" the weaker node by providing sustenance (e.g., energy) or repairs (e.g., software patches).

This is not meant to be an exhaustive list but to merely inspire thought on how models of living systems can be directly applied to the improvement of WSNs. Many of these examples require extensive cooperation among nodes in a network and among WSNs. To avoid the requirement (and all of its weaknesses) of a centralized controller, the WSNs must be designed for the emergence of desirable behavior from this cooperation.

33.7 Biomimetic Implementation Techniques

Cognitive societies of WSNs will require hardware more capable and less constrained than what is available today. As stated, these improvements are to come soon enough. As Kurzweil recognized (Richards et al., 2002), designing software systems to fully utilize these improvements for building cognitive societies is a much more difficult problem. Therefore, the design of software should not wait until improved hardware is available. Software design for the advanced WSNs can and should begin now. Tools and techniques from the Alife communities directly apply.

A major building block for the *embodied cognitive scientists* is the *intelligent agent.* Although definitions abound for the intelligent agent, Noor Malone (1999) offers a definition that is pertinent to this discussion: "All agent categories have the common feature that they, to a large extent, *independently* perform tasks on behalf of their contracting party (or user) for which specialized knowledge is needed, or which consist of many time-intensive individual steps. . . .a software/hardware agent resides in an environment, uses sensors to identify certain aspects of the environment, and executes commands that affect the environment. Intelligent . . . agents act on behalf of people, take initiatives, and make suggestions."

Agents do not have to be *intelligent* to be effective. Like bees and ants, individual agents can apply a relatively small set of known rules in the presence of environmental stimuli. The actions of many such agents can result in the emergence of more complex behavior. Consider John Conway's "Game of Life" (Gardner, 1970). The environment is a cellular automaton: a rectangular grid of square blocks or cells. The grid really depicts a torus in that movement off the grid results in the reappearance on the opposite side. Agents exist as the cells that are initially in one of two states, light or dark, and apply the following rules for each change in state of the automaton:

- If a cell is lighted:
  - If it has one or no lighted neighbors, it dies (darkens), as if by loneliness.
  - If it has four or more lighted neighbors, it dies (darkens), as if by overpopulation.
  - If it has two or three lighted neighbors, it survives (remains lighted).
- If a cell is dark:
  - If it has three lighted neighbors, it lightens.

The behavior of the system *emerges* not only by the application of these rules but by the chosen initial state of the environment (i.e., the initial position of the agents: which cells are lighted). When the changing states are displayed as an animation, objects *appear* to form and move. Groups of cells consistently lighted relative to each other begin to *glide* across the grid (called *gliders*). Sometimes, objects will appear to remain *still* (called *blinkers*). Sometimes, blinkers will appear to emit gliders (called *glider guns*). The behavior seems surprising and unpredictable.

Epstein and Axtell (1996) designed a similar system they called *Sugarscape*. Using a similar cellular automaton, they endowed the environment with simple rules: each cell was assigned an initial amount of sugar, a rate of replenishment when sugar was consumed by agents, and no two agents could simultaneously occupy a single cell. Agents had two genetic endowments:

- A vision whereby they could see a given distance in cells (e.g., some could see two cells ahead; some three cells, etc.) in each of the four horizontal and vertical directions.
- A metabolism rate: the rate at which they consume sugar as they move from cell to cell.

Agents were given one simple rule: look as far as you can see in all directions, find the cell that has the most sugar, go there, and consume the sugar. Sugar was dispersed in the environment with concentrations at certain places. Four hundred agents were initially assigned locations and the rules applied in a series of states. Not surprisingly, at the end of the simulation, agents with high metabolism rates and poor vision died (they consumed all of their sugar before they found nourishment) and those with better vision and lower metabolism survived to locate the cells richest in sugar: the system demonstrated a *living Darwinian ecosystem*. They added more and different rules to eventually simulate other living attributes such as commerce, combat, and sexual reproduction.

The question is, can agent-environment systems be engineered to elicit desired behavior and prevent undesired behavior? That is, can Kelly be correct that one cannot find the hive in the bee; or is it that, if one knew enough about the bee, the hive could be seen in it. And, can this technique be applied in the design of WSNs?

Fuzzy logic and artificial neural networks (ANN) can be applied to the development of cognitive WSNs. Introduced by Zadeh (1965), fuzzy logic was conceived to define partial truth; that is, truth values that are not completely true and not completely false. ANNs are based on the concept of an artificial animal neuron, the Threshold Logic Unit (TLU), proposed by McCulloch and Pitts (1943). Gurney (1997) defines an ANN as, "an interconnected assembly of simple processing elements, units or nodes, whose functionality is loosely based on the animal neuron. The processing ability of the network is stored in the inter-unit connection strengths, or *weights*, obtained by a process of adaptation to, or learning from, a set of training patterns." The strength of the ANN is that they can be *trained* when the weights of the connections are adjusted on the basis of input experienced (i.e., the ANN *learns*). Hashem et al. (1995) demonstrated that neural networks benefit real-time data analysis by WSNs. For the task of identifying contaminants in the environment, they demonstrated that using neural network-based analysis of data from an array of heterogeneous sensors, selectivity of the array is significantly improved using sensors that are not individually selective. The combination of a heterogeneous sensor array with an automated analysis system is described as an artificial or electronic nose and has been demonstrated for use in monitoring food and beverage odors, analyzing fuel mixtures, and environmental monitoring. The array is designed such that each sensor measures a different property of the sensed sample. Each chemical composition presented to the array produces a characteristic signature through the fusion of sensor readings. By presenting many different compositions to the array, a database of signatures can be recorded. Conventionally, the number of sensors must be at least as great as the number of analytes. The quantity and complexity of data collected can make analysis of the data intricate. Using ANN to analyze the data for pattern recognition can not only reduce the necessary computation, but can reduce the required number and type of sensors in the array. Can ANNs and/or fuzzy logic be used in other applications of WSNs to facilitate *learning*?

33.8 Engineering for Emergent Behavior

There exists a good deal of parallelism between the evolutionary development of life as presented and what will be the evolutionary development of the WSN. The evolutionary roots of WSNs may be taken as having begun with the development of the microprocessor, which allowed the development of the initial WSNs: simple sensor nodes receive environmental stimuli as input (e.g., a command to take a sensory reading, the sensed value, etc.), and respond as output (e.g., transmit the sensed data). A central processor of some kind is required to eventually aggregate the sensed data from all nodes and make some global decisions based on those results. As WSNs evolve, the sensor nodes will become more intelligent, allowing greater amounts of processing to occur locally. The architectural requirement for the network to have a central, controlling processor will be greatly reduced or eliminated (Lodding, 2004a). For some purposes, nodes will be given the ability to affect their local environment through the structural coupling of sensor and effector technologies. One example of such an application is aircraft *smartskin*, in which a sensor-affector network built from a large number of smart sensors and affectors will be directly embedded into the wing structure to dynamically alter the shape of an aircraft wing, morphing it into the most efficient shape for the current aerodynamic conditions. With a biomimetic architecture there will be no need for a central, controlling computer as the wing can *think* for itself.

Technological evolution, like its biological twin, is an ongoing process. That the WSN will over time evolve from dumb sensor nodes to more intelligent and capable sensor-affector nodes is largely self-evident. A less obvious change is that for anything but the simplest of dedicated WSNs, intelligence will migrate away from a central, controlling computer, and distribute itself fully into the network, living within and across the collection of ever more intelligent and flexible nodes. The WSN is moving from a unicellular to a multicellular architecture (Lodding, 2004b).

Sensor nodes that are capable of altering both their behavior and their relationships with other nodes will be developed. As node intelligence, the number of nodes, and the number of inter-node connections increase, the stage becomes set for a phenomenon called *circular causation* to possibly arise. This is when the actions of the parts cause the overall behavior of the organism (i.e., emergent behavior) and the overall behavior guides the action of the parts making up the organism (i.e., self-organization). A possible manifestation of this behavior might take the form of network nodes forming local assemblies, or hierarchies, of embedded *micro sensor-affector networks:* organs. The formerly limited WSN will now evolve into a responsive, smart, adaptable, self-healing sensor-affector network, what can be described as a *cognitive network*. The primary question is: *will this evolution occur in an accidental, haphazard manner, or will it be engineered*?

Today, the primary focus of WSN research has been on the unicellular level of evolutionary development. Difficult questions concerning the hardware and software aspects of implementing real-world WSNs are being addressed. Concerns about electrical power requirements, communications issues, and network topology discovery are some of the basic problems that are currently being explored. While these parts of the puzzle are both necessary and appropriate for research, there is a larger, missing piece of the puzzle that is being overlooked.

The missing piece of the puzzle is *how do we engineer emergent behavior*? How do we program individual, unreliable, autonomous units that make up a larger system to exhibit specific, desired behavior? How do we predict the high-level behavior of the group from the individual programs executing in the individuals? And just as importantly, how do we inhibit the emergence of unwanted behavior from the collective interactions of numerous individuals?

There are people outside of the sensor-network research community who have been looking at just these types of problems. The majority of these researchers are in the fields of Artificial Intelligence and Artificial Life. Mathematicians have looked at the complexities of self-organizing systems, but with little, if any, direct influence in software algorithm development. Understanding how to design and build emergent systems is for all practical purposes an unknown and unexplored problem domain. If we want to control the development of *cognitive networks*, into which it is believed the WSN is predestined to evolve, then it is critical to understand and apply this new engineering science. Furthermore, the development and application of this science needs to begin now, while the focus is on the unicellular stage of evolution. Rather than wait until nodes have the expected increased capability to begin application, it is far better to begin now to understand and plan with this new method of engineering. When these networks are endowed with the qualities and abilities that are expected, it must be guaranteed that a command such as, "Please open the pod bay door, HAL.," produces the desired result.

33.9 Concluding Remarks

WSNs are destined to become pervasive in machines and ubiquitous throughout the Earth (and beyond) as they evolve into sensor/effector systems that not only monitor the environment but affect it as well. They are also going to rapidly expand their functionality. It has been demonstrated that the evolution of these networks has, parallels in the evolution of life. Current WSNs represent the beginning stage of their "life," where they function as a community of unicellular singletons receiving stimuli (input) and reacting to those stimuli (output). A centralized authority may provide some of that input and will aggregate their output for its own gain, but this is unknown or understood by the sensor nodes. The observation here is that their evolution can be accelerated and improved by engineering for emergent behavior. This engineering should not only be developed for achieving desired emergent behavior from large numbers of increasingly functional nodes (bees and ants), but should be preparing now for nodes that are cognitive (i.e., can learn). The goal for the design of future WSNs should be a "biosphere" of WSNs interacting with each other and their environment "symbiotically" such that they are autonomous, cooperative, fault-tolerant, self-regulating, self-healing, self-sustained *and* exhibit only desirable behavior.

References

Abelson, H., Allen, D., Coore, D., Hanson, C., Homsy, G., Knight, T., Nagpal, R., Rauch, E., Sussman, G., and Weiss, R. (1999). *Amorphous Computing.* MIT Artificial Intelligence AI Memo 1665. Retrieved April 5, 2004, from http://www.swiss.ai.mit.edu/projects/amorphous/papers/ aim1665.pdf.

Abelson, H., Allen, D., Coore, D., Hanson, C., Rauch, E., Sussman, G., and Weiss, R. (2000). Amorphous computing. *Communications of the ACM*, 43, Retrieved April 5, 2004, from http://www.swiss.ai.mit.edu/projects/amorphous/cacm-2000.html.

Bonabeau, E., Dorigo, M., and Theraulaz, G. (1999). *Swarm Intelligence: From Natural to Artificial Systems.* Oxford University Press, Oxford.

Budiansky, S. (2002). Creatures of our own making. *Science*, 298: 80–86.

Centre for Pervasive Computing (2004). Retrieved April 5, 2004, from http://www.pervasive.dk.

Chen, M., Majidi, C., Doolin, D., Glaser, S., and Sitar, N. (2003). *Design and Construction of a Wildfire Instrumentation System using Networked Sensors (Poster).* Network Embedded Systems Technology (NEST) Retreat, Oakland, CA. Retrieved April 5, 2004, from http://firebug.sourceforge.net.

Clark, A. (1997). *Being There: Putting Brain, Body, and the World Together Again.* MIT Press, Cambridge, MA.

Conniff, R. (1996). *Spineless Wonders.* Henry Holt and Company, New York.

Delin, K. (2004). *NASA/JPL Sensor Webs Project.* Retrieved April 5, 2004, from http://sensorwebs.jpl.nasa.gov.

Dusenbery, D. (1996). *Life at Small Scale: The Behavior of Microbes.* Scientific American Library Series No. 61.

Dyer, F. and Seeley, T. (1991). Dance dialects and foraging range in three Asian honey bee species. *Behavioral Ecology and Sociobiology*, 28: 227–233.

Epstein, J. and Axtell, R. (1996). *Growing Artificial Societies: Social Science from the Bottom Up.* The Brookings Institution, Washington, D.C.

Gardner, M. (1970). MATHEMATICAL GAMES: The fantastic combinations of John Conway's new solitaire game "life." *Scientific American*, 223: 120–123. Retrieved April 5, 2004, from http://ddi.cs.uni-potsdam.de/HyFISCH/Produzieren/lis_projekt/proj_gamelife/ConwayScientificAmerican.htm.

Gould, J., and Gould, C. (1988). *The Honey Bee.* Scientific American Library, New York.

Gregory, R. (1997). Editorial: Brains of ants and elephants. *Perception*, 26. Retrieved April 5, 2004, from http://www.perceptionweb.com/perc0397/editorial.html.

Gurney, K. (1997). *Introduction to Neural Networks.* Routledge, an imprint of Taylor and Francis Books Lt, London.

Hashem, S., Keller, P., Kouzes, R., and Kangas, L. (1995). Neural network based data analysis for chemical sensor arrays. In *Proceedings of International Society for Optical Engineering (SPIE) AeroSense Conference*, Orlando, FL (April 17–21, 1995), in *Applications and Science of Artificial Neural Networks*, Vol. 2492, Paper #2492–05, pp. 33–40. Retrieved April 5, 2004, from http://citeseer.ist.psu.edu/519919.html.

Intel Research — Exploratory Research — Deep Networking (2004). Retrieved April 5, 2004, from http://www.intel.com/research/exploratory/deep_networking.htm.

Intel Research — Exploratory Research — Deep Networking — Heterogeneous Sensor Networks (2004). Retrieved April 5, 2004, from http://www.intel.com/research/exploratory/heterogeneous.htm

Invisible Computing: Scope (2000). Retrieved April 5, 2004, from http://invisiblecomputing.org/scope.html.

Jones, K., Wadaa, A., Olariu, S., Wilson, L., and Eltoweissy, M. (2003). Towards a new paradigm for securing wireless sensor networks. In *Proceedings New Security Paradigms Workshop 2003*, Ascona, Switzerland. August 18–21, 2003, pp. 115–122.

Kahn, J., Katz, R., and Pister, K. (1999). Next century challenges: Mobile networking for "Smart Dust". In *ACM MOBICOM Conference*, Seattle, WA. Retrieved April 5, 2004, from http://www.cs.berkeley.edu/~randy/Papers/mobicom99.pdf.

Kelly, K. (1994). *Out of Control: The New Biology of Machines, Social Systems, and the Economic World.* Perseus Books.

Lachman, R., Lachman, J., and Butterfield, E. (1979). *Cognitive Psychology and Information Processing.* Lawrence Erlbaum Assoc, Hillsdale, NJ.

Lammers, D. (2001). Embedded projects take a share of Intel's research dollars. *EE Times.* Retrieved April 5, 2004, from http://today.cs.berkeley.edu/800demo/eetimes.html.

Leveraging the Infosphere: Surveillance and Reconnaissance in 2020 (1995). Airpower Journal — Summer 1995, A SPACECAST 2020 White paper. Retrieved April 5, 2004, from http://www.airpower.maxwell.af.mil/airchronicles/apj/spacast1.html.

Lodding, K.N. (2004a). *Hitchhikers Guide to Biomorphic Software.* ACM Queue.

Lodding, K.N. (2004b). Multi-agent organisms for persistent computing. In *Proceedings of the 3rd International Joint Conference an Autonomous Agents and Multi Agent Systems* (AAMASO4). New York, NY, July 19–23.

Mainwaring, A., Polastre, J., Szewczyk, R., and Culler, D. (2002). Wireless sensor networks for habitat monitoring. (Intel Research, IRB-TR-02–006, June 10, 2002) In *ACM International Workshop on Wireless Sensor Networks and Applications.* Retrieved April 5, 2004, from http://www.greatduckisland.net.

Margulis, L. and Sagan, D. (1986). *Microcosmos: Four Billion Years of Microbial Evolution.* Simon and Schuster, New York, p. 15.

McCulloch, W. and Pitts, W. (1943). A logical calculus of the ideas immanent in nervous activity. *Bulletin of Mathematical Biophysics*, 7: 115–133.

Moravec, H. (1988). *Mind Children.* Harvard University Press.

National Research Council (2001). *Embedded, Everywhere: A Research Agenda for Systems of Embedded Computers*, Committee on Networked Systems of Embedded Computers, for the Computer Science and Telecommunications Board, Division on Engineering and Physical Sciences, Washington, DC. Retrieved April 5, 2004, from http://www.nap.edu/catalog/10193.html.

Noor, A. and Malone, J. (1999). *Intelligent Agents and Their Potential for Future Design and Synthesis Environment.* NASA CP-1999-208986.

Pfeifer, R. and Scheier, C. (2000). *Understanding Intelligence.* MIT Press, Cambridge, MA.

Primio, F., Müller, B., Lengeler. (2000). Minimal cognition in unicellular organisms. In SAB2000 *Proceedings Supplement, International Society for Adaptive Behavior.* Honolulu, HI. pp. 3–12, Retrieved April 5, 2004, from http://www.ais.fraunhofer.de/BAR/papers/diprimio-mincog.pdf.

Reinhard, J., Srinivasan, M., and Zhang, S. (2004). Olfaction: scent-triggered navigation in honeybees. *Nature,* 427: 411.

Richards, J., Gilder, G., Kurzweil, R., Searle, J., Dembski, W., Denton, M., and Ray, T. (2002). *Are We Spiritual Machines?* Discovery Institute, Seattle, WA.

Tilak, S., Abu-Ghazaleh, N., and Heinzelman, W. (2002). A taxonomy of wireless micro-sensor network models. *ACM Mobile Computing and Communications Review* (MC2R), 6. Retrieved April 5, 2004, from http://www.cs.colorado.edu/~rhan/CSCI_7143_001_Fall_2002/Papers/Tilak2002_p28-tilak.pdf.

Turner, J. (2000). *The Extended Organism: The Physiology of Animal-Built Structures.* Harvard University Press.

UCB/MLB 29 Palms UAV-Dropped Sensor Network Demo (2001). University of California, Berkeley, CA. Retrieved April 5, 2004, from http://robotics.eecs.berkeley.edu/~pister/29Palms0103.

Wadaa, A., Olariu, S., Wilson, L., Eltoweissy, M., Jones, K., and Sundaram, P. (2004a). Training a sensor network. In Special Issue of MObile NETwork (MONET) on *Algorithmic Solutions for Wireless, Mobile, Ad Hoc and Sensor Networks,* Bar-Noy, A., Bertossi, A., Pinotti, M., and Raghavendra, C. Eds. January 2004.

Wadaa, A., Olariu, S., Wilson, L., Eltoweissy, M., and Jones, K. (2004b). On providing anonymity in wireless sensor networks. In *Proceedings of the 10th International Conference on Parallel and Distributed Systems,* (ICPADS-2004). Newport Beach, CA. July 2004.

Weiser, M. (1991). The computer for the Twenty-first century. *Scientific American,* 94–10. September 1991. Retrieved April 5, 2004, from http://www.ubiq.com/hypertext/weiser/ UbiHome.html.

Weiser, M. (1996). *Ubiquitous Computing.* Retrieved April 5, 2004, from http://www.ubiq.com/hypertext/weiser/UbiHome.html.

Zadeh, L. (1965). Fuzzy sets. *Information and Control,* 8: 338–353.

Appendix A — Glossary

The following terms are pertinent to this topic:

Analyte: The object of measurement in an analytical procedure (in this case a chemical property).

Artificial intelligence: The science of simulating intelligence in a creation of humans (i.e., not natural).

Artificial life: Beyond intelligence, the science of simulating a living organism or some property thereof in a creation of humans (i.e., not natural).

Biomimetics: The study of the origin, structure, or function of biological mechanisms, processes, and materials as models for the design of artificial constructs.

Circadian: Relating to approximately a 24 h period (from Latin meaning "around the day").

Effector: A device that, in response to a stimulus (input or command), initiates an effect on or affects its environment.

Emergent Behavior: Behavior that results from:

- Innate behavioral rules applied to unpredictable environmental stimuli.
- Behavior of individuals interacting and combining to result in behavior of a group.
- Learning (i.e., in cognitive systems, learning from experiences of behavior may initiate new behavior).

Evolvable hardware: Hardware designed by the application of evolution to automate its creation and adaptation. A goal is to use these techniques in situ producing hardware that can adapt to unpredicted environmental conditions, thus improving its survivability for long durations in unknown and changing environments.

Homeostasis: The state of a relatively constant internal environment. The physical and chemical states that an organism must maintain to allow proper functioning, in maximum efficiency, of its components: cells, tissues, organs, and organ systems.

Innate behavior: Behavior that results from genetic encoding. Such behavior does not require learning and changes little in response to environmental stimuli.

Learned behavior: Behavior that results from the cognitive assessment of prior experiences to determine a new response.

Metabolism: The chemical processes (breaking down substances to provide energy or synthesis of new substances) within a living organism that are necessary for life.

Mote: Defined as a small particle, here it species the self-powered, physically independent device that contains hardware and software for sensing or effecting, computing, and communication in a wireless sensor network.

Nanoscale: Measurement on a scale of nanometers (dimensions under ~100 nm).

Nano-technology: The process used to design and build electronic circuits and devices from atoms and molecules.

Node: In computer science, this is either a terminal or hop point in a communications network. Here it specifies a mote in the context of the network.

Pervasive: Defined as the quality of being present throughout; to permeate. Pervasive computing and ubiquitous computing seem to be synonymous in computer science literature, but here, pervasive is used to describe presence throughout a system (e.g., sensors are pervasive in an automobile if they are installed and function throughout the automobile).

Photosynthesis: The process in some organisms (plants and some microbes) by which carbohydrates are synthesized from carbon dioxide and water using light as an energy source. These carbohydrates serve as energy storage and are later metabolized resulting in the release of carbon dioxide, water, and energy used by the organism.

Protozoa: Plural for protozoan collectively naming any of a large group of single-celled, microscopic organisms (e.g., amoeba, sporozoans, ciliates, flagellates, etc.). Protozoans differ from bacteria (prokaryotic) in that they, like higher plants and animals, contain cellular constructs such as nucleus, mitochondria, etc. (eukaryotic).

Sensor: A device that receives and reports a signal or stimulus.

Society: In biology, a society is defined as a colony of organisms, usually of the same species such as a society of ants. Here, a society is limited to an association of cognitive animals such as wolves or humans.

Symbiosis: A relationship between organisms that is mutually beneficial and which, over time, forms a dependence.

Ubiquitous: Defined as the quality of being or seeming to be everywhere at the same time. Pervasive computing and ubiquitous computing seem to be synonymous in computer science literature, but here, ubiquitous is used to describe presence within all systems (e.g., sensors are ubiquitous in automobiles if they are installed and function in all automobiles).

34

A Cooperative Parallel Metaheuristic Applied to the Graph Coloring Problem

Benjamin Weinberg
El-Ghazali Talbi

34.1 Introduction

Combinatorial optimization often needs a large amount of computation resource. This is particularly the case for NP-hard problems, for which no efficient algorithm is known. Parallel computers may supply this resource. However, it may be interesting to use those platforms in another way than to parallelize the computation of the objective function of the visited configurations.

Coevolutionary computation [1] leads to algorithms that can be implemented in a distributed way. We call this kind of implementation cooperative algorithms. Such an algorithm presents a low coupling between its components. In this chapter, we present a distributed design and implementation of the metaheuristic COSEARCH [2,3]. This metaheuristic is a very general model which can easily be implemented in several ways. We propose a view of this algorithm bearing in mind the need to explicitly balance diversification and intensification during the search. We evaluate our approach on the graph coloring problem.

Nevertheless, the implementation of COSEARCH needs the design of mechanisms dedicated to the studied problem. In our case, we use two new operators which are the *break* and the X_ρ operators. The former is an operator modifying a unique coloring as mutation in genetic algorithms. The latter combines two colorings as crossover.

This chapter is organized as follows. In Section 34.2, we present an overview of the parallel cooperative COSEARCH method: concepts and implementations of suggestions. In Section 34.3, we present the studied problem, which is the graph coloring problem (MGCP). In Section 34.4, we present some preliminary works on the search operators. In Section 34.5, we present an implementation of COSEARCH for MGCP. In Section 34.6, we have given some experimental results. Finally, we conclude and present some perspectives in Section 34.7.

34.2 A Cooperative Metaheuristic

In this section, we present a brief overview of a parallel metaheuristic, called COSEARCH. Some work on this metaheuristic was undertaken in Reference [2] for the quadratic assignment problem (QAP) and in reference [3] for the frequency assignment problem (FAP). This presentation is very general and this metaheuristic may be instantiated in many ways. The head of the algorithm is the use of a memory gathering information on the overall search.

This section is organized as follows: In Section 34.2.1, we present the main idea to visit the search space. In Section 34.2.2, we present the challenge to memorize information during the search. In Section 34.2.3, we present an overview of our approach.

34.2.1 First Step Search

Optimizing an NP-hard problem leads in general to the use of heuristic approaches. One of the most used approach is the local search. Those algorithms modify configurations (candidate solution) by local transformations. Generally, it is said, that we move from a configuration to one of its neighbors by a small modification.

Iterating successive modification leads to a configuration which is better than the initial configuration. But generally, the configuration found is only a local optimum; that is we cannot improve the configuration by any small modifications, we use. This is typically the case in the hill climbing algorithm. In such an algorithm, an elementary operation is called a move.

That is why the additional search mechanism is often used to escape from the local optimum and continue the search. Then, the best visited local optimum is returned at the end of the algorithm. There are plenty of such mechanisms more or less dedicated to the studied problem. We can cite the three following examples:

- Random mechanisms using biased probabilities of acceptance of the worst configurations as in simulated annealing [4].
- Memorizing of the last visited configurations or the last made moves to prevent visiting again a previously visited local optimum as in a tabu search [5].
- Multistart methods consist in the searching of the local optimum from different initial configurations [6].

Such methods may provide good results. However, in some cases, these methods may be insufficient to find the optimum. We propose to use the local search algorithm as a basic component of our algorithm. Several parallel/distributed local search algorithms may work in a cooperative way to optimize a problem. In our approach, such an algorithm is called a (search) agent.

34.2.2 Memorizing during the Search

Search agents work from different configurations of the search space. We suggest to iterate their application in order to cover the whole search space (as in multistart approach). However, for a better guidance, we have to synchronize the different agents around an adaptive memory.

The adaptive memory gathers information on the visited configurations. The main difficulty in applying this algorithm is to know what is pertinent to keep in this memory. Indeed, the search may take a long time and it would be unhelpful and inapplicable to keep every visited configuration in the memory.

The second difficulty is to answer the following question: "what do we use with this memory?" A search agent needs an initial configuration to search. The memory has to supply interesting initial configurations. For that reason, we may be able to build configurations with new or insufficiently explored characters with regard to the memory.

So, in answer to the second point, the way to surmount the main difficulties, in our view, is that the memory must keep some good configurations and notice the unexplored area of the search space.

34.2.3 Cooperative Search

Our approach consists in cooperation of several search algorithms (called agent) through the adaptive memory (see Figure 34.1).

Typically, an agent is a local search strategy (due to its efficiency), but it is not a strict rule. So, we can imagine to select any algorithm as an agent.

The different searches made by agents are totally independent of each other. So, each agent can be executed in parallel.

Furthermore, we can use heterogeneous agents. As the coupling is very low, several algorithms can be used in a complementary manner:

- Local searches with different operators
- Different metaheuristics as simulated annealing or tabu search
- Complex algorithms using two or more configurations as genetic algorithms

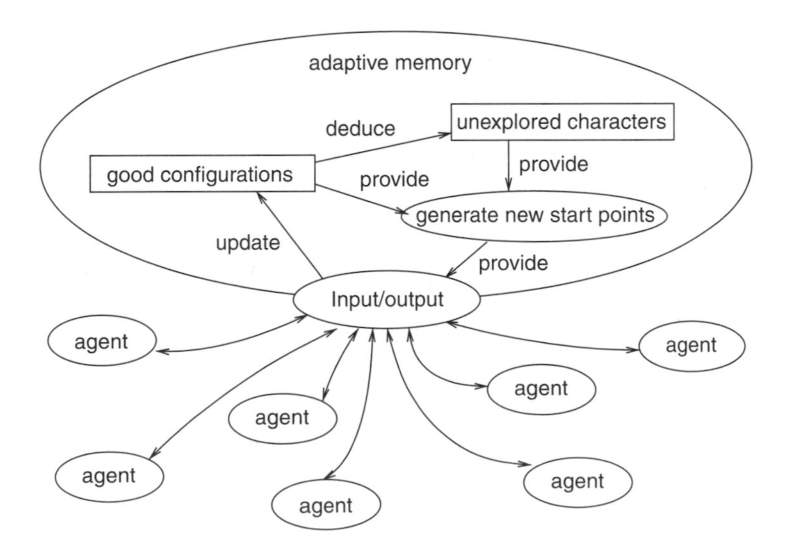

FIGURE 34.1 Global overview of our approach.
Note: In this figure, we use ellipses to indicate the presence of dynamic code and rectangles for static data.

34.3 Graph Coloring Problem

This section is about recalling the definition of the academic generic problem which is graph coloring problem (MGCP). In Section 34.3.1, we present some applications of the graph coloring problem. Then in Section 34.3.2, we present the formalizations that we use. In Section 34.3.3, we present an overview of the heuristic methods applied to this problem.

34.3.1 Applications

Graph coloring problem models a lot of real-life problems. Those problems consist in distributing objects into different groups in order to separate some pairs of objects, for example, the Register Allocation Problem (RAP) [7]. In this problem, the objects are the variables of a source code and the groups are the registers of a computer. For each variable, we must assign a register number. Two variables, having the same register number, cannot exist at the same time. Thus, we obtain a set of binary constraints of mutual exclusion on variables. The smaller the number of used registers, the more efficient is the program. The objective function is obviously to minimize the number of registers.

There are a lot of other problems, which are direct applications of MGCP. In this category, we can find:

- *Frequency assignment problem* [8]. In this problem, the objects are the channels of the transceivers and the groups are the available frequencies. Two channels covering near by areas have to communicate using different frequencies.
- *Timetabling problem* [9]. A course combines the constraints from three elements: teacher, class, and location; knowing that a teacher cannot make two lectures at the same time, idem for class and location.
- Some other problems such as pattern matching [10], air route conception [11], etc.

34.3.2 Formalization

All problems presented above can be modeled using a graph $G(V, E)$. The set of vertices V is the set of object to color (distribute or assign) and the set of edge E symbolizes the constraints between vertices: vertices connected by an edge must have different colors (be assigned with different values).

We use the following formalization for a given graph $G(V, E)$.

Definition 34.1 *We call coloring (of G) any mapping from V into the set of natural numbers* $\{1 \ldots |V|\}$.

Definition 34.2 *We call color of v with regards to a coloring $\mathcal{C}$, the image of v by $\mathcal{C}$ (i.e., $\mathcal{C}(v)$).*

Definition 34.3 *We call ith class of a coloring $\mathcal{C}$, the set of vertices colored into i (i.e., $\mathcal{C}^{-1}(i)$).*

Technically, a coloring may be trivially encoded by an array of integers indexed by the vertices. Nevertheless, for some algorithms it will be helpful to have the vertices gathered by classes. We choose a mixed encoding (see Figure 34.2). This encoding allows the visit of vertices of a class without investigating the other vertices. An array allows us to change quickly (in $O(1)$) the color of one vertex.

Definition 34.4 *We call* violation of constraint *(in a coloring $\mathcal{C}$), an edge whose extremities is colored by the same color (i.e., $\mathcal{C}^{-1}(i)$).*

Definition 34.5 *We define g the mapping defined by:*

$$g : \{colorings\} \rightarrow \mathbb{N}$$

$$\mathcal{C} \mapsto g(\mathcal{C}) = \frac{1}{2} \times \sum_{i,j,\in V} \delta_{\mathcal{C}(i),\mathcal{C}(j)} \times \mathbb{I}_E(i,j)$$

where $\delta_{i,j}$ is the Kronecker's symbol (i.e., $\delta_{i,j} = 1$ if $i = j$; 0 otherwise) and $\mathbb{I}_E(i,j)$ is the characteristic function of E (i.e., $\mathbb{I}_E(i,j) = 1$ if $(i,j) \in E$; 0 otherwise).

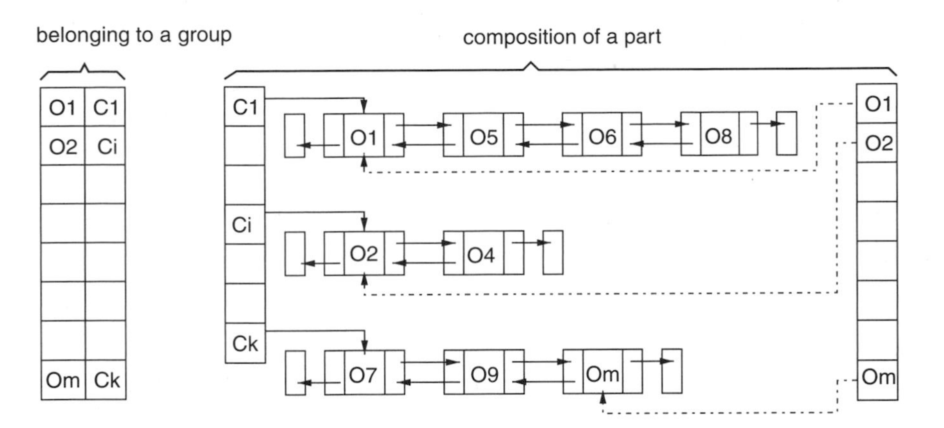

FIGURE 34.2 Example of encoding.
Note: *Technically the number $\chi(\mathcal{C})$ of non-empty parts is stored and only the first $\chi(\mathcal{C})$ labels are used.*

Definition 34.6 *We say that a coloring $\mathcal{C}$ is a* proper *coloring if there is not any violation (i.e., $g(\mathcal{C}) = 0$).*

Definition 34.7 *A coloring is a k-coloring if the used colors are less than or equal to k.*

Definition 34.8 *The chromatic number of graph is the smaller number such that the graph allows a proper k-coloring. For a given graph G, we denote $\chi(G)$ as its chromatic number.*

Definition 34.9 *Let k be a positive number. We call graph k-coloring problem (GkCP), the problem consists in finding a k-coloring such that $g(\mathcal{C})$ is minimal.*

Definition 34.10 *We call Minimal graph coloring problem (MGCP), the problem consisting in finding a coloring with the less used colors. So, we minimize the following function:*

$$f : \{\textit{proper colorings}\} \rightarrow \mathbb{N}$$
$$\mathcal{C} \mapsto f(\mathcal{C}) = \max_{v \in V} \mathcal{C}(v)$$

This is this second objective function, we want to optimize.

Both problems are known to be NP-hard [12].

In this problem, there is a symmetry. Indeed, we do not care about the effective value assigned to a vertex. The unique interest resides in the fact that vertices share or do not share the same color. So, the graph coloring problem is a partitioning problem (also called grouping problem).

In Reference [13], we propose some tools to operate on the space of partitioning without the problems due to the symmetry. We provide a format to compute in linear time if two colorings represent the same partitioning.

We also present how to compute the distance between two partitioning independently of the choice of their representations. Our study is based on the use of what we call a refining matrix. The refining matrix of two colorings $\mathcal{A}$ and $\mathcal{B}$ is a matrix $\mathcal{M}^{AB}$ defined by:

$$m_{ij}^{\mathcal{AB}} = \text{card}(\mathcal{A}(i) \cap \mathcal{B}(j)). \tag{34.1}$$

This matrix will be helpful for the design of search operators.

34.3.3 Used Methods

Minimal graph coloring problem is an old and deeply studied problem. As the problem is NP-hard, the heuristic approach is often used. There are two great families of algorithms for MGCP. The first one operates on proper coloring:

- Greedy algorithm DSATUR [14]
- Recursive Largest First RLF [15]
- Local search algorithm using the Kempe chain [16]
- Iterated Greedy, proposed by J. Culberson [17] is a kind of local search
- Memetic Algorithm using a special encoding of a coloring [18]
- Ants colony [19]
- Constraint Programming [11]

The second one operates on k-colorings. The search is decomposed into steps. At each step, the algorithm works with a fixed k and tries to find a proper k-coloring in the whole k-colorings. Then, k is decreased for a new step. When a step fails (does not find a proper coloring), the previously found proper $k+1$-coloring is returned. This strategy was used in many heuristics:

- Local search (hill climbing, simulated annealing) [16] with the change of one vertex as operator
- Local search with using "permutation-neighborhood" operator [20]
- Tabu search [21–23]
- Addition of several diversification techniques to local search procedures [24]
- Local search with variable neighborhood search [25]
- Modification of bias of neighborhood operator [26,27]
- Evolutionary Algorithm with dedicated operators [23,28,29]
- A multi-level approach [30]

Nevertheless, Clerc [31] studied a method using a mix of both strategies. He used a method working on valid colorings and k-colorings in particle swarm optimization algorithm.

In this study, our attention was focused on the first strategy.

34.3.4 Reference Graphs

We evaluate our algorithm on several graphs from the site of Trick.[1] Those graphs are issued from the files of the second DIMACS challenge. Table 34.1 presents the benchmark and some of their features.

34.4 Basic Algorithm

This study is based on the use of two algorithms modifying a coloring. Both of them change proper coloring into another proper one where the number of colors decreases or remains the same. In this section, we present those operators. Then, we experiment upon their efficiencies.

34.4.1 IG Operator

Culberson [17] proposes a manner to generate a coloring $\mathcal{C}'$, where the number of colors is less than or equal to the number of colors of the genitor $\mathcal{C}$. The new coloring is built as follow: First, all the vertices are uncolored in $\mathcal{C}'$. Second, all (uncolored in $\mathcal{C}'$) vertices of class $\mathcal{C}^{-1}(1)$ are placed into the first class of $\mathcal{C}'$. Then, the class is filled with other (uncolored in $\mathcal{C}'$) vertices (i.e., adding vertices until the class does not accept any one without inducing constraint violations). This step is repeated to build the

[1] http://mat.gsia.cmu.edu/COLOR/color.html

TABLE 34.1 Used Benchmarks

Name	nb vertices	Density
DSJC125.1	125	0.1
DSJC125.9	125	0.9
DSJC500.9	500	0.9
DSJC125.5	125	0.5
DSJC250.9	250	0.9
DSJR500.1c	500	0.1
le450_5c	450	0.1
le450_5d	450	0.1
queen8_12	96	0.3
queen9_9	81	0.3
queen10_10	100	0.3
queen11_11	121	0.3
school1	385	0.3
school1_nsh	352	0.2

second color and so on. The visit of the class of $\mathcal{C}$ are made in a disorder by manner to generate different colorings. Technically, this operator can be implemented using a unique coloring. So, we can apply again this technique on the obtained coloring.

Algorithm 34.1 presents this idea. In line 5, we complete the class without regarding the colorings $\mathcal{C}$. The uncolored vertices are shuffled to increase the diversity of the search.

Algorithm 34.1 IG Operator

Require: $\mathcal{C}$ current coloring
1: $\mathcal{C}'$ is uncolored;
2: shuffle the class of $\mathcal{C}$;
3: **for all** class i of $\mathcal{C}$ **do**
4: $\quad \mathcal{C}'^{-1}(i) = \mathcal{C}^{-1}(i) \bigcap \{$uncolored vertices of $\mathcal{C}'\}$;
5: $\quad$ Complete $\mathcal{C}'^{-1}(i)$ with uncolored vertices of $\mathcal{C}'$;
6: **end for**
7: $\mathcal{C} \leftarrow \mathcal{C}'$;

34.4.2 Break Operator

This operator operates as opposed to the previous one. Instead of filling the classes we try to empty them. For each class, we distribute its vertices into the others (nonempty) classes. In this case too, the number of used colors cannot increase.

The Algorithm 34.2 explains how we proceed to break a coloring. We notice that we use the same order to visit the classes of the coloring in line 1 and line 3.

34.4.3 Operators Efficiency

Both methods usually change the current coloring to a coloring with the same number of colors. To evaluate their performance, we iterate them at most $|V|$ times without improvement. Table 34.2 summarizes the results, we obtain.

We notice that the number of colors needed by iteration of break operator are always less than those obtained by IG. We can state that the break operator is better than IG when they are used alone.

Algorithm 34.2 Break Operator

```
Require: C the current coloring
1: shuffle the class of C;
2: for all class i of C do
3:   for all vertices s of C^{-1}(i) do
4:     for all class j of C ≠ i do
5:       if s is colorable in j then
6:         color s into j;
7:         break;
8:       end if
9:     end for
10:   end for
11: end for
```

TABLE 34.2 Operatory Efficiency

	IG				Break			
Name	min	max	avg	std dev.	min	max	avg	std dev.
DSJC125.1	6	7	6.90	0.31	6	7	6.4	0.5
DSJC125.5	19	21	20.64	0.59	19	21	20.35	0.59
DSJC125.9	45	47	46.4	0.60	44	47	45.8	0.77
DSJC250.9	78	81	79.15	0.88	76	79	77.2	0.89
DSJC500.9	141	145	143.2	1.24	137	141	138.85	1.46
DSJR500.1c	85	85	85	0	85	85	85	0
le450_5c	6	8	6.8	0.52	5	7	6.40	0.6
le450_5d	5	8	6.7	0.66	5	7	6.05	0.69
queen10_10	12	14	12.95	0.39	12	13	12.95	0.22
queen11_11	14	15	14.1	0.31	14	15	14.5	0.51
queen8_12	13	14	13.05	0.22	13	13	13	0
queen9_9	12	12	12	0	11	12	11.8	0.41
school1	14	19	14.7	1.13	14	15	14.25	0.44
school1_nsh	14	18	15.6	1.05	14	16	14.6	0.6

34.5 A Cooperative Metaheuristic

Our goal is to provide cooperative heuristic coloring graphs. In this section, we present the implementation we made of COSEARCH to MGCP.

34.5.1 Information to Memorize

In the adaptive memory, we store a set of N colorings, called *elite*. This elite present no double, for that we use the techniques presented in Reference [13]. When a coloring is received, the new coloring replaces the worst coloring (if the new coloring is better).

We also use four specific sets: one for each type of agents. More precisely, when a new coloring is obtained from an agent a, we try to add it to the elite set. If this fails, we try to add it to the set corresponding to a. The replacement strategy keeps only the best colorings in both cases.

The adaptive memory also records some additional information: for each pair of vertices, we notice the last iteration where they were in the same class, and the last iteration they were placed in different classes. Using this data, we can infer if some area of the search space is insufficiently explored.

The adaptive memory takes decision from the following list:

- *Intensify*: The memory randomly picks a coloring in elite and sends it to an agent IG or a break agent. Indeed those agents can easily modify a coloring and a better coloring can be reached by several applications of those algorithms. We call such an agent an I-agent.
- *Force a same class*: The memory randomly picks one of the best colorings. It chooses the pair of nonconnected vertices that are not colored by the same color for a long time. The memory forces both vertices to be in the same class. This may increase the number of colors. Finally, the new coloring is sent to a break agent. The agent respects always that both vertices are in the same class. It behaves as both chosen vertices form a new one. The set of neighbors of the new vertex is the union of the set of neighbors of the old ones. We call such an agent an FS-agent.
- *Force different classes*: The memory picks randomly one of the best colorings. It chooses the pair of nonconnected vertices that are always colored by the same color for a long time. The memory forces both vertices to be in different classes. This usually increases the number of colors. Finally, the new coloring is sent to a break agent. The agent respects always that the two vertices are in different classes. It behaves as if there is an additional edge between the two chosen vertices. We call such an agent an FD-agent.
- *Greedy crossover*: Memory extracts from two colorings the *commun part*. This partial coloring is sent to an agent to be completed. This completion is made in a greedy manner. Then, the break operator is applied few times. Several compressions are tested before returning the best one. We call such an agent an X_ρ-agent.

34.5.2 Partial Coloring Building

To construct a partial coloring $\mathcal{C}$ representing the commun part of two colorings, we use the refining matrix defined as below.

For two colorings $\mathcal{A}$ and $\mathcal{B}$, and their associated refining matrix $\mathcal{M}^{\mathcal{AB}}$, we build a mapping $\tau\colon \{1, \ldots, \chi(\mathcal{A})\} \to \{1, \ldots, \chi(\mathcal{B})\}$ defined by $\tau(i) = \arg\max_j m_{i,j}^{\mathcal{AB}}$.

With τ, we build $\mathcal{C}$ as follows:

$$\forall j \in \{1, \ldots, \chi(\mathcal{B})\}\ \mathcal{C}(j) = \bigcup_{i \in \tau^{-1}(j)} \mathcal{A}(i). \tag{34.2}$$

Typically, this partial coloring $\mathcal{C}$ has few nonempty classes than $\mathcal{B}$.

We remark also that such a scheme presents no constraints violations as soon as the coloring $\mathcal{B}$ is proper. Indeed classes of $\mathcal{C}$ are part of classes of $\mathcal{B}$.

34.5.3 Parallel Implementation

Our algorithm is implemented using the C-PVM environment [32]. For the synchronization of the different tasks, we use three (kind of) automata: one to control the distributed agents (see Figure 34.3), one to control a manager of a fleet of several agents of the same type[2] (see Figure 34.4) and the last one to control the adaptive memory (see Figure 34.5).

A fleet manager works as an asynchronous gate. A new job is provided with workers because of the manager since one of its agents is available.

The automaton can be in one of the three main states:

- Every worker is working, so the adaptive memory waits for one result.
- There are some computational available resources, so a policy is chosen with regard to the availabilities.

[2]Scheme, break, force different or force same.

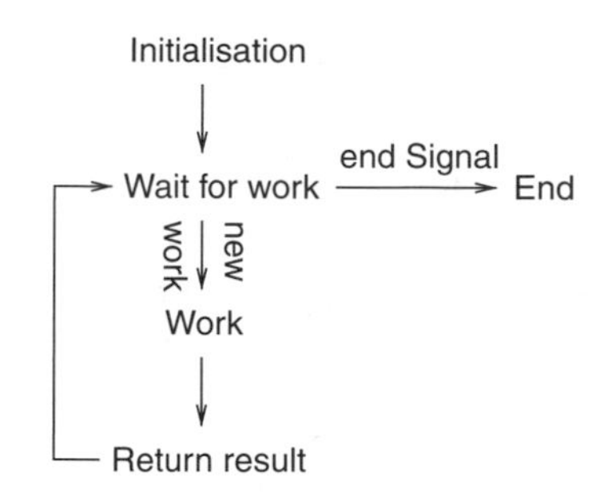

FIGURE 34.3 Automaton for the searching agents.

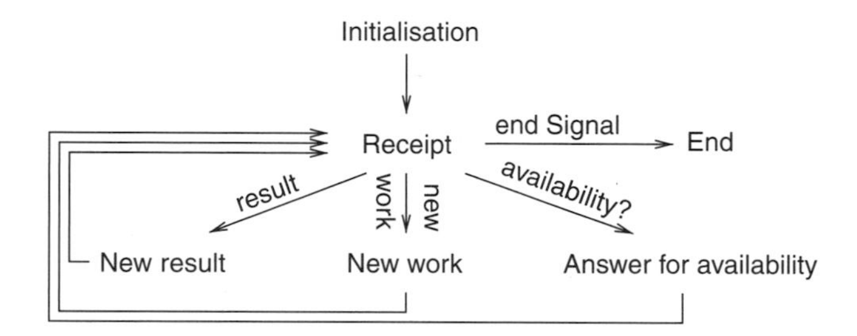

FIGURE 34.4 Automaton for the fleet managers.

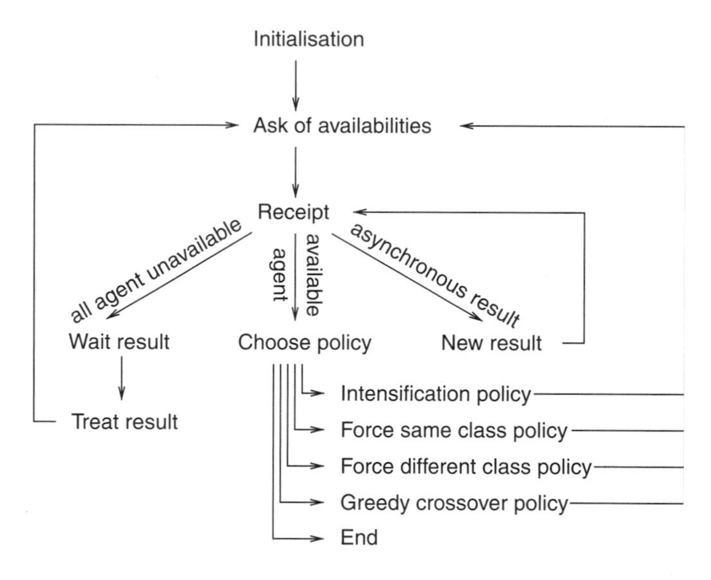

FIGURE 34.5 Automaton for the adaptive memory.

- A new result is asynchronously obtained, so the memory incorporates the result before any choice of a policy.

As suggested in Section 34.2.3, the automaton of the adaptive memory and the agents communicate with the fleet managers. A fleet manager corresponds to the Input/Output interface of the memory.

The end of the algorithm is decided by the adaptive memory. The end is chosen as a policy but it happens after a fixed number of iterations. The other policies of the memory is randomly picked from the available agents with fixed biases. For example, we may apply 60% of I-agent, 10% of FD-agent,

10% of FS-agent, and 10% of X_ρ-agent. At a given moment, if all intensification agents are working then the policy cannot be an intensification.

34.6 Experiments

Our algorithm is composed of many components. It is possible that some presented agents do not contribute well to the global quality of the search. So, we have to tune the use of each agent. This is complicated by the fact that all agents are based on the break operator that is very efficient due to its capacity to diversify. Thus, we cannot detect if we have to continue to use I-agent or operate some radical changes.

In a first time, we evaluate each agent with the I-agent. Finally, we present results of the global architecture.

34.6.1 Preliminary Tests

For our experiments, we use the following parameters:

- *Number of iterations of the global cycle*: At each iteration, adaptive memory collects a new coloring and chooses a policy. We bound experimentally this parameter to 500.
- *Number of iterations for each type of agent*: The number of application of the break operator by the I-agents, FS-agents and FD-agents is equal to the number of vertices. The X_ρ agents recombine two colorings and halve the number of vertices applications of break operator on each obtained colorings.
- *Number of available agent for each type of agent*: We use ten I-agents, four FS-agents, four FD-agents, and four X_ρ-agents. For those tests, we use six diversification agents (FS, FD, X_ρ) in parallel with ten intensification agents (break).

Table 34.3 summarizes the obtained results. For each combination, we made four runs on different hardware platforms using different architectures: Linux PCs (between 2.7 and 3 Ghz , 512 Mo and 1 Go RAM) and an IBM SP3 (with 16 processors Power3 NH2 per node running at 375 Mhz, 16 Go RAM).

We can see that the combination allows us to go further in the search. Indeed, the qualities of the overall solutions are better for any combination than for the break operator.

Incidentally, we observe that the different combinations perform more or less well on different graphs. I+FS provides the best results on DSJC250.9, I+FD provides the best results on Queen9_9 and I+X always found the same quality of results for Queen8_12.

34.6.2 Global Architecture

To profit from different diversification mechanism, place into concurrence all the agents. Results using four instances of each diversification agents and ten intensification agents are gathered in Table 34.3.

We observe that the complete combination provides results at least as good as the worst combination previously seen. But for some graphs (DSJC125.9 and DSJC250.9) results are not abreast of the best combination. Nevertheless, for one graph (DSJC125.5) interactions between components provide the best colorings than those seen before (Table 34.4).

34.7 Conclusion and Future Works

In this chapter, we present a cooperative metaheuristic called COSEARCH makes an explicit balancing of intensification and diversification in the search space. This metaheuristic is intrinsically parallel. As the model is very general, it can be implemented in many ways. We propose a manner to implement it in parallel. This implementation is hierarchical and is based on three levels. At the top the adaptive

TABLE 34.3 Summarize Experiments with one Diversification Agent and the Break Agent

	I+FS				I+FD				I+X			
Name	min	max	avg	std dev.	min	max	avg	std dev.	min	max	avg	std dev.
DSJC125.1	6	6	6	0	6	6	6	0	6	6	6	0
DSJC125.5	19	19	19	0	19	19	19	0	19	19	19	0
DSJC125.9	44	44	44	0	44	44	44	0	44	44	44	0
DSJC250.9	**73**	75	74	0.81	74	74	74	0	74	74	74	0
DSJC500.9	131	133	132	0.81	131	132	131.75	0.5	133	134	133.5	0.57
DSJR500.1c	85	85	85	0	85	85	85	0	85	85	85	0
le450_5c	5	5	5	0	5	5	5	0	5	5	5	0
le450_5d	5	5	5	5	5	5	5	0	5	5	5	0
queen10_10	12	12	12	0	12	12	12	0	12	12	12	0
queen11_11	13	14	13.5	0.57	13	14	13.75	0.5	14	14	14	0
queen8_12	12	13	12.25	0.5	12	13	12.5	0.57	12	12	**12**	**0**
queen9_9	11	11	11	0	**10**	11	10.75	0.5	11	11	11	0
school1	14	14	14	0	14	14	14	0	14	14	14	0
school1_nsh	14	14	14	0	14	14	14	0	14	14	14	0

Note: We use the following notation:

- I+FS for combination of break agent and FS-agent
- I+FD for combination of break agent and FD-agent
- I+X for combination of break agent and X_ρ-agent

TABLE 34.4 Used Benchmarks Presentation

	I+FS+FD+X			
Name	min	max	avg	std dev.
DSJC125.1	6	6	6	0
DSJC125.5	18	19	18.6	0.54
DSJC125.9	44	44	44	0
DSJC250.9	74	75	74.4	0.54
DSJC500.9	132	133	132.4	0.54
DSJR500.1c	85	85	85	0
le450_5c	5	5	5	0
le450_5d	5	5	5	0
queen10_10	12	12	12	0
queen11_11	13	14	13.25	0.5
queen8_12	12	13	12.25	0.5
queen9_9	11	11	11	0
school1	14	14	14	0
school1_nsh	14	14	14	0

memory summarizes information and decides the direction of the search: intensification or a type of diversification. Below the memory, specific agents allow to manage an asynchronous fleet of agents of the same type. At the lowest level, the workers called agents apply strategies dedicated to the problem.

We apply such a strategy to the graph coloring problem. For that, we develop dedicated operators: *Break* operator and X_ρ. We propose several ways to diversify the search forcing the presence of characters in colorings used by searching agents.

Combinations of components was tested. We saw that that diversification leads to obtain better results but no diversification worth better than all others.

The combination of all diversification techniques improves results on a graph but pays back its performance on two other graphs. This may be due to the too heavy diversification made in this case.

We are now working on new search agents as agents working with k-colorings and a way to tune to each diversification technique.

References

[1] Jan Paredis. Coevolutionary computation. *Artificial Life,* 2: 355–375, 1995.

[2] V. Bachelet. *Métaheuristique parallèle hybride: application au problème d'affectation quadratique.* Ph.D. thesis, Université des sciences et technologies de Lille, cité scientifique Villeneuve d'Ascq 59655, December 1999.

[3] B. Weinberg, V. Bachelet, and E.-G. Talbi. A co-evolutionnist meta-heuristic for the assignment of the frequencies in cellular networks. In *First European workshop on Evolutionary Computation in Combinatorial Optimization (EvoCOP),* Como, Italy, 2001. Springer-Verlag, pp. 140–149.

[4] S. Kirkpatrick, D.C. Gelatt, and M.P. Vecchi. Optimization by simulated annealing. *Science,* 220: 671–680, 1983.

[5] F. Glover and M. Laguna. Tabu search. In C. Reeves, Ed., *Modern Heuristic Techniques for Combinatorial Problems,* Blackwell Scientific Publishing, Oxford, England, 1993.

[6] T. Feo and M. Resende. Greedy randomized adaptive search procedures. *Journal of Global Optimization,* 6: 109–133, 1995.

[7] G.J. Chaitin, M. Auslander, A.K. Chandra, J. Cocke, M.E. Hopkins, and P. Markstein. *Computer Languages,* chapter Register allocation via graph coloring, IBM. T.J. Watson Research Center, 1981, pp. 47–57.

[8] A. Gamst. Some lower bounds for a class of frequency assignment problems. *IEEE Transactions of Vehicular Technology,* 1(35): 8–14, 1986.

[9] S. Miner, S. Elmohamed, and H. Yau. Optimizing timetabling solutions using graph coloring. In *NPAC REU Program, NPAC,* Syracuse University, NY, 1995.

[10] H. Ogawa. Labeled point pattern matching by delaunay triangulation and maximal cliques. In *Pattern Recognition,* number 1 in 19, 1986, pp. 35–40.

[11] N. Barnier and P. Brisset. Coloriage de graphe en programmation par contraintes. In *ROADEF 2003,* ROADéF, February 2003, pp. 348–349.

[12] R.M. Karp. *Complexity of Computer Computations,* chapter Reducibility among combinatorial problems, Plenm Press, New York, 1972, pp. 85–103.

[13] B. Weinberg and E.-G. Talbi. On symmetry of partitionning problems. In J. Gottlieb and G. Raidl Eds., *EvoCOP,* LNCS, 2004. To appear.

[14] D. Brelaz. New methods to color the vertices of a graph. *Communications of the ACM,* 22: 251–256, 1979.

[15] F.T. Leighton. A graph coulouring algorithm for large scheduling problems. *J. Res Natl Bur. Standards,* 84: 489–506, 1979.

[16] D.S. Johnson, C.R. Aragon, L.A. McGeoch, and C. Schevon. Optimization by simulated annealing: An experimental evaluation; part II, graph coloring and number partitioning. *Operations Research,* 39: 378–406, 1991.

[17] J. Culberson. Iterated Greedy Graph Coloring and the Difficulty Landscape. Technical report, University of Alberta, June 1992.

[18] S. Hurley, D. Smith, and C. Valenzuela. A permutation based genetic algorithm for minimum span frequency assignment. In T. Baeck, A. Eiben, M. Schoenauer, and H. Schwefel, Eds, *PPSN V: Proceedings of the Fifth International Conference on Parallel Problem Solving from Nature,* Vol. 1498 of *Lecture Notes in Computer Science,* Amsterdam, The Netherlands, September 1998. Springer-Verlag Publication, pp. 907–916.

[19] D. Costa and A. Hertz. Ants can colour graphs. *Journal of the Operational Research Society,* 48: 295–305, 1997.

[20] C.A. Glass and A. Prügel-Bennett. A Polynomially Searchable Exponential Neighbourhood for Graph Colouring. Technical report, Departement of Electronics and Computer Science, University of Southampton, 1998.

[21] A. Hertz and D. de Werra. Using tabu search techniques for graph coloring. *Computing*, 39: 345–351, 1987.

[22] R. Dorne. *Étude des méthodes heuristiques pour la coloration, la T-coloration et l'affectation de fréquence.* Ph.D. thesis, Université de Montpellier II Science et Technique, May 1998.

[23] P. Galinier and J-K. Hao. Hybrid evolutionary algorithms for graph coloring. *Journal of Combinatorial Optimization*, 3: 379–397, 1999.

[24] L. Paquete and T. Stützle. An experimental investigation of iterated local search for coloring graphs. In S. Cagnoni, J. Gottlieb, E. Hart, M. Middendorf, and G. Raidl, Eds, *Applications of Evolutionary Computing, Proceedings of Evo Workshops2002: EvoCOP, EvoIASP, EvoSTim*, Vol. 2279, Kinsale, Ireland, 3–4 Springer-Verlag, 2002, pp. 121–130.

[25] C. Avanthay, A. Hertz, and N. Zufferey. A variable neighborhood search for graph coloring. *European Journal of Operational Research*, 151: 379–388, 2003.

[26] A. Vesel and J. Zerovnik. How good can ants color graphs? *Journal of computing and Information Technology*, 8: 131–136, 2000.

[27] A. Petford and D. Welsh. A randomised 3-colouring algorithm. *Discrete Mathematics*, 74: 253–261, 1989.

[28] J-P. Hamiez and J-K. Hao. Scatter search for graph coloring. In *Artificial Evolution*, Le Creusot, France, October 2001, pp. 267–278.

[29] D. Fotakis, S. Likothanassis, and S. Stefanakos. An evolutionary annealing approach to graph coloring. In *Proceedings of Applications of Evolutionary Computing*, Vol. 2037 of *Lecture Notes in Computer Science*, Evo Workshops 2001, Springer-Verlag, April 2001, pp. 120–129.

[30] C. Walshaw and M.G. Everett. Multilevel Landscapes in Combinatorial Optimisation. Technical Report 02/IM/93, Comp. Math. Sci., Univ. Greenwich, London SE10 9LS, UK, April 2002.

[31] M. Clerc. Optimisation par essaim particulaire et coloriage de graphe. Technical report, France Télécom, 2001.

[32] A. Geist, A. Beguelin, J. Dongarra, W. Jiang, R. Manchek, and V. Sunderam. *PVM: Parallel Virtual Machine, A Users' Guide and a Tutorial Networked Parallel Computing.* The MIT Press, Cambridge, MA, 1994.

35

Frameworks for the Design of Reusable Parallel and Distributed Metaheuristics

N. Melab
El-Ghazali Talbi
S. Cahon

35.1 Introduction

Real-world optimization problems are often NP-hard, complex, and CPU time-consuming. Moreover, their modeling evolves continuously in terms of constraints and objectives. Therefore, their resolution requires the use of parallel/distributed hybrid metaheuristics. Unlike exact methods, metaheuristics allow to find sub-optimal solutions in a reasonable execution time. They allow to meet the resolution delays often imposed in the industrial field.

Metaheuristics fall in two categories: single solution-oriented or local search (LS) methods, and population-based or evolutionary algorithms (EAs). An LS starts with a single initial solution. At each step of the search the current solution is replaced by another (often the best) solution found in its neighborhood.

This work is a part of the current national joint grid computing project ACI-GRID DOC-G (Défis en Optimisation Combinatoire sur Grilles). It includes research teams from different laboratories: OPAC from LIFL, OPALE from PRISM and O2 and P3-ID from IMAG. The project is supported by the French government.

Very often, LS methods allow to find a local optimal solution, and so are called exploitation-oriented methods. On the other hand, evolutionary algorithms work on a randomly generated population of solutions. The initial population is enhanced through a natural evolution process. At each generation of the process, the whole or a part of the population is replaced by the newly generated individuals (often the best ones). EAs are often called exploration-oriented methods.

Although their time complexity is polynomial, metaheuristics remain insufficient for large-size problems. Therefore, parallel/distributed and concurrency tools are necessary to tackle these problems. Different parallel/distributed models have been proposed for each class of methods. They are detailed in Sections 35.2 and 35.3. In order to take benefit from the exploitation power of the LS methods and the exploration merit of EAs, their hybridization is recommended [1]. Hybrid metaheuristics allow to deliver high quality and robust solutions.

Several parallel and distributed metaheuristics and their implementations have been proposed in the literature. Most of them are available on the Internet, and can be reused and adapted to its own problems. Reusability may be defined as the ability of software components to build many different applications [2]. However, one has to rewrite the problem-specific sections of the code. Such task is tedious, error prone, energy and time-consuming. Moreover, the new developed code is harder to maintain. A better way to reuse the code of existing parallel and distributed metaheuristics is the use of libraries [3]. Their objective is twofold: they are reliable as they are often well tested and documented. In addition, they allow a better maintainability and efficiency. However, libraries do not allow the reuse of design. A better approach to reuse the design and the code at the same time is the framework-based reuse [4].

In the literature, very often the authors do not make a clear of difference between a library and a framework. In a framework, the provided code calls the user defined one according to the Hollywood property: "do not call us, we call you." Therefore, frameworks provide the full control structure of the invariant part of the algorithms, and the user has to only supply the problem-specific details. This chapter focuses on the frameworks, and aims at removing the ambiguity on their use, highlighting their characteristics, requirements, and objectives.

Most of existing frameworks related to metaheuristics for discrete optimization problems are Object-Oriented (OO) [5–14]. They include a set of classes that embody an abstract design of solution methods of a family of related problems [2]. They are based on a strong conceptual separation of the invariant (generic) part of PDM and their problem-specific part. Such characteristic allows the PDM programmer to redo very little code.

The frameworks focus only on either EA [5–9] or LS [10,11]. Only few frameworks are dedicated on the design of both EA and LS, and their hybridization [12–14]. All these frameworks are described, summarized, and compared in this chapter. The comparison is mainly based on the class of provided solution methods, the parallel/distributed models they implement, the hybridization mechanisms they allow, and some implementation choices, mainly the programming language and the communication and concurrency API. The presented overview will help the user to choose the framework corresponding to his/her needs. At our best known, such overview has never been proposed in the literature.

The rest of the chapter is organized as follows: Sections 35.2 and 35.3 present the working principles and their major parallel/distributed models of, respectively, LS methods and EA. In Section 35.4, we present the main hybrid mechanisms of metaheuristics. In Section 35.5, we propose an overview of the major frameworks dedicated to the LS methods, to the EA, and both of them. The main characteristics of each of these frameworks are summarized. Section 35.6 ends the chapter with some concluding remarks.

35.2 Parallel and Distributed Local Searches (PDLS)

In this section we first present the main existing LS methods and their working principles. Afterward, we describe the main existing parallel/distributed models of their design and implementation.

35.2.1 Principles of LS

Local search are metaheuristics dedicated to the improvement of a single solution. They are generally based on the concept of the neighborhood. They start from a solution randomly generated or provided by another metaheuristic. This solution is then updated, systematically, by replacing the current solution by another found in the neighborhood. The specific features of LS are mainly: the heuristic internal memory, the strategy used to choose the initial solution, the generator of candidate solutions, and the selection policy of the candidate moves. Three major LS stand out: Hill Climbing (HC) [15], Simulated Annealing (SA) [16], and Tabu Search (TS) [17].

A serial LS is composed of generic and specific features. Generic features include the initializing of a given movement, the exploration strategy of the neighborhood, and the computation of the fitness value of a given solution corresponding to a given movement. Specific features, such as the Tabu list involved in the TS method, allow to differentiate LS.

35.2.2 Parallel and Distributed Models of LS

There exist mainly three parallel distributed models of LS: the parallel/distributed exploration of the neighboring model, the multi-start model, and the parallel/distributed evaluation of each solution. The first two models are illustrated in Figure 35.1.

- *Parallel/distributed exploration of the neighboring*: This model is a kind of Farmer–Worker model. At the beginning of each iteration, the farmer copies the current solution on the workers. Each worker explores some neighboring candidates, and returns the results to the farmer. The model could be efficient if the evaluation of each solution is time consuming and there is a large number of candidate neighbors to evaluate. Such a model is not well suited to SA since only one candidate solution is evaluated at each iteration. On the other hand, the efficiency of the model for HC is not always guaranteed, as the number of neighboring solutions to process before finding one that improves the current objective function may be highly variable.
- *Multi-start model*: The model consists of launching in parallel several independent homo/heterogeneous LS. It allows to enhance the robustness of the execution. On the other hand, the model is well suited for low-speed networks of workstations. Moreover, its exploitation is natural and easy for the user.
- *Parallel/distributed evaluation of a single solution*: The fitness of each solution is evaluated in a parallel centralized way. That model could be efficient when the evaluation function is CPU time

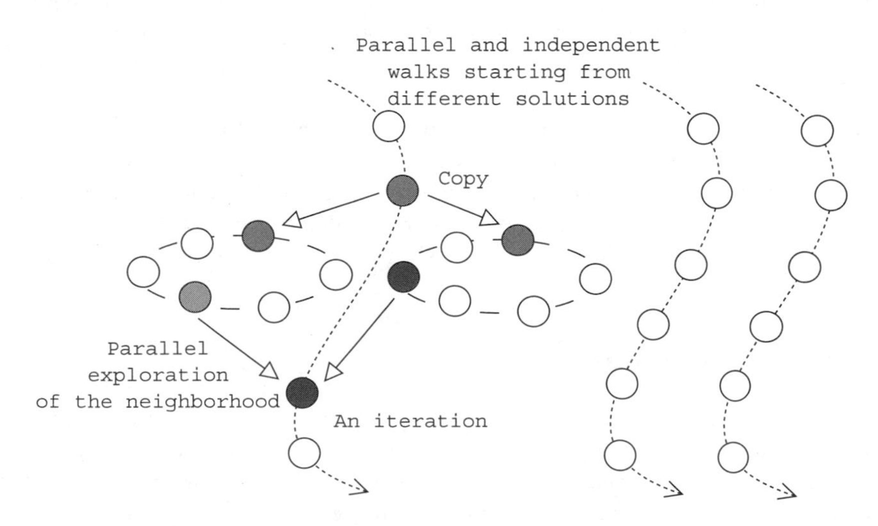

FIGURE 35.1 The first two parallel models for LS.

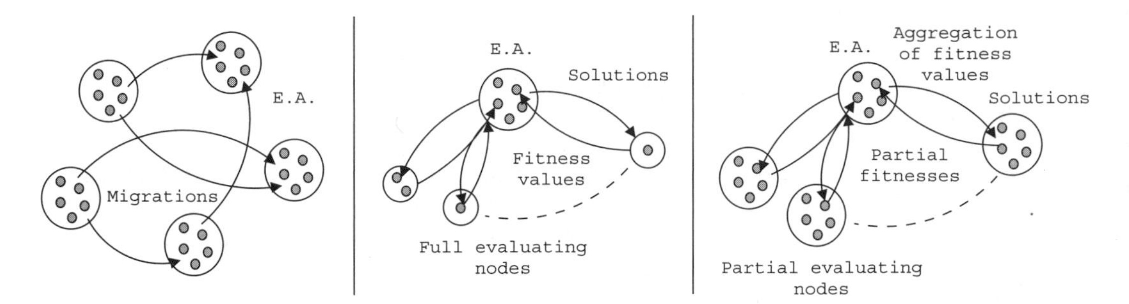

FIGURE 35.2 Three major parallel distributed models for EA.

consuming and IO intensive. The function could be viewed as an aggregation of a set of partial functions. A reduction operation is performed on the results returned by the partial functions. Consequently, for this model the user has to indicate a set of partial functions and an aggregation operator of these.

35.3 Parallel and Distributed Evolutionary Algorithms (PDEAs)

The EAs [18] are population-based metaheuristics. They improve an initial randomly generated population of solutions during a certain number of generations. At each generation, individuals are selected, paired, and recombined in order to generate new solutions that replace others. The major characteristics of EA are the way the population is initialized, the selection strategy (deterministic/stochastic) by fostering "good" solutions, the replacement strategy that withdrawn individuals, and the continuation/stopping criterion, evaluated at the end of each generation.

Three major parallel and distributed models for EA stand out in the literature (Figure 35.2): the island (a)synchronous cooperative model, the parallel/distributed evaluation of the population model, and the parallel/distributed evaluation of a single solution model. The last model is the one presented in Section 35.2.2.

In the Island model, several homo/heterogeneous EA run simultaneously and cooperate to compute better and robust solutions. They exchange genetic stuff to improve the diversity of the search. The model aims at delaying the global convergence, especially when the EA are heterogeneous regarding to the variation operators. The migration of individuals follow a policy defined by the following parameters: the migration decision criterion, the exchange topology, the number of emigrants, the emigrants selection policy, and the replacement/integration policy.

The evaluation of the population is necessary because it is, in general, the most time-consuming step of an EA. The parallel evaluation is farmer/worker oriented. The farmer applies the selection, transformation, and replacement operations as they require a global management of the population. At each generation, it distributes the new solutions among the workers, which evaluate them and return their fitness values. An efficient execution is often obtained particularly when the evaluation of each solution is costly.

35.4 Hybrid Metaheuristics

Hybridization allows to combine different metaheuristics and make them cooperate in order to solve the same optimization problem. Nowadays, hybrid metaheuristics have gained great interest [1]. They allow to solve many practical and academic problems, and to provide the best-known solutions. In Reference 1, two levels (low and high) and two modes (relay and cooperative) of hybridization have been distinguished (Figure 35.3).

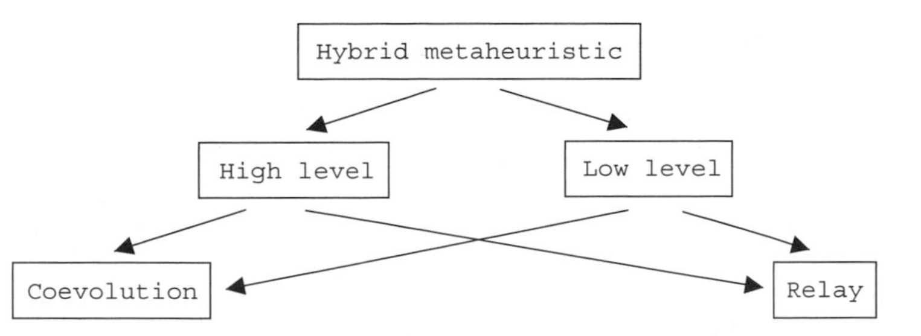

FIGURE 35.3 Hierarchical taxonomy of hybrid metaheuristics.

The low-level hybridization consists in changing the internal component of the metaheuristic on which it is performed. A given function of a given metaheuristic is replaced by another metaheuristic. For instance, the mutation operator of a given GA could be replaced by an LS method. In order to make easier this kind of hybridization the semantics of the internal function must be the same as the metaheuristic. The low-level hybridization requires to examine the internal working of the metaheuristic [14]. Conversely, in high-level hybrid algorithms the combined metaheuristics are self-containing, meaning no direct relationship to their internal working is considered.

On the other hand, in the relay hybridization mode the metaheuristics are applied in a pipeline way. The output of a given metaheuristic (except the last) is the input of its successor (except the first) in the pipeline. At the contrary, coevolutionist hybridization is a cooperative optimization model. Each metaheuristic performs a search in a solution space, and exchange good solutions with others.

35.5 Frameworks for the Design of PDM

In this section we first explain why it is important to use frameworks for the reusable design of PDM, their design requirements and objectives, and an overview and a comparison of existing frameworks.

35.5.1 Why Using Frameworks?

There are mainly three ways to reuse existing PDM: *from scratch or no reuse*, *only code reuse*, and *both design and code reuse*. The *from scratch-oriented* approach is attractive due to the apparent simplicity of the metaheuristics code. Indeed, some programmers are tempted to develop themselves their code rather than reusing a code developed by a third party. However, several problems arise: the coding effort is time and energy consuming, error prone, and difficult to maintain, etc. *Only code reuse* is the reuse of the third-party code available on the Internet either as free single programs or as libraries. Old third-party code has usually application-dependent sections that must be extracted before the new application-dependent code can be inserted. Changing these sections is often time consuming and error prone. The code reuse through libraries [3] is obviously better because these are often well tried, tested, and documented, thus, more reliable. However, libraries allow to reuse the code but they do not make easier the reuse of the complete invariant part of algorithms, particularly their whole design.

The *both code and design reuse* approach is devoted to overcome this problem, that is, to redo as little code as possible each time a new optimization problem is dealt with. The basic idea is to capture into special components the recurring (or invariant) part of solution methods to standard problems belonging to a specific domain. Frameworks provide the full control structure of the invariant part of the algorithms, and the user only has to supply the problem-specific details. In order to meet this property the design of a framework must be based on a clear conceptual separation between the solution methods and the problems they tackle.

35.5.2 Design Requirements and Objectives

A framework is normally intended to be exploited by as many users as possible. Therefore, its exploitation could be successful only if some important user criteria are satisfied. The following criteria are the major of them:

- *Maximum design and code reuse*: The framework must provide for the user a whole architecture (design) of his/her solution method. Moreover, the programmer may redo as little code as possible. This objective requires a clear and maximum conceptual separation between the solution methods and the problems to be solved, and thus a deep problem domain analysis. The user might, therefore, develop only the minimal problem-specific code.
- *Utility and extendibility*: The framework must allow the user to cover a broad range of metaheuristics, problems, parallel distributed models, hybridization mechanisms, etc. It must be possible for the user to easily add new features/metaheuristics or change existing ones without implicating other components. Furthermore, as in practice existing problems evolve and new others arise these have to be tackled by specializing/adapting to them the framework components.
- *Transparent use of parallel/distributed models and hybridization mechanisms*: In order to facilitate its use it is implemented so that the user can deploy his/her parallel algorithms in a transparent manner. Moreover, the execution of the algorithms must be robust to guarantee the reliability and the quality of the results. The hybridization mechanism allows to obtain robust and better solutions.
- *Portability*: In order to satisfy a large number of users the framework must support different material architectures and their associated operating systems.

35.5.3 An Overview of Existing Frameworks

It is quoted that a framework is mainly based on the conceptual separation between the solution methods and the problems they deal with. This separation requires a deep understanding of the application domain. In Reference 19, OO composition rather than inheritance is recommended to perform the conceptual separation. Indeed, classes are easier to reuse than individual methods. The modeling of the application domain results in a set of reusable classes with some constant and variable aspects. The implementation of the constant part is provided by the framework, and encapsulated into generic/abstract classes or skeletons [12]. The variable part is problem specific, and is specified and fixed in the framework, but it is implemented by the user. This part is a set of holes or hot spots [20] that serve to fill the skeletons provided by the framework when building specific applications.

According to the extendibility requirement, two types of frameworks can be distinguished: white- or glass-box frameworks and black-box frameworks. In black-box frameworks one can reuse components by plugging them together through static parameterization and composition, unaware of their internal working [4]. In contrast, white-box frameworks require an understanding of the internal working of the classes so that correct subclasses (inheritance-based) can be developed. Therefore, they allow more extendibility of the provided classes.

Several white-box frameworks for the reusable design of PDM are available on the Internet. This table summarizes the most important frameworks. The different frameworks are classified according to some criteria: the classes of methods they are dedicated to (LS or EA), the kind of hybridization mechanisms they provide, the parallel/distributed models they implement, the language they are developed with, and the API used for communication and concurrency. These criteria allow to evaluate the different frameworks regarding the quality and design requirements and objectives of frameworks quoted in Section 35.5.2.

	LS	EA	Hybrid	Parallel/Distributed	Language	Communication/ Concurrency
ECJ	—	+	EA/EA	Island	Java	Java threads TCP/IP Sockets
D. Beagle	—	+	EA/EA	M/S	C++	TCP/IP Sockets
Jdeal	—	+	EA/EA	M/S	Java	TCP/IP Sockets
Easylocal ++	+	—	LS/LS	—	C++	—
Localizer ++	+	—	LS/LS	—	C++	—
MAFRA	—	+	EA/EA	—	Java	—
DREAM	—	+	EA/EA	Island	Java	Java threads TCP/IP Sockets
MALLBA	+	+	EA/LS EA/EA LS/LS	all	C++	Netstream MPI
ParadisEO	+	+	EA/LS LS/LS EA/EA	all	C++	MPI PVM PThreads

According to the first two criteria, the frameworks fall in three categories that distinguish those, respectively, dedicated to only EA, only LS, and both of them. Some of the frameworks limited to only EA are the following: DREAM[1] [5], ECJ[2] [6], JDEAL[3] [7], and Distributed BEAGLE[4] [9]. These software are frameworks as they are based on a clear OO conceptual separation. They are portable as they are developed in the Java language except the last system, which is programmed in C++. However, they are limited regarding the parallel distributed models. Indeed, in DREAM and ECJ only the island model is implemented using Java threads and TCP/IP sockets. DREAM is particularly deployable on peer-to-peer platforms. Furthermore, JDEAL and Distributed BEAGLE provide only the Master–Slave (M/S) model using TCP/IP sockets. The latter also implements the synchronous migration-based island model, but deployable on only one processor.

In the LS domain, most of existing frameworks [10,11] do not allow parallel distributed implementations. Those enabling parallelism/distribution are often dedicated to only one solution method. For instance, Reference 21 provides parallel skeletons for the TS method. Two skeletons are provided and implemented in C++/MPI: independent runs (multi-start) model with search strategies, and a Master–Slave model with neighborhood partition. The two models can be exploited by the user in a transparent way.

In practice, only few frameworks available on the Internet are devoted to both PDEA and PDLS, and their hybridization. MALLBA[5] [12], MAFRA[6] [13], and ParadisEO[7] [14] are good examples of such frameworks. MAFRA is developed in Java using design patterns [22]. It is strongly hybridization oriented; however, it is very limited regarding parallelism and distribution. MALLBA and ParadisEO have numerous common characteristics. They are C++/MPI open source frameworks. They provide all the

[1] Distributed Resource Evolutionary Algorithm Machine: http://www.world-wide-dream.org
[2] Java Evolutionary Computation: http://www.cs.umd.edu/projects/plus/ec/ecj/
[3] Java Distributed Evolutionary Algorithms Library: http://laseeb.isr.ist.utl.pt/sw/jdeal/
[4] Distributed Beagle Engine Advanced Genetic Learning Environment: http://www.gel.ulaval.ca/~beagle
[5] MAlaga+La Laguna+BArcelona: http://neo.lcc.uma.es/mallba/mallba.html
[6] Java Mimetic Algorithms Framework: http://www.cs.nott.ac.uk/~nxk/MAFRA/MAFRA.html
[7] Parallel and distributed Evolving Objects: http://www.lifl.fr/~cahon/paradisEO

previously presented parallel/distributed models, and the different hybridization mechanisms. However, they are quite different as ParadisEO seems to be more flexible because the granularity of its classes is finer. Moreover, ParadisEO (an extended EO [23]) provides also the PVM-based communication layer and PThreads-based concurrency. On the other hand, MALLBA is deployable on wide area networks [12]. Communications are based on *NetStream*, an ad hoc flexible and OOP message passing service upon MPI. Furthermore, MALLBA allows the cooperation between metaheuristics and exact methods.

35.6 Concluding Remarks

In this chapter, we presented the different commonly used parallel/distributed models and hybridization mechanisms of metaheuristics. Most of these models and mechanisms have been implemented and are available on the Internet. We believe that the best way to reuse them is to encapsulate them into frameworks. Frameworks provide the invariant part of the metaheuristics; therefore the user has to just supply the specific part of his/her problem. We have also proposed an overview of existing white-box frameworks, including those dedicated either only to LS, only to EA, or to both.

The different frameworks have been compared according to some objectives: the code and design reuse, their utility and extendibility, the transparent use of parallelism/distribution and hybridization, and the portability. These objectives are evaluated according to the following criteria: the families of methods the frameworks are dedicated to, the parallel/distributed models and hybridization mechanisms they provide and to be exploited transparently by the user, the language and communication and concurrency API they are developed with.

All the presented frameworks are object oriented, and are based on a clear conceptual separation between solution methods and problems. Therefore, they allow a maximum code and design reuse. Moreover, they allow a portable deployment as they are developed in either C++ or Java. In the literature, only few frameworks provide the major parallel/distributed models. ECJ and DREAM implement the island model for cooperative EA, and Distributed Beable and Jdeal provide the Farmer/Worker model. Finally, ParadisEO and MALLBA are particularly outstanding. Indeed, they allow to achieve all the objectives quoted above. They include LSs as well as EAs. They provide all the major parallel/distributed models and the hybridization mechanisms.

References

[1] E.-G. Talbi. A taxonomy of hybrid metaheuristics. *Journal of Heuristics*, 8: 541–564, 2002.

[2] A. Fink, S. Vo, and D. Woodruff. Building reusable software components for heuristc search. In P. Kall and H.-J. Luthi (Eds.), *Operations Research Proc., 1998*, Springer-verlag, Berlin, 1999, pp. 210–219.

[3] M. Wall. GAlib: A C++ library of genetic algorithm components. http://lancet.mit.edu/ga/.

[4] R. Johnson and B. Foote. Designing reusable classes. *Journal of Object-Oriented Programming*, 1: 22–35, 1988.

[5] M.G. Arenas, P. Collet, A.E. Eiben, M. Jelasity, J.J. Merelo, B. Paechter, M. Preuß and M. Schoenauer. A framework for distributed evolutionary algorithms. In *Proceedings of PPSN VII*, September 2002.

[6] S. Luke, L. Panait, J. Bassett, R. Hubley, C. Balan, and A. Chircop. ECJ: a Java-based evolutionary computation and genetic programming research system. http://www.cs.umd.edu/projects/plus/ec/ecj/.

[7] J. Costa, N. Lopes, and P. Silva. JDEAL: The Java Distributed Evolutionary Algorithms Library. http://laseeb.isr.ist.utl.pt/sw/jdeal/home.html.

[8] E. Goodman. An Introduction to GALOPPS — The "Genetic Algorithm Optimized for Portability and Parallelism" System. Technical report, Intelligent Systems Laboratory and Case Center for Computer-Aided Engineering and Manufacturing, Michigan State University, November 1994.

[9] C. Gagné, M. Parizeau, and M. Dubreuil. Distributed BEAGLE: An environment for parallel and distributed evolutionary computations. In *Proceedings of the 17th Annual International Symposium on High Performance Computing Systems and Applications (HPCS) 2003*, May 11–14, 2003.

[10] L. Di Gaspero and A. Schaerf. Easylocal++: An object-oriented framework for the design of local search algorithms and metaheuristics. In *MIC '2001 4th Metaheuristics International Conference*, Porto, Portugal, July 2001, pp. 287–292.

[11] L. Michel and P. Van Hentenryck. Localizer++: An Open Library for Local Search. Technical report CS-01-02, Brown University, Computer Science, 2001.

[12] E. Alba and the MALLBA Group. MALLBA: A library of skeletons for combinatorial optimization. In R.F.B. Monien, Ed., *Proceedings of the Euro-Par*, Vol. 2400 of *Lecture Notes in Computer Science* Paderborn, Springer-Verlag, Heidelberg, 2002, pp. 927–932.

[13] N. Krasnogor and J. Smith. MAFRA: A java memetic algorithms framework. In Alex A. Freitas, William Hart, Natalio Krasnogor, and Jim Smith, Eds, *Data Mining with Evolutionary Algorithms*, Las Vegas, Nevada, USA, August 2000, pp. 125–131.

[14] S. Cahon, N. Melab, and E.-G. Talbi. ParadisEO: A framework for the reusable design of parallel and distributed metaheuristics. *Journal of Heuristics*, 10(3): 357–380, 2004.

[15] C.H. Papadimitriou. The Complexity of Combinatorial Optimization Problems. Master's thesis, Princeton University, 1976.

[16] S. Kirkpatrick, C.D. Gelatt, and M.P. Vecchi. Optimization by simulated annealing. *Science*, 220: 671–680, 1983.

[17] F. Glover. Tabu search, part I. *ORSA. Journal of Computing*, 1: 190–206, 1989.

[18] J.H. Holland. *Adaptation in Natural and Artificial Systems*, The University of Michigan Press, Ann Arbor, MI, 1975.

[19] D. Roberts and R. Johnson. Evolving frameworks. A pattern language for developing object-oriented frameworks. In *Proceedings of the Third Conference on Pattern Languages and Programming (PLoP '96)*, Allerton Park, Illinois, September 4–6, 1996.

[20] W. Pree, G. Pomberger, A. Schappert, and P. Sommerlad. Active guidance of framework development. *Software — Concepts and Tools*, 16: 94–103, 1995.

[21] M.J. Blesa, Ll. Hernandez, and F. Xhafa. Parallel skeletons for tabu search method. In *8th International Conference on Parallel and Distributed Systems (ICPADS'01)*, IEEE Computer Society Press, Kyongju City, Korea, 2001, pp. 23–28.

[22] E. Gamma, R. Helm, R. Johnson, and J. Vlissides. *Design Patterns, Elements of Reusable Object-Oriented Software*, Addison-Wesley, Reading, MA, 1994.

[23] M. Keijzer, J.J. Morelo, G. Romero, and M. Schoenauer. Evolving objects: A general purpose evolutionary computation library. In *Proceedings of the 5th International Conference on Artificial Evolution (EA '01)*, Le Creusot, France, October 2001.

36

Parallel Hybrid Multiobjective Metaheuristics on P2P Systems

N. Melab
El-Ghazali Talbi
M. Mezmaz
B. Wei

36.1 Introduction

Metaheuristics allow to provide near-optimal solutions of NP-hard complex problems in a reasonable time. They fall into two complementary categories: evolutionary algorithms (EAs) that have a good exploration power, and local searches (LSs) characterized by better intensification capabilities. The hybridization of the two categories permits to improve the effectiveness (quality of provided solutions) and the robustness of the metaheuristics [11]. Nevertheless, as it is CPU time consuming it is not often fully exploited in practice. Indeed, experiments with hybrid metaheuristics are often stopped before the convergence is reached. Nowadays, Peer-to-Peer (P2P) computing [8] and grid computing [5] are two powerful ways to achieve high performance on long-running scientific applications. Parallel hybrid metaheuristics used for solving real-world multiobjective problems (MOPs) are good challenges for P2P and grid computing. However, to the best of our knowledge no research work has been published on that topic.

In this chapter, we contribute with the first results on parallel hybrid multiobjective metaheuristics on P2P systems. The design and deployment of these optimization methods require a middleware that allows cooperation between parallel tasks. In addition, the traditional parallel models and hybridization mechanisms have to be re-thinked and adapted to be scaled up. Moreover, these require to be fault-tolerant to allow long-running problem resolutions. We particularly focus here on the island model and the multistart model.

Recently, few middlewares [1,4,13] allowing to exploit P2P systems have emerged. These middlewares are well suited for embarrassingly parallel applications such as multi parameter simulations. However, they are limited regarding the parallelism as they do not allow direct cross-peer (or cross-task) communication. Our contribution is to propose a Linda-like [7] coordination model and its implementation on top of XtremWeb [4]. This is a Dispatcher/Worker oriented middleware, in which the Dispatcher distributes application tasks submitted by clients to volunteer worker peers at their request. In addition, the considered middleware provides fault-tolerance mechanisms that are costly in a highly volatile P2P environment. Indeed, a work unit is re-started from scratch each time it fails. Another contribution of this chapter is to deal with the fault-tolerance issue at application level. We propose a check-pointing approach for the two parallel models quoted above.

To be validated the proposed approaches have been experimented on the Bi-criterion Permutation Flow-Shop Problem (BPFSP) [12]. The problem consists roughly to find a schedule of a set of jobs on a set of machines that minimizes the makespan and the total tardiness. Jobs must be scheduled in the same order on all machines, and each machine cannot be simultaneously assigned to two jobs. In Reference 2, a hybrid MultiObjective Metaheuristic (MOM) has been proposed to solve this problem. In this chapter, we extend this work with two P2P-based fault-tolerant parallel models: the island and multistart models. Our extended version allows to fully exploit the hybridization and provides clearly better results. This constitutes another contribution of this chapter.

This chapter is organized as follows: Section 36.2 presents briefly parallel hybrid multiobjective optimization (MOO). Section 36.3 highlights the requirements of MOO and describes the proposed coordination model and its implementation on top of XtremWeb. Section 36.4 presents the experimentation of the model and its implementation through a parallel hybrid metaheuristic applied to the BPFSP, and analyzes the preliminary experimental results. Finally, Section 36.5 concludes the chapter.

36.2 Parallel Hybrid MOMs and P2P Computing

36.2.1 Multiobjective Optimization

An MOP consists generally in optimizing a vector of nb_{obj} objective functions $F(x) = (f_1(x), \ldots, f_{nb_{\text{obj}}}(x))$, where x is an d-dimensional decision vector $x = (x_1, \ldots, x_d)$ from some universe called *decision space*. The space the objective vector belongs to is called the *objective space*. F can be defined as a cost function from the decision space to the objective space that evaluates the quality of each solution $(x_1, \ldots, x_d)$ by assigning it an objective vector $(y_1, \ldots, y_{nb_{\text{obj}}})$, called the *fitness*.

Unlike single-objective optimization problems, an MOP may have a set of solutions known as the *Pareto optimal set* rather than a unique optimal solution. The image of this set in the objective space is denoted as *Pareto front*. Graphically, a solution x is Pareto optimal if there is no other solution x' such that the point $F(x')$ is in the dominance cone of $F(x)$. This dominance cone is the box defined by $F(x)$, its projections on the axes and the origin (Figure 36.1).

36.2.2 Parallelism and Hybridization

In Reference 3, different parallel models have been distinguished: those associated with LSs and those dedicated to EAs. Three major parallel models for EAs are presented: the island (a)synchronous cooperative model, the parallel evaluation of the population, and the distributed evaluation of a single solution. The parallel models for LSs are mainly: the parallel exploration of neighboring candidate solutions and

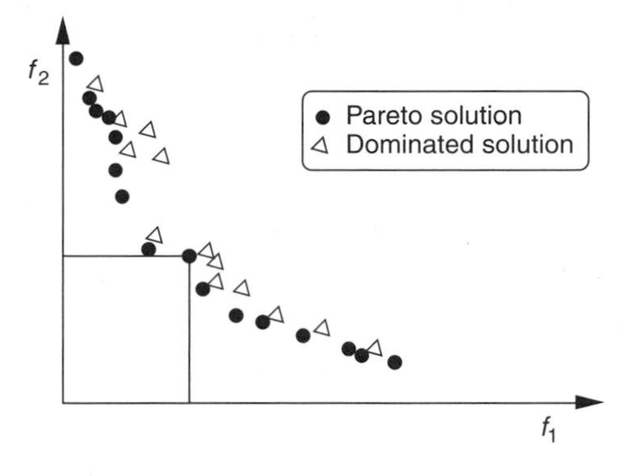

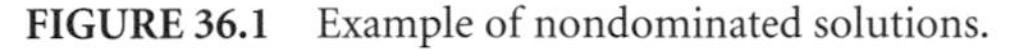
FIGURE 36.1 Example of nondominated solutions.

the multistart model. In this chapter, we focus only on the coarse-grained models: the island model and the multistart model. Due to the communication delays, fine-grained models are often inefficient when they are deployed in a large-scale network.

In the island (a)synchronous cooperative model, different EAs are simultaneously deployed and cooperate for computing better and robust solutions. They exchange, in an asynchronous way, the genetic stuff to diversify the search. The objective is to allow to delay the global convergence, especially when the EAs are heterogeneous regarding the variation operators. The migration of individuals follows a policy defined by few parameters: the migration decision criterion, the exchange topology, the number of emigrants, the emigrants selection policy, and the replacement/integration policy.

The multistart model consists in simultaneously launching several local searches. They may be heterogeneous, but no information is exchanged between them. The results would be identical as if the algorithms were sequentially run. Very often deterministic algorithms differ by the supplied initial solution and/or some other parameters. This trivial model is convenient for low-speed networks of workstations.

Combinations of different metaheuristics often provide very powerful search methods. In Reference 11, two levels and two modes of hybridization are distinguished: *Low* and *High* levels, and *Relay* and *Cooperative* modes. The low-level hybridization consists in replacing an internal function (e.g., an operator) of a given metaheuristic by another metaheuristic. In high-level hybrid algorithms, the different metaheuristics are self-containing, meaning no direct relationship to their internal working is considered. Relay hybridization means a set of metaheuristics is applied in a pipeline way. The output of a metaheuristic (except the last) is the input of the following one (except the first). Conversely, teamwork hybridization is a cooperative optimization model. Each metaheuristic performs a search in a solution space, and exchanges solutions with others. In this chapter, we address the high-level hybridization mechanism in the relay and cooperative modes.

36.2.3 P2P Computing for Parallel MO Optimization

In this chapter, we focus on Dispatcher/Worker-oriented P2P middlewares such as XtremWeb [4] and SETI@Home [1]. In such systems, clients can submit their jobs to the Dispatcher. A set of volatile workers (peers) request the jobs from the Dispatcher according to the cycle stealing model. Then, they execute the jobs and return the results to the Dispatcher to be collected by the clients. In these middlewares, even a central server (the Dispatcher) is required for controlling the peers (workers) they are considered as P2P software environments. Indeed, an important part of these systems is executed on these peers with a high autonomy.

One of the major limitations of P2P computing environments is that they are well suited for embarrassingly parallel (e.g., multiparameter) applications with independent tasks. In this case, no communication is required between the tasks, and thus peers. The deployment of parallel hybrid metaheuristics that need cross-peer/task communication is not straightforward. The programmer has the burden to manage and control the complex coordination between the workers. To deal with such problem existing middlewares must be extended with a software layer that implements a coordination model. Several interesting coordination models have been proposed in the literature [6,9]. In this chapter, we focus only on one of the most popular of them, that is, Linda [7], as our proposed model is an extension of this model.

In the Linda model, the coordination is performed through generative communications. Processes share a virtual memory space called a *tuple-space* (set of tuples). The fundamental data unit, a tuple, is an ordered vector of typed values. Processes communicate by reading, writing, and consuming these tuples. The "eval" operation is particularly useful in a P2P environment as it allows to spawn tasks to be executed on volunteer peers. A small set of four simple operations allows highly complex communication and synchronization schemes:

- *Out(tuple)*: puts *tuple* into *tuple-space.*
- *In(pattern)*: removes a (often the first) tuple matching *pattern* from *tuple-space.*
- *rd(pattern)*: is the same as *in(pattern)*, but does not remove the tuple from *tuple-space.*
- *Eval(expression)*: puts *expression* in *tuple-space* for evaluation. The evaluation result is a tuple left in *tuple-space.*

Nevertheless, Linda has several limitations regarding the design and deployment of parallel hybrid metaheuristics for P2P systems. First, it does not allow rewriting operations on the tuple space. Due to the high communication delays in a P2P system, tuple rewriting is very important as it allows to reduce the number of communications and the synchronization cost. Indeed, in Linda a rewriting operation is performed as an "in" or "rd" operation followed by a local modification and an "out" operation. The operations "in"/"rd" and "out" involve two communications and a heavy synchronization. Therefore, the model needs to be extended with a rewriting operation. Furthermore, the model does not support group operations that are useful for efficiently writing/reading Pareto sets in/from the tuple-space. Finally, nonblocking operations that are very important in a P2P context are not supported in Linda. In the next section, we propose an extension of the Linda model that allows to meet these requirements.

36.3 A Model for P2P Coordination

36.3.1 Model Description

Designing a coordination model for parallel MOO requires the specification of the content of the tuple-space, a set of coordination operations and a pattern matching mechanism. The tuple-space may be composed of a set of Pareto optimal solutions and their corresponding solutions in the objective space. For the parallel island model of the multiobjective metaheuristics, the tuple-space contains a collection of (parts of) Pareto optimal sets deposited by the islands for migration. The mathematical formulation of the tuple-space (*Pareto Space* or *PS*) is the following:

$$PS = \bigcup PO, \quad \text{with } PO = \{(x, F(x)),\ x \text{ is Pareto optimal}\}.$$

In addition to the operations provided in Linda, parallel P2P multiobjective optimization needs other operations. These operations fall into two categories: *group* operations and *nonblocking* operations. Group operations are useful to manage multiple Pareto optimal solutions. *Nonblocking* operations are necessary to take into account the volatile nature of P2P systems. In our model, the coordination primitives are defined as follows:

- *in, rd, out and eval*: These operations are the same as those of Linda defined in Section 36.2.3.
- *ing(pattern)*: Withdraws from *PS* all the solutions matching the specified pattern.
- *rdg(pattern)*: Reads from *PS* a copy of *all* the solutions matching the specified pattern.
- *outg(setOfSolutions)*: Inserts multiple solutions in *PS*.

- *update(pattern, expression)*: Updates *all* the solutions matching the specified pattern by the solutions resulting from the evaluation of *expression*.
- *inIfExist, rdIfExist, ingIfExist, and rdgIfExist*: These operations have the same syntax than respectively *in, rd, ing,* and *rdg* but they are *non blocking* probe operations.

The *update* operation allows to locally update the PS, and so to reduce the communication and synchronization cost. The pattern matching mechanism depends strongly on how the model is implemented, and in particular on how the tuple-space is stored and accessed. For instance, if the tuple-space is stored in a database the mechanism can be the request mechanism used by the database management system. More details on the pattern matching mechanism of our model are given in the next section.

36.3.2 Implementation on Top of XtremWeb

XtremWeb [4] is a Java P2P project developed at Paris-Sud University. It is intended to distribute applications over a set of peers, and is dedicated to multiparameter applications that have to be computed several times with different inputs. XtremWeb manages tasks following the Dispatcher/Worker paradigm (see Figure 36.2). Tasks are scheduled by the Dispatcher to workers only on their specific demand since they may adaptively appear (connect to the Dispatcher) and disappear (disconnect from the Dispatcher). The tasks are submitted by either a client or a worker, and in the latter case, the tasks are dynamically generated for parallel execution. The final or intermediate results returned by the workers are stored in a MySQL database. These results can be requested later by either the clients or the workers. The database stores also different information related to the workers and the deployed application tasks.

XtremWeb is well suited for embarrassingly parallel applications where no cross-peer communication occurs between workers, and these can only communicate with the Dispatcher. Yet, many parallel distributed applications particularly parallel MOMs need cooperation between workers. In order to free the user from the burden of managing himself/herself such cooperation we propose an extension of the middleware with a software layer.

The software layer is an implementation of the proposed model composed of two parts (see Figure. 36.3): a coordination API and its implementation at the worker level and a coordination request broker (CRB). The PS is a part of the MySQL database associated with the Dispatcher. Each tuple or solution of the PS is stored as a record in the database.

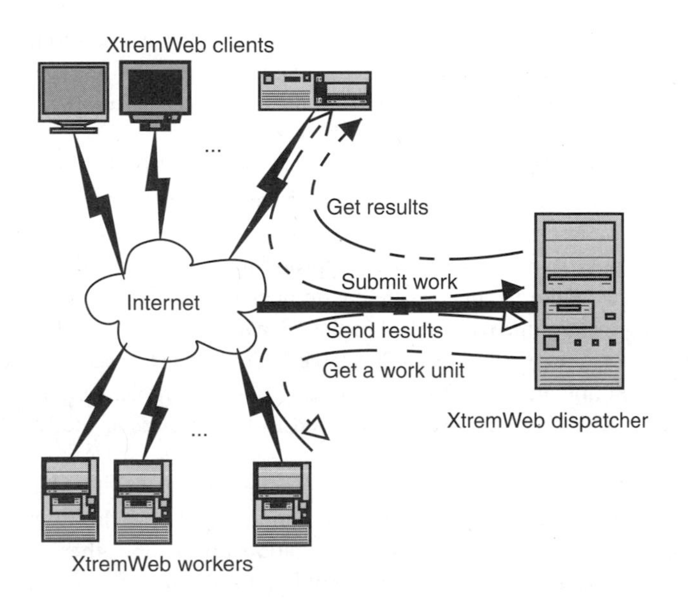

FIGURE 36.2 Global architecture of XtremWeb.

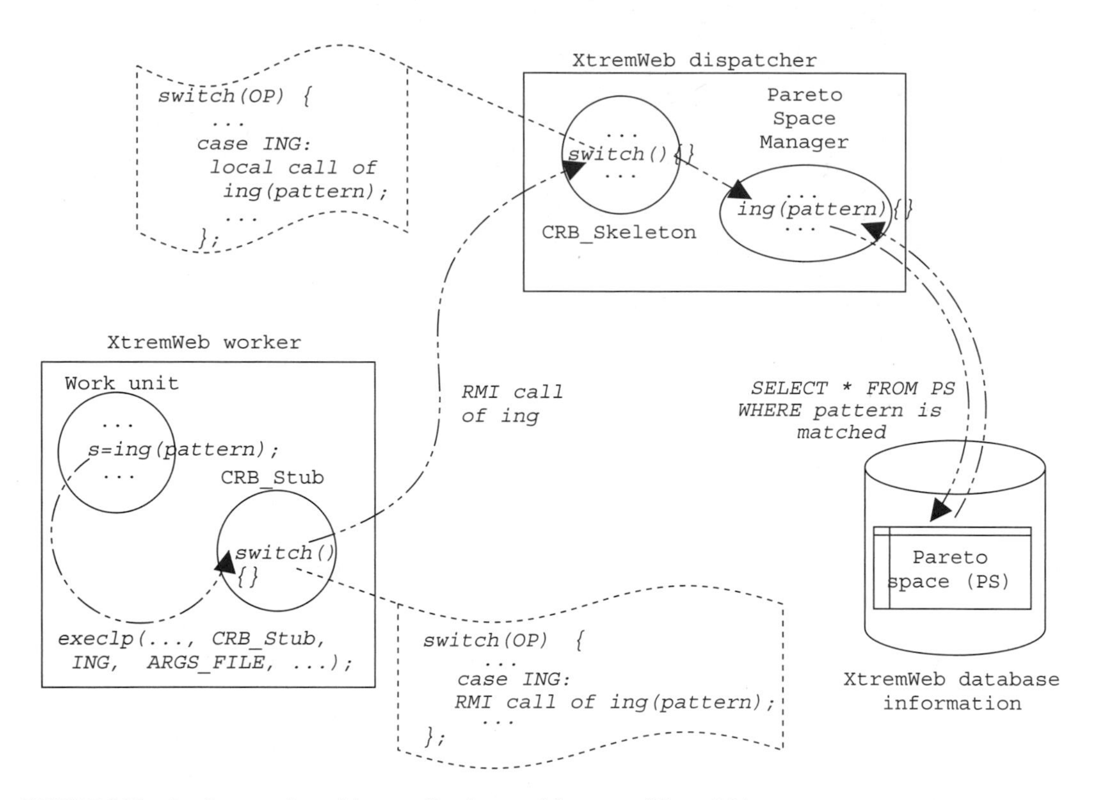

FIGURE 36.3 Implementation of the coordination model on top of XtremWeb.

From the worker side the coordination API is implemented in Java and in C/C++. The C/C++ version allows the deployment and execution of C/C++ applications with XtremWeb (written in Java). The coordination library must be included in these programmer applications. From the Dispatcher side, the coordination API is implemented in Java as a PS manager. The CRB is a software broker allowing the workers to transport their coordination operations calls to the Dispatcher, and has two components: one for the worker (CRB stub) and another for the Dispatcher (CRB skeleton). The role of the CRB stub is to transform the local calls to the coordination operations performed by the tasks executed by the worker into RMI calls. The role of the CRB skeleton is to transform these RMI calls into local calls to the coordination operations performed by the PS Manager. These local calls are translated into MySQL requests addressed to the PS.

To illustrate the implementation of the coordination layer on top of XtremWeb, let us consider the scenario presented in Figure 36.3. The work unit performed by an XtremWeb worker calls the *ing*(*template*) coordination operation. In the C++ version of the coordination API, the implementation of each coordination operation makes the system call *execlp*() with appropriate parameters to plug in the *CRB_Stub* Java object. In our scenario, the major parameters are the number *ING* designating the operation and the file *ARGS_FILE* containing the arguments specified in the *template* parameter. *CRB_Stub* translates the *ing* local call into an RMI call to the *CRB_Skeleton* Java object. This latter translates the RMI call into a local call to the *ing* operation implemented in the PS Manager class. The implementation of the coordination operation consists in a MySQL select request addressed to the PS part of the XtremWeb information database.

Note that the method declarations for the coordination operations in the PS Manager class contain the Java *synchronized* keyword. Hence, the system associates a unique lock with the instance of the PS Manager class. Whenever control enters a synchronized coordination operation, other calls to a synchronized cooperation method are blocked until the PS Manager object is unlocked. In the next section, the proposed coordination model is applied to parallel hybrid MOMs.

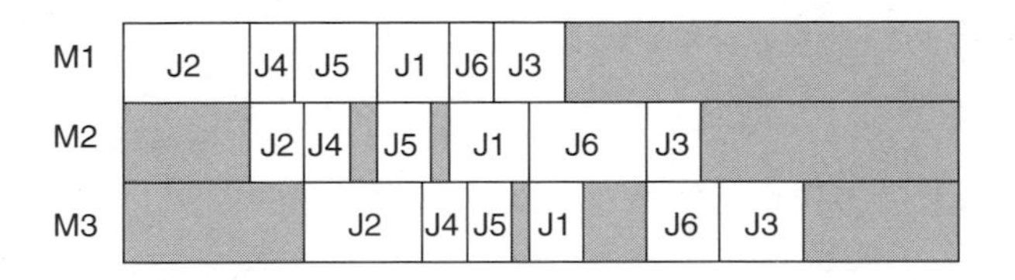

FIGURE 36.4 Example of permutation flow-shop with 6 jobs and 3 machines.

36.4 Application to BPFSP and Experimentation

36.4.1 Problem Formulation

The Flow-Shop problem is a scheduling problem [12] that has received a great attention given its importance in many industrial areas. The problem can be formulated as a set of N jobs $J_1, J_2, \ldots, J_N$ to be scheduled on M machines. The machines are critical resources as each machine cannot be simultaneously assigned to two jobs. Each job J_i is composed of M consecutive tasks $t_{i1}, \ldots, t_{iM}$, where t_{ij} represents the jth task of the job J_i requiring the machine m_j. To each task t_{ij} is associated a processing time p_{ij}, and each job J_i must be achieved before a due date d_i.

In this chapter, we focus on the *BPFSP* where jobs must be scheduled in the same order on all the machines (see Figure 36.4). Therefore, two objectives have to be minimized:

- $C_{\max}$: Makespan (Total completion time)
- T: Total tardiness

The task t_{ij} being scheduled at time s_{ij}, the two objectives can be formulated as follows:

$$f_1 = C_{\max} = \max\{s_{iM} + p_{iM} | i \in [1, \ldots, N]\},$$

$$f_2 = T = \sum\nolimits_{i=1}^{N} [\max(0, s_{iM} + p_{iM} - d_i)].$$

The Pareto front *PF* associated with *BPFSP* may be formulated as follows:

$$\forall y, \exists x \in PF, (m(x) \leq m(y)) \quad \text{or} \quad (t(x) \leq t(y)),$$

where x and y are solutions of the MOP, and $m(x)$ (respectively $t(x)$) is the value of x corresponding to the makespan (respectively tardiness) criterion.

36.4.2 A Genetic–Mimetic Algorithm for Solving BPFSP

In single objective optimization, it is well known that GAs provide better results when they are hybridized with LS algorithms. Indeed, the GA convergence is too slow to be really effective without any hybridization [10]. In Reference 2, a hybrid Genetic–Mimetic algorithm named *AGMA* has been proposed for solving BPFSP. The simplified pseudocode of the algorithm is presented in Algorithm 36.1 and illustrated in Figure 36.5.

AGMA combines a genetic algorithm (GA) and a mimetic algorithm (MA). In this chapter, we do not give the details and parameters of the two algorithms, and if needs be, the reader is referred to Reference 2. The GA uses mainly two parameters: an archive (Pareto Front) PO^* of nondominated solutions, and a progression ratio P_{PO^*} of PO^*. At each generation, these two parameters are updated. If no significant progression is noticed ($P_{PO^*} < \alpha$, where α is a fixed threshold), an intensified search process is triggered. The intensification consists in applying MA (see Algorithm 36.2) to the current population during one generation. The application of MA returns a Pareto Front $PO^{*\prime}$ that serves to update the Pareto Front PO^* of the GA.

Algorithm 36.1 AGMA Algorithm

```
Create an initial population
while run time not reached do
  Perform a GA generation with adaptive mutation
  Update PO* and P_PO*
  if P_PO* < α then
    Perform a generation of MA on the population (Algorithm 2)
    Update PO* and P_PO*
  end if
  Update selection probability of each mutation operator
end while
```

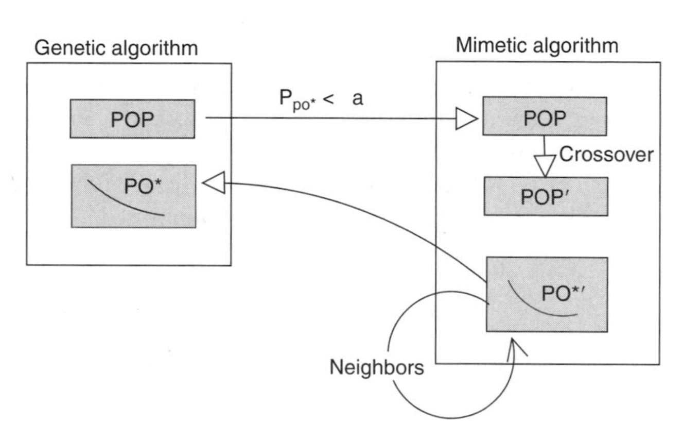

FIGURE 36.5 Illustration of AGMA.

Algorithm 36.2 MA algorithm

```
while MA run time not reached do
  Select randomly a set P of solutions from the current population
  Apply the crossover on P to generate a set P′ of new solutions
  Compute the nondominated set PO*′ from P′
  while New solutions found do
    Create the neighborhood N of each solution of PO′
    Let PO*′ be the nondominated set of N ⋃ PO*′
  end while
end while
```

Mimetic algorithm consists in selecting randomly a set of solutions from the current population of the GA. A crossover operator is then applied to these solutions and new solutions are generated. Among these new solutions only nondominated ones are maintained to constitute a new Pareto Front $PO^{*\prime}$. An LS is then applied to each solution of $PO^{*\prime}$ to compute its neighborhood. The nondominated solutions belonging to the neighborhood are inserted into $PO^{*\prime}$.

36.4.3 Parallel Hybrid AGMA

Different parallel models have been sketched and analyzed in Section 36.2. The fine-grained parallel models could not be exploited efficiently in a P2P environment due to the communication delays. In BPFSP, the model based on parallel evaluation of each solution is fine-grained and is not likely to lead to better performance. Indeed, the evaluation of each objective has a low cost. Therefore, it is useless to evaluate in parallel the two objectives and evaluate each of them in parallel. Conversely, it is useful to exploit the following parallel models: (1) the island model that consists in performing in parallel several cooperative AGMAs; (2) the parallel evaluation of the population of each AGMA; (3) the multistart model that consists in applying in parallel an LS on each solution of the Pareto Front $PO^{*\prime}$ in MA. The parallel evaluation of the neighborhood of each solution could not be efficient for the same reason as the parallel evaluation of each solution.

We have limited our implementation to the coarse-grained parallel models, that is, the island model and the multistart model. Figure 36.6 illustrates the parallel hybrid AGMA exploiting these two models.

- *The island model*: Due to its exorbitant cost in terms of CPU time on large-size instances of BPFSP, the island model has not been exploited in Reference 2. Indeed, the exploitation of the model on large-size BPFSP is possible only on large-scale P2P networks or grids. In our implementation (see Figure 36.6), the parameters of the model are the following: The different cooperative AGMA

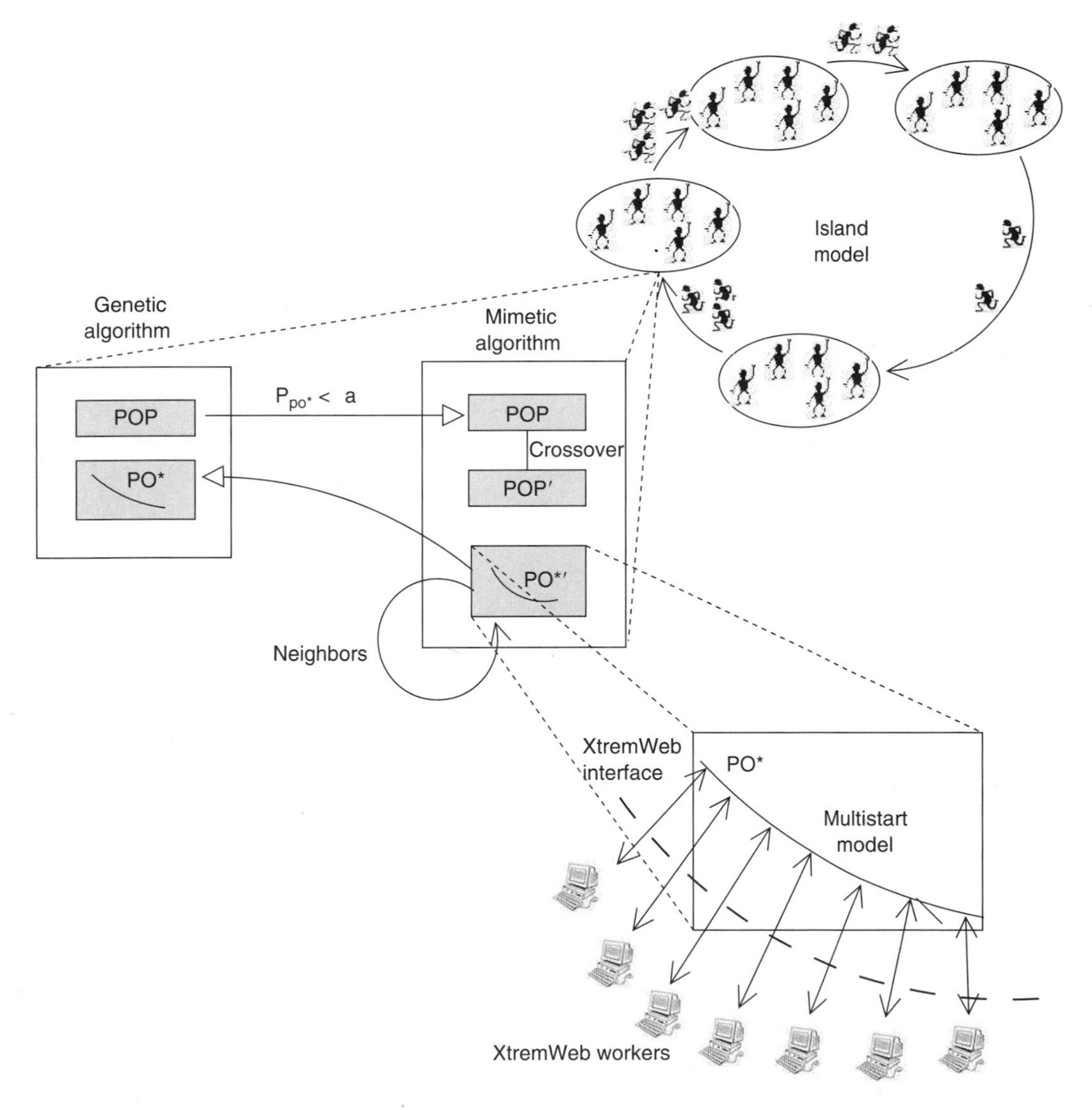

FIGURE 36.6 Illustration of parallel AGMA.

exchange their whole archives PO^*, and the number of emigrants is dynamic. At its arrival, the immigrant archive is merged with the local one. Migrations occur periodically (each a fixed number of iterations). The migration topology is the random one, meaning the destination island is selected randomly.

- *The multistart model*: The multistart model is exploited during the execution of MA. Each solution of the Pareto Front $PO^{*\prime}$ computed by the algorithm represents the initial solution of an LS method that calculates its neighborhood. The different LSs are executed in parallel according to the Master–Slave model. The master, that is, the algorithm MA merges with $PO^{*\prime}$ the neighborhoods returned by the different slaves and computes the new $PO^{*\prime}$ that contains the nondominated solutions.

36.4.4 Deployment and Experimentation

In this section, we present the deployment scheme of the different parallel hybrid models on the XtremWeb architecture, and the preliminary experimental results obtained on the application presented above.

36.4.5 Deployment and Fault Tolerance

A deployment scheme may be defined as a function that consists in embedding the different components of the parallel models on the different components of the P2P architecture. Different deployment schema of the island and multistart models on XtremWeb are possible. Indeed, the AGMA algorithms of the island model can be deployed either as XtremWeb clients or workers. For the multistart model, the master can be either a client or a worker, and the slaves are necessarily deployed as workers. For our experimentation, the deployment scheme is illustrated in Figure 36.7.

The island model is deployed on three XtremWeb clients, and each client runs the AGMA algorithm. During the hybridization phase (execution of MA), the LSs initiated on the Pareto Front PO^* are submitted as tasks to the Dispatcher that launches them on the workers at their request. The multistart model is thus deployed on a client and a set of workers.

In XtremWeb, the fault tolerance issue is tackled at Worker and Dispatcher levels. When a worker fails the work unit being executed is re-started from scratch. If the Dispatcher crashes it is re-started using its information database. The problem with such solution is that in a highly volatile environment a large

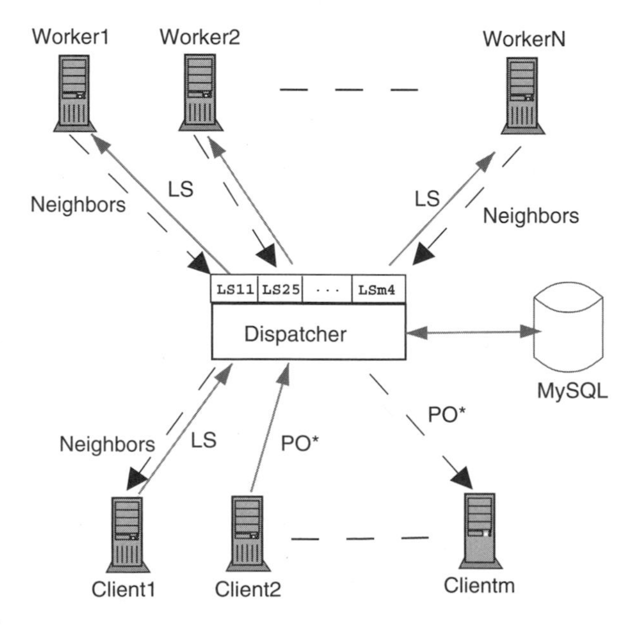

FIGURE 36.7 Deployment schema of parallel hybrid AGMA on top of XtremWeb.

amount of CPU time is wasted as the system spends its time in re-starting work units performed by the workers. Therefore, we propose a check-pointing approach at the client level that allows to solve more efficiently the fault-tolerance problem. Indeed, the problem data and intermediate results are periodically stored. If the Dispatcher fails the application is restored and re-started from the last checkpoint. In case of worker failure the work unit is re-started using the intermediate results. The check-pointing operation (storing) is performed after each LS and/or every 100 generations. The second condition is necessary when no LS has been launched during the last 100 generations. This means that a significant progression of the Pareto Front has been observed at each generation, what excludes any resort to the hybridization.

36.4.6 Experimental Results

In our experiments, we consider the BPFSP instance 200 jobs on 10 machines. The parameters of the island model are fixed as the following: migrations occur every 10 generations, the number of emigrants is fixed at each migration operation to 20 the size of the archive PO^* is upper than 20, and the whole archive otherwise, and the population size of each AGMA is 100.

The application has been deployed during working days (nondedicated environment) on the education network of the *Polytech'Lille* engineering school. The experimentation hardware platform is composed of 120 heterogeneous Linux Debian PCs. The characteristics of these PCs are presented in Table 36.1.

Three parallel hybrid versions of AGMA are experimented, evaluated and compared:

- *Version 1* is that proposed in Reference 2, and exploits only the multistart model. This means that the GA is executed on a single machine and the hybridization phase is deployed on a parallel machine (IBM-SP2) according to the Master–Slave model (*Push* mode, i.e., work distribution is initiated by the master).
- *Version 2* is the same as Version 1 except that the hybridization is deployed in a distributed way on a set of XtremWeb workers according to the cycle stealing paradigm (*Pull* mode, i.e., work distribution is initiated by the workers).
- *Version 3* is not considered in Reference 2, and is a combination of the multistart and island models. As illustrated in Figure 36.7, three AGMA algorithms are deployed on client machines and cooperate according to the island model. Each AGMA is an implementation of Version 2.

Figure 36.8 illustrates the Pareto Fronts obtained with the versions 1 and 2 after 80 LSs. The two fronts are approximately the same, but Version 2 has the advantage to be fault tolerant.

The execution of Version 1 is stopped after 80 LSs as it is not fault tolerant. Conversely, with Version 2 long-lasting executions are possible. For instance, Figure 36.9 shows that the execution goes on up to 350 LSs. The execution lasted one week, and 10 failures have been observed and as many check-pointing operations have been performed. As a result, the Pareto Front obtained with 350 LSs is clearly better than that obtained with 80 LSs using Version 1 or Version 2. One has to note that such results are possible only with a scalable and fault tolerant version of the algorithm.

TABLE 36.1 Experimentation Hardware Platform

Processor	Number
AMD Duron(tm) Processor	14
Celero (Coppermine)	14
Intel(R) Celeron(R) CPU 2.00 GHz	8
Intel(R) Celeron(R) CPU 2.20 GHz	28
Intel(R) Celeron(R) CPU 2.40 GHz	21
Intel(R) Celeron(R) CPU 1400 MHz	7
Pentium III (Katmai)	28
Total	120

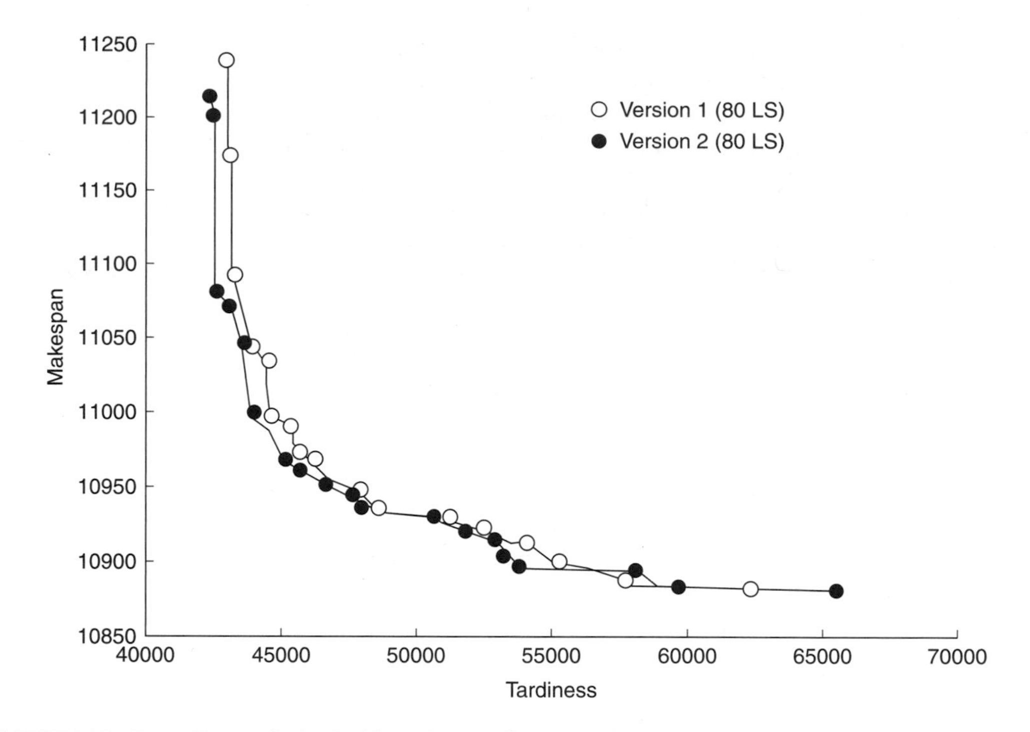

FIGURE 36.8 Pareto Fronts obtained with Version 1 and Version 2 (80 LSs).

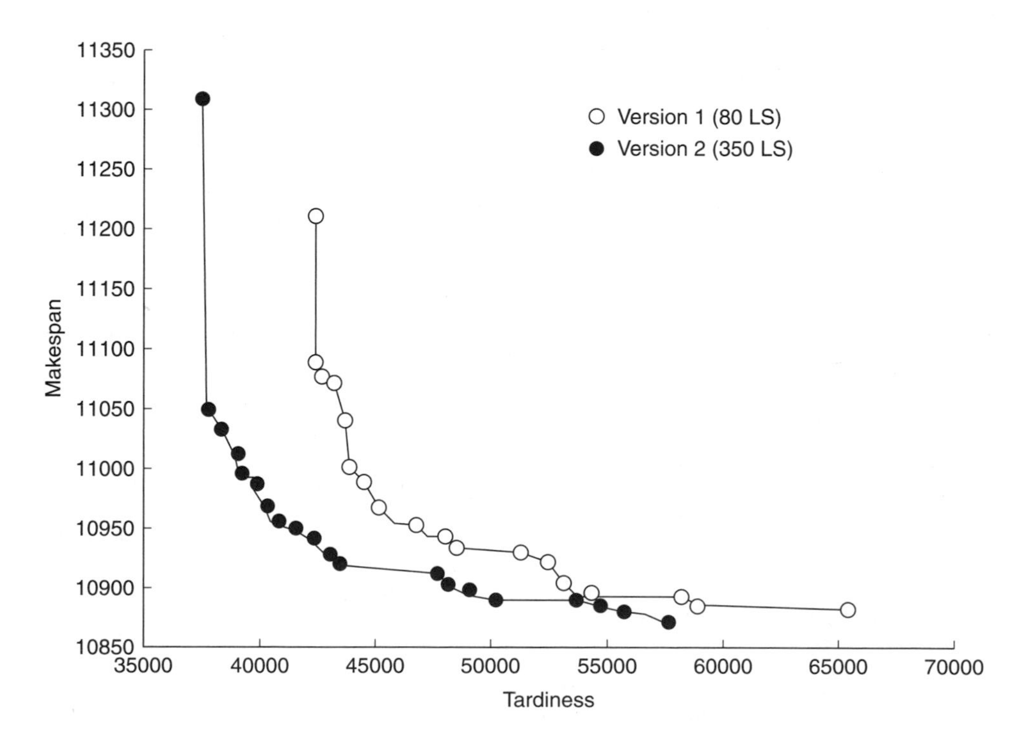

FIGURE 36.9 Pareto Fronts with Version 1 (80 LSs) and Version 2 (350 LSs).

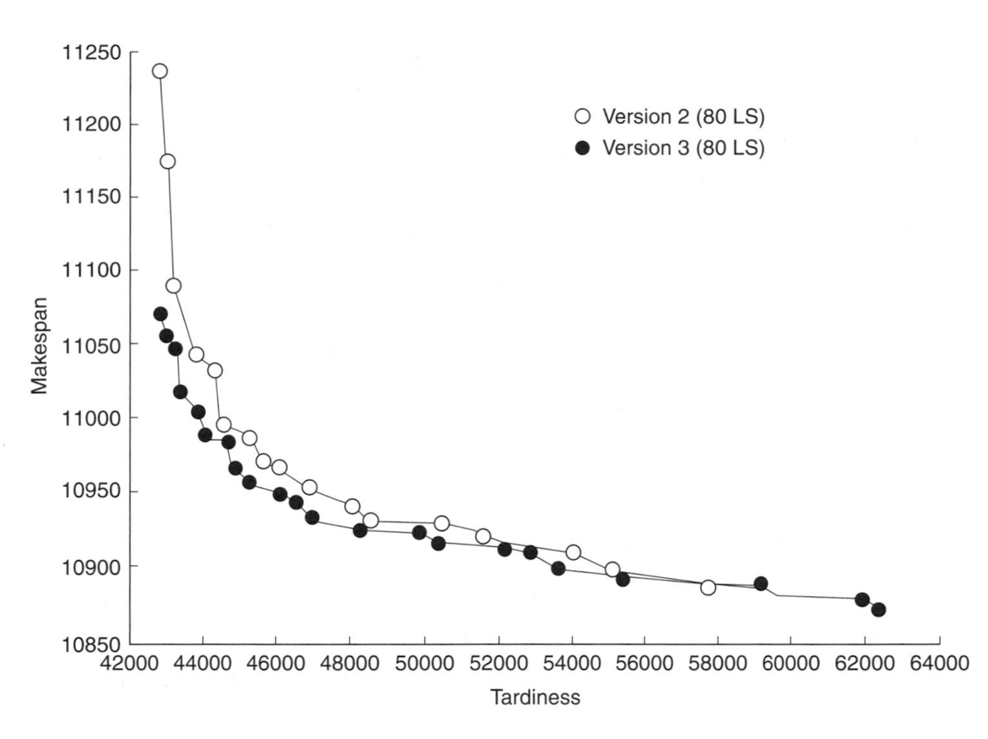

FIGURE 36.10 Pareto Fronts with Version 2 (80 LSs) and Version 3 (80 LSs).

Figure 36.10 allows to compare the Pareto Fronts obtained with Version 2 and Version 3 and to demonstrate the contribution of the island model to the effectiveness. With 80 LSs, the Pareto Front obtained using Version 3 is better than that obtained using Version 2. More experiments with more LSs are in progress.

Figure 36.11 (Part A) shows the oscillation between GA and MA (or LS) over time obtained with Version 3. Figure 36.11 (Part B) is a zoom up of Figure 36.11 (Part A) on the 50,000 first time units. It shows that the MA (thus LS) is frequently solicited and this lasts long. Figure 36.11 (Part C) illustrates the evolution in time of the number of deployed workers at the beginning of the execution (zoom out on the first 350 time-units). The maximum number of workers is 60 because during the starting phase the Pareto Front contains a small number of solutions. The number of workers decrease to 0 when the GA succeed to improve the Pareto Front without calling MA.

One has to note that the spectrum blackens with the time (from left to right). This means that the GA solicits more and more the MA, that is, the LS because it never enhances again the Pareto Front, in other words the GA converges. On the other hand, the local search lasts less and less time. Therefore, even the intensification (by LS) does not contribute to enhance the effectiveness, meaning that the AGMA converges. Through this experimentation, we have learned more on the convergence of the AGMA algorithm. Therefore, one can note that P2P computing allows to "push far" the limits in terms of computing resources to better evaluate the contribution of the hybridization but also its limitations.

36.5 Conclusion and Future Work

The hybridization of metaheuristics having complementary behaviors allows to enhance the effectiveness and robustness in combinatorial optimization [11]. However, its exploitation on industrial applications is possible only by using a great computing power. Large-scale parallelism based on the use of computational grids and/or P2P systems is recently revealed to be a good way to get at hand such computing power and exploit hybridization. To our best of knowledge, no research work has been published on parallel hybrid

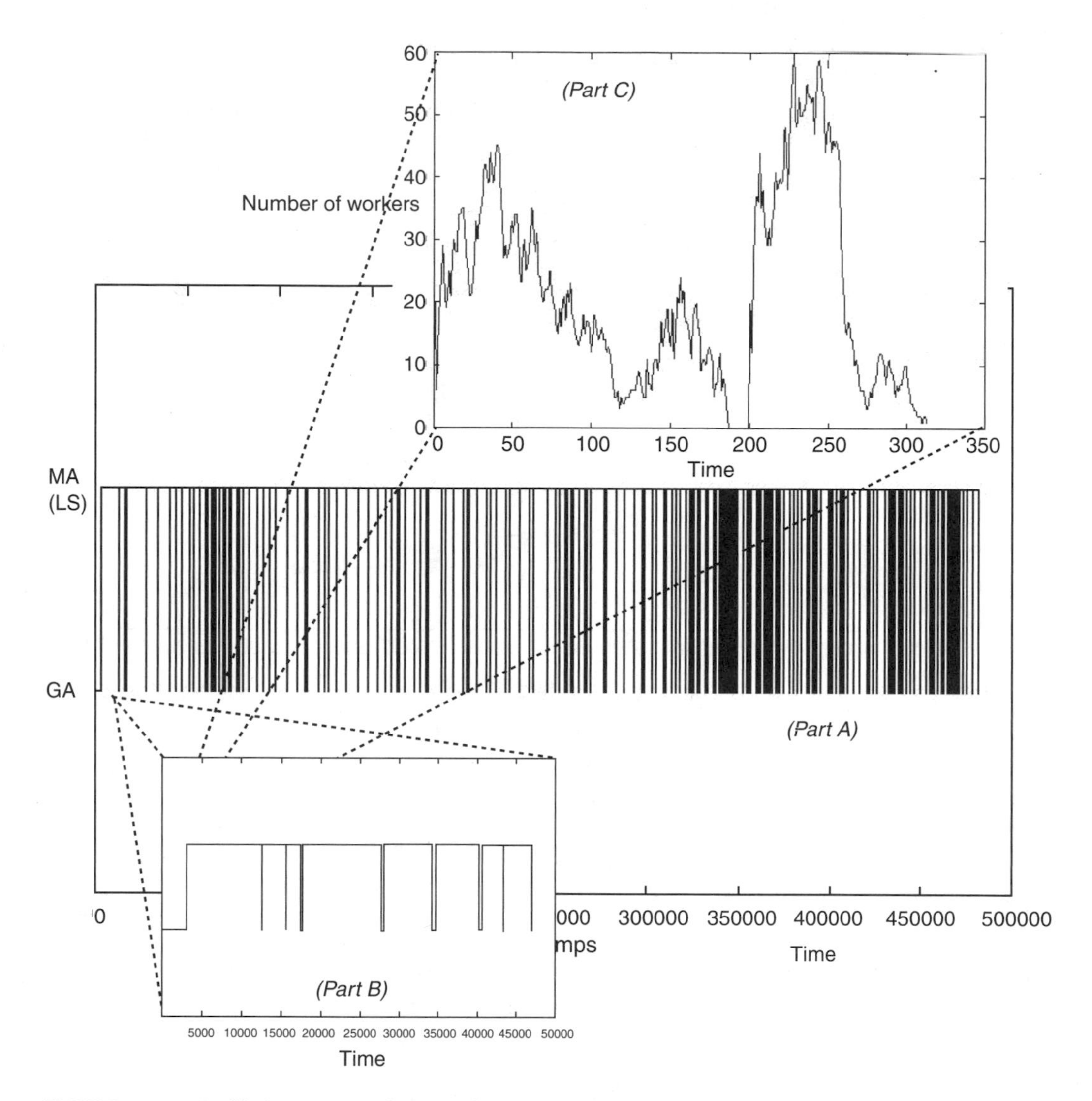

FIGURE 36.11 Oscillation spectrum between the GA and MA (thus LS).

metaheuristics on P2P systems. Nowadays, existing P2P computing middlewares are inadequate for the deployment of parallel cooperative applications. Indeed, these need to be extended with a software layer to support the cooperation. In this chapter, we have proposed a Linda-like cooperation model that has been implemented on top of XtremWeb.

In Reference 2, a hybrid metaheuristic (AGMA) has been proposed and experimented on BPFSP. The performed experiments on large-size instances such as 200 jobs on 10 machines are often stopped without the convergence is reached. The full exploitation of the hybridization needs a large amount of computational resources and the management of the fault-tolerance issue. We have proposed a fault-tolerant hybrid parallel design of the AGMA combining two parallel models: the multistart model and the island model. The algorithm has been implemented on our extended version of XtremWeb.

The first experiments have been performed on the education network of the Polytech'Lille engineering school. The network is composed of 120 heterogeneous Linux PCs. The preliminary results, obtained after several execution days, demonstrate that the use of P2P computing allows to fully exploit the benefits of hybridization. Indeed, the obtained Pareto Front is clearly better than that obtained in Reference 2.

On the other hand, the deployment of the island model allows to improve the effectiveness. Beyond the improvement of the effectiveness, the parallelism on P2P systems allows to push far the limits in terms of computational resources. As a consequence, it permits to better evaluate the benefits and limitations of the hybridization. Such result has to be confirmed again on a larger P2P network and larger instances of the problem.

References

[1] D.P. Anderson, J. Cobb, E. Korpela, M. Lepofsky, and D. Werthimer. SETI@home: An experiment in public-resource computing. *Communications of the ACM*, 45: 56–61, 2002.

[2] M. Basseur, F. Seynhaeve, and E.-G. Talbi. Adaptive mechanisms for multi-objective evolutionary algorithms. In *Congress on Engineering in System Application CESA '03*, Lille, France, 2003, pp. 72–86.

[3] S. Cahon, N. Melab, and E.-G. Talbi. ParadisEO: A framework for the reusable design of parallel and distributed metaheuristics. *Journal of Heuristics*, 10: 353–376, 2004.

[4] G. Fedak, C. Germain, V. Neri, and F. Cappello. XtremWeb: Building an experimental platform for Global Computing. *Workshop on Global Computing on Personal Devices (CCGRID2001)*, IEEE Press, May 2001.

[5] I. Foster and C. Kesselman. *The Grid: Blueprint for a New Computing Infrastructure.* Morgan Kaufmann, San Fransisco, CA, 1999.

[6] D. Gelernter and N. Carriero. Coordination languages and their significance. *Communications of the ACM*, 35: 97–107, 1992.

[7] D. Gelernter. Generative communication in Linda. *ACM Transactions on Programming Languages and Systems*, 7: 80–112, 1985.

[8] A. Oram. *Peer-to-Peer: Harnessing the Power of Disruptive Technologies.* O'Reilly & Associates, 2001.

[9] G.A. Papadopoulos and F. Arbab. Coordination models and languages. *Advances in Computers: The Engineering of Large Systems*, Academic Press, 1998, p. 46.

[10] E.-G. Talbi, M. Rahoual, M.-H. Mabed, and C. Dhaenens. A hybrid evolutionary approach for multicriteria optimization problems: Application to the Flow Shop. In E. Zitzler et al., Eds., *Evolutionary Multi-Criterion Optimization*, Vol. 1993 of *Lecture Notes in Computer Science*, Springer-Verlag, Heidelberg, 2001, pp. 416–428.

[11] E.-G. Talbi. A Taxonomy of Hybrid Metaheuristics. *Journal of Heuristics*, 8: 541–564, 2002.

[12] V. T'kindt and J.-C. Billaut. *Multicriteria Scheduling — Theory, Models and Algorithms.* Springer-Verlag, Heidelberg, 2002.

[13] J. Verbeke, N. Nadgir, G. Ruetsch, and I. Sharapov. Framework for peer-to-peer distributed computing in a heterogeneous, decentralized environment. In *Proceedings of the Third International Workshop on Grid Computing (GRID '2002)*, Baltimore, MD, January 2002, pp. 1–12.

Index